£14.56
TO

PHASE TRANSFORMATIONS
AND THE EARTH'S INTERIOR

UPPER MANTLE PROJECT SCIENTIFIC REPORT No. 26

PHASE TRANSFORMATIONS AND THE EARTH'S INTERIOR

*Proceedings of a symposium held in Canberra Australia
6–10 January 1969 by the International Upper Mantle
Committee and the Australian Academy of Sciences*

Editors:

A. E. Ringwood

D. H. Green

Australian National University

UPPER MANTLE PROJECT SCIENTIFIC REPORT No. 26

NORTH-HOLLAND PUBLISHING COMPANY – AMSTERDAM

© North-Holland Publishing Company, Amsterdam, 1970

All Rights Reserved. No part of this publication may be reproduced, stored in a retrieval system, or transmitted, in any form or by any means, electronic, mechanical, photocopying, recording or otherwise, without the prior permission of the Copyright owner.

ISBN 7204 0167 4

Publisher:
NORTH-HOLLAND PUBLISHING COMPANY – AMSTERDAM

Reprinted from

PHYSICS OF THE EARTH AND PLANETARY INTERIORS VOL. 3, 1970

PRINTED IN THE NETHERLANDS

PREFACE

Papers in this issue of Physics of the Earth and Planetary Interiors were presented at the UMC Symposium on Phase Transformations and the Earth's Interior, Canberra, 6th–10th January 1969.

The International Upper Mantle Committee (UMC) is a committee set up jointly by the International Union of Geodesy and Geophysics and the International Union of Geological Sciences to co-ordinate the Upper Mantle Project, an international program of research on the solid Earth sponsored by the International Council of Scientific Unions.

The symposium was made possible through financial support by the Australian Academy of Science, the International Union of Geodesy and Geophysics, the International Union of Geological Sciences and UNESCO. The conference was organized by

> Prof. A. E. Ringwood (Convenor)
> Dr. D. H. Green (Secretary)
> Prof. J. C. Jaeger and
> Dr. S. D. Hamann

The use of the facilities of the Australian Academy of Science and Australian National University is gratefully acknowledged.

A. E. RINGWOOD
D. H. GREEN

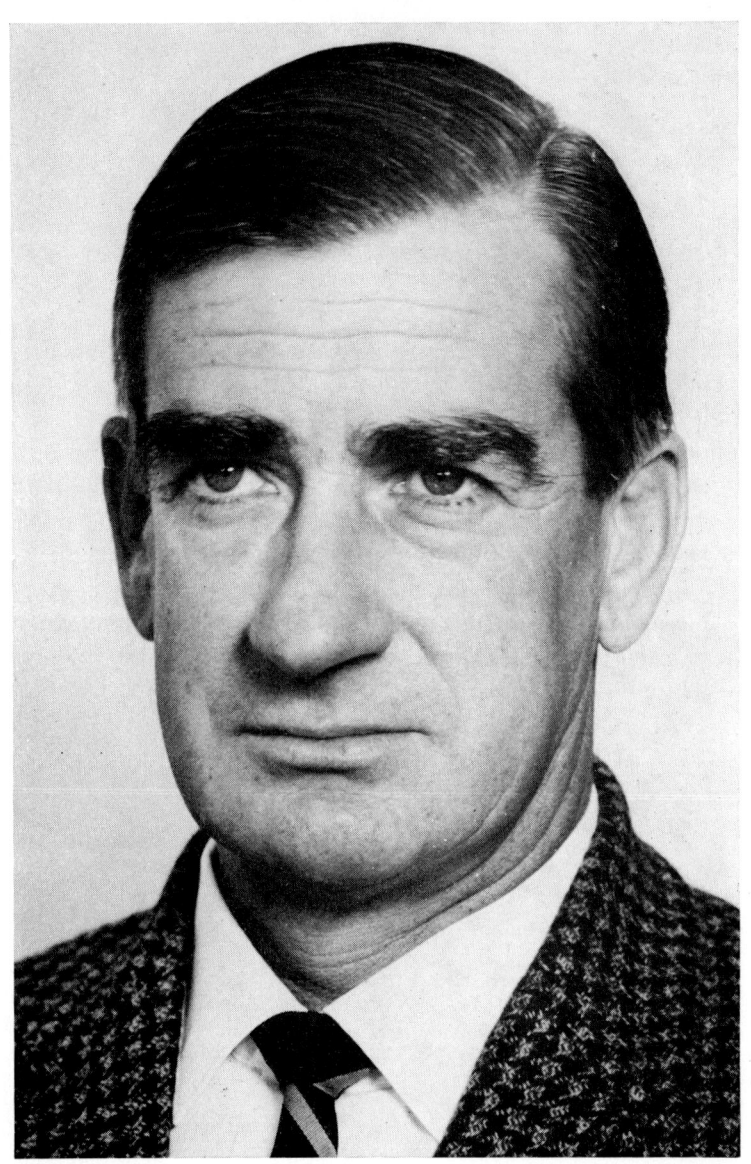

DR. A. D. WADSLEY

1918-1969

On January 6, 1969, Arthur David Wadsley introduced the opening speaker to an International Symposium in Canberra. Some few minutes later he suffered a massive coronary occlusion and died shortly afterwards.

By training and environment Wadsley should have been a metallurgist. Born in Hobart, Tasmania, on August 1, 1918, he joined the Electrolytic Zinc Company at the age of 16 and, during his subsequent university career, twice returned to work in the Company's plants at Risdon and Rosebery. In the final year of his B.Sc. course he gained a high distinction in primary metallurgy, and subsequently studied secondary metallurgy at the Hobart Technical College while completing his Honours year. His M.Sc. thesis was concerned with the production of magnesium from Tasmanian dolomite, and in 1943 he joined the CSIR Division of Industrial Chemistry to study the war-time use of Australian pyrolusite in dry cells.

It took less than two years for Australia to lose a budding metallurgist; by the end of 1944 Wadsley was seeking the property which would distinguish one manganese dioxide from another, and his life work on non-stoichiometric compounds had commenced. The influence of Byström, Hägg and Magneli can be detected in many of his early papers but by 1954 he was ready to advance his own explanations for the tremendous proliferation of complex, and seemingly irrational, structures which he and other workers were discovering. Within a year he had proposed his now famous concept of crystallographic shear, and had extended his search for the structural principles which would be the guide to the systematic design of the apparently complex, yet quite specific, structures which he was predicating. His Doctorate of Science from the University of Tasmania in 1956 was a fitting recognition of his new role in the field of chemical crystallography.

By 1958 he was confidently postulating changes in the grouping of structural polyhedra entitles; corner-sharing, edge-sharing and face-sharing became his tools of trade, and the arrival in Melbourne of Dr. Sten Andersson from Sweden was the signal for a dynamic leap forward. Their first joint publication in 1960 mounted a direct attack on the traditional concept of anion vacancies in a perovskite structure, and the subsequent description of the $K_2Ti_2O_5$ structure reflected Wadsley's search for a rational explanation of the coordination of the metal atom in a "non-stoichiometric" phase range.

It simultaneously attracted the attention of some of the larger industrial laboratories in America, and in 1962 Wadsley was invited to try out his ideas in front of an international forum. When he presented his thoughts on order and disorder in non-stoichiometric metal oxides to a symposium of the American Chemical Society, a controversy which had been simmering for several years, flared up. It was only resolved when Dr. Robert G. Roth was granted leave by the National Bureau of Standards to work with Wadsley in Australia.

Together they placed the long-postulated "block" principle on a sound experimental foundation, and presented the ordered intergrowth of phases as a logical simplification of the complex "homologous series" approach. A further visit to Australia by Andersson in 1966 paved the way for a lucid explanation of diffusion paths in terms of the fundamental concept of crystallographic shear.

By now, Wadsley had reached CSIRO's top grading of Chief Research Scientist and he devoted an increasing portion of his busy life to the encouragement of younger workers. In 1966, he was appointed Head of the Inorganic Chemistry Section, and a year later Assistant Chief of the Division of Mineral Chemistry. Around this time, Wadsley became interested in the crystallography of high pressure phases and the general problem of phase transformations in the mantle. Several geochemically significant contributions resulted from his joint work with Reid and Ringwood – particularly those on the felspar-hollandite transformation in orthoclase and the olivine to strontium plumbate transformation in Mn_2GeO_4.

Overseas requests for his participation in conferences became increasingly pressing. In 1966 he took advantage of the Werner Centenary Lectures in Switzerland to enunciate the two rules which he had used to predict the structure of previously unsuspected mixed-oxide phases of niobium and tungsten, and which were to supplement and modify the 40-year old Pauling principles for delineating the structure of crystalline solids.

Independent and adventurous, David Wadsley always enjoyed the reasoned questioning and probing of his scientific colleagues. One of his greatest pleasures was the entertainment of his international friends in his home at Kew or his beach-house at Frankston. His sudden and tragic death leaves a long list of engagements unfulfilled; his passing will be mourned in many countries.

IVAN NEWNHAM
Chief Division of Mineral Chemistry CSIRO, Melbourne

CONTENTS

Preface . . . V
Obituary to Dr. A. D. Wadsley . . . VII
Contents . . . IX

PHYSICS OF THE MANTLE

F. Press, Earth models consistent with geophysical data . . . 3
D. L. Andersen and T. Jordan, The composition of the lower mantle . . . 23
K. E. Bullen, Comparison of sources of evidence on the variation of incompressibility in the Earth's deeper interior . . . 36
D. L. Andersen and C. Sammis, Partial melting in the upper mantle . . . 41
H. K. Mao, T. Takahashi and W. A. Basset, Isothermal compression of the spinel phase of Ni_2SiO_4 up to 300 kbars at room temperature . . . 51
W. A. Basset and J. D. Barnett, Isothermal compression of stishovite and coesite up to 85 kbars at room temperature by X-ray diffraction . . . 54
O. L. Anderson and R. C. Liebermann, Equations for the elastic constants and their pressure derivatives for three cubic lattices and some geophysical applications . . . 61

PHASE TRANSFORMATIONS IN THE DEEP MANTLE

A. E. Ringwood and A. Major, The system Mg_2SiO_4–Fe_2SiO_4 at high pressures and temperatures . . . 89
A. E. Ringwood, Phase transformations and the constitution of the mantle . . . 109
R. A. Binns, $(Mg,Fe)_2SiO_4$ spinel in a meteorite . . . 156
M. Morimoto, K. Koto and M. Tokonami, Crystal structures of high pressure modifications of Mn_2GeO_4 and Co_2SiO_4 . . . 161
P. B. Moore and J. V. Smith, Crystal structure of β-Mg_2SiO_4: crystal-chemical and geophysical implications . . . 166
F. Birch, Interpretations of the low-velocity zone . . . 178
N. Kawai, S. Endo and K. Ito, Split sphere high pressure vessel and phase equilibrium relation in the system Mg_2SiO_4–Fe_2SiO_4 . . . 182
S. Akimoto and I. Syono, High-pressure decomposition of the system $FeSiO_3$–$MgSiO_3$ . . . 186
S. Akimoto, High-pressure synthesis of a "modified" spinel and some geophysical implications . . . 189
W. S. Fyfe, Lattice energies, phase transformations and volatiles in the mantle . . . 196
J. C. Jamieson, The phase behavior of simple compounds . . . 201
A. F. Reid, Crystal chemistry of high-pressure polymorphs of ABO_3, AB_2O_4 and AB_4O_8 compounds, and their possible identifications with the phases occurring in the mantle . . . 204
E. S. Gaffney and T. J. Ahrens, Stability of mantle minerals from lattice calculations and shock wave data . . . 205
C. Y. Wang, Can mantle minerals have the NiAs structure . . . 213

MAGMAS, XENOLITHS AND EXPERIMENTAL PETROLOGY

D. H. Green, A review of experimental evidence on the origin of basaltic and nephelinitic magmas . . 221

M. J. O'Hara, Upper mantle composition inferred from laboratory experiments and observations of volcanic products 236

P. W. Gast, Dispersed elements in oceanic volcanic rocks 246

D. H. Green and W. Hibberson, Experimental duplication of conditions of precipitation of high-pressure phenocrysts in a basaltic magma 247

G. J. H. McCall, Gabbroic and ultramafic nodules; High level intracrustal nodular occurrences in alkalic and associated volcanics from Kenya, described and compared with those of Hawaii . . . 255

H. Kuno and K.-I. Aoki, Chemistry of ultramafic nodules and their bearing on the origin of basaltic magmas 273

J. D. Kleeman and J. A. Cooper, Geochemical evidence for the origin of some ultramafic inclusion from Victorian Basanites 302

R. W. Nesbitt and D. L. Hamilton, Crystallisation of alkali-olivine basalts under controlled P_{O_2}, P_{H_2O} conditions 309

I. B. Lambert and P. J. Wyllie, Melting in the deep crust and upper mantle and the nature of the low-velocity layer 316

F. A. Frey, Rare Earth abundances in Alpine ultramafic rocks 323

A. L. Boettcher, Hydrothermal melting relationships in silicate-H_2O systems at vapor pressures greater than 10 kbars 331

C. W. Burnham and N. F. Davis, Thermodynamic properties of water-bearing magmas 332

C. B. Sclar, High-pressure studies in the system MgO–SiO_2–H_2O 333

N. I. Khitarov, A. B. Slutsky and V. A. Pugin, Electrical conductivity of basalts at high T–P and phase transitions under upper mantle conditions 334

A. A. Kadik and N. I. Khitarov, Influence of water on melting of silicates at high pressure 343

N. L. Dobretsov, A. A. Deribas and V. I. Maly, Shock compression of powdered SiO_2, Mg_2SiO_4, $ZrSiO_4$ and other materials 348

PETROLOGY OF THE UPPER MANTLE AND LOWER CRUST

D. H. Green and A. E. Ringwood, Mineralogy of peridotitic compositions under upper mantle conditions 359

I. D. Macgregor, The effect of CaO, Cr_2O_3 and Al_2O_3 on the stability of spinel and garnet peridotites 372

E. J. Essene, B. J. Hensen and D. H. Green, Experimental study of amphibolite and eclogite stability 378

A. Irving and D. H. Green, Experimental duplication of mineral assemblages in basic inclusions of the Delegate breccia pipes 385

J. F. Lovering, Granulitic and eclogitic inclusions from basic pipes in Eastern Australia 390

I. D. Macgregor and J. L. Carter, The chemistry of clinopyroxenes and garnets of eclogite and peridotite xenoliths from the Roberts Victor mine, South Africa 391

N. V. Sobolev, Eclogites and pyrope peridotites from the Kimberlites of Yakutia 398

S. Banno, Classification of eclogites in terms of physical conditions of their origin 405

W. Schreyer and F. Seifert, Pressure dependence of crystal structures in the system MgO–Al_2O_3–SiO_2–H_2O at pressures up to 30 kilobars 422

B. J. Hensen and D. H. Green, Experimental data on coexisting cordierite and garnet under high grade metamorphic conditions 431

T. H. Green, High pressure experimental studies on the mineralogical constitution of the lower crust . 441

Th. Ernst and R. Schwab, Stability and structural relations of (Mg,Fe) metasilicates 451

M. H. MANGHNANI, Analcite-jadeite phase boundary 456
N. L. DOBRETSOV and N. V. SOBOLEV, Eclogites from metamorphic complexes of the USSR 462
TH. R. MCGETCHIN and L. T. SILVER, A crustal upper mantle model for the Colorado Plateau based on observations of crystalline rock fragments in a kimberlite dike 471

TECTONOPHYSICS

H. KANAMORI, Mantle beneath the Japanese arc 475
G. P. WOOLLARD, Evaluation of the isostatic mechanism and role of mineralogical transformations from seismic and gravity data 484
S. J. SUBBOTIN, Phase transformations within the Earth's mantle as a cause of crustal movements and a source of crustal material 499
A. R. RITSEMA, The mechanism of mantle earthquakes in relation to phase transformation processes . 503
A. LACAM, B. A. LOMBOS and B. VODAR, Pressure effect on the rate of phase transition in mercury telluride 511
J. A. JACOBS, The evolution of the Earth's core and its magnetic field 513

AUTHOR INDEX 519

PHYSICS OF THE MANTLE

EARTH MODELS CONSISTENT WITH GEOPHYSICAL DATA

FRANK PRESS

Department of Earth and Planetary Sciences, Massachusetts Institute of Technology, Cambridge, Mass. U.S.A.

A suite of the most recently available geophysical data are inverted by an improved Monte Carlo procedure. The data are derived from surface waves for oceanic paths, eigenvibrations of the earth, elastic wave travel time and $dt/d\Delta$ data, mass and moment of inertia of the earth. A low velocity zone is required for the suboceanic mantle as is a high density lithosphere. The high density is related to eclogite fractionation from the underlying, partially molten asthenosphere in a process involving the creation and spreading of the lithosphere. If the asthenosphere is pyrolite or peridotite then an increase of mean atomic weight across the transition zone seems required. Fairborn's new $dt/d\Delta$ data for the lower mantle seem to show a higher shear velocity gradient than previously supposed. If correct, a compensatory lower density gradient is required. This may indicate a depletion of iron with depth in the lower mantle. The density at the top of the core is surprisingly well constrained to the range 9.9–10.2 g/cm^3 a value appropriate for a mixture of iron and about 15 wt% silicon.

1. Introduction

A major goal of geophysics is to uniquely specify the distribution of two elastic velocities and density with depth in the earth and to relate these distributions to variations in composition, phase, and temperature in the interior. Impediments which block these achievements are many. It has not been proved in the mathematical sense that a unique solution can be obtained, although BACKUS and GILBERT (1968) have shown that under certain circumstances stable weighted averages (over depth) can be calculated. Furthermore, the data set available for recovering earth structure is incomplete and imprecise. Finally the equations of state available for interpreting elasticity and density distributions in terms of composition, state, etc. are tentative ones based on uncertain theories and assumptions and limited laboratory data.

Despite these difficulties it may be possible even with the presently available data to make some meaningful statements about the interior. In this paper we explore this possibility using the most recent and best available data in a Monte Carlo inversion procedure. Our results and conclusions supersede those presented in earlier papers (PRESS, 1968a, b) because we are able to fit models to new, more extensive and accurate data with greater speed and better precision.

2. Method

The Monte Carlo method uses random selection to generate large numbers of models in a computer, subjecting each model to a test against geophysical data. Only those models are retained whose properties fit the data within a prescribed tolerance. The procedure offers the advantage that successful models are found without bias, preconceived ideas or uncertain assumptions of equations of state or composition. If the program is efficient so that a very large number of models are examined, the retained models can be considered as representative of the family of successful models which fit the data. When the successful models fall in a narrow band, geophysically meaningful conclusions can often be reached despite our inability to specify a single unique model. Under certain conditions (BACKUS and GILBERT, 1968) a single successful model can provide unique, local averages of density or velocity.

We have modified the Monte Carlo procedure reported last year (PRESS, 1968a) speeding up the process by 1–2 orders of magnitude. This improved efficiency enabled us to find a larger number of successful models fitting a more extensive suite of data with better precision. The flow diagram of the currently used system (fig. 1) is printed with each run of the program and provides diagnostics so that controlling constants can

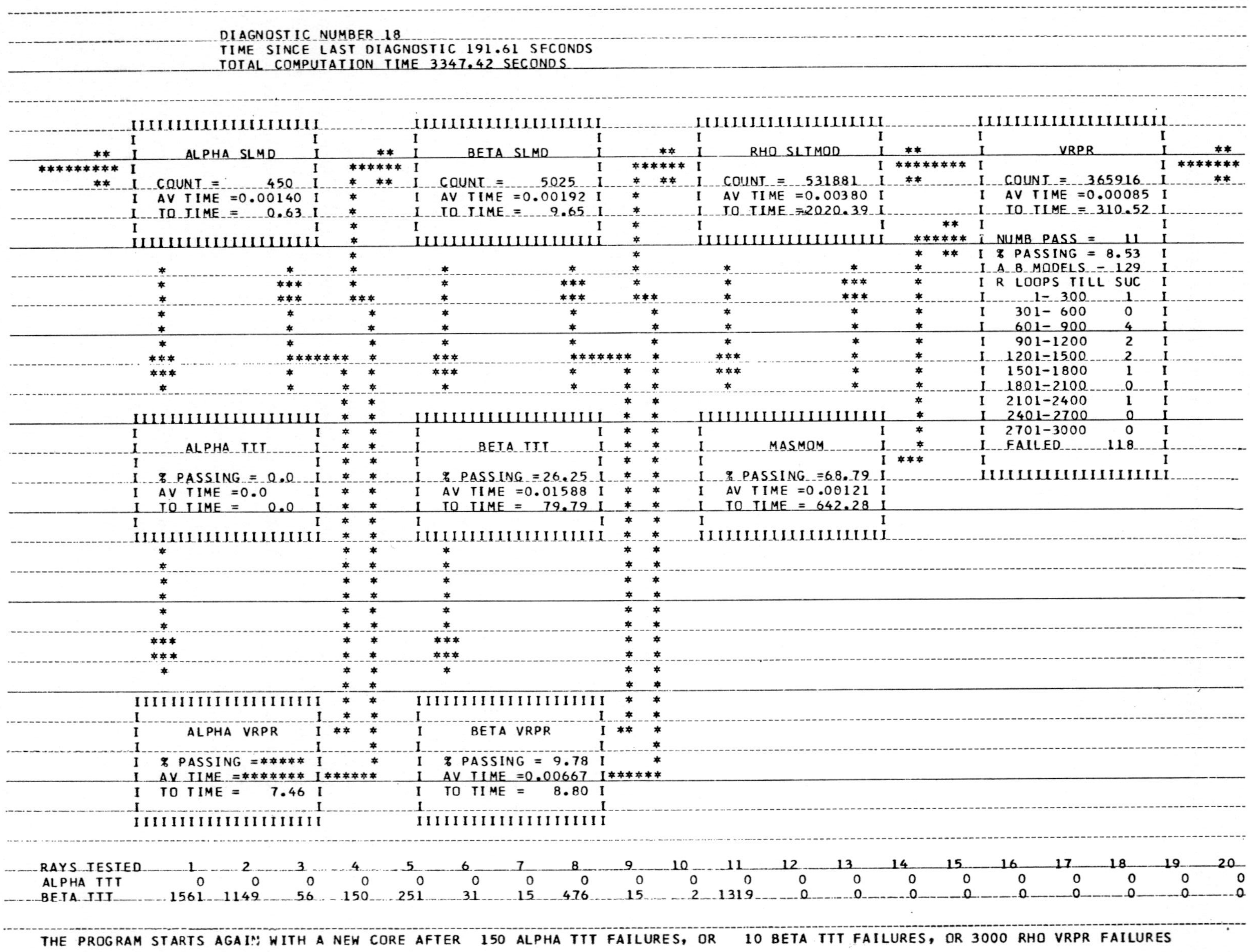

Fig. 1. Flow diagram of the Monte Carlo program during a run in which 531 881 density models and 5025 shear velocity models were tested and yielded 11 successful models.

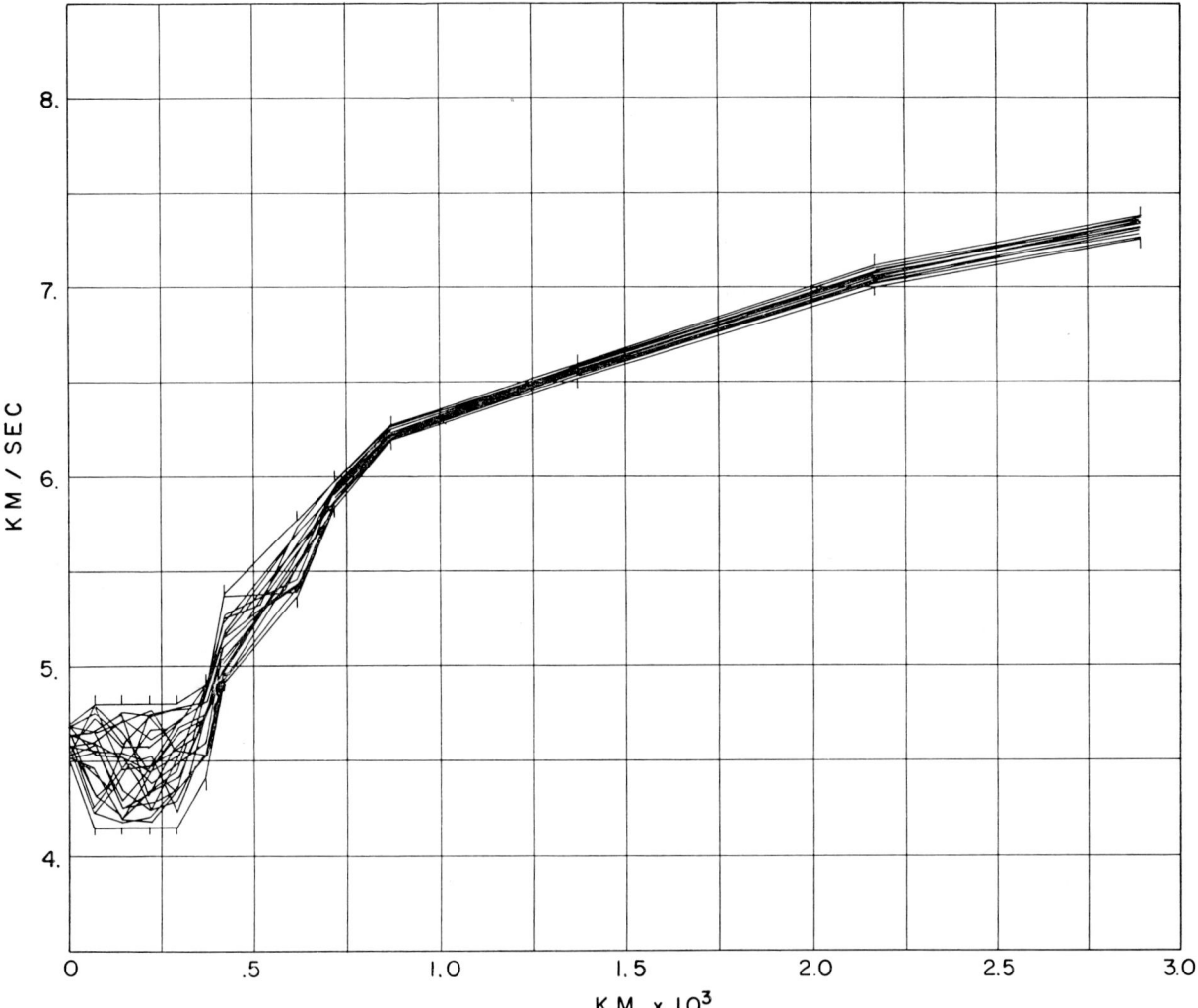

Fig. 2. Twenty-five shear velocity models of the mantle not subjected to geophysical constraints to test distribution of randomly generated models.

be set for maximum efficiency. The figure shows the diagnostics following a run of 3347 seconds on an IBM 360-65 computer which yielded 11 successful models at a cost of about $10 per successful model. SLMD is the random selection procedure fro compressional velocity (alpha), shear velocity (beta) and density (rho). TTT is a test of the model against observed travel times using BULLEN's (1961) method in which the earth is treated as a multilayered sphere, the velocity varying according to a power law within each layer. VRPR uses a table of variational parameters (WIGGINS, 1968) stored permanently in a data cell of the computer to test the perturbation of the eigenperiods due to velocity, density or core radius perturbations. This test is made after the selection of density and velocity models, the latter in order to eliminate early in the process those models which cannot be brought into agreement with eigenperiod data by density perturbations. MASMOM tests each density model against mass and moment of inertia of the earth. The flow diagram shows branching according as the several tests are passed or failed. Each box shows the number of times the corresponding step was repeated, the average time and total time for each step and the percentage of models passing. Thus the time distribution over the various components of the program is available for adjustment of input constants

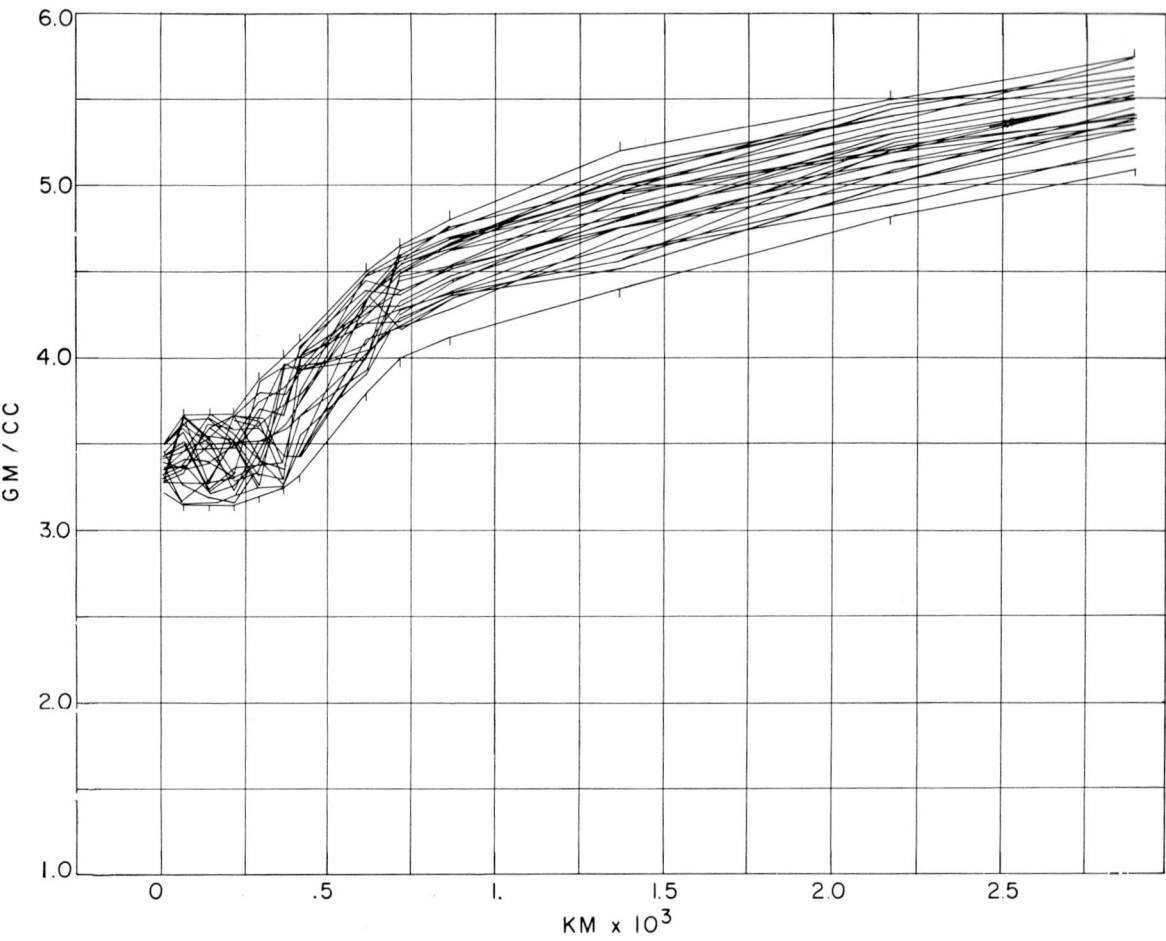

Fig. 3. Unconstrained mantle densities (see fig. 2, caption).

for maximum efficiency and insight is provided as to how the various geophysical constraints figure in the elimination of models. A key requirement of the Monte Carlo method is that the selection procedure produce an unbiased, representative suite of models for examination. Figs. 2, 3 and 4 show a run in which 25 models were generated, bypassing tests against geophysical data. It is seen that the velocity and density space between the permissible bounds is nearly uniformly filled. Since millions of earth models were generated and examined in this study it would be surprising if continued operation of the program would produce a successful model significantly different from those presented later in this paper. Some additional features of the program and procedures used are as follows:

(1) The earth is assumed to be spherically symmetrical and isotropic, with an oceanic crust-upper mantle structure. The radius of the core is selected randomly for each model in the range 3473 ± 25 km.

(2) Although α, β, ρ could be varied at 88 points in the earth, we chose the time saving device of randomly varying 19 points, (see fig. 5 section D for their location) obtaining the remaining values by linear interpolations.

(3) The fluid core was assumed to be adiabatic. The density selection procedure for the mantle below 1000 km eliminated models with extreme density gradients. The gradients in α, β, ρ were restricted to a maximum number of reversals in sign (typically 2 or 4) to restrict the complexity of models.

(4) The rigidity for the inner core only affects the mode $_0S_2$. Although zero rigidity was assumed in this

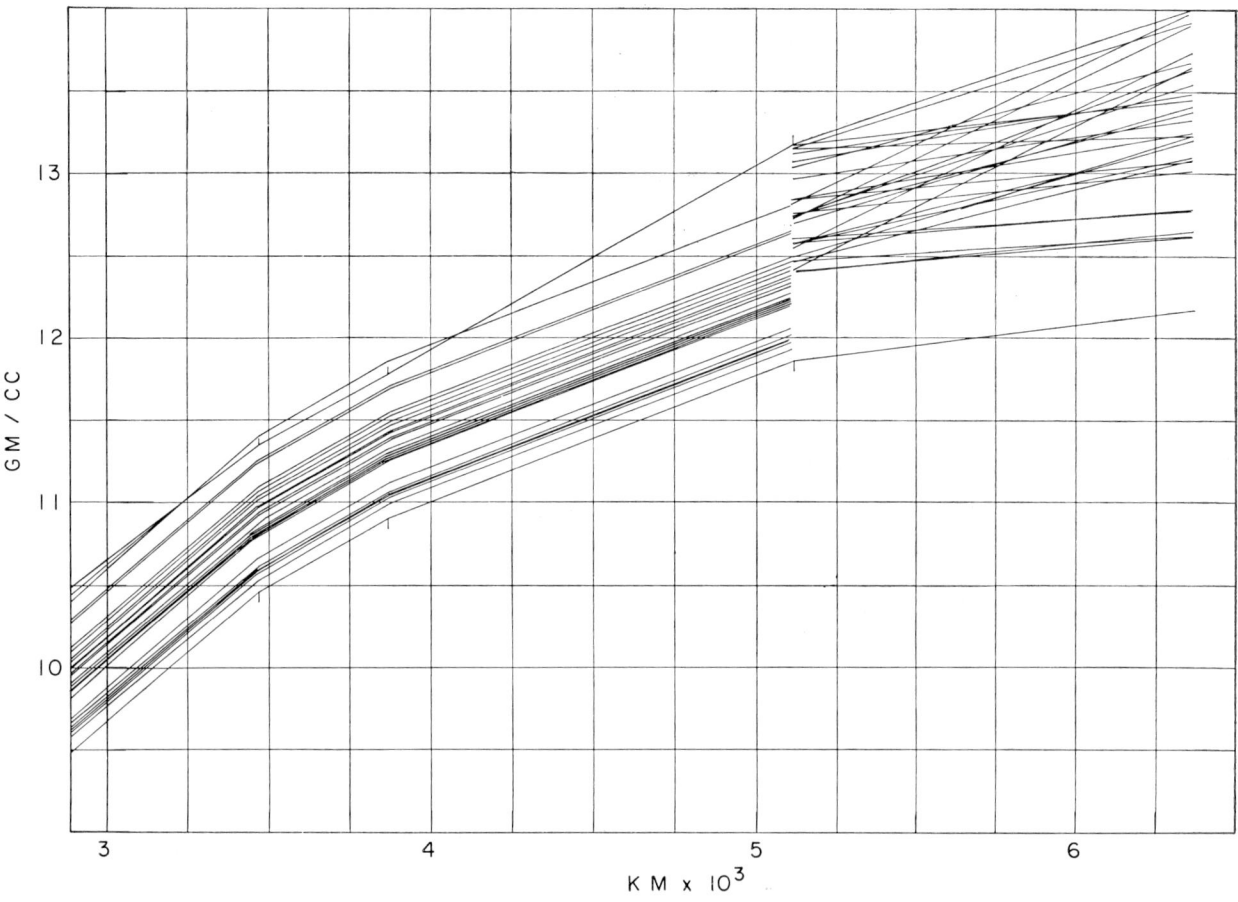

Fig. 4. Unconstrained core densities (see fig. 2, caption).

paper, the systematic, negative residuals for $_0S_2$ found in our models will be used to infer a rigidity for the inner core.

(5) Several exact calculations of eigenperiods were made to check the accuracy of the variational parameter method. The differences were small and well within the uncertainty of the data.

3. The data

Successful models were required to fit the following data:

(1) Earth mass $M = 5.976 \times 10^{27}$ g; dimensionless moment of inertia $I/Ma^2 = 0.3308$.

(2) Compressional velocity distribution in the mantle fixed very close to the models determined by JOHNSON (1969) and FAIRBORN (1969), based on $dt/d\Delta$ analyses of array data. P and PcP travel times fit the latest data (HERRIN et al., 1968) to ± 1 s. The compressional velocity distribution in the core was fixed to the recent model of HUSEBYE and TOKSÖZ (1969).

(3) Shear velocities below 800 km were restricted to lie within the narrow bounds reported by FAIRBORN (1969) who used Monte Carlo methods to interpret travel time and $dt/d\Delta$ data obtained from the Large Aperture Seismic Array (LASA). Wider bounds were used above 800 km. Travel times of S and ScS were required to fit Fairborn's data to within ± 5 s at 10 distances between 25° and 100° and a single failure was sufficient to reject a model. These travel time data primarily constrain the mantle below 800 km.

(4) Eigenperiods tested were $_0S_0$, $_0S_2$, through $_0S_{22}$, $_1S_2$, $_1S_3$, $_1S_5$, $_1S_6$, $_1S_8$, $_1S_{12}$, $_2S_4$, $_2S_6$, $_2S_{10}$; toroidal oscillations tested were $_0T_3-_0T_{21}$, $_0T_2$ not being used because of its uncertain value; models were also required to fit surface wave phase velocities for predominantly oceanic paths as follows: Rayleigh waves in

```
-6

MODEL    1    MASS= 5.97600   I= 8.02400   I/MA2= 0.33080   M.C.D.= 11.1859   IC/MA2C= 0.38854
 0.0    0.0     937.70    935.80     1.90         16.6   24.22    579.83    575.59    4.23         15.5   32.49    713.51    709.28    4.23
14.4   47.02    931.97    926.75     5.22         13.3   55.22   1045.72   1039.71    6.01         12.2   65.22   1173.89   1168.21    5.68
11.1   72.63   1260.32   1256.00     4.32         10.0   83.69   1375.71   1374.27    1.44          8.9   94.14   1474.78   1470.89    3.89
 7.8   68.09   1264.38   1260.27     4.11
0 S        2 3221.5 -7.5     4 1543.4 -3.6     6  963.7 -0.2    14  449.3  0.9    30  261.8 -0.7    50  178.7  0.5    70  134.0 -0.2
           3 2129.6 -4.4     5 1189.4  0.4     7  813.0  1.3     8  709.2  1.6     9  635.7  1.7    10  581.4  1.4    11  538.9  1.4
        * 12  504.2  2.2    13  474.7  1.6    15  427.0  0.8    16  407.3  0.8    17  389.8  0.4    18  374.1  0.7    20  347.1 -0.3
          21  335.4 -0.4    22  324.7  0.7    24  305.7 -0.0    26  289.2 -0.2    28  274.7 -0.3    32  250.2 -0.6    34  239.6 -0.4
          36  229.9 -0.1    39  216.8  0.1    42  205.0  0.3    45  194.4  0.5    55  165.2  0.6    60  153.5  0.4    65  143.1 -0.1
          75  125.8 -0.5    19  360.0 -0.6
0 T        4 1308.4  4.4     8  738.2  1.2    11  575.9  0.2    40  201.1 -0.2    80  106.5  0.0    17  409.8  1.5     3 1708.5  0.5
           5 1079.3  3.3     6  928.5  3.1     7  820.3  2.4     9  673.2  1.2    10  620.1  0.7    15  452.3 -0.0    12  538.2  1.3
          13  505.7  1.0    14  477.3  0.3    16  429.9  0.3     2 2639.3 -0.7    18  391.7  1.5    19  375.2  0.7    20  360.1 -0.5
          21  346.2 -0.9    22  333.4 -0.9    23  321.6  0.1    25  300.3  0.1    27  281.7  0.1    30  257.8 -0.1    35  225.9 -0.3
          45  181.1 -0.2    50  164.7 -0.2    60  139.4 -0.1    70  120.8 -0.0    90   95.2  0.0   105   82.1  0.0
0 R        1 1226.7 -0.9
1 S        2 1473.5  1.5     3 1067.4  3.4  *  5  734.2  5.5     6  662.2  2.2     8  559.3  3.9  * 12  393.4 -3.4
1 T
1 R
2 S   *    2  921.8 16.3  *  4  730.7  5.8     6  595.9  1.3    10  415.9  0.7
2 T
2 R
 1 6371.0  1.52  0.0   1.03      2 6368.0  1.52  0.0   1.03*     3 6368.0  6.55  3.73  2.84      4 6361.0  6.55  3.73  2.84*
 5 6361.0  8.00  4.62  3.44      6 6338.0  8.06  4.58  3.48      7 6338.0  8.06  4.58  3.48      8 6318.0  8.11  4.53  3.53
 9 6300.0  8.16  4.50  3.56     10 6275.0  8.03  4.45  3.58     11 6250.0  7.89  4.41  3.60     12 6225.0  7.76  4.36  3.62
13 6200.0  8.00  4.36  3.59     14 6175.0  8.23  4.35  3.56     15 6150.0  8.47  4.34  3.53     16 6125.0  8.51  4.36  3.54
17 6100.0  8.54  4.38  3.55     18 6075.0  8.58  4.40  3.56     19 6050.0  8.61  4.51  3.55     20 6025.0  8.65  4.61  3.54
21 6000.0  8.68  4.71  3.53     22 5975.0  9.17  5.01  3.46     23 5950.0  9.67  5.31  3.39     24 5925.0  9.71  5.34  3.48
25 5900.0  9.75  5.37  3.57     26 5875.0  9.79  5.41  3.66     27 5850.0  9.83  5.44  3.75     28 5825.0  9.88  5.47  3.84
29 5800.0  9.92  5.50  3.93     30 5775.0  9.96  5.53  4.02     31 5750.0 10.00  5.56  4.11     32 5725.0 10.24  5.67  4.20
33 5700.0 10.48  5.77  4.29     34 5675.0 10.73  5.87  4.38     35 5650.0 10.97  5.98  4.47     36 5625.0 11.01  6.02  4.49
37 5600.0 11.05  6.06  4.50     38 5550.0 11.13  6.15  4.52     39 5500.0 11.21  6.23  4.55     40 5400.0 11.37  6.30  4.58
41 5300.0 11.53  6.37  4.62     42 5200.0 11.70  6.44  4.65     43 5100.0 11.86  6.50  4.68     44 5000.0 12.02  6.57  4.72
45 4800.0 12.27  6.69  4.79     46 4600.0 12.52  6.81  4.87     47 4400.0 12.78  6.94  4.95     48 4200.0 13.03  7.06  5.03
49 4000.0 13.23  7.12  5.12     50 3800.0 13.44  7.19  5.22     51 3700.0 13.54  7.22  5.27     52 3650.0 13.59  7.24  5.29
53 3600.0 13.64  7.25  5.32     54 3575.0 13.67  7.26  5.33     55 3550.0 13.70  7.27  5.34     56 3525.0 13.72  7.28  5.36
57 3500.0 13.75  7.28  5.37     58 3476.7 13.77  7.29  5.38     59 3476.7  7.96  0.0  10.08     60 3450.0  8.01  0.0  10.12
61 3425.0  8.05  0.0  10.16     62 3400.0  8.10  0.0  10.20     63 3350.0  8.19  0.0  10.28     64 3300.0  8.28  0.0  10.36
65 3100.0  8.64  0.0  10.69     66 2900.0  9.00  0.0  11.02     67 2700.0  9.30  0.0  11.24     68 2500.0  9.60  0.0  11.45
69 2300.0  9.69  0.0  11.60     70 2100.0  9.78  0.0  11.74     71 1900.0  9.86  0.0  11.89     72 1800.0  9.91  0.0  11.96
73 1700.0  9.95  0.0  12.03     74 1600.0 10.00  0.0  12.10     75 1500.0 10.04  0.0  12.17     76 1450.0 10.06  0.0  12.21
77 1400.0 10.09  0.0  12.25     78 1350.0 10.11  0.0  12.28     79 1300.0 10.13  0.0  12.32     80 1253.0 10.15  0.0  12.35
81 1253.0 11.15  0.0  12.88     82 1100.0 11.15  0.0  12.95     83  900.0 11.16  0.0  13.05     84  700.0 11.17  0.0  13.14
85  500.0 11.17  0.0  13.24     86  300.0 11.18  0.0  13.33     87  150.0 11.19  0.0  13.40     88    0.0 11.19  0.0  13.47
 4 ALPHA REVERSALS BEFORE LAYERS      10    13    59    63
 2 BETA REVERSALS BEFORE LAYERS        6    19
 2 RHO REVERSALS BEFORE LAYERS        22    24
       5     10.   22.6       23    421.   19.3        58   2898.   24.0
       9     71.   22.6       31    621.   23.0        59   2898.   54.8
      12    146.   23.9       35    721.   23.3        66   3471.   56.0
      15    221.   21.3       39    871.   23.6        68   3871.   56.2
      18    296.   21.3       44   1371.   23.2        80   5118.   58.8
      21    371.   21.4       48   2171.   23.4        81   5118.   58.2
                                                   88   6371.   60.8
```

Fig. 5. Results for a typical model. Section A line 1: mass and moment of inertia; lines 2–4: p, Δ, theoretical times, model times and residuals for S and ScS. Section B: model eigenperiods and residuals against observed eigen periods. Section C: model printout. Section D: depths at which parameters were varied and corresponding m values.

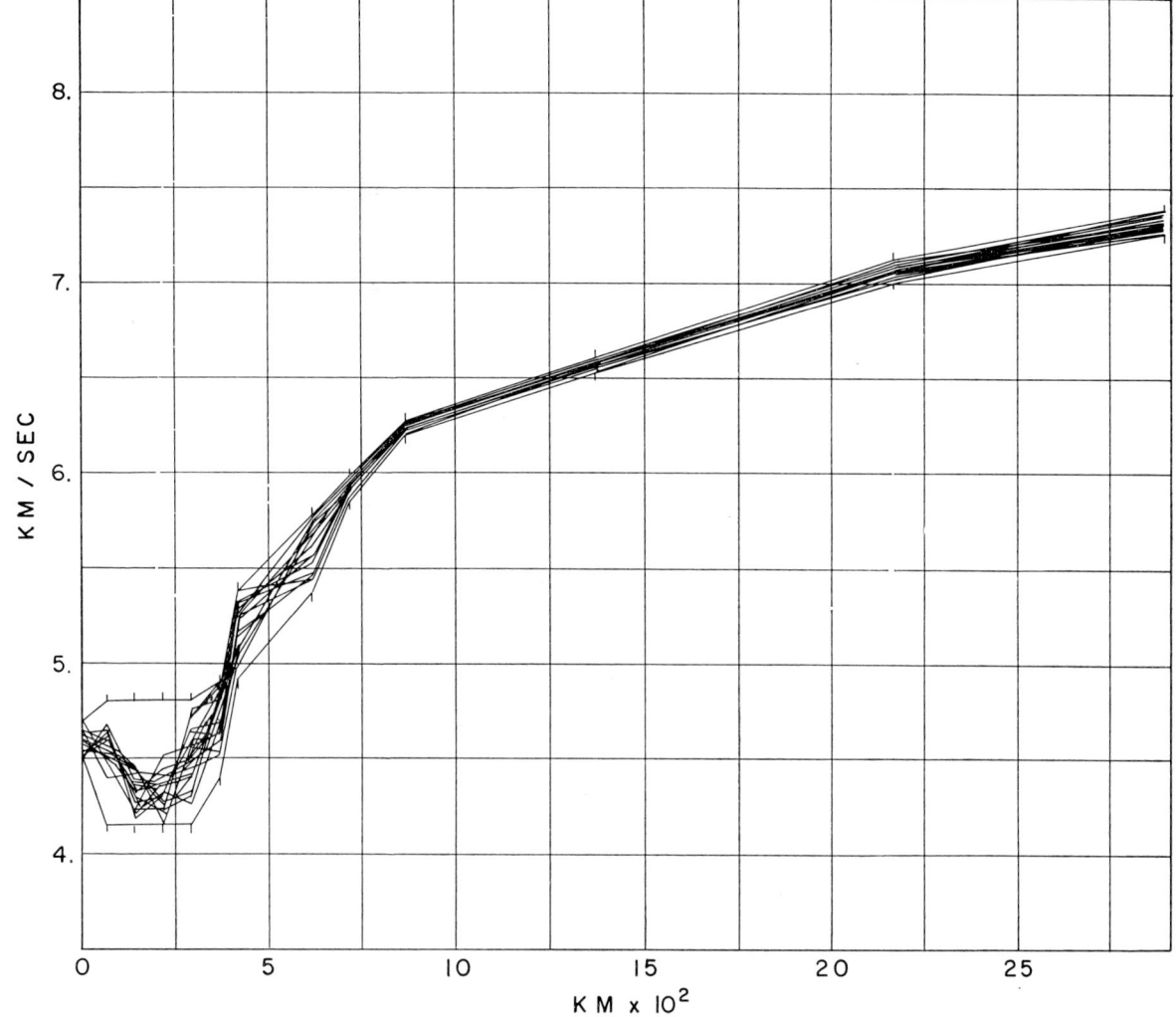

Fig. 6. Twenty-seven successful shear velocity models for the mantle. Ticks on upper and lower bounds show where parameter was randomly varied.

the period range 125–325 s (BEN-MENAHEM, 1965); Love waves in the period range 80–340 s (TOKSÖZ and ANDERSON, 1966). We used the eigenperiod data as reviewed and summarized by DERR (1969), except as follows: $_0S_2 = 3229.0$ s, $_0S_{11} = 537.5$ s. The uncertainty in the eigenperiod and dispersion data was taken to be $\pm 0.4\%$ due to asphericity, rotational splitting and experimental errors (DAHLEN, 1968). An error analysis of the oceanic surface wave data indicates that an accuracy better than 1% was achieved. Comparison with phase velocities for other oceanic paths verified this for Love and Rayleigh waves. The fit of $_0S_0$ was required to be within $\pm 0.1\%$. Actually the final models fit most of the data to about half these tolerances.

Fig. 5 shows the computed eigenperiods and the residuals for a typical model.

4. Results

The results reported here supersede our earlier conclusions (PRESS, 1968) because of the new and more extensive data set inverted in this paper. The effects of lateral variation were reduced by deriving higher mode data from oceanic surface wave phase velocities. Moreover the new procedure enabled us to find a much larger number of successful models and therefore a more representative selection from the set of successful models.

The shear velocity and density distribution are plotted

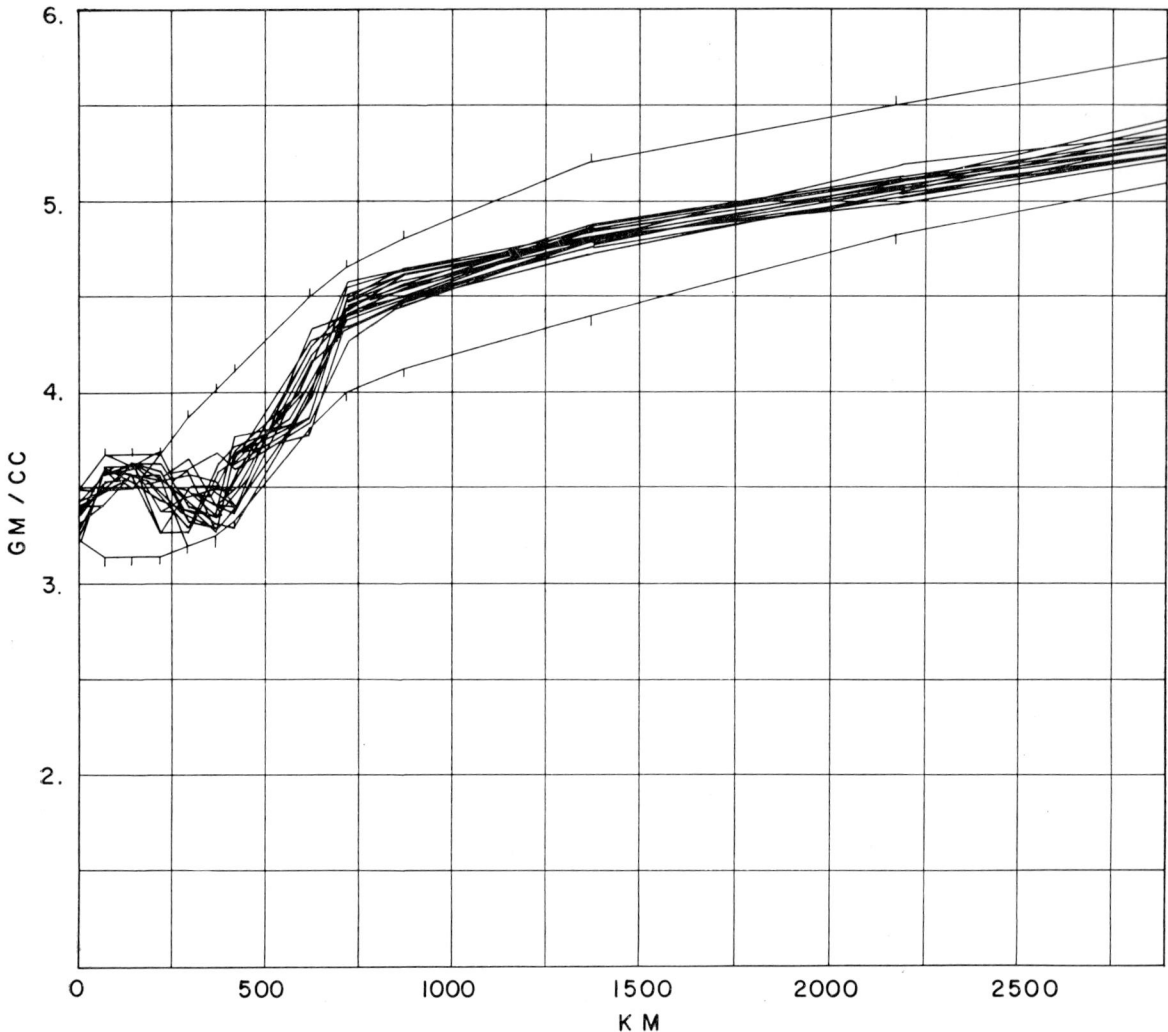

Fig. 7. Density in the mantle (see fig. 6, caption).

in figs. 6, 7 and 8 and are also tabulated (together with the fixed compressional velocity distribution) in figs. 9–13.

5. The upper mantle under oceans

Without exception every successful model contains a low velocity zone for shear waves which centers at depths between 150 and 250 km. If the lid of this zone is characterized by $\beta > 4.5$ km/s, then its thickness is 50–100 km. We failed to find a single model without a low velocity zone despite a special search in which 162 000 monotonic shear velocity models were examined. A low velocity zone seems required by our data since essentially every possible model without it was examined and eliminated. Nevertheless, HADDON and BULLEN (1969) reported a successful monotonic model, probably because: (1) they only use modes through $n = 44$, whereas our data go to $n = 105$; (2) our Love wave phase velocities tend towards lower values than the HB data (see fig. 14).

The several mechanisms which might account for the low velocity channel are reviewed by BIRCH (1970). We favor the hypothesis of partial (grain boundary) melting for the following reasons:

(1) Shear velocity and Q are sensitive to the presence of small amounts of melt along grain boundaries.

(2) Data presented at this conference by several investigators show that the temperature at which melting

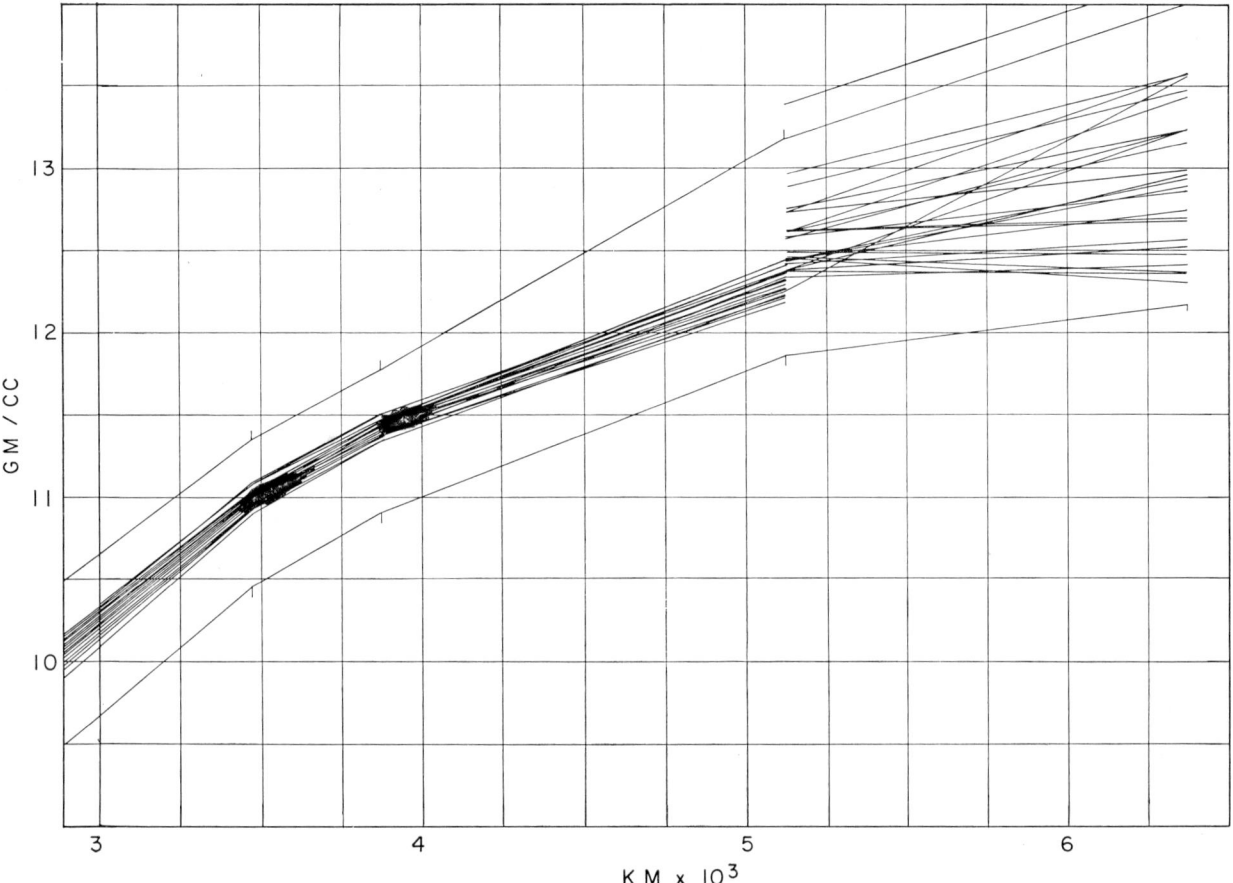

Fig. 8. Density in the core (see fig. 6, caption).

begins in the wet state for candidate upper mantle mineral assemblages is sufficiently low to be reached by the geotherm for most thermal models of the earth.

(3) The partial melting product of candidate mineral assemblages can account for basaltic vulcanism.

(4) A partially molten, low strength zone would serve to mechanically decouple the lithosphere from the underlying mantle as is required by some proposed mechanisms for the spreading sea floor.

The density values shown in figs. 7 and 15 fill the entire permissible range at the M-discontinuity indicating a lack of constraint by the geophysical data. However, the initial density gradients are all positive and in the vicinity of 100 km all the values fall in the narrow band 3.5–3.6 g/cm³ in the upper part of the permissible range. As a check on this result a special search was made without success to find models with densities be- low 3.4 g/cm³ in this depth range. For additional confirmation of this result we applied the BACKUS and GILBERT (1968) δ-ness criterion using weighting functions computed for our data by WIGGINS (1969). According to Backus and Gilbert, if the weighting functions are concentrated over narrow depth intervals, a stable local average can be obtained from a single model. Using this procedure every one of the models yielded an average density in the range 3.5–3.6 g/cm³ for the depth interval 75–125 km. Presumably the average density near 100 km is uniquely determined in the sense that any model computed from our data set should give the same value.

Unfortunately the density resolution deteriorates below 100 km as can be seen by the wider band of Monte Carlo solutions. At 300 km the resolving length inferred from the δ-ness criterion is 200 km. One might argue

10.	8.000	4.576	3.412	36.1	1.72	0.257
71.	8.160	4.534	3.491	39.2	1.91	0.277
146.	7.760	4.311	3.621	35.4	1.91	0.277
221.	8.470	4.376	3.263	46.2	2.41	0.318
296.	8.580	4.556	3.504	45.9	2.21	0.304
371.	8.680	4.527	3.503	48.0	2.34	0.313
421.	9.670	5.208	3.476	57.3	2.11	0.296
621.	10.000	5.622	4.145	57.9	1.83	0.269
721.	10.970	5.860	4.332	74.6	2.17	0.300
871.	11.210	6.263	4.488	73.4	1.87	0.273
1371.	12.020	6.597	4.869	86.5	1.99	0.284
2171.	13.030	7.029	5.107	103.9	2.10	0.295
2898.	13.770	7.375	5.264	117.1	2.15	0.299
2898.	7.960	0.0	9.956			
3471.	9.000	0.0	10.925			
3871.	9.600	0.0	11.379			
5118.	10.150	0.0	12.334			
5118.	11.150	0.0	12.723	5.70		
6371.	11.190	0.0	13.577			

10.	8.000	4.530	3.222	36.6	1.79	0.264
71.	8.160	4.499	3.575	39.6	1.96	0.282
146.	7.760	4.343	3.492	35.1	1.86	0.272
221.	8.470	4.350	3.472	46.5	2.46	0.321
296.	8.580	4.598	3.446	45.4	2.15	0.299
371.	8.680	4.620	3.408	46.9	2.20	0.302
421.	9.670	5.070	3.355	59.2	2.30	0.310
621.	10.000	5.701	4.220	56.7	1.74	0.259
721.	10.970	5.914	4.415	73.7	2.11	0.295
871.	11.210	6.258	4.605	73.4	1.88	0.274
1371.	12.020	6.521	4.775	87.8	2.06	0.291
2171.	13.030	7.085	5.098	102.9	2.05	0.290
2898.	13.770	7.340	5.293	117.8	2.19	0.302
2898.	7.960	0.0	10.031			
3471.	9.000	0.0	10.965			
3871.	9.600	0.0	11.403			
5118.	10.150	0.0	12.325			
5118.	11.150	0.0	12.374	7.43		
6371.	11.190	0.0	13.236			

10.	8.000	4.587	3.432	35.9	1.71	0.255
71.	8.160	4.504	3.464	39.5	1.95	0.281
146.	7.760	4.338	3.620	35.1	1.87	0.273
221.	8.470	4.345	3.579	46.6	2.47	0.321
296.	8.580	4.522	3.366	46.4	2.27	0.308
371.	8.680	4.683	3.335	46.1	2.10	0.295
421.	9.670	5.088	3.522	59.0	2.28	0.309
621.	10.000	5.759	4.092	55.8	1.68	0.252
721.	10.970	5.951	4.366	73.1	2.06	0.291
871.	11.210	6.231	4.487	73.9	1.90	0.276
1371.	12.020	6.559	4.862	87.1	2.03	0.288
2171.	13.030	7.052	5.054	103.5	2.08	0.293
2898.	13.770	7.371	5.242	117.2	2.16	0.299
2898.	7.960	0.0	10.150			
3471.	9.000	0.0	11.037			
3871.	9.600	0.0	11.436			
5118.	10.150	0.0	12.219			
5118.	11.150	0.0	12.459	6.17		
6371.	11.190	0.0	12.361			

10.	8.000	4.573	3.345	36.1	1.73	0.257
71.	8.160	4.540	3.611	39.1	1.90	0.276
146.	7.760	4.285	3.617	35.7	1.95	0.281
221.	8.470	4.243	3.503	47.7	2.65	0.332
296.	8.580	4.733	3.478	43.7	1.95	0.281
371.	8.680	4.822	3.348	44.3	1.91	0.277
421.	9.670	5.081	3.449	59.1	2.29	0.309
621.	10.000	5.694	4.012	56.8	1.75	0.260
721.	10.970	5.921	4.435	73.6	2.10	0.294
871.	11.210	6.195	4.563	74.5	1.94	0.280
1371.	12.020	6.600	4.802	86.4	1.98	0.284
2171.	13.030	7.005	5.047	104.4	2.13	0.297
2898.	13.770	7.373	5.388	117.1	2.15	0.299
2898.	7.960	0.0	10.087			
3471.	9.000	0.0	11.002			
3871.	9.600	0.0	11.424			
5118.	10.150	0.0	12.291			
5118.	11.150	0.0	12.443	1.64		
6371.	11.190	0.0	12.872			

10.	8.000	4.517	3.247	36.8	1.80	0.266
71.	8.160	4.612	3.514	38.2	1.80	0.265
146.	7.760	4.216	3.614	36.5	2.05	0.291
221.	8.470	4.301	3.532	47.1	2.54	0.326
296.	8.580	4.656	3.189	44.7	2.06	0.291
371.	8.680	4.689	3.237	46.0	2.09	0.294
421.	9.670	5.306	3.619	56.0	1.99	0.285
621.	10.000	5.467	4.014	60.1	2.01	0.287
721.	10.970	5.849	4.397	74.7	2.18	0.301
871.	11.210	6.235	4.555	73.8	1.90	0.276
1371.	12.020	6.590	4.864	86.6	1.99	0.285
2171.	13.030	7.050	5.086	103.5	2.08	0.293
2898.	13.770	7.318	5.422	118.2	2.21	0.303
2898.	7.960	0.0	9.979			
3471.	9.000	0.0	10.908			
3871.	9.600	0.0	11.338			
5118.	10.150	0.0	12.227			
5118.	11.150	0.0	12.244	10.33		
6371.	11.190	0.0	13.557			

10.	8.000	4.625	3.437	35.5	1.66	0.249
71.	8.160	4.647	3.466	37.8	1.75	0.260
146.	7.760	4.191	3.567	36.8	2.09	0.294
221.	8.470	4.384	3.578	46.1	2.40	0.317
296.	8.580	4.424	3.342	47.5	2.43	0.319
371.	8.680	4.860	3.474	43.8	1.86	0.272
421.	9.670	5.281	3.671	56.3	2.07	0.287
621.	10.000	5.416	3.866	60.9	2.08	0.292
721.	10.970	5.958	4.465	73.0	2.06	0.291
871.	11.210	6.212	4.571	74.2	1.92	0.278
1371.	12.020	6.548	4.792	87.3	2.04	0.289
2171.	13.030	7.101	5.070	102.5	2.03	0.289
2898.	13.770	7.349	5.283	117.6	2.18	0.301
2898.	7.960	0.0	10.005			
3471.	9.000	0.0	10.973			
3871.	9.600	0.0	11.430			
5118.	10.150	0.0	12.404			
5118.	11.150	0.0	12.613	9.71		
6371.	11.190	0.0	12.683			

Fig. 9. Tabulated parameters of successful models. Change in core radius shown in lower right corner. Each model has fixed crustal layers for depth, alpha, beta and rho as follows: 0., 1.52, 0., 1.03; 3., 6.55, 3.73, 2.84; 10, 6.55, 3.73, 2.84.

Fig. 10. See fig. 9.

Fig. 11. See fig. 9.

DEPTH	ALPHA	BETA	RHO	φ	K/μ	σ
10.	8.000	4.595	3.462	35.8	1.70	0.254
71.	8.160	4.490	3.559	39.7	1.97	0.263
146.	7.760	4.373	3.634	34.7	1.82	0.267
221.	8.470	4.748	3.370	47.7	2.44	0.332
296.	8.580	4.501	3.457	45.5	2.16	0.289
371.	8.680	4.834	3.489	44.2	1.89	0.275
421.	9.670	5.055	3.491	59.4	2.33	0.312
621.	10.000	5.734	4.168	56.2	1.71	0.255
721.	10.970	5.883	4.414	74.2	2.14	0.298
871.	11.210	6.722	4.496	44.0	1.91	0.277
1371.	12.020	6.576	4.753	86.8	2.01	0.286
2171.	13.030	7.040	5.117	103.7	2.09	0.294
2898.	13.770	7.346	5.253	117.7	2.19	0.301
2898.	7.960	0.0	10.138			
3471.	9.000	0.0	11.036			
3871.	9.600	0.0	11.446		1.07	
5118.	10.150	0.0	12.272			
5118.	11.150	0.0	12.429			
6371.	11.190	0.0	12.893			

DEPTH	ALPHA	BETA	RHO	φ	K/μ	σ
10.	8.000	4.510	3.482	36.9	1.81	0.267
71.	8.160	4.613	3.493	38.2	1.80	0.265
146.	7.760	4.329	3.610	35.2	1.88	0.274
221.	8.470	4.349	3.461	46.5	2.46	0.333
296.	8.580	4.476	3.283	46.9	2.34	0.313
371.	8.680	4.703	3.606	45.9	2.07	0.292
421.	9.670	5.080	3.714	59.1	2.29	0.309
621.	10.000	5.755	3.857	55.8	1.69	0.252
721.	10.970	5.878	4.315	74.3	2.15	0.299
871.	11.210	6.736	4.502	73.8	1.90	0.276
1371.	12.020	6.598	4.856	86.4	2.00	0.284
2171.	13.030	7.050	5.102	103.5	2.08	0.293
2898.	13.770	7.262	5.311	119.3	2.26	0.307
2898.	7.960	0.0	10.067			
3471.	9.000	0.0	10.967			
3871.	9.600	0.0	11.374		5.18	
5118.	10.150	0.0	12.182			
5118.	11.150	0.0	12.609			
6371.	11.190	0.0	13.432			

DEPTH	ALPHA	BETA	RHO	φ	K/μ	σ
10.	8.000	4.628	3.359	35.4	1.65	0.249
71.	8.160	4.388	3.620	40.9	2.12	0.287
146.	7.760	4.422	3.523	34.1	1.75	0.260
221.	8.470	4.397	3.424	46.0	2.38	0.316
296.	8.580	4.484	3.374	46.8	2.33	0.312
371.	8.680	4.603	3.338	47.1	2.22	0.304
421.	9.670	5.210	3.535	57.4	2.11	0.295
621.	10.000	5.655	4.124	57.4	1.79	0.265
721.	10.970	5.906	4.456	73.2	2.12	0.296
871.	11.210	6.193	4.583	74.5	1.94	0.280
1371.	12.020	6.587	4.786	86.6	2.00	0.285
2171.	13.030	7.039	5.066	103.7	2.09	0.294
2898.	13.770	7.367	5.198	117.3	2.17	0.300
2898.	7.960	0.0	10.097			
3471.	9.000	0.0	11.033			
3871.	9.600	0.0	11.466		3.15	
5118.	10.150	0.0	12.367			
5118.	11.150	0.0	12.382			
6371.	11.190	0.0	12.958			

DEPTH	ALPHA	BETA	RHO	φ	K/μ	σ
10.	8.000	4.502	3.359	37.0	1.82	0.268
71.	8.160	4.632	3.575	38.0	1.77	0.262
146.	7.760	4.294	3.567	35.6	1.93	0.279
221.	8.470	4.208	3.494	48.1	2.72	0.336
296.	8.580	4.761	3.325	43.4	1.91	0.278
371.	8.680	4.807	3.578	44.5	1.93	0.279
421.	9.670	4.975	3.618	60.5	2.44	0.320
621.	10.000	5.720	3.860	56.4	1.72	0.257
721.	10.970	5.905	4.448	73.8	2.12	0.296
871.	11.210	6.227	4.615	74.0	1.91	0.277
1371.	12.020	6.577	4.795	86.9	2.01	0.287
2171.	13.030	7.040	5.036	103.7	2.09	0.294
2898.	13.770	7.288	5.374	118.8	2.24	0.305
2898.	7.960	0.0	10.097			
3471.	9.000	0.0	11.036			
3871.	9.600	0.0	11.473		6.92	
5118.	10.150	0.0	12.289			
5118.	11.150	0.0	12.382			
6371.	11.190	0.0	12.523			

DEPTH	ALPHA	BETA	RHO	φ	K/μ	σ
10.	8.000	4.576	3.349	36.1	1.72	0.257
71.	8.160	4.488	3.563	39.7	1.97	0.263
146.	7.760	4.446	3.593	33.9	1.71	0.256
221.	8.470	4.238	3.370	47.8	2.66	0.333
296.	8.580	4.538	3.384	46.2	2.24	0.306
371.	8.680	4.874	3.277	43.7	1.84	0.270
421.	9.670	5.058	3.672	59.4	2.32	0.312
621.	10.000	5.654	3.938	57.4	1.79	0.265
721.	10.970	5.907	4.441	73.8	2.12	0.296
871.	11.210	6.573	4.536	74.4	1.94	0.280
1371.	12.020	6.573	4.835	86.9	2.01	0.287
2171.	13.030	7.053	5.119	103.5	2.08	0.293
2898.	13.770	7.307	5.294	118.4	2.27	0.304
2898.	7.960	0.0	10.057			
3471.	9.000	0.0	10.977			
3871.	9.600	0.0	11.399		3.62	
5118.	10.150	0.0	17.258			
5118.	11.150	0.0	12.338			
6371.	11.190	0.0	12.415			

DEPTH	ALPHA	BETA	RHO	φ	K/μ	σ
10.	8.000	4.526	3.400	36.7	1.78	0.265
71.	8.160	4.685	3.474	37.3	1.70	0.254
146.	7.760	4.313	3.493	37.0	2.12	0.296
221.	8.470	4.637	3.666	46.9	2.52	0.325
296.	8.580	4.613	3.396	44.9	2.09	0.294
371.	8.680	5.374	3.406	47.0	2.21	0.303
421.	9.670	5.440	3.390	55.0	1.90	0.277
621.	10.000	5.840	3.994	60.5	2.05	0.302
721.	10.970	6.243	4.542	74.2	2.20	0.302
871.	11.210	6.554	4.632	73.7	1.89	0.275
1371.	12.020	6.554	4.790	87.2	2.03	0.288
2171.	13.030	7.033	5.057	103.8	2.10	0.294
2898.	13.770	7.333	5.241	117.9	2.19	0.302
2898.	7.960	0.0	10.133			
3471.	9.000	0.0	11.020			
3871.	9.600	0.0	11.419		7.48	
5118.	10.150	0.0	12.765			10464
5118.	11.150	0.0	12.567			
6371.	11.190	0.0	13.220			

Fig. 12. See fig. 9.

10.	8.000	4.610	3.363	35.7	1.69	0.251
71.	8.160	4.558	3.527	38.9	1.87	0.273
146.	7.760	4.428	3.559	34.1	1.74	0.259
221.	8.470	4.151	3.580	48.8	2.83	0.342
296.	8.580	4.594	3.323	45.5	2.15	0.299
371.	8.680	4.770	3.382	45.0	1.98	0.284
421.	9.670	5.050	3.762	59.5	2.33	0.312
621.	10.000	5.740	3.849	56.1	1.70	0.254
721.	10.970	5.930	4.471	73.5	2.09	0.294
871.	11.210	6.253	4.607	73.5	1.88	0.274
1371.	12.020	6.545	4.799	87.4	2.04	0.289
2171.	13.030	7.025	5.007	104.0	2.11	0.295
2898.	13.770	7.377	5.238	117.1	2.15	0.299
2898.	7.960	0.0	10.155			
3471.	9.000	0.0	11.076			
3871.	9.600	0.0	11.501			
5118.	10.150	0.0	12.373	−3.41		-----
5118.	11.150	0.0	12.483			
6371.	11.190	0.0	12.471			
10.	8.000	4.552	3.514	36.4	1.76	0.261
71.	8.160	4.536	3.659	39.2	1.90	0.276
146.	7.760	4.332	3.598	35.2	1.88	0.274
221.	8.470	4.377	3.572	46.2	2.41	0.318
296.	8.580	4.413	3.373	47.7	2.45	0.320
371.	8.680	4.840	3.477	44.1	1.88	0.274
421.	9.670	5.318	3.684	55.8	1.97	0.283
621.	10.000	5.515	4.041	59.4	1.95	0.281
721.	10.970	5.933	4.308	73.4	2.09	0.293
871.	11.210	6.263	4.470	73.4	1.87	0.273
1371.	12.020	6.587	4.718	86.6	2.00	0.285
2171.	13.030	7.048	5.086	103.5	2.08	0.293
2898.	13.770	7.343	5.341	117.7	2.18	0.301
2898.	7.960	0.0	10.192			
3471.	9.000	0.0	11.076			
3871.	9.600	0.0	11.476			
5118.	10.150	0.0	12.261	3.33		-----
5118.	11.150	0.0	12.449			
6371.	11.190	0.0	12.300			
10.	8.000	4.601	3.289	35.8	1.69	0.253
71.	8.160	4.557	3.542	38.9	1.87	0.273
146.	7.760	4.316	3.513	35.4	1.90	0.276
221.	8.470	4.402	3.515	45.9	2.37	0.315
296.	8.580	4.451	3.583	47.2	2.38	0.316
371.	8.680	4.522	3.274	48.1	2.35	0.314
421.	9.670	5.252	3.686	56.7	2.06	0.291
621.	10.000	5.660	3.769	57.3	1.79	0.264
721.	10.970	5.891	4.484	74.1	2.13	0.297
871.	11.210	6.202	4.644	74.4	1.93	0.279
1371.	12.020	6.598	4.807	86.4	1.99	0.284
2171.	13.030	7.047	5.051	103.6	2.09	0.293
2898.	13.770	7.281	5.218	118.9	2.24	0.306
2898.	7.960	0.0	10.041			
3471.	9.000	0.0	10.996			
3871.	9.600	0.0	11.450			
5118.	10.150	0.0	12.426	4.28		-----
5118.	11.150	0.0	12.578			
6371.	11.190	0.0	12.853			

Fig. 13. See fig. 9.

on physical grounds that the lower density solutions should be favored below 150 km because of the low shear velocities. This implies a density reversal from the lithosphere to the asthenosphere (3.5–3.6 g/cm³ at 100 km to 3.3–3.5 g/cm³ at 300 km).

More complex models were found involving two low velocity or two low density zones in the upper mantle. However, these models yield the same indication of high density near 100 km.

The indicated density for the lithosphere near 100 km is so high as to narrow the range of its possible composition to an eclogitic facies. This follows if the selection is made from the current petrologic hypotheses for the constitution of the upper mantle. In fig. 15 densities computed by CLARK and RINGWOOD (1964) for a mantle composed of pyrolite (peridotite or dunite would give about the same values) and eclogite. Only the eclogite model is consistent with our results between 80 and 150 km. Either model is acceptable above this region and the pyrolite model is weakly favored near 300 km.

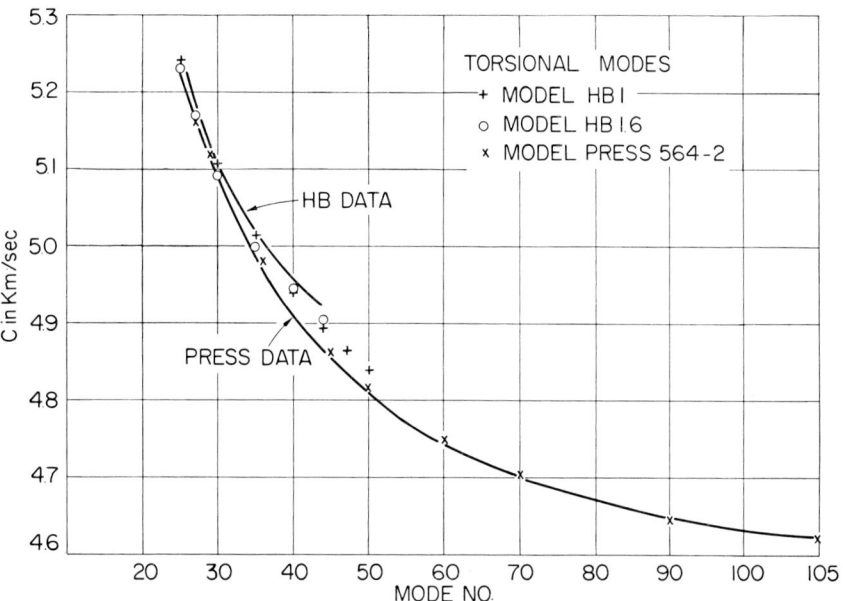

Fig. 14. Differences in Love wave phase velocity data used by Hadden and Bullen and by Press which accounts for latter's requirement of low velocity zone. Points show how models fit the data.

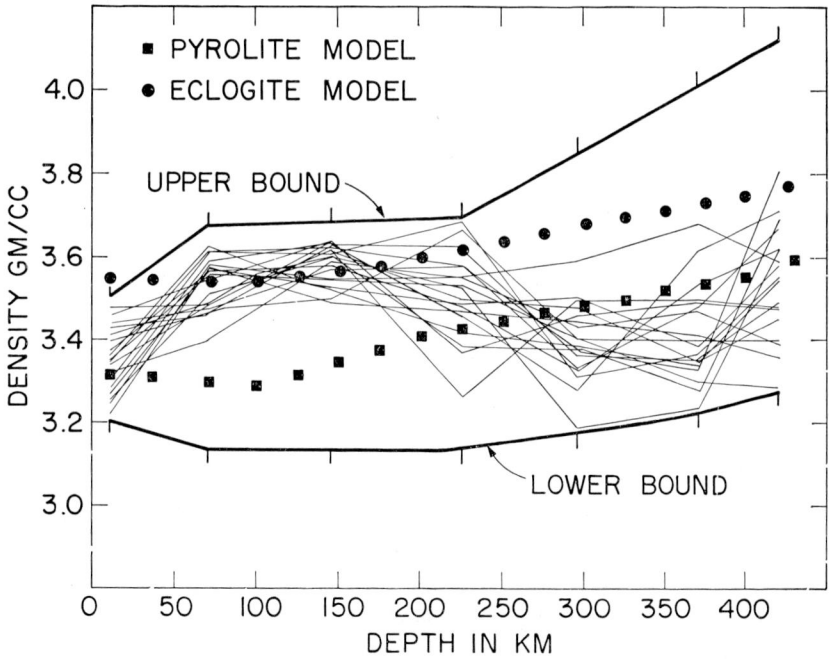

Fig. 15. Successful density models for the upper mantle plotted together with Clark and Ringwood models for pyrolite and eclogite.

A more extended discussion of these results can be found in another paper (PRESS, 1969) where a hypothesis is proposed in which eclogite fractionation from the underlying, partially molten asthenosphere is involved in the creation and spreading of the sub-oceanic, rigid, lithospheric plate. BIRCH (1970) also interpreted these results to imply an eclogitic composition.

6. The transition zone

Seismic array data have been used recently to establish rapid velocity changes near 400 and 700 km (see for example JOHNSON, 1967). These results have been incorporated in our models by fixing the compressional velocity and narrowing the range of permissible shear velocities at these depths to conform to the rapid increases, as seen in fig. 6. Although no such restrictions were placed on the density values the rapid increase in density across the transition zone is evident on all models in fig. 7. This increase is due to compression, phase changes and possibly to composition changes. Phase transitions are also inferred from the laboratory verification of the olivine–spinel phase change at pressures corresponding to depths near 400 km and by the theoretical and experimental indications for a post-spinel phase transformation. (See for example D. L. ANDERSON, 1967 or H. FUJISAWA, 1968).

The occurrence of composition changes in the transition zone are more difficult to establish. BIRCH (1961) used the velocity change $\Delta \alpha$, and the density change $\Delta \rho$ to separately estimate the effects of phase and composition change. Using $\Delta \alpha$ and $\Delta \rho$ values for each model between 333 and 871 km, allowing 0.36 g/cm^3 for compression and using Birch's values for $(\partial \alpha/\partial \rho)_m$, $(\partial \rho/\partial m)_{T,p}$, $(\partial \alpha/\partial m)_\rho$, the change in mean atomic weight Δm was computed across the transition zone for each model. Those models with reduced densities in the asthenosphere ($\rho < 3.4$ g/cm^3) showed an increase of 1–2 units in m. Thus for an asthenosphere with $m \approx 21$, and Fe/(Fe+Mg) ≈ 0.1, as would be the case for peridotite or pyrolite, the Fe/(Fe+Mg) ratio would increase to 0.2 or 0.3 across the transition zone. On the other hand, no increase in m was found for those models with a high density asthenosphere ($\rho > 3.5$ g/cm^3). If the entire upper mantle is closer to eclogite in its iron content no increase in the Fe/(Fe+Mg) ratio seems to be required across the transition zone. In a recent paper

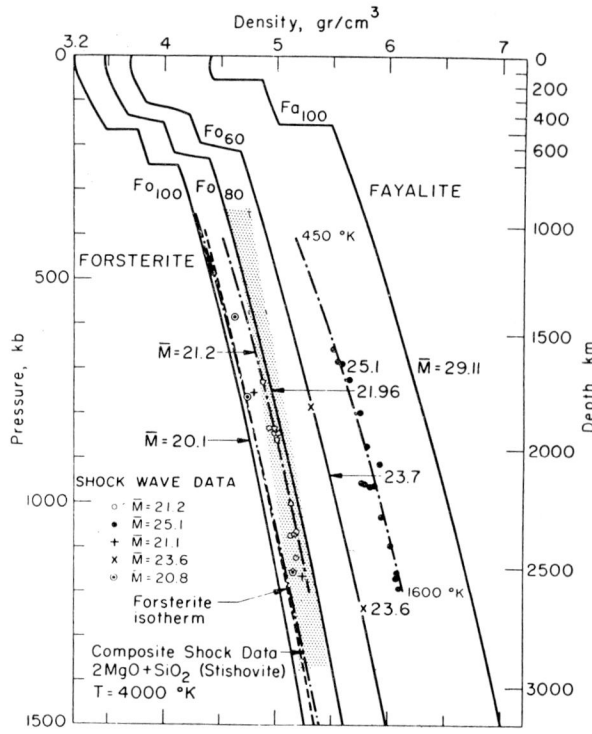

Fig. 16. D. L. Anderson's theoretical models and his summary of experimental data plotted with our density solutions shown by the shaded band.

D. L. ANDERSON (1968) proposed that $\Delta m \approx 1.5$, and BIRCH (1961) gave $\Delta m \approx 1.0$ for one model.

7. The lower mantle

Our results for the lower mantle rest heavily on Fairborn's independent determinations of a band of shear velocity distributions consistent with $dt/d\Delta$ and travel time data obtained at LASA. The range of shear velocities permitted by Fairborn's results is quite narrow as can be seen in fig. 6. This enables us to use eigenperiod data to constrain the density in the lower mantle to a greater degree than was possible before. Fairborn's shear velocity envelope shows a higher gradient than has usually been assumed (e.g. when compared to the Gutenberg model) and this requires a compensatory reduction in the density gradient in order to fit the spheroidal eigenperiod data. The results are shown in figs. 6 and 7. The density is constrained surprisingly well, to within about 0.2 g/cm^3 for most of the lower mantle. The density gradient is less than the adiabatic

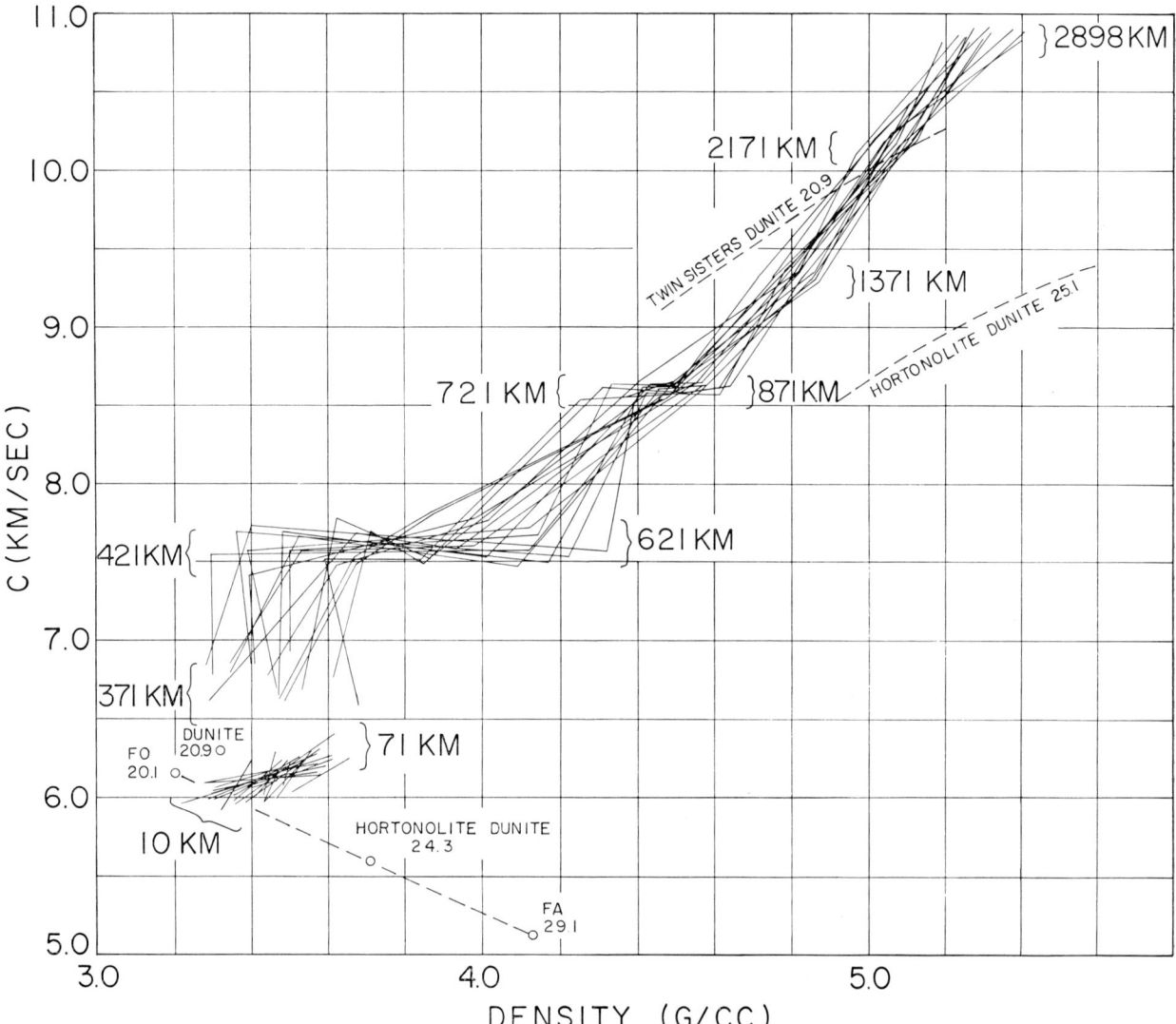

Fig. 17. Bulk sound velocity–density plot for successful models of the mantle together with static and shock wave data for dunite and forsterite–fayalite.

gradient as can be seen by comparison with the lower bound which approximates an adiabatic gradient. Fig. 16 shows our band of solutions plotted against D. L. ANDERSON's (1968) theoretical calculations for density of the solid solution series forsterite–fayalite and his summary of shock wave data. The band of density solutions is discordant with respect to profiles of constant composition, suggesting a change of mean atomic weight from 22–23 at the top of the lower mantle to 20–22 at the bottom of the mantle. This implies a depletion of iron with depth with the Fe/(Fe+Mg) ratio going from 0.2–0.3 to 0.1–0.2. Although superadiabatic temperature gradients might also account for the smaller density gradient, the augmented shear velocity gradient argues against this.

This can also be seen in fig. 17 where the bulk sound velocity and density values for each model are plotted. The figure also shows the shock wave values for Twin Sisters dunite ($m = 20.9$) and hortonolite dunite ($m = 25.1$), as reduced by AHRENS et al. (1969). Although the data are scanty, a reduction in mean atomic weight from 22–23 at 871 km to 20–21 at 2898 km is indicated.

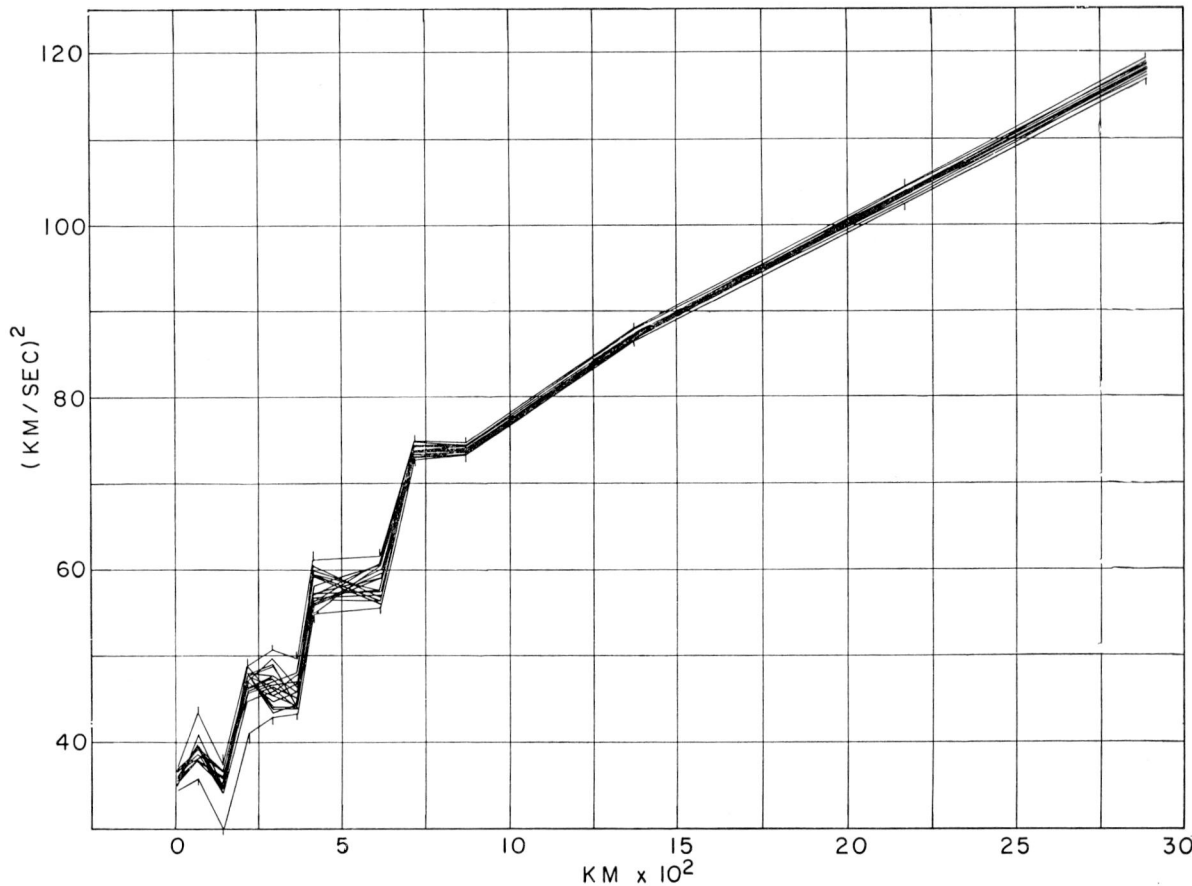

Fig. 18. Seismic parameter ϕ for the mantle obtained from successful models.

The discordance with lines of constant composition seems too large to be accounted for by superadiabatic temperature gradients. WANG (1969) also suggested that these data might indicate decreasing m in the lower mantle. If subsequent studies do not establish Fairborn's shear velocity distribution as a world wide phenomenon, this conclusion will have to be changed.

8. The mantle

The c–ρ graph illustrates the main features of the mantle discussed earlier. The olivine–spinel phase transformation is evidenced between 371 km and 421 km by models with increasing c and ρ. Between 421 km and 621 km models with large increases in ρ and with little change in c could be interpreted as the result of compression and increasing iron content, the two effects having the same sign for ρ and opposite signs for c. The increase in c and ρ between 621 km and 721 km implies a phase change as the major feature. Decomposition of the ferro–magnesium–aluminium silicate to close packed simple oxides, or transformations to structures such as ilmenite or perovskite have been suggested for this region (BIRCH, 1952, D. L. ANDERSON, 1967, RINGWOOD, 1970).

The distributions for ϕ and k/μ in the mantle are given in figs. 18 and 19.

9. The core

Results for the core are shown in figs. 8 and 20. The assumption of adiabaticity in the fluid core and the constraints imposed by the data prescribe the densities to the surprisingly small range of 0.25 g/cm³. The δ-ness criterion also indicates high resolving power for density at the top of the core. Using the shock wave data of MCQUEEN and MARSH (1966) and BALCHAN and COWAN (1966) we see that iron alloyed with a miscible, abun-

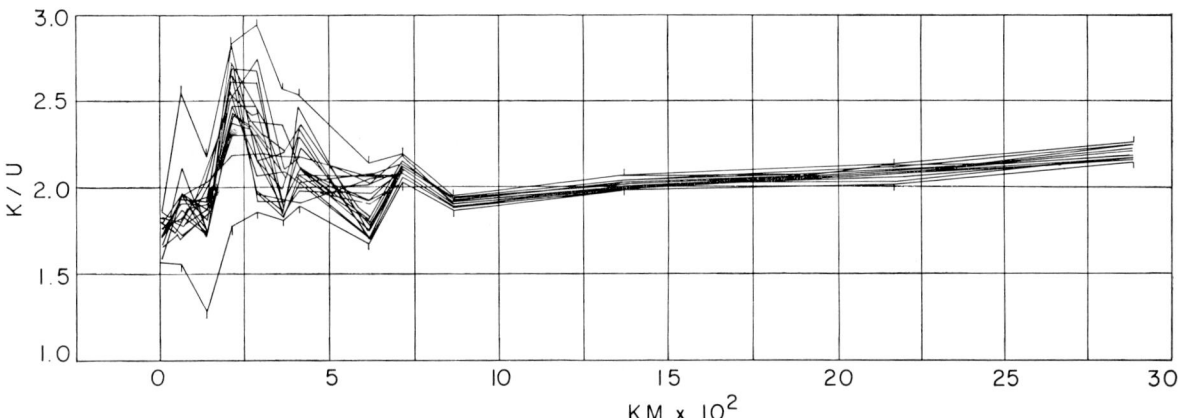

Fig. 19. Ratio k/μ for the mantle obtained from successful models.

dant element such as silicon (15 wt%) would account for the core densities. There is no control on the density in the inner core with the data used here. Changes in the core radius ranged from -3 km to $+10$ km.

The requirement for finite rigidity of the inner core was evidenced in an interesting way. Our procedure neglects core rigidity and the only mode affected by this assumption, $_0S_2$, showed negative residuals for every model. Using ALSOP's (1963) correction for rigidity resulted in a reduction of the residuals for $_0S_2$, the largest discrepancy being 0.15%. This agrees with D. L. ANDERSON's conclusions concerning rigidity in the inner core (1970).

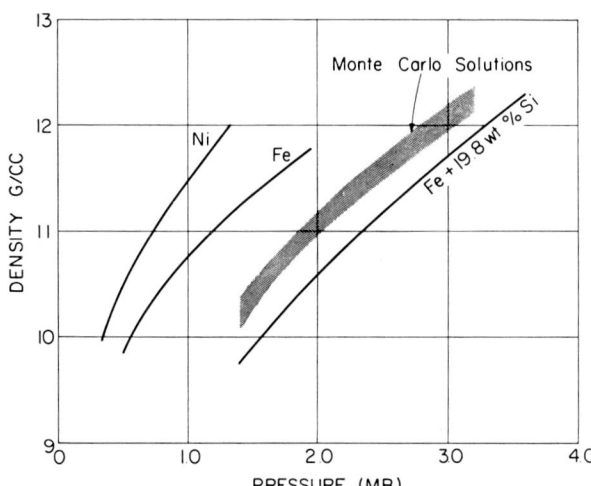

Fig. 20. Band of core densities from successful models together with shock wave density data for Fe, Ni and Fe+19.8 wt.% Si.

10. Discussion

The question of uniqueness arises in all discussions of internal earth structure. The Backus–Gilbert δ-ness criterion demonstrates how stable local averages can be computed for α, β and ρ using eigenperiod data. However, with currently available data the resolving power is adequate at too few places in the earth. Also the procedure does not yet allow for travel time or $dt/d\Delta$ data which under certain circumstances have high resolving power, nor does it consider errors in data. The band of solutions provided by the Monte Carlo method, if sufficiently narrow and if derived without bias from a large and representative selection of models, can under certain circumstances lead to meaningful conclusions. Unfortunately one is never quite sure that continued search would not reveal models significantly different from those already found, vitiating the conclusions. The use of physical arguments, laboratory experiments, and theoretical-empirical equations of state will eventually provide powerful constraints. However, the poor state of knowledge of the behavior of materials at internal earth pressures and temperatures, though improving rapidly, is a severe, current limitation.

With regard to the present paper and the other studies cited, we believe the following results are firm if the assumptions and data are correct:

(1) The low velocity zone for shear waves in the suboceanic upper mantle. The parameterization included sufficiently few elements so that all possible models

without a low velocity zone could be tested and eliminated.

(2) The high density for the lithosphere near 100 km. The δ-ness criterion has high resolving power for density at this depth and the narrow spread of Monte Carlo solutions indicate that errors in surface wave data of about 1% do not weaken the constraint. As mentioned earlier, we believe this accuracy was achieved.

(3) The rapid velocity increase near 400 km and its association with the olivine–spinel transformation: directly obtainable from $dt/d\Delta$ data (with minor depth uncertainty); the phase transformation was experimentally verified in the laboratory and olivine is almost certainly a major constituent of the upper mantle.

(4) The rapid velocity increase near 700 km: directly obtainable from $dt/d\Delta$ data (with some depth uncertainty).

(5) The density at the top of the core is between 9.9 and 10.2 g/cm^3. The δ-ness criterion shows high resolving power and the spread of Monte Carlo solutions is small.

11. Implications for mass transport in the mantle

The vertical heterogeneity of the mantle below the transition zone indicated by the superadiabatic density gradient is destabilizing and would facilitate convection. If the transition zone is a region of phase change and increasing mean atomic weight (as our preferred solutions imply), then it could inhibit penetration of deep mantle convection currents to the upper mantle. The dense lithosphere is also destabilizing and would foster gravitational instability mechanisms for sea floor spreading and continental drift.

Acknowledgements

Joel Karnofsky was an invaluable research assistant in this project. R. A. Wiggins kindly made his δ-ness weighting functions available in advance of publication. D. L. Anderson, F. Birch, C. Wang, J. Fairborn, K. E. Bullen, J. Derr, M. N. Toksöz made manuscripts available in advance of publication. A grant from the Australian National University made it possible for me to participate in this Symposium. The technique was developed as part of an inversion procedure for lunar data with support from the National Aeronautics and Space Administration under grant NGR 22-009-123. The application to the earth was supported by the Advanced Research Projects Agency and monitored by the Air Force Office of Scientific Research under contract AF 49(638)-1632.

References

AHRENS, T. J., D. L. ANDERSON and A. E. RINGWOOD (1969) in press.
ALSOP, L. E. (1963) Bull. Seismol. Soc. Am. **53**, 503.
ANDERSON, D. L. (1967) Science **157**, 1165.
ANDERSON, D. L. (1968) Earth Planet. Sci. Letters **5**, 89.
ANDERSON, D. L. (1970) Phys. Earth Planet. Interiors **3**, 41.
BACKUS, G. and F. GILBERT (1968) Geophys. J. **16**, 169.
BALCHAN, A. S. and G. R. COWAN (1966) J. Geophys. Res. **71**, 3577.
BEN-MENAHEM, A. (1965) J. Geophys. Res. **70**, 4641.
BIRCH, F. (1952) J. Geophys. Res. **57**, 227.
BIRCH, F. (1961) Geophys. J. **4**, 295.
BIRCH, F. (1970) Phys. Earth Planet. Interiors **3**, 178.
BULLEN, K. E. (1961) Geophys. J. **4**, 93.
CLARK, S. P. and A. E. RINGWOOD (1964) Rev. Geophys. **2**, 35.
DAHLEN, F. A. (1968) Geophys. J. **16**, 329.
DERR, J. (1969) Bull. Seismol. Soc. Am., in press.
FAIRBORN, J. W. (1969) Bull. Seismol. Soc. Am., in press.
FUJISAWA, H. (1968) J. Geophys. Res. **73**, 3281.
HADDON, R. A. W. and K. E. BULLEN (1969) in press.
HERRIN, E., W. TUCKER, J. TAGGART, D. W. GORDON and J. L. LOBDELL (1968) Bull. Seismol. Soc. Am. **58**, 1273.
HUSEBYE, E. S. and M. N. TOKSÖZ (1969) J. Geophys. Res., in press.
JOHNSON, L. R. (1967) J. Geophys. Res. **72**, 6309.
JOHNSON, L. R. (1969) in press.
MCQUEEN, R. G. and S. P. MARSH (1966) J. Geophys. Res. **71**, 1751.
PRESS, F. (1968a) J. Geophys. Res. **73**, 5223.
PRESS, F. (1968b) Science **160**, 1218.
PRESS, F. (1969) Science **165**, 174.
RINGWOOD, A. E. (1970), Phys. Earth Planet. Interiors **3**, 109.
TOKSÖZ, M. N. and D. L. ANDERSON (1966) J. Geophys. Res. **71**, 1649.
WANG, C. (1969) J. Geophys. Res., in press.
WIGGINS, R. A. (1968) Phys. Earth Planet. Interiors **1**, 201.
WIGGINS, R. A. (1969) in preparation.

THE COMPOSITION OF THE LOWER MANTLE*

DON L. ANDERSON and T. JORDAN

Seismological Laboratory, California Institute of Technology, Pasadena, California, U.S.A.

The equation of state of the lower mantle is found by requiring that both the density and the seismic parameter $\Phi = (\partial P/\partial \rho)_s = V^2 - (4/3)V_s^2$ be satisfied. The zero-pressure density and Φ of the lower mantle can then be found and the mean atomic weight can be inferred. For most published density models for the Earth the FeO content of the lower mantle is greater than the upper mantle. This indicates that the mantle is not homogeneous in composition as is usually assumed.

1. Introduction

The region of the lower mantle between about 1000 and 2700 km, Bullen's region D', seems to be relatively uniform. Recent solutions show some minor structure in this region but departures from a smooth curve amount to less than 1%. BIRCH (1952), CLARK and RINGWOOD (1964) and others have proposed that below 1000 km the silicates have collapsed to a close packed oxide structure. It is usually assumed that the rapid change in properties in the C-region of the mantle can be fully accounted for by solid-solid phase changes and that the composition of the lower mantle is the same as the upper mantle.

In this paper we determine an equation of state for the lower mantle which is consistent with both the density and the seismic parameter Φ. In general, it is possible to do this by assuming that the temperature gradient in the lower mantle is adiabatic. We take eight recently proposed density models and three models for the elastic ratio, Φ, for the lower mantle and with the use of the finite-strain and seismic equations of state we have determined ρ_0, Φ_0 and the mean atomic weight, $\overline{M}$, implied by these models. For most models the mean atomic weight of the lower mantle is higher than is appropriate for the composition which has been inferred for the upper mantle from petrological considerations.

* Contribution 1623, Division of Geological Sciences, California Institute of Technology, Pasadena, California 91109, U.S.A.

2. Discussion of the density models

The eight density models considered in this paper represent a variety of assumptions concerning the lower mantle. Briefly these assumptions may be outlined as follows:

a) Birch's Models I and II (BIRCH, 1964) were constructed by using Birch's empirical relationship $\rho = a + bV_p$ (a and b are constants) between density, ρ, and compressional velocity V_p, to integrate through regions B and C, and by using the Adams-Williamson equation to obtain densities in the lower mantle and core. Jeffrey's velocities were used in the Adams-Williamson integration. The two models differ in the choice of starting density for the upper mantle.

b) The Pyrolite and Eclogite Models (CLARK and RINGWOOD, 1964) were constructed assuming specific compositions for the upper mantle. These compositions were based on mineralogical considerations of supposedly mantle derived material and utilize Gutenberg's velocities in the Adams-Williamson integration.

c) Problem 1 of MCQUEEN et al. (1964) was obtained by fitting a shock wave equation of state to Jeffrey's seismic velocities.

d) Model BH_1 (BULLEN and HADDON, 1967a, b) was constructed using the method of variation of parameters on Bullen's Model A" to obtain a density distribution which agreed with the observed periods of the low-order modes of the free oscillation of the Earth. The core radius used in this model is now believed to be unacceptably large.

e) Model G1 (GILBERT and BACKUS, 1968) was derived from free oscillation data. It was not tested against travel time data.

f) CIT Model 200204 (ANDERSON and SMITH, 1968) is a tentative model incorporating the P-wave velocities of JOHNSON (1967); the shear velocity and density distributions were adjusted in order to satisfy observed free oscillation data.

The Birch and Clark-Ringwood models assumed an adiabatic temperature gradient in the lower mantle. The McQueen model is for a particular assumed temperature gradient which is slightly superadiabatic. The other models required no assumptions about the temperatures in the Earth.

The only model inconsistent with the equation of state analysis was G1. The density structure in region D for this model implies departure from the assumption of a homogeneous, self-compressed solid. For each of the other models (correcting McQueen, Problem 1 for a superadiabatic gradient) good agreement with observed seismic velocities could be obtained.

3. Equation of state fit to pressure-density data

The maximum pressure in the lower mantle is about 1.4 megabars and the maximum linear strain is about 12 % (CLARK and RINGWOOD, 1964). BIRCH (1952) has formulated a theory based on finite strain considerations which has proved successful under these conditions. MURNAGHAN (1937) showed that, if f is the negative strain, then

$$f = (\tfrac{1}{2})[(\rho/\rho_0)^{\tfrac{2}{3}} - 1],$$

where ρ_0 is the zero pressure density. To third order Birch's results are

$$P = 3K_0 f(1+2f)^{\tfrac{5}{2}}(1-2\xi f),$$

$$K = K_0(1+2f)^{\tfrac{5}{2}}[1+7f-2\xi f(2+9f)],$$

where K_0 is the zero pressure bulk modulus and ξ is a dimensionless constant equal to $(3/4)(4-K_0')$, where K_0' is the pressure derivative of K at $P = 0$. Substituting, we obtain

$$P = (\tfrac{3}{2})K_0\{(\rho/\rho_0)^{\tfrac{7}{3}} - (\rho/\rho_0)^{\tfrac{5}{3}}\}\{1 - \xi[(\rho/\rho_0)^{\tfrac{2}{3}} - 1]\},$$

which is called the Birch-Murnaghan equation of state. This equation was originally derived by Birch as an isothermal equation of state, but the substitution of adiabatic constants, K_{0s} and ξ_s, transform the equation into an adiabatic equation of the same form (CLARK, 1962).

In this paper it is assumed that the temperature gradient approximates the adiabat in the lower mantle. A direct least-squares fit of the pressure-density data to Birch's third-order equation is performed and the corresponding adiabatic constants are found. Two regions of the lower mantle were investigated. The first, designated region 1, covered the depth range from about 1000 km to 2800 km depth. The second, designated region 2, included a narrower range of data varying in depth from about 1000 km to 2200 km. The exact ranges used for each model are given in table 1. The effect of varying this range is minor and will be discussed in a later section.

TABLE 1

Range of R/R_M used in equation of state fit to model data

Model	Region 1	Region 2
Birch I	0.840–0.560	0.840–0.660
Birch II	0.840–0.560	0.840–0.660
McQueen 1	0.840–0.600	0.840–0.660
Bullen–Haddon 1	0.840–0.560	0.840–0.660
G 1	0.840–0.560	No fit attempted
Pyrolite	0.847–0.545	0.847–0.658
Eclogite	0.847–0.545	0.847–0.658
CIT 200204	0.842–0.558	0.842–0.653

$R_M = 6338$ km.

In order to fit the pressure-density data to the equation of state it is necessary to assume an initial density. Since this value is unknown we obtained fits for a range of initial densities. The choice of the initial density had little effect upon the goodness of the fit, as illustrated in table 5, which gives K_0, ξ and the standard deviations for a range of ρ_0 for Birch model I. For this model a ρ_0 of 3.90 gm/cm^3 gives a fit that is good to 0.08 %. With the exception of model G1, all fitted curves deviated from the assumed density distribution by less than 0.025 (relative standard deviation). Model G1 was found to be incompatable with a Birch-Murnaghan equation of state fit, and had relative deviations nowhere less than 14 %.

The zero-pressure density was estimated by requiring

TABLE 2

Seismic parameter in the mantle: Jeffreys*

R/R_M	Depth (km)	$\Phi = V_p^2 - (4/3)V_s^2$ (km/sec)^2
1.00	33	34.80
0.98	160	38.63
0.96	287	42.96
0.94	413	47.63
0.92	540	58.42
0.90	667	65.67
0.88	794	72.75
0.86	920	74.63
0.84	1047	77.72
0.82	1174	80.15
0.80	1301	82.97
0.78	1427	85.99
0.76	1554	88.88
0.74	1681	91.91
0.72	1808	95.00
0.70	1934	98.19
0.68	2061	101.18
0.66	2188	103.96
0.64	2315	106.46
0.62	2441	109.51
0.60	2568	112.07
0.58	2695	114.72
0.56	2822	114.72
0.55	2885	114.92

* Adapted from the tabulation of PRESS (1966), $R_M = 6338$ km.

TABLE 3

Seismic parameter in the mantle: Gutenberg*

R/R_M	Depth (km)	$\Phi = V_p^2 - (4/3)V_s^2$ (km/sec)^2
0.996	60	38.21
0.989	100	38.19
0.982	150	36.39
0.974	200	38.99
0.958	300	44.04
0.942	400	48.33
0.926	500	54.71
0.911	600	60.20
0.895	700	63.84
0.879	800	68.38
0.863	900	74.77**
0.847	1000	76.20
0.816	1200	82.91
0.784	1400	87.12
0.753	1600	90.54
0.721	1800	94.94
0.690	2000	99.44
0.658	2200	103.67
0.627	2400	107.03
0.595	2600	111.78
0.563	2800	117.61
0.548	2900	118.57
0.545	2920	117.20

* Adapted from the tabulation of PRESS (1966), $R_M = 6338$ km.
** Discontinuity in slope.

TABLE 4

Seismic parameter in the mantle: Lane R. JOHNSON (1968) and ANDERSON and JULIAN (1969)

R/R_M	Depth (km)	$\Phi = V_p^2 - (4/3)V_s^2$ (km/sec)^2
0.974	200	40.5
0.958	300	45.3
0.926	500	55.6
0.911	600	63.7
0.879	800	70.7
0.863	900	74.1
0.847	1000	77.3
0.816	1200	83.0
0.784	1400	87.3
0.753	1600	91.7
0.721	1800	95.4
0.690	2000	100.0
0.658	2200	105.0

that the seismic velocity data also be satisfied. By differentiating the Birch-Murnaghan equation an expression can be obtained for the seismic parameter Φ as a function of ρ/ρ_0. We required that the computed elastic ratio agree with the value obtained directly from seismic observations. For this criteria to be valid the observed values of the elastic ratio must correspond to an adiabatic temperature gradient.

Three velocity distributions were used. The two standard distributions, those of Jeffrey and Gutenberg, were obtained from the tabulation of PRESS (1966). A third distribution is based on the P-wave velocities of JOHNSON (1967) and S-wave velocities obtained by ANDERSON and JULIAN (1969) from data of IBRAHIM and NUTTLI (1968). This distribution was used only with data from region 2. These velocities are tabulated in tables 2, 3, and 4.

In tables 6 through 9 the calculated seismic parameter Φ for a range of initial densities is compared with the above three Φ distributions for the lower mantle. Although comparisons have been made for all possible combinations of density and velocity models, internal consistency requires that the models of Birch, Clark and Ringwood and McQueen et al. be compared with the velocity models from which they were derived. However, the spread of inferred initial densities for a given density model due to differences in the Φ models

TABLE 5

Equation of state parameters and deviation of fit from pressure data for Birch model I

ρ_0 (gm/cm^3)	Region 1				Region 2			
	K_0 (mb)	ξ	std dev (kb)	rel dev	K_0 (mb)	ξ	std dev (kb)	rel dev
3.88	1.901	−0.040	1.16	0.0018	1.928	−0.088	0.59	0.0011
3.89	1.949	0.010	0.87	0.0012	1.940	−0.019	0.51	0.0008
3.90	1.999	0.059	0.67	0.0008	1.995	0.049	0.53	0.0008
3.91	2.049	0.109	0.68	0.0010	2.052	0.118	0.66	0.0011
3.92	2.101	0.159	0.93	0.0016	2.110	0.186	0.85	0.0014
3.93	2.153	0.209	1.30	0.0024	2.170	0.255	1.08	0.0019

TABLE 6

Deviation of observed phi from phi computed from fit to Earth models for various initial densities

ρ_0 (gm/cm^3)	Region 1				Region 2					
	Jeffrey		Gutenberg		Jeffrey		Gutenberg		LRJ–BRJ	
	std (km/sec)^2	rel	std (km/sec)^2	rel	std (km/sec)^2	rel	std (km/sec)^2	rel	std (km/sec)^2	rel
Birch Model 1:										
3.88	0.670	0.008	1.538	0.017	0.637	0.008	1.502	0.018	1.764	0.021
3.89	0.450	0.005	1.285	0.014	0.418	0.005	1.283	0.014	1.596	0.019
3.90	0.310	0.004	1.027	0.012	0.306	0.004	1.082	0.013	1.455	0.017
3.91	0.382	0.004	0.773	0.009	0.427	0.005	0.920	0.011	1.356	0.015
3.92	0.609	0.007	0.545	0.006	0.681	0.007	0.835	0.010	1.319	0.014
3.93	0.888	0.010	0.413	0.005	0.982	0.011	0.863	0.010	1.356	0.014
3.94	1.194	0.013	0.492	0.006	1.308	0.014	1.009	0.011	1.471	0.015
Birch model 2:										
3.95	0.921	0.009	1.711	0.018	1.531	0.016	1.753	0.018	1.550	0.017
3.96	0.805	0.008	1.501	0.015	1.386	0.015	1.507	0.016	1.283	0.014
3.97	0.770	0.008	1.306	0.013	1.280	0.014	1.263	0.013	1.008	0.011
3.98	0.837	0.009	1.140	0.011	1.232	0.014	1.031	0.011	0.730	0.008
3.99	0.994	0.011	1.029	0.010	1.259	0.014	0.836	0.009	0.465	0.005
4.00	1.214	0.014	0.999	0.010	1.364	0.016	0.723	0.008	0.305	0.003
4.01	1.478	0.018	1.067	0.012	1.541	0.018	0.753	0.009	0.434	0.004
4.02	1.773	0.021	1.227	0.014	1.778	0.021	0.927	0.012	0.730	0.008

is usually less than 0.03 gm/cm^3. The actual zero pressure density of the lower mantle is taken to be that which gives the best fit to the observed Φ of the lower mantle.

The "best" ρ_0 and Φ_0 for each density model and the Birch–Murnaghan parameter ξ are tabulated in tables 10, 11 and 12 for the three velocity distributions.

Differences between fits to region 1 and fits to region 2 are small and reflect the smooth behavior throughout the entire lower mantle of the density distributions used. In region 2, from approximately 1000 km to 2200 km depth, the compressional wave velocities of JOHNSON (CIT 204, 1967) when combined with the shear wave velocities of ANDERSON and JULIAN (1969) give a new Φ distribution (JAJ) which is in close agreement with that of Gutenberg. The models considered best fits to these curves give zero-pressure densities 0.01 to 0.03 gm/cm^3 higher than the best fits to Jeffrey's Φ distribution.

The computed initial densities ranged from 3.90 gm/cm^3 for the eclogite model to 4.13 gm/cm^3 for CIT model 200204. For the pyrolite and eclogite models, CLARK and RINGWOOD (1964) performed an equation of state extrapolation to zero-pressure using Birch's third order equation and derived the same initial densities as this study. BIRCH (1964) used his second order equation on his two models and calculated zero-pressure densities of 3.90 gm/cm^3 for model I and 3.96 gm/

TABLE 7

Deviation of observed phi from phi computed from fit to Earth models for various initial densities

ρ_0 (gm/cm³)	Region 1					Region 2						
	Jeffrey		Gutenberg		Jeffrey		Gutenberg		LRJ–BRJ	
	std (km/sec)²	rel	std (km/sec)²	rel	std (km/sec)²	rel	std (km/sec)²	rel	std (km/sec)²	rel
Pyrolite:										
4.06	0.835	0.010	1.713	0.019	1.241	0.014	1.760	0.020	1.705	0.020
4.07	0.568	0.007	1.452	0.016	0.978	0.010	1.474	0.016	1.435	0.017
4.08	0.334	0.004	1.196	0.013	0.745	0.008	1.191	0.013	1.174	0.014
4.09	0.285	0.003	0.958	0.010	0.601	0.007	0.925	0.010	0.946	0.011
4.10	0.504	0.006	0.767	0.008	0.638	0.008	0.716	0.008	0.796	0.009
4.11	0.806	0.010	0.683	0.008	0.851	0.010	0.651	0.008	0.793	0.008
4.12	1.139	0.013	0.759	0.009	1.161	0.014	0.790	0.010	0.956	0.010
4.13	1.489	0.018	0.970	0.012	1.524	0.018	1.073	0.013	1.237	0.013
Eclogite:										
3.87	0.487	0.006	1.332	0.015	0.672	0.008	1.263	0.014	1.345	0.016
3.88	0.303	0.004	1.083	0.012	0.471	0.005	1.010	0.012	1.144	0.014
3.89	0.351	0.004	0.855	0.010	0.441	0.005	0.795	0.009	0.999	0.012
3.90	0.593	0.007	0.684	0.008	0.631	0.007	0.678	0.008	0.952	0.010
3.91	—	—	—	—	0.930	0.011	0.729	0.009	1.030	0.011
3.92	1.219	0.014	0.754	0.010	1.277	0.015	0.940	0.011	1.224	0.012

TABLE 8

Deviation of observed phi from phi computed from fit to Earth models for various initial densities

ρ_0 (gm/cm³)	Region 1					Region 2						
	Jeffrey		Gutenberg		Jeffrey		Gutenberg		LRJ–BRJ	
	std (km/sec)²	rel	std (km/sec)²	rel	std (km/sec)²	rel	std (km/sec)²	rel	std (km/sec)²	rel
McQueen et al., problem 1										
3.94	3.802	0.039	4.003	0.039	5.349	0.057	4.949	0.051	4.380	0.045
3.95	3.772	0.039	3.879	0.038	5.284	0.057	4.835	0.050	4.256	0.043
3.96	3.756	0.040	3.780	0.038	5.235	0.057	4.735	0.050	4.148	0.043
3.97	3.776	0.041	3.711	0.038	5.206	0.057	4.653	0.049	4.061	0.043
3.98	3.835	0.042	3.679	0.039	5.200	0.058	4.593	0.050	3.999	0.043
3.99	—	—	—	—	5.221	0.059	4.561	0.050	3.969	0.043
Model BH$_1$:										
3.91	1.078	0.012	1.491	0.017	0.545	0.007	1.308	0.015	1.595	0.019
3.92	1.031	0.010	1.280	0.015	0.392	0.005	1.087	0.013	1.436	0.017
3.93	1.058	0.010	1.090	0.012	0.415	0.005	0.893	0.010	1.314	0.015
3.94	1.162	0.011	0.945	0.010	0.614	0.007	0.762	0.009	1.249	0.014
3.95	1.341	0.013	0.890	0.009	0.889	0.010	0.745	0.008	1.258	0.013
3.96	1.552	0.016	0.920	0.009	1.200	0.013	0.863	0.009	1.315	0.014
3.97	1.811	0.018	1.065	0.011	1.535	0.017	1.087	0.012	1.521	0.015

cm³ for model II. These are slightly lower than the initial densities given in table 12. McQueen et al. (1964) using a Hugoniot equation of state, obtained a ρ_0 of 4.04 gm/cm³, 0.06 gm/cm³ greater than the results of the calculation in table 12.

The zero pressure ρ_0 and Φ_0 tabulated refer to a temperature on the adiabat which passes through the lower mantle. Clark and Ringwood (1964) estimate that this temperature is about 1600 °C and consider that the density at 20 °C will be 4 to 5 % greater. From the seismic equation of state (Anderson, 1967) and simple solid state physics we have

$$(\partial \ln \Phi / \partial \ln \rho)_p \sim (\partial \ln \Phi / \partial \ln \rho)_T - 1 \sim 2$$

TABLE 9

Deviation of observed phi from phi computed from fit to Earth models for various initial densities

ρ_0 (gm/cm³)	Region 1				Region 2					
	Jeffrey		Gutenberg		Jeffrey		Gutenberg		LRJ–BRJ	
	std (km/sec)²	rel	std (km/sec)²	rel	std (km/sec)²	rel	std (km/sec)²	rel	std (km/sec)²	rel
Model G 1:										
4.10	14.94	0.150	15.09	0.149	—	—	—	—	—	—
4.20	15.31	0.158	15.34	0.156	—	—	—	—	—	—
4.30	16.53	0.179	16.39	0.174	—	—	—	—	—	—
4.40	19.53	0.221	19.19	0.214	—	—	—	—	—	—
CIT 200204*:										
4.09	3.257	0.034	3.364	0.037	0.957	0.012	1.355	0.016	1.685	0.020
4.10	3.233	0.033	3.265	0.035	0.838	0.010	1.105	0.013	1.516	0.018
4.11	3.239	0.032	3.191	0.033	0.826	0.009	0.884	0.011	1.392	0.016
4.12	3.278	0.032	3.145	0.032	0.939	0.010	0.736	0.008	1.335	0.015
4.13	3.352	0.032	3.134	0.031	1.154	0.012	0.730	0.008	1.363	0.014
4.14	3.462	0.032	3.160	0.031	1.436	0.015	0.885	0.009	1.483	0.015

* Fit to region 1 made using pressures from Birch 1.

TABLE 10

Equation of state parameters for best fit to Jeffrey's velocity distribution

Model	$\rho_0/\overline{M}_1 = 0.048\ \Phi^{0.323}$ Region 1				$\rho_0/\overline{M}_2 = 0.0492\ \Phi^{\frac{1}{3}}$ Region 2				
	ρ_0 (gm/cm³)	Φ_0 (km/sec)²	ξ	$\overline{M}_1$	ρ_0 (gm/cm³)	Φ_0 (km/sec)²	ξ	$\overline{M}_1$	$\overline{M}_2$
Birch 1*	3.90	51.25	0.059	22.8	3.90	51.16	0.049	22.8	21.4
Birch 2*	3.97	51.58	0.020	23.1	3.98	52.44	−0.035	23.1	21.6
Pyrolite	4.09	53.08	0.086	23.6	4.09	52.74	0.047	23.7	2.22
Eclogite	3.88	51.20	0.063	22.7	3.89	52.15	0.086	22.6	21.2
McQueen 1*	3.9	52.72	−0.080	22.8	3.98	52.72	−0.184	22.9	21.6
BH₁**	3.93	52.57	0.146	22.8	3.93	52.09	0.089	22.8	21.4
200204**	4.12	55.10	0.282	23.5	4.11	55.33	0.403	23.5	21.9

* Self consistent comparison.
** Density models not directly determined from a velocity distribution.

TABLE 11

Equation of state parameters for best fit to Gutenberg's velocity distribution

Model	$\rho_0/\overline{M}_1 = 0.048\ \Phi^{0.323}$ Region 1				$\rho_0/\overline{M}_2 = 0.0492\ \Phi^{\frac{1}{3}}$ Region 2				
	ρ_0 (gm/cm³)	Φ_0 (km/sec)²	ξ	$\overline{M}_1$	ρ_0 (gm/cm)³	Φ_0 (km/sec)²	ξ	$\overline{M}_1$	$\overline{M}_2$
Birch 1	3.93	54.80	0.209	22.5	3.92	53.82	0.186	22.5	21.1
Birch 2	3.99	53.95	0.123	22.9	4.00	55.21	0.179	22.8	21.4
Pyrolite*	4.11	55.45	0.186	23.4	4.11	55.54	0.196	23.4	21.9
Eclogite*	3.88	51.20	0.625	22.7	3.90	53.52	0.157	22.5	21.0
McQueen 1	3.96	55.41	0.050	22.6	3.97	55.71	−0.017	22.6	21.1
BH₁**	3.95	54.94	0.244	22.6	3.95	54.80	0.228	22.6	21.1
200204**	4.14	57.54	0.381	23.3	4.13	56.73	0.536	23.3	21.9

* Self consistent comparison.
** Density models not directly determined from a velocity distribution.

4. The composition of the lower mantle

Birch demonstrated that the mean atomic weight $\overline{M}$ is an appropriate measure of the composition when discussing the elastic properties of most common rocks. ANDERSON (1967) proposed a seismic equation of state of the form

$$\frac{\rho}{\overline{M}_1} = A\Phi^n,$$

which explicitly relates the mean atomic weight to density and the elastic ratio. He fit this equation to thirty-one selected minerals and rocks and obtained values of A and n of 0.048 and 0.323, respectively. Thus, if the zero-pressure density and zero-pressure elastic ratio are known in the lower mantle, the mean atomic weight can be estimated. The results of the calculations are summarized in tables 10, 11, and 12.

The seismic equation of state was based on many rocks and minerals, but primarily on relatively low-density crustal and upper mantle material. Since the minerals in the lower mantle are much more closely packed ANDERSON (1969) proposed that an equation of the form $\rho/\overline{M}_2 = 0.0492\ \Phi^{\frac{1}{3}}$ be used in interpreting lower mantle data. This equation satisfies data for the close packed oxides MgO, Al_2O_3 and SiO_2 (stishovite). Use of this equation leads to the values for mean atomic weight labelled $\overline{M}_2$ in tables 10, 11, and 12. These estimates for mean atomic weight are about 1.4 units less than computed from the seismic equation of state and indicate the degree of uncertainty involved in estimating this parameter for the lower mantle.

In computing the composition of the lower mantle we assume that the region is homogeneous throughout, both in phase and in composition. To facilitate the later analysis it is also assumed that the composition corresponds to mixtures or compounds of the three primary oxides MgO, FeO and SiO_2 (stishovite).

The density models that seem most satisfactory are Birch model I, Birch model II, pyrolite and eclogite. All of these models can be fit to one of the seismic Φ distributions with an error in Φ of 0.4% or less. The range in inferred zero-pressure densities for these models is from 3.88 gm/cm³ for the eclogite model to 4.11 gm/cm³ for the pyrolite model. CLARK and RINGWOOD (1964) estimated zero-pressure densities of 3.90 and 4.11 gm/cm³ for their eclogite and pyrolite models,

TABLE 12

Equation of state parameters for best fit to JAJ velocity distribution

$\rho_0/\overline{M}_1 = 0.048\ \Phi^{0.323}$
Region 2

Model	ρ_0 (gm/cm³)	Φ_0 (km/sec)²	ξ	$\overline{M}_1$	$\overline{M}_2$
Birch 1	3.93	55.21	0.255	22.4	21.0
Birch 2	4.00	55.21	0.179	22.8	21.4
Pyrolite	4.11	55.54	0.196	23.4	21.9
Eclogite	3.90	53.52	0.157	22.5	21.0
McQueen 1	3.98	57.26	−0.067	22.4	21.0
BH_1*	3.95	54.80	0.228	22.6	21.1
200204*	4.13	56.73	0.536	23.3	21.9

* Density models not directly determined from a velocity distribution.

TABLE 13

Parameters of lower mantle at 20 °C

Model	ρ_0 (gm/cm³)	Φ_0 (km/sec)²	M_1	M_2
Birch 1	4.08	59.6	22.7	21.2
Birch 2	4.16	59.6	23.0	21.6
Pyrolite	4.27	60.0	23.6	22.2
Eclogite	4.20	57.8	23.5	22.1
McQueen 1	4.14	61.8	22.7	21.3
BH_1	4.11	59.2	22.8	21.4
200204	4.30	61.3	23.6	22.2

(see also CLARK and RINGWOOD, 1964). The relative change in Φ is therefore twice the relative change in ρ. Taking the temperature corrected density to be 4% greater than those in table 6 we obtain the room temperature values in table 13.

In terms of their ability to match the observed Φ structure of the Earth the most satisfactory models are those of BIRCH (1964) and CLARK and RINGWOOD (1964). For these models the range of temperature corrected values for ρ_0 are 4.08 to 4.27 and for Φ_0 are 59.6 to 60.0. These can be compared with the room temperature values of 4.0 to 4.3 gm/cm³ for the density and 60–62 (km/sec)² for Φ_0 determined by CLARK and RINGWOOD (1964) for their models.

The model 200204 (ANDERSON and SMITH, 1968) is the most geophysically plausible model since it satisfies the periods of free oscillation as well as the mass and moment of inertia. The density and Φ at ambient conditions for this model are 4.30 gm/cm³ and 61.3 (km/sec)² respectively.

respectively. Their estimates were based solely on the goodness of fit of a Birch–Murnaghan equation of state to lower mantle densities derived from the Adams–Williamson method. The latter method involves an integration of the seismic Φ so the derived density should be consistent with this parameter. The range in mean atomic weight, using the original form of the seismic equation of state, for the above models is from 22.4 to 23.7. This can be compared with the values 22.5 and 22.7 which were found by BIRCH (1961) and MCQUEEN et al. (1964). The modified form of the seismic equation of state gives 21.0 to 22.3 as the range of mean atomic weight in the lower mantle.

5. Superadiabatic temperature gradient

In the previous sections it was assumed that the temperature gradient in the lower mantle is adiabatic. Satisfactory agreement was obtained for all models except problem 1 of McQueen et al. which had minimum standard deviations exceeding 1% (table 12). In all other cases the assumption of a superadiabatic temperature gradient gives poorer fits than those reported here. MCQUEEN et al. (1964) explicitly assumed a temperature gradient for the lower mantle which is probably slightly superadiabatic.

For purposes of discussion it is assumed that the temperature gradient departs from the adiabat below 600 km depth. The coefficient of thermal expansion at zero-pressure was taken as 10.6×10^{-6} °C^{-1}, a value considered appropriate for the post-spinel phase of Mg_2SiO_4. The effect of pressure on this quantity was computed using Birch's equation:

$$\alpha/\alpha_0 = 1 - \frac{nP}{K} \quad \text{(BIRCH, 1952, 1968)}.$$

Here P is the pressure and n is the exponent in the Murnaghan equation of state, taken as 3.096 (ANDERSON, 1967). K is the bulk modulus computed from $K = K_0(\rho/\rho_0)^n$.

Assuming a superadiabatic gradient, the density was corrected to an adiabat using the above equation. The Birch–Murnaghan equation of state was fitted to the results and, a new Φ, appropriate to the temperature corrected model, was generated. The observed JAJ Φ distribution was also corrected to the adiabat using an equation of the form

$$\left(\frac{d \ln \Phi}{d \ln \rho}\right)_P = R \quad R \text{ constant}.$$

The deviations of the temperature corrected computed Φ from the temperature corrected observed Φ are given in table 14 for superadiabatic gradients of 0.75 °C/km and 1 °C/km and values of R equal to one and two. Relative deviations of less than 1% were obtained assuming a gradient 1 °C/km above the adiabat. This is consistent with the actual temperature gradient assumed by McQueen et al.

6. Composition of the upper mantle

There have been several recent attempts to estimate the composition of the upper mantle. RINGWOOD (1966) constructed a model which is a 1:3 mixture of basalt:peridotite. He was led to this model by requiring that the upper mantle yield basalt magma upon fractional melting. WHITE (1967) analyzed olivine modules and peridotites to establish the amount of the major constituents which he felt were representative of undepleted upper mantle. NICHOLLS (1966) synthesized a composition for the upper mantle by combining a basaltic fraction with a residual fraction. He considers olivine modules and various peridotites and serpentinites to represent the residual fraction rather than the total mantle composition. NELSON et al. (1967) compiled data on the composition of the St. Paul's rocks

TABLE 14

Standard deviations and relative standard deviations of temperature corrected φ distributions for McQueen problem 1 from the temperature corrected φ distribution

ρ_0 gm/cm^3	φ_0 (km/sec)^2	R = 1		R = 2	
		Standard deviation	Relative deviation	Standard deviation	Relative deviation
Super-adiabatic gradient of 0.75 °C/km					
3.96	53.39	1.745	0.018	1.384	0.014
3.97	54.84	1.627	0.017	1.259	0.013
3.98	56.33	1.573	0.017	1.217	0.013
3.99	57.86	1.597	0.018	1.277	0.014
4.00	59.43	1.708	0.020	1.439	0.017
Super-adiabatic gradient of 1 °C/km					
3.96	53.09	0.937	0.010	0.624	0.007
3.97	54.52	0.755	0.008	0.418	0.004
3.98	55.99	0.693	0.007	0.410	0.004
3.99	57.49	0.800	0.009	0.649	0.007
4.00	59.04	1.042	0.012	0.982	0.011

which have been postulated to be an exposure of the suboceanic mantle. Using a partial fusion model CARTER (1967) estimated the chemical composition of the primary upper mantle under Kilbourne Hole, New Mexico. ITO and KENNEDY (1967) report on the chemical composition of a garnet lherzolite nodule in a South African kimberlite which they considered to be a representative of rock constituting the upper mantle. There is remarkably little difference among these various estimates of the composition of the upper mantle. The mean atomic weights range from 20.96 to 21.26 and the average value is 21.1. This can be compared with 21.2 for a typical chrondrite.

By contrast our best estimates of $\overline{M}_1$ (seismic equation of state) for the lower mantle range from about 22.4 to 23.4 (table 12). These are 1.3 to 2.3 units greater than the upper mantle. This is in agreement with ANDERSON (1968) who concluded that the $\overline{M}$ of the lower mantle was 1.5 to 1.6 units greater than the upper mantle. This conclusion was based on the relative increase of density, compressional velocity and seismic parameter Φ across the transition region of the upper mantle. Using the modified form of the seismic equation of state, thought to be more appropriate for close-packed minerals, the $\overline{M}$ of the lower mantle corresponding to the above range is 21.0 to 22.0. If the latter values are preferable the mean atomic weight of the lower mantle is very close to that of the upper mantle, differing at most by about 1 unit. Because of the uncertainty in the ρ_0, $\overline{M}$, Φ_0 relationships the conclusions regarding $\overline{M}$ of the lower mantle are much weaker than the conclusions regarding ρ_0 and Φ_0. Since MgO and SiO_2 have nearly the same mean atomic weight this parameter is pertinent to the amount of iron or iron oxides. We will show in a later section an alternate way to get at the iron content of the lower mantle.

BIRCH (1961), using his velocity-density correlation, found an increase in the mean atomic weight through the transition region of 1 unit, but, considering the uncertainties he concluded that "there is little change of iron content through the transition layer, phase changes accounting for nearly all of the anomalous increase of density". CLARK and RINGWOOD (1964) found that their estimates of the zero pressure density of the lower mantle were about 5% greater than the probable assemblage of upper mantle composition at lower mantle pressures. They offered three possible explanations:

a) The discrepancy may not be real and may be a result of their treatment of the transition region or of errors in the lower mantle velocities.

b) Phase transformations in the transition region may result in assemblages denser than those considered.

c) The abundance of iron in the lower mantle may be appreciably higher than assumed for pyrolite.

The treatment given here and in ANDERSON (1968) are independent of details in the transition region and the conclusions regarding the mean atomic weight of the lower mantle are independent of the velocity model chosen and are consistent with the most recent determinations of velocity. Explanation a) can therefore be ruled out.

ANDERSON (1968) showed that not only was the zero-pressure density of the lower mantle greater than the upper mantle but the seismic velocities (V_p, Φ) were less. If it is assumed that a collapse to a denser phase also increases the elastic properties then it is expected that phase changes will increase the density and the seismic velocities. Likewise temperature will affect density and velocity in the same direction. Explanation b) can also be ruled out. In silicates and oxides iron substitution increases the density and decreases the seismic velocities and the seismic parameter Φ_0.

Any increase in the mean atomic weight is most probably due to an increase in the iron content of the lower mantle relative to the upper mantle. The $FeO + Fe_2O_3$ weight per cent of proposed upper mantle rocks varies from about 8 to 10%. The range of possible lower mantle compositions, found above from $\overline{M}_1$, corresponds to $FeO + Fe_2O_3$ of about 20 to 31 weight per cent. The lower values of $\overline{M}_2$ correspond to 7 to 18 weight per cent $FeO + Fe_2O_3$ for the lower mantle.

Strictly speaking the seismic equation of state should only be applied to room temperature data. The mean atomic weight calculated from the two variants of the seismic equation of state are given in table 13 for the 7 models. The preferred models, Birch 2, Pyrolite and 200204 give $\overline{M}_1$ of 23.0, 23.6 and 23.6 respectively. For the alternate form of the seismic equation of state the corresponding $\overline{M}_2$ are 21.6, 22.2 and 22.2 respectively. These values are only about 0.2 units greater than those derived from the temperature uncorrected data. The inferred weight per cent $FeO + Fe_2O_3$ is correspondingly raised by about 1.8%. The lower estimate of this percentage for the lower mantle is therefore raised to

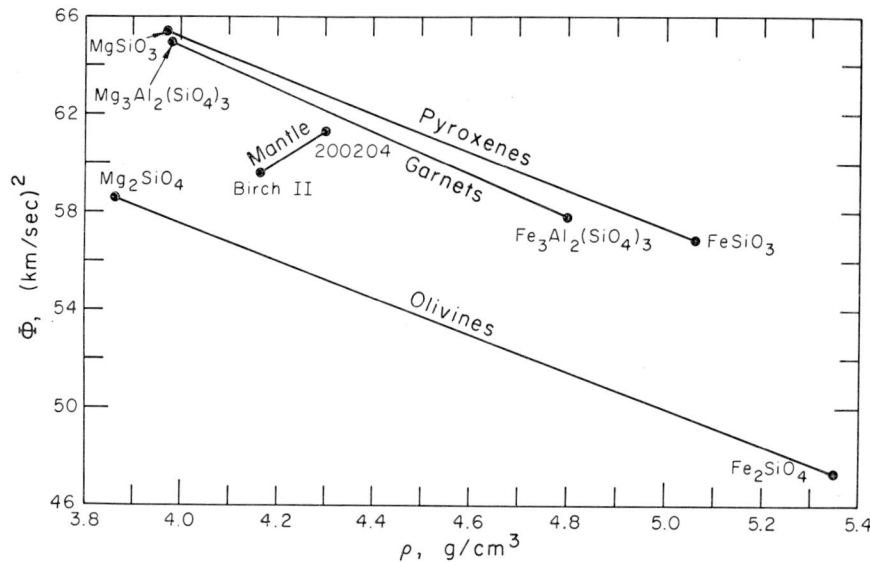

Fig. 1. Φ versus ρ for olivines, pyroxenes and garnets assuming that both molar volumes and seismic ratios are molar averageable. The composition of the lower mantle can be inferred from this graph.

about 10%, or close to the upper bound for upper mantle rocks.

Based on the seismic equation of state the lower mantle is clearly enriched in FeO relative to the upper mantle. Using the modified seismic equation of state the situation is marginal. An alternate approach, not involving the mean atomic weight, is given in the next section.

7. Alternate approaches

If the molar volumes and elastic ratios of deep mantle minerals can be considered to be molar averages of the oxides (ANDERSON, 1969) then both the composition and mineralogy of the lower mantle can be discussed by comparing the density and Φ_0 inferred in this paper with those predicted for the dense forms of olivines, pyroxenes and garnets. On a plot of ρ_0 versus Φ_0 (fig. 1) the values of Φ_0 for the mantle fall between the curves for olivine and the curves for pyroxenes and garnets. The inferred zero pressure values of density for the lower mantle are more than 5% greater than the density of the magnesium rich end member of any of the above minerals. We assume that the coexisting minerals in the lower mantle all have the same mole percentage of the iron-rich end member. For a garnet free lower mantle (i.e. no aluminum) the olivine content varies between about 20 and 57% and the molar percentage of Fe_2SiO_4 or $FeSiO_3$ lies between about 19 and 30. By combining the values of ρ_0 and Φ_0, inferred for the lower mantle, we obtain the estimates for mole per cent olivine, pyroxene (or garnet), SiO_2, MgO and FeO for Birch model 2 and Anderson–Smith model 200204. These are given in table 15. The ranges for

TABLE 15

Composition of lower mantle (in mole fraction)

Model	olivine	pyroxene	garnet	SiO_2	MgO	FeO
Birch 2	0.57	0.43		0.39	0.49	0.12
Birch 2	0.50		0.50			
200204	0.20	0.80		0.46	0.38	0.16
200204	0.10		0.90			

these models are 39–46 mole percent SiO_2, 38–49 mole per cent MgO and 12–16 mole per cent FeO. These can be compared with the corresponding values for ultrabasic rocks (WHITE, 1967) of 38% SiO_2, 52% MgO and 6% [$FeO+\frac{1}{2}Fe_2O_3$], all in mole per cent. These values were estimated from White's histograms in which dunites were excluded.

This approach, which is independent of the previous methods based on mean atomic weight, suggests more strongly that the lower mantle is enriched in FeO relative to the upper mantle.

A third approach, and the most direct, makes use of

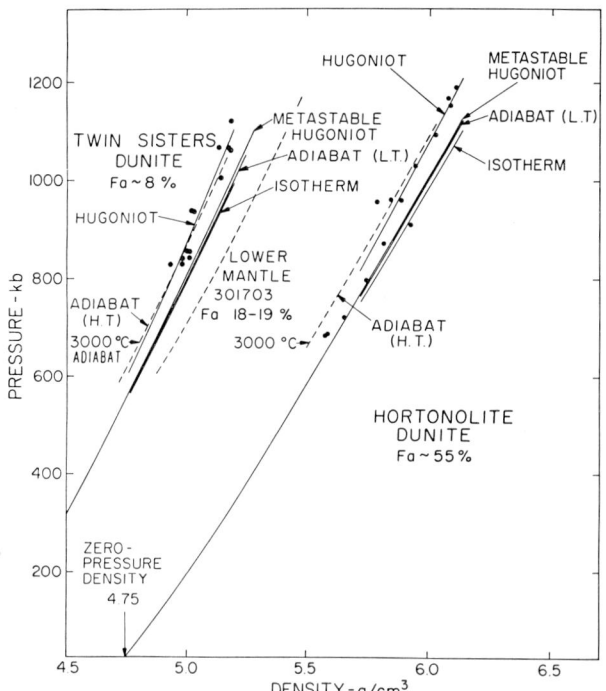

Fig. 2. Density versus pressure for two dunites (MCQUEEN et al., 1967, AHRENS et al., 1969). The high temperature (H.T.) adiabat is computed for temperatures thought to be appropriate for the lower mantle. The curve labelled lower mantle 301703 is from inversion of free oscillation and seismic travel time data (ANDERSON et al., 1969).

shock wave data. Two dunites have been shocked to pressure of the order of 2 megabar by MCQUEEN et al. (1967). Their data are shown in fig. 2. The Twin Sister's dunite contains about 8 mole per cent Fe_2SiO_4; the Hortonolite dunite contains about 55 mole per cent Fe_2SiO_4. Also shown are derived metastable hugoniots, and low-temperature adiabats and isotherms (AHRENS et al., 1969). The dashed curves labeled adiabat (H.T.) are adiabats calculated for temperatures thought to be appropriate for the ower mantle. The curve for the lower mantle is for an Earth model (301703) derived from free oscillation data (ANDERSON et al., 1969). It differs only slightly in this region from model 200204. The inferred fayalite content, assuming a purely dunitic lower mantle, is 18–19 mole per cent which compares with the 12–16 mole per cent found by the previous method. This argument is based strictly on a comparison of the densities. Since the densities of the high-pressure forms of olivines, pyroxenes and garnets containing approximately 20 mole per cent of their iron rich end members are similar the argument will be only slightly affected by including these other possible constituents.

8. Rationale for an iron-rich lower mantle

If the lower mantle is enriched in iron this condition was probably established early in the history of the Earth, possibly at the time of the differentiation of the crust and the core. There are several possibilities; among the most obvious are:

1) The mantle, as well as the whole Earth, has been extensively differentiated and the chemical zoning reflects gravitational stratification. In magnesium rich olivine rocks, intruded near the surface, the melt phase is rich in iron but the volume change upon melting and, more importantly, the concentration of volatiles and other light low-melting components in the melt, more than offsets the increased iron content so the iron rich magmas rise and magnesium rich crystals settle. Thus the iron content of such intrusives decrease with depth. It is likely that the density of the melt increases more rapidly with depth than the density of the solid phase so it is possible that the above relationship is reversed at depth. Once the volatiles and lighter constituents are removed from the mantle partial melting will generate iron-rich melts which are denser than the residual crystals.

From the phase diagram for melting in the system forsterite-fayalite (BOWEN and SCHAIR, 1935) at atmospheric pressure, olivine of 20 weight percent Fe_2SiO_4 will first yield a melt composition of about 50 weight percent Fe_2SiO_4. The density of an olivine containing 50 weight percent fayalite is about 3.7 gm/cm^3 and even if this is reduced by 5% to allow for the volume increase upon melting the density of the melt will be about 3.5 gm/cm^3 which is appreciably more than the density of 3.38 gm/cm^3 of the olivine containing 20% weight percent fayalite. Once the volatiles and other low density, low-melting components have been removed from the mantle to form the hydrosphere and crust melting of olivine will generate an iron rich melt which is denser than the remaining crystals and it will tend to sink.

Recent thermal history calculations (HANKS and ANDERSON, 1969) suggest that a large part of the mantle was partially molten shortly after accretion. It is even possible that the outer parts of the Earth were partially molten during most of the time of accretion and differentiation was therefore occurring during the accretion process. The present low-velocity zone is possibly a relic of this situation.

2) If partial melting occurs at great depth (~ 200 kb) the melting of olivine in the spinel structure probably gives a magnesium rich fluid which will rise.

3) The sinking of free iron or iron rich compounds through the lower mantle during the core formation process may have enriched this region in iron. This possibility is difficult to evaluate.

4) It is well established that the iron-magnesium ratio of crustal rocks is much greater than the iron content of the upper mantle.

For example estimates of FeO, Fe_2O_3 and MgO for the continental crust are (MASON, 1958) 3.9, 3.1 and 3.6 weight per cent, respectively. Corresponding estimates for the upper mantle are about 8, 2 and 40 for the same compounds. There is therefore roughly an enrichment of 8 in the FeO/MgO ratio of the continental crust as compared with the upper mantle. If the continental crust has been derived from the upper 400 km of the mantle the FeO/MgO ratio of the primitive upper mantle will be increased by about 0.17. If the difference in mean atomic weight between the upper and lower mantle is taken as 1.5, this implies a difference in the FeO/MgO ratio (weight per cent) of about 0.5 between these two regions or 3 times the above. The oceanic crust is primarily basalt and the FeO, Fe_2O_3 and MgO contents of oceanic basalt are about 9, 1 and 10 so that the FeO/MgO ratio is about unity. The maximum thickness of the basalt layer is about 10 km in normal oceanic areas but is presumably much thicker under the mid-oceanic rises.

The high pressure chemical equivalent of basalt is eclogite which, according to the moho phase hypothesis, underlies the oceanic crust. The total thickness of oceanic crust and upper mantle of basaltic composition that would be required to increase the FeO/MgO content of the upper mantle by a further 0.33 is 155 km. It would be correspondingly less if the difference in mean atomic weight between the upper and lower mantle were less than 1.5.

Although the FeO/MgO ratio is much greater in the crust than in the upper mantle the FeO content of the crust is slightly less than the upper mantle. Therefore, in order to match the primitive upper mantle with the present lower mantle one must keep the crust enriched in SiO_2, Al_2O_3 and alkalies and speculate that these compounds represented an even earlier process of differentiation. Otherwise mixing of the present crust into the present upper mantle will not appreciably raise its mean atomic weight.

It appears probable that the crust-upper mantle system is depleted in iron relative to the lower mantle. We envisage a two-stage process. The upper mantle (above 400–600 km) was differentiated from a mantle of composition more primitive than pyrolite. The crust was then differentiated from the upper mantle for which pyrolite is probably a reasonable model.

An independent estimate of the change in composition through the transition region can be obtained by comparing the zero pressure densities and seismic parameters of the upper and lower mantle. Unfortunately the density and Φ_0 of the upper mantle are not known precisely. Estimates of density range from about 3.31 to over 3.4 gm/cm^3. For illustrative purposes we will adopt the conventional value of 3.32 gm/cm^3 which is close to the value for pyroxene pyrolite (GREEN and RINGWOOD, 1963) and the value adopted by BIRCH (1964) for his solution II. Estimates for Φ_0 of the upper mantle range from 33 to 39 (km/sec)^2. We will adopt the value 37.4 (km/sec)^2 which corresponds to $V_p = 8.1$ km/sec and $V_s = 4.6$ km/sec, typical values for the top of the mantle. The effects of temperature and pressure on ρ and Φ roughly cancel in the upper parts of the Earth and any correction would be less than the uncertainty. Accepting these values for the upper mantle and adopting the temperature corrected values of the Birch I model, $\rho_0 = 4.05$ gm/cm^3 and $\Phi_0 = 55.4$ (km/sec)^2 we can estimate the change in mean atomic weight and the change in density due to phase changes through the transition region. The pertinent equations are

$$d \ln \rho_0 = (\partial \ln \Phi_0 / \partial \ln \overline{M})_x \, d \ln \overline{M} + d \ln x,$$

$$d \ln \Phi_0 = (\partial \ln \Phi_0 / \partial \ln \overline{M})_{\rho_0} + \\
+ (\partial \ln \rho_0 / \partial \ln \overline{M})_x (\partial \ln \Phi_0 / \partial \ln \rho_0)_{\overline{M}} \, d \ln \overline{M} \\
+ (\partial \ln \Phi_0 / \partial \ln \rho_0)_{\overline{M}} \, d \ln x,$$

where $d \ln x$ is the relative density increase due to phase changes and the subscript x refers to constant phase. Taking $(\partial \ln \rho / \partial \ln \overline{M})_x = 0.85$ (BIRCH, 1961), $(\partial \ln \rho_0 / \partial \ln \overline{M})_\rho = -3$ and $(\partial \ln \Phi_0 / \partial \ln \rho_0)_{\overline{M}} = +3$ (ANDERson, 1967, 1968) we have $d\overline{M}/\overline{M}_{av} = +0.056$ and $d\rho/\rho_{av} = 0.15$. The former value yields a $d\overline{M}$ of about 1.18 if $\overline{M}_{av} = 21$ and the latter value can be compared with the relative increase of density between the low pressure and high pressure assemblages of pyrolite (0.20) (CLARK and RINGWOOD, 1964), forsterite (0.18), fayalite (0.19) (ANDERSON, 1967), and dunites (0.22) (AHRENS et al., 1969). Except for the pyrolite model the models of BIRCH (1961) and CLARK and RINGWOOD (1964) all have relative increases of density, referred to zero pressure, of less than 0.22 so that these models, having only the mass and moment of inertia as constraints, can be compatible with little or no composition change through the transition region. Taking $d \ln x = 0.20$ the pyrolite model, with $\Delta\rho_0/\rho_{av} = 0.256$ gives a $d\overline{M}$ of 1.4 through the transition region and a Φ_0 for the upper mantle of about 33 $(km/sec)^2$ which is low but not unreasonable.

Acknowledgments

This research was supported by National Science Foundation Grant GA 1003.

References

AHRENS, T. J., DON L. ANDERSON and A. E. RINGWOOD (1969) Rev. Geophys. (in press).
ANDERSON, DON L. (1967) Geophys. J. **13**, 9.
ANDERSON, DON L. (1968) Earth Planet. Sci. Letters **5**. 89.
ANDERSON, DON L. (1969) J. Geophys. Res. (in press).
ANDERSON, DON L. and M. SMITH (1968) Abstract, Trans. Am. Geophys. U.
ANDERSON, DON L. and B. R. JULIAN (1969) J. Geophys. Res. (in press).
ANDERSON, DON L., T. JORDAN and M. SMITH (1969) in preparation.
BIRCH, F. (1952) J. Geophys. Res. **57**, 227.
BIRCH, F. (1961) Geophys. J. **4**, 295.
BIRCH, F. (1964) J. Geophys. Res. **69**, 4377.
BIRCH, F. (1968) J. Geophys. Res. **73**, 817.
BULLEN, K. E. and R. A. HADDON (1967a) Nature 574.
BULLEN, K. E. and R. A. HADDON (1967b) Proc. Nat. Acad. Sci. **58**.
BOWEN, N. L. and J. F. SCHAIR (1935) Am. J. Sci. **29**, 151.
CARTER, J. L. (1967), Southwest Center for Advanced Studies, Dallas, Annual Report, 24.
CLARK, S. P., Jr. (1962) *Research in Geochemistry*, ed. P. H. Abelson (Wiley, New York).
CLARK, S. P. and A. E. RINGWOOD (1964) Rev. Geophys. **2**, 35.
GILBERT, J. T. and G. E. BACKUS (1968) Bull. Seismol. Soc. Am. **58**, 103.
GREEN, D. H. and A. E. RINGWOOD (1963) J. Geophys. Res. **68**, 937.
HANKS, T. and DON L. ANDERSON (1969) Phys. Earth Planet. Interiors **2**, 19.
IBRAHIM, A. K. and O. W. NUTTLI (1968) Bull. Seismol. Soc. Am. **58**, 339.
ITO, K. and G. KENNEDY (1967) Am. J. Sci. **265**, 519.
JOHNSON, L. R. (1967) J. Geophys. Res. **72**, 6309.
MASON, B. (1958) *Principles of Geochemistry* (Wiley, New York).
MCQUEEN, R. G., J. N. FRITZ and S. P. MARSH (1964) J. Geophys. Res. **69**, 2947.
MCQUEEN, R. G., S. P. MARSH and J. N. FRITZ (1967) J. Geophys. Res. **72**, 4999.
NELSON, W. G., E. VAROSEWICH, V. BOWEN and G. THOMPSON (1967) Science **155**, 1532.
MURNAGHAN, F. D. (1937) Am. J. Math. **59**, 235.
NICHOLLS, G. D. (1966) in: *Mantles of the Earth and Terrestrial Planets*, ed. S. K. Runcorn (Interscience Publishers, London).
PRESS, F. (1966) in: *Handbook of Physical Constants*, ed. S. P. Clark, Memoir 97, G.S.A. 215.
RINGWOOD, A. E. (1966) in: *Advances in Earth Sciences*, ed. P. M. Hurley (M.I.T. Press, Cambridge, Massachusetts).
WHITE, I. G. (1967) Earth Planet. Sci. Letters **3**, 11.

COMPARISON OF SOURCES OF EVIDENCE ON THE VARIATION OF INCOMPRESSIBILITY IN THE EARTH'S DEEPER INTERIOR

K. E. BULLEN

Department of Applied Mathematics, University of Sydney, Australia

A comparison is made of two series of estimates of the variation of the incompressibility k below 1000 km depth in the Earth. The estimates are those derived by Birch using an adaptation of Murnaghan's finite-strain theory and those derived by the writer using seismic data and the relation (6) below.

The comparison shows several discordancies, chiefly in estimates of the changes in k at the mantle-core boundary N and in estimates of dk/dp (p = pressure) inside the core. For the discordancies at N to be reconcilable, a combination of the following circumstances would be required: (a) the mean temperature gradient between 1000 and 2900 km depth is greater than expected; (b) the mean density in the outermost 200–300 km of the mantle is greater than expected; (c) the materials on the two sides of N are much less different in composition than is currently supposed. Reconciliation through (a) alone would require the temperature gradient to be at least (about) 1.7 deg/km (this value would have to be greater if there should be significant inhomogeneity in this part of the Earth). Reconciliation is not possible through (b) alone even should the immediate sub-crustal density be as high as 3.7 g/cm³. The analysis shows that, failing such a combination of circumstances, Birch's finite-strain theory needs significant amendment at pressures of the order of a million atmospheres. The discordancies in estimates of dk/dp in the core are irreconcilable with Birch's theory unless the seismic data used are more seriously in error than expected. The importance of making a new check on the variation of the P velocity inside the outer core is stressed.

Details in the calculations lead to the further suggestion that the region E_1 (2900 to 3600 km depth) of the outer core is mildly inhomogeneous, the inhomogeneity being possibly associated with continuous phase changes. The region E_2 (3600 to 4500 km) is likely to be one of uniform chemical composition and phase, with $dk/dp \approx 3.5$. Below E_2, the evidence on k is more meagre, but it is suggested that k is continuous to good approximation, with dk/dp lying between about 3.5 and 4.2.

1. Introduction

The incompressibility k is an important parameter in investigating fine changes of property inside the Earth at depths z greater than about 1000 km. Evidence on k comes from several sources, especially seismology, finite-strain theory and laboratory experiments on rocks and other materials. A main section of the seismological evidence is brought to bear through the relation

$$\alpha^2 - 4\beta^2/3 = k/\rho = \phi \text{ (say)}, \tag{1}$$

where α and β are the P and S velocities, resulting in fairly well determined values of k/ρ throughout much of the Earth. Birch's application of Murnaghan's finite-strain theory has yielded, for chemically homogeneous regions, neglecting certain temperature effects, equations such as

$$p = 3k_0 f(1+2f)^{\frac{5}{2}}, \tag{2}$$

$$k = k_0(1+2f)^{\frac{5}{2}}(1+7f), \tag{3}$$

$$3\frac{dk}{dp} = 7 + \frac{5}{1+7f}, \tag{4}$$

where p is the pressure, f is the compression, the subscript zero relates to zero pressure (assuming no phase changes as the pressure is reduced), and isothermal conditions are assumed.

The existence of independent approaches to estimating k inside the Earth enables important checks to be made. Such a check made by BIRCH (1952) as part of a wider survey showed agreement, at least to a useful first approximation, in results derived from his finite-strain theory and the seismological evidence available at the time. The writer has examined the matter further using more recent data, and while confirming broad agreement in the first approximation, finds that there are, however, some significant discrepancies. The present paper is concerned with exploring possible ways of resolving the discrepancies and with certain interesting results that have emerged on the way. Experimental results, for example those of Orson ANDERSON (1968), are brought to bear where relevant.

Throughout the paper, discussion will be principally confined to the lower mantle and core, which will be treated as spherically symmetrical, and k will be taken to be the adiabatic incompressibility. In the light of pertinent investigations by Birch, differences between adiabatic and isothermal conditions will be treated as negligible in the present context. Throughout the paper dk/dp means, strictly, $(dk/dz)/(dp/dz)$.

Reference will be made to the relations (BULLEN, 1963, 1967)

$$dp/dz = \eta g \rho/\phi, \qquad (5)$$

$$dk/dp = \eta + g^{-1} d\phi/dz, \qquad (6)$$

where ρ and g denote density and gravitational intensity and η is an index which allows for effects, such as continuous variation of chemical composition and phase, additional to the effect of pure compression.

2. Compression in the Earth

Before eqs. (2), (3) and (4) can be applied to the Earth, assessments of the compression f are required. BIRCH (1952) gave values for f in the mantle, and the writer, by a simpler procedure using the relation

$$f = p/(3k - 7p) \qquad (7)$$

(BULLEN, 1968a), gave values of f throughout the Earth. The estimates of f by both Birch and the writer are minimal in that no allowances are made for possible phase changes. Any needed amendments are, however, likely to be small unless there are large phase changes such as might occur at the mantle-core boundary. The writer's procedure involving eq. (7) applies inside chemically inhomogeneous as well as homogeneous regions of the Earth.

Currently preferred results (1968a) give f increasing steadily with z throughout the mantle to 0.13 at the base, and give for the core $f = 0.13, 0.17, 0.19, 0.19$ at $z = 2900, 3600, 4500, 6370$ km, respectively.

3. Trends on the variation of k with p

Consider the application of eqs. (2) and (3) to two different materials M_1 and M_2. Let the subscripts 1 and 2 relate to properties of M_1 and M_2, respectively, at a common pressure p, and the subscript zero to values at zero pressure. Let $|k_2/k_1 - 1| = \theta$. The quantity θ_0 is an index of the compositional difference between M_1 and M_2, and the variation of θ with p indicates the variation of the dependence of $k_2 - k_1$ on compositional difference.

A specimen calculation of Birch postulating $f_1 = 0.10$ and $f_2 = 0.15$ at some common pressure p gave θ diminishing from 0.83 at zero pressure to 0.24 at p. The results in section 2 show that these values postulated for f_1, f_2 are roughly of the order of the values at the Earth's mantle-core boundary N.

A formal study (BULLEN, 1968b) generalizing this section of Birch's work gave inter alia values of θ in terms of series of assigned values of θ_0 and f_1. Results of interest to the present discussion are $\theta = (0.12, 0.24)$, $(0.06, 0.13)$, $(0.05, 0.09)$ for $\theta_0 = (0.20, 0.40)$ and $f_1 = 0.05, 0.14, 0.20$, respectively.

The same study incidentally gave a useful approximate formula

$$\theta = \theta_0(1 + 2f_1)(1 + 7f_1)^{-2} \qquad (8)$$

for estimating θ.

Now let the subscripts 1, 2 apply in particular to the materials at the base of the mantle and top of the core, respectively, and let $\Delta k = k_2 - k_1$. Assuming the materials to be principally composed of the equivalent of ultrabasic rock and iron, respectively, θ_0 would be expected to be about 0.40, which would entail $\Delta k/k_1 = +0.13$. Allowance for such possible effects as compositional changes inside the region C ($400 < z < 1000$ km, approx.) and alloying inside the outer core E could reduce these estimates of θ_0 and $\Delta k/k_1$ somewhat, but most improbably below 0.20 and $+0.06$ unless the changes at N are principally phase changes.

4. Comparison of estimates of $\Delta k/k_1$

Section 3 shows that, on currently preferred views on the compositions of the Earth's mantle and core, Birch's theory yields $\Delta k/k_1 \geq +6$ per cent. In contrast, the writer's original Model A calculations (see BULLEN, 1965a) gave $\Delta k/k_1 = -5$ per cent. The writer has since sought to find how far this gap between $+6$ and -5 per cent might be reduced.

In calculations taking account of the recently revised estimate of the Earth's moment of inertia (which supplied the biggest correction, equal to $+4.2$ per cent), abnormal density gradient in the region D'' (2700 <

$z < 2900$ km), recent work on the lower core (BULLEN, 1965b) and evidence from Earth oscillations (BULLEN and HADDON, 1967a), the writer's earlier result of -5 per cent was amended (BULLEN, 1968c, 1968d) to about -0.5 ± 1 per cent.

In a further paper, BULLEN and HADDON (1969a) have sought to set an upper bound to $\Delta k/k_1$ by postulating the values $0, +1, +2, \ldots$ per cent for $\Delta k/k_1$, and otherwise following the writer's general procedures for constructing Earth models based on seismic data, and then examining the consequences on density distribution. It was initially assumed (a) that the coefficient η is not significantly less than unity in the lower mantle and (b) that the mean density in the outermost 200–300 km of the mantle is not significantly greater than that in Earth models such as, for example, Model A, Model HB_1 (BULLEN and HADDON, 1967a) or recent models of B type (BULLEN and HADDON, 1967b, 1969b). On these assumptions, it was found to be unlikely that $\Delta k/k_1$ could exceed $+1$ per cent, though an unusual combination of circumstances might raise the value to an extreme of $+2$ per cent; the favoured value remains insignificantly different from zero.

There is therefore an irreducible gap of at least 4 per cent (and probably appreciably more) unless one or both of the assumptions (a) and (b) is error, or (c) the chemical compositions of the lower mantle and outer core are less different than most investigators, including Birch, currently think likely, or (d) Birch's theory is less reliable than other approaches to the determination of k in the Earth's deeper interior. It may be noted that even though the percentages in question may seem small, the calculation details show that it is very difficult to sustain a significantly positive value of Δk, which is an almost essential requirement of Birch's theory.

As regards (a), HADDON and the writer have carried out an auxiliary calculation (1969a) which formally shows that the upper bound to $\Delta k/k_1$ could be raised to $+6$ per cent if the mean value of η in the range $1000 < z < 2900$ km could be reduced to 0.8 or less. But this would require a mean temperature gradient of at least 1.7 ± 0.5 deg/km in this part of the Earth if chemically homogeneous. (The gradient would be greater still if there were significant departures from homogeneity.) The actual temperature gradient is generally held to be significantly smaller than this value. As regards (b), Professor F. Press at the 1969 Canberra Symposium quoted some evidence pointing to possibly higher densities just below the crust than have lately been considered. However, it can be inferred from the recent work of HADDON and the writer (1969a) that even raising the immediate sub-crustal density to 3.7 g/cm^3, which is approximately the extreme permissible value, would raise the estimated upper bound to $\Delta k/k_1$ only to about $+0.4$ per cent.

An analysis of estimates of $\Delta k/k_1$ therefore leads to a strong suggestion that Birch's finite-strain theory needs some small but significant amendments in applications at pressures of the order of a million atmospheres.

5. dk/dp in the outer core

Further differences between results on Birch's and the writer's approaches emerge from an analysis of dk/dp in the Earth's outer core.

Birch's equation (4) gives dk/dp as determined solely in terms of f, regardless of composition. Eq. (4) requires dk/dp to be always less than the value 4 which applies at zero pressure, and to decrease steadily as f increases. Using the values of f in section 2, and noting that these are minimum values, eq. (4) gives $dk/dp \leq 3.21, 3.09, 3.05$ for $z = 2900, 3600, 4500$ km, respectively.

In contrast, the writer's procedures give dk/dp on the whole increasing with p throughout the lower mantle and outer core. The empirical relation

$$k = 2.34 + 3.00p + 0.10p^2 \qquad (9)$$

(BULLEN, 1968e), where the units are 10^{12} dyn/cm^2, represents within 1 per cent the writer's most recent results for $1000 < z < 4500$ km. (The best fitting linear formulae for the lower mantle and outer core separately give dk/dp greater in the outer core than in the mantle by about 0.3.)

Using the equivalent of eq. (6) with η taken equal to unity, and using data of Gutenberg on $d\phi/dz$, BIRCH (1952) inferred a mean value of 3.0 for dk/dp and thence reasonable agreement with his results derived through eq. (4). He noted at the same time that the data of Jeffreys on $d\phi/dz$ did not give satisfactory agreement but ascribed this to uncertainties in the seismic data. At the time when Birch drew these conclusions, it had been reasonable to assume $\eta \approx 1$ in the outer core down to nearly 5000 km depth. But this assumption can no longer be relied on between 4500 and

5000 km depth, a range inside which η may exceed unity substantially (BULLEN, 1965b).

The writer has therefore re-examined the seismic evidence relating to η and dk/dp in the outer core. The re-examination, details of which are being published elsewhere (BULLEN, 1969), removes much of the seismic support for eq. (4) and has brought to light some new features of interest.

It will now be convenient to refer to the regions for which $2900 < z < 3600$ km, $3600 < z < 4500$ km, as E_1, E_2, respectively. Throughout the whole of E_1 and E_2, the Jeffreys and Gutenberg curves of $g^{-1}d\phi/dz$ against z run nearly parallel; the Jeffreys values are greater than the Gutenberg throughout, but only by about 0.2. Below E_2, the differences are much greater. Inside E_1, the Jeffreys and Gutenberg data both give $g^{-1}d\phi/dz$ increasing fairly steeply, the total increase being about 0.8 in both cases. Inside E_2, $g^{-1}d\phi/dz$ is given as approximately constant, equal to 2.4–2.6. The suggestion thus arises that the two parts E_1 and E_2 of the outer core are in slightly different states.

In interpreting the results, account has been taken of experimental data such as those of Orson ANDERSON (1968). Inside E_2, the overall evidence indicates fairly firmly that η and dk/dp are both fairly constant at about 1.0 (corresponding to chemical homogeneity) and 3.5, respectively. The most probable interpretation for E_1 gives η decreasing from about 1.4 at the top to 1.0 at the bottom, and dk/dp increasing from about 3.1 to 3.5 (sharply contrary to the trend indicated by eq. (4)). At the top of E_1, there is some flexibility in the assessments: while $dk/dp - \eta$ is well determined in terms of $g^{-1}d\phi/dz$, there is no definite evidence enabling the separate values of η and dk/dp to be closely determined. The interpretation taken gives the least departures from simple behaviour inside E_1 and also happens to give dk/dp continuous at N. Decreasing the preferred estimate of η at the top of E_1, i.e. moving towards homogeneity in E_1, would decrease the estimate of dk/dp at the top of E_1 and thereby accentuate the discordancy with eq. (4).

Quite apart from the apparent trend of dk/dp inside E_1 being against the indication of eq. (4), it is to be noted that the more definitely determined value of about 3.5 inside E_2 exceeds that indicated by eq. (4) by about 0.5; for $f \approx 0.18$, the right side of eq. (4) would need a positive correction of about one-sixth to produce agreement. Looking at the matter the other way round, putting $dk/dp = 3.5$ in eq. (4) would give $f = 0.06$, a value very much less than the minimum compression which is expected to exist anywhere inside the Earth's core. Hence the suggestion strongly arises that Birch's finite-strain theory needs significant amendment on this count also. At the same time it needs to be pointed out that much depends on the degree of reliability that can be attached to the available seismic estimates of $d\phi/dz$. It is very desirable that new checks should be made on the variation of α in the outer core.

Although the analysis suggests no departures from uniform composition and phase inside E_2, it indicates the presence of mild departures inside E_1. There is a suggestion, but no definite evidence, that these departures may be associated with phase change, a point which it may interest geochemists to examine.

The whole range of depth from about 2700 to 3600 km thus now appears to be slightly abnormal. Should the abnormalities inside D'' ($2700 < z < 2900$ km) and E_1 be substantially connected with phase changes, there incidentally arises the question as to whether some modified form of the mantle-core phase-change hypothesis earlier considered by RAMSEY (1948) and the writer (1949), involving a measure of continous phase change, might be relevant, at least as a contributing factor, to the changes in density which take place in this part of the Earth.

6. dk/dp below the outer core

Below the region E_2, the observational evidence is insufficient to yield closely reliable estimates of dk/dp through eq. (6). Immediately below E_2 for some distance, η is likely to reach 3 units or so (BULLEN, 1965b) but the value of η here is appreciably uncertain. Inside the inner core G, where β is likely to be significant, direct evidence on β is lacking, so that $d\phi/dz$ is much less certainly determined than in the outer core.

Direct application of Birch's eq. (4), using the data on f in section 2, would give $dk/dp \leq 3.05$ inside G, but the discussion in section 5 indicates that this assessment cannot be relied on.

Values of $\partial k/\partial p$ for particular materials at pressures equal to those inside G can be inferred from curves presented by BIRCH (1963) using data from shock-wave experiments. But $\partial k/\partial p$ here relates to changes of pressure under Hugoniot thermodynamical conditions. For iron, which is likely to be sufficiently representative of the inner core composition, the curves yield $\partial k/\partial p \approx 6$. It is a question as to how far the corresponding adiabatic or isothermal value of dk/dp would be less than this value of 6, and whether a useful assessment of dk/dp inside G is possible by this route.

In most of the writer's Earth models, dk/dp has been taken as continuous throughout the whole core, as a consequence of the evidence of near continuity of dk/dp throughout a wide range of depth which includes most of the Earth between the upper mantle and lower core. The value usually taken for dk/dp inside G has been around 4. Taking note of the results in section 5, it seems reasonable to suppose on current evidence that, below E_2, dk/dp lies between about 3.5 and 4.2. In constructing Earth models, it continues to be appropriate to assume that, to good approximation, k is continuous throughout the core.

7. Concluding remarks

The above analysis shows that estimates made by Birch and the writer of k and dk/dp in the Earth's deeper interior, while agreeing in order of magnitude, are discrepant in important details. There is a strong suggestion that Birch's finite-strain theory needs some amendment before it can be closely relied on in respect of results at pressures of the order of a million atmospheres. The only alternative to this appears to be that errors in the available seismic data for the outer core are greater than expected and that there are greater departures than expected from assessments of certain mantle properties, for example, the mean density in the outer part of the mantle and the mean temperature gradient in the lower mantle. Before too definite a conclusion can be drawn, however, it is desirable that new checks should be made on the P velocity variation in the outer core.

Subject again to the degree of reliability of the seismic data for the outer core, there is a strong suggestion of mild inhomogeneity, which may be due to continuous phase changes, in the outermost 700 km of the core.

Various sources of evidence relating to dk/dp in the lower core are discussed but do not lead to closely determined direct estimates of dk/dp. Suggestions are made concerning the most suitable values to take for k and dk/dp in the lower core.

References

ANDERSON, O. L. (1968) Phys. Earth Planet. Interiors **1**, 169.
BIRCH, F. (1952) J. Geophys. Res. **57**, 227.
BIRCH, F. (1963) in *Solids under Pressure* (McGraw-Hill, New York) ch. 6, pp. 137–162.
BULLEN, K. E. (1949) Monthly Notices Roy. Astron. Soc. **5**, 355.
BULLEN, K. E. (1963) Geophys. J. **7**, 584.
BULLEN, K. E. (1965a) *Introduction to the Theory of Seismology*, 3rd ed. (Cambridge University Press).
BULLEN, K. E. (1965b) Geophys. J. **9**, 233.
BULLEN, K. E. (1967) Geophys. J. **13**, 459.
BULLEN, K. E. (1968a) Geophys. J. **16**, 31.
BULLEN, K. E. (1968b) Phys. Earth Planet. Interiors **1**, 297.
BULLEN, K. E. (1968c) Nature **218**, 262.
BULLEN, K. E. (1968d) Proc. Nat. Acad. Sci. U.S. **60**, 752.
BULLEN, K. E. (1968e) Geophys. J. **16**, 235.
BULLEN, K. E. (1969) "Compressibility-pressure-gradient and the constitution of the Earth's outer core", Geophys. J., in course of publication.
BULLEN, K. E. and R. A. W. HADDON (1967a) Proc. Nat. Acad. Sci. U.S. **58**, 846.
BULLEN, K. E. and R. A. W. HADDON (1967b) Phys. Earth Planet. Interiors **1**, 1.
BULLEN, K. E. and R. A. W. HADDON (1969a) "Upper bound to change in incompressibility at the Earth's mantle-core boundary", Geophys. J. **17**, 179.
BULLEN, K. E. and R. A. W. HADDON (1969b) "Corrections to three Earth models", Phys. Earth Planet. Interiors **2**, 35.
RAMSEY, W. H. (1948) Monthly Notices Roy. Astron. Soc. **4**, 498.

PARTIAL MELTING IN THE UPPER MANTLE*

DON L. ANDERSON and CHARLES SAMMIS

Seismological Laboratory, California Institute of Technology, Pasadena, Calif., U.S.A.

The low velocity zone in tectonic and oceanic regions is too pronounced to be the effect of high temperature gradients alone. Partial melting is consistent with the low velocity, low Q and abrupt boundaries of this region of the upper mantle and is also consistent with measured heat flow values. The inferred low melting temperatures seem to indicate that the water pressure is sufficiently high to lower the solidus about 200 °C to 400 °C below laboratory determinations of the melting point of anhydrous silicates.

The mechanical instability of a partially molten layer in the upper mantle is probably an important source of tectonic energy. The top of the low-velocity zone can be considered a self-lubricated surface upon which the top of the mantle and the crust can slide with very little friction. Lateral motion of the crust and upper mantle away from oceanic rises is counterbalanced by the flow of molten material in the low-velocity layer toward the rise where it eventually emerges as new crust. If this lateral flow of molten material is not as efficient as the upward removal of magma, then regions of extrusion, such as oceanic rises, will migrate.

1. Introduction

The low velocity zone is now a well established feature of the upper mantle, particularly in tectonic and oceanic regions. This region of the mantle also attenuates seismic waves more rapidly than adjacent regions. The boundaries of the low velocity zone are near 60 and 150 km and seem to be relatively sharp. The total decrease in velocity, amounts to about 3 to 5%; the attenuation increases by a factor of 3 or more.

A velocity reversal may be due to a high temperature gradient, a change in composition, solid–solid phase changes or to the onset of partial melting. Ultrasonic data indicate that a temperature gradient of the order of 6 to 10 °C/km can cancel out the effect of pressure for compressional waves. The critical gradient for shear waves is of the order of 2.5 to 4 °C/km. Thus, low or negative velocity gradients can be expected in the upper mantle, particularly for shear waves. The current general concensus seems to be that high temperature gradients are an adequate explanation of low velocity zones (MACDONALD and NESS, 1961; LUBIMOVA, 1967; VALLE, 1956). RINGWOOD (1962a) pointed out what he considered serious thermal difficulties associated with this interpretation and he and CLARK and RINGWOOD (1964) proposed a change in mineralogy. MAGNITSKIY (1965), after considering several possibilities, concluded that "amorphization" was responsible for the decrease in velocity. PRESS (1959) attributed the low velocity zone to a "state near the melting point". SHIMOZURU (1963) emphasized the possible role of partial melting. It is the purpose of this paper to investigate in greater detail the origin and implications of the low velocity zone.

We combine ultrasonic data with recent seismic data in order to estimate the thermal gradients in the upper mantle. In the first model considered we assume that the upper mantle is homogeneous in composition and phase. In a second model we include changes in mineralogy such as have been proposed by GREEN and RINGWOOD (1967).

2. Seismic results

The difficulties associated with detecting low velocity zones with body wave data are well known. For sources above the zone all the evidence is indirect and involves an anomalous decrease of amplitudes with distance (the shadow zone effect) and a delay between the P_n (or S_n) branch of the travel time curve and the branch associated with waves being refracted upward from below the low velocity zone. However, with improvements in

* Contribution 1624, Division of Geological Sciences, California Institute of Technology, Pasadena, Calif. 91109.

timing and in amplitude calibration these methods can now be used with some assurance in detecting the presence of a velocity reversal and in determining the top of the zone and the total delay through it. The thickness and velocity of the zone can, to some extent, be traded off against each other so the details of this region cannot be unambiguously determined with these techniques. Surface wave studies and the method of vertical travel times, applicable in some tectonic regions give a more direct determination. Over a large part of the Earth the only pertinent data come from the study of surface waves which are mainly sensitive to the shear velocity structure.

If high temperature gradients are responsible for the low velocity zone it should begin and end gradually as in the original model of GUTENBERG (1959) and in the model of BROOKS (1962). Recent models have the top and bottom of this zone relatively abrupt.

Figure 1 shows recent body wave solutions for the western United States (JOHNSON, 1967; JULIAN and ANDERSON, 1968). The region between about 60 and 150 km, the low velocity zone is the subject of the present paper. Also shown are three theoretical mantle models which indicate the effect of temperature, pressure and phase changes on compressional velocities in purely olivine mantles, and in a mantle which contains 20% (molar) pyroxene. Note the shallow low-velocity zone centered near 100 km which results from the relatively high temperature gradient in this region of the mantle. Clearly the actual low velocity zone is much more pronounced than for these particularly theoretical homogeneous models.

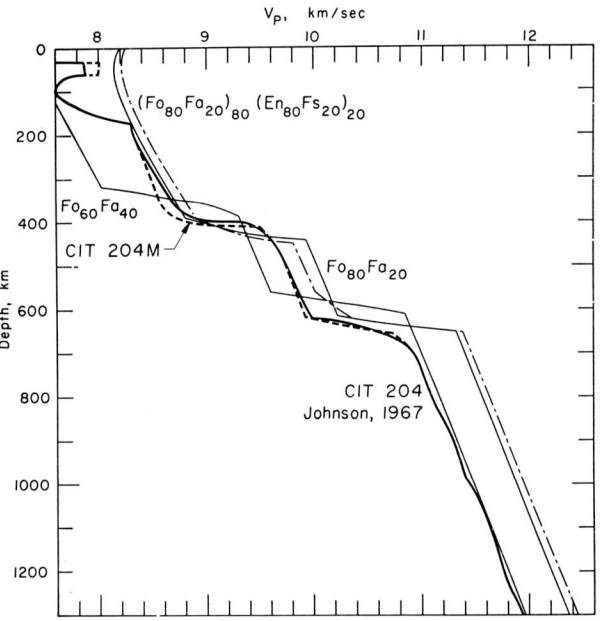

Fig. 1. Recent compressional velocity profiles for western United States (JOHNSON, 1968; JULIAN and ANDERSON, 1968). Theoretical curves for olivine mantles modified from ANDERSON (1967).

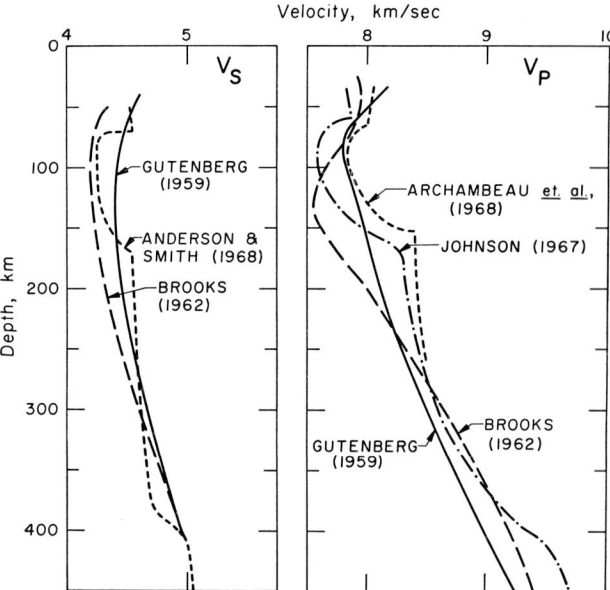

Fig. 2. Summary of upper mantle structures used to compute temperature gradients in the uppermost mantle and in the low velocity zone.

Figure 2 summarizes some of the published solutions for the upper mantle. These are primarily oceanic and tectonic models and the following discussion is for these regions of the Earth. The structures used for the present calculations are due to JOHNSON (1967), ARCHAMBEAU et al. (1968), GUTENBERG (1969), BROOKS (1962), and D. L. ANDERSON and SMITH (1968). The important parameters of these models are tabulated in table 1. Since the details of the low velocity zone are not well determined the total decrease in velocity or the velocity gradient through this region is poorly defined, but the total decrease in velocity for these structures is about 2.4% to 5.4%. If the velocity in this region is continuous the gradient is about -0.003 to -0.01 (km/s)/km. As we shall show these changes in

TABLE 1

Velocity gradients in the top of the low velocity zone

Model	Bottom of lid		Minimum		$\partial V/\partial z$ (km/s km) ($\times 10^3$)	ΔV (%)
	depth (km)	velocity (km/s)	depth (km)	velocity (km/s)		
Gutenberg (P)[2]	60	8.15	150	7.85	−3.33	3.7
Gutenberg (S)[2]	60	5.6	150	4.35	−2.78	3.4
Brooks (P)[1]	50	7.92	130	7.5	−5.25	5.3
Brooks (S)[1]	70	4.35	125	4.20	−2.7	3.4
Johnson (P)[3]	70	8.05	100	7.85	−6.67	2.5
200202 (S)[4]	70	4.39	100	4.26	−4.33	3.0
Archambeau et al. (P)[5]						
Bilby SE	60	7.93	90	7.66	−9.00	3.4
Bilby NE	60	8.02	80	7.83	−9.50	2.4
Shoal Fallon NE	60	7.93	95	7.70	−6.57	2.9
Shoal Fallon SE	No lid					

1. BROOKS (1962).
2. GUTENBERG (1959).
3. JOHNSON (1967).
4. D. L. ANDERSON and SMITH (1968).
5. ARCHAMBEAU et al. (1968).

velocity, if attributed to effects of temperature and pressure alone, give unreasonably high thermal gradients and heat flow values. While it is fairly easy to obtain a low velocity zone with reasonable thermal gradients, it is quite difficult to obtain the observed decreases in velocity unless partial melting in the upper mantle is allowed.

3. Ultrasonic data

Laboratory data on the temperature and pressure derivatives of velocity exist only for relatively low temperatures and pressures. The most accurate data are for single crystals and for powdered ceramics which are sintered to near theoretical density. In natural rock specimens the problems of porosity and cracks persist to at least 10 kb and heterogeneity and large grain size reduce the precision of measurement, particularly of the derivatives. Table 2 summarizes most of the relevant laboratory data on elastic properties of silicates and oxides. Unfortunately, the two most pertinent materials for discussions of the upper mantle, olivine and pyroxene, were measured on high porosity specimens. Once the effects of porosity are removed the velocity derivatives probably increase with temperature and decrease with pressure and this further complicates discussions of temperature gradients in the mantle from seismic data. The pressures and temperatures in the vicinity of the low-velocity zone of the mantle are now

TABLE 2

Velocity derivatives

Parameter	Olivine[1]	Pyroxene	Garnet[1]	Spinel[1]	Al_2O_3[1]	MgO[1]	MgO[1]
ρ (g/cm^3)	3.021	3.279[3]	4.160	3.619	3.972	3.580	3.583
V_P (km/s)	7.59	7.64[3]	8.53	9.91	10.845	9.661	9.692
V_S (km/s)	4.36	4.59[4]	4.76	5.65	6.373	5.9974	6.041
$(\partial V_P/\partial P)_T$ (10^{-3} km/s kb)	10.3	19[3]	7.84	4.9	5.18	8.66	8.35
$(\partial V_P/\partial T)_P$ (10^{-4} km/s °C)	−4.1	−6.4[2]	−3.9	−3.1	−3.6	5.0	5.2
$(\partial V_S/\partial P)_T$ (10^{-3} km/s kb)	2.45	7[4]	2.17	0.43	2.21	4.23	4.02
$(\partial V_S/\partial T)_P$ (10^{-4} km/s °C)	−2.9	−6.0[2]	−2.2	−2.2	−3.1	−4.8	−4.4
$(\partial T/\partial z)_C$ P wave	7.6	9.5	8.2	5.6	5.7	6.13	5.7
$(\partial T/\partial z)_C$ S wave	2.5	3.9	4.1	0.69	3.9	3.12	3.2

1. O. L. ANDERSON et al. (1968).
2. HUGHES and NISHITAKE (1963).
3. BIRCH (1960).
4. SIMMONS (1964).

accessible in the laboratory but the measurement of the velocity derivatives at high temperature and pressure remains one of the most important unexplored areas of geophysics.

The variation of velocity with depth in homogeneous regions of the mantle depends, of course, on both the temperature and pressure gradient. These opposing effects are conveniently summarized in the critical gradient, $(\partial T/\partial z)_c$, the temperature gradient which leads to constant velocity with depth. For compressional waves, of main concern here, this parameter only varies by 15% for materials important in the upper mantle (olivine, pyroxene and garnet) although it varies by 52% when all the materials in table 2 are considered.

Note that pressure increases the compressional velocity more effectively than it increases the shear velocity, but the effects of temperature are roughly comparable. In terms of relative changes in velocity, pressure affects the compressional velocity most and temperature affects the shear velocity most. These conditions combine to make it easier to generate a low shear velocity zone. For an upper mantle composed primarily of olivine and pyroxene the critical temperature gradient, i.e. the condition for constant velocity with depth, is 7.6 to 9.5 °C/km for compressional waves and 2.5 to 3.9 °C/km for shear waves. Larger temperature gradients are required to decrease the velocity. Note that the critical gradient for shear waves in spinel is almost negligible. If this is typical of the spinel structure we can expect a negative shear velocity gradient in the spinel region of the mantle, between about 450 and 600 km.

Most of the temperature and pressure derivatives in table 2 were measured at Lamont Geological Observatory by O. L. Anderson, Schrieber and Soga (O. L. ANDERSON et al., 1969). Pyroxenes have been measured by BIRCH (1960), SIMMONS (1964) and HUGHES and NISHITAKE (1963) and we have estimated the derivatives from their data. Both the temperature and pressure derivatives for pyroxene are much higher than for the other materials and the uncertainty is much greater. Unfortunately, the data for olivine were obtained on a porous ceramic specimen. There is still some uncertainty in the velocities of olivine. VERMA (1960) measured the elastic constants of a gem quality single crystal. The P and S velocities which he computed by the Voigt averaging technique were 13% higher than those given by O. L. ANDERSON et al. (1968) for the porous polycrystalline sample. Anderson's velocities are given in table 2 and were used for the computations. If the velocity in olivine is higher, as suggested by Verma's work, the following conclusions are strengthened.

SOGA and O. L. ANDERSON (1967) also measured the effect of temperature on the Young's modulus and the shear modulus for porous forsterite and porous protoenstatite. The effect of temperature was less effective in decreasing the moduli for $MgSiO_3$ than for Mg_2SiO_4 which is not in agreement with the results in table 2 which were obtained in different laboratories.

4. Interpretation of seismic results using ultrasonic data

The effects of temperature, pressure, composition and phase change on the seismic velocity gradient may be expressed as

$$\frac{dV}{dz} = \left(\frac{\partial V}{\partial P}\right)\frac{dP}{dz} + \left(\frac{\partial V}{\partial T}\right)\frac{dT}{dz} + \left(\frac{\partial V}{\partial C}\right)\frac{dC}{dz} + \left(\frac{\partial V}{\partial \phi}\right)\frac{d\phi}{dz}, \quad (1)$$

where z = depth, and the independent variables are P = pressure, T = temperature, C = composition, ϕ = phase, V = seismic velocity. In homogeneous regions of the Earth this expression may be solved for the temperature gradient

$$\frac{dT}{dz} = \left[\frac{dV}{dz} - \left(\frac{\partial V}{\partial P}\right)_T \frac{dP}{dz} - \left(\frac{\partial V}{\partial \phi}\right)\frac{d\phi}{dz}\right] \bigg/ \left(\frac{\partial V}{\partial T}\right)_P. \quad (2)$$

In solving this equation in the next few sections the derivative dV/dz is taken directly from the seismic profiles and dP/dz is taken to be a hydrostatic pressure gradient of 0.32 kb/km. The ultrasonically determined temperature and pressure derivatives of the velocity were used for $(\partial V/\partial P)_T$ and $(\partial V/\partial T)_P$.

Two sets of temperature derivatives were calculated using eq. (2). The first set of calculations assumed constant phase, $\partial \phi/\partial z = 0$; the second set of calculations included the effects of the changes in mineralogy proposed by GREEN and RINGWOOD (1967).

5. Temperature gradients in the uppermost mantle

The temperature gradient in the upper mantle, above the low-velocity zone, in the absence of partial melting, probably lies between about 7 and 10 °C/km (RINGWOOD, 1966). The melting point gradient for most silicates is about 3 to 4 °C/km so the geotherm must level off at greater depth if melting is to be avoided. If the geotherm intersects the melting point curve the latent heat of melting and the possibility of heat transfer by motion of the fluid phase will tend to make the melting point gradient the limiting gradient so we should expect a gradient of 4 °C/km or less in partically molten regions of the upper mantle.

The thermal conductivity of forsterite at 30 km and 700–1100 °K is 0.010 to 0.014 cal/cm °K s (FUJISAWA et al. 1968). CLARK and RINGWOOD (1964) estimate a value of 0.006 cal/cm °K s for the upper mantle. The conducted heat flow from the mantle would therefore be of the order 0.4 to 1.4 μcal/cm^{2}s for combinations of parameters thought to be appropriate. Estimates of heat flow from the mantle vary between about 0.3 to 0.6 μcal/cm^{2}s (CLARK and RINGWOOD, 1954; BIRCH, 1966).

Since olivine and pyroxene probably make up more than 90% of the upper mantle, the velocity derivatives with respect to temperature and pressure will probably be controlled by these components. The ultrasonic data for olivine and pyroxene were used in eq. (2) to estimate the temperature gradients in the upper mantle above the low velocity zone. These estimated gradients for four different seismic profiles are given in table 3. The

TABLE 3

Temperature gradients in the lid constant composition

Profile	Depth interval (km)	$\partial V_P/\partial z$ (km/s km)	$(\partial T/\partial z)$ Olivine (°C/km)	$(\partial T/\partial z)$ Pyroxene (°C/km)
Johnson CIT204	40–60	+1.0	5.5	7.8
Bilby SE	37–60	−1.2	10.9	11.3
Bilby NE	35–60	−1.2	10.9	11.3
Shoal Fallon NE	38–60	−1.1	10.6	11.1

values range from about 5 °C/km to 11 °C/km. There is uncertainty in both seismic gradients and in the ultrasonically determined derivatives but the above estimates seem reasonable on other grounds. Surface gradients are on the order of 20 °C/km and the gradient should decrease with depth. With a conductivity of 6×10^{-3} cal/cm °K s the implied heat flow is of the order of 0.3 to 0.7 μcal/cm^{2}s through the base of the crust. Based on the above estimates we would expect the temperature gradient to be less than 11 °C/km below about 50 km depth. In the next section we show that with available seismic and ultrasonic data the tem-

perature gradients required to create the low velocity zone are much greater than this.

6. Temperature gradient in the low velocity zone

Seismic velocity gradients are uncertain in the low velocity zone. The difference between the velocity at the bottom of the lid and the minimum velocity was used to define the gradient. If the velocity drop is more abrupt the arguments in the following section are strengthened. If the minimum in velocity occurs deeper then the problem of terminating the low velocity zone arises. In any case current velocity structures between about 50 and 200 km cannot be explained by the effects of pressure and temperature alone.

6.1. Model A: Constant composition and phase

The velocity gradients dV/dz from table 1 were used in eq. (2) with $d\phi/dz$ taken to be zero. The results of the calculations are summarized in table 4 for the region of decreasing velocity. These temperature gradients for an olivine mantle are plotted in fig. 3.

TABLE 4

Uniform upper mantle.
Temperature gradients required to generate low velocity zone

Model	Olivine $\partial T/\partial z$ (°C/km)	Pyroxene $\partial T/\partial z$ (°C/km)	Q* (μcal/cm^{2}s)
Gutenberg (P)	16	15	1.0–0.9
Gutenberg (S)	12	8	0.7–0.5
Brooks (P)	21	18	1.3–1.1
Brooks (S)	12	8	0.7–0.5
Johnson (P)	25	20	1.5–1.2
200202[1] (S)	18	11	1.1–0.7
Bilby SE[2] (P)	32	25	1.9–1.5
Bilby NE[2] (P)	31	24	1.9–1.4
Shoal Fallon NE[2] (P)	26	21	1.6–1.3

* Heat flow with thermal conductivity $k = 0.006$ cal/cm °K s.
1. D. L. ANDERSON and SMITH (1968).
2. ARCHAMBEAU et al. (1968).

Temperature gradients of the order of 15 to 32 °C/km are required to generate the observed low velocity zone for compressional waves. Most of the models require a gradient of 20 °C/km or greater. The shear wave data requires gradients of 8 to 18 °C/km. For a given region the implied thermal gradient for shear waves is less than for compressional waves, i.e. the same thermal gradient apparently cannot explain both sets of data. This in itself is an argument against the conventional

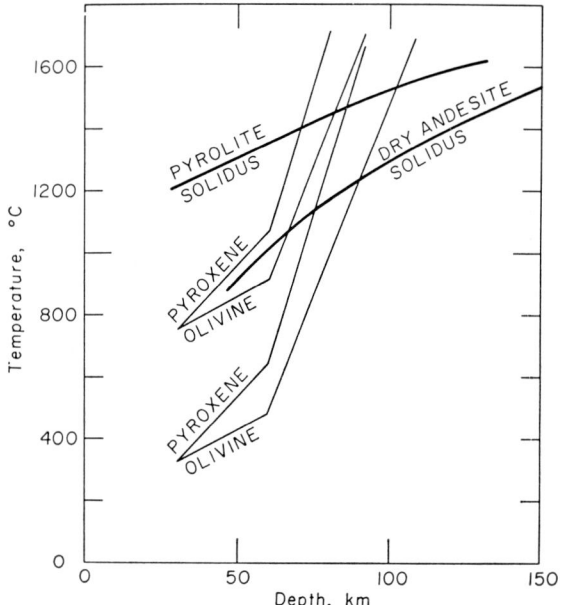

Fig. 3. Geotherms in a pure olivine and pure pyroxene mantle which correspond to the low velocity zone of JOHNSON (1967). Assuming temperatures of 330 °C and 750 °C at the base of the crust, partial melting is suggested below 100 km, even in a dry mantle.

interpretation of low velocity zones in terms of high thermal gradients.

The temperature at the base of the crust probably varies from about 300 °C in continental regions of low heat flow to about 750 °C in regions of high heat flow such as in the Basin and Range province of the western United States (ROY and BLACKWELL, personal communication). Computed temperatures are shown for two assumed temperatures at 30 km. The upper curves are based on the Bilby SE profile of ARCHAMBEAU et al. (1968). The lower curves were computed from JOHNSON's (1967) model CIT204. The melting curves for pyrolite and dry andesite are from GREEN and RINGWOOD (1967) and GREEN and RINGWOOD (1966). These curves probably bracket the possibilities for a dry upper mantle. Slight amounts of water can reduce the solidus by several hundred degrees. Even for dry rock partial melting is suggested at depths shallower than 100 km. In the Basin and Range partial melting is implied at the base of the crust.

The temperature gradients in the vicinity of the low velocity zone give a heat flow through this region of the mantle of 1 to 2 $\mu cal/cm^2 s$. This is two to four times larger than the 0.5 $\mu cal/cm^2 s$ which CLARK and RINGWOOD (1964) estimate should be coming from the mantle and is inconsistent with the values determined in the previous section for the uppermost mantle.

There are regions of the Earth where high surface heat flow values do imply higher heat flow from the mantle. For example ROY et al. (1968) estimate that 1.1 $\mu cal/cm^2 s$ is the heat flow from the mantle in the Basin and Range province of the western United States, and estimate that the temperature at 30 km is about 750 °C. They interpret their results in terms of the upwelling of partially molten material to near the base of the crust. Their temperature gradient is consistent with partial melting in the upper 100 km even for a dry refractory mantle. It is yet to be established whether low-velocity zones of the type discussed here occur only in regions of high heat flow. If they do, then partial melting is implied without having to invoke any seismological data. One of the main points of the present paper is that pronounced low-velocity zones are incompatible with regions of normal heat flow.

6.2. Model B: *Constant composition, variable mineralogy*

GREEN and RINGWOOD (1967) presented a mineralogical model for the upper mantle based on the stability fields of aluminous pyroxene peridotite and garnet peridotite. We now wish to see if the solid–solid phase changes in this model can overcome the objections to the constant phase model just investigated.

In order to maximize the effect of phase change in reducing the velocity between 60 and 100 kilometers one would phase out all the relatively high velocity spinel over this interval. This leads to a velocity decrease of about 1 % if one assumes that about 5 % spinel phases out while 5 % garnet phases in between the top and the minimum of the low velocity zone.

Taking this decrease into account temperature gradients in the region of decreasing velocity were recalculated and are summarized in table 5. The proposed change in mineralogy does not lower the velocity enough to give a reasonable heat flow, and melting is still predicted for the upper mantle.

7. Bottom of low velocity zone

The extremely high temperature gradients which are required to generate the observed velocity decrease in the upper part of the low velocity zone lead to further

TABLE 5

Temperature gradients required to generate low velocity zone in a pyrolite upper mantle taking solid–solid phase changes into account

Model	Olivine $\partial T/\partial z$ (°C/km)	Pyroxene $\partial T/\partial z$ (°C/km)	Q^* (μcal/cm^{2}s)
Gutenberg (P)	14	14	0.8
Gutenberg (S)	11	8	0.7–0.5
Brooks (P)	19	16	1.1–1.0
Brooks (S)	9	7	0.5–0.4
Johnson (P)	19	16	1.1–1.0
200202[1] (S)	13	8	0.8–0.5
Bilby SE[2] (P)	24	20	1.4–1.2
Bilby NE[2] (P)	23	19	1.4–1.1
Shoal Fallon Ne[2] (P)	19	17	1.1–1.0

* Heat flow with thermal conductivity $k = 0.006$ cal/cm s°C.
1. D. L. ANDERSON and SMITH (1968).
2. ARCHAMBEAU et al. (1968).

difficulties if one attempts to explain the termination of this zone by the same mechanism. If the upper mantle is homogeneous, then large *negative* temperature gradients are required to increase the velocity in the bottom part of the low velocity zone. This being unacceptable a change in phase, either solid–solid or liquid–solid, or a change in composition is required in order to terminate the low velocity zone.

If the upper mantle contains small amounts of water and if $P_{H_2O} < P_{total}$ then the upper and lower boundaries of the low-velocity zone can represent the intersection of the geotherm with the solidus. The upper boundary of the low-velocity zone could represent the onset of instability of hydrous phases and the first appearance, with depth, of free water. If $P_{H_2O} = P_{total}$ then the solidus is lowered so much that it is difficult to terminate the partial melt zone unless the water has migrated to the top of the upper mantle; the bottom would then represent the absence of free water and a consequent increase of the solidus temperature. Another possibility for terminating the low-velocity zone would be the appearance of the high pressure hydroxylated phases discussed by SCLAR (1968). In all the above cases the top and the bottom of the low-velocity would be relatively sharp.

It does not appear possible to explain the increase in velocity in terms of a phase change in a pyrolite upper mantle. The major change which occurs between 90 and 170 km is the increase in garnet from about 5% to about 15%. This would correspond to a velocity increase of about 1%.

On the basis of the seismic velocity data, then, it appears that the low velocity zone must be a region of different composition than the surrounding mantle or else it is partially molten. The fact that the attenuation increases in this zone strongly suggests the latter possibility although it is not out of the question that a compositional change could lower the elasticity and increase the anelasticity. The decrease in velocity of 2 to 5% could be accounted for by adding about 10% basalt. However, basalt is unstable below about 50 kilometers and the transformation to garnet granulite and eclogite would increase the velocity just where we require a velocity decrease. The presence of hydrated minerals is also unlikely at the temperatures and pressures in the vicinity of the low velocity zone. In fact, the instability and consequent dehydration of the low pressure, low temperature hydrous phases which may occur immediately below the Moho would serve to decrease the melting point and may facilitate the intersection of the geotherm with the solidus.

Very little data exists on the elastic properties of partially molten systems. SPETZLER and D. L. ANDERSON (1968) studied the effects of grain boundary melting in the NaCl–ice system. As the sample was warmed, both the compressional and shear velocities dropped abruptly at the eutectic temperature, i.e. at the onset of partial melting. For a dilute solution containing 3.3% melt at the eutectic temperature the compressional and shear velocities were 9.5% and 13.5% respectively, less than in the unmelted solid. The attenuation increased at the same time by 37 to 48%. Thus, a small amount of melt can have significant effects on the elastic and anelastic properties of a material. Other mechanical properties such as strength and viscosity can be expected to behave in a similar fashion.

8. Implications of a partial molten upper mantle

The latent heat associated with partial melting will tend to stabilize the temperature at the melting point gradient unless melting is so extensive that heat transfer by mass transport is allowed. In this case the temperature gradient would attempt to fall to the adiabat. The efficient removal of heat by convection would lower the temperature below the solidus but the relatively slow process of conduction would permit radioactivity to heat the material back up to the melting point so that the actual temperature gradient would probably oscillate between some temperature above the solidus and the adiabat. It will stay above the adiabat since any tendency for the temperature to decrease in the upper mantle will be counteracted by the more efficient heat flow from below allowed by the steepened gradient at the base of the partially molten zone.

Completely free vertical convection is constrained since we are discussing a penetrative convective system with perhaps only about 1% melt. In fact, the competition between heating up, further melting and heat transfer by fluid motions will stabilize the region at some, probably small, amount of partial melt.

The melting point gradient of many volcanic rocks is 10 °C/kb or about 3 °C/km. If the amount of melt is so slight that vertical convection is inhibited the heat flow through an upper mantle with this gradient will be about 0.3 μcal/cm^{2}s, very similar to previous estimates based on observed surface heat flow values and radioactivities.

The solidus temperatures of calc-alkaline and basic rocks range from about 1250 °C for dry quartz tholeiite to near 1400 °C for dry pyrolite (GREEN and RINGWOOD, 1966). These temperatures are well above those estimated for this depth in the mantle. A small amount of water can reduce the melting point by about 200 °C at these pressures. A small amount of water in the mantle seems to be required if melting is to begin near 60 km.

Figure 4 summarizes the pressure dependence of the solidus as determined by GREEN and RINGWOOD (1966) on dry calc-alkali rocks and on wet basalts (YODER and TILLEY, 1962; GENSHAFT et al., 1967). Also are shown estimates of geotherms by CLARK and RINGWOOD (1964). Comparing the solidus of the lowest melting point dry material with the oceanic geotherm, which is probably an upper bound, we see that melting is not possible above about 90 km. Small amounts of water decrease the solidus temperature considerably but it is not yet clear how much water is required to depress the solidus by the 200 to 400 °C that is required to generate a partial melt zone starting near 60 km. Presumably, trace amounts of H$_2$O, certainly less than 1%, would be adequate.

Once the minimum melting point, or solidus, of a region of the mantle is exceeded it will stay partially molten unless the removal of heat is more efficient than the generation of heat. In general the heat conducted upwards is replaced by heat conducted from below so a volume element will continue to heat up, or to supply heat for further melting, until the melt content is so

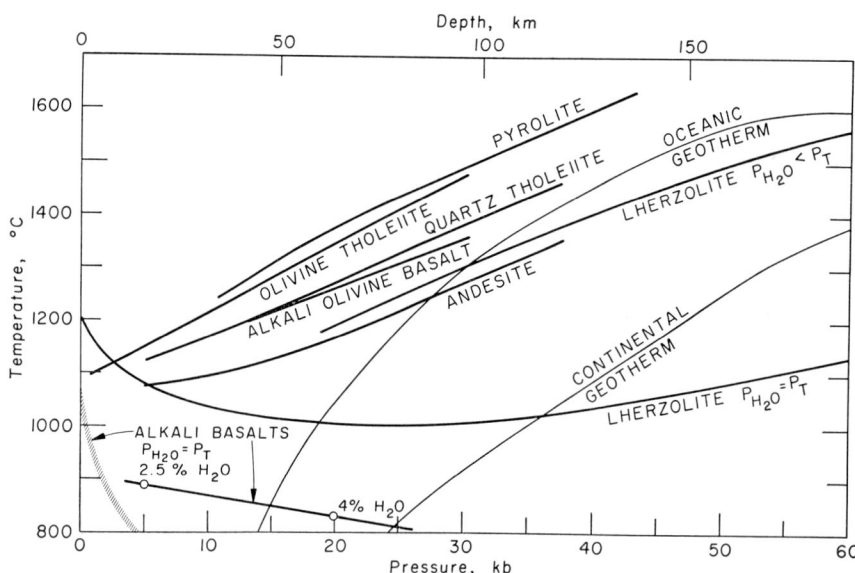

Fig. 4. Solidus curves of various rocks. The presence of a small amount of water significantly lowers the solidus. The pyrolite solidus is from GREEN and RINGWOOD (1967). The dry olivine tholeiite, dry alkali olivine basalt, dry quartz tholeiite and dry andesite are from GREEN and RINGWOOD (1966). The wet alkali basalts are from YODER and TILLEY (1962) and from GENSHAFT et al. (1967). The lherzolite curves are from KUSHIRO et al. (1968).

large that efficient removal of heat upwards is allowed by the mass transport of the molten phase. The critical amount of melt is perhaps 5%. An interval of igneous activity and volcanism may then occur until the melt content of the upper mantle is reduced below the threshold value. If we take the heat production rate of the upper mantle as 0.3×10^{-6} cal/g yr and the latent heat of fusion as 100 cal/g then 5% of the upper mantle can be melted 70 million years after the temperature initially reached the solidus. If the subcritical concentration is 2% then 42 million years will be the interval between volcanic episodes in a closed system. The material lost to the surface will be some fraction of the available melt and it can presumably be replenished from adjacent portions of the mantle, particularly the low velocity zone. The newly generated crust, and possibly upper mantle, will be less dense than surrounding columns and the lateral pressure gradient will facilitate the flow of molten material, at depth, toward the region being depleted. This lateral flow of material can be partially compensated for by flow from regions still further away from the region being depleted and partially by subsidence of the overlying crust and upper mantle. If lateral transport of magma at depth is less efficient than removal of magma to the surface, then seccessive eruptions will be adjacent to the depleted region and zones of volcanic and igneous activity will tend to migrate.

9. Alternate explanation

Hydrated minerals are generally unstable at temperatures in excess of about 500 °C. Dehydration of these minerals will probably lead to phenomena similar to partial melting. Concentration of H_2O vapor at grain boundaries will probably decrease seismic velocities and increase attenuation even though the dehydrated matrix itself has good mechanical properties. In either case, partial melting or dehydration, the existence of H_2O in the upper mantle is implied.

10. Tectonic implications

The concept of a wide spread partially molten layer in the upper mantle has important tectonic implications. Such a layer will be less dense than the surrounding mantle and will be mechanically unstable. Any perturbations in such a layer will tend to grow with time. For a uniform low density layer buoyancy forces will deform the surface of the layer and the optimal wavelength is about 6.3 times the thickness of the layer (RAMBERG, 1967). For the dimensions of the low velocity zone this optimal wavelength is about 1000 km. Accidental perturbations in the thickness and state of stress in the crust and upper mantle above the low velocity zone, however, will probably control the location and spacing of upwarpings and downbucklings. In regions of crustal tension, which may themselves be due to the attempted rise of the light magma the instability can be relieved by the extrusion of the molten material. In regions of crustal and upper mantle compression the instability is more likely to be relieved by sinking of the crust and upper mantle into the low density layer. In any event, if partial melting has proceeded to a sufficiently advanced stage the magma and associated volatiles in the low velocity zone will tend to rise to the top of the zone and may effectively lubricate the boundary between the base of the lid and the top of the partially molten layer, around 50 to 60 km. This may help explain the extreme mobility of the oceanic crust which apparently can slide around on the upper part of the low velocity zone with very low friction.

11. Conclusions

It does not appear possible to obtain a self-consistent interpretation of pronounced low velocity zones in terms of high temperature gradients. The thermal gradients required imply partial melting in the upper 100 km of the Earth and imply unacceptably high values for heat flow through the upper mantle. In addition thermal gradients required to satisfy the P velocity structures do not satisfy the S velocity structures and are not consistent with gradients estimated from velocities in the lid of the low velocity zone. A small amount of partial melting is consistent with the drop in velocity, the more pronounced effect on shear waves, and the increase in absorption. If the effects of partial melting in the NaCl ice system can be used as a guide only about 1% melt is required. Furthermore, the melting point gradient is consistent, if the melting is not too extensive, with the heat flow through the base of the crust. If large amounts of melt are generated, upward transport of heat by convection will periodically give rise to large heat flow values and, possibly to intrusive and extrusive igneous activity.

Acknowledgments

This research was partially supported by the Air Force Office of Scientific Research, Office of Aerospace Research, United States Air Force, under AFOSR contract number AF-49(638)-1337, and National Science Foundation grant GA 1003.

References

ANDERSON, D. L. (1967) Science **157**, 1165.
ANDERSON, D. L. and M. SMITH (1968) Trans. Am. Geophys. Union **49**, 283.
ANDERSON, O. L., E. SCHREIBER, R. C. LIEBERMANN and N. SOGA (1968) Rev. Geophys. **6**, 491.
ARCHAMBEAU, C. B., R. ROY, D. BLACKWELL, D. L. ANDERSON, L. JOHNSON and B. JULIAN (1968) Trans. Am. Geophys. Union **49**, 328.
BIRCH, F. (1960) J. Geophys. Res. **65**, 1083.
BIRCH, F. (1966) *Advances in Earth science*, P. M. Hurley, ed. (M.I.T. Press, Cambridge, Mass.) 403.
BROOKS, J. A. (1962) Geophys. Monograph **6**, G. A. MacDonald and Hisashikuno, eds., 2.
CLARK, S. P. and A. E. RINGWOOD (1964) Rev. Geophys. **2**, 35.
FUJISAWA, H., N. FUJII, H. MIZUTANI, H. KANAMORI and S. AKIMOTO (1968) Tech. Rept. ISSP Tokyo **A298**.
GENSHAFT, YU S., V. V. NASEDKIN, YU. N. RYABININ and V. P. PETROV (1967), Izv. Acad. Sci. USSR Ged. Ser. English Transl., Phys. Solid Earth **9**, 567.
GREEN, D. H. and A. E. RINGWOOD (1967) Earth Planet. Sci. Letters **3**, 151.
GREEN, T. H. and A. E. RINGWOOD (1966) Earth Planet. Sci. Letters **1**, 307.
GUTENBERG, B. (1959), Ann. Geofis. Rome **12**, 439.
HUGHES, D. S. and T. NISHITAKE (1963) Geophys. Papers dedicated to Prof. K. Sassa.
JOHNSON, L. R. (1967) J. Geophys. Res. **72**, 6309.
JULIAN, B. R. and D. L. ANDERSON (1968) Bull. Seismol. Soc. Am. **58**, 339.
KUSHIRO, I., Y SYONO and S. AKIMOTO (1968) J. Geophys. Res., in press.
LUBIMOVA, E. A. (1967) in: *The Earth's mantle* (Academic Press, London) 213.
MACDONALD, G. J. F. and N. F. NESS (1961) J. Geophys. Res. **66**, 1865.
MAGNITSKIY, V. A. (1967) Izdatel'stro "Nedra" (1965), NASA Tech. Transl. TT F-395.
PRESS, F. (1959) J. Geophys. Res. **64**, 565.
RAMBERG, H. (1967) *Gravity and deformation of the Earth's crust, as studied by centrifuged models* (Academic Press, London, New York).
RINGWOOD, A. E. (1962a) J. Geophys. Res. **67**, 857.
RINGWOOD, A. E. (1962b) J. Geophys. Res. **67**, 4473.
RINGWOOD, A. E. (1966) Mineralogy of the mantle, in: P. M. Hurley, ed., *Advances in Earth science*, (MIT Press, Cambridge, Mass.) 357.
SCLAR, C. B., L. C. CARRISON and O. M. STEWART (1967) Trans. Am. Geophys. Union **48**, 226.
SHIMOZURU, D. (1963) J. Phys. Earth **11**, 19.
SIMMONS, G. (1964) J. Geophys. Res. **69**, 1117.
SOGA, N. and O. L. ANDERSON (1967) J. Am. Ceram. Soc. **50**, 239.
SPETZLER, H. and D. L. ANDERSON (1968) J. Geophys. Res. **73**, 6051.
VALLE, P. E. (1956) Ann. Geofis. Rome **9**, 371.
VERMA, R. K. (1960) J. Geophys. Res. **65**, 757.
YODER, H. S. and C. E. TILLEY (1962) J. Petrol. **3**, 342.

ISOTHERMAL COMPRESSION OF THE SPINEL PHASE OF Ni_2SiO_4 UP TO 300 KILOBARS AT ROOM TEMPERATURE

H. K. MAO*, TARO TAKAHASHI and WILLIAM A. BASSETT

Department of Geological Sciences and Space Science Center, University of Rochester, Rochester, N.Y. 14627, U.S.A.

The effect of pressure on the lattice parameter, hence the volume, for the high-pressure spinel phase of Ni_2SiO_4 has been determined at room temperature up to 300 kb by means of the X-ray diffraction method employing a diamond-anvil high-pressure cell. The isothermal bulk modulus was calculated to be 2.14 ± 0.07 Mb by a least squares fit of the experimental data to the Birch–Murnaghan equation. This value is consistent with the Anderson–Nafe relationship, but deviates from the seismic equation of state of D. L. Anderson.

1. Introduction

A knowledge of the relationships between the chemical composition, crystal structure and elastic properties of rock forming minerals is essential for the study of the composition and state of the Earth's interior by means of seismic data. O. L. ANDERSON and NAFE (1965) have systematized the effect of volume on the bulk and shear moduli for a number of crystalline solids including oxides, silicates and halides. D. L. ANDERSON (1965) has found a similar relationship between density and bulk modulus for oxides and silicates, and called it a seismic equation of state. This seismic equation of state, however, is a special case of the Anderson–Nafe relationship. We have been investigating systematically the effect of phase changes on bulk modulus in order to provide basic data needed for further extension and evaluation of the existing theories relating the elastic properties to crystallo-chemical parameters.

The olivine–spinel phase transformation in Ni_2SiO_4 was first observed by RINGWOOD (1962), and the temperature–pressure stability field for each phase was determined by AKIMOTO et al. (1965). The purpose of this paper is to report new measurements of the effect of pressure on the lattice parameter, hence the volume, of the spinel phase of Ni_2SiO_4 up to 300 kb at room temperature.

* Presently at the Geophysical Laboratory, Washington, D.C., U.S.A.

2. Experimental method

A high-pressure X-ray diffraction camera designed by BASSETT et al. (1967) was used in this study. The high-pressure cell of this camera consists of two gem quality diamond anvils ($\frac{1}{8}$ carat brilliant – cut with a 0.3 mm diameter culet) driven by a screw assembly. A sample is subjected to pressure between the anvil faces, while exposed to a beam of X-rays. A finely collimated X-ray beam (approximately 50 μ in diameter) of molybdenum $K_{\alpha 1}$ radiation, monochromatized with a curved quartz crystal, passes through one of the anvils and impinges on the central portion of the sample area. When a polycrystalline sample is compressed between the anvil faces, a maximum pressure and a minimum pressure gradient are produced at the center of the sample area. Thus, a small X-ray beam directed at the center of the sample area produces a high-resolution X-ray diffraction pattern. Diffracted rays pass out through the other anvil to a cylindrical film having a radius of 50 mm and a maximum 2θ angle of 45°. The X-ray geometry is essentially that of the Debye–Scherrer camera.

The samples consisted of a mixture of powdered Ni_2SiO_4 spinel crystals and sodium chloride, which serves as an internal pressure standard. The X-ray diffraction pattern therefore consists of the lines for the specimen and for NaCl. For calculations of the lattice parameter the diffraction lines (111), (200), (311), (222), (400), (333), (511) and (440) for the spinel phase of

Ni$_2$SiO$_4$, and lines (200), (220) and (222) for NaCl were used. The lattice parameter values thus calculated are accurate to $\pm 0.1\%$, and the volumes to $\pm 0.3\%$.

The pressures to which the samples were subjected were calculated from the pressure–volume relation for NaCl, which was computed with the vibrational formulation of the Hildebrand equation by WEAVER et al. (1969) and MAO et al. (1969). On the basis of the uncertainties in the parameters used for the computation, the accuracy for the pressure–volume relation for NaCl has been estimated to be $\pm 2.5\%$ in pressure for a given value of compression.

The sample of the spinel phase of Ni$_2$SiO$_4$ used for this work is a quench product of a high pressure-temperature synthesis at about 32 kb and 900 °C, and was kindly provided by Dr. S. Akimoto of the University of Tokyo.

3. Results

The results of the measurements at 23 ± 3 °C are listed in table 1, and also plotted in fig. 1. The isothermal bulk modulus at zero pressure K_0 was calculated by a least squares fit of the volume data to the first order Birch–Murnaghan equation

$$P = \tfrac{3}{2} K_0 \left\{ \left(\frac{V}{V_0}\right)^{-\tfrac{7}{3}} - \left(\frac{V}{V_0}\right)^{-\tfrac{5}{3}} \right\}.$$

Since the precision of the data did not warrant a reliable determination of the second order term (i.e. the ξ-parameter), the ξ-parameter was assumed to be zero (or equivalently $(\partial K/\partial P)_T$ at $P = 0$ is equal to 4). In the least squares procedure the uncertainty in all the lattice parameter measurements (ΔL) was taken to be 0.1%,

TABLE 1

Experimental data for the spinel phase of Ni$_2$SiO$_4$ at 23 ± 3 °C

Run no.	NaCl		P(kb)	Ni$_2$SiO$_4$ (spinel)		
	a (Å)	V/V$_0$		a (Å)	V(cm^3/mole)	V/V$_0$
DS-1	—	—	0.0	8.043	39.174	1.000
5LD28	5.636	0.998	0.6	8.044	39.19	1.000
3LD88	5.592	0.975	6.5	8.029	38.97	0.995
5LD34	5.246	0.805	87.2	7.965	38.05	0.971
5LD37	5.171	0.771	115.9	7.894	37.04	0.946
3LD71	5.153	0.763	123.5	7.919	37.39	0.954
3LD69	5.044	0.715	178.5	7.851	36.43	0.930
3LD70	4.999	0.693	210.0	7.820	36.01	0.919
3LD75	4.940	0.672	244.5	7.791	35.60	0.909
3LD86	4.857	0.639	309.7	7.733	34.81	0.889

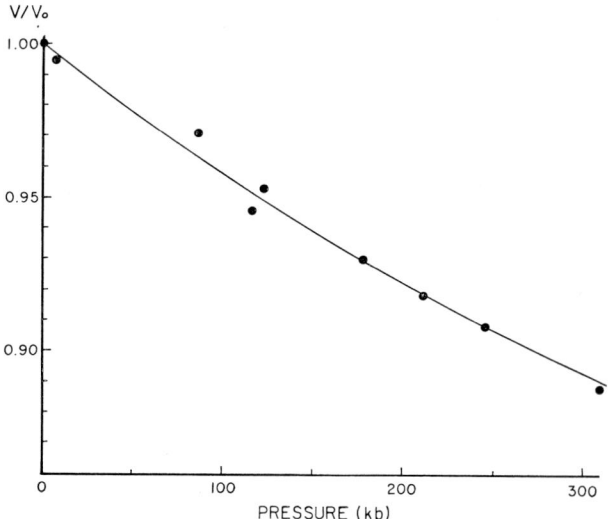

Fig. 1. The effect of pressure on the molar volume of the spinel phase of Ni$_2$SiO$_4$ at 23 ± 3 °C. The curve indicates the first order Birch–Murnaghan equation with a K_0 value of 2.14 Mb.

and the relationship between the bulk modulus at zero pressure and the lattice parameter uncertainty was obtained through the Birch–Murnaghan equation

$$\Delta L = \Delta \left(\frac{l}{l_0}\right) \approx -\left(\frac{dL}{dP}\right) \Delta P =$$
$$= \frac{2L^8}{3K_0(7-5L^2)} \cdot P - \frac{(1-L^2)L}{7-5L^2},$$

where l and l_0 are respectively the lattice dimensions at pressure P and zero pressure, and L is the linear compression l/l_0. The least-squares operation on this equation yields a relation

$$\frac{1}{K_0} = \beta_0 = \frac{\sum\limits_{i=1}^{n} A_i B_i}{\sum\limits_{i=1}^{n} A_i^2},$$

where

$$A_i = \frac{2L_i^8 P}{3(7-5L_i^2)}$$

and

$$B_i = \frac{(1-L_i^2)L_i}{7-5L_i^2}.$$

The standard deviation for the values of K_0 and β_0 thus calculated can be calculated by

$$\{\sigma(\beta_0)\}^2 = \sum_{i=1}^{n} \left(\frac{\partial \beta_0}{\partial L_i}\right)^2 \sigma(L_i)^2,$$

where $\sigma(\beta_0)$ is the standard deviation for β_0 and $\sigma(L_i)$ the standard deviation for the lattice parameter measurements

$$\sigma(L_i) = \left[\sum_{i=1}^{n} \frac{(A_i\beta_0 - B_i)^2}{n-1}\right]^{\frac{1}{2}},$$

where n is the number of measurements. Since $\sigma(L_i)$ is set to be constant for all measurements, the equation above may be altered to

$$\{\sigma(\beta_0)\}^2 = \{\sigma(L_i)\}^2 \sum_{i=1}^{n} \left(\frac{\partial \beta_0}{\partial L_i}\right)^2 =$$

$$= \left[\sum_{i=1}^{n} \frac{(A_i\beta_0 - B_i)^2}{n-1}\right] \times$$

$$\times \sum_{j=1}^{n} \left\{\frac{B_j \Sigma A_i^2 - 2A_j \Sigma A_i B_i}{(\Sigma A_i^2)^2}\left(\frac{\partial A_j}{\partial L_j}\right) + \frac{A_j}{\Sigma A_i^2}\left(\frac{\partial B_j}{\partial L_j}\right)\right\}$$

The value of K_0 ($= 1/\beta_0$) with the uncertainty at 50% confidence level thus calculated is 2.136 ± 0.045 Mb. In addition to the error caused by the random scatter of the measurements, the absolute uncertainty in the NaCl pressure scale, i.e. $\pm 2.5\%$ in pressure, would introduce an uncertainty of ± 0.05 Mb. Since these sources of errors are independent, the total uncertainty for the K_0 value is estimated to be ± 0.07 Mb.

4. Discussion

The K_0 value thus obtained for the spinel phase of Ni_2SiO_4 may be compared with the K_0 values for other silicates and oxides. The parameters required for such a comparison are listed in table 2. O. L. ANDERSON

TABLE 2

The parameters for the spinel phase of Ni_2SiO_4 at room temperature and 1 b pressure

Molar volume	= 39.17 cm³/mole
Formula weight (M)	= 209.51 g/mole
Mean atomic weight ($\bar{M}$)	= 29.93 g
Density (ρ)	= 5.35 g/cm³
Bulk modulus (K_0)	= 2.14 Mb
$2\bar{M}/\rho$	= 11.19 cm³
$\rho/\bar{M}$	= 0.18 cm⁻³
ϕ ($= K_0/\rho$)	= 40.0 (km/s)²

and NAFE (1965) plotted $\ln K_0$ versus $\ln (2\bar{M}/\rho)$ for a number of silicates and oxides, where $\bar{M}$ is the mean atomic weight and ρ is the density of the compound, and showed that practically all the points for these compounds fall on a straight line having a slope of -4. We have plotted the data of the spinel phase of Ni_2SiO_4 and also of the spinel phase of Fe_2SiO_4 (MAO et al., 1969), and found that the points fall on the same straight line.

On the other hand, D. L. ANDERSON and KANAMORI (1968) plotted the bulk modulus and density data of various oxides and silicates on a $\ln \phi$ versus $\ln (\rho/\bar{M})$ diagram, where ϕ is the seismic parameter K_0/ρ, and found that a straight line having a slope of approximately $\frac{1}{3}$ is consistent with most of the data points. In this plot, however, the point for Ni_2SiO_4 spinel fails to fall on the straight line, whereas the point for Fe_2SiO_4 spinel does fall on the straight line. It should also be pointed out that the points for the olivine phase of Mg_2SiO_4(forsterite) and Fe_2SiO_4(fayalite) fall on the line, and that the point for FeO(wüstite) fails to fall on the line, while the point for MgO (periclase), which is isostructural to FeO, is on the straight line. Each pair of these isostructural compounds has SiO^{2-} radicals or O^{2-} ions in common, and differs only in the cation species. Therefore the observed deviation of the data points for the spinel phase of Ni_2SiO_4 from D. L. Anderson's seismic equation of state

$$\ln (\rho/\bar{M}) = \tfrac{1}{3} \ln \phi + \text{const.},$$

is probably due to the difference in the nature of bondings of the Ni and Fe atoms in the spinel structure.

Acknowledgments

We are grateful to Dr. S. Akimoto of the University of Tokyo for providing a sample of the spinel phase of Ni_2SiO_4, and to L. G. Liu of our laboratory for beneficial discussions. This work was supported by a National Science Foundation research grant, GP-5304.

References

AKIMOTO, S., H. FUJISAWA and T. KATSURA (1965) J. Geophys. Res. **70**, 1969.
ANDERSON, D. L. (1967) Geophys. J. **13**, 9.
ANDERSON, D. L. and H. KANAMORI (1968) J. Geophys. Res. **73**, 6477.
ANDERSON, O. L. and J. E. NAFE (1965) J. Geophys. Res. **70**, 3951.
BASSETT, W. A., T. TAKAHASHI and P. W. STOOK (1967) Rev. Sci. Instr. **38**, 37.
MAO, H. K., T. TAKAHASHI, W. A. BASSETT, J. S. WEAVER and S. AKIMOTO (1969) J. Geophys. Res. **74**, 1061.
RINGWOOD, A. E. (1962) Geochim. Cosmochim. Acta **26**, 457.
WEAVER, J. S., T. TAKAHASHI and W. A. BASSETT (1969) E. Lloyd, ed., Symp. Characterization of High Pressure Environments (Natl. Bur. Std., Washington, D.C.) in press.

ISOTHERMAL COMPRESSION OF STISHOVITE AND COESITE UP TO 85 KILOBARS AT ROOM TEMPERATURE BY X-RAY DIFFRACTION

WILLIAM A. BASSETT* and J. DEAN BARNETT

Department of Physics, Brigham Young University, Provo, Utah 84601, U.S.A.

X-ray diffraction (MoK_α) was used to obtain compression data on stishovite and coesite subjected to pressures up to 85 kb at 25 °C in a tetrahedral press. Pressure was calculated from the lattice parameter of NaCl dispersed along with the sample in a plastic (Durez) which served as a pressure transmitting medium. Bulk moduli calculated from these data are 3.0 ± 0.3 Mb and 1.47 ± 0.15 Mb for stishovite and coesite respectively. The compression along the a axis ($\Delta a/a_0$) is approximately twice the compression along the c axis ($\Delta c/c_0$) in stishovite with both axes decreasing monotonically.

1. Introduction

Coesite is a monoclinic phase of silica having silicon atoms in four-fold coordination and is about 10% denser than α-quartz. Stishovite is a tetragonal phase of silica with silicon atoms in six-fold coordination and has a density approximately 60% greater than that of α-quartz. Since the first synthesis of coesite by COES (1953) and stishovite by STISHOV and POPOVA (1961) under conditions of high pressure and high temperature, there has been much interest among geophysicists in these two substances.

BIRCH (1952) pointed out that the observed increase in seismic velocities with depth from 200 to 900 kilometers in the Earth can only be accounted for by phase transformations probably involving an increase in coordination number of silicon. Furthermore, it has been demonstrated that the pressure-temperature conditions in the Earth's mantle coincide with the stability fields of coesite and stishovite. Therefore, knowledge of the properties of these substances is vitally important for the study of the Earth's mantle.

2. Previous work

The stability fields of the high pressure polymorphs of silica have been investigated by MACDONALD (1956), DACHILLE and ROY (1959), BOYD and ENGLAND (1960), TAKAHASHI (1963), KITAHARA and KENNEDY (1964), COHEN and KLEMENT (1967), BOETTCHER and WYLLIE (1968), and AKIMOTO and SYONO (1968). The crystal structure of coesite has been worked out by RAMSDELL (1955), and ZOLTAI and BUERGER (1959), and for stishovite by CHAO et al. (1962) and STISHOV (1964). The optical properties of both phases have been studied by SCLAR et al. (1962a, 1962b). Thermodynamic properties have been measured by HOLM et al. (1967).

WACKERLE (1962) obtained compression data for SiO_2 by shock loading quartz crystals and fused quartz. McQUEEN et al. (1963) and ANDERSON and KANAMORI (1968) calculated the zero-pressure bulk modulus of stishovite from Wackerle's data thus obtaining values of 4.35 Mb and 3.92 Mb respectively. The compression of stishovite at room temperature has also been measured by X-ray diffraction by IDA et al. (1967) with a resulting zero-pressure bulk modulus of 7.15 Mb. These values are in serious disagreement with each other. The purpose of this paper is to report new X-ray diffraction measurements for the effect of pressure on the lattice parameters, hence molar volume, for stishovite up to 85 kb at room temperature and to evaluate the earlier measurements. Compression data for coesite are reported for the first time.

3. Experimental techniques

The adaptation of the tetrahedral press for X-ray diffraction studies of samples under pressure has been described by BARNETT and HALL (1964). Filtered MoK_α radiation is collimated so as to pass between two of

* On leave from the University of Rochester, Rochester, New York 14627, U.S.A.

the anvils, through the pressure transmitting plastic-boron mixture, and impinge upon the sample. The direct beam as well as the diffracted rays then exit between the other two anvils. This was designated geometry B by Barnett and Hall. The diffracted rays are detected by two scintillation counters which scan a 2θ range of $50°$ on each side of the direct beam thus giving a symmetrical diffraction pattern.

The stishovite sample was synthesized from silicic acid in a tetrahedral press at a pressure of approximately 80 kb and a temperature of approximately 900 °C. Coesite was synthesized in a tetrahedral press at approximately 50 kb and 900 °C. The lattice parameters for the stishovite sample calculated from a Debye-Scherrer pattern are $a_0 = 4.178 \pm 0.001$ Å and $c_0 = 2.663 \pm 0.001$ Å. These are in reasonable agreement with those reported by CHAO et al. (1962), $a_0 = 4.1790$ Å and $c_0 = 2.6649$ Å. The lattice parameters obtained for the coesite sample are $a_0 = b_0 = 7.143 \pm 0.007$ Å, $c_0 = 12.36 \pm 0.01$ Å, $\gamma = 120°$. These are in satisfactory agreement with the values reported by SCLAR et al. (1962b), $a_0 = 7.14$ Å, $b_0 = 7.14$ Å, $c_0 = 12.37$ Å.

The stishovite and coesite samples were lightly ground under water in a mullite mortar and pestle. They were then mixed with NaCl in ratios ranging from 2:1 to 5:1 by weight. Each of these mixtures was then thoroughly mixed with approximately equal amounts of plastic (Durez) by volume. These in turn were pelletized to cylindrical samples 3 mm in diameter and approximately 2 mm thick. In each case the amount of sample was carefully controlled for optimum diffraction efficiency.

A sample pellet produced in this way was then centered within a tetrahedron consisting of a boron-epoxy or boron-Durez mixture. The tetrahedron was 7 mm on the edge and had preformed gaskets 2.5 mm thick and extending 2.5 mm from the edge. The sample pellet was oriented within the tetrahedron so that the X-ray beam would pass along the cylinder axis.

This type of sample preparation by dispersion in a plastic of low shear strength was used to maximize the isotropy and uniformity of pressure on the grains of both the SiO_2 sample and the NaCl calibrant.

This method was able to produce diffraction patterns containing six reflections from NaCl: (200), (220), (222) (400), (420) and (422); and seven reflections from stishovite: (110) (101), (111), (210), (211), (220) and (301).

The coesite diffraction pattern consists of nine peaks, but most of these can be indexed with more than one set of indices. Each set of brackets represents a single peak in the coesite pattern: [(111), (103)], [(200), ($\bar{2}$12), (004)], [($\bar{2}$14)], [(115), (123)], [(303)], [(220), ($\bar{2}$16)], [(131), (206), (107)], [(400), (224), (008)], [(226), (141), (323), (307), (109)].

The assignment of the indices to these peaks was based upon a theoretical powder pattern calculated using a computer program by D. K. SMITH (1963) from the structure given by ZOLTAI and BUERGER (1959). Indices were assigned only when calculated intensities warranted it.

Interference between sample reflections and those of NaCl required that one or two reflections from each pattern be rejected. Hence, lattice parameter calculations were based generally upon four or five NaCl reflections, five or six stishovite reflections, and seven or eight coesite reflections.

The change in lattice parameter, hence molar volume, of NaCl with pressure was used to determine the pressure to which the sample was being subjected. The relationship used for this calculation was derived by DECKER (personal communication) and is based upon the Mie–Gruneisen equation of state employing the best available zero-pressure compressibility 4.22×10^{-3} kb^{-1} (CHANG (1965), SLAGLE and McKINSTRY (1967), DRABBLE and STRATHEN (1967)). At a compression of $V/V_0 = 0.80$ this scale yields a pressure of 90.5 kb. At the same compression Decker's earlier scale (DECKER (1966)) also based on the Mie–Gruneisen equation of state but using older values for the zero-pressure compressibility of NaCl yields a pressure of 88.2 kb. A pressure–volume relationship based upon the vibrational formulation of the Hildebrand equation by WEAVER et al. (1969) and MAO et al. (1969) yields a pressure of 91.1 kb for the same compression.

Lattice parameters were calculated from the powder diffraction patterns by means of a least squares computer program. The accuracy of the lattice parameter measurements is estimated to be $\pm 0.1\%$ and of volume measurements $\pm 0.3\%$. The accuracy of the pressure determinations by the NaCl method is estimated to be $\pm 2\%$ for a given value of compression.

4. Results

The compression data for stishovite are presented in

Table 1
Compression data for stishovite

Run	NaCl		stishovite				
	a/a_0	Pressure kb	a/a_0	c/c_0	c/a	V/V_0	Number of reflections
B13	0.994	4.2	1.000	1.001	0.639	1.001	6
B2	0.987	10.1	0.998	1.006	0.643	1.002	5
A2	0.986	10.9	0.999	0.998	0.638	0.995	5
B3	0.972	25.4	0.996	1.004	0.643	0.996	5
A3	0.971	25.7	0.995	1.005	0.646	0.994	6
B4	0.957	44.4	0.996	0.998	0.640	0.989	5
B5	0.949	54.1	0.992	0.996	0.641	0.980	6
B6	0.946	58.9	0.992	0.997	0.642	0.980	5
B7	0.941	67.4	0.992	0.999	0.643	0.984	5
B9	0.936	75.8	0.990	0.992	0.639	0.972	5
B8	0.936	76.3	0.989	0.997	0.644	0.974	5
B11	0.933	82.1	0.990	9.992	0.640	0.972	6
B10	0.932	83.2	0.988	0.996	0.644	0.972	6
B12	0.931	84.6	0.986	0.996	0.645	0.969	6

table 1 and fig. 1, for coesite in table 2 and fig. 2. In the stishovite the compression along the a axis is more than twice as great as the compression along the c axis:

$$\frac{1}{a_0}\frac{\Delta a}{\Delta P} = -1.38 \times 10^{-4} \text{ kb}^{-1},$$

$$\frac{1}{c_0}\frac{\Delta c}{\Delta P} = -0.57 \times 10^{-4} \text{ kb}^{-1}.$$

The data are not sufficiently precise to indicate a variation in these values up to 85 kb.

In the coesite diffraction patterns the multiple indexing and the 2θ resolution of $0.5°$ combine to make it virtually impossible to detect differences in rates of compression along different crystallographic axes or changes in interfacial angles. The indexing is such that most of the 2θ measurements from a coesite pattern contribute strongly to the a, b and c-axis determinations. Routine least squares calculations of lattice parameters therefore indicate essentially no pressure dependence of the axial ratios. The multiple indexing of reflections in the coesite diffraction pattern however does not severely jeopardize the molar volume calculations since it has the effect of averaging compression rates in the different directions.

Relative volumes calculated from lattice parameters for both stishovite and coesite are plotted as a function of pressure in figs. 1 and 2. In the pressure range 0–85 kb a Birch–Murnaghan equation for bulk modulus greater than 1.5 Mb is essentially a straight line. There-

Table 2
Compression data for coesite

Run	NaCl		Coesite				
	a/a_0	Pressure kb	a/a_0	c/c_0	c/a	V/V_0	Number of reflections
A3	0.989	8.4	0.998	0.998	1.730	0.993	20
C2	0.987	9.9	0.996	0.996	1.731	0.987	19
C4	0.964	33.8	0.991	0.991	1.731	0.974	22
A5	0.957	42.9	0.993	0.994	1.731	0.980	19
C5	0.957	43.0	0.991	0.990	1.730	0.971	18
A6	0.952	49.4	0.991	0.990	1.728	0.972	19
C6	0.946	58.7	0.988	0.987	1.730	0.962	18
C7	0.941	68.0	0.982	0.981	1.730	0.947	18
B2	0.940	68.7	0.985	0.984	1.729	0.954	18
B3	0.936	75.8	0.982	0.981	1.729	0.946	18
C8	0.934	80.6	0.980	0.978	1.728	0.940	18

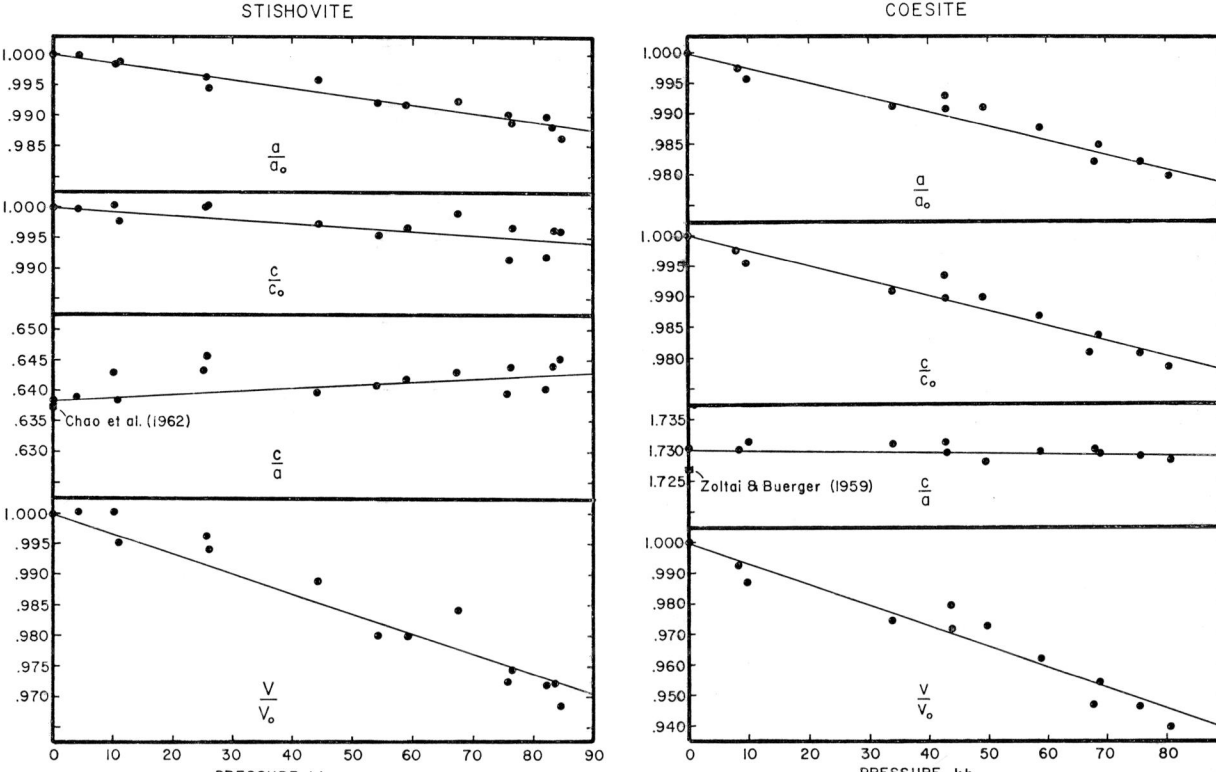

Fig. 1. Effect of pressure on lattice parameters and molar volume of stishovite at room temperature.

Fig. 2. Effect of pressure on lattice parameters and molar volume of coesite at room temperature.

fore the zero-pressure bulk moduli for stishovite and coesite were determined by fitting with a straight line, and values of 3.0 ± 0.3 Mb and 1.47 ± 0.16 Mb respectively were obtained.

5. Discussion

The bulk modulus for stishovite reported in this paper is significantly lower than those values based on Wackerle's high pressure shock data and the zero-pressure density of quenched stishovite ($\rho = 4.29$ g/cm^3). MCQUEEN et al. (1963) obtained $K_0 = 4.35$ Mb and ANDERSON and KANAMORI (1968) obtained $K_0 = 3.92$ Mb. However, the bulk modulus reported in this paper agrees well with the value of 3.0 Mb ($\phi = 75$ km^2/sec^2) which Anderson and Kanamori obtained by fitting Wackerle's data with a Birch–Murnaghan equation and solving for both the zero-pressure density and bulk modulus. The zero-pressure density they obtained by this treatment is 4.06 g/cm^3.

This low value for the density may be an indication of a real difference in the nature of the sample under shock loading. Rather than being a completely transformed crystalline stishovite, it may be a mixture of fine grained stishovite with short-range order glass. MCQUEEN et al. (1963) suggest that the high pressure phase may be "a dense silica glass, which in its short-range ordering is stishovite". DECARLI and MILTON (1965) examined shocked quartz samples, and found that they contained a low density short-range ordered material with traces of stishovite. On the basis of these observations they concluded that a dense six-fold coordinated phase in which long-range order had not been attained existed at high pressure but reverted rapidly to four-fold coordinated short-range ordered material upon release.

If shock loaded silica is in fact not crystalline stishovite, but is in part a glass having short-range ordering, it might well have a density less than that of stishovite under the same pressure-temperature conditions. In that case the density of crystalline stishovite at zero pressure should not be used along with the shock Hugoniot data for calculating the bulk modulus since crystalline stishovite would be physically inconsistent with the SiO₂ under shock loading. This may explain why the bulk modulus based entirely upon the behavior of crystalline stishovite (reported in this paper) is in better agreement with the value calculated entirely from the shock Hugoniot (reported by Anderson and Kanamori) than with the values based upon the shock Hugoniot assuming a zero-pressure density equal to that of crystalline stishovite.

The zero-pressure bulk modulus reported in this paper is also very much lower than the value of 7.15 Mb calculated by IDA et al. (1967) from high pressure X-ray diffraction measurements. Before an explanation for this discrepancy can be attempted, it is necessary to describe the compression along each crystallographic axis separately. Ida et al. report that the a axis dimension decreases gradually with increasing pressure while the c axis dimension increases with increasing pressure, passes through a maximum at about 90 kb and then decreases. The results of our work, however, indicate a monotonic decrease of both the a and the c axis dimensions with increasing pressure. These data can be fit with straight lines as shown in fig. 1. The a axis and c axis compressions derived from the diagram given by IDA et al. (1967) for the pressure range 0–90 kb are:

$$\frac{1}{a_0}\frac{\Delta a}{\Delta P} = (-1.8 \pm 0.2) \times 10^{-4} \text{ kb}^{-1}$$

$$\frac{1}{c_0}\frac{\Delta c}{\Delta P} = (+2.3 \pm 0.4) \times 10^{-4} \text{ kb}^{-1}$$

When these are compared with our a and c axis compressions, it should be noted that the a axis compressions resulting from both sets of data are in reasonable agreement, whereas the c axis compressions are in serious disagreement.

In the determinations by Ida et al. the "change of the lattice parameters a and c with pressure was calculated by means of the least squares method from shifts of the (110), (111), (211) and (301) lines". They point out that the "anomalous behavior (of the c axis dimension) is clearly revealed by the fact that the (111) line hardly shifts with increasing pressure". Of the reflections they used, the (111) has the strongest influence on the c axis determination. If it does not move during compression, the decreasing a axis dimension based on the other reflections does indeed demand that the c axis increase. However, the (220) reflection of NaCl ($I/I_{100} = 55$) seriously interferes with the (111) reflection of stishovite ($I/I_{100} = 35$) in the pressure range below 50 kb. Unless the 2θ resolution is very good this interference prevents the measurement of the (111) stishovite reflection at zero pressure and possibly up to a pressure as high as 50 kb. Ida et al. do not discuss how this problem was handled. Extrapolation of the high pressure measurements of the (111) line to zero pressure could lead to an erroneous judgement of the shift of the line. If the shift was determined by comparison of the high pressure d-values with a zero-pressure d-value calculated from a separate run in which NaCl did not interfere, a difference in instrumental geometry between the runs might lead to an erroneous indication of the magnitude of the shift.

In our lattice parameter determinations the effects of these sources of error were minimized by the use of three additional stishovite reflections. Also lattice parameters were calculated at each pressure from all of the available reflections in the diffraction pattern made at that pressure rather than by attempting to measure the shift of each reflection.

It is of geophysical interest to compare the bulk moduli for stishovite and coesite with other silicates and oxides. This can be done by means of the relationships derived by O. L. ANDERSON and NAFE (1965) and D. L. ANDERSON and KANAMORI (1968). Anderson and Nafe plot bulk modulus (K_0) versus $2\overline{M}/\rho$ where $\overline{M}$ is the mean atomic number and ρ is the density. In this plot the stishovite and coesite points (table 3) fall on a straight line of slope -4, which passes through points for quartz and other oxides.

Anderson and Kanamori plot $\ln(\rho/\overline{M})$ versus $\ln \phi$ where ϕ, the seismic parameter, is equal to K_0/ρ_0. In

TABLE 3

Data for coesite and stishovite at room temperature and 1b pressure

	coesite	stishovite
Molar volume (V)	20.56 cm³/mole	14.00 cm³/mole
Formula weight (M)	60.09 g/mole	60.09 g/mole
Mean atomic weight ($\overline{M}$)	20.03 g/at. mole	20.03 g/at. mole
Density (ρ)	2.92 g/cm³	4.291 g/cm³
Bulk modulus (K_0)	1.47 Mb	3.0 Mb
$2\overline{M}/\rho$	13.7	9.336
$\rho/\overline{M}$	0.146	0.2142
$\phi (= K_0/\rho)$	50.3 (km/sec)²	69.9 (km/sec)²

their plot a straight line having a slope of approximately $\frac{1}{3}$ (i.e. $\ln(\rho/\overline{M}) = \frac{1}{3}\ln\phi + \text{const.}$) is consistent with most of the data points for common oxides and silicates. However, the coesite point based on the data reported in this paper (table 3) falls to the right and below the straight line (i.e. it has a higher ϕ and lower $\rho/\overline{M}$ than the straight line). The stishovite point on the other hand lies to the left and above the straight line (i.e. it has a lower ϕ and higher $\rho/\overline{M}$ than the straight line). Thus the change of coordination of silicon in this pair of isochemical substances appears to lead to an inconsistency in the relationship described by ANDERSON and KANAMORI (1968) but not the one described by ANDERSON and NAFE (1965).

Acknowledgments

The authors are indebted to H. Tracy Hall for making many of his laboratory facilities available for the synthesis of the samples and for other purposes. They wish to thank Taro Takahashi for his many valuable suggestions and critical reading of the manuscript. Daniel Decker and Leo Merrill provided a great deal of assistance in collecting and interpreting the data. The work was supported in part by a grant from the National Science Foundation.

References

AKIMOTO, S. and Y. SYONO (1968) Preprint, *Symposium on accurate characterization of the high-pressure environment* (Gaithersburg, Md).
ANDERSON, D. L. and H. KANAMORI (1968) J. Geophys. Res. **73**, 6477.
ANDERSON, O. L. and J. E. NAFE (1965) J. Geophys. Res. **70**, 3951.
BARNETT, J. D. and H. T. HALL (1964) Rev. Sci. Instr. **35**, 175.
BIRCH, F. (1952) J. Geophys. Res. **57**, 227.
BOETTCHER, A. L. and P. J. WYLLIE (1968) Contr. Mineral. Petrol. **17**, 224.
BOYD, F. R. and J. L. ENGLAND (1960) J. Geophys. Res. **65**, 749.
CHAO, E. C. T., J. J. FAHEY, J. LITTLER and D. J. MILTON (1962) J. Geophys. Res. **67**, 419.
CHANG, Z. P. (1965) Phys. Rev. **140**, A1788.
COES, L. (1953) Science **118**, 131.
COHEN, L. H. and W. KLEMENT JR. (1967) J. Geophys. Res. **72**, 4245.
DACHILLE, F. D. and R. ROY (1959) Z. Krist. **111**, 451.
DECARLI, P. S. and D. J. MILTON (1965) Science **147**, 144.
DECKER, D. L. (1966) J. Appl. Phys. **37**, 5012.
DRABBLE, J. R. and R. E. B. STRATHEN (1967) Proc. Phys. Soc. **92**, 1090.
HOLM, J. L., O. J. KLEPPA and E. F. WESTRUM JR. (1967) Geochim. Cosmochim. Acta **31**, 2289.
IDA, Y., Y. SYONO and S. AKIMOTO (1967) Earth Planet. Sci. Letters **3**, 216.
KITAHARA, S. and G. C. KENNEDY (1964) J. Geophys. Res. **69**, 5395.
MAO, H. K., T. TAKAHASHI, W. A. BASSETT, J. S. WEAVER and S. AKIMOTO (1969) J. Geophys. Res. **74**, 1061.
MACDONALD, G. J. F. (1956) Am. J. Sci. **254**, 713.
MCQUEEN, R. G., J. N. FRITZ and S. P. MARSH (1963) J. Geophys. Res. **68**, 2319.
RAMSDELL, L. S. (1955) Am. Mineralogist **40**, 975.
SCLAR, C. B., A. P. YOUNG, L. C. CARRISON and C. M. SCHWARTZ (1962a) J. Geophys. Res. **67**, 4049.
SCLAR, C. B., L. C. CARRISON and C. M. SCHWARTZ (1962b) Am. Mineralogist **47**, 1292.
SLAGLE, O. D. and H. A. MCKINSTRY (1967) J. Appl. Phys. **38**, 437.
SMITH, D. K. (1963) A FORTRAN program for calculating X-ray powder diffraction patterns, U.C.R.L. Rept. 7196.
STISHOV, S. M. (1964) Tectonophysics **1**, 223.
STISHOV, S. M. and S. V. POPOVA (1961) Geokhimiya **10**, 837.
TAKAHASHI, T. (1963) High pressure measurement, in: A. A. Giardini and E. C. Lloyd eds (Butterworth, London) 240.
WACKERLE, J. (1962) J. Appl. Phys. **33**, 922.
WEAVER, J. S., T. TAKAHASHI and W. A. BASSETT (1969) *Symposium on the characterization of high pressure environments*, C. Becket and E. Lloyd eds. (National Bureau of Standards, Washington, D.C.) in press.
ZOLTAI, T. and M. J. BUERGER (1959) Z. Krist. **111**, 129.

Note added in proof

Since the manuscript for this paper was submitted for publication, two pieces of work bearing on the results reported here have been brought to the attention of the authors.

MANGHNANI (1969) has used ultrasonic interferometry to measure the effects of pressure and temperature on the elastic constants of rutile (TiO_2) which is isostructural with stishovite and has nearly the same density. He reports $K_s|_{P=0} = 2155$ kb $dK_s/dP|_{P=0} = 6.75$, where K_s is the adiabatic bulk modulus and P is the pressure in kilobars.

AHRENS *et al.* (in press) have calculated the following parameters for stishovite from shock Hugoniot data for SiO_2 samples of varying starting densities: $K_s|_{P=0} = 3.00$ Mb, $dK_s/dP|_{P=0} = 6.9$, $K_T|_{P=0} = 2.98$ Mb, and $dK_T/dP|_{P=0} = 7.0$, where K_T is the isothermal bulk modulus.

These values for the pressure derivative of the bulk modulus are considerably higher than for other common oxides and would appreciably lower the bulk modulus value derived from the X-ray diffraction measurements if the X-ray diffraction measurements

Table

Zero pressure bulk moduli for stishovite and coesite calculated from the Birch-Murnaghan equation assuming different values for $dK_T/dP \vert_{P=0}$

Stishovite		Coesite	
$K_T \vert_{P=0}$	$dK_T/dP \vert_{P=0}$	$K_T \vert_{P=0}$	$dK_T/dP \vert_{P=0}$
2.62 Mb	8	1.14 Mb	8
2.69	6	1.19	6
2.76	4	1.26	4
2.83	2	1.32	2
2.92	0	1.40	0

are fitted to a Birch-Murnaghan equation. For this reason we are adding a table of bulk moduli calculated from the X-ray measurements by fitting to the Birch-Murnaghan equation with values of $dK_T/dP \vert_{P=0}$ ranging from 0 to 8. Although we have no information on the pressure dependence of the bulk modulus for coesite, a similar table is given here for future reference.

MANGHNANI, M. H. (1969) J. Geophys. Res. **74**, 4317.
AHRENS, T. J., T. TAKAHASHI and G. F. DAVIES (in press) J. Geophys. Res.

EQUATIONS FOR THE ELASTIC CONSTANTS AND THEIR PRESSURE DERIVATIVES FOR THREE CUBIC LATTICES AND SOME GEOPHYSICAL APPLICATIONS*

ORSON L. ANDERSON and ROBERT C. LIEBERMANN

Lamont–Doherty Geological Observatory of Columbia University, Palisades, N.Y. 10964, U.S.A.

Lattice dynamical considerations and a Born repulsive potential between atoms lead to equations for the elastic constants of the cubic NaCl, CsCl and ZnS lattices as a function of compression. The NaCl lattice is unstable for $n < 4.6$ and the CsCl lattice is unstable for $n < 7.2$, where n is the exponent of the repulsive power law. The pressure derivatives of the shear constants (c', c_{44}) show a strong dependence upon crystallographic structure; for the ZnS lattice both dc'/dP and dc_{44}/dP are negative for all reasonable values of n. The theoretical values of dc_{ij}/dP compare favorably with experimental results.
For the isotropic shear modulus (μ), $d\mu/dP$ is determined by Poisson's ratio and the coordination of the ions in the lattice. Low (and possibly negative) values of $d\mu/dP$ are likely for several materials of importance to geophysics; such values would make low-velocity zones possible and interpretation of shock-wave data difficult. Vanishing of a shear constant predicts phase transitions at compression (V/V_0) of 0.95 for the ZnS lattice, 0.75 for the NaCl lattice, and 0.60 for the CsCl lattice. The CsCl transition is predicted in spite of the fact that none of the elastic constants have a negative pressure derivative at zero pressure. The equation of state parameters, K_0 and $(dK/dP)_0$ (where K is the bulk modulus), are almost independent of crystallographic structure. K_0 arises entirely from the Laplacian of the repulsive potential. $(dK/dP)_0 = \frac{1}{3}(n+7)$ so that the stability limits on n make it unlikely that $(dK/dP)_0 < 3.5$ for the NaCl lattice or less than 5 for the CsCl lattice.

1. Introduction

The shear wave velocity decreases with pressure for isotropic ZnO (SOGA and ANDERSON, 1967) and isotropic α-SiO$_2$ (SOGA, 1968). This behavior has been correlated to BLACKMAN's (1958) lattice dynamic theory of ZnS by O. L. ANDERSON (1968b). Anderson concluded, following the earlier suggestion of Blackman, that compounds with both a low relative shear modulus and a low coordination number should be expected to have a low value, and possibly a negative value, of $d\mu/dP$ where μ is the isotropic shear modulus. The ZnS lattice is a prototype of the structures for which $d\mu/dP$ is anomalous. (In all derivatives in this paper, we are dealing with the adiabatic elastic constants and their isothermal pressure derivatives.)

The suggestion that certain crystallographic structures will propagate shear waves with a velocity largely insensitive to pressure, is considered in detail in this paper. We will derive and present the equations for dc_{11}/dP, dc_{12}/dP, and dc_{44}/dP (where c_{11}, c_{12}, c_{44} are the usual symbols for the elastic constants) for three

* Lamont–Doherty Geolog. Observ. Contrib. no. 1393.

cubic structures: NaCl, CsCl, and ZnS, and indicate why certain structural features of the ZnS lattice give rise to large negative values of dc_{44}/dP and dc'/dP, where $c' = \frac{1}{2}(c_{11}-c_{12})$. Further we will show why these features are absent in the other cubic lattices. From these conclusions, we will generalize rules for structures of lower symmetry.

From the first, it is important to point out that no claim of priority is made for most of the derivations. Most of the equations we will present are implicit in BLACKMAN's (1958) publication, who based his theory on BORN's (1923) earlier treatise in the field. However, the critical equations we wish to examine are missing from the published literature on lattice dynamics, even though they are implicit in these works. In order to focus on the essential applications to geophysics, we believe it is worthwhile to present certain equations that are skipped over in the lattice dynamics papers. Our goals are different from those of most lattice dynamic calculations. Here we are mostly concerned with the elastic constants and their pressure derivatives and these quantities are important enough to examine in considerable detail. The goals of lattice dynamics generally involve the frequency spectrum, and the elastic constants are used as numerical parameters to derive the frequency spectrum and associated properties.

There are four applications of the results of this paper to geophysical problems. The first concerns the low-velocity zone of the earth's mantle. A low value of dv_s/dP allows a low-velocity zone to be compatible with a small thermal gradient. It is important, therefore, to determine what crystallographic structures are most likely to have a low value of dv_s/dP. The second concerns the problem of phase transitions. Materials with a sufficiently low value of dc'/dP will undergo a phase transition at higher pressure due to an instability in the elastic energy of the lattice. The third concerns the problem of computing the temperatures in a material which has been shocked to a high compression. This calculation requires that the thermal energy be computed, which in turn depends upon the evaluation of the Grüneisen parameter, γ, and its pressure derivative. The calculation of γ and $d\gamma/dP$ for materials with a low value of dv_s/dP is fraught with uncertainty (THOMSEN and O. L. ANDERSON, 1969; KNOPOFF and SHAPIRO, 1969). The fourth application to geophysics discusses the significance of our calculations of the parameters which affect equations of state in the earth.

We shall return to these geophysical problems in the last sections of the paper, but the major portion will deal with the lattice dynamical expressions themselves. An outline of the derivations themselves now follows.

The beginning point of the derivation is the expressions for the elastic constants of a cubic lattice from Born's lattice theory, written in the form appropriate to the central force approximation as presented by BLACKMAN (1958). This means, of course, that for the case of the two lattices, NaCl and CsCl, we cannot examine the causes for departures from the Cauchy relation. By this simplification, however, we are able to keep the derivation in a simple form, and are thereby able to easily expose the considerable differences in the pressure derivatives of the elastic constants arising from the separate three cubic lattices. The second important assumption is that we take only nearest neighbors in the lattice sum for the repulsive potential. By this simplification we are able to define the repulsive contribution of the elastic constants in terms of the Madelung constant by invoking the condition of equilibrium. The third important assumption is that we assume a power law for the repulsion and thereby are able to derive explicit and dimensionless equations for the pressure derivatives of the elastic constants in terms of n, the exponent of the power law. The three assumptions have been used many times before (e.g. BARRON, 1955; BLACKMAN, 1957, 1958; REDDY and RUOFF, 1965). We believe the explicit expressions for dc_{ij}/dP in terms of n are new, and we have not been able to find the details of the derivation of CsCl lattice in the literature, so it may be new.

The assumptions used will necessarily affect the result, but probably not by much. The results derived from this simple model are in fairly good agreement with the experimental results for the NaCl lattice and the CsCl lattice. Data on the ZnS lattice is missing although, as pointed out by O. L. ANDERSON (1968b), Blackman's equations for ZnS predict the result found for polycrystalline ZnO. Calculations using a non-central potential, and other calculations using second-nearest neighbors have been made, but they show that only minor adjustments are made in the computed results for dc_{ij}/dP, and that the qualitative conclusions are not affected. These refinements will not be presented.

The derivations are made side-by-side for the three lattices. The equations are presented in a number of different forms because each of these forms has a useful interpretation. For example, in the derivation of the dc_{ij}/dP, it is important to carefully separate the coulombic and the repulsive contribution in c_{ij}. On the other hand, in determining the conditions under which the elastic constants vanish, an equation showing c_{ij} as a linear equation in n is more useful.

In deriving these equations, whose beginning and end points were fixed by eq. (1), (4) and (6) in BLACKMAN (1958), eq. (14) of BARRON (1957) and eq. (11) and (12) of BLACKMAN (1957), we found that the chief difficulty was in properly accounting for the appropriate Madelung constant, and distinguishing between the two space variables (r, the interatomic spacing constant, and a, representing half the lattice constant). Matching lattice sums of the coulombic and repulsive components crucially depend upon the proper choices. We think it is worthwhile to repeat a caution made by Blackman: it is important that the equilibrium condition of the lattice be invoked at the appropriate point in the derivation; otherwise the volume dependence of elastic constants is lost.

2. Lattice dynamic calculations for diatomic ionic cubic crystals

2.1. Central force approximations

The derivatives of the potential energy ϕ for central forces are most conveniently expressed in spherical coordinates. The transformation from rectangular to spherical coordinates is given by

$$\left(\frac{\partial \phi}{\partial \alpha}\right)_r = \alpha P, \quad (1)$$

$$\left(\frac{\partial^2 \phi}{\partial \alpha \partial \beta}\right)_r = \delta_{\alpha\beta} P + \alpha\beta Q, \quad (2)$$

where

$$P = \left(\frac{1}{r}\frac{\partial \phi}{\partial r}\right), \quad (3)$$

$$Q = \left[\frac{1}{r}\frac{\partial}{\partial r}\left(\frac{1}{r}\frac{\partial \phi}{\partial r}\right)\right]_r, \quad (4)$$

$$\delta_{\alpha\beta} = \begin{cases} 1 \text{ for } \alpha = \beta, \\ 0 \text{ for } \alpha \neq \beta. \end{cases}$$

The equations for the elastic constants are then defined by (BLACKMAN, 1958) (where Δ is the volume per cell (ion pair), not the crystallographic cell):

$$c_{11} = (c_{xx,xx}) = \frac{1}{\Delta}\sum_l \{Q(x^l)^4 + P(x^l)^2\}, \quad (5)$$

$$c_{12} = (c_{xx,yy}) = \frac{1}{\Delta}\sum_l \{Q(x^l)^2(y^l)^2 - P(x^l)^2\}, \quad (6)$$

$$c^*_{44} = (c^*_{xy,xy}) = \frac{1}{\Delta}\sum_l \{Q(x^l)^2(y^l)^2 + P(x^l)^2\}, \quad (7)$$

and

$$c_{44} = c^*_{44} - \frac{C^2}{D}, \quad (8)$$

where

$$C = \frac{1}{\Delta}\sum_l x^l y^l z^l Q, \quad (9)$$

$$D = \frac{1}{\Delta}\sum_l \{P + Q(x^l)^2\}. \quad (10)$$

In the above there is a built-in assumption that there are two Bravais lattices, $k = 1$ and $k = 2$. Further, the summation over k has been accomplished.

The summation over l yields results which clearly reflect the distinctive features of the crystallographic lattice under consideration. The ratios $x/r, y/r, z/r$ give the relative positions of lattice sites in the crystallographic cell. Of especial interest is C (eq. 9), which is an odd function of the lattice positions. If all the atoms are on sites where one of the three variables is zero (such as on planes lying on the coordinate axes), C will be zero, or if all the atoms lie on central symmetric positions, the sum over l will add out. The constant D is a spring constant and related in an important way to optical properties.

2.2. Separating the long-range component of the potential

Since forces are additive, the derivatives of the potential in the equations of motion are additive, and the elastic constants given above can be broken down into sums:

$$(c_{\alpha\beta,\gamma\lambda}) = (c_{\alpha\beta,\gamma\lambda})^e + (c_{\alpha\beta,\gamma\lambda})^R, \quad (11)$$

$$C = C^e + C^R, \quad (12)$$

$$D = D^e + D^R. \quad (13)$$

The reason these constants are broken down into two parts is that the summation is taken to infinity for the e component (standing for electrostatic or coulombic), while the summation is taken only over nearest neighbors for the R component (standing for repulsive). This separation corresponds to the separate parts of the lattice potential arising between the k and k' positions which is appropriate to define the energy per unit cell:

$$\phi_{kk'} = -\frac{A_r Z_k Z_{k'} e^2}{r_{kk'}} + M_k v(r), \quad (14)$$

where M_k is the coordination number around the k position, A_r is the Madelung constant of the lattice, Z_k and $Z_{k'}$ are the valence numbers and e is the electronic charge, and $v(r)$ is the repulsive potential.

Designate the potential above by the symbols R and e corresponding to the two terms in (14)

$$\phi = \phi^e + \phi^R. \quad (15)$$

Now designate the operators P and Q by the same symbols so that, for example,

$$P^R = \frac{1}{r}\left(\frac{\partial \phi^R}{\partial r}\right)_r; \quad Q^R = \left[\frac{1}{r}\frac{\partial}{\partial r}\left(\frac{1}{r}\frac{\partial \phi^R}{\partial r}\right)\right]_r. \quad (16)$$

Then the summation on l to the elastic constants proceeds as follows (using c_{11}, for example)

$$c_{11} = c_{11}^e + c_{11}^R, \tag{17}$$

$$c_{11}^e = \frac{1}{\Delta} \sum_{l=1}^{\infty} \{Q^e(x^l)^4 + P^e(x^l)^2\}, \tag{18}$$

$$c_{11}^R = \frac{1}{\Delta} \sum_{l=1}^{M} \{Q^R(x^l)^4 + P^R(x^l)^2\}. \tag{19}$$

The summation of l on c_{11}^e proceeds to infinity because the function $1/r$ in ϕ^e converges very slowly for large r. The summation of l on c_{11}^R is not taken beyond the M nearest neighbors because the repulsive potential $\phi(r)$ between adjacent atoms varies rapidly with r and it is assumed that it can be ignored at distances $2r$ and greater.

The summation for the coulombic part of the lattice extends to infinity, and involves mathematical problems in convergence. It has been worked out by Born, Evjen, Kellerman, Cowley, Tosi and others. The series which arises directly from equations such as eq. (18) is not absolutely convergent, and it must be transformed into other series which correspond to the physical conditions of a neutrally charged cell, and which are quickly convergent.

These mathematical problems of the convergence of the coulombic part of the potential have been worked out, and the results for several simple cubic solids will be reproduced in the table below. We use the numbers given by COWLEY (1962).

Caution must be used with the definitions used for the convergent sums. The Madelung constants can be defined in two ways: a constant A_r corresponding to the nearest-neighbor distance r, or A_a corresponding to half distance a of the unit cell. For comparison purposes, $A_r = A_a$ for NaCl, $A_a = \frac{1}{3}A_r\sqrt{3}$ for CsCl, $A_a = \frac{2}{3}A_r\sqrt{3}$ for ZnS. Similarly, $r = a$ for NaCl, $a = \frac{1}{3}r\sqrt{3}$ for CsCl, and $a = \frac{2}{3}r\sqrt{3}$ for ZnS.

These lattice sum calculations have certain internal cross-checks. It is required that $c_{11}^e = -2c_{44}^{*e}$, and that the sum $c_{11}^e + 2c_{12}^e$ is related to the Madelung constant:

$$(c_{11}^e + 2c_{12}^e)\frac{\Delta a}{Z^2 e^2} = -\tfrac{4}{3}A_a. \tag{20}$$

2.3. The short-range component of the potential

The calculation of c_{11}^R, c_{12}^R, etc., are of especial interest because the assumptions of the physical model enter strongly into the results. The assumptions center around the upper limit of the sum and the nature of the repulsive potential, as well as the assumptions already implicit in this paper, namely, central forces. In eq. (14), for example, there is the assumption that the repulsive potential is radially symmetric, that only nearest neighbors interactions need be considered, and that the potential falls off so rapidly that the sum of the nearest neighbors is terminated by the nearest-neighbor limit. For the case of NaCl, M is 6; for ZnS, M is 4; and for CsCl, M is 8.

We now proceed to evaluate the repulsive portion of eqs. (5)–(10) for the three crystal structures by using table 2.

Of particular interest is the sum $\sum_{l=1}^{M} x^l y^l z^l$, which is zero for NaCl because every term is zero, is zero for CsCl because there are equal numbers of positive and negative terms, but is $+\frac{4}{9}r^3\sqrt{3}$ for ZnS because all terms have the same sign.

The sums over l reduce to the expressions of table 3 for the operators P^R and Q^R, and for the three lattices.

We now express the operators P and Q in terms of the derivatives of the repulsive potential with respect to r. In the sum corresponding to eq. (14), we sum on $v(r)$ over l up to the coordination number for the individual constants. It is readily shown that

$$Q^R = \frac{1}{r^2}\frac{d^2 v}{dr^2} - \frac{1}{r^3}\frac{dv}{dr}, \tag{21}$$

$$P^R = \frac{1}{r}\frac{dv}{dr}. \tag{22}$$

In tabular form, the results are given in table 4. Two new constants are added: $c' = \frac{1}{2}(c_{11} - c_{12})$ a shear

TABLE 1

Values of lattice sums for the Coulombic potential. The units of c_{ij} are $Z^2 e^2 / \Delta a$ where Δ is the volume of the cell

	NaCl	CsCl	ZnS
A_a	1.74756	1.01768	1.89147
A_r	1.74756	1.76267	1.63805
c_{11}^e†	−2.55604	1.40179	0.12381
c_{12}^e†	0.11298	−1.37936	−1.32296
c_{44}^{*e}	1.27802	−0.70089	−0.06190
C^e	0	0	$-2.51/r$
D^e	$-4.189/r^2$		$-2.0953/r^2$

† The values for c_{ij}^e refer to lattice sums taken appropriate to a, or the Madelung constant A_a.

TABLE 2

A detailed table showing the contributions of the lattice components to the sum in eqs. 5–10 affecting c_{11}, c_{12} and c_{44}^*

Atom position NaCl	x	y	z	x^2	y^2	x^2y^2	x^4	xyz
	in units of r			in units of r^2		in units of r^4		in units of r^3
1	1	0	0	1	0	0	1	0
2	0	1	0	0	1	0	0	0
3	-1	0	0	1	0	0	1	0
4	0	-1	0	0	1	0	0	0
5	0	0	1	0	0	0	0	0
6	0	0	-1	0	0	0	0	0
CsCl	in units of $\tfrac{1}{3}r\sqrt{3}$			in units of $\tfrac{1}{3}r^2$		in units of $\tfrac{1}{9}r^4$		in units of $\tfrac{1}{9}r^3\sqrt{3}$
1	1	1	1	1	1	1	1	1
2	-1	-1	-1	1	1	1	1	-1
3	-1	-1	1	1	1	1	1	1
4	1	1	-1	1	1	1	1	-1
5	1	-1	1	1	1	1	1	-1
6	-1	1	-1	1	1	1	1	1
7	-1	1	1	1	1	1	1	-1
8	1	-1	-1	1	1	1	1	1
ZnS								
1	1	1	1	1	1	1	1	1
2	-1	-1	1	1	1	1	1	1
3	1	-1	-1	1	1	1	1	1
4	-1	1	-1	1	1	1	1	1

TABLE 3

	NaCl	CsCl	ZnS
Δc_{11}^R	$2Q^R r^4 + 2P^R r^2$	$\tfrac{8}{9}(Q^R r^4 + 3P^R r^2)$	$\tfrac{4}{9}(Q^R r^4 + 3P^R r^2)$
Δc_{12}^R	$-2P^R r^2$	$\tfrac{8}{9}(Q^R r^4 - 3P^R r^2)$	$\tfrac{4}{9}(Q^R r^4 - 3P^R r^2)$
Δc_{44}^{*R}	$2P^R r^2$	$\tfrac{8}{9}(Q^R r^4 + 3P^R r^2)$	$\tfrac{4}{9}(Q^R r^4 + 3P^R r^2)$
ΔC^R	0	0	$\tfrac{4}{9}Q^R r^3 \sqrt{3}$
ΔD^R	$8P^R + 2Q^R r^2$	$8(P^R + \tfrac{1}{3}Q^R r^2)$	$4(P^R + \tfrac{1}{3}Q^R r^2)$

constant, and $K = (\tfrac{1}{3}c_{11} + 2c_{12})$ the bulk modulus.

There are two ways to proceed from this point. One way is to represent the operators d^2v/dr^2 and dv/dr at the equilibrium condition $r = r_0$ as pure numbers and then evaluate the numbers from experiments on the elastic constants. The second way is to assume some function for $v(r)$ and make the evaluation of this function through the operators. We will proceed with the second alternate because it allows us to retain the elastic constants as functions of r.

Let us take for the repulsive potential

$$v(r) = \frac{b}{r^n}, \quad (23)$$

where b is a constant and n is the index of the repulsion. Using eq. (23) in table 4, we have evaluated the repulsive term in the elastic constants in terms of this potential in table 5.

The above constants can be evaluated providing the constant b is known. This is not ordinarily known, but it can be expressed in terms of the Madelung constant by requiring that the derivative of eq. (14) vanishes at equilibrium, $r = r_0$. Thus

$$\frac{nb}{r_0^n} = \frac{A_r Z^2 e^2}{M r_0}, \quad (24)$$

and

TABLE 4

The elastic constant parameters in terms of the derivatives of the repulsive potential [$v(r)$]

	NaCl	CsCl	ZnS
Δc_{11}^R	$2r^2 \dfrac{d^2v}{dr^2}$	$\dfrac{8}{9}\left(r^2 \dfrac{d^2v}{dr^2} + 2r \dfrac{dv}{dr}\right)$	$\dfrac{4}{9}\left(r^2 \dfrac{d^2v}{dr^2} + 2r \dfrac{dv}{dr}\right)$
Δc_{12}^R	$-2r \dfrac{dv}{dr}$	$\dfrac{8}{9}\left(r^2 \dfrac{d^2v}{dr^2} - 4r \dfrac{dv}{dr}\right)$	$\dfrac{4}{9}\left(r^2 \dfrac{d^2v}{dr^2} - 4r \dfrac{dv}{dr}\right)$
Δc_{44}^{*R}	$2r \dfrac{dv}{dr}$	$\dfrac{8}{9}\left(r^2 \dfrac{d^2v}{dr^2} + 2r \dfrac{dv}{dr}\right)$	$\dfrac{4}{9}\left(r^2 \dfrac{d^2v}{dr^2} + 2r \dfrac{dv}{dr}\right)$
ΔC^R	0	0	$\dfrac{4\sqrt{3}}{9}\left(r \dfrac{d^2v}{dr^2} - \dfrac{dv}{dr}\right)$
ΔD^R	$8 \dfrac{d^2v}{dr^2} + \dfrac{6}{r} \dfrac{dv}{dr}$	$\dfrac{8}{3}\left(\dfrac{d^2v}{dr^2} + \dfrac{2}{r} \dfrac{dv}{dr}\right)$	$\dfrac{4}{3}\left(\dfrac{d^2v}{dr^2} + \dfrac{2}{r} \dfrac{dv}{dr}\right)$
$\Delta c'^R$	$r^2 \dfrac{d^2v}{dr^2} + r \dfrac{dv}{dr}$	$+\dfrac{8}{3} r \dfrac{dv}{dr}$	$+\dfrac{4}{3} r \dfrac{dv}{dr}$
ΔK^R	$\dfrac{2}{3}\left(r^2 \dfrac{d^2v}{dr^2} - 2r \dfrac{dv}{dr}\right)$	$\dfrac{8}{9}\left(r^2 \dfrac{d^2v}{dr^2} - 2r \dfrac{dv}{dr}\right)$	$\dfrac{4}{9}\left(r^2 \dfrac{d^2v}{dr^2} - 2r \dfrac{dv}{dr}\right)$

TABLE 5

The elastic constant parameters in terms of the Born repulsive potential

	NaCl	CsCl	ZnS
c_{11}^R	$\dfrac{2nb}{\Delta r^n}(n+1)$	$\dfrac{8nb}{\Delta r^n} \dfrac{1}{9}(n-1)$	$\dfrac{4nb}{\Delta r^n} \dfrac{1}{9}(n-1)$
c_{12}^R	$\dfrac{2nb}{\Delta r^n}$	$\dfrac{8nb}{\Delta r^n} \dfrac{1}{9}(n+5)$	$\dfrac{4nb}{\Delta r^n} \dfrac{1}{9}(n+5)$
c_{44}^{*R}	$-\dfrac{2nb}{\Delta r^n}$	$\dfrac{8nb}{\Delta r^n} \dfrac{1}{9}(n-1)$	$\dfrac{4nb}{\Delta r^n} \dfrac{1}{9}(n-1)$
C^R	0	0	$+\dfrac{4nb}{\Delta r^n}\left(\dfrac{1}{r}\right)\dfrac{1}{9}(n+2)\sqrt{3}$
D^R	$\dfrac{2nb}{\Delta r^n}\left(\dfrac{1}{r^2}\right)(n-2)$	$\dfrac{8nb}{\Delta r^n}\left(\dfrac{1}{r^2}\right)\dfrac{1}{3}(n-1)$	$\dfrac{4nb}{\Delta r^n}\left(\dfrac{1}{r^2}\right)\dfrac{1}{3}(n-1)$
c'^R	$\dfrac{nb}{\Delta r^n} n$	$-\dfrac{8nb}{\Delta r^n} \cdot \dfrac{1}{3}$	$-\dfrac{4nb}{\Delta r^n} \cdot \dfrac{1}{3}$
K^R	$\dfrac{2nb}{\Delta r^n} \dfrac{1}{3}(n+3)$	$\dfrac{8nb}{\Delta r^n} \dfrac{1}{9}(n+3)$	$\dfrac{4nb}{\Delta r^n} \dfrac{1}{9}(n+3)$

TABLE 6

The repulsive part of elastic parameters in units of $Z^2e^2/\Delta_0 a_0$

	NaCl	CsCl	ZnS
c_{11}^R	$\frac{1}{3} A_a (n+1) \left(\frac{a_0}{a}\right)^{n+3}$	$\frac{1}{9} A_a (n-1) \left(\frac{a_0}{a}\right)^{n+3}$	$\frac{1}{9} A_a (n-1) \left(\frac{a_0}{a}\right)^{n+3}$
c_{12}^R	$\frac{1}{3} A_a \left(\frac{a_0}{a}\right)^{n+3}$	$\frac{1}{9} A_a (n+5) \left(\frac{a_0}{a}\right)^{n+3}$	$\frac{1}{9} A_a (n+5) \left(\frac{a_0}{a}\right)^{n+3}$
c_{44}^{*R}	$-\frac{1}{3} A_a \left(\frac{a_0}{a}\right)^{n+3}$	$\frac{1}{9} A_a (n-1) \left(\frac{a_0}{a}\right)^{n+3}$	$\frac{1}{9} A_a (n-1) \left(\frac{a_0}{a}\right)^{n+3}$
c'^R	$\frac{1}{6} A_a n \left(\frac{a_0}{a}\right)^{n+3}$	$-\frac{1}{3} A_a \left(\frac{a_0}{a}\right)^{n+3}$	$-\frac{1}{3} A_a \left(\frac{a_0}{a}\right)^{n+3}$
K^R	$\frac{1}{9} A_a (n+3) \left(\frac{a_0}{a}\right)^{n+3}$	$\frac{1}{9} A_a (n+3) \left(\frac{a_0}{a}\right)^{n+3}$	$\frac{1}{9} A_a (n+3) \left(\frac{a_0}{a}\right)^{n+3}$
D^R	$\frac{1}{3} A_a \left(\frac{n-2}{a_0^2}\right) \left(\frac{a_0}{a}\right)^{n+5}$	$\frac{1}{9} A_a \left(\frac{n-1}{a_0^2}\right) \left(\frac{a_0}{a}\right)^{n+5}$	$\frac{4}{9} A_a \left(\frac{n-1}{a_0^2}\right) \left(\frac{a_0}{a}\right)^{n+5}$
C^R	0	0	$\frac{2}{9} A_a \left(\frac{n+2}{a_0}\right) \left(\frac{a_0}{a}\right)^{n+4}$

$$\frac{nb}{r^n} = \frac{A_r Z^2 e^2}{M r_0} \left(\frac{r_0}{r}\right)^n. \quad (25)$$

Further, since $1/\Delta = (1/\Delta_0)(r_0/r)^3$:

$$\frac{nb}{\Delta r^n} = \frac{Z^2 e^2}{\Delta_0 r_0} \frac{A_r}{M} \left(\frac{r_0}{r}\right)^{n+3}. \quad (26)$$

The elastic parameters, for the repulsive components, are displayed in table 6, where the multiplying factor $(Z^2 e^2/\Delta_0 a_0)$ is assumed. The values of the coordination number M put into eq. (26): 6, 8, 4 for the NaCl, CsCl, and ZnS structures, respectively. We now wish to use eq. (26) to evaluate b in table 5, and we wish further to transform the equations for the repulsive elastic constants from the space variable r (interatomic separation) to the space variable a (half the lattice constant). The purpose of this transformation is to change the Madelung constant in eq. (26) from A_r to A_a, because the latter is compatible with the coulombic contributions of the elastic constants given in table 1. In this transformation we use $A_r/r_0 = A_a/a_0$, $r_0 = a_0 \sqrt{3}$ for CsCl, and $r_0 = \frac{1}{2} a_0 \sqrt{3}$ for ZnS.

2.4. The elastic constants

The elastic constant expressions can be found by combining the results shown in table 1 with those in table 6. Using the values for the lattice sums found in table 1, the full expressions for the elastic constants are obtained.

NaCl structure:

Repulsive   Coulombic

$$c_{11} = \frac{Z^2 e^2}{\Delta_0 a_0} \left\{ 0.58252(n+1) \left(\frac{a_0}{a}\right)^{n+3} - 2.55604 \left(\frac{a_0}{a}\right)^4 \right\} \quad (27)$$

$$c_{12} = \frac{Z^2 e^2}{\Delta_0 a_0} \left\{ 0.58252 \left(\frac{a_0}{a}\right)^{n+3} + 0.11298 \left(\frac{a_0}{a}\right)^4 \right\} \quad (28)$$

NaCl structure:

Repulsive   Coulombic

$$c_{44} = \frac{Z^2 e^2}{\Delta_0 a_0} \left\{ -0.58252 \left(\frac{a_0}{a}\right)^{n+3} + 1.27802 \left(\frac{a_0}{a}\right)^4 \right\} \tag{29}$$

$$c' = \frac{Z^2 e^2}{\Delta_0 a_0} \left\{ 0.29126 n \left(\frac{a_0}{a}\right)^{n+3} - 1.33451 \left(\frac{a_0}{a}\right)^4 \right\} \tag{30}$$

$$K = \frac{Z^2 e^2}{\Delta_0 a_0} \left\{ 0.19417(n+3) \left(\frac{a_0}{a}\right)^{n+3} - 0.7767 \left(\frac{a_0}{a}\right)^4 \right\} \tag{31}$$

$$D = \frac{Z^2 e^2}{\Delta_0 a_0} \left(\frac{1}{a_0^2}\right) \left\{ 0.58252(n-2) \left(\frac{a_0}{a}\right)^{n+5} - 4.1890 \left(\frac{a_0}{a}\right)^6 \right\} \tag{32}$$

$$C = 0 \tag{33}$$

CsCl structure:

$$c_{11} = \frac{Z^2 e^2}{\Delta_0 a_0} \left\{ 0.113075(n-1) \left(\frac{a_0}{a}\right)^{n+3} + 1.40179 \left(\frac{a_0}{a}\right)^4 \right\} \tag{34}$$

$$c_{12} = \frac{Z^2 e^2}{\Delta_0 a_0} \left\{ 0.113075(n+5) \left(\frac{a_0}{a}\right)^{n+3} - 1.37935 \left(\frac{a_0}{a}\right)^4 \right\} \tag{35}$$

$$c_{44}^* = \frac{Z^2 e^2}{\Delta_0 a_0} \left\{ 0.113075(n-1) \left(\frac{a_0}{a}\right)^{n+3} - 0.70089 \left(\frac{a_0}{a}\right)^4 \right\} \tag{36}$$

$$c_{44} = c_{44}^* \tag{37}$$

$$c' = \frac{Z^2 e^2}{\Delta_0 a_0} \left\{ -0.33923 \left(\frac{a_0}{a}\right)^{n+3} + 1.39057 \left(\frac{a_0}{a}\right)^4 \right\} \tag{38}$$

$$K = \frac{Z^2 e^2}{\Delta_0 a_0} \left\{ 0.113075(n+3) \left(\frac{a_0}{a}\right)^{n+3} - 0.452303 \left(\frac{a_0}{a}\right)^4 \right\} \tag{39}$$

$$C = 0 \tag{40}$$

ZnS structure:

$$c_{11} = \frac{Z^2 e^2}{\Delta_0 a_0} \left\{ 0.210163(n-1) \left(\frac{a_0}{a}\right)^{n+3} + 0.12381 \left(\frac{a_0}{a}\right)^4 \right\} \tag{41}$$

$$c_{12} = \frac{Z^2 e^2}{\Delta_0 a_0} \left\{ 0.210163(n+5) \left(\frac{a_0}{a}\right)^{n+3} - 1.32296 \left(\frac{a_0}{a}\right)^4 \right\} \tag{42}$$

$$c_{44}^* = \frac{Z^2 e^2}{\Delta_0 a_0} \left\{ 0.210163(n-1) \left(\frac{a_0}{a}\right)^{n+3} - 0.06190 \left(\frac{a_0}{a}\right)^4 \right\} \tag{43}$$

$$c' = \frac{Z^2 e^2}{\Delta_0 a_0} \left\{ -0.63049 \left(\frac{a_0}{a}\right)^{n+3} + 0.723385 \left(\frac{a_0}{a}\right)^4 \right\} \tag{44}$$

$$K = \frac{Z^2 e^2}{\Delta_0 a_0} \left\{ 0.210163(n+3) \left(\frac{a_0}{a}\right)^{n+3} - 0.84070 \left(\frac{a_0}{a}\right)^4 \right\} \tag{45}$$

ZnS structure:

Repulsive — Coulombic

$$D = \frac{Z^2 e^2}{\Delta_0 a_0}\left(\frac{1}{a_0^2}\right)\left\{0.84065(n-1)\left(\frac{a_0}{a}\right)^{n+5} - 2.095\left(\frac{a_0}{a}\right)^6\right\} \tag{46}$$

$$C = \frac{Z^2 e^2}{\Delta_0 a_0}\left(\frac{1}{a_0}\right)\left\{0.420326(n+2)\left(\frac{a_0}{a}\right)^{n+4} - 2.51\left(\frac{a_0}{a}\right)^5\right\} \tag{47}$$

$$c_{44} = c_{44}^* - C^2/D$$

At zero pressure ($r = r_0$, $a = a_0$) the values of the elastic constants may be expressed as simple functions of n. The value of n appropriate to different crystallographic structures may then be estimated by comparison with experimental values of the elastic constants. This is a common approach in lattice dynamical calculations. We prefer, however, to retain the equations in the form of table 6 for the moment and to only comment briefly on several interesting properties of the elastic constants at zero pressure.

It is interesting to note that the formula for K_0 is the same for all three structures as shown by eq. (20) and table 6:

$$K_0 = K_0^R + K_0^e = \frac{Z^2 e^2}{\Delta_0 a_0} A_a(n-1). \tag{48}$$

We see that for NaCl, c_{12} and c_{44} are independent of the repulsion law chosen, while the shear modulus c' is linear with n. For the CsCl and ZnS structure, c_{12} and c_{44}^* are linear with n, while c' is independent of the repulsion law chosen.

It can be seen that for the CsCl structure n must be large in order for c_{12} and c_{44} not to be negative numbers. In fact, n must be greater than 7.2. On the other hand, no such restriction is imposed upon the NaCl lattice. In the case of the ZnS lattice, n need only be larger than 1.3 for the lattice to have positive values of all elastic constants. For the NaCl lattice, c' is negative for n less than 4.6.

The ratio of c_{11} to c_{12} is always over unity, and is greater in the NaCl lattice, next greatest in the CsCl lattice, and least in the ZnS lattice.

For the case of NaCl, it is possible to choose a value of n so that at equilibrium isotropy occurs for about $n = 7$: isotropy occurs when $c' = c_{44}$. The condition of isotropy never occurs for ZnS lattice but it occurs for the CsCl structure at about $n = 11$.

It can be seen that for both NaCl and CsCl, the Cauchy relation holds ($c_{12} = c_{44}$). However, this equality only holds for equilibrium conditions $a = a_0$. At higher pressures (or a higher value of (a_0/a)), we have $c_{12} > c_{44}$. For NaCl we find

$$c_{12} - c_{44} = \frac{Z^2 e^2}{2a_0^4}\left(\frac{a_0}{a}\right)^4 1.16504\left\{\left(\frac{a_0}{a}\right)^{n-1} - 1\right\}. \tag{49}$$

2.5. Pressure derivatives of the elastic constants

Using the relationships

$$K = -V\frac{dP}{dV}, \tag{50}$$

and

$$\frac{3r}{r}\frac{dP}{dV} = \frac{dP}{dr}, \tag{51}$$

it is easily shown that

$$\frac{dc_{ij}}{dP} = -\frac{r}{3K}\frac{dc_{ij}}{dr} = -\frac{a}{3K}\frac{dc_{ij}}{da}. \tag{52}$$

The pressure derivatives of the elastic constants are evaluated at zero pressure by applying the eq. (52) above to the general formula

$$c_{ij}(a) = (c_{ij}^R)_0\left(\frac{a_0}{a}\right)^{n+3} + (c_{ij}^e)_0\left(\frac{a_0}{a}\right)^4, \tag{53}$$

with the result that

$$\left(\frac{dc_{ij}}{dP}\right)_0 = \tfrac{1}{3}(n+3)\left(\frac{c_{ij}^R}{K}\right)_0 + \tfrac{4}{3}\left(\frac{c_{ij}^e}{K}\right)_0, \tag{54}$$

$$\left(\frac{dC}{dP}\right)_0 = \tfrac{1}{3}(n+4)\left(\frac{C^R}{K}\right)_0 + \tfrac{5}{3}\left(\frac{C^e}{K}\right)_0. \tag{55}$$

$$\left(\frac{dD}{dP}\right)_0 = \tfrac{1}{3}(n+5)\left(\frac{D^R}{K}\right)_0 + 2\left(\frac{D^e}{K}\right)_0. \tag{56}$$

Of special interest is the case of the relationship between K and c_{44}^{*R} in the CsCl and ZnS structure, which we shall now consider. From eq. (48) and table 6 we see that for ZnS and CsCl

$$K_0 = (c_{11}^R)_0 = (c_{44}^{*R})_0. \tag{57}$$

Immediately then, we have from eq. (54)

$$\left(\frac{dc_{11}}{dP}\right)_0 = \tfrac{1}{3}(n+3) + \tfrac{4}{3}\left(\frac{c_{11}^e}{K}\right)_0, \tag{58}$$

$$\left(\frac{dc_{44}^*}{dP}\right)_0 = \tfrac{1}{3}(n+3) + \tfrac{4}{3}\left(\frac{c_{44}^{e*}}{K}\right)_0. \tag{59}$$

There is also a simple formula for $(dK/dP)_0$. Applying the eq. (58)–(59) above for K_0^e and K_0^R to eq. (54) we find, after some rearranging, that

$$\left(\frac{dK}{dP}\right)_0 = \tfrac{1}{3}(n+7). \tag{60}$$

The importance of eq. (60) arises in the fact that this formula is independent of the three cubic structures we have considered. Therefore, we conclude that only the repulsive law and not the crystallographic structure affects the values of $(dK/dP)_0$.

However, the structure, as well as the repulsive law, affects the pressure derivatives of shear constants, as will become evident. Returning to eq. (59) for the ZnS structure

$$\frac{dc_{44}^*}{dP} = \tfrac{1}{3}(n+3) - \frac{0.393}{(n-1)}. \tag{61}$$

To reduce eq. (61) to (dc_{44}/dP), which is a quantity that can be measured, we use eq. (8), and must therefore evaluate

$$\frac{d}{dP}\left(\frac{C^2}{D}\right) = \frac{C^2}{D}\left(\frac{2}{C}\frac{dC}{dP} - \frac{1}{D}\frac{dD}{dP}\right). \tag{62}$$

Using eqs. (55) and (56)

$$\frac{d}{dP}\left(\frac{C^2}{D}\right) = \frac{C^2}{3DK}\left[2(n+4)\frac{C^R}{C} + 10\frac{C^e}{C} - (n+5)\frac{D^R}{D} - 6\frac{D^e}{D}\right]. \tag{63}$$

After some arranging

$$\frac{d}{dP}\left(\frac{C^2}{D}\right) = \frac{C^2}{DK}\left[1 + \frac{n}{3}\left\{2\left(\frac{C^R}{C}\right) - \frac{D^R}{D}\right\} + \tfrac{1}{3}\left\{2\left(\frac{C^e}{C}\right) - \frac{D^e}{D}\right\}\right]. \tag{64}$$

Thus we have for the ZnS structure

$$\left(\frac{dc_{44}}{dP}\right)_0 = \tfrac{1}{3}(n+3) - \frac{0.391}{n-1} - \frac{C^2}{DK} \times$$
$$\times \left\{1 + \tfrac{1}{3}n\left[2\frac{C^R}{C} - \frac{D^R}{D}\right] + \tfrac{1}{3}\left[2\frac{C^e}{C} - \frac{D^e}{D}\right]\right\}. \tag{65}$$

From eq. (47)

$$Ca_0 = \left(\frac{Z^2e^2}{\varDelta_0 a_0}\right)\tfrac{2}{9}A_a\left\{(n+2) - \frac{2.51 \times 9}{2A_a}\right\} = \tag{66}$$

$$= \frac{Z^2e^2}{\varDelta_0 a_0}\tfrac{2}{9}A_a(n - 3.9715). \tag{67}$$

Similarly, from eq. (46)

$$Da_0^2 = \frac{Z^2e^2}{\varDelta_0 a_0}\tfrac{4}{9}A_a(n - 3.4921). \tag{68}$$

Using the above equations with eq. (48) we find

$$\frac{C^2}{KD} = \left(\frac{1}{n-1}\right)\cdot\frac{(n-3.9715)^2}{(n-3.4921)}. \tag{69}$$

Also

$$\frac{C^e}{C} = \frac{-2.51}{0.4203(n+2)-2.51}; \quad \frac{C^R}{C} = \frac{0.4203(n+2)}{0.4203(n+2)-2.51}; \tag{70}$$

$$\frac{D^e}{D} = \frac{-2.095}{0.84065(n-1)-2.0953};$$
$$\frac{D^R}{D} = \frac{0.84065(n-1)}{0.84065(n-1)-2.0953}. \tag{71}$$

Evaluating eq. (65)

$$\left(\frac{dc_{44}}{dP}\right)_0 \begin{cases} = -0.740, & \text{for } n = 6, \\ = -1.29, & \text{for } n = 9. \end{cases}$$

This is to be compared with

$$\left(\frac{dc_{44}}{dP}\right)_0 = \tfrac{1}{3}(n+3) - \frac{8.264}{n-1} \tag{72}$$

TABLE 7

Pressure derivatives of the elastic constants for arbitrary n in the Born repulsive potential

Pressure derivative of	NaCl	CsCl	ZnS
c_{11}	$\dfrac{(n+3)(n+1)-17.55}{n-1}$	$\tfrac{1}{3}(n+3)+\dfrac{16.53}{n-1}$	$\tfrac{1}{3}(n+3)+\dfrac{0.785}{n-1}$
c_{12}	$\dfrac{n+3.776}{n-1}$	$\dfrac{(n+3)(n+5)-48.78}{3(n-1)}$	$\dfrac{(n+3)(n+5)-25.17}{3(n-1)}$
c_{44}^{*}	$\dfrac{5.776-n}{n-1}$	$\tfrac{1}{3}(n+3)-\dfrac{8.264}{n-1}$	$\tfrac{1}{3}(n+3)-\dfrac{0.391}{n-1}$
C^2/D	0	0	See eq. (65)
c_{44}	$\dfrac{dc_{44}^{*}}{dP}$	$\dfrac{dc_{44}^{*}}{dP}$	$\dfrac{dc_{44}^{*}}{dP}-\dfrac{d}{dP}\left(\dfrac{C^2}{D}\right)$
c'	$\dfrac{n(n+3)-18.327}{2(n-1)}$	$\dfrac{13.396-n}{n-1}$	$\dfrac{1.589-n}{n-1}$
K	$\tfrac{1}{3}(n+7)$	$\tfrac{1}{3}(n+7)$	$\tfrac{1}{3}(n+7)$

for CsCl, and

$$\left(\frac{dc_{44}}{dP}\right)_0 = \tfrac{1}{3}(n+3)\frac{c_{44}^R}{K}+\tfrac{4}{3}\frac{c_{44}^e}{K}=\frac{5.776-n}{n-1} \quad (73)$$

for NaCl. Thus $(dc_{44}/dP)_0$ is negative for all reasonable values of n, and increases in magnitude as n increases for the ZnS structure; it is always positive and increases steadily with n for the CsCl structure; and it is small and changes sign at about $n = 6$ for the NaCl structure.

These great differences in the c_{44} behavior arise because of lattice sums arising from the different cubic structures; the shear constants stand in contrast to the case of K and $(dK/dP)_0$ which do not depend upon structure according to this model.

Similar differences can be found for dc'/dP. The formulas for this and the remaining derivatives are given in table 7 based on eqs. (54)–(56).

2.6. *Comparison of calculated and measured pressure derivatives of elastic constants*

The pressure derivatives of the elastic constants for the NaCl and CsCl lattices are calculated from the equations in table 7 for $n = 6$ and $n = 9$ and are listed in table 8. For comparison we have also given the measured derivatives for LiF, NaF, NaCl, KCl, KBr, KI and RbBr (all NaCl structures) and for CsCl, CsBr, and CsI (CsCl structures); these measured values are taken from the tabulation by BARSCH and CHANG (1967), and are also listed in complete form in table 9. Examination of table 8 clearly shows that the Born model of the potential between adjacent atoms, given by eq. (14) accounts for the experimental values of the pressure derivatives of the elastic constants for the materials with the NaCl and CsCl structures. For the NaCl structure, LiF appears to be in better agreement with the $n = 6$ calculation, while the $n = 9$ case is more compatible with the data for the other materials. For the CsCl structure, the data are more closely approximated by the $n = 9$ calculation. The agreement is not perfect but it is quite good enough for qualitative purposes.

The effect of n upon the pressure derivatives of the elastic constants for the ZnS structure is investigated in table 10. It is quite clear that the ZnS structure is characterized by negative values of dc_{44}/dP and dc'/dP

TABLE 8

Comparison of calculated and measured pressure derivatives for NaCl and CsCl structures

Pressure derivative of	NaCl structure									CsCl structure				
	Calculated		Measured							Calculated		Measured		
	$n=6$	$n=9$	LiF	NaF	NaCl	KCl	KBr	KI	RbBr	$n=6^*$	$n=9$	CsCl	CsBr	CsI
c_{11}	9.09	12.81	9.92	11.59	11.71	12.77	13.47	14.56	13.53	6.31	6.07	6.82	6.30	6.46
c_{12}	1.96	1.60	2.72	1.99	2.06	1.61	1.61	2.45	3.05	3.35	4.97	5.05	4.93	4.78
c_{44}	−0.04	−0.40	1.38	0.21	0.37	−0.39	−0.30	−0.20	−0.36	1.35	2.97	3.56	3.68	3.74
c'	3.57	5.60	3.60	4.80	4.83	5.58	5.93	6.05	5.24	1.48	0.55	1.35	0.68	0.84
K	4.33	5.33	5.12	5.19	5.28	5.33	5.56	6.49	6.54	4.33	5.33	5.64	5.39	5.34

* This condition is unstable according to eq. (36).

TABLE 9

Adiabatic elastic constants (c_{ij}, in kb) and their isothermal pressure derivatives [$(\partial c_{ij}/\partial P)_T = c_{ij}'$] for the alkali halides and other materials

Material (1 b, 298 °K)	c_{11}	c_{12}	c_{44}	c_{11}'	c_{12}'	c_{44}'	Reference
LiF	1137	476	637	9.92	2.72	1.38	1
NaF	970	238	282	11.59	1.99	0.205	1
NaCl	490	126	127	11.71	2.06	0.372	1, 2
KCl	405	70	63	12.77	1.61	−0.392	1
KBr	342	52	51	13.47	1.61	−0.296	1
KI	268	41	37	14.56	2.45	−0.204	1
RbBr	316	49	38	13.53	3.05	−0.361	1
CsCl	368	89	82	6.82	5.05	3.56	1
CsBr	308	83	76	6.30	4.93	3.68	1
CsI	246	66	64	6.46	4.78	3.74	1
CuZn	1290	1097	824	3.08	3.33	1.89	1
CaF_2	1642	440	337	6.04	4.35	1.30	3, 4
BaF_2	920	416	257	4.82	5.17	0.777	4

1. Barsch and Chang (1967).
2. Bartels and Schuele (1965).
3. Wong and Schuele (1967).
4. Wong and Schuele (1968).

(the latter largely insensitive to n), while the NaCl structure is characterized by a large positive value of dc'/dP and a small (sometimes negative value) for dc_{44}/dP. In the CsCl structure, no pressure derivatives are negative, although dc'/dP is somewhat low for $n = 9$.

The elastic behavior of an isotropic material may be completely described by only two elastic constants, the bulk modulus (K) and the shear modulus (μ). For the three cubic structures we have considered (eq. 60) indicates that only the repulsive law and not the crystallographic structure affects the values of $(dK/dP)_0$. On the other hand, the effect of crystallographic structure appears to enter into the value of $d\mu/dP$ more than it does in dK/dP. This empirical result is verified by the

TABLE 10

Pressure derivatives of the elastic constants for the ZnS structure (calculated) from the equations in table 7

Pressure derivative of	$n=6$	$n=7$	$n=8$	$n=9$
c_{11}	3.16	3.42	3.78	4.10
c_{12}	4.92	5.27	5.61	5.95
c_{44}^*	2.922	3.268	3.611	3.951
C^2/D	3.662	4.271	4.780	5.238
c_{44}	−0.74	−1.00	−1.17	−1.29
c'	−0.88	−0.90	−0.92	−0.93
K	4.33	4.67	5.00	5.33

formulas in table 7 for the pressure derivatives of the shear constants c' and c_{44}; we shall return to this point in a later paragraph.

O. L. Anderson (1968b) has recently pointed out that the crystallographic dependence of $d\mu/dP$ for a number of oxides and silicates may be characterized by two parameters: the first parameter is the coordination number for the cation-anion pairs in the lattice; the second is the ratio of the shear modulus to the bulk modulus (μ/K). Anderson found that structures with the lowest coordination numbers and the lowest μ/K will have the lowest (and negative) values of $d\mu/dP$; this behavior was explained by means of an earlier theory for ZnS, which was originally proposed by Blackman (1958). It is very important to establish firmly whether or not Blackman's rule can be extended to oxides and silicates in general.

In this analysis the ultrasonic data from O. L. Anderson et al. (1968) for oxides and silicates and the data for other materials tabulated by Barsch and Chang (1967) and Wong and Schuele (1967, 1968), are used to evaluate Blackman's rule. For reference, the adiabatic elastic constants c_{ij} for the alkali halides

TABLE 11

Isotropic elastic moduli and their isothermal pressure derivatives for the alkali halides and other materials

Material	μ (kb)			K_S (kb)	μ/K_S	σ_S	$(\partial\mu/\partial P)_T$			E	$(\partial K_S/\partial P)_T$
	Voigt	Reuss	Hill				Voigt	Reuss	Hill		
LiF	514	465	489	696	0.703	0.215	2.27	3.40	2.84	0.11	5.12
NaF	316	311	313	482	0.649	0.233	2.04	1.55	1.79	0.02	5.19
NaCl	149	144	147	247	0.594	0.252	2.15	1.52	1.84	0.03	5.28
KCl	105	84	94	182	0.520	0.280	2.00	0.31	1.16	0.17	5.33
KBr	88	67	79	149	0.528	0.276	2.19	0.39	1.29	0.19	5.56
KI	68	51	59	116	0.508	0.282	2.30	0.46	1.38	0.21	6.49
RbBr	76	53	65	138	0.468	0.296	1.88	0.15	1.02	0.24	6.54
CsCl	105	98	101	182	0.556	0.266	2.49	3.30	2.90	0.06	5.64
CsBr	91	87	89	158	0.563	0.263	2.48	3.11	2.80	0.03	5.39
CsI	75	73	74	126	0.585	0.255	2.58	4.05	3.32	0.03	5.34
CuZn	533	205	369	1161	0.318	0.356	1.08	0.69	0.89	0.85	3.24
CaF$_2$	443	409	426	841	0.506	0.283	1.12	1.48	1.30	0.07	4.91
BaF$_2$	255	255	255	584	0.437	0.309	0.40	0.18	0.29	0.00	5.05

and their isothermal pressure derivatives $(\partial c_{ij}/\partial P)_T$ are given in table 9. From these data the isotropic elastic moduli and their isothermal pressure derivatives are determined in table 11 from the Voigt–Reuss–Hill approximation (CHUNG, 1963; O. L. ANDERSON, 1963, 1965). The equations for the moduli are well-known. The equations for the pressure derivatives of a polycrystalline aggregate in the Voigt–Reuss–Hill approximation have been presented for cubic crystals by CHUNG (1967) and BARSCH (1968). The Reuss pressure derivative given in BARSCH (1968) is, however, preferred (all derivatives are at isothermal conditions):

$$K^V = K^R = K^{VRH} = \tfrac{1}{3}(c_{11}+2c_{12}), \quad (74)$$

$$(dK/dP)^{VRH} = \tfrac{1}{3}(dc_{11}/dP + 2\,dc_{12}/dP), \quad (75)$$

$$\mu^V = \tfrac{1}{5}(2c'+3c_{44}), \quad (76)$$

$$(d\mu/dP)^V = \tfrac{1}{5}[2(dc'/dP)+3(dc_{44}/dP)], \quad (77)$$

$$\mu^R = \frac{10\,c'c_{44}}{[4c_{44}+6c']}, \quad (78)$$

$$(d\mu/dP)^R = \frac{(\mu^R)^2}{5}\,[2(c')^{-2}(dc'/dP) + 3(c_{44})^{-2}(dc_{44}/dP)] + E, \quad (79)$$

$$E = 6\left(\frac{c'-c_{44}}{3c'+2c_{44}}\right)^2 = 0.24\,A^2(1+\tfrac{3}{5}A), \quad (80)$$

$$A = \frac{c'-c_{44}}{c_{44}}, \quad (81)$$

$$(d\mu/dP)^{VRH} = \tfrac{1}{2}[(d\mu/dP)^V + (d\mu/dP)^R]. \quad (82)$$

The correction term (E) of BARSCH (1968) is only important in the calculation of $(\partial\mu/\mu P)^R$ when the elastic anisotropy A of the second-order elastic constants c_{ij} is large. O. L. ANDERSON (1963) has shown that A is directly related to the difference between the VOIGT (1928) and REUSS (1929) values for K and μ so that it is possible to evaluate the importance of E for non-cubic crystals. This ambiguity in the definition of $(\partial\mu/\partial P)_T$ may be eliminated entirely by using the difference method suggested by SOGA (1968):

$$dK/dP = [K^{VRH}(P) - K^{VRH}(O)]/\Delta P, \quad (83)$$

$$d\mu/dP = [\mu^{VRH}(P) - \mu^{VRH}(O)]/\Delta P. \quad (84)$$

In this manner (eqs. 83–84), the pressure derivatives of the effective elastic moduli are found directly.

In fig. 1, $d\mu/dP$ for the oxides and silicates is plotted as a function of Poisson's ratio (σ), which is simply related to μ/K:

$$\sigma = \frac{3-2(\mu/K)}{2[3+2(\mu/K)]}, \quad (85)$$

so that low values of μ/K correspond to large values of σ. The major points discussed by O. L. ANDERSON (1968b) remain valid when all the available data are included. For equal values of σ, the lower coordinated compounds have lower values of $d\mu/dP$. Note, especially for $\sigma \approx 0.2$–0.3, that $d\mu/dP$ decreases systematically as the coordination numbers decrease from 6–6 (MgO–CaO) to 6–4 (α-Al$_2$O$_3$–α-Fe$_2$O$_3$) to 6–4–4 (spi-

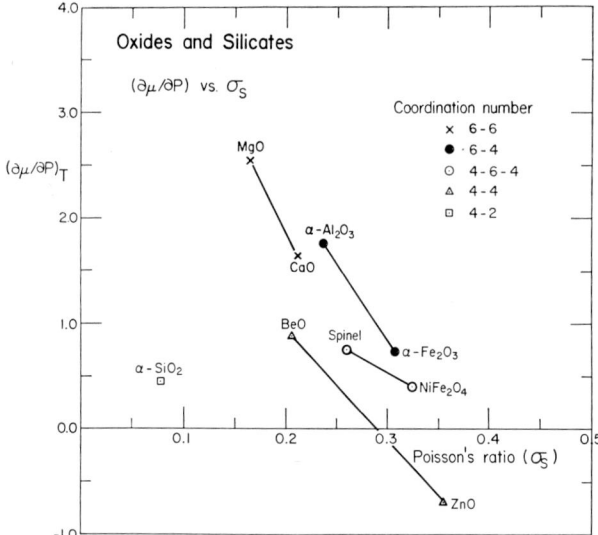

Fig. 1. $(\partial\mu/\partial P)_T$ versus Poisson's ratio for oxides and silicates of various coordination numbers (from data of O. L. ANDERSON et al., 1968 and LIEBERMANN, 1969).

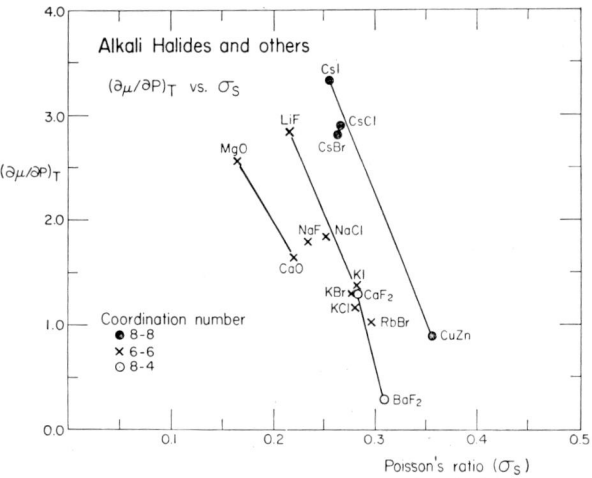

Fig. 2. $(\partial\mu/\partial P)_T$ versus Poisson's ratio for alkali halides and other materials of various coordination numbers. (From data given in table 11.)

nel–$NiFe_2O_4$) to 4–4 (BeO–ZnO). Within groups of the same coordination, $d\mu/dP$ decreases as σ increases in a similar manner for all groups. The position of α-SiO_2 in this diagram is consistent with its low coordination (4–2) and its low value of σ (0.077).

The corresponding data for the alkali halides and other materials (CaF_2, BaF_2, CuZn) are presented in fig. 2; the data for MgO and CaO from the previous figure are included for reference. Note again the systematic tendency in all coordination groups for $d\mu/dP$ to decrease as σ increases, with about the same slope for all groups. For $\sigma \approx$ constant $d\mu/dP$ decreases with decreasing average coordination number.

The details of crystallographic structure, especially coordination number, thus appear to be significantly related to the value of $d\mu/dP$ for a wide variety of materials. O. L. ANDERSON (1968b) showed that BLACKMAN's (1958) model for the ZnS structure yields results for $d\mu/dP$ that are quite close to the actual values measured for the related structure ZnO (SOGA and O. L. ANDERSON, 1967). With the more explicit formulas derived in this paper, it is possible to evaluate $d\mu/dP$ for the three structures we have considered. For convenience we use the Voigt approximation for $d\mu/dP$ (eq. 77) and obtain from table 8

	NaCl	CsCl	ZnS
$(d\mu/dP)^V$ $n = 6$	1.40	1.40	-0.80
$n = 9$	2.00	2.00	-1.15

The important result here is that polycrystalline aggregates of a NaCl or CsCl structure have positive values of $d\mu/dP$ while polycrystalline aggregates of a ZnS structure have negative values of $d\mu/dP$.

It is obvious that many structures will have values of $d\mu/dP$ between those of NaCl or CsCl and ZnS, and we should like to comment on that point by examining the details of the lattice sums which produce small or negative values of dc_{44}/dP and $d\mu/dP$.

It is apparent that when C, given by eq. (9), is non-zero, dc_{44}/dP is smaller than dc_{44}^*/dP. Thus, when C is large, dc_{44}/dP will be small or negative.

Using eqs. (5) and (6), it is seen that

$$c' = \frac{1}{2\Delta} \sum_l \{Q(x^l)^4 - Q(x^l)^2(y^l)^2 + 2P(x^l)^2\}. \quad (86)$$

Now, in the NaCl structure, the second term on the right is missing, so that

$$c' = \frac{1}{2\Delta} \sum_l \{Q(x^l)^4 + 2P(x^l)^2\}; \text{ NaCl}, \quad (87)$$

but in the CsCl and ZnS structure, $\sum_l Q(x^l)^4 = \sum_l Q(x^l)^2(y^l)^2$, so that

$$c' = \frac{1}{\Delta} \sum_l P(x^l)^2 = \frac{1}{\Delta} \sum_l \frac{\partial v(r)}{\partial r}(r)^l. \quad (88)$$

This difference shows up in table 4, where C'^R is negative for CsCl and ZnS, but positive for NaCl.

It seems clear that dc_{44}/dP will be small and possibly negative in structures where

$$\sum_{l=1}^{M} Q(x^l)(y^l)(z^l) \tag{89}$$

is non-zero and that dc'/dP will be small and possibly negative in structures where

$$\sum_{l=1}^{M} Q(x^l)^2(y^l)^2 \tag{90}$$

is non-zero.

In order for expression (89) to be non-zero, the structure must not be *centrosymmetric*. In order for expression (90) to be non-zero, certain atoms in the cell must be in positions which do not lie on the axial planes (001), (100), (010).

For materials which have structures more complicated than diatomic cubic, generalized equations replacing the two above are required. If the structure is triatomic or greater, we must deal with more than closest nearest neighbors. In this case, the repulsion potential is given by

$$\phi_{kk'}(r) = M_1 \frac{b_{kk'}^1}{r^p} + M_2 \frac{b_{kk'}^2}{r^p} + \ldots .$$

The question on the size of dc_{44}/dP will depend upon whether

$$\sum_{k}\sum_{k'}\sum_{l} Q_{kk'}^l (x_{kk'}^l)(y_{kk'}^l)(z_{kk'}^l) \tag{91}$$

is non-zero. The question on the size of dc'/dP will depend upon whether

$$\sum_{k}\sum_{k'}\sum_{l} Q_{kk'}^l (x_{kk'}^l)^2(y_{kk'}^l)^2 \tag{92}$$

is non-zero.

The case of second-nearest neighbors for the NaCl structure will be considered in a subsequent paper. However, it is apparent that expression (91) will be non-zero for the rutile and α-quartz lattice, so it is not surprising that dc_{44}/dP is small for these lattices. Expression (92) is non-zero for spinel and corundum lattices (due to positions in the cell having finite values of x and y), so that it is not surprising that dc'/dP is small for these lattices.

3. Geophysical applications of lattice dynamical calculations

3.1. *Low-velocity zone*

One plausible explanation for the existence of low-velocity layers and/or low-density layers in the upper mantle is that the thermal gradient becomes sufficiently large to cause the effect of a velocity decrease with increasing temperature to overcome the effect of a velocity increase with increasing pressure. This critical thermal gradient is determined by parameters which can be derived from the ultrasonic laboratory data (SCHREIBER and ANDERSON, 1966; following a suggestion apparently first made by BIRCH and BANCROFT, 1938).

For a variable which is a function only of pressure and temperature, we may write

$$f = f(P, T),$$

$$df = \left(\frac{\partial f}{\partial P}\right)_T dP + \left(\frac{\partial f}{\partial T}\right)_P dT. \tag{93}$$

The critical thermal gradient which must be exceeded to produce a *decrease* in f with an *increase* in P is

$$(\partial T/\partial P)_f = -\frac{(\partial f/\partial P)_T}{(\partial f/\partial T)_P}. \tag{94}$$

A low value of $(\partial v_s/\partial P)_T$ thus allows a low-velocity zone for shear waves to be compatible with a small thermal gradient. The fact that certain crystallographic structures yield low values of $d\mu/dP$ and dv_s/dP is therefore very important in interpretations of the velocity structure of the Earth's mantle. A more detailed discussion of these critical thermal gradients and the low-velocity zones in the mantle is given in a separate publication (LIEBERMANN, 1969). It is sufficient to point out here that compounds with a low value of $d\mu/dP$ have low values of coordination as shown in figs. 1 and 2.

3.2. *Phase transitions*

It is ordinarily assumed that the vanishing of certain of the elastic constants will result in a lattice instability, and be manifested by a phase transition. Using the equations of section 2, the compression at which an elastic constant vanishes can be computed. From eqs.

(30), (38) and (44) the constant c' can be written as a function of density:

$$c' = \frac{Z^2 e^2}{\Delta_0 a_0} 1.33451 \left(\frac{\rho}{\rho_0}\right)^{\frac{4}{3}} \left[0.21825 \, n \left(\frac{\rho}{\rho_0}\right)^{\frac{1}{3}(n-1)} - 1\right]$$

for NaCl, (95)

$$= \frac{Z^2 e^2}{\Delta_0 a_0} 1.39057 \left(\frac{\rho}{\rho_0}\right)^{\frac{4}{3}} \left[1 - 0.24395 \left(\frac{\rho}{\rho_0}\right)^{\frac{1}{3}(n-1)}\right]$$

for CsCl, (96)

$$= \frac{Z^2 e^2}{\Delta_0 a_0} 0.72339 \left(\frac{\rho}{\rho_0}\right)^{\frac{4}{3}} \left[1 - 0.87158 \left(\frac{\rho}{\rho_0}\right)^{\frac{1}{3}(n-1)}\right]$$

for ZnS. (97)

The corresponding equations for c_{44} are

$$c_{44} = \frac{Z^2 e^2}{\Delta_0 a_0} 1.27802 \left(\frac{\rho}{\rho_0}\right)^{\frac{4}{3}} \left[1 - 0.45580 \left(\frac{\rho}{\rho_0}\right)^{\frac{1}{3}(n-1)}\right]$$

for NaCl, (98)

$$= \frac{Z^2 e^2}{\Delta_0 a_0} 0.70089 \left(\frac{\rho}{\rho_0}\right)^{\frac{4}{3}} \left[0.16133 (n-1) \left(\frac{\rho}{\rho_0}\right)^{\frac{1}{3}(n-1)} - 1\right]$$

for CsCl. (99)

For the more complicated case of ZnS

$$c_{44} = c_{44}^* - C^2/D,$$

$$c_{44}^* = \frac{Z^2 e^2}{\Delta_0 a_0} 0.06190 \left(\frac{\rho}{\rho_0}\right)^{\frac{4}{3}} \left[3.395(n-1)\left(\frac{\rho}{\rho_0}\right)^{\frac{1}{3}(n-1)} - 1\right],$$
(100)

$$C = \frac{Z^2 e^2}{\Delta_0 a_0} 2.51 \left(\frac{\rho}{\rho_0}\right)^{\frac{5}{3}} \left[0.16746(n+2)\left(\frac{\rho}{\rho_0}\right)^{\frac{1}{3}(n-1)} - 1\right],$$
(101)

$$D = \frac{Z^2 e^2}{\Delta_0 a_0} 2.095 \left(\frac{\rho}{\rho_0}\right)^{2} \left[0.40126(n-1)\left(\frac{\rho}{\rho_0}\right)^{\frac{1}{3}(n-1)} - 1\right].$$
(102)

These equations have been evaluated for the case $n = 9$ and are plotted in fig. 3. Here it is shown that according to the model c' vanishes at $\rho/\rho_0 = 1.05$, and c_{44} van-

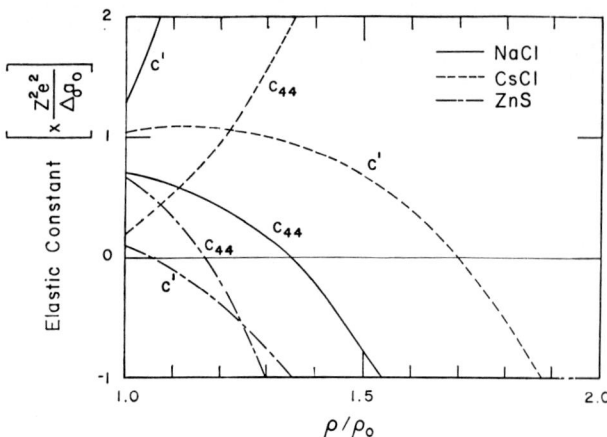

Fig. 3. Elastic constants (c' and c_{44}) as a function of compression for NaCl, CsCl and ZnS lattices.

ishes at $\rho/\rho_0 = 1.17$ for the ZnS lattice. The next highest vanishing point is for NaCl for which c_{44} vanishes at $\rho/\rho_0 = 1.34$; and the highest vanishing point occurs for CsCl for which c' vanishes at $\rho/\rho_0 = 1.7$.

Thus our model accounts for the well-known sequence of transitions in cubic lattices

$$ZnS \rightarrow NaCl \rightarrow CsCl,$$

It is possible to compute the pressures at which the transitions occur if we assume an equation of state. Let us take the MURNAGHAN (1944) equation to approximate the transition pressure

$$\frac{P}{K_0} = \frac{1}{K_0'}\left[\left(\frac{\rho}{\rho_0}\right)^{K_0'} - 1\right]. \quad (103)$$

The value of K_0' is 5.333 for $n = 9$, our choice of the parameter. Thus we compute from the numbers listed above

$$\frac{P^*}{K_0} = 0.07 \text{ for vanishing } c', \quad \text{ZnS}$$

$$= 0.25 \text{ for vanishing } c_{44}, \quad \text{ZnS}$$
$$= 0.74 \text{ for vanishing } c_{44}, \quad \text{NaCl}$$
$$= 3.00 \text{ for vanishing } c', \quad \text{CsCl}.$$

These pressures are only approximate, of course, since the equation of state which we have assumed affects the resulting calculation. Nevertheless, the order of magnitude is probably correct; thus we should expect phase transitions in the region of 60–80 kb for ZnS lattices, 200 kb for NaCl lattices (750 kb for MgO) and several megabars for CsCl lattices.

The elastic constants c_{11}, c_{12} and K all increase monotonically with pressure. It is of some interest to determine if the elastic constants which vanish at high compressions decrease approximately linearly with pressure up to the transition pressure (P^*). For this argument we assume

$$c_{ij}(P) = c_{ij}(0) + \left(\frac{dc_{ij}}{dP}\right)_0 P, \qquad (104)$$

then

$$\frac{P^*}{K_0} = -\frac{c_{ij}(0)/K_0}{\left(\frac{dc_{ij}}{dP}\right)}. \qquad (105)$$

Using previous tables we compute for the transition pressure from eq. (105)

$$\frac{P^*}{K_0} = 0.06 \text{ for vanishing } c', \quad \text{ZnS}$$

$$= 0.30 \text{ for vanishing } c_{44}, \quad \text{ZnS}$$

$$= 1.92 \text{ for vanishing } c_{44}, \quad \text{NaCl}$$

$$= \text{negative for vanishing } c', \quad \text{CsCl}.$$

It is seen that only in the case of small transition pressures eq. (105) is very helpful in computing pressures at which the elastic constants vanish. In the case of CsCl, eq. (105) fails to predict any positive transition pressure, even though the lattice dynamic calculation, given by eq. (96), demands a transition at $\rho/\rho_0 \approx 1.7$. The answer to this apparent contradiction can be found in a more formal theory of lattice instability. It turns out that a phase transition is required for any solid in which either c' or c_{44} is less than (positive) unity, if one stipulates eq. (104).

From the point of view of lattice dynamics, for a lattice to be stable, the energy density at equilibrium must be a positive definite quadratic form so that the energy is raised by the imposition of any small strains. In terms of the elastic constants, at one atmosphere pressure, the energy density, or the strain-energy function, is

$$U = \tfrac{1}{2}\sum_i \sum_j c_{ij}\varepsilon_i\varepsilon_j, \quad i,j = 1, 2, \ldots, 6 \qquad (106)$$

where ε_i and ε_j are the six strains and the number of independent values of c_{ij} depends upon the crystal class. For the triclinic class, the elastic constants are given by the following matrix:

$$\begin{bmatrix} c_{11} & c_{12} & c_{13} & c_{14} & c_{15} & c_{16} \\ c_{12} & c_{22} & c_{23} & c_{24} & c_{25} & c_{26} \\ c_{13} & c_{23} & c_{33} & c_{34} & c_{35} & c_{36} \\ c_{14} & c_{24} & c_{34} & c_{44} & c_{45} & c_{46} \\ c_{15} & c_{25} & c_{35} & c_{45} & c_{55} & c_{56} \\ c_{16} & c_{26} & c_{36} & c_{46} & c_{56} & c_{66} \end{bmatrix} \qquad (107)$$

As pointed out by BORN and HUANG (1954, p. 141), a well-known theorem in algebra shows that the above quadratic form (eq. 106) will be positive definite if the determinants of the principal minors of the matrix (107) are all positive. If we start from the lower right in the matrix (107) we see that these determinants are:

$$|c_{66}|, \quad \begin{vmatrix} c_{55} & c_{56} \\ c_{56} & c_{66} \end{vmatrix}, \quad \begin{vmatrix} c_{44} & c_{45} & c_{46} \\ c_{45} & c_{55} & c_{56} \\ c_{46} & c_{56} & c_{66} \end{vmatrix}, \ldots.$$

For solids in the class of cubic symmetry, the matrix shown in (107) is much simpler:

$$\begin{vmatrix} c_{11} & c_{12} & c_{12} & 0 & 0 & 0 \\ c_{12} & c_{11} & c_{12} & 0 & 0 & 0 \\ c_{12} & c_{12} & c_{11} & 0 & 0 & 0 \\ 0 & 0 & 0 & c_{44} & 0 & 0 \\ 0 & 0 & 0 & 0 & c_{44} & 0 \\ 0 & 0 & 0 & 0 & 0 & c_{44} \end{vmatrix} \qquad (108)$$

In order for all the determinants of matrices of successive orders to be zero, the elastic constants must satisfy the condition

$$c_{44} > 0, \qquad c_{11}^2 - c_{12}^2 > 0,$$
$$c_{11} > 0, \qquad c_{11} + 2c_{12} > 0.$$

Noting that since $c_{11} > 0$ and $c_{11} + 2c_{12} > 0$, it must necessarily follow that $c_{11} + c_{12} > 0$; as a consequence $c_{11}^2 - c_{12}^2 > 0$ in turn requires that $c_{11} - c_{12} > 0$. This is important because it means that the shear constant $\tfrac{1}{2}(c_{11} - c_{12}) = c'$ cannot be zero or negative if the cubic lattice is to be stable. In lattices of lower symmetry than cubic, the constant c_{66} always arises, which in many symmetries is equal to c'. Thus a common feature of all crystal lattices at zero pressure is that c_{66} or c' be greater than zero in order to insure mechanical stability.

For hexagonal symmetry, the elastic constant matrix is

$$\begin{bmatrix} c_{11} & c_{12} & c_{13} & 0 & 0 & 0 \\ 0 & c_{11} & c_{13} & 0 & 0 & 0 \\ 0 & 0 & c_{33} & 0 & 0 & 0 \\ 0 & 0 & 0 & c_{44} & 0 & 0 \\ 0 & 0 & 0 & 0 & c_{44} & 0 \\ 0 & 0 & 0 & 0 & 0 & c' \end{bmatrix} \quad (109)$$

where

$$c' = \tfrac{1}{2}(c_{11} - c_{12}). \quad (110)$$

In order for the determinants of matrices of successive orders of (109) to be zero, the elastic constants must satisfy the condition

$$\begin{aligned} & c' > 0, & & c_{11}^2 - c_{12}^2 > 0, \\ & c_{44} > 0, & & c_{11} + 2c_{12} > 0. \\ & c_{33} > 0, & & \end{aligned} \quad (111)$$

Stability of the crystal itself must be carefully distinguished from the stability criteria of the energy density described above. It is clear that if the elastic constants fail to satisfy the above conditions, the lattice is unstable. Thus these conditions are a sufficient condition for a phase transition. Phase transitions may occur for a variety of reasons even if the above conditions on the elastic constants are satisfied. Thus, these conditions are not *necessary* for a phase transition. Three exceptions will be given. First, the conditions on the energy density criteria represent homogeneous deformations (a deformation which leaves the resulting structure a perfect lattice). A lattice may be unstable under local deformations which are not reflected in the elastic constants. Second, a lattice may change to another phase because the Gibbs free energy of the second phase is lower than that of the first phase at some external condition. Third, instability of a lattice may set in because one of the frequencies of the normal modes becomes negative. For example, a ferroelectric transition may take place because one of the frequencies of the optical modes is zero (CHOCHRAN, 1960); this condition having little connection to the elastic constants. Both the ferroelectric transition and the mechanical stability reflected in (111) are cases in lattice dynamics of solutions for normal modes at vanishing wave number. An instability may arise for a value of the wave number other than zero, in particular at the boundary of the Brillouin zone. Consequently, the conditions under which the critical elastic constants vanish, represent limits on the conditions of mechanical stability.

If the lattice is compressed isothermally, the conditions of equilibrium change, and even the symmetry changes. At a higher pressure, the elastic constant matrix changes in two ways. First, the elastic constants change as a function of pressure, and second, terms in the pressure itself are added to the elements of the matrix (THURSTON, 1965). The elastic constant matrix for the hexagonal case at high pressure is given as follows:

$$\begin{bmatrix} -P + \dfrac{\lambda_1^3}{\lambda_1 \lambda_3} c_{11} & \dfrac{\lambda_1 \lambda_2}{\lambda_3} c_{12} & \dfrac{\lambda_1 \lambda_3}{\lambda_2} c_{13} & 0 & 0 & 0 \\ 0 & -P + \dfrac{\lambda_2^2}{\lambda_1 \lambda_2} c_{11} & \dfrac{\lambda_2 \lambda_3}{\lambda_1} c_{12} & 0 & 0 & 0 \\ 0 & 0 & -P + \dfrac{\lambda_3^2}{\lambda_1 \lambda_2} c_{33} & 0 & 0 & 0 \\ 0 & 0 & 0 & -P + \dfrac{\lambda_2 \lambda_3}{\lambda_1} c_{44} & 0 & 0 \\ 0 & 0 & 0 & 0 & -P + \dfrac{\lambda_1 \lambda_3}{\lambda_2} c_{44} & 0 \\ 0 & 0 & 0 & 0 & 0 & -P + \dfrac{\lambda_1 \lambda_2}{\lambda_3} c' \end{bmatrix} \quad (112)$$

where λ_1, λ_2, λ_3 are the principal stretches resulting from the deformation and are elements of the Jacobian relating the undeformed axes to the deformed axes. The stability of the hexagonal lattice under pressure depends upon the conditions

$$-P + \frac{\lambda_1 \lambda_2}{\lambda_3} c > 0; \qquad -P + \frac{\lambda_2 \lambda_3}{\lambda_1} c_{44} > 0, \text{etc.}$$

For the remainder of this discussion we will focus upon the condition that $-P + (\lambda_1 \lambda_2/\lambda_3)c' > 0$, since this condition appears to be the most critical for lattice stability.

The question we wish to examine is: at what pressure does the generalized elastic constant, given by the lowest element of the matrix (107) vanish. As a first approximation we might assume that the quantity $(\lambda_2 \lambda_1 \lambda_3)c'$ is linear with pressure; that is

$$\frac{\lambda_2 \lambda_1}{\lambda_3} c' = (c')_0 + \left(\frac{\partial c'}{\partial P}\right)_T P. \qquad (113)$$

Thus the lattice is mechanically unstable when

$$c' + \left(\frac{\mathrm{d}c'}{\mathrm{d}P} - 1\right) P^* = 0. \qquad (114)$$

Now note that when $\mathrm{d}c'/\mathrm{d}P$ is less than unity, there exists a positive value of pressure for which eq. (114) is satisfied, namely,

$$P^* = \frac{c'}{1 - (\mathrm{d}c'/\mathrm{d}P)} \qquad (115)$$

where P^* is the expected transition pressure. Comparing eq. (115) to eq. (105), we see that eq. (115) predicts a phase transition at finite pressure for any solid in which $\mathrm{d}c'/\mathrm{d}P$ is less than unity, whereas eq. (105) predicts a transition only when $\mathrm{d}c'/\mathrm{d}P$ is less than zero. Thus, the apparent contradiction in the vanishing of c' in the CsCl lattice is resolved and eq. (104) is wrong for extrapolation to the transition pressure.

We may attempt to predict the transition pressure of a number of solids using eq. (115). In α-quartz, $c' = c_{66} = 398$ kb and $\mathrm{d}c'/\mathrm{d}P = 2.67$ (McSkimin et al. 1965). Thus, the expected P^* at which the effective shear modulus makes the lattice unstable, is 108 kb. Indeed, there is a phase transition in α-quartz in the vicinity of 120 kb (see O. L. Anderson, 1966, fig. 15). The equation (115) works well for α-quartz but tends to overestimate the transition pressure for other compounds.

For CdS (ZnS structure), $c_{66} = 162$ kb and $\mathrm{d}c_{66}/\mathrm{d}P = -0.805$ (Corll, 1967), which gives $P^* = 90$ kb; this pressure is higher than the observed transition which occurs near 20–25 kb. (Calculation of P^* from eqs. (97) and (103) gives $P^* = 50$ kb, which is in better agreement with the experimental evidence.) For polycrystalline ZnO, which is structurally similar to ZnS, $\mu = 442$ kb and $\mathrm{d}\mu/\mathrm{d}P = -0.69$; thus, eq. (115) predicts a transition at $P^* = 261$ kb, while eqs. (97) and (103) give an estimate of $P^* = 90$ kb; there is a transition near 90 kb (John Jameison, private communication). A transition in calcite occurs at about 15 kb; if we take $c_{66} = 433$ kb (Dandekar, 1968) the value of $\mathrm{d}c_{66}/\mathrm{d}P$ would have to be -30 in order to predict this transition; this is hardly reasonable, especially since Dandekar (1968) measured $\mathrm{d}c_{66}/\mathrm{d}P = 0.35$. This transition may arise from a lattice instability which is not associated with an acoustic mode.

In summary, we find the computation of the transition pressure from the formulas presented to be marginally satisfactory. The reason arises because an equation of state must be assumed, and the pressure dependence of the elastic constant must be assumed. A more reliable theoretical approach (eqs. 95–102) is to find the compression at which the elastic constant vanishes. These compression equations arise in a straightforward manner from lattice dynamical calculations such as those presented in section 2 of this paper.

3.3. Grüneisen parameter and the interpretation of shock-wave data

The parameter $(\mathrm{d}\mu/\mathrm{d}P)_T$ exerts a dominant influence in the evaluation of the acoustic Grüneisen parameters at low temperatures and at high temperatures and thus dominates the thermal pressure. It is interesting to ascertain what values of $(\mathrm{d}\mu/\mathrm{d}P)_T$ would be predicted if the acoustic γ's are required to be equal to the thermal $\gamma_\mathrm{th} = \alpha K_S/\rho C_P$. At low temperature $\gamma_{LT} \approx \gamma_s$. The form of γ_s has been suggested by Smith (1958):

$$\gamma_s = -\tfrac{1}{6} + \frac{K_T}{\mu}\left(\frac{\partial \mu}{\partial P}\right)_T. \qquad (116)$$

If $\gamma_{LT} = \gamma_\mathrm{th}$ were required, then

$$(\partial \mu / \partial P)_T = \frac{\mu}{K_T}(2\gamma_\mathrm{th} + \tfrac{1}{3}) \qquad (117)$$

TABLE 12

Comparison of measured and predicted $(\partial \mu/\partial P)_T$ values

Material	Measured	Predicted			
		Slater $(\mu/K_T)(\partial K_T/\partial P)_T$	Dugdale-MacDonald $\frac{1}{3}(\mu/K_T)$	Acoustic (LT) $(\mu/K_T)(2\gamma_{th}+\frac{1}{3})$	Acoustic (HT) $(\mu/K_T)(3\gamma_{th}+\frac{1}{3}-\gamma_p)$
α-SiO₂	0.45	7.54	0.40	2.06	1.50
ZnO	−0.69	1.52	0.11	0.62	0.51
BeO	0.88	4.06	0.25	2.13	1.94
α-Fe₂O₃	0.73	1.99	0.15	1.94	2.14
α-Al₂O₃	1.76	2.55	0.21	1.91	1.78
CaO	1.65	3.76	0.25	1.97	1.59
MgO (single crystal)	2.62	3.63	0.27	2.79	2.65
CsCl	2.90	3.14	0.20	2.64	2.44

would be predicted. At high temperatures if $\gamma_{HT} = \gamma_{th}$ where required, then

$$(\partial \mu / \partial P)_T = \frac{\mu}{K_T}(3\gamma_{th} + \tfrac{1}{3} - \gamma_p) \qquad (118)$$

would be predicted, since $\gamma_{HT} = \frac{1}{2}(\gamma_p + 2\gamma_s)$. In fig. 4 the deviation (Δ) of the measured value of $(\partial \mu/\partial P)_T$ from that predicted by eqs. (117) and (118) is illustrated

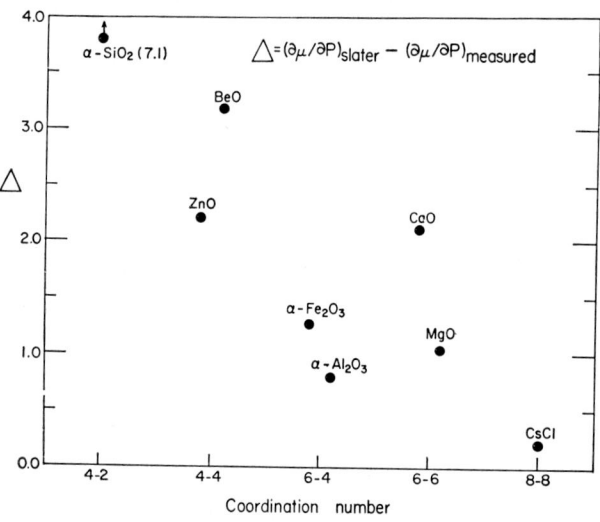

Fig. 4. Deviations for different coordination numbers of measured $(\partial \mu/\partial P)_T$ from the value predicted by requiring the thermal γ (γ_{th}) to equal the acoustic γ (γ_{ac}). Dots are for $\gamma_{ac} = \gamma_{LT} = \gamma_{th}$; arrows for $\gamma_{ac} = \gamma_{HT} = \gamma_{th}$. (From data in table 12.)

for materials of different coordination numbers (see table 12 for results of calculations for eqs. (117) and (118)). The solid dots represent Δ for $(\partial \mu/\partial P)_T$ predicted by eq. (117) and the arrows indicate the shift in Δ if $(\partial \mu/\partial P)_T$ is predicted from eq. (118). Note the tendency

for the acoustic γ ($\gamma_{ac} = \gamma_{LT}$ or γ_{HT}) to deviate from γ_{th} as the coordination number decreases, especially for α-SiO₂, ZnO and BeO. For α-Al₂O₃, CaO MgO and CsCl, $\gamma_{ac} = \gamma_{th}$ is not a bad approximation to the data. The Δ for α-Fe₂O₃ is somewhat larger than might have been anticipated and may be related to the anomalous behavior of the thermal expansion just below room temperature.

In the analysis of shock-wave data via the Mie-Grüneisen equation of state (GRÜNEISEN, 1926), it is necessary to specify the volume dependence of γ (see O. L. ANDERSON, 1968a). The formulas for γ of SLATER (1939) and of DUGDALE and MACDONALD (1953) are often employed by various authors (e.g. TAKEUCHI and KANAMORI, 1966) in the reduction of shock-wave Hugoniots to isotherms. Slater's formula has also been used to determine from the values of $(\partial K_T/\partial P)_T$ obtained by fitting a Birch-Murnaghan equation to the shock-wave data for the high-pressure region (D. L. ANDERSON and KANAMORI, 1968; AHRENS et al., 1969). Since both the Slater (γ_{SL}) and the Dugdale-MacDonald (γ_{DM}) determinations for the Grüneisen parameter involve simplifying assumptions, it is of interest to investigate the implication of these assumptions for the parameter $(\partial \mu/\partial P)_T$. KNOPOFF and SHAPIRO (1969) have recently presented an authoritative discussion of these implications; the format chosen in this work is another way of illustrating the points that they emphasized but was chosen to provide a clearer picture of the dependence of $(\partial \mu/\partial P)_T$ upon crystallographic structure.

As pointed out by KNOPOFF and SHAPIRO (1969), the important implications of γ_{SL} and γ_{DM} are that the shear modulus is assumed to vary with pressure in a known way. Inspection of fig. 3 shows how wrong this assumption is for cubic solids. The Slater definition of γ requires that $(\partial\sigma/\partial P)_T = 0$. Differentiation of eq. (85) with respect to pressure yields

$$\left(\frac{\partial\sigma}{\partial P}\right)_T = \frac{9\mu K}{(3K+\mu)^2}\left[\left(\frac{\partial \ln K}{\partial P}\right)_T - \left(\frac{\partial \ln \mu}{\partial P}\right)_T\right]. \quad (119)$$

$\left(\frac{\partial\sigma}{\partial P_T}\right) = 0$ implies that

$$\left(\frac{\partial \ln K_T}{\partial P}\right)_T = \left(\frac{\partial \ln \mu}{\partial P}\right)_T, \quad (120)$$

or

$$\left(\frac{\partial \mu}{\partial P}\right)_T = \frac{\mu}{K_T}\left(\frac{\partial K_T}{\partial P}\right)_T. \quad (121)$$

This relationship (121) can never be true for a ZnS structure, is unlikely for a CsCl structure and is only occasionally true for a NaCl structure. From eq. (119) it is clear that γ_{SL} implies that all of the acoustic modes vary in the same way with pressure (or volume). The results of the calculations of $(\partial\mu/\partial P)_T$ from Slater (eq. 121) are given in table 12, and the deviation Δ of the measured $(\partial\mu/\partial P)_T$ from these values are plotted in fig. 5 as a function of coordination number. The Slater approximation (eq. 121) is observed to be the least valid for the compounds of low coordination number (α-SiO$_2$, BeO, ZnO). Only for CsCl (8-8 coordination) does eq. (121) provide a reasonable estimate of $(\partial\mu/\partial P)_T$. For all other materials $\Delta > 0.8$.

DUGDALE and MACDONALD (1953) (see O. L. ANDERSON, 1968a) have defined an ideal harmonic solid which may be considered as one of primitive cubic symmetry in which nearest neighbors are connected with simple springs and for which γ is defined by

$$\gamma_{DM} = -\tfrac{1}{3}-\tfrac{1}{2}V\left[\frac{\partial(PV^{\frac{2}{3}})}{\partial V}\right]^{-1}\left[\frac{\partial^2(PV^{\frac{2}{3}})}{\partial V^2}\right], \quad (122)$$

or

$$\gamma_{DM} = -\tfrac{1}{2}+\tfrac{1}{2}\left(\frac{\partial K_T}{\partial P}\right)_T. \quad (123)$$

PASTINE (1965) pointed out that γ_{DM} may be viewed as being based only on modes which propagate parallel to the principal axes of such an ideal solid; he showed that along these axes only compressional modes are possible, which implies that

$$\gamma_s = 0, \quad (124)$$

$$\gamma_p = \gamma_{SL} = \frac{V}{1-\sigma^2}\left(\frac{d\sigma}{dV}\right), \quad (125)$$

where σ is now allowed to vary with volume. It follows from conservation of particles that $\gamma_{DM} = \gamma_p$ as defined by eq. (125) (O. L. ANDERSON, 1968a). Combining eqs. (116) and (124), it is seen that the Dugdale–MacDonald formulation of γ implies that

$$(\partial\mu/\partial P)_T = \tfrac{1}{3}(\mu/K_T). \quad (126)$$

This result (126) is compatible with Blackman's rule in the sense that materials with low values of μ/K_T will have low values of $(\partial\mu/\partial P)_T$ (see data for α-SiO$_2$ in table 12). However, the Dugdale–MacDonald prediction eq. (126) fails to satisfy the data for other materials given in table 11, and fails to account for the effect of coordination which we have demonstrated in figs. 1 and 2.

The deviation (Δ) of the measured $(\partial\mu/\partial P)_T$ from the Slater and Dugdale–MacDonald estimates is very significant in attempts to predict the behavior of materials at high pressures. AHRENS et al. (1969) have analyzed the extant shock-wave data to obtain the equation of state parameters of the shock-induced high-pressure phases; from these parameters they have calculated γ_{SL} and γ_{DM}. The probable crystallographic structures of these high-pressure phase materials inferred from the classical laws of crystal chemistry and, in some cases, from static high-pressure recovery experiments on analog compounds by AHRENS et al. (1968) are given in table 13. From the coordination numbers of the high-pressure phases and the deviations in fig. 5, it is clear that the Slater γ is a reasonable approximation only for the perovskite (12-6-6) and calcium ferrite (8-6-5) structures; for all of the other structures $(\partial\mu/\partial P)_T$ is probably smaller than the value implicit in the Slater formulation of γ.

3.4. Equation of state parameters: K_0 and $K_0' = (\partial K/\partial P)_0$

The quantities K_0 and $(dK/dP)_0$ are important parameters in the equations of state which have been commonly employed in geophysical discussions in the past (e.g MURNAGHAN, 1944; BIRCH, 1938, 1947, 1952); these parameters retain their importance in the fourth-order anharmonic equation of state recently proposed by THOMSEN (1969). Thus, it is of interest to emphasize the significance of our formulas for these parameters to geophysics.

TABEL 13

Coordination in oxide and silicate phase transitions (after AHRENS et al., 1969)

Compound	Low-pressure phase		High-pressure phase	
	Structure	Coordination	Structure	Coordination
MgO	NaCl	6-6[1]	NaCl	6-6[1]
Al_2O_3	corundum	6-4[1]	corundum	6-4[1]
β-MnO_2	rutile	6-3[1]	rutile	6-3[1]
α-SiO_2	α-quartz	4-2[1]	rutile	6-3[1]
TiO_2	rutile	6-3[1]	fluorite	8-4[1]
Fe_2O_3	corundum	6-4[1]	perovskite ($YFeO_3$)	12-6-6[2]
			B rare earth (Sc_2O_3)	
$MgAl_2O_4$	normal spinel	4-6-4[1]	calcium ferrite	8-6-5[2]
$Fe^{2+}Fe_2^{3+}O_4$	inverse spinel	6-(4,6)-4[1]	calcium ferrite	8-6-5[2]
			$FeO+Fe_2O_3$*	6-6[1]+12-6-6[2]
			$Fe+Fe_2O_3$*	8-8[1]+12-6-6[2]
Mg_2SiO_4	chrysoberyl	6-4-4[1]	K_2NiF_4	9-6-6
			pseudobrookite (Fe_2TiO_5)	
(Mg, Fe)$_2SiO_4$	chrysoberyl	6-4-4[1]	ilmenite+NaCl	6-6-4[1]+6-6[1]
(>10% FeO)			Sr_2PbO_4	
(Mg, Fe)SiO_3	orthopyroxene	8-4-4[3]	$MgAl_2O_4$*+SiO_2*	8-6-5[2]+6-3[1]
(>10% FeO)			ilmenite	6-4-4[1]
R(Al, Si)$_4O_8$	feldspar	8-4-4	hollandite (RM_4O_8)	8-6-4
Al_2SiO_5	andalusite	(6,5)-4-3[1]	corundum+SiO_2*	6-4[1]+6-3[1]

1 WYCKHOFF (1963).
2 AHRENS et al. (1969).
3 BRAGG et al. (1965).
* High-pressure phase.

As noted in section 2 the formula for K_0 is the same for all three structures considered:

$$K_0 = \frac{Z^2 e^2}{\Delta_0 a_0} \frac{1}{9} A_a (n-1). \qquad (48)$$

A similar result has been derived by GRÜNEISEN (1912, eq. 28; 1926, eq. 6) from his theory of solids and independently by BRIDGMAN (1923, p. 234); these authors also used the Born–Mie potential but with arbitrary m in the attractive term. As GRÜNEISEN (1926, p. 12) has pointed out, this formula (48) is valid for zero pressure and absolute zero temperature. From eq. (48) K_0 is inversely proportional to $\Delta_0^{\frac{4}{3}}$ (where Δ_0 = volume per ion pair). The effect of crystallographic structure upon K_0 is in the Madelung constant A_a. For a more complete discussion of this power law dependence of K_0 upon Δ_0 and an indication of the compounds to which it applies, see O. L. ANDERSON and NAFE (1965).

In section 2 we have discussed the fact that our calculations of $(dK/dP)_0$ for the three cubic structures considered indicate that this parameter depends only upon the exponent in the repulsive potential (eq. 23) and not upon the details of the crystallographic structure:

$$(dK/dP)_0 = K_0' = \tfrac{1}{3}(n+7). \qquad (60)$$

This result was obtained by separating the interatomic potential (eq. 14) into an attractive term ($\sim 1/r^m$) and a repulsive term ($\sim 1/r^n$). In our treatment we have assumed a coulombic attraction ($m = 1$) and taken the summation of the attractive term over all atoms. For

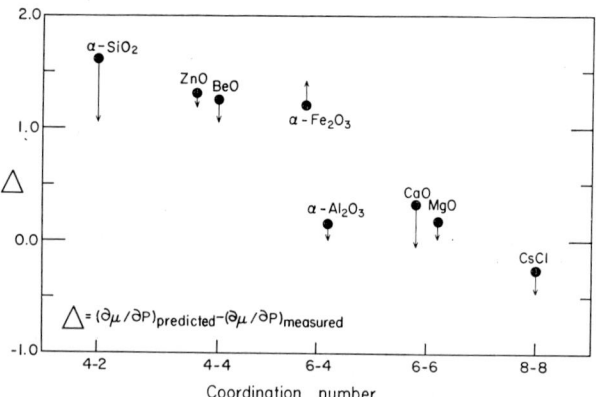

Fig. 5. Deviations of measured $(\partial \mu / \partial P)_T$ from the value predicted by the Slater formula for γ for materials with different coordination numbers. (From data in table 12.)

the repulsive term, the summation is taken only over nearest neighbors. It is instructive to compare our result (60) with that obtained by previous investigators.

GRÜNEISEN (1912, 1926) assumed a Born–Mie potential (MIE, 1903) of the type that we have adopted

$$\phi = -\frac{a}{r^m} + \frac{b}{r^n}, \qquad (127)$$

where a and b are constants and r is again the interatomic distance; by invoking the condition of equilibrium in the Mie–Grüneisen equation of state, he obtained:

$$P = -\left(\frac{\partial \phi}{\partial V}\right)_T. \qquad (128)$$

With his definition of $K_T = -V_0(\partial P/\partial V)_T$, a double differentiation of (128) yields

$$(\partial K_T/\partial P) = \tfrac{1}{3}(m+n+9) \qquad (129)$$

at zero pressure. This relationship (129) has also been derived by BRIDGMAN (1923) and by FÜRTH (1944). With K_T defined by $K_T = -V(\partial P/\partial V)_T$ eq. (129) becomes

$$(\partial K_T/\partial P)_T = \tfrac{1}{3}(m+n+6). \qquad (130)$$

The difference between eqs. (129) and (130) arises solely from the definition of K_T and has been a source of confusion in the past as noted by BIRCH (1952; p. 246 footnote) and D. L. ANDERSON (1967).

A more general form of eq. (130) has been given by O. L. ANDERSON (1968a):

$$(\partial K_T/\partial P)_T = (\tfrac{1}{3}n+1) + $$
$$+ (\tfrac{1}{3}m+1)(n-m)\left[(n+3)\left(\frac{V_0}{V}\right)^{n-\tfrac{1}{3}m} - (m+3)\right]^{-1} \qquad (131)$$

At zero pressure ($V = V_0$) this reduces to eq. (130) while for large compressions [(V_0/V) large] eq. (131) becomes $(\partial K_T/\partial P)_T = \tfrac{1}{3}(n+3)$. The fact that at large compressions $(\partial K_T/\partial P)_T$ is controlled entirely by the repulsive potential was first pointed out by RAMSEY (1950) and was used by him to construct theoretical Earth models.

Thus we see that our result for K_0' is equivalent (for $m = 1$) to that derived by Grüneisen directly from the Born–Mie potential (127). This result, $K_0' = \tfrac{1}{3}(n+7)$ provides a strong argument against values of $K_0' < 3.5$, since such values would require $n < 4$. Such values of n require a very weak repulsion between adjacent atoms. Furthermore, the NaCl lattice is unstable for $n < 4.6$ (c' is then negative) and the CsCl lattice is unstable for $n < 7.2$ (c_{44} is then negative). From our acoustic measurements on oxides and silicates, we find the value of K_0' to be between the limits of 6.4 for α-quartz and 4.0 for corundum (O. L. ANDERSON et al., 1968). To our knowledge the lowest value of K_0' for a solid, measured acoustically, is that for tantalum, $K_0' = 3.20$ (CHECHILE, 1967). On the other hand, the highest values for solids, measured acoustically, are those for cadmium, $K_0' = 6.77$ (CORLL, 1967), and rutile, $K_0' = 6.8$ (MANGHNANI, 1969). These higher values would require $n = 13$ on the basis of eq. (60). Thus the evidence from acoustics is against values of K_0' much less than 3.5 or larger than 7 for crystalline solids.

We strongly recommend that values of $K_0' < 3.5$, which are obtained by a parametric fit of an equation of state to static pressure-volume data, be ignored until confirmed by a reliable acoustic method. We believe that such low values of K_0' violate the condition of lattice stability for most solids.

Not only are the equations for K_0 and K_0' substantially independent of structure (except for insignificant differences arising from (A_r/Δ_0) for K_0), but their derivation always leads to the same result regardless of the degree of sophistication in the theoretical approach. This means, we believe, that the equations may be safely applied to structures of lower symmetry than we have considered.

The formula for K_0 for all three cubic structures is given by eq. (48). Furthermore, eq. (57) shows that $K_0 = (c_{11}^R)_0$ for CsCl and ZnS. Thus we can identify the operators for K_0 by comparing the entry for c_{11}^R in table 6 with the corresponding entry in table 4. The bulk modulus arises simply from the operator $r^2\nabla^2$ [$r^2\nabla^2 = (r^2(d/dr^2) + 2r(d/dr)]$ acting upon $\tfrac{4}{9}[v(r)/\Delta_0]$:

$$K = \frac{4r^2}{9\Delta_0}\nabla^2 v(r) = \left(\frac{4a^2}{9\Delta_0}\right)\nabla^2 v(a) \qquad (132)$$

where ∇^2 is the Laplacian operator. Note that $r^2\nabla^2 v$ is identical to $r^2\nabla^2\phi$, since the Laplacian of $(1/r)$, proportional to the coulombic potential, vanishes identically. Since the bulk modulus is the Laplacian of the total potential, or equivalently, the Laplacian of the repulsive potential, the details and the refinements of lattice sums which are required for all other elastic constants never need enter into considerations of the bulk modulus. This, we believe, has accounted, on the one hand, for the success of simple-minded approaches to considerations of the bulk modulus, and on the other hand, for the lack of progress in understanding the shear modulus.

Acknowledgments

The authors owe a great debt to Prof. M. Blackman, FRS, with regard to his published works and his personal correspondence on the intricacies of lattice sums.

This research was supported by the Air Force Office of Scientific Research under contract AF-F44620-68-C-0079.

References

AHRENS, T. J., D. L. ANDERSON and A. E. RINGWOOD (1969) Equations of state and crystal structure of high pressure phases of shocked silicates and oxides, Rev. Geophys., in press.

ANDERSON, D. L. (1967) A seismic equation of state, Geophys. J. **13**, 9–30.

ANDERSON, D. L. and H. KANAMORI (1968) Shock wave equations of state for rocks and minerals, J. Geophys. Res. **73**, 6477–6502.

ANDERSON, O. L. (1963) A simplified method for calculating the Debye temperature from elastic constants, J. Phys. Chem. Solids **24**, 909–917.

ANDERSON, O. L. (1965) Determination and some uses of isotropic elastic constants of polycrystalline aggregates, using single crystal data, in W. P. Mason, ed., *Physical acoustics* IIIB (Academic Press, New York) 43–95.

ANDERSON, O. L. (1966) The use of ultrasonic measurements under modest pressure to estimate compression at high pressure, J. Phys. Chem. Solids **27**, 547–565.

ANDERSON, O. L. (1968a) Some remarks on the volume dependence of the Grüneisen parameter, J. Geophys. Res. **73**, 5187–5194.

ANDERSON, O. L. (1968b) Comments on the negative pressure dependence of the shear modulus found in some oxides, J. Geophys. Res. **73**, 7707–7712.

ANDERSON, O. L., E. SCHREIBER, R. C. LIEBERMANN and N. SOGA (1968) Some elastic constant data on minerals relevant to geophysics, Rev. Geophys. **6**, 491–524.

ANDERSON, O. L. and J. E. NAFE (1965) The bulk modulus–volume relationship for oxide compounds and related geophysical problems, J. Geophys. Res. **70**, 3951–3963.

BARRON, T. H. K. (1955) On the thermal expansion of solids at low temperatures, Phil. Mag. Ser. 7, **46**, 720–734.

BARRON, T. H. K. (1957) Grüneisen parameters for the equation of state of solids, Ann. Phys. N.Y. **1**, 77–90.

BARSCH, G. R. (1968) Relation between third-order elastic constants of single crystals and polycrystals, J. Appl. Phys. **39**, 3780–3793.

BARSCH, G. R. and Z. P. CHANG (1967) Adiabatic, isothermal, and intermediate pressure derivatives of the elastic constants for cubic symmetry II: numerical results for 25 materials, Solid State Phys. **19**, 139–151.

BARTELS, R. A. and D. E. SCHUELE (1965) Pressure derivatives of the elastic constants of NaCl and KCl at 295 °K and 195 °K, Phys. Chem. Solids, **26**, 537–549.

BIRCH, F. (1938) The effect of pressure upon the elastic parameters of isotropic solids, according to Murnaghan's theory of finite strain, J. Appl. Phys. **9**, 279–288.

BIRCH, F. (1947) Finite elastic strain of cubic crystals, Phys. Rev. **71**, 809–824.

BIRCH, F. (1952) Elasticity and constitution of the earth's interior, J. Geophys. Res. **57**, 227–286.

BIRCH, F. and D. BANCROFT (1938) The effect of pressure on the rigidity of rocks II, J. Geol. **4**$_i$, 113–141.

BLACKMAN, M. (1957) On the thermal expansion of solids, Proc. Phys. Soc. London Ser. B **70**, 827–832.

BLACKMAN, M. (1958) On negative volume expansion coefficients, Phil. Mag. Ser. 8, **3**, 831–838.

BORN, M. (1923) *Atomtheorie des festen Zustandes* (Teubner, Leipzig).

BORN, M. and K. HUANG (1954) *Dynamical theory of crystal lattices* (Oxford Univ. Press, London).

BRAGG, L., G. F. CLARINGBULL and W. H. TAYLOR (1965) *The crystalline state* IV: *Crystal structure of minerals* (Cornell Univ. Press, Ithaca, N.Y.) 409 p.

BRIDGMAN, P. W. (1923) The compressibility of thirty metals as a function of pressure and temperature, Proc. Am. Acad. Arts Sci. **58**, 165–242.

CHECHILE, R. A. (1967) Ultrasonic equations of state of tantalum, Case Inst. Tech. ONR Tech. Rept. **10**, May 1967.

CHOCHRAN, W. (1960) Crystal stability and the theory of ferroelectricity, Phil. Mag. Ser. 8, **5**, 387–423.

CHUNG, D. H. (1963) Elastic moduli of single crystal and polycrystalline MgO, Phil. Mag. Ser. 8, **8**, 833–841.

CHUNG, D. H. (1967) First pressure derivatives of polycrystalline elastic moduli: Their relation to single-crystal acoustic data and thermodynamic relations, J. Appl. Phys. **38**, 5104–5113.

CORLL, J. A. (1967) Effect of pressure on the elastic parameters and structure of CdS, Phys. Rev. **57**, 623–626.

COWLEY, R. A. (1962) The elastic and dielectric properties of crystals with polarizable atoms, Proc. Roy. Soc. London Ser. A. **268**, 121–144.

DANDEKAR, D. P. (1968) Variation in the elastic constants of calcite with pressure, J. Appl. Phys. **39**, 3694–3699.

DUGDALE, J. S. and D. K. C. MACDONALD (1953) The thermal expansion of solids, Phys. Rev. **89**, 832–834.

FÜRTH, R. (1944) On the equation of state for solids, Proc. Roy. Soc. London Ser. A **183**, 87–110.

GRÜNEISEN, E. (1912) Theorie des festen Zustandes einatomiger elemente, Ann. Phys. Berlin **39**, 257–306.

GRÜNEISEN, E. (1926) Die Zustandsänderungen fester Körpers, Handbuch der Phys. **10**, 1–59. [English transl.: The state of a solid body, NASA Republ. RE2-18-59W, Feb. 1959.] 76 pp.

KNOPOFF, L. and J. N. SHAPIRO (1969) Comments on the interrelationships between Grüneisen's parameter and shock and isothermal equations of state, J. Geophys. Res. **74**, 1439–1450.

LIEBERMANN, R. C. (1969) Effect of iron content upon the elastic properties of oxides and some applications to geophysics, Ph.D. Thesis, Columbia Univ., New York.

MANGHNANI, M. H. (1969) Elastic constants of single-crystal rutile under pressures to 7.5 kb, J. Geophys. Res., in press.

MCSKIMIN, H. J., P. ANDREATCH jr. and R. N. THURSTON (1965) Elastic moduli of quartz versus hydrostatic pressure at 25 °C and −195.8 °C, J. Appl. Phys. **36**, 1624–1632.

MIE, G. (1903) Zur kinetischen Theorie der einatomigen Körper Ann. Phys. Berlin **11**, 657–697.

MURNAGHAN, F. D. (1944) The compressibility of media under extreme pressures, Proc. Natl. Acad. Sci. **30**, 244–247.

PASTINE, D. J. (1965) Formulation of the Grüneisen parameter for monatomic cubic crystals, Phys. Rev. **138A**, 767–770.

RAMSEY, W. H. (1950) On the compressibility of the earth, Monthly Notices Roy. Astron. Soc. Geophys. Suppl. **6**, 42–59.

REDDY, P. J. and A. L. RUOFF (1965) Pressure derivatives of the elastic constants in some alkali halides, in C. T. Tomizuka and R. M. Emrick, eds., *Physics of solids at high pressures* (Academic Press, New York) 510–523.

REUSS, A. (1929) Berechnung de Fliessgrenze von Mischkristallen auf Grund der Plastizitätsbedingung für Einkristalle, Z. Angew. Math. Mech. **9**, 49–58.

SCHREIBER, E. and O. L. ANDERSON (1966) Temperature dependence of the velocity derivatives of periclase, J. Geophys. Res. **71**, 3007–3012.

SLATER, J. C. (1939) *Introduction to chemical physics* (McGraw-Hill, New York).

SMITH, C. C. (1958) The relationship between the pressure derivatives of the elastic constants and Grüneisen's gamma (mimeographed notes passed out at the 1958 Gordon Conf. High Pressure, Meriden, N.H.), private correspondence dated July 16, 1958.

SOGA, N. (1968) The temperature and pressure derivatives of isotropic sound velocities of α-quartz, J. Geophys. Res. **73**, 827–829.

SOGA, N. and O. L. ANDERSON (1967) Anomalous behavior of the shear-sound velocity under pressure for polycrystalline ZnO, J. Appl. Phys. **38**, 2985–2988.

TAKEUCHI, H. and H. KANAMORI (1966) Equations of state of matter from shock-wave experiments, J. Geophys. Res. **71**, 3985–3994.

THOMSEN, L. (1969) On the fourth-order anharmonic equation of state of solids, Ph.D. Thesis, Columbia Univ., New York.

THOMSEN, L. and O. L. ANDERSON (1969) On the high-temperature equation of state of solids, J. Geophys. Res. **74**, 981–991.

THURSTON, R. N. (1965) Effective elastic coefficients for wave propagation in crystals under stress, J. Acoust. Soc. Am. **37**, 348–356.

VOIGT, W. (1928) *Lehrbuch der Kristallphysik* (Teubner, Leipzig).

WONG, C. and D. E. SCHUELE (1967) The pressure derivatives of the elastic constants of CaF$_2$, J. Phys. Chem. Solids **28**, 1225–1231.

WONG, C. and D. E. SCHUELE (1968) Pressure and temperature derivatives of the elastic constants of CaF$_2$ and BaF$_2$, J. Phys. Chem. Solids **29**, 1309–1330.

WYCKOFF, R. W. G. (1963) *Crystal structures* 1–3, 2nd ed. (Wiley, New York).

Note added in proof

Mineo Kumazawa and Don L. Anderson have both pointed out to the authors that the Murnaghan equation of state, eq. (103), is inconsistent with the other elastic constants and should not, therefore, be used to calculate the pressure P^* for vanishing elastic constants. The consistent equation for P is found by integrating the expression for the bulk modulus, eqs. (31), (39), and (45). The result for each of the cubic solids is

$$\frac{P}{K_0} = \frac{3}{(n-1)} \left[\left(\frac{\rho}{\rho_0}\right)^{\frac{1}{3}(n+3)} - \left(\frac{\rho}{\rho_0}\right)^{\frac{4}{3}} \right],$$

which is a classic isothermal equation of state employed by numerous authors for many years [for example, FÜRTH (1944) and BIRCH (1952)]. The calculation of transition pressures for the vanishing elastic constants given by eq. (103) should be ignored. We are grateful to Dr's. Kumazawa and Anderson.

The authors now believe that the arguments between eq. (112) and eq. (115) are not helpful in calculating transition pressures, and should be ignored. Eq. (115) implies a linearity between the shear constant and P (although displaced by unity value in P) and is inconsistent with the main arguments previously presented. The authors have recently discovered an elegant proof by M. Born (On the stability of crystal lattices, Proc. Cambridge Phil. Soc. **36** (1940) 160–165), which shows why the vanishing of c_{44}, c', or K results in a lattice instability.

PHASE TRANSFORMATIONS IN THE DEEP MANTLE

THE SYSTEM Mg$_2$SiO$_4$-Fe$_2$SiO$_4$ AT HIGH PRESSURES AND TEMPERATURES

A. E. RINGWOOD and A. MAJOR

Department of Geophysics and Geochemistry, Australian National University, Canberra

Previous investigations on the olivine-spinel transformation are reviewed. The present investigation was carried out on synthetic (MgFe)$_2$SiO$_4$ olivine solid solutions using a Bridgman-anvil apparatus equipped with an internal heater. The experimental method and methods of calibration are described and the precision evaluated. More than 140 runs have been carried out in the pressure range 50–200 kb and approximately at 1000 °C. After quenching, the phases produced were examined by X-ray diffraction and optical methods.

A continuous series of spinel solid solutions from Fe$_2$SiO$_4$ to (Mg$_{0.8}$Fe$_{0.2}$)$_2$SiO$_4$ was synthesized, and lattice parameters and refractive indices determined. The density of pure Mg$_2$SiO$_4$ was determined by extrapolation to be 3.56 g/cm^3, some 10.6% denser than forsterite. A phase diagram for the Fe$_2$SiO$_4$–Mg$_2$SiO$_4$ system was constructed on the basis of the experimental results. In compositions between Fe$_2$SiO$_4$ and (Mg$_{0.8}$Fe$_{0.2}$)$_2$SiO$_4$, olivines transform to spinels with a large two-phase loop of coexisting olivine plus spinel. Between (Mg$_{0.8}$Fe$_{0.2}$)$_2$SiO$_4$ and pure Mg$_2$SiO$_4$, olivines transform to a new orthorhombic phase, β-Mg$_2$SiO$_4$ which is 8% denser than forsterite. This phase is believed to be stable in its synthesis field. The olivine to beta phase transformation at 1000 °C occurs at approximately 120 kb. Reconnaissance investigations were also carried out on the Co$_2$SiO$_4$–Mg$_2$SiO$_4$ system, in which the beta phase possesses a more extensive field of occurrence.

The application of the experimental phase diagram to the constitution and seismic structure of the mantle is discussed. Assuming that the olivine of the mantle has an Mg/(Mg+Fe) ratio of 0.89, olivine would partially transform with increasing depth first to spinel. At a greater depth spinel reacts to produce beta phase, and at still greater depths, the remaining olivine transforms to beta phase. The depth interval over which these transformations occur is 27 km and contains a first order density discontinuity at the spinel–beta phase reaction point. Assuming a temperature in the mantle of 1600 °C at a depth of about 400 km, and a gradient for the transformations of 30 b/°C, the olivine–spinel–beta phase transformations would occur over an interval of about 27 km, with a median depth of 397 km and with a first order discontinuity at 403 km. This agrees closely with the depth of a major seismic discontinuity in the mantle. The experimental data thus provide a satisfactory explanation of this important geophysical feature of the mantle.

1. Introduction

It is widely believed that olivine (MgFe)$_2$SiO$_4$, is the most abundant mineral in the upper mantle. Accordingly it is necessary to determine the stability and phase relationships of olivine at high pressures and temperatures in order to understand the properties and constitution of deeper regions of the mantle. It was suggested by BERNAL (1936) that common olivine might transform in the mantle under a sufficienctly high pressure to a new polymorph possessing the spinel structure, which would be about 9% denser than the olivine. This suggestion was based upon a previous observation by GOLDSCHMIDT (1931) that the analogous compound Mg$_2$GeO$_4$ was dimorphous, displaying both olivine and spinel polymorphs at atmospheric pressure, the spinel being 9% denser. Bernal's suggestion was adopted by JEFFREYS (1937) as the basis for an explanation of a rapid increase in seismic velocity which was believed to occur near 400 km – "the 20 degree discontinuity".

This hypothesis was submerged in controversy for some years. The occurrence of a seismic discontinuity near 400 km was disputed by seismologists, particularly GUTENBERG (1959), who believed that the curvature in the travel-time curve formerly associated with the 20 degree discontinuity arose from the structure of the "low-velocity zone" at much shallower depths. BIRCH (1939, 1952) carried out a classic investigation on the elastic properties of the mantle, and concluded that the region between 300 and 900 km was characterized by a series of major phase transformations, one of which might be the olivine–spinel transformation. Birch's hypothesis has subsequently been proven correct. Nevertheless this was not immediately clear and it was opposed vigorously by VERHOOGEN (1954), GRIGGS (1954), EVERNDEN (1958) and others. Bernal's specific suggestion of an olivine–spinel transformation was the subject of extensive theoretical investigations by WADA (1960), MIKI (1955) and SHIMAZU (1958) who concluded that the transformation was unlikely or impossible.

Clearly, for an adequate appraisal of the olivine–spinel phase change hypothesis, it would be necessary to obtain relevant experimental data. This presented problems, however, since the pressure at 400 km is about 130 kb and the temperature probably in the vicinity of 1600 °C. Until recently P–T conditions in this range were well beyond the range of available equipment. Since this circumstance prevented a direct test of the hypothesis, it was necessary to revert to indirect experimental methods.

RINGWOOD (1956, 1958a) approached this problem by an experimental investigation of solid solubility relations between Ni_2GeO_4 spinel and Mg_2SiO_4 olivine at 1500 °C and atmospheric pressure. These measurements yielded a value for ΔG_0, the free energy of transformation of Mg_2SiO_4 from the olivine to the hypothetical spinel structure at zero pressure. From a well known thermodynamic relationship, ΔG_0 is approximately equal to $P \cdot \Delta V$ where ΔV is the difference in molar volume between olivine and spinel polymorphs and P is the equilibrium pressure needed to transform Mg_2SiO_4 from the olivine to the spinel structure at 1500 °C. Using this method, with ΔV derived from crystal chemical measurements, Ringwood obtained a value for P of 175 ± 55 kb, and a density change of $11 \pm 3\%$. This pressure range is equivalent to a depth interval of 520 ± 180 km, which overlaps the depth of the 20 degree discontinuity as originally obtained by Jeffreys. This was the first quantitative evidence supporting the Jeffreys–Bernal hypothesis.

Another method of using germanate-silicate phase equilibria to predict pressures at which transformations should occur in silicate end-members is by determining the solid solubility boundaries in a given system over the available range of pressures, and then extrapolating the phase boundaries into higher pressure regions. RINGWOOD (1958b) extrapolated the solubility of Mg_2SiO_4 in Ni_2GeO_4 over the range 0–90 kb at 600 °C, and obtained a value of about 125 kb for the olivine–spinel transition in pure Mg_2SiO_4. A more comprehensive investigation of this system by RINGWOOD and SEABROOK (1962, fig. 1) led to the estimate of 130 ± 20

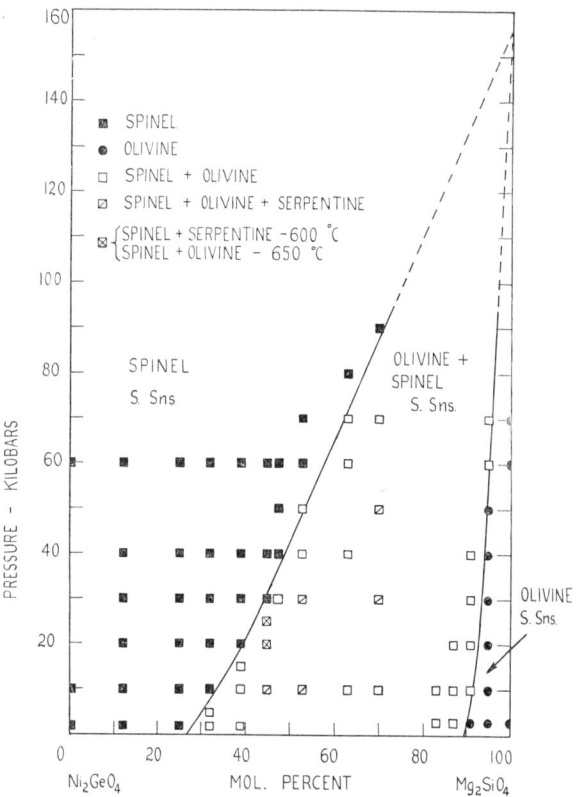

Fig. 1. The system Ni_2GeO_4–Mg_2SiO_4 at 600 °C and 0–90 kb. (The transition pressure in pure Mg_2SiO_4 obtained by direct extrapolation is 155 kb. However this is somewhat high because of pressure gradients in the squeezer. The corrected transition pressure was estimated to be 130 ± 20 kb.) After RINGWOOD and SEABROOK (1962).

kb and a density change $\geq 9\%$ for the transition. An analogous investigation of the system Mg_2GeO_4–Mg_2SiO_4 over the range 0–60 kb by DACHILLE and ROY (1960) led to an estimate of 100 ± 15 kb at 530 °C and a density change of 4.7%.

The general plausibility of the transformation hypothesis for common olivine in the mantle was considerably enhanced by the discovery (RINGWOOD, 1958b, 1962, 1963) that the olivines Fe_2SiO_4, Ni_2SiO_4 and Co_2SiO_4 transformed to spinel structures at pressures between 20 and 70 kb (700 °C) accompanied by average density increases of 10%. This removed any lingering doubts that silicates could crystallize in the spinel structure. Furthermore, the fact that Fe_2SiO_4 was a significant component of natural olivine in the mantle was particularly relevant in this context.

Further examples of olivine–spinel transformations have been uncovered in recent years. BLASSE (1963) found that LiMgVO$_4$ displays an olivine–spinel transformation at high pressures. RINGWOOD and REID (1969) have observed this transformation in MgMnGeO$_4$, FeMnGeO$_4$ and CoMnGeO$_4$. RINGWOOD and REID (1968) also showed that a related compound LiAlSiO$_4$ (eucryptite) which possesses a phenacite structure at low pressures, transforms to a spinel at very high pressure. A list of olivine–spinel transformations is given in table 1.

because of difficulty in attaining equilibrium at this temperature, a better fit to the remaining data points is possible.

The experimental investigations of olivine–spinel transformations which have been reviewed above sufficed to eliminate earlier objections which had been expressed against the possibility that phase transformations in general might be important in the earths mantle, and against the specific objections which had

TABLE 1

Parameters of olivine–spinel transformations

Compound	Ref.	Transition pressure (kb)	Temperature (°C)	$V_{olivine}$† (cm^3)	V_{spinel}† (cm^3)	Density increase (%)	$\dfrac{dP}{dT}$ (b/°C)
Mg$_2$GeO$_4$	1, 2	0	820	45.87	42.35	8.3	40
Fe$_2$SiO$_4$	3, 4, 5	49	1000	46.39	42.03	10.4	28
Ni$_2$SiO$_4$	6, 7, 5	31	1000	42.51	39.17	8.5	16
Co$_2$SiO$_4$	8, 9	70	900	44.56	40.58	9.8	32
Mg$_2$SiO$_4$‡	5	125	1000	43.79	39.58	10.6	30
MgMnGeO$_4$	10	(35)*	1100	—	—	—	—
FeMnGeO$_4$	10	(35)*	1100	—	—	—	—
CoMnGeO$_4$	10	(35)*	1100	—	—	—	—
LiMgVO$_4$	11	—	—	44.71	42.58	5.0	—

* Brackets denote synthesis pressure for spinel. Equilibrium pressures are probably smaller.
† V denotes molar volume.
‡ Data obtained by extrapolation and refer to metastable Mg$_2$SiO$_4$.

References: 1. GOLDSCHMIDT (1931).
2. DACHILLE and ROY (1960).
3. RINGWOOD (1958b).
4. AKIMOTO, KOMADA and KUSHIRO (1967).
5. This paper.
6. RINGWOOD (1962).
7. AKIMOTO, FUJISAWA and KATSURA (1965).
8. RINGWOOD (1963).
9. AKIMOTO and SATO (1968).
10. RINGWOOD and REID (1969).
11. BLASSE (1963).

The olivine–spinel transformations in Fe$_2$SiO$_4$, Ni$_2$SiO$_4$ and Co$_2$SiO$_4$ have been restudied in greater detail by AKIMOTO et al. (1965), AKIMOTO et al. (1967), and AKIMOTO and SATO (1968). These studies have provided valuable data on the gradients (dP/dT) of the transitions which are given in table 1. These have been corrected to be consistent with the pressure scale discussed in section 3. The gradient for the olivine–spinel transformation in Ni$_2$SiO$_4$ given in table 1 (16 b/°C) is higher than is given by AKIMOTO et al. (1965). If one of their data points at the lowest temperature (750 °C) is ignored as likely to be in greater error than the others

been raised against the possibility of an olivine–spinel transformation. Recent advances in seismology, on the other hand, have finally confirmed the existence of the rapid increase in seismic velocity near 400 km earlier inferred by Jeffreys. Investigations by ANDERSON and TOKSOZ (1963), NIAZI and ANDERSON (1965), JOHNSON (1967), KANAMORI (1967), HALES et al. (1960), JULIAN and ANDERSON (1968), and ARCHAMBEAU et al. (1969) agree in requiring an increase of seismic velocity of 0.6 to 1.0 km/sec over an interval of about 50 km in the vicinity of 400 km (fig. 14).

2. Previous high-pressure investigations on the system Fe_2SiO_4–Mg_2SiO_4

Because of apparatus limitations, studies of olivine–spinel transformations (reviewed above) were mostly carried out on systems other than the key mantle system Fe_2SiO_4–Mg_2SiO_4 in the years prior to 1966. However BOYD and ENGLAND (1960) reported the synthesis of a spinel in the latter system containing 4% of Mg_2SiO_4 at 75 kb and 1300 °C. During 1966, the development of new types of high pressure–high temperature apparatus in several laboratories resulted in major progress in the system Fe_2SiO_4–Mg_2SiO_4. SCLAR and CARRISON (1966, March) reported the synthesis of a series of $(FeMg)_2SiO_4$ spinel solid solutions containing up to 35 mol% of Mg_2SiO_4. RINGWOOD and MAJOR (1966, July) described the synthesis at 180 kb and 900 °C of a continuous series of spinel solid solutions between pure Fe_2SiO_4 and a magnesia-rich spinel containing 80 mol% of Mg_2SiO_4*. These were the first syntheses of spinels with Mg/Fe ratios close to those in the earth's mantle and thus confirmed to a considerable degree, the predictions which had been made earlier. The magnesia-rich spinels were found to be 10.6% denser than the corresponding olivines. These results were in close agreement with RINGWOOD's (1956, 1958a) prediction of a density change of $11 \pm 3\%$, and showed that DACHILLE and ROY's (1960) estimate of 4.6% was in error.

Also in July, 1966, Akimoto and Fujisawa published the preliminary results of a comprehensive investigation of equilibria in the Fe_2SiO_4–Mg_2SiO_4 system at pressures up to 95 kb and at 800 °C. This work was definitive in establishing the nature and width of the two phase field at the iron-rich end of the series (spinels containing up to 30% of Mg_2SiO_4 were synthesized) and permitted an extrapolation to the Mg-rich end of the series. More recently, AKIMOTO and FUJISAWA (1968) have greatly extended their results, giving detailed isothermal sections of the system at 800 °C, 1000 °C and 1200 °C. Their results at 800 °C are shown in fig. 2. A homogeneous series of spinels containing up to 40% of Mg_2SiO_4 was synthesized, the lattice

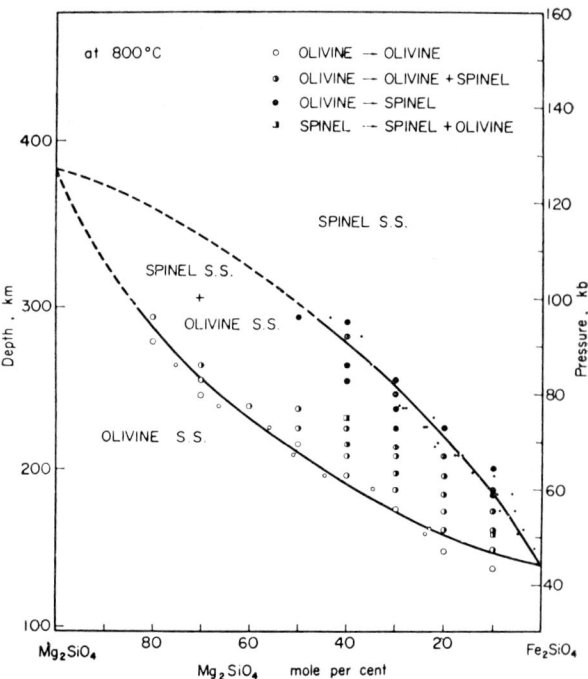

Fig. 2. The system Mg_2SiO_4–Fe_2SiO_4 at 800 °C and 0–95 kb. After AKIMOTO and FUJISAWA (1968).

parameters of which agreed well with the more extensive measurements by Ringwood and Major. Akimoto and Fujisawa's experiments were carried out using a tetrahedral anvil press.

The nature of the high pressure transformations in compositions close to pure Mg_2SiO_4 has been the subject of some confusion in the past, and a principal objective of the present investigation was to resolve this confusion. RINGWOOD and MAJOR (1966a) found that at pressures greater than 150 kb† at about 1000 °C, pure Mg_2SiO_4 transformed completely to a birefringent phase with a complex X-ray diffraction pattern, which possessed some resemblance to a spinel (fig. 3), but which had many extra lines and was clearly of a symmetry lower than cubic. When transformed under completely dry conditions, the crystal size of the new phase was sufficiently small so that it appeared pseudo-isotropic, with a mean refractive index of 1.702 ± 0.005 (RINGWOOD and MAJOR, 1966b). Based upon the Gladstone–Dale rule, this indicates a density of 3.46 g/cm³, which is 8% greater than that of forsterite (RINGWOOD,

---

* RINGWOOD and MAJOR (1966a) estimated that the maximum Mg_2SiO_4 content of the spinels which they synthesized was 85 mol%. Our later more extensive investigations (figs. 7, 11) indicate that the composition was closer to 80% Mg_2SiO_4.

† The pressure given in the cited publication was 170 kb. This value has been corrected downwards in accord with subsequent revisions of the pressure scale – see section 3.

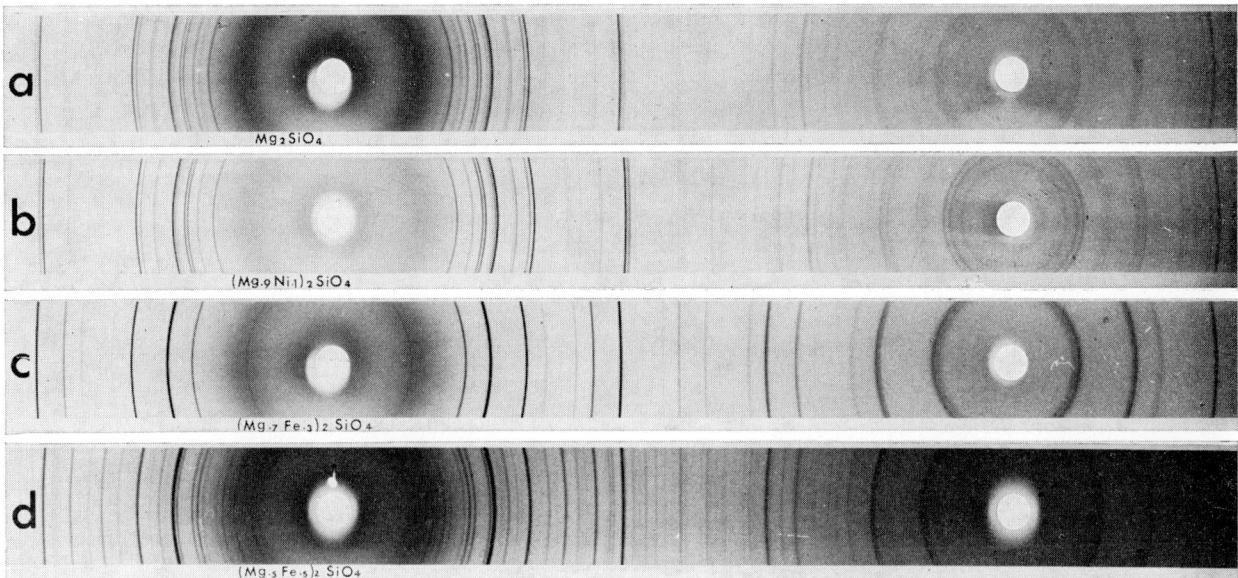

Fig. 3. X-ray diffraction photographs of phases produced in the present experiments. All photographs were taken with cobalt radiation and iron filter using 11 cm powder camera. (a) β-Mg_2SiO_4 synthesized at 148 kb from anhydrous fine grained forsterite. (b) β-$(Mg_{0.9}Ni_{0.1})_2SiO_4$ synthesized at 180 kb from olivine solid solution. Moore and Smith (1969, 1970) indexed 103 reflexions belonging to this specimen in their structure determination. Note sharply resolved back diffraction lines. (c) $(Mg_{0.7}Fe_{0.3})_2SiO_4$ spinel synthesized from olivine solid solution at 160 kb. (d) Olivine and spinel in $(\alpha+\gamma)$ field of fig. 11. Run was carried out at 110kb on sample of composition $(Mg_{0.5}Fe_{0.5})_2SiO_4$. Note sharply resolved back diffraction lines of spinel in equilibrium with olivine. The lattice parameter of the spinel 8.182 Å implies a composition $(Mg_{0.32}Fe_{0.68})_2SiO_4$.

1968). This may be compared with the density of ideal Mg_2SiO_4 spinel which is calculated to be 3.56 g/cm^3 (section 3). The X-ray diffraction pattern suggested that the new phase might have some kind of distorted spinel structure.

This phase was also found in some runs at the $(Mg_{0.85}Fe_{0.15})_2SiO_4$ composition, but its occurrence seemed to depend upon the quenching procedure which was followed in an individual run. Ringwood and Major (1966a) tentatively concluded from this behaviour that the phase which was actually stable under high pressure was probably the spinel, but that when pressure was released, it was impossible to preserve the spinel, which transformed retrogressively to a distorted modification. Evidence of such retrogressive transformations was found about the same time in other systems. Subsequently, more extensive investigations were carried out which cast doubt upon this interpretation and suggested that the new phase, "β-Mg_2SiO_4", was thermodynamically stable in its synthesis field (Ringwood, 1968).

Two months after publication of Ringwood and Major's (1966a) results on the transformation of Mg_2SiO_4, Akimoto and Ida (1966, September) claimed to have synthesized a true Mg_2SiO_4 spinel. This claim was repeated in subsequent papers (Akimoto and Fujisawa, 1968; Fujisawa, 1968). Without wishing to detract from the major contributions by these workers to the study of olivine–spinel transformations, this particular claim must be subjected to critical scrutiny since it has been widely quoted in subsequent literature.

Akimoto and Ida (1966) carried out 9 runs on pure Mg_2SiO_4 at pressures between 140 and 230 kb (nominal) and temperatures of 700–1100 °C, using an apparatus of the kind described by Minomura et al. (1964). In runs at pressures of 155 kb and above a new phase was observed. The maximum degree of conversion of the olivine was estimated to be 50%, and only 6 X-ray reflexions attributed to the new phase were identified, 4 of which were classed as weak. (In contrast, Ringwood and Major obtained 100% transformation of olivine and recorded 21 reflexions. In a later run, 103 reflexions were identified (fig. 3).)

Although Akimoto and Ida clearly succeeded in partially transforming forsterite into a high pressure phase, there is considerable doubt as to whether the high pressure phase was a true spinel as claimed rather than the β-Mg$_2$SiO$_4$ phase found by Ringwood and Major. Thus the refractive index of Akimoto and Ida's phase (1.70 ± 0.01) agrees with that of β-Mg$_2$SiO$_4$ (1.702 ± 0.005) but is significantly smaller than would be expected for Mg$_2$SiO$_4$ spinel (1.720) from the Gladstone–Dale density-refractive index rule. Of the 6 new X-ray lines observed by Akimoto and Ida, 4, including the two strongest, agreed closely with the *d* spacings of β-Mg$_2$SiO$_4$ whilst one of the remaining two lines at 2.33 Å is very close to olivine reflexions at 2.32 and 2.35 Å and may have been confused. The intensities of the X-ray reflexions do not match well with those expected for Mg$_2$SiO$_4$ spinel. Thus the 222 reflexion is expected to be very weak in Mg$_2$SiO$_4$ spinel whereas 220 would be quite strong. Yet in the new phase, the former reflexion was identified but not the latter. Furthermore, for a true Mg$_2$SiO$_4$ spinel the intensity of 440 is expected to exceed that of 400, which is the reverse of that observed for the new phase. Finally, the lattice parameters calculated from the observed *d* spacings range from 8.025 to 8.084 Å. This variation would normally be considered excessive for a correctly indexed spinel phase.

In view of these considerations, Akimoto and Ida's claim to have synthesized a true Mg$_2$SiO$_4$ spinel is judged to be premature. Certainly, a considerable amount of further detailed evidence must be produced before the claim can be accepted.

Another claim to have synthesized a true Mg$_2$SiO$_4$ spinel in 3 runs at 150 kb and above and at 800 °C was made by Kawai et al. (1966), and has since been widely acknowledged in the literature. The claimed synthesis was achieved by reacting MgO and silicic acid together in an ingenious segmented sphere high pressure apparatus (Kawai, 1966). The X-ray diffraction pattern of the run product is shown in fig. 4. It is seen that out of more than 20 recognisable and unidentified reflexions 5 have been selected and assigned as spinel reflexions. Many of the reflexions which were arbitrarily ignored are stronger than some which were selected as spinel reflexions. The lattice parameters have been calculated for each of the chosen reflexions and are given in table 2.

It is seen that these values cover a range of 0.22 Å. This is more than 20 times larger than can be considered acceptable for the correct measurement and

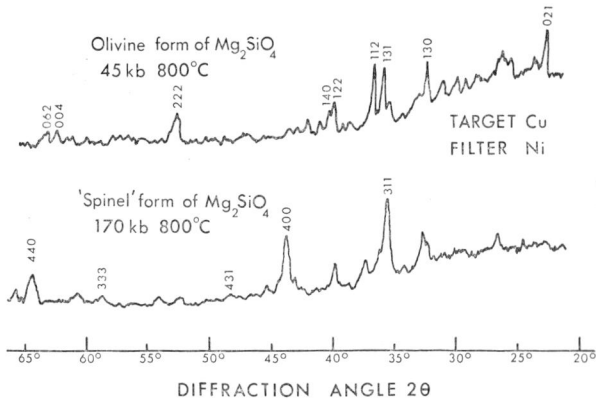

Fig. 4. X-ray diffraction chart of phases produced by reacting MgO and silica at 180 kb, 800 °C. After Kawai et al. (1966).

assignment of spinel *d* spacings in the circumstances pertaining. Errors of this magnitude simply mean that the indexed reflexions do not belong to a spinel phase. Moreover the correct lattice parameter of Mg$_2$SiO$_4$ spinel is 8.071 ± 0.005 Å (section 3, figs. 7, 8). This figure is so far below the calculated values of Kawai et al. (table 2) as to be irreconcilable. It appears certain that the phase described by Kawai et al. (1966) was neither Mg$_2$SiO$_4$ spinel nor β-Mg$_2$SiO$_4$. It appears most likely that the phase was a hydrated magnesium silicate. Ringwood and Major (1966a, 1967) observed that when magnesium oxide and silicic acid are reacted together at pressures greater than 120 kb and at 1000 °C a hydrated phase of low density is formed. These reaction conditions are apparently similar to those used by Kawai et al., except that the temperatures in the former work are somewhat higher. The X-ray diffraction pattern of this phase resembles fig. 4 in several respects, and appears to be identical with the hydroxylated pyroxene described by Sclar et al. (1967) and prepared similarly.

During 1968 several important advances in the study

TABLE 2

X-ray diffraction data on Mg$_2$SiO$_4$ "spinel"
(after Kawai et al. (1966), fig. 2)

Plane	*d* (Å)	Calculated lattice parameter (Å)
311	2.52	8.36
400	2.065	8.26
331	1.88	8.19
333	1.57	8.17
440	1.44	8.14

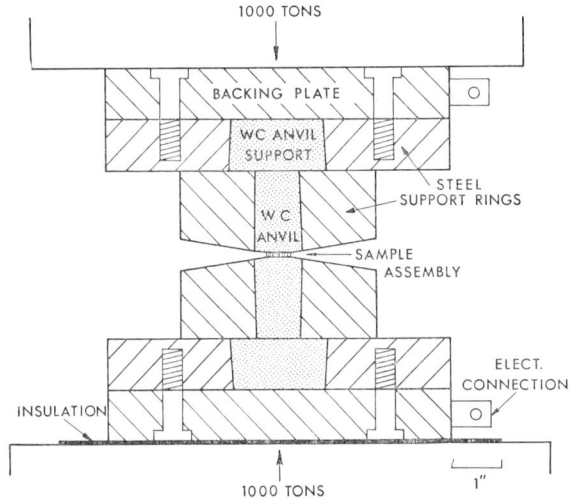

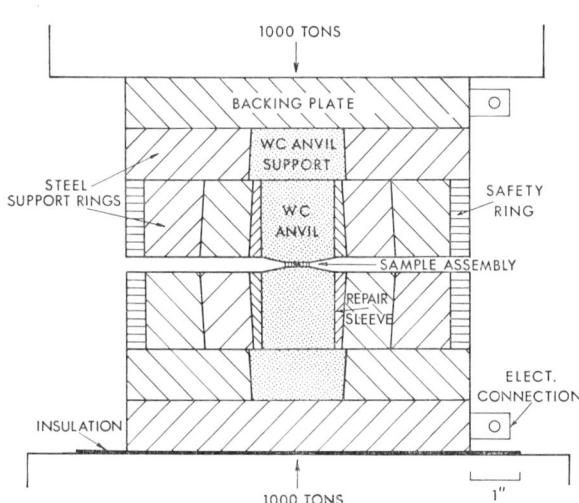

Fig. 5. Bridgman anvils used in present experimental investigations. (a) single support ring, (b) large anvil with compound support rings.

of the olivine–spinel transformation were made and are reported by the workers concerned in other parts of this volume. AKIMOTO and SATO (1968) discovered new high pressure polymorphs of Co_2SiO_4 and Mn_2GeO_4 which were isostructural with β-Mg_2SiO_4. It was demonstrated that the new phases were thermodynamically stable. MORIMOTO et al. (1968) successfully determined the unit-cell dimensions and space groups of these phases. Using these data MORIMOTO et al. (1970) and MOORE and SMITH (1969, 1970) independently solved the crystal structures of β-Mn_2GeO_4 and β-Mg_2SiO_4.

A milestone was reached when BINNS (1970) and BINNS et al. (1969) discovered the first natural occurrence of $(MgFe)_2SiO_4$ spinel in a chondritic meteorite where it had been formed under shock wave conditions. Finally KAWAI et al. (1970) report the partial transformation of natural olivine to a spinel-like phase.

3. Experimental investigations

3.1. Apparatus and procedure

We describe now the experimental methods used in the present investigation of the system Mg_2SiO_4–Fe_2SiO_4. The high pressure apparatus used (RINGWOOD and MAJOR, 1968) consisted of a pair of Bridgman anvils (fig. 5) with a pressure cell designed for internal heating of the sample (fig. 6). The cell is designed to function as a pressure intensifier, so that the pressure on the specimen is about 60 kb *higher* than the mean pressure across the anvils. After assembly of the cell as in fig. 6 it is placed in position between the anvils and pressure is slowly applied by a hydraulic press. When the desired load has been applied, the sample is heated by passing an electric current through the nickel strip heater whilst the anvils are kept cool by a blast of air. The power input is usually 15–18 W, the exact value used depending upon the attainment of a characteristic inflexion in the resistivity of the nickel heater which may correspond to some sort of transformation. This power input heats the specimen in the central portion of the strip to a temperature of about 1000 °C. After heating the specimen for the desired period (usually 3 minutes) it is quenched by terminating the power supply. The specimen remains under full pressure for a further 10 minutes whilst air-cooling of the anvils continues. Then the pressure is released, and the pressure cell removed. The nickel heater containing the specimen is extracted and dissected under a binocular microscope. The samples which are recovered for X-ray diffraction and optical study are taken from the hot spot within the heater. This is easily recognisable from the diffusion of gold into the nickel (fig. 6). In the cooler regions of the furnace, the gold foil does not diffuse. By selecting the samples to be examined from such a small region, the effects of temperature gradients and pressure gradients are minimised. That these are often remarkably small can be seen from fig. 3d. Notice the sharply resolved $\alpha_1\alpha_2$ back diffraction lines of the

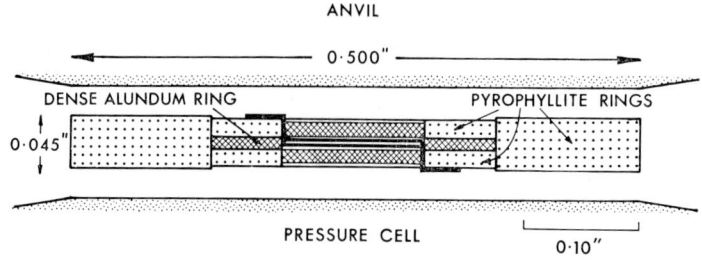
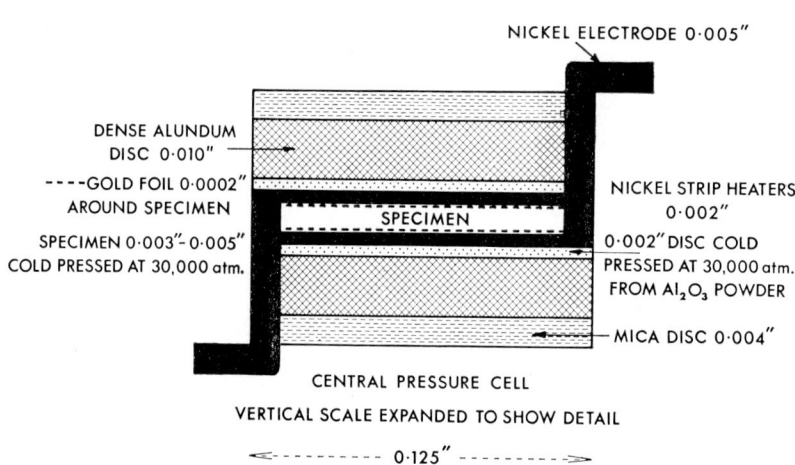

Fig. 6. Internally heated pressure cell used in conjunction with Bridgman anvils.

spinel phase in equilibrium with the olivine. Within the two-phase region of this system, the composition of the spinel phase and hence the width and resolution of its back diffraction lines, are sensitively dependent upon both temperature and pressure (fig. 11, see also AKIMOTO and FUJISAWA, 1968). The sharpness of the $\alpha_1\alpha_2$ doublet implies pressure variation smaller than ± 1 kb and temperature variations smaller than $\pm 35\,°C$. The temperature attained in the hot spot is estimated by comparing the crystallization and transformations of a number of substances in an externally heated squeezer apparatus (KENNEDY et al., 1956) where the temperature is accurately known, with their corresponding behaviour in the present apparatus. It is also checked by inserting a thermocouple in dummy runs, although this is only possible in the lower pressure range. From these observations, the temperature of the central region of the furnace is found to be about 1000 °C under the stated operating conditions. The possible error in individual runs is estimated to be less than 200 °C. Although this is rather large, it must be remembered that temperature is not employed as an independent variable in these studies – rather, its role is to activate the transformation under investigation so that it proceeds to completion within an acceptable interval. Temperature errors are of a random nature so that they are minimised by carrying out large numbers of runs and averaging results, as in the present investigation.

The average length of experimental runs during which simultaneous pressure and temperature are applied to the specimen is 3 to 5 minutes. Because of the strain energy and dislocations introduced in to the sample when pressure is first applied, transformations are observed to proceed to completion in much shorter times and at lower temperatures than are required for the same transformations in apparatus where the initial pressure distribution is much more hydrostatic.

3.2. Pressure calibration

Pressure calibration is ultimately dependent upon pressures assigned to certain key phase transformations – particularly those occurring in bismuth, thallium, barium, tin and iron. These pressures have fluctuated alarmingly during recent years, but there are now heartening signs of convergence towards values for transformations below 100 kb which are unlikely to be seriously in error. In the present investigation we use values recommended at the *Symposium on the accurate characterization of the high pressure environment Natl. Bur. St. U.S.*, Oct. 1968) as follows:

Bi_{I-II}	25.5 kb	Bi_{III-V}	77 kb
Tl_{II-III}	36.7 kb	Sn	97 kb
Ba_{I-II}	55 kb	Fe	126 kb

Some of these values are significantly higher than those obtained by JEFFREY et al. (1966) and which have been previously used by us and by Akimoto and co-workers. Apart from the iron transition, calibration points for pressures above 100 kb are highly uncertain and this should be realized when considering results from high pressure apparatus which have been "calibrated" using the lead, high barium and other transitions. Hopefully, the cross-checking of pressures for solid-solid transitions against a widely accepted continuous pressure-density relationship for sodium chloride will alleviate these difficulties in the future.

The measurement of pressures in internally heated apparatus of small volume poses an additional problem. Use of the "standard" transitions determined at low temperature is of little avail since the operation of the heater in the very small volume of the pressure cell has a substantial effect upon pressure distribution. This problem is encountered particularly by the present apparatus, by the high compression belt and by the apparatus of MINOMURA et al. (1964). Accordingly it is necessary to use a series of secondary standards – transformations determined in large-volume apparatus at temperatures similar to those which are used in the small-volume apparatus. We use the coesite–stishovite transformation as our principal secondary standard. During 1967–68, we took this as occurring at 92–95 kb at 1000 °C. Following the latest investigation of this transition by AKIMOTO and SYONO (1968) and the revisions of the pressure scale noted above, we now assume this transition to occur at 98 kb, 1000 °C. We have also made some use of the olivine–spinel transformations in Fe_2SiO_4 and Co_2SiO_4 (AKIMOTO et al., 1967; AKIMOTO and SATO, 1968) referred to the latest N.B.S. pressure scale as secondary standards – these transitions at 1000 °C are taken as 52 kb (Fe_2SiO_4) and 73 kb (Co_2SiO_4).

In an experiment, the pressure on the sample is given by the sum of two terms, P – the mean pressure over the anvil faces and p – the difference between the central pressure and the mean pressure, caused by the existence of a pressure gradient across the cell. P is usually the larger term and is determined by knowledge of the press load and area of piston face with an accuracy better than one percent. The second term p must be determined by an independent calibration. This is achieved by determining the *mean* pressure P at which the coesite–stishovite transition at 1000 °C occurs in the apparatus. This varies according to the state of dishing of the piston but is in the range 30–37 kb. Since this transition is known to occur at 98 kb, this means that the pressure on the specimen in the centre of the cell is between 60 and 68 kb *higher* than the mean pressure P. This pressure intensification effect is an important aspect of the design of the cell, and enables the generation of extremely high total pressures.

It is assumed that this pressure gradient remains constant at higher pressures up to the limit where plastic deformation of the piston becomes significant (approximately 150 kb total pressure). This assumption is not strictly correct but the errors introduced within the limited range over which it is employed (80–150 kb) are not believed to be large. The pressure gradient p is probably a slowly varying function of mean pressure P, for a limited range of P. Most of the critical experiments which define the phase diagram (fig. 11) were carried out between 90 and 120 kb and within this interval, experimental pressures are closely tied to the coesite–stishovite calibration point at 98 kb. Errors introduced by this assumption are applicable only to the term p, which is smaller than P in the higher pressure range where the errors would be most significant. Hence the proportional error in the *total* pressure on the specimen is reduced by a factor of 2 to 4. Whilst this calibration procedure is not ideal, it is the only one available, since no suitable calibration points at 1000 °C exist above 100 kb. After consideration of the various sources of uncertainty and extensive experience with the operation and behaviour of the apparatus (800 runs above 100 kb) we believe that the pressures in individual runs between 80 and 150 kb are likely to be correct within 10%, assuming that the pressure assigned to the coesite-stishovite transition is correct. Most of the pressure error is of a random nature, caused by minor differences in the properties and dimensions of the components of the pressure cell, which are unavoidable. Hence by carrying out large numbers of runs, the pressure error can be reduced. Thus the pressure-error in the phase equilibrium boundaries determined in the present investigation (fig. 11) is expected to be in the vicinity of 5%.

Below 80 kb, the pressure–load relationship is complex and highly non-linear because of differential "take-up" of components of the pressure cell, and the behaviour also becomes increasingly erratic at pressures below 70 kb. Calibrations in this range were based upon the olivine–spinel transformations in Co_2SiO_4 and Fe_2SiO_4 as discussed previously. Above 150 kb, continuous plastic deformation of the anvils during runs becomes increasingly serious, and it is probable that the effective surface area of the pistons increases whilst the pressure differential p decreases. Accordingly, pressure errors become large and total pressure tends to be overestimated. Nominally, using the anvils in fig. 5b, a mean pressure of 250 kb can be applied to the cell and, with a pressure differential in the cell of 60 kb, this would imply a total pressure of 310 kb. Because of the effects mentioned, the true pressure is probably closer to 200 kb.

In carrying out an extended series of runs, as in the present investigation, the pistons gradually become dished and the pressure required for the coesite–stishovite transformation increases. Starting with new pistons, this is located reproducibly at 30 ± 1 kb. After carrying out 5 to 10 experimental runs, the mean pressure for the coesite–stishovite transformation is re-determined and may be found to have increased to 32 kb. Then another series of runs is carried out, followed by recalibration. The coesite–stishovite transformation is located reproducibly, within ±1 to ±2 kb, for pistons at similar stages of deformation.

3.3. *Starting materials*

The starting materials used in most runs consisted of previously synthesized $(MgFe)_2SiO_4$ olivine solid solutions of known composition. These were prepared by intimately mixing MgO, Fe_2O_3, metallic iron powder and dehydrated silicic acid in the required ratios, pressing the mixture into a pellet which was wrapped in platinum foil, sealed in an evacuated silica tube, and heated at 950–1000 °C for 24 hours. After cooling the sample was X-rayed to check for complete conversion into an olivine solid solution. In some cases, where some inhomogeneity remained, the sample was reground and the procedure repeated. Olivines prepared by this method were very finely crystalline and reactive starting materials.

A number of experiments were also carried out on other olivines in which varying amounts of Co, Ni and Ge had been substituted for the Mg and Si. These were prepared by mixing the components (cobalt from Co_3O_4) intimately, pressing into pellets and firing in air at appropriate temperatures (RINGWOOD, 1958a). A few runs were also conducted on oxide mixtures of periclase-wüstite solid solutions of known composition with silicic acid in the orthosilicate ratio. The (MgFe)O solid solutions were prepared analogously to the MgFe olivines. RINGWOOD and MAJOR (1966a) employed similar mixtures in their syntheses of $(MgFe)_2SiO_4$ spinels.

We found that pure Mg_2SiO_4 synthesized by solid state reaction at about 1000 °C was not very reactive, and frequently required much higher pressures to transform than were indicated by other data. Accordingly, a more reactive sample of Mg_2SiO_4 was prepared from an evaporated mixture of magnesium nitrate and silicic acid. The mixture was heated at 700 °C for 5 days, ground and mixed thoroughly, and heated for a further 7 days at 700 °C. This resulted in about 80–85% transformation into a very fine grained, reactive forsterite, displaying pronounced line-broadening. In most runs, this material was mixed with 5% of brucite $(Mg(OH)_2)$ as a mineralizer. Material produced in this way readily and completely transformed to β-Mg_2SiO_4 in the appropriate pressure interval. Complete transformation was also obtained on material to which brucite had not been added.

Samples so prepared were pressed into small wafers at 30 kb, and placed in the pressure cell. After being subjected to the desired pressure for 3–5 minutes at 1000 °C as described above, the samples were removed from the pressure cell and investigated by X-ray diffraction and optical methods. The nature and proportions of phases produced were determined from the X-ray powder diffraction photographs and the lattice parameters of the spinels were measured. The refractive indices of the spinels were determined in immersion liquids.

4. Experimental results

4.1. *Properties of high-pressure phases*

A series of spinel solid solutions at 10 mol% intervals ranging from pure Fe_2SiO_4 to $(Mg_{0.8}Fe_{0.2})_2SiO_4$ was synthesized at 170 kb. Conversion of olivines to spinels was complete (e.g. figs. 3, 11) except at $(Mg_{0.8}Fe_{0.2})_2$-SiO_4 where a small amount of β-$(MgFe)_2SiO_4$ also occurred. Lattice parameters of the spinels are shown in fig. 7, which extrapolates to a value of 8.075 Å for a pure Mg_2SiO_4 spinel. Lattice parameters were also determined upon a series of Mg_2SiO_4–Co_2SiO_4 spinels and upon a spinel composition Mg_2SiO_4 (85) Ni_2GeO_4 (15) (fig. 8). Extrapolation of these results gives 8.067 Å for pure Mg_2SiO_4 spinel. Accordingly, the lattice parameter of pure Mg_2SiO_4 spinel may be taken as 8.071 ± 0.005 Å from which a molar volume of 39.58 cm^3 and density of 3.56 g/cm^3 are calculated. Mg_2SiO_4 spinel is accordingly 10.6% denser than forsterite.

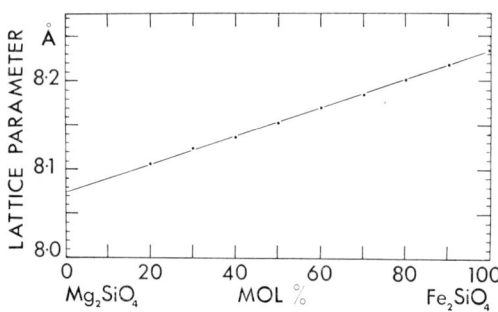

Fig. 7. Lattice parameters of Mg_2SiO_4–Fe_2SiO_4 spinel solid solutions. Synthesized from olivine solid solutions at 170 kb, 1000 °C.

Following the discovery by AKIMOTO and SATO (1968) of a phase, β-Co_2SiO_4*, (fig. 9) possessing a powder pattern similar to that of β-Mg_2SiO_4 (fig. 3), Morimoto and co-workers carried out a single crystal analysis and successfully determined the space group and unit cell dimensions of β-Co_2SiO_4, showing that this phase was orthorhombic with space group *Ibmm*. On the basis of this space group, they were able to index the d spacings of β-Mg_2SiO_4 published by RINGWOOD and MAJOR (1966a) and demonstrated that the two phases were indeed isostructural. Unit cell dimensions and density calculated for β-Mg_2SiO_4 were

$$a = 8.248 \pm 0.009 \text{ Å},$$
$$b = 11.45 \pm 0.02 \text{ Å},$$
$$c = 5.710 \pm 0.004 \text{ Å},$$
$$\rho = 3.47 \text{ g/cm}^3.$$

Thus β-Mg_2SiO_4 was found to be 7.9% denser than forsterite. This was in close agreement with the 8% difference given by RINGWOOD (1968) on the basis of refractive index data and the Gladstone–Dale relationship.

In May 1968 we sent a sample of a particularly well crystallized β phase, composition $(Mg_{0.9}Ni_{0.1})_2SiO_4$, to Drs. P. Moore and J. V. Smith of the University of Chicago. The powder pattern of this sample, contain-

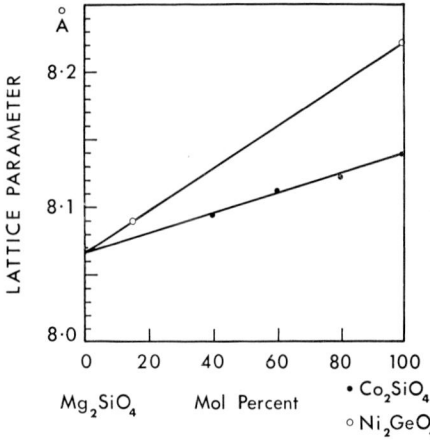

Fig. 8. Lattice parameters of Co_2SiO_4–Mg_2SiO_4 and Ni_2GeO_4–Mg_2SiO_4 spinel solid solutions.

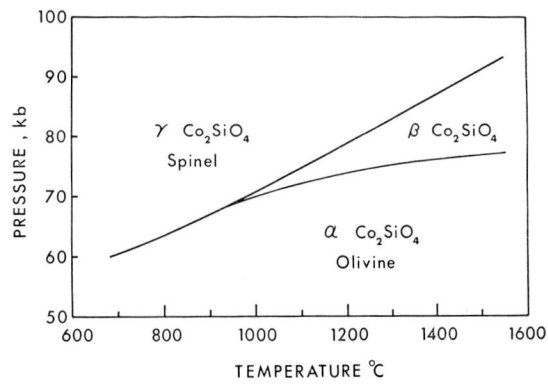

Fig. 9. Relations between the three polymorphs of Co_2SiO_4. After AKIMOTO and SATO (1968).

ing some 103 β-phase reflexions is shown in fig. 3. Lacking single crystal data, Moore and Smith were unable to establish the unit-cell dimensions and space group. However, later in 1968, when Morimoto (reported by AKIMOTO and SATO, 1968) had determined

---

* Initially this phase was called Co_2SiO_4 II. Subsequently it has been agreed to change the terminology to β-Co_2SiO_4.

the space group of β-Co$_2$SiO$_4$ from a single crystal, Moore and Smith indexed all 103 reflexions of β-(Mg$_{0.9}$Ni$_{0.1}$)$_2$SiO$_4$ (fig. 3) on the basis of this space group, and were successful in determining the crystal structure of this phase (MOORE and SMITH, 1969, 1970). In the meantime Akimoto had discovered a third example of a beta phase occurring as a high pressure modification of Mn$_2$GeO$_4$. A full structure analysis on a single crystal of this phase was carried out (MORIMOTO et al., 1970). The results of these two independent investigations were first reported at the 1969 Symposium on phase transformations in Canberra. The structures were found to be in close agreement. The oxygen anions in the structure are approximately in cubic close packing as in the spinel structure. The Mg^{2+}, Co^{2+} and Mn^{2+} cations are surrounded by octahedra of oxygen anions, whilst the Si^{4+} and Ge^{4+} cations are tetrahedrally coordinated with respect to oxygen. However, whereas the MO$_4$ tetrahedra in spinel are separated, the tetrahedra in the beta phase share one of their oxygen ions, resulting in M$_2$O$_7$ sorosilicate doublets and corresponding oxygen ions which are not bonded to Si^{4+} or Ge^{4+}.

Dale rule is shown. There is good agreement between theoretical and observed values for the spinels considering the uncertainties. However the observed indices of the beta phases fall significantly below the theoretical value for spinels, as would be expected from the fact that the beta-phases are two percent less dense than the corresponding spinels.

If the beta phase is synthesized by transforming forsterite in the absence of water, it is sometimes pseudo-isotropic, because of the very small crystal size. A sample of this material was found to possess mean refractive index of 1.702 ± 0.005. However by transforming pure finely crystallized Mg$_2$SiO$_4$ in the presence of 5% of brucite (Mg(OH)$_2$) which acts as a mineralizer, the crystal size is substantially increased and it was possible to determine the alpha and gamma refractive indices

$$\alpha = 1.689 \pm .003,$$
$$\gamma = 1.704 \pm .003.$$

4.2. Phase diagram for the system Mg$_2$SiO$_4$–Fe$_2$SiO$_4$

The nature and proportions of phases observed in the experiments are shown in fig. 11. The experiments clearly define three fields of olivine (α) solid solutions, spinel (γ) solid solutions and β phase solid solutions, and two phase regions of ($\alpha+\gamma$), ($\gamma+\beta$) and ($\alpha+\beta$). Lattice parameters of spinel solid solutions in the two phase ($\alpha+\gamma$) field were determined and used, together with fig. 7, to determine the compositions of spinels and accordingly the spinel solvus. These are shown as dots in fig. 11.

The phase relationships and lattice parameters closely determine the location of the spinel solvus between 80 and 116 kb. Because of apparatus limitations, few runs have been carried out below 80 kb, and the results of AKIMOTO and FUJISAWA (1968) remain definitive in this region. In the 80 to 95 kb region where our results overlap with those of Akimoto and Fujisawa, agreement is within experimental error (when reduced to the same pressure scale). However our results show a stronger curvature of the spinel solvus.

In magnesia-rich compositions, complete transformations of olivines to the equilibrium high-pressure assemblages above 120 kb were sometimes not attained because of kinetic difficulties. Another non-equilibrium feature is seen in fig. 11, where a few runs show olivine

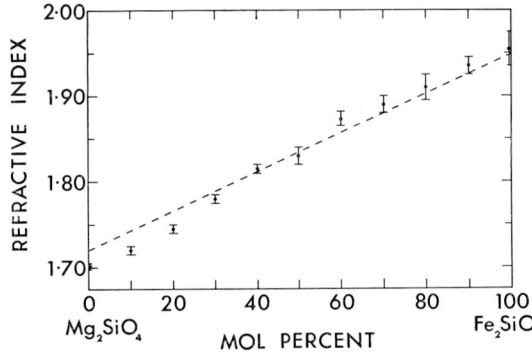

Fig. 10. Refractive indices of spinel solid solutions between pure Fe$_2$SiO$_4$ and (Mg$_{0.7}$Fe$_{0.3}$)$_2$SiO$_4$ and of β-(MgFe)$_2$SiO$_4$ solid solutions between (Mg$_{0.8}$Fe$_{0.2}$)SiO$_4$ and pure Mg$_2$SiO$_4$. Broken line joins the theoretical refractive indices of Mg$_2$SiO$_4$ spinel and Fe$_2$SiO$_4$ as given by the Gladstone–Dale rule.

The refractive indices of spinels and the mean refractive indices of β phases synthesized in the present work are shown in fig. 10. The precision of measurement in some of the iron-rich varieties is not high owing to the presence of fine inclusions. Also in fig. 10 the calculated refractive indices of spinels obtained from their densities by application of the Gladstone–

alone within the $(\alpha+\gamma)$ field. In these runs, nucleation and growth of spinels from olivines within the time of the runs has apparently required a substantial overpressure.

The transformations in fig. 11 are unreversed because of apparatus limitations. Nevertheless the general consistency of the results, the agreement in the position of the spinel solvus obtained by the disappearing-phase method and from the compositions of spinels in the two-phase field as determined from their lattice parameters, and the broad agreement of the field boundaries below 100 kb with those determined by AKIMOTO and FUJISAWA (1968), who succeeded in reversing their boundaries, leave little doubt that the olivine-spinel boundaries in fig. 11 represent a close approach to equilibrium.

Within the $(\gamma+\beta)$ field of fig. 11, the occurrences of phases are extremely complex and erratic. Samples of $(Mg_{0.8}Fe_{0.2})_2SiO_4$ have been converted to spinel and/or to the beta phase, whilst in some runs a substantial amount of residual untransformed olivine is present. In most runs, mixtures of all three phases in varying proportions are recovered, and there is no systematic effect with increasing pressure.

The occurrence of olivine in this field is clearly explicable in terms of non-equilibrium and kinetics. The relations between spinel and β phase presented some problems in interpretation. When we first observed the erratic behaviour at this composition (RINGWOOD and MAJOR, 1966a) we suspected that the beta phase which we called "distorted spinel" represented a retrogressive metastable transformation product of a true spinel which had been synthesized at pressure. This type of behaviour is not uncommon in high-pressure experimentation. However as we carried out many more runs above 120 kb in compositions richer in magnesia than $Mg_{80}Fe_{20}$, spinel was never encountered. The only high pressure phase occurring in this region was β-$(MgFe)_2SiO_4$. These experiments indicated a clearly defined field of occurrence of the beta phase and were consistent with the assumption, although not proving, that the beta phase was stable in its synthesis field (RINGWOOD, 1968).

The accumulation of later evidence has strongly supported this assumption. The facts that isostructural beta phases occur in the Co_2SiO_4 and Mn_2GeO_4 systems and have been demonstrated by Akimoto and co-workers to occur as *equilibrium* transformation products of olivines, suggests that the beta phase in the present system is not a freak, as first suspected, but rather, is likely to occupy an analogous stability field. The X-ray diffraction film of the beta-phase (fig. 3b) shows very sharply resolved $\alpha_1\alpha_2$ back diffraction lines. If the beta phase represented a retrogressive transformation product of a stable high-pressure spinel, strong broadening, or perhaps, complete disappearance of these lines would be expected. In cases of retrogressive reconstructive transformation at room temperature from a high pressure phase on release of pressure, the crystallite size of the low-pressure phase is nearly always very small, and the crystals frequently strained, resulting in line-broadening and/or loss of high angle reflexions. More recently we have synthesized relatively large (5 to 10 μ) unstrained crystals of beta Mg_2SiO_4 possessing sharp optical extinctions. It is improbable that such crystals should form by retrogressive transformation. Finally, studies of high-pressure equilibria in the system Co_2SiO_4–Mg_2SiO_4 discussed later in this section further support the view that the beta phase is thermodynamically stable in its synthesis field.

Accepting this interpretation, a possible explanation of the erratic behaviour within the $(\gamma+\beta)$ field becomes apparent. As is seen in fig. 11 the width of this field is comparatively small. It will be recalled that the experimental conditions employed were not strictly isothermal and that individual runs may have been carried out as much as ± 200 °C from the median temperature of 1000 °C. Since the spinel to beta phase equilibrium is probably temperature sensitive, as demonstrated for Co_2SiO_4 (fig. 9) the boundaries of the $(\gamma+\beta)$ field would be significantly different from run to run, so that for a fixed composition $(Mg_{80}Fe_{20})$ the specimen might lie either in the γ, β or $(\gamma+\beta)$ fields according to the temperature reached in the particular run.

The transformation of pure forsterite to the beta phase occurs at 120 kb* and 1000 °C (fig. 11). By extrapolating the olivine–spinel phase boundaries in fig. 11 the corresponding pressure for the metastable olivine–spinel transformation in pure Mg_2SiO_4 is 125 kb. This small difference in pressures, combined with the small

---

* In one run at 118 kb, a sample of forsterite was completely converted to β-Mg_2SiO_4 and it is possible that the transformation pressure is closer to 115 kb than to 120 kb. The latter value however was preferred in constructing fig. 11 in order to be more consistent with some results on other compositions.

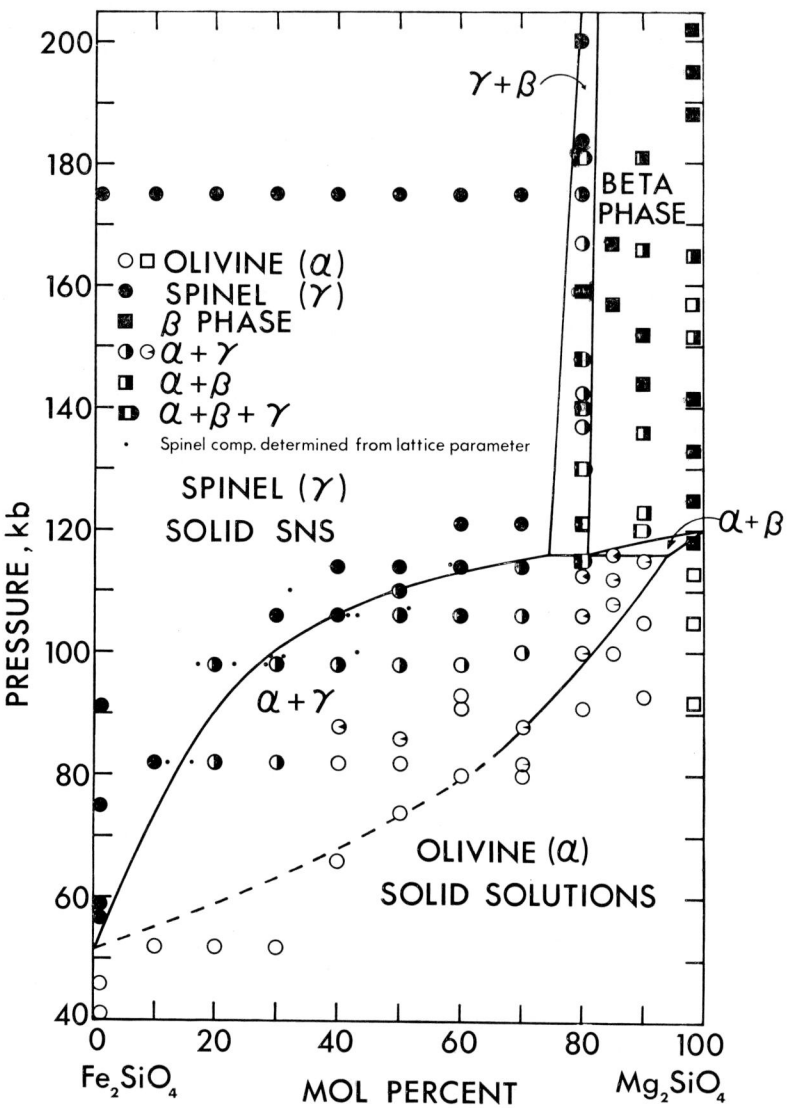

Fig. 11. Phase relationships in the system Mg_2SiO_4–Fe_2SiO_4 at 50–200 kb and at 1000 °C.

To conserve space, detailed tables showing results in individual runs have not been prepared. The nature of phases observed in individual runs is shown by symbols and the proportions of phases present can usually be depicted with an accuracy comparable to the approximate visual estimates made from the X-ray diffraction photographs. Additional details on particular runs are given below.

$Mg_{80}Fe_{20}$ composition:
  Run at 184 kb contained about 10% of beta phase.
  Runs at 148 and 159 kb were carried out on mixtures of periclase–wustite solid solutions with silicic acid.

$Mg_{85}Fe_{15}$ composition:
  Runs at 157 and 167 kb were carried out on mixtures of periclase–wustite solid solutions with silicic acid.

Pure Mg_2SiO_4:
  The runs at 110, 114, 118, 125 and 133 kb were carried out on highly reactive forsterite in presence of 5% brucite as described in text. The run at 141 kb was carried out on similar material except that brucity was absent. Remaining runs were carried out on samples of coarse grained synthetic forsterite.

General:
  A substantial number of runs was carried out in the beta phase field without obtaining any transformation of olivine. This is attributed to kinetic difficulties and the runs have not been recorded in fig. 11.
  In most runs two percent of water was added to the charge as a mineralizer to facilitate reaction. This was not done with the run on pure Mg_2SiO_4 at 141 kb.
  This diagram contains the results of further runs carried out since an earlier publication of this diagram (RINGWOOD, 1969b) and some of the boundaries at the Mg-rich end of the system have been changed slightly.

difference in molar volumes of the two polymorphs in turn implies that the free energy difference between the two high pressure polymorphs of Mg_2SiO_4 is small. Accordingly the positions of the $(\gamma-\beta)$ phase boundaries may be rather sensitively dependent upon temperature and possibly, in the more complex natural system in the mantle, upon the presence of other components which form solid solutions preferentially with one of these phases, thus increasing its relative stability. These circumstances complicate the application of the phase relations in the simple system to the mantle. The small difference in free energies between the gamma and beta phases is also reflected in the narrowness of the $(\gamma+\beta)$ field.

The $(\gamma+\beta)$ field boundaries in fig. 11 are almost vertical. At atmospheric pressure, the spinel is about 2% denser than the beta phase and accordingly it might be expected that the $(\gamma+\beta)$ boundaries should have a positive slope and should intersect the Mg_2SiO_4 axis, corresponding to a transformation of β-Mg_2SiO_4 to Mg_2SiO_4 spinel at higher pressures. This is observed in the Co_2SiO_4 system (AKIMOTO and SATO, 1968). The present experimental data give no indication of this however. It appears that the compressibility of the beta phase may be slightly greater than that of spinel, so that at a pressure of about 100 kb, the density difference between the two polymorphs has been reduced to zero, thus causing the vertical slope of the phase boundaries.

Following the discovery of the beta modification of Co_2SiO_4 (AKIMOTO and SATO, 1968; fig. 9) it appeared that an investigation of the system Co_2SiO_4–Mg_2SiO_4 might throw further light upon spinel–beta phase relationships. The results of our experimental studies of this system are shown in fig. 12. The system is qualitatively similar to Fe_2SiO_4–Mg_2SiO_4, with fields of α, β and γ phases. However the extent of the spinel field relative to β and $(\beta+\gamma)$ fields is substantially reduced. This is consistent with behaviour to be expected under equilibrium conditions in this system. The free energy difference between spinel and beta modifications of Co_2SiO_4 is very small near 1000 °C (fig. 9), whereas the absence of a beta modification of Fe_2SiO_4 in the very wide P–T field over which this compound has been investigated indicates that the free energy difference between the spinel and hypothetical beta modifications of Fe_2SiO_4 is probably much greater than for Co_2SiO_4.

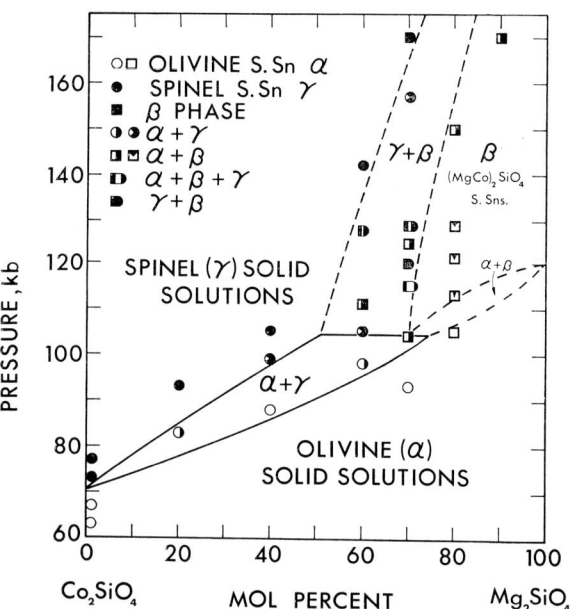

Fig. 12. Phase relationships in the system Mg_2SiO_4–Co_2SiO_4 at 60–170 kb, 1000 °C.

Hence the addition of Co_2SiO_4 to Mg_2SiO_4 extends the stability field of the beta phase more than the equivalent addition of Fe_2SiO_4 (figs. 11, 12).

The width of the olivine-spinel two phase field in the Co_2SiO_4–Mg_2SiO_4 system is substantially smaller than for Fe_2SiO_4–Mg_2SiO_4. The olivine–spinel phase boundaries in the Co_2SiO_4–Mg_2SiO_4 system can be extrapolated to give a pressure of 125 kb for the metastable olivine–spinel transformation in pure Mg_2SiO_4. This value agrees with that obtained by extrapolation in the Fe_2SiO_4–Mg_2SiO_4 system. The spinel–beta phase boundaries in the Co_2SiO_4–Mg_2SiO_4 system (fig. 12) appear to have a positive slope suggesting that for these compositions, over the pressure range investigated, the spinel phase is still slightly denser than the beta phase. Similar indications of a positive slope have been obtained for the same boundary in the Ni_2GeO_4–Mg_2SiO_4 system (unpublished observations); however the experimental evidence in both of these systems is too sparse to be definitive.

A few experiments have been carried out to study the influence of other elements upon the transformation. Olivines with compositions $Mg_2(Si_{0.9}Ge_{0.1})O_4$ and $(Mg_{0.9}Ni_{0.1})_2SiO_4$ transformed to beta phases. An olivine of composition $85Mg_2SiO_4 \cdot 15Ni_2GeO_4$ transformed dominantly to a spinel, with less than 10%

of beta phase. However at the composition $90Mg_2SiO_4 \cdot 10Ni_2GeO_4$ a run at 120 kb produced complete transformation to beta phase, whereas a run at 150 kb produced mainly spinel, with about 10% of beta phase and 10% of untransformed olivine. Evidently the spinel stability field is more extensive in the system Ni_2GeO_4–Mg_2SiO_4 than in the Fe_2SiO_4–Mg_2SiO_4 system.

The gradients dP/dT of the transformations are important for the application of the experimental results to the mantle. AKIMOTO et al. (1967) obtained a gradient of 33 b/°C for the olivine–spinel transformation in pure fayalite. This value was obtained on the assumption that the $Ba_{II\ III}$ transition occurs at 59 kb. This transition is now believed to occur at 55 kb (section 3). This would reduce the gradient for the olivine–spinel transition in Fe_2SiO_4 to 28 b/°C – a value which is used here.

Gradients for the olivine–spinel transformation in more magnesia-rich compositions may be estimated from the 800 °C, 1000 °C and 1200 °C isothermal sections (fig. 13) of AKIMOTO and FUJISAWA (1968). These authors estimate a value of 62 b/°C for pure Mg_2SiO_4 "spinel". However this high value is obtained from a calculated thermodynamic extrapolation of their primary data. This extrapolation contains large possible errors and it is preferable to estimate the gradient directly from the experimental observations over a limited range of compositions (fig. 13).

Bearing in mind the several substantial sources of experimental uncertainty implicit in these boundaries, we believe that the only conclusion which can reasonably be drawn from the experimental data as they stand, is that the gradient in the magnesia-rich compositions is probably similar to that for pure Fe_2SiO_4 (fig. 13). A gradient of about 30 b/°C for the transition in magnesia-rich compositions is indicated.

5. The transformation of olivine in the mantle

Recent detailed seismic investigations by ANDERSON and TOKSOZ (1963), NIAZI and ANDERSON (1965), JOHNSON (1967), KANAMORI (1967), HALES et al. (1968), JULIAN and ANDERSON (1968) and ARCHAMBEAU et al. (1969) have re-established the occurrence of the major seismic discontinuity near 400 km, earlier inferred by Jeffreys. The detailed properties of the discontinuity have not yet been finally determined. However most seismic solutions indicate that it occurs between about 370 and 420 km (fig. 14) and that a total increase in velocity of 0.6 to 1.0 km/s is involved. The nature of the velocity distribution within these limits is not known and fine structure may be present (BOLT et al., 1968).

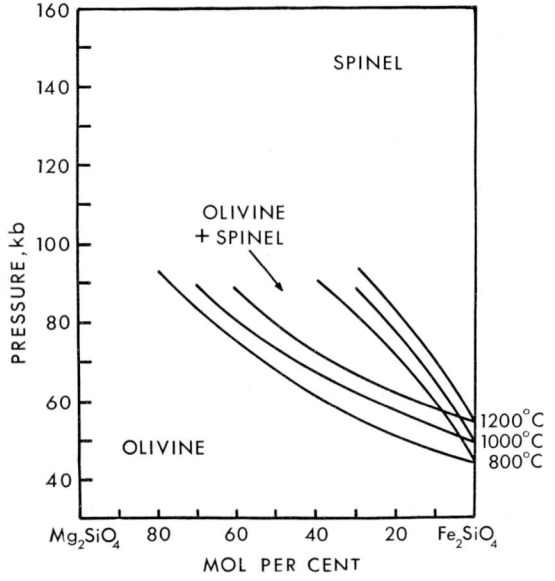

Fig. 13. Isothermal sections at 800 °C, 1000 °C and 1200 °C through the system Mg_2SiO_4–Fe_2SiO_4 in the pressure range 0–95 kb. After AKIMOTO and FUJISAWA (1968). The phase boundaries apply only to those regions of the system which have been experimentally studied.

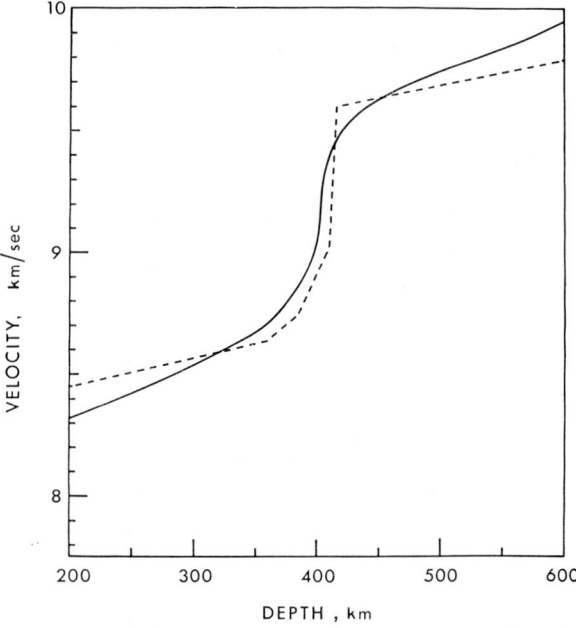

Fig. 14. Distribution of P wave velocities between depths of 200 and 600 km in the mantle. Solid line – JOHNSON (1967). Broken line – ARCHAMBEAU et al. (1969).

This discontinuity is thus comparable in magnitude with the Mohorovicic discontinuity and is one of the two major discontinuities within the mantle. The seismic discontinuity is doubtless accompanied by corresponding changes in other important physical properties – e.g. electrical conductivity and rheological properties, in addition to density and elastic properties.

It is generally agreed that olivine is a major constituent of the upper mantle. The chemical and mineralogical constitution of this region has been discussed extensively by CLARK and RINGWOOD (1964) and RINGWOOD (1966a,b, 1969a), and it was concluded that the Mg/(Fe+Mg) (molecular) ratio was close to 0.89 down to depths of 200 km or so. It will be assumed that this ratio also holds at greater depths in the mantle. The extent to which transformations in mantle olivine of this composition can account for the seismic discontinuity is now examined.

We have determined above, the phase diagram for the system Fe_2SiO_4–Mg_2SiO_4 at 1000 °C. Application of this diagram to the mantle may be complicated by the presence of other components capable of forming solid solutions and by the higher temperatures in the mantle. We consider first the effect of other components.

Previous investigations (RINGWOOD, 1967, 1969b) have shown that the principal phase co-existing with the spinel and beta phases around 400 km is a complex garnet solid solution containing some octahedral silicon – $A_3B_2Si_3O_{12}$ – $A_3(ASi)Si_3O_{12}$ where A^{2+} = Mg, Fe or Ca and B^{3+} = Al, Cr or Fe. Of the divalent ions which are available to substitute in the spinel–β phase structures, Ca^{2+} would be too large and is accommodated in the garnet. Site preference considerations show that Ni^{2+} would preferentially substitute in the octahedral positions of the spinel and beta structures, but the relatively low abundance of nickel in the mantle ($\approx 0.3\%$) combined with our studies of the behaviour of this component show that it would have a negligible effect upon the α–β–γ equilibria. The same can be said for Co^{2+} and probably for Mn^{2+}.

The only geochemically abundant element which appears capable of substituting for Si^{4+} in the tetrahedral sites of the spinel or beta phases is Al^{3+}. This would imply entry as an $(MgAl)^{VI}Al^{IV}O_4$* component, the

* Note that this component is an *inverse* spinel whereas $MgAl_2O_4$ is *normal* with Mg^{2+} in 4-fold coordination.

superscripts representing the coordinations with respect to oxygen. (If this substitution were possible, Cr^{3+} and Fe^{3+} would probably replace some of the Al^{3+} in the octahedral sites.) However the substitution of Al^{3+} for silicon in four-fold coordination would be anomalous under these very high pressures. It is well known that at pressures of a few tens of kilobars, most aluminosilicates possessing Al^{3+} replacing Si^{4+} in tetrahedral coordination transform to structures in which the Al^{3+} is octahedrally coordinated and not replacing Si^{4+}. In the present case, where we are considering the partition of Al^{3+} between tetrahedral sites in a spinel and octahedral sites in a coexisting garnet, it appears likely that the partition will strongly favour the garnet.

These considerations suggest that the experimental equilibria in the simple system are not likely to be seriously modified by other components present in the mantle. The fact that the A^{2+} and B^{4+} cations in olivine, spinel and beta phases possess similar oxygen coordinations is also significant in limiting strong differential solid solution effects. Since additional components are likely to be more soluble in the spinel and beta phases than in the olivine, the most probable effect of other components in the mantle would be a slight reduction of the pressures required to transform olivine into spinel or beta phases. Because of the small free energy difference between these latter phases, the phase boundaries between them are perhaps likely to be the most sensitive to solid solution formation, although, following previous discussion, major displacements are not expected. Nevertheless, because these boundaries occur at Mg/(Mg+Fe) compositions believed to be close to those in the mantle, even relatively small displacements may be significant in influencing the mineralogy of the mantle.

The temperature in the mantle at a depth of about 400 km is probably much higher than 1000 °C. Accordingly, this may substantially expand or diminish the primary fields of α, β and γ phases relative to one another. In the Co_2SiO_4 system, the stability field of the beta phase expanded relatively to spinel with increasing temperature (fig. 9). However, in view of the very small free energy difference between the phases, it it not certain whether this relationship would hold in the present system. Application of the experimental equilibria to the mantle requires knowledge of the gradients dP/dT of the olivine–spinel and olivine–beta phase boundaries particularly. The former was taken as 30 b/°C in the previous section, with an uncertainty of perhaps 10 b/°C. The latter is unknown. However thermodynamic considerations imply that if the latter gradient is significantly higher than the former, the extent of the beta phase field will be decreased relatively to the spinel field as temperature increases, and the beta-phase will therefore not play an important role in the mantle. Accordingly the olivine–spinel gradient of about 30 b/°C forms an effective *upper limit* for the application of this diagram at higher temperatures in the mantle. A further source of uncertainty in applying the experimental results to the mantle is introduced by possible experimental errors in fixing the phase boundaries in fig. 11, as discussed earlier.

Future experimental investigations may be expected to narrow these sources of uncertainty. They have been stressed above since some workers (e.g. AKIMOTO and FUJISAWA, 1968; FUJISAWA, 1968; ANDERSON, 1967) have attempted to determine within rather narrow limits, the temperature at 400 km and the Fe/(Fe+Mg) ratio in this region by varying these quantities until the best agreement between seismic properties and the assumed Fe_2SiO_4–Mg_2SiO_4 phase diagram have been obtained. When a realistic assessment is made of uncertainties, particularly the probability that the gradients used by these workers were incorrect by a factor of two, and that the phase diagram is very different from that assumed, these attempts would appear premature. Nevertheless, in the future, as our knowledge of the phase relationships, particularly the temperature gradients, in this fundamental system improves, it will be possible to repeat these exercises with reasonable expectation of success.

For the present, as a first approximation to the problem we will assume that the phase diagram for the system Mg_2SiO_4–Fe_2SiO_4 at 1000 °C as determined experimentally is applicable to the mantle as it stands, except for the appropriate temperature correction. The latter we will take to be 30 b/°C. The Mg/(Mg+Fe) ratio in the mantle is taken as 0.89 as discussed previously.

Referring to fig. 11, we see that for Mg/(Fe+Mg) = 0.89, as pressure increases, at 109 kb, olivine first begins to transform to a spinel of composition $(Mg_{0.46}Fe_{0.54})_2$-SiO_4. As pressure increases the amount of spinel increases whilst the spinel becomes richer in magnesia, reaching $(Mg_{0.75}Fe_{0.25})_2SiO_4$ at 116 kb. At this point the spinel phase completely reacts to form a beta phase of composition $(Mg_{.8}Fe_{.2})_2SiO_4$. Above this pressure, spinel is absent and we are in the field of olivine+beta phase. With a further slight increase of pressure, olivine finally transforms to the beta phase by 118 kb. The total transformation is thus spread over an interval of 9 kb with a median value of 114 kb.

Now let us assume that the temperature near 400 km is 1600 °C, which was obtained in an earlier study by CLARK and RINGWOOD (1964). With a gradient of 30 b/°C, the entire phase diagram would be raised in pressure by 18 kb and the median pressure for the transition would be 132 kb. Thus the transformation of olivine through spinel into the beta phase would occur at a median depth of 397 km, with a width of about 27 km. This is in excellent agreement with the depth of the seismic discontinuity (fig. 14) considering the uncertainties in the seismic data. Although the phase transformations extend over an interval of 27 km the existence of the spinel–beta phase reaction point at 116 kb (134 kb or 403 km at 1600 °C) causes a discontinuous change in mineralogy and density at this depth, and a further rapid increase of density between 403 and 410 km. These effects will produce a first order seismic discontinuity, rather than a second order discontinuity as was suggested in earlier investigations. The reflections observed by BOLT et al. (1968) may well have been produced by this feature.

It is most satisfactory that when we take an Mg/(Fe+Mg) ratio for the upper mantle which is strongly supported by petrological and geochemical considerations, and a temperature at 400 km which is consistent with geothermal considerations, the laboratory investigations are so closely consistent with seismic observations. This agreement suggests that the possible uncertainties in the experimental work are not very serious – nevertheless this will have to be confirmed in the future.

The principal experimental uncertainties apply to the relationships between spinel and beta phases. It will be necessary to establish these relationships since they will have a bearing upon the thermodynamic properties of the mantle and upon its elastic and rheological behaviour, among other things. Nevertheless, as far as the broad seismological implications are concerned, the spinel phase and the beta phase are essentially identical since we have concluded that above 100 kb, their densities are identical, and from the velocity–density relationship of BIRCH (1961) we can expect that the seismic velocities are similar. RINGWOOD (1969b) has shown furthermore, that when the influence of other minerals besides olivine which are believed present in the mantle is taken into account, the magnitude of the velocity change near 400 km (fig. 14) is explained quantitatively by the transformations which have been discovered in the laboratory. It appears then that a rather complete and satisfying explanation of a major geophysical feature of the mantle has been accomplished.

Although β-(MgFe)$_2$SiO$_4$ has not yet been discovered in nature, there is little doubt that it exists as a major phase in the earth's mantle. We suggest that it would be appropriate to name this mineral Wadsleyite in honour of the late Dr. A. D. Wadsley whose outstanding crystallographic investigations on minerals and inorganic compounds are renowned to all within his field and to many outside. As a matter of record he was the first crystallographer to take an interest in β-Mg$_2$SiO$_4$, and also suggested (with Reid and Ringwood) that it would probably transform to the strontium plumbate structure at even higher pressures in the mantle. This transformation is probably responsible for the second major discontinuity within the mantle, occurring at a depth of about 650 km.

References

AKIMOTO, S. and I. SYONO (1970) High-pressure decomposition of the system FeSiO$_3$–MgSiO$_3$, Phys. Earth Planet. Interiors **3**, 186–188.

AKIMOTO, S. and H. FUJISAWA (1966) Olivine–spinel transition in system Mg$_2$SiO$_4$–Fe$_2$SiO$_4$ at 800 °C, Earth Planet. Sci. Letters **1**, 237–240.

AKIMOTO, S. and H. FUJISAWA (1968) Olivine–spinel solid solution equilibria in the system Mg$_2$SiO$_4$–Fe$_2$SiO$_4$, J. Geophys. Res. **73**, 1467–1479.

AKIMOTO, S. and Y. IDA (1966) High-pressure synthesis of Mg$_2$SiO$_4$ spinel, Earth Planet. Sci. Letters **1**, 358–359.

AKIMOTO, S. and Y. SATO (1968) High pressure transformation in Co$_2$SiO$_4$ olivine and some geophysical implications, Tech. Rept. Inst. Solid State Phys. Univ. Tokyo Ser. A **328**.

AKIMOTO, S. and Y. SYONO (1968) The coesite-stishovite transition, Tech. Rept. Inst. Solid State Phys. Univ. Tokyo Ser. A **327**.

AKIMOTO, S., H. FUJISAWA and T. KATSURA (1965) The olivine–spinel transition in Fe$_2$SiO$_4$ and Ni$_2$SiO$_4$, J. Geophys. Res. **70**, 1969–1977.

AKIMOTO, S., E. KOMADA and I. KUSHIRO (1967) Effect of pressure on the melting of olivine and spinel polymorphs of Fe$_2$SiO$_4$, J. Geophys. Res. **72**, 679–686.

ANDERSON, D. L. (1967) Phase changes in the upper mantle, Science **157**, 1165–1173.

ANDERSON, D. L. and M. N. TOKSOZ (1963) Surface waves on a spherical earth I, Upper mantle structure from Love waves, J. Geophys. Res. **68**, 3483–3500.

ARCHAMBEAU, C. B., E. A. FLINN and D. G. LAMBERT (1969), in press.

BERNAL, J. D. (1936) Discussion, Observatory **59**, 268.

BIRCH, F. (1939) The variation of seismic velocities within a simplified earth model in accordance with the theory of finite strain, Bull. Seismol. Soc. Am. **29**, 463–479.

BIRCH, F. (1952) Elasticity and constitution of the Earth's interior, J. Geophys. Res. **57**, 227–286.

BIRCH, F. (1961) The velocity of compressional waves in rocks to 10 kilobars 2, J. Geophys. Res. **66**, 2199–2224.

BINNS, R. A. (1970) (Mg,Fe)$_2$SiO$_4$ spinel in a meteorite, Phys. Earth Planet. Interiors **3**, 156–160.

BINNS, R. A., R. J. DAVIS and S. B. J. REED (1969) Ringwoodite, natural (MgFe)$_2$SiO$_4$ spinel in the Tenham meteorite, Nature **221**, 943–944.

BLASSE, G. (1963) Crystal structures of some compounds of the type LiMe^{3+}Me^{4+}O$_4$ and LiMe^{2+}Me^{5+}O$_4$, J. Inorg. Nucl. Chem. **25**, 230–231.

BOLT, B. A., M. O'NEILL and A. QAMAR (1968) Seismic waves near 110°: is structure in core or upper mantle responsible? Geophys. J. **16**, 475–487.

BOYD, F. R. and J. L. ENGLAND (1960) Minerals of the mantle, Carnegie Inst. Wash. Yearbook **59**, 48–52.

CLARK, S. P. and A. E. RINGWOOD (1964) Density distribution and constitution of the mantle, Rev. Geophys. **2**, 35–88.

DACHILLE, F. and R. ROY (1960) High pressure studies of the system Mg$_2$GeO$_4$–Mg$_2$SiO$_4$ with special reference to the olivine–spinel transition, Am. J. Sci. **258**, 225–246.

EVERNDEN, J. (1958) Finite strain theory and the earth's interior, Geophys. J. **1**, 1–8.

FUJISAWA, H. (1968) Temperature and discontinuities in the transition layer within the earth's mantle: Geophysical application of the olivine–spinel transition in the Mg$_2$SiO$_4$–Fe$_2$SiO$_4$ system, J. Geophys. Res. **73**, 3281–3294.

GOLDSCHMIDT, V. M. (1931) Zur Kristallchemie des Germaniums, Nachr. Akad. Wiss. Göttingen Math. Physik. Kl. **1** no. 2, 184–190.

GRIGGS, D. (1954) The earth's mantle: discussion, Trans. Am. Geophys. Union **35**, 93–96.

GRIGGS, D. T. and G. C. KENNEDY (1956) A simple apparatus for high pressures and temperatures, Am. J. Sci. **254**, 722–735.

GUTENBERG, B. (1959) *Physics of the Earth's interior* 1, Intern. Geophys. Series, J. V. Mieghem ed. (Academic Press, New York).

HALES, A. L., J. CLEARY, H. DOYLE, R. GREEN and J. ROBERTS (1968) P-wave station anomalies and the structure of the upper mantle, J. Geophys. Res. **73**, 3885–3896.

JEFFEREY, R. N., J. D. BARNETT, H. B. VANFLEET and H. T. HALL (1966) Pressure calibration to 100 kb based upon the compression of NaCl, J. Appl. Phys. **37**, 3172–3180.

JEFFREYS, H. (1937) On the materials and density of the earth's crust, Monthly Notices Roy. Astron. Soc. Geophys. Suppl. **4**, 50–61.

JOHNSON, L. (1967) Array measurements of P velocities in the upper mantle, J. Geophys. Res. **72**, 6309–6325.

JULIAN, B. R. and D. L. ANDERSON (1968) Bull. Seismol. Soc. Am. **58**, 339–366.

KANAMORI, H. (1967) Upper mantle structure from apparent velocities of P waves recorded at Wakayama micro-earthquake observatory (1967) Bull. Earthquake Res. Inst. Tokyo Univ. **45**, 657–678.

KAWAI, N. (1966) A static high-pressure apparatus with tapering multi-pistons forming a sphere I., Proc. Japan. Acad. **42**, 385–388.

KAWAI, N., S. ENDO and K. ITO (1970) Split sphere high pressure vessel and phase equilibrium relation in the system Mg_2SiO_4–Fe_2SiO_4, Phys. Earth Planet. Interiors **3**, 182–185.

KAWAI, N., S. ENDOH and S. SAKATA (1966) Synthesis of Mg_2SiO_4 with spinel structure. Proc. Japan. Acad. **42**, 626–628.

MIKI, H. (1955) Is the C-layer (413–984 km) inhomogeneous? J. Phys. Earth. (Tokyo) **3**, 1–6.

MINOMURA, S., K. ITO and B. OKAI (1964) Pressure and temperature measurements in Drickhamer's resistance cell up to 161 kb and 4000 °C, Am. Soc. Mech. Eng. Symp. High pressure technology ASME Publ. **64-WA/PT6**, 1–4.

MORIMOTO, N., S. AKIMOTO, K. KOTO and M. TOKONAMI (1970) Crystal structures of high pressure modifications of Mn_2GeO_4 and Co_2SiO_4, Phys. Earth Planet. Interiors **3**, 161–165.

MOORE, P. B. and J. V. SMITH (1969) High pressure modification of Mg_2SiO_4: Crystal structure and crystallochemical and geophysical implications, Nature **221**, 653–655.

MOORE, P. B. and J. V. SMITH (1970) Crystal structure of β-Mg_2SiO_4: crystal-chemical and geophysical implications, Phys. Earth Planet. Interiors **3**, 166–177.

NIAZI, M and D. L. ANDERSON (1965) Upper mantle structure of western North America from apparent velocities of P waves, J. Geophys. Res. **70**, 4633–4640.

RINGWOOD, A. E. (1956) The olivine–spinel transition in the earth's mantle, Nature **178**, 1303–1304.

RINGWOOD, A. E. (1958a) The constitution of the mantle I; Thermodynamics of the olivine–spinel transition, Geochim. Cosmochim. Acta **13**, 303–321.

RINGWOOD, A. E. (1958b) The constitution of the mantle II; Further data on the olivine–spinel transition. Geochim. Cosmochim. Acta **15**, 18–29.

RINGWOOD, A. E. (1962) Prediction and confirmation of olivine–spinel transition in Ni_2SiO_4, Geochim. Cosmochim. Acta **26**, 457–469.

RINGWOOD, A. E. 1963) Olivine–spinel transformation in cobalt orthosilicate, Nature **198**, 79–80.

RINGWOOD, A. E. (1966a) The chemical composition and origin of the earth, in: P. Hurley ed., *Advances in Earth Science* (M.I.T. Press, Boston) 287–356.

RINGWOOD, A. E. (1966b) Mineralogy of the mantle, in: P. Hurley, ed. *Advances in Earth Science* (M.I.T. Press, Boston) 357–399.

RINGWOOD, A. E.(1967) The pyroxene-garnet transformation in the earth's mantle, Earth Planet. Sci. Letters **2**, 255–263.

RINGWOOD, A. E. (1968) High pressure transformations in A_2BO_4 compounds (abstract), Trans. Am. Geophys. Union **49**, 355.

RINGWOOD, A. E. (1969a) Composition and evolution of the upper mantle, in: *The Earths Crust and Upper Mantle*, Am. Geophys. Union Monograph **13**, 1–17.

RINGWOOD, A. E. (1969b) Phase transformations in the mantle, Earth Planet. Sci. Letters **5**, 401–412.

RINGWOOD, A. E. and A. MAJOR (1966a) Synthesis of Mg_2SiO_4–Fe_2SiO_4 solid solutions, Earth Planet. Sci. Letters **1**, 241–245.

RINGWOOD, A. E. and A. MAJOR (1966b) Some high-pressure transformations in olivines and pyroxenes, J. Geophys. Res. **71**, 4448–4449.

RINGWOOD, A. E. and A. MAJOR (1967) High pressure reconnaissance investigations in the system Mg_2SiO_4–MgO–H_2O, Earth Planet. Sci. Letters **2**, 130–133.

RINGWOOD, A. E. and A. MAJOR (1968) Apparatus for phase transformation studies at high pressures and temperatures, Phys. Earth Planet. Interiors **1**, 164–168.

RINGWOOD, A. E. and A. F. REID (1968) High pressure transformations of spinels I, Earth Planet. Sci. Letters **5**, 245–250.

RINGWOOD, A. E. and A. F. REID (1969) New high pressure olivine–spinel transformations, in press.

RINGWOOD, A. E. and M. SEABROOK (1962) Olivine-spinel equilibria at high pressure in the system Ni_2GeO_4–Mg_2SiO_4, J. Geophys. Res. **67**, 1975–1985.

SCLAR, C. B. and L. C. CARRISON (1966) High pressure synthesis of a spinel on the join M_2SiO_4–Fe_2SiO_4 (abstract), Trans. Am. Geophys. Union **41**, 207.

SCLAR, C. B., L. C. CARRISON and O. M. STEWART (1967) High-pressure synthesis of a new hydroxylated pyroxene in the system MgO–SiO_2–H_2O (abstract), Trans. Am. Geophys. Union **48**, 226.

SHIMAZU, Y. (1958) A chemical phase transition hypothesis and the origin of the C-layer within the mantle of the earth, J. Earth Sci. Nagoya Univ. **6**, 12–30.

VERHOOGEN, J. (1954) Elasticity of olivine and constitution of the earth's mantle, J. Geophys. Res. **58**, 337–346.

WADA, T. (1960) On the physical properties within the B-layer deduced from olivine-model and on the possibility of polymorphic transition from olivine to spinel at the 20° discontinuity, Bull. Disaster Prevent. Inst. Kyoto Univ. **37**, 1–20.

PHASE TRANSFORMATIONS AND THE CONSTITUTION OF THE MANTLE

A. E. RINGWOOD

Department of Geophysics and Geochemistry, Australian National University, Canberra

This paper presents a review of the constitution of the mantle between depths of 200 and 1200 km in the light of recent high pressure experimental investigations on silicates and on their germanate analogues. These studies indicate that the rapid increase of seismic velocity around 350–450 km is caused mainly by the transformation of olivine into a spinel-like structure and by the transformation of pyroxenes into a new kind of garnet structure characterised by octahedral coordination of a substantial proportion of silicon atoms. Current understanding of the constitution of the mantle below 600 km rests heavily upon interpretation of phase transformations in germanate analogue systems, upon solid solubility relationships of silicates in high pressure germanate phases, and upon shock wave investigations. These show that the Ca-rich component of garnets is likely to transform to the perovskite structure (12% increase in density) whereas Mg-rich garnets probably transform to the ilmenite structure (8% increase in density). A systematic study of transformations in spinels leads to the conclusion that the "post-spinel" transformation of Mg_2SiO_4 will probably be to a strontium-plumbate structure or to a mixture of $MgSiO_3$ (ilmenite) plus MgO. The transformations of some aluminosilicates and aluminogermanates are also discussed.

The transformations of "spinels" and garnets to new phases possessing strontium plumbate, ilmenite and perovskite structures probably occur around depths of 600–700 km and are responsible for an inferred major seismic discontinuity in this region. The densities and elastic properties of these high-pressure phases closely approach those of an isochemical mixture of oxides $(MgO + FeO + CaO + Al_2O_3 + SiO_2$ (as stishovite)). At greater depths, a further set of transformations to new phases may occur, causing the "zero-pressure density" of the mantle to attain values about 5% higher than an isochemical mixture of oxides. These may involve the transformation of Mg_2SiO_4 to the calcium ferrite or K_2NiF_4 structures, of $MgSiO_3$ to the perovskite structure, and of $MgAl_2O_4$ to the calcium ferrite structure. The phase transformations which have been directly observed or inferred, provide a quantitative explanation of the seismic velocity and density distribution between 200 and 1200 km. An increase of iron content with depth is not necessitated by presently available data.

CONTENTS

1. Introduction — 110
2. Study of phase transitions in mantle minerals by indirect methods — 111
3. High pressure transformations in A_2BO_4 compounds — 115
   - 3.1. Further transformation of $A_2^{2+}B^{4+}O_4$ spinels and beta phases — 115
     - 3.1.1. Transformation into single dense phases — 115
     - 3.1.2. Disproportionation of spinels into a binary ABO_3 compound plus an oxide AO — 116
     - 3.1.3. Complete disproportionation into constituent simple oxides — 117
     - 3.1.4. Discussion — 117
   - 3.2. Transformations in $A^{2+}B_2^{3+}O_4$ spinels — 118
   - 3.3. Further transformations in olivines — 118
   - 3.4. Possible transformations of Mg_2SiO_4 to structures denser than isochemical mixtures of MgO plus SiO_2 — 119
4. High pressure hydrated magnesium silicates — 120
5. High pressure transformations in ABO_3 compounds — 120
   - 5.1. Disproportionation of pyroxenes into spinel + rutile structures — 121
   - 5.2. The pyroxene–ilmenite transformation — 122
   - 5.3. The pyroxene–garnet transformation — 124
   - 5.4. Transformation of $MgSiO_3$ — 128
   - 5.5. The garnet–ilmenite transformation — 129
   - 5.6. The garnet–perovskite transformation — 130
6. Transformations in alkali aluminosilicates and aluminogermanates — 132
   - 6.1. The feldspar–hollandite transformation — 132
   - 6.2. Transformation of nepheline and jadeite — 134
7. Mineralogical constitution of the mantle — 134
   - 7.1. Chemical composition — 134
   - 7.2. Seismic velocity distributions — 138
   - 7.3. Mineralogy in a pyrolite mantle as a function of depth — 139
     - 7.3.1. The region between depths of approximately 150–350 km — 139
     - 7.3.2. Seismic discontinuity and high velocity gradients between approximately 360 and 420 km — 140

7.3.3. The region between 420 and 600 km   140
7.3.4. The 650 km "discontinuity"   141
7.3.5. The region between 700 and 1050 km   141
7.3.6. Phase transitions in the lower mantle 142
7.4. Magnitude of velocity changes at seismic discontinuities 142
7.5. Density of the lower mantle   144
7.6. Elasticity of the lower mantle   148

8. Conclusions   151

References   152

1. Introduction

The mantle may be divided into three major regions on the basis of the seismic velocity-depth distributions (fig. 1). In the *upper mantle*, extending down to about 400 km, velocity gradients are generally small, except possibly in the vicinity of the low-velocity zone. In the *transition zone*, between about 400 and 1050 km, velocity gradients are high on the average. Recent observations (D. L. ANDERSON and TOKSOZ, 1963; NIAZI and D. L. ANDERSON, 1965; JOHNSON, 1967; KANAMORI, 1967; JULIAN and D. L. ANDERSON, 1968; ARCHAMBEAU et al., 1969; HALES et al., 1968) show that most of the velocity increase in this region is concentrated in two or three narrow zones around 400, 650 and (perhaps) 1050 km. The velocity gradients in these zones are so high that they are often regarded as representing seismic discontinuities. Below 1050 km and extending to the core at 2900 km lies the *lower mantle*, where velocity gradients are relatively small and uniform. This distribution of velocities clearly reflects a corresponding variation of other important physical properties and must ultimately be explained in terms of the nature and properties of the mineral phases which are stable in the various regions of the mantle.

The first clue to the nature of the transition zone arose from the classical studies on the density of the mantle by BULLEN (1936) who demonstrated that this region is chemically inhomogeneous. It could not be established whether the inhomogeneity was due to chemical changes or phase changes. BERNAL (1936) and JEFFREYS (1937) suggested that the beginning of the transition zone (the so called "20 degree discontinuity") might be caused by a phase transformation of olivine to the spinel structure. A detailed study of the nature of the inhomogeneity was carried out by BIRCH (1939, 1952) using an equation of state based upon finite strain theory. Birch concluded that the properties of the upper mantle were consistent with this region being composed of familiar minerals such as olivines, pyroxenes and garnets. However the elastic properties of

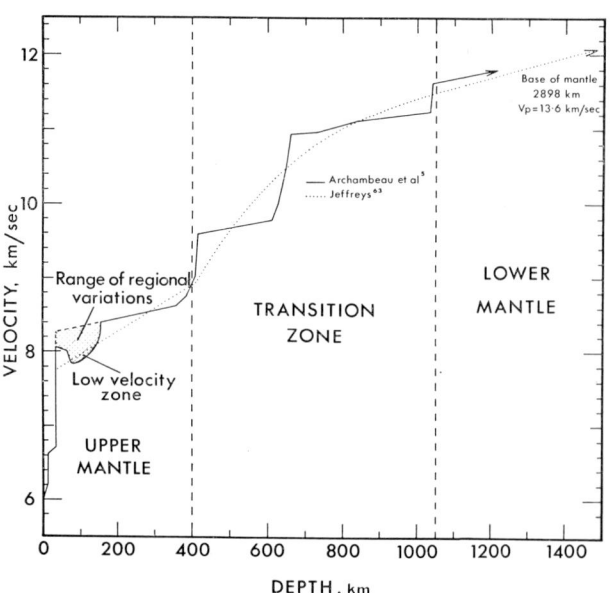

Fig. 1. Distribution of seismic P wave velocities in the outer 1200 km of the mantle according to JEFFREYS (1937) and ARCHAMBEAU et al. (1969). The latter distribution is for the Bilby NE profile in the USA. Several other groups (refs. see text) have recently reported generally similar velocity distributions in other regions.

the lower mantle were very different and resembled those possessed by relatively closely packed oxides such as corundum, periclase, rutile and spinel. Birch proposed that the transition zone was characterized by a series of major phase transformations resulting from the instability of olivines, pyroxenes and garnets at high pressure. He argued that these minerals transformed into a new assemblage of close-packed polymorphs, that the transformations were complete by about 1000 km, and that between 1000 km and 2900 km, no further transformations occurred, this region being essentially homogeneous.

Since Birch's hypothesis was based upon an equation of state which was not universally accepted (VERHOOGEN, 1953; GRIGGS, 1954; EVERNDEN, 1958) experimental verification was essential. It would be necessary to demonstrate that olivines, pyroxenes and garnets were indeed unstable at high pressure, and would transform to new polymorphs at P–T conditions equivalent to those between 400 and 1000 km, and further, that the elastic properties and densities of these new polymorphs were capable of explaining quantitatively the inferred properties of the lower mantle. The difficulty about testing the hypothesis experimentally was that until 1963, available static high pressure–temperature apparatus was incapable of reproducing the P–T conditions in the earth at depths greater than about 300 km. Accordingly, in the preceding period it was necessary to employ *indirect* experimental methods to investigate the possibility of phase transitions in mantle silicates.

2. Study of phase transitions in mantle minerals by indirect methods

The indirect methods make use of comparative crystal chemistry, thermodynamics, and particularly, upon the study of germanate isotypes of silicates. Crystal chemical relationships between germanates and silicates were first elucidated by GOLDSCHMIDT (1931). Both silicon and germanium readily form tetravalent ions which possess similar outer electronic structures and radii (Si^{4+} 0.42 Å, Ge^{4+} 0.49 Å [AHRENS (1952) personal communication]). Accordingly, the crystal chemistry of silicates is very closely related to that of germanates, particularly for oxy-compounds. Corresponding silicates and germanates are usually isostructural with one another and capable of forming continuous solid solutions. Over 60 examples are known of complex silicates which possess isostructural germanate analogues, and many more remain to be discovered. In general terms, it appears that if a germanate with a new structure should be synthesized, there would be a reasonable probability that under some appropriate P–T conditions a corresponding isostructural silicate would be stable.

A second important relationship between germanates and silicates has also emerged: it appears that germanates often behave as high pressure models for the corresponding silicates. If a germanate is found to

TABLE 1

Comparative stabilities of isostructural germanate and silicate phases

Structure type	Germanate compound	Stability conditions	Silicate compound	Stability conditions	Ref.
Rutile	GeO_2	$P = 0$ $T < 1007$ °C	SiO_2	$P > 100$ kb $T = 1000$ °C	1, 2, 3
Garnet	$Ca_3Al_2Ge_3O_{12}$	$P = 0$ $T \approx 1000$ °C	$Ca_3Al_2Si_3O_{12}$	$P > 15$ kb $T = 800$–1000 °C	4, 5
Spinel	Ni_2GeO_4	$P = 0$ $T = 650$ °C	Ni_2SiO_4	$P > 18$ kb $T = 650$ °C	6
Spinel	Co_2GeO_4	$P = 0$ $T = 700$ °C	Co_2SiO_4	$P > 70$ kb $T = 700$ °C	7
Spinel	Fe_2GeO_4	$P = 0$ $T = 600$ °C	Fe_2SiO_4	$P > 38$ kb $T = 600$ °C	8
Spinel	$(Mg_{0.8}Fe_{0.2})GeO_4$	$P = 0$ $T = 800$ °C	$(Mg_{0.8}Fe_{0.2})SiO_4$	$P > 110$ kb $T = 800$ °C	9, 10
Spinel	$LiAlGeO_4$	$P = 10$ kb $T \approx 1000$ °C	$LiAlSiO_4$	$P > 120$ kb $T = 1000$ °C	11, 12
Kyanite	Al_2GeO_5	$P = 15$ kb $T = 1200$ °C	Al_2SiO_5	$P > 20$ kb $T = 1200$ °C	13, 14
Scheelite	$ZrGeO_4$	$P = 0$ $T = 1000$ °C	$ZrSiO_4$	$P > 100$ kb $T = 1000$ °C	15, 16
Hollandite	$KAlGe_3O_{12}$	$P = 35$ kb $T = 1100$ °C	$KAlSi_3O_8$	$P > 100$ kb $T = 1000$ °C	17, 18
Jadeite	$NaAlGe_2O_6$	$P > 13$ kb $T = 1000$ °C	$NaAlSi_2O_6$	$P > 20$ kb $T = 1000$ °C	19, 20

1. ROBBINS and LEVIN (1959)
2. STISHOV and POPOVA (1961)
3. AKIMOTO and SYONO (1968)
4. TAUBER *et al.* (1958)
5. PISTORIUS and KENNEDY (1960)
6. RINGWOOD (1962a)
7. RINGWOOD (1963)
8. RINGWOOD (1958b)
9. DACHILLE and ROY (1960)
10. RINGWOOD and MAJOR (1970)
11. ROOYMANS (1967)
12. RINGWOOD and REID (1968b)
13. CLARK (1961)
14. RINGWOOD and REID (1968b)
15. WYCKOFF (1965)
16. REID and RINGWOOD (1969c)
17. KUME *et al.* (1966)
18. RINGWOOD *et al.* (1967a)
19. ROBERTSON *et al.* (1957)
20. RINGWOOD, unpublished observations

display a given phase transformation at a particular pressure, the corresponding silicate often displays the same transformation but at a much higher pressure (table 1). The reverse of this relationship has not been observed. A similar relationship is found for disproportionation reactions (table 2). Alternatively a germanate may crystallize at atmospheric pressure in a structure which is only attained by the corresponding silicate at very high pressure.

TABLE 2

Comparison of germanate and silicate disproportionation reactions

		P (kb)	T (°C)	Ref.
$2FeGeO_3$	$\rightarrow Fe_2GeO_4 + GeO_2$	10	700	1
$2FeSiO_3$	$\rightarrow Fe_2SiO_4 + SiO_2$	100	1000	1
(pyroxene)	$\rightarrow$ (spinel) + (rutile)			
$2CoGeO_3$	$\rightarrow Co_2GeO_4 + GeO_2$	10	700	1
$2CoSiO_3$	$\rightarrow Co_2SiO_4 + SiO_2$	100	1000	1
(pyroxene)	$\rightarrow$ (spinel) + (rutile)			
$NaAlGe_3O_8$	$\rightarrow NaAlGe_2O_6 + GeO_2$	15	1100	2
$NaAlSi_3O_8$	$\rightarrow NaAlSi_2O_6 + SiO_2$	28	1100	3
(albite)	$\rightarrow$ (jadeite) + (rutile, quartz)			
$3KAlGe_2O_6$	$\rightarrow 2KAlGe_3O_8 + KAlO_2$	90	1000	2
$3KAlSi_2O_6$	$\rightarrow 2KAlSi_3O_8 + KAlO_2$	100	1000	2

1. This paper.
2. Ringwood et al. (1967a, b).
3. Birch and le Comte (1960).

This behaviour can be understood in terms of crystal chemical principles. The structures of ionic compounds are largely governed by ionic radius ratios of constituent ions. This applies particularly to oxide compounds in which the critical parameters are usually the radius ratios between the small cations and the large oxygen anions. When a silicate or germanate is subjected to high pressure, the large oxygen anions (radius 1.40 Å) tend to contract* relatively more than the small Si^{4+} and Ge^{4+} cations, hence the radius ratios R_{Si}/R_O and R_{Ge}/R_O increase. Transformation into a new high pressure phase occurs when these radius ratios attain some critical value. Since the zero-pressure radius of Ge^{4+} (0.48 Å) is already slightly larger than that of Si^{4+} (0.42 Å), germanates require smaller pressures to achieve the critical radius ratios required for given transitions than the corresponding silicates do. Alternatively, because of their initially higher radius ratios, germanates may crystallize at zero pressure in a structure which is only attained by the silicate at high pressure.

For these reasons, the study of germanates as high pressure models of silicates offers us the possibility of obtaining useful information about phase transformations which may occur in silicates at pressures beyond the range of currently available experimental techniques. There are several ways in which these relationships may be taken advantage of. The first is purely qualitative. A systematic study at high pressure is made of the stabilities of germanate isotypes of the principal mantle minerals. In nearly all cases investigated in my laboratory, the germanates have been found to transform to denser structures. From the considerations advanced above, there is a fairly high probability that silicates will transform to these structures at much higher pressures. This is a most valuable aid in exploration. Furthermore, from analogy with germanates and from crystallographic considerations, various important properties of the high pressure silicates, particularly density and elasticity, can be estimated, and thus used to aid in the interpretation of the properties of the deep mantle.

A note of caution should be injected here. Whilst the correspondence between germanates and silicates is very close, it is not of universal applicability, and a significant number of cases is known where a silicate does not possess a germanate analogue (e.g. coesite) or vice versa. Accordingly the above relationships must be used with due regard to the probabilities involved.

---

* The contraction is a more complex phenomenon than simple differential compression. It is probably connected both with deformation of the ions and with pressure-induced changes in the nature of the chemical bonding. The polarisability of the O^{-2} ion is 3.1×10^{-24} cm^3 compared to 0.43×10^{-24} cm^3 for Si^{4+} (Born and Heisenberg, 1924). Thus the oxygen ion is much more readily deformed into non-spherical configurations, resulting in a reduction of its "effective radius" and an increase of the "effective radius ratio" of cation to oxygen anion. It has also been pointed out that compression may cause an increase in the covalent component of the bond in oxides (Magnitsky and Kalinin, 1959; Wada, 1960; Drickhamer, 1963). This is manifested by an increased electron density between ions and a contraction of bond length, e.g. the ionic Mg^{+2}–O^{-2} bond length is 2.10 Å compared to 1.95 Å for the covalent Mg–O bond (Magnitsky and Kalinin, 1959). More importantly, the position of the electron-density minimum which marks the nominal point of contact of the ions or atoms moves closer to the oxygen atom as the covalent contribution to the bond increases. This causes an increase in the cation to anion radius ratio. Thus the Mg^{2+}/O^{2-} ionic radius ratio is 0.5 compared to 2.7 for the corresponding atomic radius ratio. Clearly, a change in the nature of the bond has a large influence on the effective radius ratio.

These methods do not offer a comparable substitute for the direct investigation of stabilities of silicates at high pressure. Nevertheless, when this is not possible because of technical difficulties, the indirect methods are of considerable value. At present, it is possible to reproduce the P–T conditions in the mantle at depths down to about 600 km, so that stabilities of silicate phases can be investigated directly over the corresponding pressure range. However, below 600 km, our understanding of the probable nature of the mineral phases present depends heavily upon studies of germanates and their relationships with silicates.

It is also possible to take advantage of solid solubility relationships between silicates and germanates in systems in which the germanate forms a high pressure phase but not the silicate. Thus, by measuring the solid solubility of a silicate in a high pressure germanate structure over an experimentally accessible range of pressures, the pressure required for the transition of the silicate to the new structure may be estimated by extrapolation. Examples of the use of this technique were cited in the accompanying paper (RINGWOOD and MAJOR, 1970). Finally, the solid solubility relationships between silicates and germanates provide data on which thermodynamic calculations of the pressure required for a given transformation in a silicate can be made. Where a low pressure silicate phase (A) with a relatively open structure displays substantial solid solubility in a closer packed germanate phase (B), the pressure (P) required to transform the silicate into the structure possessed by the germanate is approximately

$$P = -\frac{RT}{\Delta v} \ln \frac{a_2}{a_1}$$

(RINGWOOD, 1958a), where a_1 represents the activity of the silicate component in phase B, a_2 represents the activity of the silicate component of phase A in equilibrium with B, and Δv is the difference in molar volume between the pure silicate in its low pressure structure (A) and its hypothetical high pressure structure (B). Because of the similarity between the germanium and silicon ions, it appears probable that germanate-silicate solid solutions will behave approximately ideally at elevated temperatures, so that the activities are obtained from the compositions of the equilibrium solid solutions which are measured directly. The use of this prediction method in a specific instance is discussed in the accompanying paper. More generally, the observation of a substantial solid solubility of a silicate in a closer packed germanate under high pressure conditions is an extremely valuable indicator as to the probable transformation mode and to the approximate pressure of transformation of the silicate. We will discuss several examples of the use of this relationship in later sections.

As experimental observations on high pressure transformations have expanded during recent years, it has become possible to make some wider crystal-chemical generalisations than those reached previously. An important result is the strong tendency of high pressure phases to crystallise in structures which are already known. Of about 70 high pressure transformations discovered in my laboratory, only 8 have been to structures which were unknown, or not closely related to known structures. It is likely that current structural investigations will reveal that some of these exceptions belong to known structure-types. Bearing in mind the large number and complexity of existing silicate and germanate structures, this is a fortunate circumstance which could not have been confidently predicted before the advent of high pressure experimentation. The probability that transformations deep in the mantle will involve presently known structure types greatly simplifies the interpretation of the mineralogy of this region. Furthermore, when transformations occur from one structure to another, a clearly defined trend emerges. Previously this was noted just for silicates and germanates, but it turns out to be much more general. Goldschmidt and his colleagues demonstrated the decisive role which the radius ratio $R_{\text{cation}}/R_{\text{oxygen}}$ plays in determining the structures of oxides. Thus, in ternary oxide compounds $A_xB_yO_z$ of given stoichiometry, when the radius ratios R_A/R_O and R_B/R_O are plotted against one another it is found that structures of a given type fall into well defined fields which can be interpreted in terms of simple geometric packing of anions around cations. When high pressure transformations occur, almost invariably, the new structure is found to be one which is characterized by higher cation to oxygen radius ratios than the original structure. This is to be understood in terms of the previous discussion on silicate-germanate relationships, whereby under pressure, the "effective radius" of the oxygen ions decrease more than those of the cations, so that the radius ratio of cation to anion increases. This systematic behaviour allows a broader use to be made of model relationships in interpreting the transformations in silicates and other compounds over very wide pressure ranges. Thus we can regard titanates and stannates (Ti^{4+} 0.68 Å, Sr^{4+} 0.71 Å) as indicating model structures which might ultimately be adopted by silicates at pressures beyond those at which silicates adopt germanate structures, or as providing high pressure models for germanates. For example, the compound $KAlTi_3O_8$ has a hollandite structure at atmospheric pressure, whereas $KAlGe_3O_8$ and $KAlSi_3O_8$ have felspar structures. $KAlGe_3O_8$ transforms to the hollandite structure at 35 kb, 1100 °C whereas $KAlSi_3O_8$ transforms to this structure at about 100 kb (RINGWOOD et al., 1967a). These considerations will be found helpful in proposing structures for mineral phases in the deepest regions of the mantle, where even the germanate models may be of restricted utility.

Summary of results obtained by indirect methods, 1956–63

As available apparatus was not capable of generating P–T conditions equivalent to deeper than about 300 km until 1963, the experimental testing of Birch's hypothesis during this period necessarily involved the use of indirect methods as discussed earlier. The results of this phase of investigation were reviewed by RINGWOOD (1966b). Studies of the system Ni_2GeO_4–Mg_2SiO_4 provided the first quantitative evidence that Mg_2SiO_4 olivine would be expected to transform to a spinel structure in the transition zone (RINGWOOD, 1956, 1958a). This was supported by the discovery that the silicate olivines Fe_2SiO_4, Co_2SiO_4 and Ni_2SiO_4 could be transformed to spinel structures at 20–70 kb and that the spinels were about 10% denser than the corresponding olivines (RINGWOOD, 1958b, 1963). The latter transformation was of interest since it was first predicted on the basis of a thermodynamic study of the system Ni_2GeO_4–Ni_2SiO_4, and subsequently discovered near the predicted pressure (RINGWOOD, 1962a). Fayalite Fe_2SiO_4 is a significant component of mantle olivine. The results further supported the probability that magnesia-rich olivines would transform at higher pressure. All of the germanate olivines and pyroxenes which were studied in the pressure range 0–90 kb were found to be unstable at high pressures and transformed to denser phases, suggesting strongly that the corresponding silicates would transform similarly at higher pressures (RINGWOOD, 1966b). Whenever quantitative estimates of the pressures required for the transformation of upper mantle silicates were made by the methods described above, the transformation pressures were found to be within the pressure range in the transition zone (RINGWOOD, 1958a, 1962b; RINGWOOD and SEABROOK 1962a). An important discovery was made by STISHOV and POPOVA (1961) during this period who showed that at about 100 kb (1600 °C) silica could be transformed to the rutile structure. (GeO_2 displays this same transformation at atmospheric pressure.) The new polymorph "stishovite" possessed a density of 4.28 g/cm^3. The demonstration that the coordination of silicon (like germanium) could change from 4 to 6 under pressure greatly extended the range of possible transformation structures for silicates.

These results provided powerful qualitative support for Birch's hypothesis. It was possible to construct a model series of transformations in mantle silicates based upon the above indirect evidence which yielded densities and elastic properties in good agreement with those inferred for the lower mantle (RINGWOOD, 1958c, 1962b; CLARK and RINGWOOD, 1964). Nevertheless, there remained a strong incentive to construct apparatus which could develop pressures and temperatures equivalent to those in the transition zone so that the predicted new silicate polymorphs could be synthesized directly and their properties measured, leading to an understanding of the fine structure of the mantle in this region. Also, the possibility of discovery of completely new and unpredicted phases remained. The indirect methods do not predict with certainty, and are inapplicable in cases (e.g. coesite) where a silicate does not have a germanate isotype. Also, it sometimes happens that the indirect methods indicate more than one possible transformation for a silicate at high pressures. Which of these transformations will actually be displayed can only be discovered by direct experiment.

An apparatus capable of developing pressures higher than 200 kb simultaneously with high temperatures was developed by RINGWOOD and MAJOR (1966c, 1968a) and is further described in the accompanying paper (RINGWOOD and MAJOR, 1970). This is equivalent to a depth of 600 km. Using this apparatus, many new phase transformations of importance in the mantle have been discovered in silicates, germanates and related compounds. Other workers have also reported the operation of apparatus capable of pressures in 100–200 kb range (AKIMOTO and IDA, 1966; KAWAI, 1966; BUNDY 1963) but the results of application of these apparati to silicate phase transformations have so far been rather limited.

Another comparatively recent development has been the application of shock wave techniques capable of developing transient pressures in the megabar range to materials of geophysical interest (MCQUEEN et al., 1967). This technique has revealed a number of phase transformations in silicates and has provided valuable information on the equations of state of the high pressure phases (AHRENS et al., 1969). Shock wave techniques have the advantage of generating much higher pressures than static methods but these are maintained only for a few microseconds. Consequently, phase transformations in silicates particularly, rarely if ever occur under equilibrium conditions. Furthermore, shock wave experiments in themselves do not provide direct information on the nature and structures of the high pressure phases. The static and dynamic (shock wave) experimental methods are complementary rather than competitive. An interpretation of transformations achieved under shock wave conditions by MCQUEEN et al. (1967) in terms of static transformations in silicates and model systems is given by AHRENS et al. (1969).

In the following sections, experimental results on phase transformations on silicates, germanates and related model systems are reviewed.

3. High pressure transformations in A_2BO_4 compounds

3.1. Further transformations of $A_2^{2+}B^{4+}O_4$ spinels and beta phases

The singular role of olivine in the upper mantle and the importance of transformations of olivine to the spinel and beta-phase structures in causing the seismic discontinuity near 400 km are discussed in detail in the accompanying paper (RINGWOOD and MAJOR, 1970). In both the spinel and beta phases, the Si^{4+} ions are tetrahedrally coordinated. The observation that at high pressures, the coordination of silicon in SiO_2 changes from 4 to 6 accompanied by a large increase in density (STISHOV and POPOVA, 1961) and the observations of many such coordination changes in germanate analogues, suggest that silicate spinels and beta phases are likely to transform to denser structures characterized by octahedral coordination of silicon (RINGWOOD, 1962b; STISHOV, 1962). This is confirmed by the results of shock wave experiments on $(MgFe)_2SiO_4$ olivines (MCQUEEN et al., 1967) at pressures in the megabar range. These experiments indicated that olivines had been shocked directly into modifications which considerably exceeded the densities anticipated for spinels. Under the non-equilibrium conditions which prevailed, the "spinel" field was apparently overshot, and the field of "post-spinel" transformation products was reached directly. WANG (1967) and AHRENS et al. (1969) have reduced the shock wave data to obtain the densities of the high pressure phases extrapolated to atmospheric pressure. It was found that these phases possessed densities very similar to those of isochemical mixed oxides: $MgO + FeO + SiO_2$ (as stishovite).

Direct attempts to synthesize the post spinel phase at pressures up to 200 kb were unsuccessful (RINGWOOD and MAJOR, 1970). Accordingly, studies of analogue compounds were indicated. Three major classes of transformations in oxide spinels and related compounds have been observed (table 3).

3.1.1. Transformations into single dense phases.
The olivine Mn_2GeO_4 is one such possible analogue. MORIMOTO et al. (1969) showed that Mn_2GeO_4 olivine transforms to a beta phase isostructural with β-Mg_2SiO_4 at high pressure. Previously, WADSLEY et al. (1968) demonstrated that at still higher pressures, Mn_2GeO_4 transforms to the strontium plumbate (Sr_2PbO_4) structure and suggested that Mn_2GeO_4 might represent a high pressure model for Mg_2SiO_4. The radius ratio Mn^{2+}/Ge^{4+} is very similar to that of Mg^{2+}/Si^{4+}. If, as discussed earlier, we regard the principal effect of pressure as shrinking the large oxygen anions relatively to the small cations, then it appears possible that $(MgFe)_2SiO_4$ solid solutions may be capable of crystallizing in this structure at high pressure.

RINGWOOD and REID (1968b, 1969) have investigated the transformation of a large number of spinels at high pressure in an attempt to further clarify the systematic crystal chemical and thermodynamic factors governing spinel transformations. Several examples of transformations into single dense phases were observed.

The compound $FeMnGeO_4$ possesses an olivine structure at atmospheric pressure. At 35 kb, 1100 °C it transformed to a spinel structure. In runs above 100 kb,

TABLE 3*

High pressure transformations in oxide spinels and closely related structures

	Ref.
(A) Transformation into single dense phases	
$\Delta \rho/\rho \approx 10\%$	
Mn$_2$GeO$_4$ ('beta' phase) transforms to Sr$_2$PbO$_4$ str.	1, 6
FeMnGeO$_4$ spinel ,, ,, ,, ,, ,,	2
Mn$_2$SnO$_4$,, ,, ,, ,, ,, ,,	3, 7
Mn$_3$O$_4$ (tetragonal spinel) transforms to calcium manganite structure	3, 4
(B) Disproportionation into oxide plus ilmenite structure	
$\Delta \rho/\rho \approx 8\%$	
A$_2$BO$_4$ (spinel) → ABO$_3$ (ilmenite) + AO (rocksalt)	
Mg$_2$TiO$_4$, Co$_2$TiO$_4$, Fe$_2$TiO$_4$	5
MgZnTiO$_4$	3
(C) Complete disproportionation into oxides	
$\Delta \rho/\rho \approx 10\%$	
A$_2^{2+}$B^{4+}O$_4$ (spinel) → 2AO (rocksalt) + BO$_2$ (rutile)	3
Mg$_2$SnO$_4$, Co$_2$SnO$_4$	
A$^{2+} \cdot$ B$_2^{3+}$O$_4$ (spinel) → AO (rocksalt) + B$_2$O$_3$ (cor.)	
MnAl$_2$O$_4$, FeAl$_2$O$_4$, NiAl$_2$O$_4$, CoAl$_2$O$_4$	3

1. Wadsley et al. (1968)   5. Akimoto and Syono (1967)
2. Ringwood and Reid (1969)   6. Morimoto et al. (1970)
3. Ringwood and Reid (1968b)   7. Syono et al. (1969)
4. Reid and Ringwood (1969)

the spinel transformed to the strontium plumbate structure. This compound may also behave as a model for (MgFe)$_2$SiO$_4$. The spinel Mn$_2$SnO$_4$ also transformed into the Sr$_2$PbO$_4$ structure. Efforts to transform Mg$_2$GeO$_4$, Fe$_2$GeO$_4$, Co$_2$GeO$_4$ and Ni$_2$GeO$_4$ spinels were unsuccessful.

In the strontium plumbate modification of Mn$_2$GeO$_4$, the Mn^{2+} ions are surrounded by six oxygens at the corners of a trigonal prism, whilst the germanium is surrounded by 6 oxygens at the corners of an octahedron. (The trigonal prism coordination for Mg^{2+} is also known in MgSc$_2$O$_4$ so that this arrangement is stereochemically possible.) All known strontium plum-

* In a preliminary investigation (Ringwood and Reid, 1968b), transformations were also observed in Mn$_2$TiO$_4$, Zn$_2$SnO$_4$ and Zn$_2$TiO$_4$. The powder patterns of the high pressure phases showed a qualitative resemblance to those of Sr$_2$PbO$_4$ structures and it was suggested that the new phases might be related to this structure. Subsequent investigations have not confirmed this interpretation.

bate isotypes are formed between end-members possessing rocksalt and rutile structures. It is instructive to compare the molar volumes of Sr$_2$PbO$_4$ structures with those of their isochemical rocksalt and rutile type end-members (table 4). It is seen that all members of this

TABLE 4

Molar volumes of compounds possessing Sr$_2$PbO$_4$ structure* compared to the volumes of the isochemical rocksalt plus rutile-type mixtures

Compound	Molar volume V (cm^3)	Isochemical oxide volume v (cm^3)	$\frac{\Delta v}{v}$ (%)**
Sr$_2$PbO$_4$	65.23	66.40	−1.7
Ca$_2$PbO$_4$	57.83	58.54	1.2
Ca$_2$SnO$_4$	54.75	55.07	−0.5
Cd$_2$SnO$_4$	52.89	52.73	0.3
Mn$_2$GeO$_4$	43.32	43.10	0.5

* After Ringwood and Reid (1968a)
** $\Delta v = V - v$.

group are characterized by molar volumes which are practically identical with the mixed oxides. This is an interesting and important property in view of the observation previously noted that (MgFe)$_2$SiO$_4$ olivines transform under shock to a phase also possessing a molar volume equal to that of the isochemical mixed oxides. Furthermore, unpublished investigations by Ringwood and Reid have shown that a substantial amount of Mg^{2+} and Si^{4+} enters into solid solution in Mn$_2$GeO$_4$ at very high pressures. Thermodynamic calculations of the type earlier discussed suggest that Mg$_2$SiO$_4$ would transform from the beta structure to the strontium plumbate structure at pressures of 200–300 kb. It may be concluded then, that this structure is very plausible for the post-spinel phase of Mg$_2$SiO$_4$.

3.1.2. *Disproportionation of spinels into a binary ABO$_3$ compound plus an oxide AO.* A further possibility, originally noticed by Birch (1952) and further considered by Ringwood (1962b, 1966b) is that Mg$_2$SiO$_4$ might disproportionate in the deep mantle according to the reaction

$$\underset{\text{(spinel)}}{A_2BO_4} \rightarrow \underset{\text{(ilmenite str.)}}{ABO_3} + \underset{\text{(rocksalt str.)}}{AO}$$

Studies of the MgGeO$_3$–MgSiO$_3$ system (section 5.2) have shown that an ilmenite form of MgSiO$_3$ will probably become stable between 200 and 300 kb, so that the above reaction must be seriously considered for Mg_2SiO_4.

Experimental examples of these equilibria have been provided by AKIMOTO and SYONO (1967) who showed that the spinels Mg_2TiO_4, Co_2TiO_4 and Fe_2TiO_4 disproportionated into ilmenite phases plus oxides at pressures of a few tens of kilobars. RINGWOOD and REID (1968b) showed that $MgZnTiO_4$ spinel transformed similarly. The densities of the ilmenite-oxide mixtures are usually 1 to 4% smaller than the mean densities of the isochemical component oxides.

3.1.3. Complete disproportionation into constituent simple oxides.

It has been suggested by several workers (e.g. BIRCH, 1952, 1964; MACDONALD, 1956; SHIMAZU, 1958; MACQUEEN et al., 1964; AHRENS and SYONO, 1967) that under the high pressures existing in the lower mantle, silicate minerals may actually disproportionate into physical mixtures of their constituent oxides; principally SiO_2 (as stishovite), MgO, FeO, CaO and Al_2O_3. Such reactions are theoretically possible if the density of the oxide mixture is greater than that of the mineral or compound. The free energy change ΔG in these reactions is equal to $\Delta G_0 + \int_0^P \Delta v \, dP$ where ΔG_0 is the free energy of formation of the compound from its constituent oxides at atmospheric pressure and Δv is the difference in molar volume at pressure P between the compound and its isochemical oxide mixture. The condition for disproportionation is for ΔG to be zero, hence $\int_0^P \Delta v \, dP = -\Delta G_0$. If we consider Δv to be constant (a fair approximation in many cases) then $P \Delta v = -\Delta G_0$ and hence $P = -\Delta G_0/\Delta v$. Many calculations of the pressures at which Mg_2SiO_4 "spinel" might be expected to disproportionate into oxide components have been carried out (RINGWOOD, 1962; MACDONALD, 1962; STISHOV, 1962; AHRENS and SYONO, 1967; D. L. ANDERSON, 1967). The calculated pressures are in the vicinity of 300 kb at 1000–2000 °C.

Examples of these transformations have been discovered by RINGWOOD and REID (1968b) who observed that the spinels Mg_2SnO_4 and Co_2SnO_4 disproportionated into oxide mixtures at high pressures. The $A^{2+}B_2^{3+}O_4$ spinels ($MnAl_2O_4$, $FeAl_2O_4$, $NiAl_2O_4$ and $CoAl_2O_4$) were similarly found to disproportionate into constituent oxides at high pressures. Another compound which disproportionated into oxides was Al_2GeO_5 (kyanite).

3.1.4. Discussion.

We have considered three possible modes of transformation for spinels and related structures. All transformations lead to densities similar to the mixed oxides and involve a change of coordination of one cation from fourfold in spinel to sixfold in the post-spinel phases.

The bearing of these observations on the probable transformation mode of Mg_2SiO_4 has been discussed by RINGWOOD and REID (1968b). The nature of the post-spinel transformation is strongly influenced by (i) the free energy ΔG_0 of formation of the spinel from constituent oxides and (ii) by stereochemical factors controlling the stability of dense binary phases such as ABO_3 ilmenite and A_2BO_4 (strontium plumbate). It is readily seen that where ΔG_0 is relatively small, a correspondingly small pressure may be sufficient to cause dissociation according to the relationship $P = -\Delta G_0/\Delta v$. On the other hand, where ΔG_0 is large (i.e. a strong compound-forming tendency exists between the oxides), a correspondingly higher pressure will be required to cause dissociation into oxides, and transformation, if it occurs, is more likely to result in the formation of a new single phase (providing that this is permitted by stereochemical considerations).

The experimental results are generally in accord with these expectations. The spinels which disproportionated into oxides, e.g. some of the stannates and $A^{2+}B_2^{3+}O_4$ spinels, were characterized by relatively low ΔG_0 values. On the other hand, none of the silicate and germanate spinels, possessing higher ΔG_0 values (RINGWOOD and REID, 1968b) were observed to disproportionate into simple oxides. Where transformations were observed, they were to isochemical denser phases possessing A_2BO_4 stoichiometry. The titanate spinels, which would be expected to possess ΔG_0 values intermediate between the stannates on the one hand, and the silicates and germanates on the other, tended to display an intermediate transformation behaviour, mostly disproportionating into mixtures of ABO_3 (ilmenite) + AO (rocksalt). The energy required for these transformations is much smaller than that for complete disproportionation into oxides.

In view of the strong compound forming tendency of MgO and SiO_2, as shown by the comparatively high free energy of formation of Mg_2SiO_4 and the evidence previously cited which makes it probable that dense Mg_2SiO_4 and $MgSiO_3$ compounds possessing strontium plumbate and ilmenite structures are stable under high pressures, it appears most unlikely that Mg_2SiO_4 will disproportionate completely into simple oxides at very high pressures. The evidence suggests that the most probable mode of transformation of Mg_2SiO_4 will be to the strontium plumbate structure. However, disproportionation into a mixture of $MgSiO_3$ (ilmenite) plus MgO remains a distinct possibility.

3.2. Transformations in $A^{2+}B_2^{3+}O_4$ spinels

Another important class of spinels include those which are formed between oxides of divalent elements (usually rocksalt structure) and trivalent elements (usually corundum structure). Transformations are known in these spinels and may be of importance in determining the behaviour of trivalent oxides in the lower mantle (section 7).

We have already mentioned the observed disproportionation of the spinels $FeAl_2O_4$, $CoAl_2O_4$, $MnAl_2O_4$ and $NiAl_2O_4$ into constituent oxides at high pressures (RINGWOOD and REID, 1968b). However at even higher pressures there is evidence that these spinels may be capable of transforming to structures which are substantially denser than the mixed oxides. Thus, MCQUEEN and MARSH (1966) observed transformations of $MgAl_2O_4$ and Fe_3O_4 under shock. AHRENS et al. (1969) estimated the zero-pressure density of the high pressure $MgAl_2O_4$ to be 4.19 g/cm³ compared to 3.86 g/cm³ for the density of an isochemical mixture of $MgO+Al_2O_3$. Similarly magnetite is shocked to a phase with an estimated zero pressure density of 6.3 g/cm³ compared to a density of 5.54 for a mixture of $FeO+Fe_2O_3$.

Possible high pressure structures for these spinels are calcium ferrite $CaFe_2^{3+}O_4$, or the closely related calcium titanite $CaTi_2^{3+}O_4$ and calcium manganite $CaMn^{3+}O_4$ (REID and RINGWOOD, 1969c). These structures are characteristically about 4–6% denser than the mixed oxides. In the calcium ferrite structure the calcium atoms are in eightfold coordination whereas the trivalent atoms are octahedrally coordinated.

REID and RINGWOOD (1969c) have shown that Mn_3O_4 which possesses a hausmanite structure (i.e. a tetragonal spinel) transforms to the $CaMn_2O_4$ structure at high pressure with a ten percent increase in density. This suggests that the transformation observed under shock wave conditions in Fe_3O_4 may be similar. The inferred zero pressure density for the high pressure phase of Fe_3O_4 is somewhat high for this transformation, but considering the uncertainties in deriving this zero pressure density from the shock data, the discrepancy may not be significant. The calcium ferrite structure would provide a satisfactory explanation of the density of shocked $MgAl_2O_4$. In view of the facts that $MgSc_2O_4$ has a calcium ferrite structure at atmospheric pressure (MÜLLER-BUSCHBAUM, 1966) and that $CaAl_2O_4$ transforms to the calcium ferrite structure at high pressure (REID and RINGWOOD, 1969c) it appears likely that the observed shock transformation of $MgAl_2O_4$ is to this or to a closely related structure. RINGWOOD and REID (1968b) have also observed that the spinels $CdCr_2O_4$, $CdFe_2O_4$ and $LiFeTiO_4$ transform to denser phases under high pressure. Structural investigations are in progress. Perhaps some of these phases will be found to be related to the calcium ferrite structure.

3.3. Further transformations in olivines

The olivine Ca_2GeO_4 transforms to the K_2NiF_4 structure at high pressure (RINGWOOD and REID, 1968a). This transformation is accompanied by a density increase of 25%. The K_2NiF_4 structure is commonly formed between end members possessing rocksalt and rutile structures, and is between 2 and 7% *denser* than isochemical mixtures of these end members – e.g. Ca_2MnO_4, Sr_2TiO_4, Sr_2SnO_4, Ba_2PbO_4 (table 5). Such compounds will therefore be stabilized relatively

TABLE 5

Molar volumes of compounds possessing K_2NiF_4 structures compared to the volumes of the isochemical rocksalt plus rutile-type mixtures*

Compound	Molar volume V (cm³)	Isochemical oxide volume v (cm³)	$\dfrac{\Delta v}{v}$ (%)**
Ba_2PbO_4	74.08	75.76	−2.2
Ba_2SnO_4	68.62	72.29	−5.1
Sr_2SnO_4	61.50	62.93	−2.3
Sr_2MoO_4	59.42	61.23	−3.0
Sr_2IrO_4	58.88	60.59	−2.8
Sr_2RuO_4	57.46	60.42	−4.9
Sr_2TiO_4	57.14	60.18	−4.9
Sr_2MnO_4	53.77	57.99	−7.3
Ca_2MnO_4	49.00	50.13	−2.3
Ca_2GeO_4	48.98	50.18	−2.4
Ca_2SiO_4	45.7***	47.55	−4***

* After RINGWOOD and REID (1968a).
** $\Delta v = V-v$.
*** Estimated on basis of average $\Delta v/v$.

to the isochemical oxide mixture by pressure. The K$_2$NiF$_4$ structure consists of alternate stacking of sub-units possessing rock-salt and perovskite structures. The large divalent cation is surrounded by 9 oxygen atoms whereas the small tetravalent cation is octahedrally coordinated. Most of the oxides capable of crystallizing in the K$_2$NiF$_4$ structure are also capable of crystallising in the perovskite structure when reacted in 1:1 molecular ratios. Perovskites are also substantially denser than their isochemical oxide mixtures (section 5.6). From the germanate–silicate relationships discussed previously, it appears likely that Ca$_2$SiO$_4$ would be capable of adopting the K$_2$NiF$_4$ structure under very high pressure. The possibility that Mg$_2$SiO$_4$ might ultimately be capable of adopting this structure is discussed in a later section.

In contrast to Ca$_2$GeO$_4$, the closely related olivine Cd$_2$GeO$_4$ disproportionates under high pressure (RINGWOOD and REID, 1968a)

$$\underset{\text{(olivine)}}{Cd_2GeO_4} \rightarrow \underset{\text{(perovskite)}}{CdGeO_3} + \underset{\text{(rocksalt)}}{CdO}$$

Analogous reactions are displayed by MgCaGeO$_4$ and MnCaGeO$_4$.

$$\underset{\text{(olivine)}}{CaMgGeO_4} \rightarrow \underset{\text{(perovskite)}}{CaGeO_3} + MgO$$

$$\underset{\text{(olivine)}}{CaMnGeO_4} \rightarrow \underset{\text{(perovskite)}}{CaGeO_3} + MnO$$

The density increases are about 25%.

The difference in behaviour between Ca$_2$GeO$_4$ and Cd$_2$GeO$_4$ is readily attributed to the much higher free energies of reaction of CaO with given oxides than the corresponding free energies of combination of CdO. This tends to inhibit disproportionation in the former case, leading to transformation to a single dense phase. However, Cd$_2$GeO$_4$, with a lower free energy of formation displays disproportionation. In the cases of CaMgGeO$_4$ and CaMnGeO$_4$, disproportionation may be only partly due to the lower free energy of combination of MgO and MnO with other oxides as compared with CaO. An additional and perhaps a more important cause may be the mismatch between MgO, MnO subunits and CaGeO$_3$ subunits owing to differences between Ca–O, Mg–O and Mn–O bond lengths.

3.4. Possible transformations of Mg$_2$SiO$_4$ to structures denser than isochemical mixtures of MgO plus SiO$_2$

We have already discussed in some detail the transformations by which A$_2^{2+}$B^{4+}O$_4$ spinels may attain densities similar to those of the isochemical mixtures of rocksalt and rutile end members. There is some evidence that in the deep mantle, silicates may be about 5% denser than the equivalent oxide mixtures. Furthermore, the density of the high pressure polymorph of pure forsterite produced in shock wave experiments (MCQUEEN and MARSH, 1966) appears to be several percent higher than that of the mixed oxides (AHRENS et al., 1969). This is in contrast to the behaviour of (MgFe)$_2$SiO$_4$ olivines which attain densities similar to the mixed oxides. The reason for this contrasting behaviour is not established. It may be connected with the fact that the forsterite used in the shock experiments possessed a substantial porosity, so that it reached much higher temperatures during the shock than the (MgFe)$_2$SiO$_4$ olivines. This might have influenced the transformation kinetics and permitted transformation into a denser phase. (Alternatively, there is a possibility that errors exist in the experimental measurements upon pure Mg$_2$SiO$_4$ (MCQUEEN, personal communication) and that the high inferred density is not real.)

In either case it is important to consider possible transformations for Mg$_2$SiO$_4$ into states which are denser than the mixed oxides.

One possibility is that Mg$_2$SiO$_4$ might ultimately transform to the K$_2$NiF$_4$ structure, displayed by the high pressure modification of Ca$_2$GeO$_4$. This structure is between 2 and 7% denser than the isochemical mixed oxides (table 5). We noted earlier that the structure consists of alternate sub-units possessing perovskite and rocksalt structures. In order to form the perovskite layers, a substantial change in the nature of Mg–O bond under pressure towards a more covalent type would probably be necessary (sections 2, 5.6). Alternatively, Mg$_2$SiO$_4$ might disproportionate under extreme pressures into a mixture of MgSiO$_3$ (perovskite) + MgO. The conditions under which MgSiO$_3$ might exist in the perovskite structure are discussed in section 5.6. Such an assemblage would be about 5% denser than the isochemical mixed oxides. Possible models for these transformations are the olivines MgCaGeO$_4$, Cd$_2$GeO$_4$ and MnCaGeO$_4$, which transform to Ca-

GeO$_3$ (perovskite) plus MgO, CdO, MnO (rocksalt). A further increase in density would be attained if the rocksalt structures transformed to the caesium chloride structure.

A third dense structure to which Mg$_2$SiO$_4$ might ultimately transform is the calcium ferrite structure or the related CaTi$_2$O$_4$ and CaMn$_2$O$_4$ structures (REID and RINGWOOD, 1969c). This structure contains two non-equivalent octahedral sites and an 8-fold coordinated site. It would be possible for one Mg^{2+} cation to occupy the 8-fold site as in pyrope garnet, whereas the octahedral sites might be shared by Mg^{2+} and Si^{4+} as in the Mg$_3$(MgSi)Si$_3$O$_{12}$ high pressure garnet end member (section 5.3). The calcium ferrite structure is stable for a wide variety of ions possessing different charges and radii, and extensive solid solutions are possible at high temperature. The formation of a calcium ferrite polymorph of Mg$_2$SiO$_4$ would not require an important change in the nature of the Mg–O bond as in the cases discussed previously. The density of Mg$_2$SiO$_4$ in this structure would be about 5% higher than an isochemical mixture of MgO plus stishovite.

4. High-pressure hydrated magnesium silicates

RINGWOOD and MAJOR (1967b) carried out some reconnaissance work on the system MgO–SiO$_2$–H$_2$O in compositions where the MgO/SiO$_2$ ratio varied from 1 to 5 at pressures between 100 and 170 kb and at 1000 °C. Starting materials consisted of hydrous gels containing 2 to 10% H$_2$O and mixtures of periclase with silicic acid (5% H$_2$O). In the high pressure regime, familiar phases such as talc and serpentine were not encountered and were replaced by a new series of hydrated phases. The principal X-ray d-spacings of these phases were established but it was not possible to characterize these phases completely. Accordingly they were named phases A, B and C. Phase A had a mean refractive index of about 1.65 and a low to medium birefringence. It was believed to be hydrated, with a MgO/SiO$_2$ ratio of 1.5 to 2. This phase was not formed from runs at pressures below 100 kb, its place being taken by forsterite.

Phase B was almost ubiquitous in runs above 110 kb but always occurred in association with other phases including phases A, C, MgO and stishovite. This rendered optical identification and characterization difficult. It appeared to correspond to one of two optically identifiable phases possessing mean refractive indices of about 1.71 and 1.77 respectively. From the preferred synthesis field, it appeared that phase B was hydrated and that the MgO/SiO$_2$ ratio was between 2 and 3. The refractive indices indicate that this phase possesses a high density, probably between 3.5 and 3.8 g/cm^3.

Phase C occurred sporadically in runs on water-rich gels with MgO/SiO$_2$ ratios between 3 and 5 and was believed to be a layer-lattice mineral, perhaps related to the chondrodite–humite series.

The importance of these phases, as noted by Ringwood and Major, lies in the possibility that they may serve as hosts for OH^{-1} ions in the mantle. Phases A and B are both stable at relatively high temperatures and phase B in particular possess a high density.

SCLAR et al. (1967a, b), SCLAR (1970) have carried out extensive investigations of the system MgO–SiO$_2$–H$_2$O with MgO/SiO$_2$ in the 2/1 ratio. Their results thus far have been reported in abstract form. Three phases at this composition were recognised, and are believed to represent different polymorphs of hydroxylated pyroxenes, isostructural with orthopyroxene and clinopyroxene. The deduced formula was MgSi$_{0.5}$(H$_4$)$_{0.5}$O$_3$ in which only half the tetrahedral sites are occupied by silicon, charge compensation occurring by substitution of (H$_4$O$_4$)$^{-4}$ for SiO$_4^{-4}$. The principal X-ray "d"-spacings for the phase published by SCLAR et al. (1967a) show that it is identical with the phase A of RINGWOOD and MAJOR (1967a). Sclar et al. also draw attention to the geochemical significance of these hydroxylated pyroxenes for the storage and release of large amounts of water in the upper mantle, according to the reaction

$$Mg_2SiO_4 + 2H_2O \rightleftharpoons 2MgSi_{0.5}(H_4)_{0.5}O_3$$

However confirmation that this phase indeed possesses the structure of a hydroxylated pyroxene is desirable. If this inference is correct it will be very important to investigate the solid solubility relations of the hydroxylated pyroxenes with normal pyroxenes. It may be that normal mantle pyroxenes are able to take significant quantities of OH^{-1} into their structure, and that these "normal" phases represent the reservoirs for water in the upper mantle rather than the pure, completely hydroxylated end members so far synthesized.

5. High pressure transformations in ABO$_3$ compounds

Next to olivine, the major mineral constituents of

the upper mantle are pyroxenes and garnets. The fundamental chemical formulae of these two mineral families are similar, having 2 cations to 3 oxygen anions. Although characterized by complex compositions and the formation of extensive solid solutions, the end members of these groups usually have simple formulae of the type $A^{2+}B^{4+}O_3$ and $C_2^{3+}O_3$. Because of the abundance of these minerals in the upper mantle, their behaviour at high pressures is of vital importance to an understanding of the constitution of the deep mantle. As we shall see, they display a wide variety of transformations, most of which have been discovered only very recently. We shall commence with a description of transformations in pyroxenes and pyroxenoids.

5.1. *Disproportionation of pyroxenes into spinel+rutile structures*

A number of pyroxenes are observed to disproportionate at high pressures into a mixture of spinel+rutile-type phases according to the following reaction (table 6)

$$2ABO_3 \rightarrow A_2BO_4 + BO_2.$$
$$\text{(pyroxene)} \quad \text{(spinel)} \quad \text{(rutile)}$$

Transformations of this type were discovered by RINGWOOD and SEABROOK (1963) in the germanate pyroxenes $FeGeO_3$, $CoGeO_3$ and $(Mg_{0.75}Ni_{0.25})GeO_3$. They occurred at pressures between 10 and 25 kb at 700 °C and were accompanied by density increases of about 15% (table 6).

More recently RINGWOOD and MAJOR (1966a, b, d) found that the pyroxenes $FeSiO_3$ and $CoSiO_3$ likewise

TABLE 6

Disproportionation transformations displayed by pyroxenes at 700–1000 °C*

$2ABO_3 \rightarrow A_2BO_4 + BO_2$
(pyroxene) → (spinel)+(rutile)
$\Delta \rho / \rho \approx 15\%$

Pyroxene	Disproportionation pressure (approx.) (kb)
$FeGeO_3$	10
$CoGeO_3$	10
$(MgNi)GeO_3$	10
$FeSiO_3$	100
$CoSiO_3$	100
$FeSiO_3$–$MgSiO_3$ solid solutions as far as $(Mg_{0.43}Fe_{0.57})SiO_3$	100–200

* References: see text.

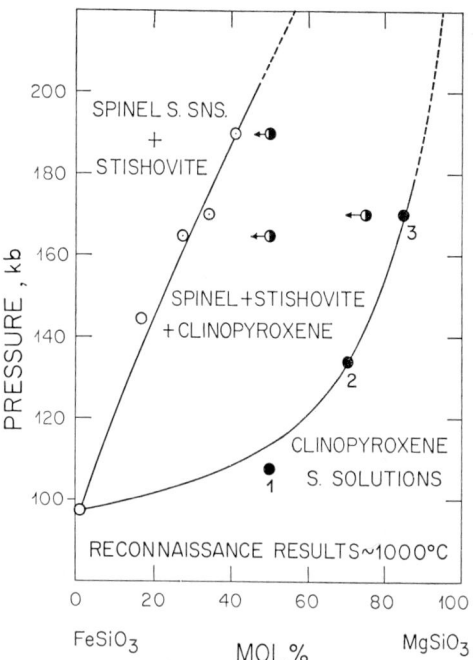

Fig. 2. Phases observed in the system $FeSiO_3$–$MgSiO_3$ at pressures up to 200 kb and at approximately 1000 °C. Symbols as follows: ○ complete transformation to spinel+stishovite, ◉ Boundary determined by spinel composition obtained from lattice parameter, ←◐ Experimental run in 3 phase field in which composition of spinel was determined, ●$_{1,2,3}$ inferred points on boundary of pyroxene stability field. After RINGWOOD and MAJOR (1968b).

transformed to Fe_2SiO_4 and Co_2SiO_4 spinels plus SiO_2 (stishovite) respectively at pressure close to 100 kb and around 1000 °C (The pressures cited are somewhat smaller than those given in the original references because of revisions in the pressure scale.) It is interesting to observe the similarity in the modes of transformation of $FeSiO_3$ and $FeGeO_3$ on one hand and $CoSiO_3$ and $CoGeO_3$ on the other, and also that the germanates transform at only 10 kb compared to 100 kb for the silicates.

Having obtained a transformation in $FeSiO_3$, a component of natural pyroxenes, the next step was to attempt to extend this towards more magnesian compositions by means of a study of the system $FeSiO_3$–$MgSiO_3$. Reconnaissance results in this system are shown in fig. 2. The position of the spinel boundary is reasonably well established from measurements of the lattice parameters of the spinels; however the boundary for initial breakdown of pyroxene is not fixed so well because of problems in obtaining equilibrium during crystallization of $MgSiO_3$–$FeSiO_3$ glasses. Nevertheless the general nature of the phase diagram is clearly established. As pressure increases from 100 to 190 kb, the composition of the pyroxenes which can be completely transformed to $(MgFe)_2SiO_4$ (spinel) + SiO_2 (stishovite) rises from $FeSiO_3$ to $(Mg_{0.43}Fe_{0.57})SiO_3$. The field of spinel, stishovite and untransformed pyroxene is wide so that a broad spectrum of pyroxene compositions ranging to about $(Mg_{0.8}Fe_{0.2})SiO_3$ can be transformed partially according to this equilibrium. An interesting feature is that at the same Fe/Mg ratio, pyroxenes are stable to much higher pressures than the corresponding olivines. Assuming that no other phases become stable it would be possible to extrapolate the phase boundaries of fig. 2 to obtain the transformation pressure for pure $MgSiO_3$. The uncertainties in the location of the present phase boundaries in fig. 2 make extrapolation hazardous. Nevertheless it appears reasonable to infer that $MgSiO_3$ would transform to β-Mg_2SiO_4 plus SiO_2 (stishovite) at some pressure between about 200 and 300 kb, *providing no other major phase transformation intervenes*. This pressure is surprisingly high, but is supported by direct experiments (unpublished results) which have failed to transform pure $MgSiO_3$ at pressure substantially exceeding 200 kb.

5.2. *The pyroxene–ilmenite transformation*

The possibility that $MgSiO_3$ might ultimately transform to the corundum structure (essentially a disordered ilmenite structure) was first suggested by J. B. Thompson (BIRCH, 1952, p. 234). The corundum and ilmenite structures are based upon hexagonal close packing of oxygen ions with the octahedral interstices occupied by the cations. Such structures are commonly formed between end members possessing rutile and rocksalt structures (table 7).

Experimental evidence supporting this suggestion was obtained by RINGWOOD and SEABROOK (1962c, 1963) who showed that the orthopyroxenes $MgGeO_3$ and $MnGeO_3$ were transformed to ilmenite structures at 25–30 kb, 700 °C accompanied by density increases of 15 and 18% respectively. These large density increases were due to the fact that the germanium ion changed from 4- to 6-fold coordination through the transition. In view of the model relationships between germanates and silicates previously discussed, it was concluded that ultimate transformation of $MgSiO_3$ to the ilmenite

TABLE 7

Comparative molar volumes of compounds possessing ilmenite structures and formed from end-members possessing rocksalt and rutile structures

Compound	Molar volume V (cm³)	Isochemical oxide volume* v (cm³)	$\frac{\Delta v^*}{v}$ (%)
$CdTiO_3$	35.69	30.39	3.8
$CoTiO_3$	31.05	30.44	2.0
$MgGeO_3$	29.14	27.91	4.4
$MgTiO_3$	30.86	30.05	2.7
$MnGeO_3$	31.28	29.88	4.7
$MnTiO_3$	32.76	32.02	2.3
$FeTiO_3$	31.69	30.84	2.8
$NiTiO_3$	30.56	29.77	2.7
$NiMnO_3$	28.42	27.58	3.0
$CoMnO_3$	29.43	28.25	4.2
$MgSiO_3$	26.40	25.26	4.5 (assumed)

* $\Delta v = V - v$.

structure was probable. Here then, was an alternative transformation to the disproportionation inferred in the previous section.

Further information was sought via a study of the system $MgGeO_3$–$MgSiO_3$ at high pressure. RINGWOOD and SEABROOK (1963) working at 700 °C and 0–50 kb obtained evidence in some runs that $Mg(GeSi)O_3$ solid solutions disproportionated into a mixture of spinel, rutile, plus residual pyroxene. In only three runs was spinel definitely observed; in most runs the inference that partial disproportionation was occurring according to the reaction

$$2Mg(GeSi)O_3 \rightarrow Mg_2(GeSi)O_4 + (GeSi)O_2$$
(pyroxene) (spinel) (rutile)

was based upon the occurrence of the rutile phase in high pressure runs. The problem with identification of spinel was that most of its lines overlapped those of pyroxene. Only when an extensive degree of transformation occurred, was it possible to identify spinel. At higher pressures, 50–100 kb, there was evidence that at the composition $Mg(Ge_{0.9}Si_{0.1})O_3$, spinel plus rutile recombined to form an ilmenite phase.

Accordingly it was concluded from these experiments that pure $MgSiO_3$ would probably transform initially to a spinel-stishovite mixture and that at higher pressures, recombination to denser $MgSiO_3$ ilmenite would occur.

More recent work on this system by RINGWOOD and MAJOR (1966b, 1968b) required a modification of these conclusions. In the course of a large number of runs at temperatures in the vicinity of 1000 °C, no sign of disproportionation into spinel + rutile was found. Some evidence indicated that Mg(GeSi)O$_3$ pyroxenes at high temperature may be non-stoichiometric, and that this may be responsible for precipitation of the rutile phase frequently observed by Ringwood and Seabrook. Accordingly it was concluded that the P–T conditions under which disproportionation into spinel occurred were very restricted and in view of the fact that experiments were conducted near the lowest devitrification temperature of the glass starting materials, the appearance of spinel in the 3 runs may indeed have been a non-equilibrium effect.

A large number of runs has now been carried out at higher temperatures and pressures in this system. All evidence indicates that above 40 kb it is essentially a binary system (ignoring the possibility of non-stoichiometric pyroxenes) involving pyroxene and ilmenite solid solutions.

The results of reconnaissance work in this system by RINGWOOD and MAJOR (1966b, 1968b) at 1000 °C are shown in fig. 3. MgSiO$_3$ is observed to display substantial solid solubility in MgGeO$_3$ ilmenite and the degree of solid solubility increases with pressure reaching 25 ± 5 % at 170 kb. The solid solubility of MgGeO$_3$ in the ilmenite decreases regularly with pressure.

The observation of a substantial solid solubility of MgSiO$_3$ in the ilmenite structure at 170 kb is important in showing that the free energy difference between the ilmenite and pyroxene forms of MgSiO$_3$ is of small magnitude (≈ 2.6 kcal/mol) at this pressure (section 2). A formal calculation along the lines discussed in section 2 leads to an estimated pressure of 213 kb for the transformation pressure in pure MgSiO$_3$. Because of the possibility that the departures from ideality of the solid solutions may increase at high pressure, this estimate has a large margin of uncertainty. Nevertheless it is indicative.

Alternatively, one may simply extrapolate the phase diagram (fig. 3) across to the MgSiO$_3$ side in which case it would be reasonable to conclude that MgSiO$_3$ is likely to transform to an ilmenite structure at some pressure between 200 and 300 kb, *providing that transformation to other phases does not intervene.*

Incorporating the results of the previous section we see that two alternative modes of transformation be-

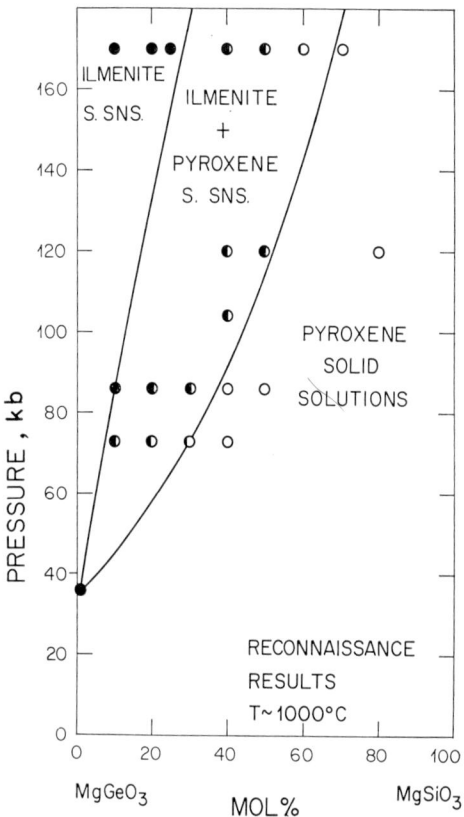

Fig. 3. Phases present in the system MgGeO$_3$–MgSiO$_3$ at pressures up to 170 kb and at approximately 1000 °C. Symbols as follows: ● homogeneous ilmenite type solid solutions, ◐ two phase field of ilmenite + pyroxene solid solutions, ○ pyroxene solid solutions. After RINGWOOD and MAJOR (1968b).

tween pressures of 200 and 300 kb are indicated by the experimental data. Enstatite may either transform directly to an ilmenite structure, or alternatively, the stability field of β-Mg$_2$SiO$_4$ + SiO$_2$ (st.) would be reached at a lower pressure, in which case disproportionation to this assemblage will intervene. The uncertainties in the experimental data (figs. 2 and 3) are such that it is not reasonable to prefer one mode of transformation over the other. The important result is the inferred instability of enstatite at high pressure and the conclusion that it may transform either to ilmenite or to "spinel"-stishovite between 200 and 300 kb.

The density of the ilmenite form of MgSiO$_3$ is a matter of some importance. The molar volumes of ilmenite compounds compared to the volumes of their component rocksalt and rutile-structure end members are given in table 7.

The densities of the ilmenite compounds are observed

to be between 2.0 and 4.7% smaller than the isochemical oxide mixtures. This is apparently caused by the slight distortion involved in introducing two cations of differing size into the octahedral interstices of a hexagonal close packing of oxygen anions. The density differential of $MgSiO_3$ ilmenite compared with its oxide components is probably close to that of $MgGeO_3$ (4.4%) and $MnGeO_3$ (4.7%). Assuming a figure of 4.5%, the density of $MgSiO_3$ (ilmenite) would be 3.80 g/cm³. An alternative method based upon extrapolation of bond lengths for A_2O_3 compounds gives a value of 3.76 g/cm₃ for $MgSiO_3$ ilmenite (REID and RINGWOOD, 1969a).

5.3. The pyroxene–garnet transformation

A Goldschmidt diagram showing the structure fields of some ABO_3 and $A_3B_5O_{12}$ compounds in relation to the ionic radii of A and B cations is shown in fig. 4. The central and transitional position of the garnet field, relative to the pyroxene, ilmenite and perovskite fields is to be noted. This suggests the possibility that garnet structures may represent a third mode of transformation from pyroxenes and pyroxenoids.

Until about twenty years ago, the garnet group was thought to be essentially restricted to a rather small number of rock-forming silicates $A_3B_2Si_3O_{12}$ where A = Mg, Ca, Fe or Mn and B = Al, Fe or Cr. Since then, systematic studies of the crystal chemistry of garnets have revealed that a remarkable range of elements are capable of entering this structure and an extremely complex series of ionic substitutions and charge balances are possible. See GELLER (1968) for a recent review of this subject. The general formula is $A_3^{VIII}B_2^{VI}C_3^{IV}O_{12}$ where the superscripts denote the co-ordination of the specified cation by oxygen. In table 8 the cations which are capable of entering the A, B and C sites are shown, together with examples of various general and specific types of ionic substitutions. Notice particularly that it is possible to replace the characteristic trivalent cations in the B position (e.g. Al^3) by a combination of a divalent and tetravalent atom e.g. $Mg^2 + Ti^4$. The most general case of this replacement would be where the B atoms were replaced by an A + C couple (table 8). In such a case the garnet has the formula $A_3(AC)C_3O_{12}$ or more simply $(ACO_3)_4$ where A is divalent and C is tetravalent.

The first example of this type of structure was discovered by RINGWOOD and SEABROOK (1963) who found that $CaGeO_3$ which has a wollastonite-like structure at atmospheric pressure was transformed com-

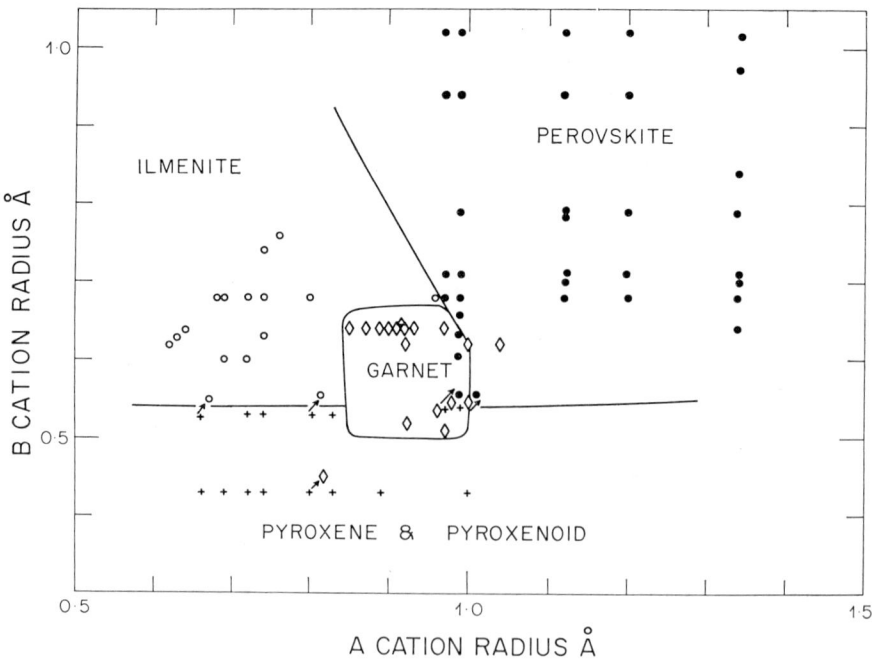

Fig. 4. Plot of A and B ionic radii for $A^{2+}B^{4+}O_3$ compounds possessing pyroxene, ilmenite, perovskite and garnet structures and for $A_3^{3+}B_5^{3+}O_{12}$ garnets. Arrows indicate high pressure transformations.

TABLE 8

Structural formulae, ionic substitutions and examples of garnets*

Structural formulae	Ionic substitutions†			Examples
$A_3^{VIII}B_2^{VI}C_3^{IV}O_{12}$	A cations	1+	Na, K	$Mg_3Al_2Si_3O_{12}$
		2+	Ca, Mg, Fe, Mn, Cd, Co, Cu, Zn	
$A_3^{VIII}C_2^{VI}C_3^{IV}O_{12}$		3+	Rare earths, Bi, Y	$Ca_3(TiMg)Ge_3O_{12}$
				$(CdGd_2)Mn_2Ge_3O_{12}$
$A_3^{VIII}(AC)\ C_3^{IV}O_{12}$	B cations	2+	Mn, Co, Mg, Ni, Zn	$(NaCa_2)Ni_2V_3O_{12}$
		3+	Fe, Ga, Al, Cr, V	
		4+	Sn, Ge, Si, Ti, Zr, Hf	$Na_3Al_2Li_3F_{12}$
		5+	Nb	$(CaNa_2)Ti_2Ge_3O_{12}$
	C cations	3+	Fe, Ga, Al	$Y_3Al_2Al_3O_{12}$
		4+	Si, Ge, Sn, Ti	
		5+	As, V	$Ca_3(CaGe)Ge_3O_{12}$
		1+	Li	$Mn_3(MnSi)Si_3O_{12}$
	Anions		O^{-2}, F^{-1}, OH^{-1}	

* General references: WYCKOFF (1965), GELLER (1968).   † For $A_3B_2C_3O_{12}$ type garnets.

pletely into a garnet structure at 40 kb and 700 °C. In some runs the presence of line splitting indicated that the garnet was slightly distorted with a symmetry lower than cubic. The same workers found that $CdGeO_3$ which has a complex structure of low density at atmospheric pressure transformed to a distorted garnet structure at pressures greater than about 10 kb, 700 °C. Detailed studies of the structures of these germanate garnets have been undertaken by PREWITT and SLEIGHT (1969). The first such transformation in a silicate was observed by RINGWOOD and MAJOR (1967a) in rhodonite $MnSiO_3$. This mineral has a pyroxenoid structure at atmospheric pressure. At pressures of about 60 kb, 700 °C, transformation to a true pyroxene structure was observed. Finally in runs between 120 and 170 kb (1000 °C) transformation into a garnet structure occurred. This had a density of 4.27 g/cm³ compared to 3.71 g/cm³ for rhodonite. The structural formula of the garnet can be expressed as $Mn_3^{VIII}(MnSi)^{VI}Si_3^{IV}O_{12}$ where the superscripts denote coordination numbers. The fact that two cations varying in radius by as much as Mn^{2+} and Si^{4+} share the octahedral site almost certainly implies ordering at this site. It is notable that this inferred transformation implies a change in coordination of one quarter of the silicon atoms from fourfold to sixfold.

The occurrence of these transformations suggests the possibility that $MgSiO_3$ and $CaSiO_3$ might perhaps transform to garnet structures at high pressure. This would represent a third mode of transformation for $MgSiO_3$ in addition to those previously considered.

RINGWOOD and MAJOR (1966b) and RINGWOOD (1967) crystallized a glass of composition $MgSiO_3 \cdot 10\% Al_2O_3$ (wt%) at various pressures at about 1000 °C. This composition is equivalent to a mixture of 60% $MgSiO_3$ (enstatite) and 40% $Mg_3Al_2Si_3O_{12}$ (pyrope). Results are shown in figs. 5 and 6. At pressures up to 90 kb, the

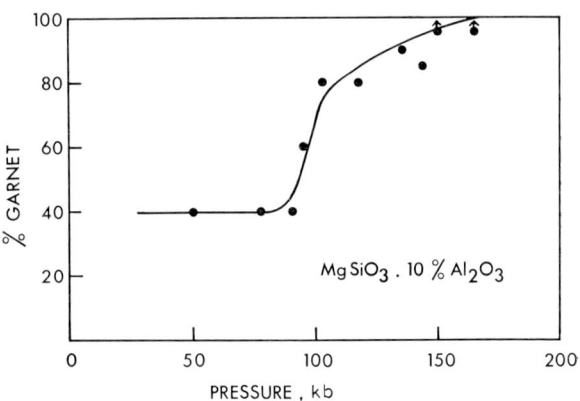

Fig. 5. Proportions of garnet crystallizing from an $MgSiO_3$. 10% Al_2O_3 glass as a function of pressure at approximately 1000 °C. After RINGWOOD (1967).

glass crystallized to a mixture of clinoenstatite 40 and pyrope 60 as expected. However between 90 and 100 kb, a dramatic increase in the proportion of garnet present occurred and by 150 kb the glass had transformed completely to a garnet. The composition of this garnet would be, according to previous discussion and interpretation

$$[Mg_3(MgSi)Si_3O_{12}]_{60} \quad [Mg_3Al_2Si_3O_{12}]_{40}$$

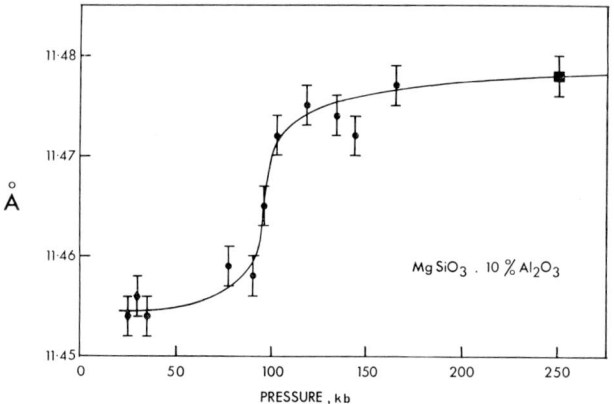

Fig. 6. Lattice parameters of garnets crystallized from a glass of composition $MgSiO_3 \cdot 10\% \ Al_2O_3$ as a function of pressure at (approx.) 1000 °C. Run at 250 kb (square symbol) was on an $MgSiO_3 \cdot 5\% \ Al_2O_3$ glass. Runs at 25, 30 kb (diamond symbols) were on a glass of pyrope composition. After RINGWOOD (1967).

Similar results have been obtained on glasses with compositions $MgSiO_3 \cdot 13.5\% \ Al_2O_3$ and $MgSiO_3 \cdot 5\% \ Al_2O_3$. In the latter case complete transformation to garnet was not obtained, the lowest Al_2O_3 content of a homogeneous garnet at the limit of experimental pressures being about 7%. However the experimental results imply that at even higher pressures garnets with still smaller amounts of Al_2O_3 would be stable, and ultimately, at very high pressures, a pure $MgSiO_3$ garnet may occur.

These results show that at high pressures, enstatite is capable of dissolving extensively in the garnet structure. The fact that solid solubility increases so rapidly at about 100 kb suggests that this is not a normal solid solution but may be more closely related to compound formation. The composition of the garnet formed at about 100 kb is very nearly $Mg_3(AlMg_{0.5}Si_{0.5})Si_3O_{12}$. It appears that ordering of Al, Mg and Si in the 2:1:1 proportion in the B sites may in fact define a compound which displays solid solubility with pure $MgSiO_3$ at higher pressures (figs. 5 and 6). The transformation involves an effective increase in density of about 10% in the $MgSiO_3$ component as it transforms from pyroxene to garnet. In view of the fact that this phase was synthesized from a glass, the question of possible metastability arises. RINGWOOD (1967) discussed this question in detail and concluded that the new garnets were almost certainly stable in their synthesis fields. The principal evidence was that they could be transformed from crystallized aluminous enstatite and that the Ca-

GeO_3, $CdGeO_3$ and $MnSiO_3$ garnets had been transformed from crystallized phases, not glasses. Nevertheless, the former experiment, although it provides very strong evidence of stability does not constitute a strict reversal of the equilibrium in question and it is desirable that this should be accomplished in order to settle the matter beyond all doubt. Because of the sluggishness of the reaction, a reversal will require apparatus capable of generating higher temperatures than can be reached in the internally heated Bridgman anvils.

RINGWOOD and MAJOR (1966) and RINGWOOD (1967, unpublished) carried out an analogous series of experiments upon glasses of composition $CaMgSi_2O_6 \cdot 10\% \ Al_2O_3$, $CaSiO_3 \cdot 10\% \ Al_2O_3$ and $FeSiO_3 \cdot 10\% \ Al_2O_3$ These glasses also crystallized to form homogeneous garnet phases at high pressures, and these are inferred to possess similar structures to that of $MgSiO_3 \cdot 10\% \ Al_2O_3$. The X-ray reflexions of the $CaSiO_3 \cdot 10\% \ Al_2O_3$ garnet were split, indicating lower symmetry than cubic. A series of runs on the aluminous diopside glass yielded similar quantitative results to the runs on $MgSiO_3 \cdot 10\% \ Al_2O_3$. Solid solubility of diopside in garnet was negligible or low up to about 90 kb, beyond which it increased rapidly. At 120 kb, the glass crystallized to garnet 80%, clinopyroxene 20%, whilst at 150 kb the proportion was garnet 90%, clinopyroxene 10%. The composition of the garnet in the latter run was approximately 50% $(CaMg)_3Al_2Si_3O_{12}$ to 50% $CaMgSi_2O_6$. In some of the runs the X-ray reflexions of the garnet were split, indicating lower symmetry. A list of pyroxene-garnet transitions is given in table 9.

From these studies on Mg, Fe and Ca bearing pyroxenes it appears that the pyroxene-garnet transformation in the presence of Al_2O_3 is of rather general occurrence. This was confirmed by some experiments (RINGWOOD, 1967) upon more complex, eclogitic systems. Compositions of glass starting materials are given in table 10. The results of high pressure experiments on the "alkali-poor olivine tholeiite" are given in fig. 7. This composition crystallizes to an eclogite containing about 50% of garnet and 50% of pyroxene at pressures over 20 kb (RINGWOOD and GREEN, 1966) and possesses a density of 3.55 g/cm³. The behaviour of this complex eclogitic system at higher pressures is seen to be very similar to the simple systems. There is a drastic increase in the solubility of pyroxene in the garnet structure at about 100 kb. At a pressure of 120 kb, crys-

TABLE 9

Transformations of pyroxenes and related structures to garnets $\Delta\rho/\rho \approx 10\%$*

	P (kb) (approx.)
$CaGeO_3$	35
$CdGeO_3$	10
$MnSiO_3$	100
$(Mg_3 \overline{MgSi} Si_3O_{12})_{70}(Mg_3Al_2Si_3O_{12})_{30}$	200
$(Fe_3 \overline{FeSi} Si_3O_{12})_{50}(Fe_3Al_2Si_3O_{12})_{50}$	100–150
$(Ca_3 \overline{CaSi} Si_3O_{12})_{50}(Ca_3Al_2Si_3O_{12})_{50}$	100–150
$\{(CaMg)_3 \overline{Ca_{0.5}Mg_{0.5}Si} Si_3O_{12}\}_{50}\{(CaMg)_3\text{-}Al_2Si_3O_{12}\}_{50}$	100–150
Eclogite → "Garnetite"	100–150
$CaMgGe_2O_6 \rightarrow CaGeO_3 + MgGeO_3$ (diopside) (garnet) (ilmenite)	70
$MgSiO_3$ (inferred from presence of garnet in chondritic meteorite)	>200

* References: see text.

TABLE 10

Compositions of eclogitic (i.e. basaltic) glasses used in studies of eclogite–garnetite transformations

	Alkali-poor olivine tholeiite	Quartz tholeiite
SiO_2	46.23	52.16
TiO_2	0.07	1.86
Al_2O_3	14.52	14.60
Fe_2O_3	0.54	2.46
FeO	11.80	8.39
MnO	0.30	0.14
MgO	12.45	7.36
CaO	13.00	9.44
Na_2O	0.81	2.68
K_2O	0.01	0.73
P_2O_5	0.03	0.18
Cr_2O_3	0.24	—
	100.00	100.00

tallization of the glass to a "garnetite" possessing a density of 3.72 g/cm³ has occurred. The glass starting material was formed by the fusion of pyrope garnet and omphacite extracted from a kimberlite xenolith and accordingly it may be held to represent the composition of a typical mantle eclogite xenolith.

Analogous behaviour was observed in the case of the quartz tholeiite. At 23 kb this glass crystallized to clinopyroxene 55%, garnet 40% and quartz 5%. At 170 kb, the mineral assemblage was garnet 70%, clinopyroxene 25% and stishovite 5%. The larger proportion of pyroxene which persisted to higher pressures in this run is caused by the larger amount of alkalis in the quartz tholeiite. These do not enter the garnet and remain as jadeite which is stable to extremely high pressures. The jadeite stabilizes a further amount of Ca-rich clinopyroxene.

The pyroxene–garnet transformation is probably of considerable importance in the earth's mantle. The above experimental results imply that in the presence of 8–10% Al_2O_3 (relative to pyroxene) the entire pyroxene component of the mantle would ultimately be transformed to the garnet structure. An alternative statement is that the pyroxene component will form extensive solid solutions with existing normal garnet and that this reaction will set in suddenly at a pressure of about 100 kb, 1000 °C, and will be extended to somewhat higher pressures by solid solution effects. The effective increase in density of the pyroxene as it enters the garnet structure is about 10%. Since pyroxenes are abundant components of the upper mantle and almost any geochemically reasonable model of the mantle (section 7) contains sufficient R_2O_3 ($Al_2O_3 + Cr_2O_3 + Fe_2O_3$) in relation to pyroxene for this transformation to proceed essentially to completion, it is probably of considerable geophysical importance and will strongly influence the seismic velocity and density distribution between about 300 and 400 km depth, depending upon the temperature gradient which is not yet known. It is of interest that at 1000 °C the transformation proceeds nearly to completion at a slightly

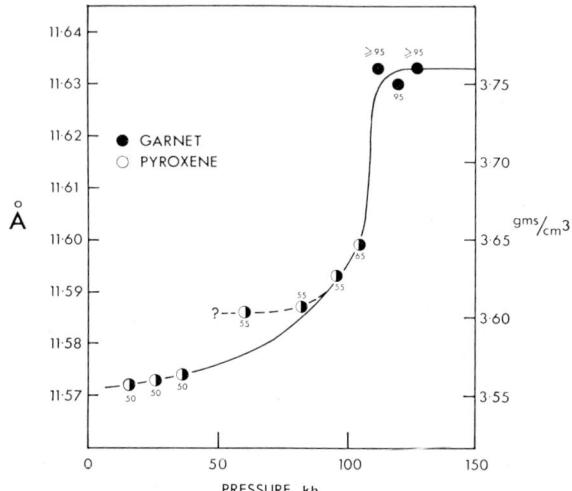

Fig. 7. Lattice parameters of garnets, proportions of garnets and mean densities for an eclogite of alkali-poor olivine–tholeiite composition (table 10) in the pressure range 16–130 kb, 1000 °C.

smaller pressure than is required for the transformation of mantle olivine to spinel and beta phase (RINGWOOD and MAJOR 1970; fig. 11).

Two further transformations of a pyroxene and pyroxenoid might be noted. RINGWOOD and SEABROOK (1963) showed that germanium diopside $CaMgGe_2O_6$ disproportionates at ≈ 70 kb, 700 °C into a mixture of $CaGeO_3$ (garnet)+$MgGeO_3$ (ilmenite). RINGWOOD and MAJOR (1967a) crystallized $CaSiO_3$ glass at several pressures. Two new polymorphs of wollastonite were found. The higher pressure modification had a refractive index of 1.745 indicating a density of about 3.5 g/cm^3 compared to 2.91 for wollastonite. Only a minute amount of material was recovered and the X-ray pattern was of poor quality. It appeared complex, and could not be matched with that of a known structure.

5.4. Transformation of $MgSiO_3$

The static high pressure experiments described previously indicate three possible modes of transformation of pure $MgSiO_3$:
(i) disproportionation to β-Mg_2SiO_4+stishovite,
(ii) transformation to ilmenite structure,
(iii) transformation to garnet structure.

The pressures estimated for these transformations were between 200 and 300 kb. However the static experiments were unable to establish which of the three transformation modes was the most probable. Since the density of $MgSiO_3$ ilmenite was greater than the garnet and β-Mg_2SiO_4–SiO_2 mixture, it would be possible for $MgSiO_3$ to transform first to one of the latter, and subsequently, in a further transformation at a higher pressure, to transform ultimately to the ilmenite structure.

Under shock wave conditions, major phase transformations were inferred in two bronzitites (($Mg_{0.9}$-$Fe_{0.1}$)SiO_3) by MCQUEEN et al. (1967). The zero pressure density of the high pressure phase was estimated by AHRENS et al. (1969) to be 3.74 g/cm^3 and by WANG (1967) to be within the range 3.71–3.84 g/cm^3. The densities of possible high pressure modifications of bronzite for this composition are: β-Mg_2SiO_4+SiO_2—3.77, garnet—3.64 and ilmenite—3.87 g/cm^3 respectively. The inferred shock density agrees best with that of an isochemical β-Mg_2SiO_4–stishovite mixture, and WANG (1967) suggested that transformation into this assemblage had occurred during the shock. However this appears unlikely. A large amount of information has been obtained on shocked olivines and in no case is there evidence of transformation to a phase with the density of spinel[*]. It appears that in all cases, above about 400 kb, the olivines transformed into a phase which was much denser than spinel (AHRENS et al. 1969). There is no reason to expect the bronzitite to transform partially to a spinel in a pressure range where olivines do not display this transformation.

This leaves the garnet and ilmenite structures as possible candidates. The inferred low-pressure density of the shocked phases are intermediate between these alternatives. The shock data display evidence of an abnormal compressibility, suggesting that complete transformation may not have been obtained. If so, this would imply that the calculated density is on the low side so that the ilmenite structure would be the most probable in the highest pressure shock regime. The density of bronzitite at 1000 kb along the Hugoniot is indeed similar to the probable density of an ilmenite form of bronzite under the same P–T conditions.

If we accept the abnormal compressibility of the bronzitites along the Hugoniot as evidence of partial transformation, then it is possible that a garnet form of $(MgFe)SiO_3$ was formed in the intermediate pressure shock regime (300–700 kb) and that the degree of transformation of garnet to ilmenite increased with pressure. This speculation is strengthened by recent observations on the occurrence of garnet in heavily shocked regions of chondritic meteorites. The first such occurrence of garnet was discovered by MASON et al. (1968) in a shock-produced vein of the Coorara chondrite. Electron-probe microanalysis showed the absence of Al and indicated an olivine composition, and Mason et al. suggested that olivine in the meteorite had transformed to a garnet phase under shock conditions. This interpretation did not appear probable since garnet does not possess olivine stoichiometry and in addition the observed lattice parameter was not consistent with the suggested composition (GELLER, 1969).

Recently BINNS (1970) and BINNS et al. (1969) have discovered garnet in heavily shocked regions of the Tenham chondrite and suggest that this may be the

---

[*] This statement applies to the extensive shock experiments carried out by MCQUEEN et al. (1967). However spinel has been formed under natural shock wave conditions in a meteorite (BINNS et al., 1969).

high pressure polymorph of pyroxene as indicated by the work of RINGWOOD and MAJOR (1966b) and RINGWOOD (1967). Although these workers were unable to establish the composition, the garnet did not appear to contain aluminium. At about the same time SMITH and MASON (1969) re-examined the Coorara garnets and obtained evidence that they were indeed of pyroxene composition. Hence it now seems clear that the meteoritic garnets indeed represent high pressure polymorphs of pyroxene with the approximate composition $(Mg_{0.75}Fe_{0.25})SiO_3$.

The shock wave evidence must be treated with some caution since it does not necessarily pertain to conditions of chemical equilibrium. Nevertheless, when combined with the interpretation based on static high pressure experimentation, it appears probable that the principal initial transformation of Mg-rich pyroxenes is to the garnet structure and that at still higher pressures, transformation of garnet to the denser ilmenite structure may occur.

5.5. *The garnet–ilmenite transition*

In section 5.2, we concluded that an ilmenite polymorph of $MgSiO_3$ would be stable relative to $MgSiO_3$ pyroxene at a pressure between 200 and 300 kb. However, in the previous section we found that $MgSiO_3 \cdot xAl_2O_3$ compositions transformed initially to a garnet structure and that there was also strong evidence that pure $MgSiO_3$ would transform first to this structure. The garnet structure has most of its silicon atoms in 4-fold coordination and is less dense than the ilmenite structure in which all cations are octahedrally coordinated. Consequently it appears likely that $MgSiO_3$ and $MgSiO_3 \cdot xAl_2O_3$ garnets may transform to the ilmenite structure at even higher pressures. This would essentially be a solid solution between $MgSiO_3$ (ilmenite) and Al_2O_3 (which already has the corundum (ilmenite) structure) as suggested by RINGWOOD (1962b, 1966b). CLARK et al. (1962) and BOYD (1964) have made the more specific suggestion that pyrope garnet might transform to an ilmenite-type structure at high pressures.

As attempts to transform silicate garnets at high pressure were unsuccessful, recourse was made to germanate analogue methods described earlier. RINGWOOD and MAJOR (1967c) investigated the system $Mg_3Al_2Ge_3O_{12}$–$Mg_3Al_2Si_3O_{12}$ at 0–170 kb and 1000 °C. Results are shown in fig. 8. Above 70 kb the system is

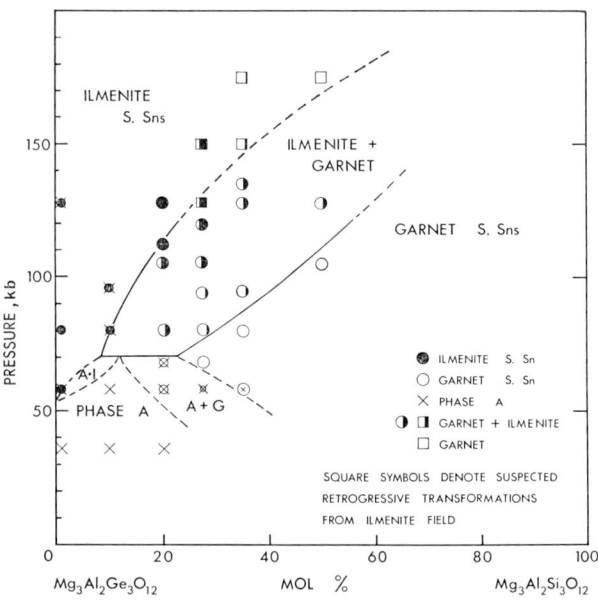

Fig. 8. Phases present in the system $Mg_3Al_2Ge_3O_{12}$–$Mg_3Al_2Si_3O_{12}$ as a function of pressure at about 1000 °C. After RINGWOOD and MAJOR (1967c).

binary and it is clear that Ge–Si pyrope garnets indeed transform to ilmenite structures at high pressure. Some of the points in fig. 8 (squares) illustrate a problem which is likely to arise increasingly in experiments at the highest pressures using quenching methods. Exsolution and retrogressive transformation of ilmenite solid solutions has apparently occurred when pressure was released.

At pressures above 70 kb at 1000 °C the garnet composition $Mg_3Al_2Ge_3O_{12}$ crystallizes as an ilmenite-type solid solution. In view of the model relationships between germanate and silicate systems this result supports the suggestion that silicate garnets may transform to the ilmenite structure deep within the mantle. The existence of a homogeneous series of ilmenite solid solutions containing up to at least 20 mol % $Mg_3Al_2Si_3O_{12}$ is demonstrated. It is possible also that homogeneous ilmenite solid solutions containing up to 50 % of $Mg_3Al_2Si_3O_{12}$ were synthesized during the experiment but were retrogressively transformed on release of pressure. The extensive degree of solid solution of pyrope in the germanium ilmenite structure at high pressure further supports the view that pure pyrope will ultimately transform to an ilmenite polymorph at sufficiently high pressure (following the discussion in section 2). Extrapolation of phase boundaries in fig. 8 suggests that pressures of 200–300 kb would be required to transform pure Si pyrope to an ilmenite structure. Calculations based upon the solid solution relationships (section 2) indicate a transition pressure of only 180 kb. This is probably too low because of non-ideality of the solid solutions at high pressure. Nevertheless the result is considered indicative. The ilmenite structure is about 8% denser than the corresponding garnet. The density of $Mg_3Al_2Si_3O_{12}$ ilmenite is expected to be about 3.85 g/cm^3.

5.6. *The garnet–perovskite transformation*

It was concluded in section 5.3 that calcium bearing pyroxenes in the upper mantle would transform to garnet structures at depths in the mantle of 300–400 km. It is of interest therefore to study the behaviour of the calcium-rich component of natural mantle garnets at high pressure. Since no direct transformation has been found at pressures up to 200 kb, we turn to a study of germanium analogues.

RINGWOOD and MAJOR (1967a) showed that $CaGeO_3$ garnet which is stable at moderate pressures is further transformed to a perovskite structure above 100 kb at 1000 °C. Similarly, $CdGeO_3$ garnet was transformed into a distorted perovskite structure. These results suggest that $CaSiO_3$ in the mantle is likely ultimately to transform to a perovskite structure. Studies of the system $CaGeO_3$–$CaSiO_3$ at high pressure strongly support this. In a series of runs at 170 kb, RINGWOOD and MAJOR (1967a) found that $CaGeO_3$ perovskite will take at least 35 mol% of $CaSiO_3$ into solid solution (fig. 9).

Almost certainly the amount of $CaSiO_3$ dissolved substantially exceeded 35%. However, on releasing pressure, exsolution of the solid solution apparently occurred and proceeded nearly to completion in some cases. In fig. 9, it is seen that in the two phase region, the $CaSiO_3$ content of the perovskite solid solution is much smaller than in the single phase homogeneous region at saturation. This inability to quench a high pressure equilibrium will probably be a serious limitation to further extension of the quenching method.

The extensive solid solubility of $CaSiO_3$ in the perovskite structure implies a small free energy difference between the low pressure form of $CaSiO_3$ and the perovskite form. Calculations using the method outlined in section 2 indicate that a pressure of 200–250 kb should be sufficient to cause $CaSiO_3$ to transform to a perovskite structure. Thus it appears highly probable that in the deep mantle $CaSiO_3$ occurs in this form. In section 3.2 it was noted that pure $CaSiO_3$ transformed in a run above 200 kb to a phase (not perovskite) with a density of about 3.5. The new phase possessed a peculiar undulatory extinction and a poor X-ray diffraction pattern indicative of strain and fine grain size. This phase may represent a retrogressive transformation product from a denser phase on release of pressure. It is possible that the phase actually synthesized in the apparatus was $CaSiO_3$ perovskite.

The perovskite structure is based upon a cubic close packing of the large cations (e.g. Ca^{2+}) and oxygen anions, with the smaller cations, e.g. Ti^{4+}, Ge^{4+}, Si^{4+} occurring in interstices in octahedral coordination with

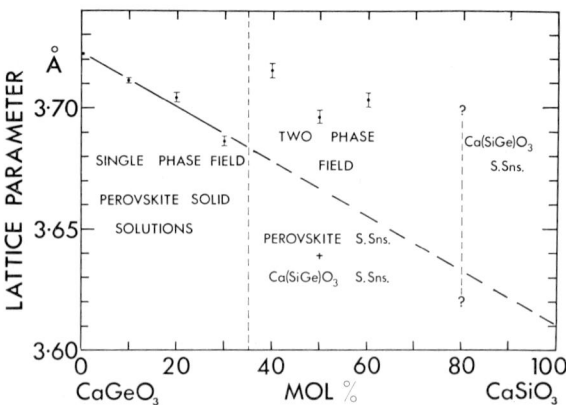

Fig. 9. Solid solubility relations in the system $CaGeO_3$–$CaSiO_3$ at about 170 kb and 1000 °C. Notice the anomalously high lattice parameters in the two phase field, probably due to exsolution of $CaSiO_3$ from perovskite solid solution on release of pressure.

TABLE 11

List of oxides forming binary compounds with $A^{2+}B^{4+}O_3$ perovskite structures, where

AO is usually a rock salt structure,
BO_2 is a rutile or fluorite structure.

AO	BO_2 (rutile)	BO_2 (fluorite)
BaO	GeO_2 (HP)	CeO_2
CaO	MnO_2	HfO_2
CdO	MoO_2	PbO_2 (HP)
(PbO)*	PbO_2	PrO_2
(SnO)*	SnO_2	PuO_2
SrO	(SiO_2) (HP solid solution)	ThO_2
EuO	TaO_2	UO_2
	TiO_2	ZrO_2
	VO_2	TiO_2 (HP)

* Non rock-salt structure.
HP = High pressure phase.

respect to oxygen. This structure is formed by a large number of binary ABO_3 compounds including many of those where AO has a rocksalt structure and BO_2 has a rutile structure (table 11). The large A atoms are surrounded by 12 oxygen anions in the ideal perovskite structure. However many perovskites show some degree of structural distortion, often accompanied by superlattice formation, so that their true symmetry is lower than cubic. The coordination numbers of the cations accordingly may often be smaller than 12.

Where perovskites are formed between rocksalt and rutile type oxides, their densities are characteristically between 5 and 10% *greater* than the isochemical mixed oxides (table 12). Thus the structure is relatively close-

TABLE 12

Molar volumes of perovskites formed from rocksalt and rutile-type end members

Compound	Molar volume V (cm³)	Isochemical oxide volume v (cm³)	$\frac{\Delta v^*}{v}$ (%)
$BaSnO_3$	42.03	46.92	−10.4
$SrSnO_3$	39.52	42.24	−6.4
$CaSnO_3$	36.28	38.31	−5.3
$CdSnO_3$	35.73	37.14	−3.8
$BaMoO_3$	39.73	45.22	−12.1
$SrMoO_3$	37.83	40.54	−6.7
$BaTiO_3$	38.89	44.17	−12.0
$SrTiO_3$	35.87	39.49	−9.2
$EuTiO_3$	35.65	39.29	−9.3
$CaTiO_3$	34.11	35.56	−4.1
$CdTiO_3$	31.76	34.39	−7.6
$CaMnO_3$	31.32	33.37	−6.1
$CaVO_3$	32.02	34.64	−7.6
$BaPbO_3$	46.99	50.39	−6.7
$CdGeO_3$	30.44	32.26	−5.6
$CaGeO_3$	31.08	33.43	−7.0
$CaSiO_3$	28.34	30.77	−7.9

* $\Delta v = V − v$.

packed. The density of $CaGeO_3$ perovskite is 5.17 g/cm³ compared to 4.81 g/cm³ for an isochemical mixture of $CaO + GeO_2$. For comparison the densities of $CaGeO_3$ wollastonite and garnet are 3.98 and 4.44 g/cm³ respectively. The density of $CaSiO_3$ perovskite may be obtained from the extrapolated lattice parameter (3.61 ± 0.01 Å, fig. 9) and is found to be 4.10 ± 0.04 g/cm³. This is 8.6% denser than an isochemical mixture of CaO + stishovite.

The calcium-rich component of natural garnet will generally have the formula $CaSiO_3 \cdot xAl_2O_3$ where for grossularite $x = 25$ wt%. In a perovskite dominantly composed of Ca^{2+} and Si^{4+}, it would be difficult to incorporate much aluminium which is obliged to enter as an $RAlO_3$ component, where R is a large trivalent cation (e.g. Y^{3+}). The low abundance of such large trivalent cations will accordingly severely restrict the amount of aluminium which can enter a $CaSiO_3$ perovskite structure in the mantle. Thus, the transformation of Ca garnet in the mantle into a perovskite will also involve disproportionation according to the equation

$$\underset{\text{(garnet)}}{CaSiO_3 \cdot xAl_2O_3} \rightarrow \underset{\text{(perovskite)}}{CaSiO_3} + xAl_2O_3$$

This transformation has been demonstrated in a germanium analogue system. At low pressures, germanium anorthite $CaAl_2Ge_2O_8$ is found to transform to a mixture of $Ca_3Al_2Ge_3O_{12}$ garnet, Al_2GeO_5 kyanite and GeO_2 rutile. However at much higher pressures the assemblage observed was $CaGeO_3$ perovskite + Al_2O_3 + GeO_2 (unpublished observations). This implies a transformation of germanium grossularite according to the reaction

$$\underset{\text{(garnet)}}{Ca_3Al_2Ge_3O_{12}} \rightarrow \underset{\text{(perovskite)}}{3CaGeO_3} + \underset{\text{(corundum)}}{Al_2O_3}.$$

Attempts to obtain this transformation directly in pure germanium grossularite were unsuccessful, presumably owing to kinetic difficulties. Analogous transformations have been reported by MAREZIO *et al.* (1966) who found that the garnets $Y_3Al_5O_{12}$ and $Y_3Fe_5O_{12}$ disproportionated into mixtures of perovskite and corundum type phases at high pressures e.g.

$$\underset{\text{(garnet)}}{Y_3Al_5O_{12}} \rightarrow \underset{\text{(perovskite)}}{3YAlO_3} + \underset{\text{(corundum)}}{Al_2O_3}.$$

The shock wave results on pure $MgSiO_3$ (MCQUEEN and MARSH, 1966) indicated that it had transformed to a structure with an estimated zero-pressure density of 4.20 g/cm³, compared to 3.98 g/cm³ for an isochemical mixture of $MgO + SiO_2$ (stishovite). Thus the pure enstatite appears to be shocked to a phase with a substantially higher density than the state attained by the shocked bronzite. (Analogous behaviour was displayed by pure forsterite and the cause may be similar—secstion 3.4). RINGWOOD (1962b, 1966b) and REID and RINGWOOD (1969a) suggested that $MgSiO_3$ might transform ultimately to the perovskite structure. From table 12 we see that perovskites are on the average about

7.5% denser than isochemical mixtures of rutile and rocksalt type end members. A perovskite modification of $MgSiO_3$ would therefore be expected to possess a density of about 4.3 g/cm^3, which is in good agreement with the density derived from shock data. Thus, if the shock data for pure $MgSiO_3$ can be taken at face value it appears likely that a perosvkite modification of $MgSiO_3$ may indeed exist at pressures greater than 400 kb.

In the perovskite structure, the larger cations and oxygen anions form a close packed lattice whilst the smaller cations occupy octahedral interstices. The ionic radii of Mg^{2+} (0.66 Å) and O^{2-} (1.40 Å) differ by too much to permit close packing and magnesium is not a constituent of $A^{2+}B^{4+}O_3$ perovskites under normal conditions (table 12). The smallest divalent cations to form a perovskite are Cd^{2+} (0.97 Å) and Ca^{2+} (0.99 Å). The transformation of $MgSiO_3$ to the perovskite structure accordingly requires a large increase in the effective Mg^{2+}/O^{2-} radius ratio so that the effective radii of these ions become more nearly equal. Following the discussion in section 2, this change in effective radius ratio would require a substantial increase in the covalent component of the Mg–O bound under pressure.

The compounds $CdTiO_3$ and $CdSnO_3$ are both dimorphous, displaying perovskite and ilmenite polymorphs at atmospheric pressure, the perovskites being about 8% denser than the ilmenite modifications (ROOYMANS, 1967). The P–T relations for the ilmenite–perovskite transition in $CdTiO_3$ have been studied by LIEBERTZ and ROOYMANS (1965). A notable property is the negative slope (dP/dT) of this transition. If $MgSiO_3$ displays an ilmenite to perovskite transformation in the earth's interior, the occurrence of an analogous negative slope would lead to some interesting geophysical properties.

An ilmenite-perovskite transformation in the mantle would involve the disproportionation of a complex solid solution between $MgSiO_3$ and $(Al, Cr, Fe)_2O_3$.

$$MgSiO_3 \cdot x(Al, Cr, Fe)_2O_3 \rightarrow$$
(ilmenite)

$$MgSiO_3 + x(Al, Cr, Fe)_2O_3.$$
(perovskite) (corundum)

It is possible that the corundum phase might react with Mg_2SiO_4 to form $MgAl_2O_4$ in the dense structure (calcium ferrite?) observed in shock waves (section 3.2) plus more $MgSiO_3$ (perovskite) i.e.

$$Al_2O_3 + Mg_2SiO_4 \rightarrow MgAl_2O_4 + MgSiO_3.$$
(corundum) (strontium plumbate str.) (calcium ferrite) (perovskite)

6. Transformations in alkali aluminosilicates and aluminogermanates

The geochemistry of the alkali metals in the crust has been the subject of a vast amount of research. In contrast, much less is known about the behaviour and distribution of these elements in the mantle. In the crust, the alkali metals occur dominantly as feldspar and other aluminosilicates. Since these open structures are unstable under high pressures, it is clear that the geochemistry of alkali metals in the mantle will be strongly influenced by the nature of the phases into which alkali aluminosilicates transform at depth. A well-known example is the transformation of albite into jadeite plus quartz (BIRCH and LE COMTE (1960)). Another more general example is of the transformation of feldspars to the hollandite structure.

6.1. *The felspar–hollandite transformation*

KUME et al. (1966) showed that at 35 kb, and 1100 °C the germanium analogue of orthoclase, $KAlGe_3O_8$, transformed to a hollandite structure. RINGWOOD et al. (1967a) later succeeded in transforming natural orthoclase to the hollandite structure. The transformation pressure is about 100 kb at 1000 °C. The new phase has a density of 3.84 g/cm^3 and is thus about 50% denser than orthoclase. This large density change is caused by the circumstance that all of the silicon and aluminium atoms are in octahedral coordination.

RINGWOOD et al. (1967b) observed this transformation in the felspars $NaAlGe_3O_8$ and $RbAlGe_3O_8$. Germanium albite first transformed at 15 kb, 1100 °C according to the reaction

$$NaAlGe_3O_8 \rightarrow NaAlGe_2O_6 + GeO_2$$
(albite) (jadeite) (rutile)

At pressures of 25 kb and above (1100 °C), the jadeite and rutile phases recombined to form the hollandite. Analogous behaviour might be expected for common albite which is known to transform to jadeite+quartz at relatively low pressures. In runs up to 150 kb, albite crystallises to jadeite+stishovite. However at still higher pressures, these phases might be expected to recombine to form $NaAlSi_3O_8$ hollandite. Evidence for this is provided by the shock wave studies of MCQUEEN and MARSH (1966) on an albitite, which yielded evidence of major phase transformation. AHRENS et al. (1969) obtained a zero-pressure density of 3.80 for the shocked phase. This is in good agreement with the density expected for a hollandite phase and it appears probable that albite was indeed transformed to hollandite in the shock.

The hollandite structure is tetragonal and based upon the α-MnO_2 structure. This consists of an open framework of MnO_6 octahedra with channels and holes along the c axis, in which are situated the large cations. The structure is closely related to that of β-MnO_2 which has the rutile structure and is characterised by a closer packing of the MnO_6 octahedra.

The general formula of the hollandite group is $A_xB_yC_{8-y}O_{16}$ where A, B and C represent the following ions (BYSTROM and BYSTROM, 1950; BAYER and HOFFMAN, 1966; WADSLEY, 1964):

A	Na, K, Rb, Cs, Ba;
B^{2+}	Mg, Co, Ni, Cu, Zn;
B^{3+}	Al, Ti, Cr, Fe, Ga, In;
C^{4+}	Mn, Ti, Ge, Si, Sn.

There is a considerable degree of flexibility in the ionic substitutions and charge balances permitted and in addition a substantial degree of non-stoichiometry often exists corresponding to deficiencies of A and B cations compared to an ideal formula represented for example by $K_2Al_2Ti_6O_{16}$. (This particular compound represents the case of a titanium analogue serving as a high pressure model for the corresponding germanate and silicate.) A list of felspar-hollandite transformations is given in table 13.

TABLE 13

High pressure hollandite phases
$\Delta\rho/\rho \approx 50\%$*

(a) Obtained under static high pressures
$K_2Al_2Si_6O_{16}$
$K_2Al_2Ge_6O_{16}$
$Rb_2Al_2Ge_6O_{16}$
$Na_2Al_2Ge_6O_{16}$
$BaAl_2Si_6O_{16}$
$SrAl_2Si_6O_{16}$

(c) Inferred to occur under shock conditions
$Na_2Al_2Si_6O_{16}$
$K_2Al_2Si_6O_{16}$
$NaCaAl_3Si_5O_{16}$

* References: see text.

The flexibility in ionic substitutions and the observation that Ba is capable of entering the hollandite structure suggests the possibility that Ba, Sr and Ca felspars may be capable of forming hollandites at high pressure. Under static conditions, anorthite is known to break down to a mixture of grossularite, kyanite and quartz (BOYD and ENGLAND, 1961). This type of breakdown continues to 150 kb except that stishovite occurs instead of quartz (unpublished observations). However, under shock conditions (McQUEEN and MARSH, 1966) an anorthosite of composition $Ab_{51}An_{49}$ transformed to a phase with an estimated zero-pressure density of 3.72 g/cm^3 (AHRENS et al., 1969). It is possible that transformation to a hollandite structure has occurred. Such a transformation would be favoured kinetically during the few microseconds duration of the shock rather than breakdown into a mixture of $NaAlSi_3O_8$ hollandite, grossularite, kyanite and stishovite, involving complex nucleation processes and diffusion.

Support for this interpretation comes from studies of barium and strontium felspars by REID and RINGWOOD (1969b). When subjected to pressures of 100–150 kb at 1000 °C, $BaAl_2Si_2O_8$ and $SrAl_2Si_2O_8$ were observed to disproportionate into hollandite phases plus unknown phases. Under similar conditions glasses with compositions $Ba_{0.5}AlSi_3O_8$ and $Sr_{0.5}AlSi_3O_8$ (compare $KAlSi_3O_8$) crystallized to single phase hollandite. However an attempt to synthesize $Ca_{0.5}AlSi_3O_8$ hollandite was unsuccessful. This behaviour may be interpreted in terms of solid solution formation in the series

$$Ba_2Al_4Si_4O_{16} - BaAl_2Si_6O_{16} - Ba_0Si_8O_{16}.$$

Entry of a barium (or strontium) ion is accompanied by the replacement of two silicon ions by aluminium in order to preserve electroneutrality. It is found that the homogeneous hollandite composition extends as far as about $Ba_{1.5}Al_3Si_5O_{16}$ for barium and $SrAl_2Si_6O_{16}$ for strontium, whilst for calcium, the formation of a hollandite does not occur. The contrasting behaviour apparently arises from the repulsion energies of the large cations in the hollandite structure. These cations occur as neighbours occupying channels in the structure and the number of cations which can be accommodated is evidently governed both by the space available and by the repulsion potential between these ions. For this reason, the hollandite field does not extend as far as $Ba_2Al_4Si_4O_{16}$. The conclusion that anorthosite $Ab_{51}An_{49}$ transformed to a hollandite structure under shock conditions is consistent with this interpretation. This composition could be written $CaNaAl_3Si_5O_{16}$ and is analogous to the barium-rich hollandite $Ba_{1.5}Al_3Si_5O_{16}$ which was synthesized.

RINGWOOD et al. (1967a, b) have also observed several other disproportionation reactions involving hollandite phases. Leucite disproportionated at high pressure into hollandite plus potassium aluminate at 100 kb, 1000 °C. Germanium leucite and germanium kalsilite displayed analogous transformations. The fact that at high pressure, a silica-rich phase such as $KAlSi_3O_8$ hollandite is stable in association with the extremely basic $KAlO_2$ shows that K hollandite would be stable in the conditions of silica saturation which occur in the mantle.

$$3KAlSi_2O_6 \rightarrow 2KAlSi_3O_8 + KAlO_2$$
$$3KAlGe_2O_6 \rightarrow 2KAlGe_3O_8 + KAlO_2$$
$$\text{(leucite)} \qquad \text{(hollandite)}$$

$$3KAlGeO_4 \rightarrow KAlGe_3O_8 + 2KAlO_2$$
$$\text{(kalsilite)} \qquad \text{(hollandite)}$$

It is possible that $KAlSi_3O_8$ hollandite will be discovered as a naturally occurring mineral in impactites associated with meteorite craters. The conditions under which stishovite was formed should also have resulted in the transformation of potassium felspar grains present in the parent rock. However since $KAlSi_3O_8$ hollandite is probably soluble in hydrofluoric acid, the solution method used to recover stishovite may not be successful in the case of $KAlSi_3O_8$.

6.2. Transformation of nepheline and jadeite

RINGWOOD and MAJOR (1967a) observed that $NaAlGeO_4$, which has a nepheline-related structure, transformed at high pressure into a new dense phase. REID et al. (1967) showed that this phase possessed the calcium ferrite structure. The sodium atoms were in eightfold coordination whilst the aluminium atoms were in sixfold coordination. This structure is displayed by a large number of $NaA^{3+}B^{4+}O_4$ compounds and it appears likely that common nepheline $NaAlSiO_4$ will ultimately transform to a calcium ferrite structure which would have a density of about 3.9 g/cm³ (REID et al., 1967). Efforts to synthesize this phase directly were unsuccessful, since nepheline was found to disproportionate initially into jadeite plus a new high pressure phase of $NaAlO_2$ possessing the sodium ferrite structure (REID and RINGWOOD, 1968):

$$2NaAlSiO_4 \rightarrow NaAlSi_2O_6 + NaAlO_2.$$
$$\text{(nepheline)} \qquad \text{(jadeite)}$$

Jadeite is remarkably stable at high pressures. Nevertheless it appears probable that at pressures above 200 kb, the jadeite and sodium aluminate will recombine to form $NaAlGeO_4$ (calcium ferrite structure).

Experiments were carried out on the germanium analogue of jadeite. This was found to disproportionate at high pressures as follows

$$NaAlGe_2O_6 \rightarrow NaAlGeO_4 + GeO_2.$$
$$\text{(jadeite)} \qquad \text{(calcium ferrite)} \qquad \text{(rutile)}$$

It appears plausible that natural jadeite may ultimately disproportionate analogously into $NaAlSiO_4$ (calcium ferrite) $+ SiO_2$ (stishovite).

The fact that sodium is able to form high pressure phases of the type $NaAlSi_2O_6$, $NaAlGe_2O_6$, $NaAlGeO_4$ and probably $NaAlSiO_4$, whereas potassium appears unable to form these phases leads to a significant difference in the geochemical behaviour of these elements in the deep mantle. Potassium probably occurs as the hollandite phase $KAlSi_3O_8$, which, as was noted, was stable in extremely Si-undersaturated environments. However, the sodium hollandite $NaAlSiO_8$ would probably be unstable under the Si-saturation conditions of the mantle because of the preferential formation of $NaAlSi_2O_6$ and at the highest pressures, of $NaAlSiO_4$ (calcium ferrite). Thus sodium and potassium probably reside in different phases, and accordingly may become fractionated from each other.

7. Mineralogical constitution of the mantle

7.1. Chemical composition

The seismic P wave velocity in most regions of the mantle immediately beneath the Mohorovicic discontinuity is usually within the range 8.2 ± 0.2 km/s. This property, together with certain broad petrological and chemical boundary conditions effectively limits the mineralogical composition of the upper mantle to some combination of olivine, pyroxene(s), garnet and, in restricted regions, amphibole. The two principal rock types carrying these minerals are peridotite (olivine–

pyroxene) and eclogite (pyroxene–garnet). Complete mineralogical transitions between these two major rock types are uncommon and usually of local significance only when they occur. A number of petrological and geochemical considerations point to ultramafic rocks as the principal component of the upper mantle, with eclogite a widely distributed accessory component, which however, may be more abundant in some restricted regions. This conclusion is based upon studies of the abundances and origin of mantle-derived xenoliths in kimberlites and alkali basalts, and from an interpretation of the occurrence and origin of alpine peridotites. The relevant arguments are derived from many sources and have been extensively discussed by RINGWOOD (1958c; 1962b, c, d; 1966a, b; 1969b), RINGWOOD and GREEN (1966), GREEN and RINGWOOD (1963; 1967a, b, c) and GREEN (1970), among others.

Although the ultramafic rocks occurring as xenoliths in kimberlites and alkali basalts and as alpine ultramafic intrusions appear to be representative of a widespread layer of the upper-most mantle, detailed studies of the chemical and isotopic compositions of these rocks show that they have been subjected to complex fractionation processes and are not representative of the entire upper mantle (e.g. GREEN and RINGWOOD, 1963, 1967c; STEUBER and MURTHY, 1968; LEGGO and HUTCHINSON, 1968; KLEEMAN and COOPER, 1970; KLEEMAN et al., 1969). One of the most important properties of large volumes of the upper mantle is the capacity to produce basaltic magmas by partial melting processes. Basaltic magmas have been erupted intermittently from the mantle throughout geological time and over all regions of the earth, continental and oceanic. Studies of the major element, trace element and isotopic compositions (refs. see above) of alpine ultramafics and of ultramafic xenoliths in kimberlites and alkali basalts reveal that few, if any, of these rocks would be capable of yielding basaltic magmas by partial melting. These studies suggest that the observed and accessible ultramafic rocks in the categories mentioned above are not primary, parental upper mantle but rather, are residual in nature, representing the unmelted, refractory component which has remained after basaltic magmas have been extracted, following partial melting.

This concept leads to a chemically zoned upper mantle with a layer of fractionated residual ultramafic rocks of varying thickness (up to perhaps 200 km) occurring immediately beneath the Mohorovicic discontinuity, and a more primitive rock-type from which the basaltic magma has not yet been extracted by partial melting, occurring below the fractionated layer (RINGWOOD, 1958c). It follows that this primitive rock type must possess a composition lying somewhere between those of basalt and peridotite. The interpretation of basalt as a *partial* melting product of this primitive rock, which can be justified on many grounds, implies that the composition of this primary rock lies closer to that of peridotite than to basalt. A number of geochemical and petrological considerations indicated a composition lying between basalt:peridotite ratios of 1:1 to 1:4. Since the chemical (including trace element and isotopic) composition of the parental rock required to produce basalt on partial melting is rarely if ever matched by naturally occurring ultramafic rocks, a new name for this hypothetical parental rock type appeared desirable and it was called "pyrolite"*, the name being

---

* It should be noted that the definition of pyrolite is based upon *chemical composition* and *not upon mineralogy*. Furthermore the composition of pyrolite is *defined* by the property that it is *required* to produce a basaltic magma upon partial melting, leaving behind a residual, refractory peridotite. This emphasis is necessitated by recent articles by O'HARA (1968, 1970) who has apparently failed to understand the basis of the pyrolite hypothesis. O'Hara incorrectly asserts that pyrolite is defined by a particular mineralogical composition (olivine+aluminous pyroxenes) and by a unique chemical composition, and attempts to prove that this composition and mineralogy cannot be the source-rock of basaltic magma. Irrespective of his detailed arguments, which I reject, this misses the whole point that the pyrolite composition is defined so as to be able to produce basalt, and if a particular assumed model composition were shown to require for example a minor alteration to the CaO/Al_2O_3 ratio (as argued by O'Hara) in order to achieve this, then this change is perfectly in order and the model composition of pyrolite may be modified accordingly.

I would also take this opportunity to refute another criticism voiced by O'Hara with respect to GREEN and RINGWOOD's (1967a) study of basalt genesis in a pyrolite upper mantle. O'Hara states that because the olivine tholeiite investigated by Green and Ringwood did not have olivine on its liquidus at 15–17 kb, it could not have been formed by partial melting in the mantle and its composition is therefore irrelevant for considerations of basalt genesis. O'Hara's discussion is worded so as to give the impression that a new argument is being introduced, and one which had been overlooked by Green and Ringwood. It is to be regretted that he did not refer to page 166 of Green and Ringwood's paper where this point was considered and its importance evaluated. It was shown there that the liquid forming by partial melting of pyrolite at 15–17 kb would have contained a few percent more normative olivine than the composition actually investigated, but otherwise, the fractionation trends would be unaltered. This has since been fully verified by additional experiments (GREEN, 1970).

derived from the predominant pyroxene–olivine mineralogy which rocks within this composition range would adopt in the mantle.

The composition of pyrolite depends upon the assumed ratio of basalt to peridotite (within the 1:1–1:4 limits) and upon the actual compositions of the basaltic and peridotitic end members chosen. Thus there is a good deal of flexibility in the concept and it is probable that the composition of pyrolite varies significantly throughout the upper mantle. In order to test the model quantitatively it is necessary to set up specific models. A number of these have been constructed during the last few years as understanding of the various equilibria involved in basalt genesis has improved. The latest and most satisfactory of these (RINGWOOD, 1966a) is derived from a mixture of 75% alpine peridotite (20% orthopyroxene) and 25% primitive Hawaiian basalt (table 14). Alternatively, the latter might be replaced by oceanic tholeiite. An improved understanding of the geochemical relationships between basalts and peridotites has strongly suggested that the peridotite:basalt ratio for pyrolite is greater than 1:1 and is probably not far from 3:1. Nevertheless there remains some flexibility in this ratio and revised model compositions may be derived in the future as additional constraints are uncovered.

The pyrolite model composition (table 14) is based upon interpretation of evidence obtained from the upper mantle, and accordingly cannot be applied to the composition of the entire mantle without further justification. Recent advances in knowledge of the abundances of elements in the sun and in meteorites, and of the origin of the earth by accretion processes in a primitive solar nebula provide important constraints upon the composition of the entire mantle.

Studies of chondritic meteorites have revealed a class, the type I carbonaceous chondrites, which have had a relatively simple chemical and thermal history, and may possess a chemical composition which is similar in many respects to that of the dust particles in the primordial solar nebula. The abundances of elements in these chondrites are plotted against the solar photosphere abundances in fig. 10 (after RINGWOOD, 1966a). When due allowance is made for experimental uncertainties, the agreement is seen to be remarkably good. On these, and on other grounds it appears reasonable to assume that the earth has been ultimately derived from material with the type I carbonaceous chondrite abundances. Furthermore there are grounds for believing that strong fractionation of the less-volatile elements during this process has not taken place. Accordingly, the chemical composition of the entire mantle (about $\frac{2}{3}$ the mass of the earth) is probably rather similar to that of the carbonaceous chondrites for non-volatile elements, after allowance has been made for the partition of elements in the earths core (RINGWOOD, 1960, 1962e, 1966a, b, c). The general pattern of chemical abundances in the sun, in all classes of chondritic meteorites, in cosmic rays, and also arising out of nucleosynthesis considerations implies that in the mantle, the elements Mg, Si and O are about an order of magnitude more abundant than the next class Ca, Al, Na and these in turn are 5 to 10 times as abundant as the next group K, Cr, Mn, P, Ti. The abundance of iron in the sun is the subject of some controversy, nevertheless, geophysical limitations show clearly that its abundance in the mantle is intermediate between those of (O, Mg, Si) and (Ca, Al, Na) and closer to the latter group.

The mineralogy of the mantle is largely controlled by this gross abundance pattern. The system SiO_2–MgO–FeO is fundamental to the entire mantle and to a first approximation, the mineralogy can be considered in terms of phase stability fields in this system. The influence of the next most abundant components Ca, Al, Na can then be treated as a further refinement.

It is instructive to compare the pyrolite composition of the upper mantle with the composition of the entire mantle as obtained from the primordial abundances,

TABLE 14

Model pyrolite composition after RINGWOOD (1966a)

	I
SiO_2	45.16
MgO	37.49
FeO	8.04
Fe_2O_3	0.46
Al_2O_3	3.54
CaO	3.08
Na_2O	0.57
K_2O	0.13
Cr_2O_3	0.43
NiO	0.20
CoO	0.01
TiO_2	0.71
MnO	0.14
P_2O_5	0.06
	100.00

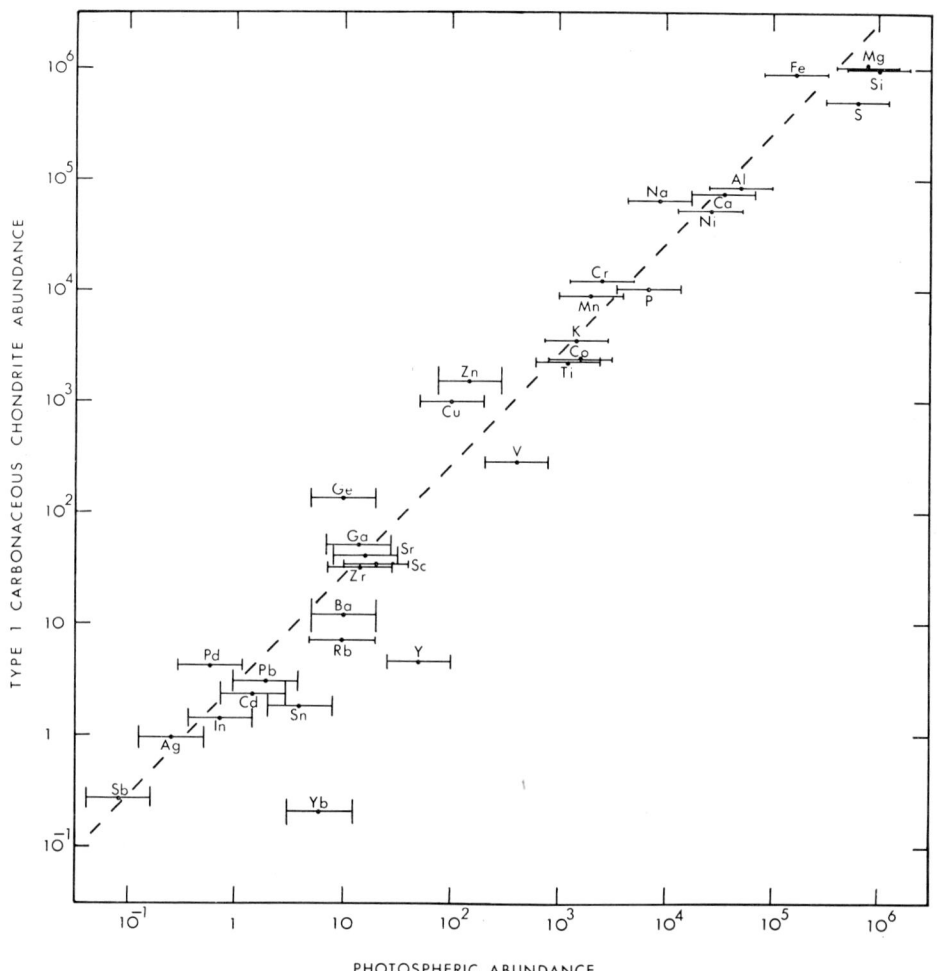

Fig. 10. Comparison of elemental abundances in type I carbonaceous chondrites with those in the solar photosphere. All abundances normalized on the basis of Si = 10^6. Horizontal bars correspond to a twofold uncertainty in the solar abundance. Vertical bars represent conservative estimates of uncertainties in the type I carbonaceous chondrite abundances. After RINGWOOD (1966c).

for the more abundant elements (table 15). It is evident that there is a strong similarity in the gross abundance patterns between the composition of the upper mantle* and the composition of the entire mantle. This is of considerable genetic significance and suggests that the mantle has been well stirred and mixed, perhaps by convection during and subsequent to core formation (RINGWOOD, 1966a).

* In the earliest formulation of the pyrolite model, the basalt to peridotite ratio was chosen so as to maximize this agreement, which was not therefore surprising. However the accumulation of subsequent petrologic and geochemical evidence points towards a peridotite/basalt ratio of about 3/1 and the formulation of the model pyrolite composition is entirely independent of chondritic abundances.

TABLE 15

Relative atomic abundances in pyrolite, type I carbonaceous chondrites and the solar photosphere

Element	Pyrolite	chondrite (carb. I)	Solar photosphere
Si	100	100	100
Mg	124	105	58
(Fe)	(16)	(89)	(12)
Al	9.5	8.5	6.8
Ca	7.3	7.4	7.2
Na	2.4	6.4	6.0
Ti	1.2	0.2	0.1
Cr	0.8	1.2	0.5
K	0.4	0.4	0.2
Mn	0.3	0.1	0.2
P	0.1	1.1	3.2

The similarity between pyrolite and primordial abundance patterns suggests that the pyrolite model composition for major elements may be applicable to the entire mantle. We will explore this hypothesis by investigating the sequence of phase changes, which, in the light of evidence reviewed earlier, would be expected to occur in material of pyrolite composition with increasing depth in the mantle. We will then enquire to what extent phase changes in pyrolite are capable of explaining the observed and inferred distribution of physical properties with depth. We will, however, treat the $Mg/(Mg+Fe)$ ratio (atomic proportions) as an independent variable in order to discuss the possibility of changes in this ratio with depth as have been suggested by several authors. The $Mg/(Mg+Fe)$ ratio of pyrolite is 0.89 (atomic). Initially, we will assume this ratio to be constant throughout the mantle. The effects of varying this ratio will then be explored. Before commencing, however, a brief discussion of seismic velocity distributions in the mantle is required.

7.2. *Seismic velocity distributions*

The seismic velocity distributions of JEFFREYS (1937, 1939) and GUTENBERG (1958, 1959) were based upon first arrivals and upon smoothed travel time data, and hence the derived velocity distributions between 400 and 1000 km were also smooth. This was held by some to be an objection against the proposal that phase changes occurred in this region (VERHOOGEN, 1953). My earlier attempts to explain this velocity distribution in terms of phase transformations attempted to use the smearing-out effects associated with solid solution formation (RINGWOOD, 1958c, 1962b, 1966b). Attempts in this direction were also made by MEIJERING and ROOYMANS (1958). In the light of evidence then available, this explanation did not appear unreasonable. However, subsequent experimental investigations in relevant systems (AKIMOTO and FUJISAWA, 1968; RINGWOOD and MAJOR, 1970; this paper) have shown that the two phase loops were generally narrower than anticipated, and that several important phase transitions in the mantle would occur within limited depth intervals producing relatively sharp velocity increases, and in some cases, discontinuities.

Fortunately, parallel developments in seismology have demonstrated that the velocity distribution between 400 and 1000 km is not smooth but that two or more seismic discontinuities are probably present, as shown for example in fig. 1. The first hint of this structure arose from the surface wave investigations of D. L. ANDERSON and TOKSÖZ (1963), D. L. ANDERSON (1964) and KOVACH and D. L. ANDERSON (1964), which indicated the presence of abnormally large increases in velocity concentrated in two relatively narrow zones 50 to 100 km thick around 400 km and 700–800 km. When these results were announced, RINGWOOD (1966b, p. 390) suggested that the high velocity gradient between 400 and 500 km might be caused by the olivine–spinel transformation and by the transformation of pyroxene, whilst the high gradients around 800–900 km might be caused by the further transformation of spinel into ilmenite and oxide structures.

The velocity distributions proposed by Anderson and co-workers have been confirmed in general by a number of subsequent investigations of mantle body waves (NIAZI and D. L. ANDERSON, 1965; JOHNSON, 1967; KANAMORI, 1967; HALES et al., 1968; JULIAN and D. L. ANDERSON, 1968; ARCHAMBEAU et al., 1969). There is general agreement concerning the presence of a major seismic discontinuity near 400 km and of a further discontinuity, or alternatively a zone of high velocity gradient, near 650 km. More recently, BOLT et al. (1968) and BOLT (1969) have obtained evidence suggesting the presence of two possibly first order discontinuities near 400 km. D. L. ANDERSON (1967) has proposed an explanation for the discontinuities similar to that advanced by RINGWOOD (1966b), i.e. the 400 km discontinuity caused dominantly by the olivine–spinel transition whereas the 650 km discontinuity is caused by a "post-spinel" transition into a state with properties resembling the mixed oxides.

The presence of seismic discontinuities or regions of anomalously high velocity gradient in the mantle below 800 km is also indicated by the investigations of ARCHAMBEAU et al. (1969) and JOHNSON (1969). The former authors suggest that a substantial increase in velocity (0.4 km/s) may occur near 1050 km (fig. 1). This feature is not regarded as uniquely established: "a single change or perhaps several small but rapid changes in velocity gradient occur in the range from 1000 km to 1200 km" (ARCHAMBEAU et al., 1969). They believe that the region of uniform gradient (fig. 1) between 700 km and 1000 km is reliably determined from the combined data of all their profiles.

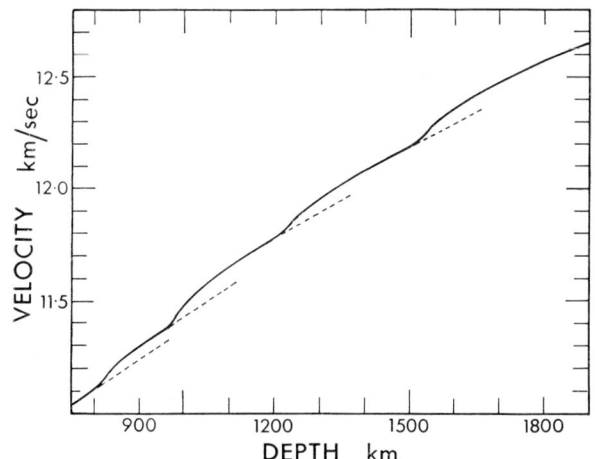

Fig. 11. P wave velocity distribution in the lower mantle. After JOHNSON (1969).

JOHNSON (1969) has carried out a detailed investigation of P velocity in the lower mantle using array data. He inferred the presence of anomalously large velocity gradients near 830, 1000, 1230, 1540 and 1910 km (fig. 11). The high-gradient regions are spread over depth intervals of about 50 km. These regions account for an integrated velocity increase of 2.7% which is about 17% of the total velocity increase between 800 and 2300 km (21% of total velocity increase between 800 and 1700 km, see fig. 11). Johnson suggests the possibility that phase changes might be responsible for these features. There appears to be a significant difference between the Johnson and Archambeau et al. solutions in the region between 800 and 1200 km. The latter authors obtain a much larger increase of velocity from discontinuities or anomalously high gradients (≈ 0.4 km/s) in this region than does Johnson (≈ 0.2 km/s). This is seen to result from the higher average gradient for "normal" regions between 700 and 1000 km obtained by Johnson, as compared with the gradient for this interval by Archambeau et al. and considered to be well established. This region warrants further study. KANAMORI (1967) using an array also finds evidence of an abnormally high gradient around 950 km.

Although further refinements to the velocity distributions appear desirable, the general pattern of the Johnson and Archambeau et al. velocity distributions is strongly suggestive of the presence of further phase transformations to denser and seismically faster structures occurring below 800 km, although the combined magnitude of these phase changes is 5 to 10 times smaller than the major phase changes occurring between 360 and 700 km. (A possible alternative explanation of these abnormal velocity gradients in terms of sudden decreases in the iron content of the mantle can be dismissed on the basis of arguments concerning the density of the lower mantle which are developed in later sections.) If the phase change interpretation is confirmed by future investigations it implies that the zero pressure density of mineral phases in the lower mantle will be significantly higher (perhaps by 5% after allowing for normal self compression) than the density of the isochemical mixed oxides.

7.3. Mineralogy in a pyrolite mantle as a function of depth

We now proceed to set up a model which depicts the most probable variation of mineralogy with depth for the pyrolite composition (table 14) based on the experimental data described in previous chapters. It is emphasized that we are dealing with a *model* subject to future changes as new and improved experimental data become available. The mineralogy down to 600 km is based directly upon the results of high pressure experiments which reproduce the P–T conditions down to this depth. Below 600 km, the model is based largely upon indirect evidence such as germanate analogue systems.

7.3.1. The region between depths of approximately 150–350 km.
The P velocity distributions of ARCHAMBEAU et al. (1969) (see fig. 1) and JOHNSON (1967) show a uniform and relatively small rate of increase of velocity with depth in this region. The observed velocity gradients, in the light of BIRCH's (1952) analysis suggest that this region is essentially homogeneous. This inference is supported by experimental data. High-pressure investigations show that throughout the P–T conditions in the region, pyrolite would crystallize to a mineral assemblage comprised of olivine, orthopyroxene, clinopyroxene (both low in alumina) and garnet. No further phase transformations which might cause velocity anomalies have yet been discovered despite intensive search.

The detailed mineralogy for the pyrolite composition is as follows:

		Wt. %
Olivine (Fo$_{89}$)	(Mg,Fe)$_2$SiO$_4$	57
Orthopyroxene	(Mg,Fe)SiO$_3$	17
Omphacitic clinopyroxene		
	(Ca,Mg,Fe)$_2$Si$_2$O$_6$–NaAlSi$_2$O$_6$ s.sn.	12
Pyrope-rich garnet	(Mg,Fe,Ca)$_3$(Al, Cr)$_2$Si$_3$O$_{12}$	14

The density (ρ_0) of this mineral assemblage reduced to the value which it would display at atmospheric pressure and temperature, is 3.38 g/cm^3. The minerals are characterized by 4-fold coordinated Si and by 6- and 8-fold coordinated Mg, Fe and Ca (with respect to oxygen).

7.3.2. Seismic discontinuity and high velocity gradients between approximately 360 and 420 km. This depth interval contains one of the two major seismic discontinuities within the mantle (fig. 1). The feature is clearly recognised on all recent studies of seismic velocity distribution with depth. The experimental investigations reviewed previously demonstrated that two major transformations would occur in pyrolite near this pressure range:

(i) the pyroxene–garnet transformation

and

(ii) the olivine–spinel–β-Mg$_2$SiO$_4$ transformation.

The former involves the solid solution of pyroxene components in the garnet structure, which would set in rather suddenly at about 105 kb, 1000 °C (figs. 5 and 6). The temperature gradient of this transition is not known. If it is normal, with a value of about 20–30 b/°C the pyroxene–garnet transformation would occur between 350 and 400 km. It involves an effective density increase of about 10% for the pyroxene component which comprises about 25% of pyrolite, and results in the formation of a complex garnet solid solution characterized by partial octahedral coordination of silicon as discussed in section 5.3. The garnet solid solutions possess the general formula M$_3$Al$_2$Si$_3$O$_{12}$–M$_3$(MSi)Si$_3$O$_{12}$ where M = Mg, Fe, Ca.

If the gradient dP/dT of the pyroxene–garnet transformation is normal, it will slightly precede the next major transformation which is of olivine through the spinel into the β-Mg$_2$SiO$_4$ structure. This transformation involves a density increase of about 8%, but since it affects a major component of the mantle (olivine 57%) a large increase of velocity is caused. This transformation, first suggested by JEFFREYS (1937) and BERNAL (1936), was discussed in detail in the accompanying paper (RINGWOOD and MAJOR, 1970). It occurs over a depth interval of about 27 km centered around a depth of 397 km (for T ≈ 1600 °C as suggested by CLARK and RINGWOOD, 1964). Within this depth interval, a reaction point between spinel and β-Mg$_2$SiO$_4$ produces a discontinuous change in mineralogy and density, and accordingly, a first order seismic discontinuity will be caused.

The experimental data thus indicate the presence of two seismic discontinuities in the 360–420 km region, together with boundary regions of high velocity gradient. Recent investigations by BOLT et al. (1968) and BOLT (1969) have obtained evidence for seismic reflections from 400 km indicating a first order discontinuity, together with the possibility of one or more first order discontinuities at shallower depth. The overall agreement between seismological observations and experimental results for a pyrolite model is most impressive.

7.3.3. The region between 420 and 600 km. The P velocity gradients in this region increase at a constant and relatively small rate with depth (ARCHAMBEAU et al., 1969; fig. 1; JOHNSON, 1967). The rate of velocity increase suggests that this region is homogeneous, following the arguments of BIRCH (1952). Extensive investigations of the stabilities of silicate garnets, silicate spinels and beta phases have been carried out in my laboratory up to pressures equivalent to depths of 600 km. No further phase transformations have been discovered. The experimental investigations therefore imply that a mantle of pyrolite composition would be homogeneous between 420–600 km in agreement with the seismic observations. The mineralogy of this region would be as follows.

	Wt. %
Spinel-like phase – β-(Mg,Fe)$_2$SiO$_4$	57
Complex garnet solid solution	
M$_3$(Al,Cr,Fe)$_2$Si$_3$O$_{12}$–M$_3$(MSi)Si$_3$O$_{12}$	39
where M = Mg, Fe, Ca	
Jadeite NaAlSi$_2$O$_6$	4

The density (ρ_0) of this mineral assemblage, reduced to atmospheric pressure and temperature would be 3.62

g/cm³. Coordination of silicon atoms by oxygen is mainly fourfold, with some Si^{VI} in the garnet.

7.3.4. The 650 km "discontinuity".

Fig. 1 shows a major seismic discontinuity occurring close to 650 km. This feature is also shown by most other recent seismic studies (e.g. JOHNSON, 1967). The pressure at a depth of 650 km is about 230 kb, which is just beyond the capabilities of the experimental apparatus used in the author's laboratory. Accordingly the study of possible phase transformations in this region must be conducted by indirect methods or by shock wave methods. The success of indirect methods used prior to 1966 in predicting the occurrence of phase transformations in olivines and pyroxenes at depths less than 600 km, provides added confidence in their applicability at depths greater than 600 km.

The germanate model studies suggested that garnets and beta-phases would transform at higher pressures into new phases possessing ilmenite, strontium plumbate and perovskite structures. The densities of this assemblage of phases would be similar to those of the isochemical mixed oxides. Thermodynamic investigations based on the observed solid solubility of silicates in the high pressure germanate structures yielded calculated pressures in the range 200–300 kb for the above transformations in silicates. These estimates, although admittedly imprecise, are in good agreement with the pressure at a depth of 650 ($\approx$ 230 kb) when the various sources of uncertainty are taken into account. Accordingly it may be concluded that the pyrolite model is capable of explaining the presence of the major seismic discontinuity near 650 km. The principal phase transformations which appear to be involved are as follows:

1. β-$(Mg,Fe)_2SiO_4$ transforms to the strontium plumbate structure.
2. Pyrope-rich component of garnet transforms to the ilmenite structure.
3. Calcium-rich component of garnet transforms to the perovskite structure.
4. Jadeite disproportionates to yield the calcium ferrite structure.

An alternative to (1) which is considered slightly less probable would be for β-$(Mg,Fe)_2SiO_4$ to disproportionate to an $(Mg,Fe)SiO_3$ ilmenite phase plus free $(Mg,Fe)O$.

It is unlikely that these transformations would occur at exactly the same depth. By far the most important transformation would be of beta $(Mg,Fe)_2SiO_4$ to the strontium plumbate structure. This transformation may well occur within a relatively narrow depth range, and is probably primarily responsible for the occurrence of the 650 km seismic discontinuity. The transformations of garnets to ilmenite and perovskite structures involve the formation of very complex solid solutions. They will probably be smeared out more than the strontium plumbate transformation and it would be coincidental if they were to occur at exactly the same pressure. It appears that they may contribute to a rather complex fine structure in this region which has yet to be resolved by seismology.

7.3.5. The region between 700 and 1050 km.

Assuming that the above transformations are complete by about 700 km, the mineralogy below this depth would consist of:

	Wt.%
1. Strontium plumbate structure $(Mg,Fe)_2SiO_4$ (alternatively $(Mg,Fe)SiO_3$ (ilmenite)+$(Mg,Fe)O$)	55
2. Ilmenite-type solid solution $(Mg,Fe)SiO_3$–$(Al,Cr,Fe)AlO_3$	36
3. Perovskite $CaSiO_3$	6.5
4. Calcium ferrite $NaAlSiO_4$	2.5

The zero-pressure density of this mineral assemblage is 3.99 g/cm³. It is probable that there would be some solid solubility of $FeSiO_3$ and minor components such as $MnSiO_3$ in the $CaSiO_3$ perovskite. Allowance for this effect could bring the zero-pressure density of this assemblage up to about 4.03 g/cm³. This mineral assemblage is thus characterized by 6-fold coordination of Mg, Fe and Si. As discussed in sections 3 and 5 the mean zero-pressure density is almost identical to the mean density of an isochemical mixture of component oxides (SiO_2 as stishovite).

The seismic velocity-depth gradients of ARCHAMBEAU et al. (1969) (fig. 1) in the region between 700 km and about 1000 km are consistent with homogeneity following the arguments of BIRCH (1952). If this is confirmed, this region may be interpreted as consisting of the mineral assemblage discussed above. JOHNSON's (1969) gradient in this region is higher but may nevertheless be consistent with homogeneity except for a small anomaly near 800 km (fig. 11).

7.3.6. Phase transitions in the lower mantle. The general nature of the phase transformations responsible for increasing the zero-pressure density of pyrolite up to that of the equivalent isochemical mixed oxides appears to be fairly well understood. If we accept the above interpretation that these transformations are complete by about 700 km, then the possibility of further transformations to denser states at greater depths must be considered. The seismic velocity distribution (fig. 1) shows a sharp increase of 0.4 km/s at 1050 km. If verified, this may well represent a further phase transformation. The regions of anomalously high velocity gradient located by JOHNSON (1969) (fig. 11) are likewise readily interpreted in terms of phase transformations, as previously noted. It should be pointed out on completely general grounds that further phase transformations in the deep mantle over the pressure range 300 to 1300 kb are not unlikely, when considered in relation to the number of transformations which occur at pressures less than 300 kb. Perhaps the most surprising aspect is the small total magnitude of the density changes which is permitted for possible further transformations by constraints arising from the density distribution and elasticity of the deep mantle.

Possible transformations leading to densities higher than the isochemical mixed oxides have already been discussed in previous sections. These result in mineralogies characterised by Mg and Fe coordinations higher than 6. A possible mineral assemblage for pyrolite composition which is suggested by analogue studies is as follows:

$(Ca,Mg,Fe)SiO_3$	perovskite structure
$NaAlSiO_4$	calcium ferrite structure
$(Mg,Fe)(Al,Cr,Fe)_2O_4$	calcium ferrite structure
$(MgFe)_2SiO_4$	calcium ferrite or K_2NiF_4 structure.

Alternatively, $(Mg,Fe)_2SiO_4$ may disproportionate to form $(Mg,Fe)SiO_3$ perovskite plus $(Mg,Fe)O$. The zero-pressure density of this assemblage would be about 5% higher than that of the mixed oxides.

7.4. Magnitude of velocity changes at seismic discontinuities

We have seen that phase transformations which are known or expected to occur on the basis of results from static high pressure experimentation in a mantle of pyrolite composition provide an excellent qualitative explanation of the principal features of the seismic velocity distribution in the mantle. Specifically, they predict the occurrence of major phase transformations and accompanying large velocity increases near 400 km, and between 600 and 900 km and also the presence of homogeneous regions between 150–360 km and 420–600 km. We now examine whether the inferred phase changes are capable of providing a quantitative explanation of the magnitudes of the velocity increases associated with the seismic discontinuities.

The basis for this comparison rests upon empirical relationships between seismic velocity V_P, density (ρ) and mean atomic weight for many silicates and oxides, established by BIRCH (1961a). Birch showed that for constant mean atomic weight, the simple relationship $V_P = a+b\rho$ was widely applicable, where the constant of proportionality b varied from 2.5 to 4.5, but was most often close to 3.2. The most relevant solution for the present purposes is given on table 15, no. 3 in BIRCH's (1961a) paper and is based upon accurately measured single crystal values. This gives $b = 3.16$ (km/s)/(g/cm^3). These empirical relationships have been developed by O. L. ANDERSON and NAFE (1965) and D. L. ANDERSON (1967b) to give simple equations connecting bulk modulus K_S and elastic ratio ϕ (equal to $K_S/\rho = V_P^2 - \frac{4}{3}V_S^2$) with density ρ.

In the previous section, the zero-pressure densities of the mineral assemblages believed to occur in different regions of the mantle were calculated. Using Birch's empirical relationship, the density changes caused by transformations from one mineral assemblage to another can be converted into equivalent changes in seismic velocity. Before these can be compared with the changes in seismic velocity inferred at the discontinuities, a correction for normal self compression within the mantle must be applied, so that the density changes occurring under the P–T conditions existing in the mantle at the discontinuities may be estimated.

This correction may be applied using the Birch–Murnaghan equation

$$P = \tfrac{3}{2}K_0\left[\left(\frac{\rho}{\rho_0}\right)^{\frac{7}{3}} - \left(\frac{\rho}{\rho_0}\right)^{\frac{5}{3}}\right]$$

where P is the pressure and K_0 is the bulk modulus of

the particular mineral assemblage at zero pressure. Values of K_0 for the various mineral assemblages can be estimated if the zero-pressure densities are known, using the BIRCH (1961a) compressional velocity–density relationship, which has been extended by WANG (1967) to a corresponding relationship between density and hydrodynamical sound velocity C. Since C is equal to $\sqrt{(K/\rho)}$ and is proportional to ρ, we have K proportional to ρ^3. Known bulk moduli are therefore plotted against the cube of density for a variety of compounds and minerals of constant mean atomic weight and the constant of proportionality determined. This relationship then serves to determine the bulk modulus of an unknown structure if the density is known. Alternative empirical methods such as those used by O. L. ANDERSON and NAFE (1965) and D. L. ANDERSON (1967b) may be employed with essentially similar results. The estimated zero pressure bulk moduli and densities obtained in this manner for different pyrolite mineral assemblages are given in table 16. Using these values the variation of density with pressure for each assemblage was calculated using the Birch–Murnaghan equation, thus yielding the density difference between assemblages as a function of pressure. This was converted into an equivalent velocity change by means of the Birch V–ρ relationship. In addition a small correction to the density changes to allow for the high temperature of the mantle was applied, using the procedure of CLARK and RINGWOOD (1964).

TABLE 16

Estimated zero pressure densities, bulk moduli and elastic ratios (ϕ) for pyrolite mineral assemblages

	ρ_0 (g/cm^3)	K_0 (kb)	$\phi_0 = K_0/\rho_0$ (km/s)^2
Olivine+pyroxenes+garnet	3.38	1300	38
β-(MgFe)$_2$SiO$_4$+garnet	3.62	1700	47
(MgFe)$_2$SiO$_4$ strontium plumbate str.+(MgFe)SiO$_3$·(AlCrFe)$_2$O$_3$ ilmenite str.+CaSiO$_3$ perovskite str.+NaAlSiO$_4$ calcium ferrite str.	3.99	2260	57
(MgFe)$_2$SiO$_4$–NaAlSiO$_4$–Mg(AlCrFe)$_2$O$_4$ calcium ferrite s.sns.+(MgFeCa)SiO$_3$ perovskite str.	4.20	2800	67

Results were as follows:
The increase in density due to phase changes near 400 km was found to be 0.22 g/cm^3 or 6%. This corresponds to a velocity change of 0.7 km/s, which may be compared with 0.65 km/s found by JOHNSON (1967) for the 400 km discontinuity and 0.9 km/s obtained by ARCHAMBEAU et al. (1969) (fig. 1). Considering the uncertainties, the agreement is satisfactory. At 640 km the transformation of the garnet–β-Mg$_2$SiO$_4$ assemblage to the "post-spinel" assemblage is accompanied by a density increase of 0.31 g/cm^3 which is equivalent to a velocity increase of 1.0 km/s (BIRCH, 1961a, solution 3). This compares with 0.9 km/s found by Johnson and 1.1 km/s found by Archambeau et al. Again the agreement is satisfactory considering the uncertainties. If we consider the total density increase due to phase changes at 400 and 640 km, extrapolated so as to have all the transformations occurring at an effective depth of 640 km, the density increase as garnet pyrolite transforms to the post-spinel assemblage is 0.48 g/cm^3 equivalent to 1.5 km/s. This compares with 1.5 km/s extrapolated from Johnson's velocity distribution and 2.0 km/s from the Archambeau et al. velocity distribution.

The conclusion following from the previous discussion is that increases of velocity arising out of phase transformations between 360 and 700 km in a mantle of pyrolite composition are *quantitatively consistent* with the density changes caused by the transformations. In particular, they do not demand any increase in the iron to magnesium ratio with depth in order to maintain consistency. This differs from a conclusion reached by D. L. ANDERSON (1967a, 1968) and PRESS (1968) that a twofold increase in Fe/(Fe+Mg) ratio is required through the transition zone if the changes in seismic velocities are to be reconciled with density changes arising out of phase transformations.

It is emphasized that the present arguments, to this stage, do not preclude the possibility of a small change in iron content with depth. The situation is that a change in iron content is not *required* in order to achieve consistency between phase changes and accompanying velocity changes in a pyrolite mantle when adequate allowance is made for the various sources of uncertainty arising from the data and the simplifying assumptions made in treating this data. The sources of uncertainty include:

(1) The magnitude of the seismic velocity change between 360 and 700 km which is caused by phase changes. The velocity changes obtained from the John-

son and Archambeau et al. velocity distributions differ substantially.

(2) The uncertainty in the proportional factor b in the Birch velocity–density relationship which varies between 2.7 and 3.6 for common rocks and minerals at atmospheric pressure and temperature* (granites and plagioclases fall well outside this range).

(3) The unknown effect of high temperature in the mantle upon b.

(4) Paucity of knowledge concerning the probable b values for closely packed silicate structures characterized by octahedrally coordinated silicon.

In the light of these uncertainties, the case for an increase in iron content with depth, based upon velocity-density relationships, appears decidedly weak.

A second argument supporting an increase in $Fe/(Fe+Mg)$ ratio near 400 km arose out of attempts to match an extrapolated phase diagram for the system Mg_2SiO_4–Fe_2SiO_4 (AKIMOTO and FUJISAWA, 1968) with the velocity distribution and probable temperature near 400 km (D. L. ANDERSON, 1967a). The argument possessed some merit as long as the extrapolated phase diagram was assumed to be correct. However a direct determination of equilibria in the Mg-rich portion of this system (RINGWOOD and MAJOR, 1970) showed that the postulated diagram was in error. RINGWOOD and MAJOR (1970) showed that when the new phase diagram for this system is used, a satisfactory explanation of the 400 km discontinuity emerges without any need to alter the $Fe/(Mg+Fe)$ ratio in order to achieve consistency.

Finally we consider the magnitude of density increases which are associated with discontinuities or abnormally high velocity gradients in the deep mantle.

---

* ANDERSON (1967b) and WANG (1967) have shown that the relationships of hydrodynamical sound velocity C and elastic ratio ϕ (equal to C^2) to density for oxide systems of similar mean atomic weight are much more regular than the corresponding Birch relationship between compressional velocity and density. This arises because the former relationship involves only the bulk modulus, through $\phi = K/\rho$, and it is found that K is comparatively "well behaved" (O. L. ANDERSON and NAFE, 1965) and rather simply related to repulsion potentials between atoms. However compressional wave velocity V_P involves both the bulk modulus and shear modulus μ through $V_P = \sqrt{(K+\frac{4}{3}\mu/\rho)}$. The shear modulus and its derivatives vary in a complex and erratic fashion and are structure-sensitive (O. L. ANDERSON and LIEBERMANN, 1970). The erratic behaviour of μ thus contributes to a rather wide variation of the parameter b in the Birch velocity-density relationship.

Between 800 km and 1700 km, these abnormal regions account for a velocity change of 0.284 km/s which is 21% of the total velocity increase (JOHNSON, 1969; fig. 11) through this region. From the Birch velocity–density relation, the corresponding density increase is 0.09 g/cm^3. Making a correction for differential self compression of the two states, taking the 0.09 g/cm^3 differential to apply at an average depth of 1200 km, and taking the bulk modulus of the dense assemblage as 2800 kb (table 11), this leads to an initial density ρ_0 which is 0.16 g/cm^3 higher than that of isochemical mixed oxides. If these velocity increases are attributed to phase transformations, as seems reasonable, this implies that the stable mineral assemblage in the deep regions of the mantle is about 4% denser than the isochemical mixed oxides.

It was pointed out previously that a difference exists between the solutions of JOHNSON (1969) and ARCHAMBEAU et al. (1969) for the region between 800 and 1200 km. The latter results which extend only as far as 1200 km imply a larger component of velocity-increase caused by phase changes. If the abnormal velocity increase found by Archambeau et al. for this depth interval is combined with Johnson's measurements of abnormal velocity increases at greater depths, a total velocity increase of 0.55 km/s due to phase changes is indicated. This corresponds to a density change of 0.17 g/cm^3 equivalent to about 0.3 g/cm^3 at atmospheric pressure. This implies a mineral assemblage about 7% denser than the isochemical mixed oxides. This inference that the mineral assemblage occurring in the deep mantle may be 4–7% denser than isochemical mixed oxides is consistent with the nature of the possible transformations to states denser than the mixed oxides as discussed in earlier sections.

7.5. Density of the lower mantle

The assumption of a chemical and mineralogical model for the upper mantle determines the mean density and moment of inertia of this region, and these in turn, constrain the density distribution in the lower mantle. CLARK and RINGWOOD (1964) discussed the density distribution within the earth using two chemical-petrological models for the upper mantle, namely, the pyrolite model with a mean density of 3.40 in the upper 400 km, and the eclogite model with a mean density of 3.63 in this region. For assumed upper mantle com-

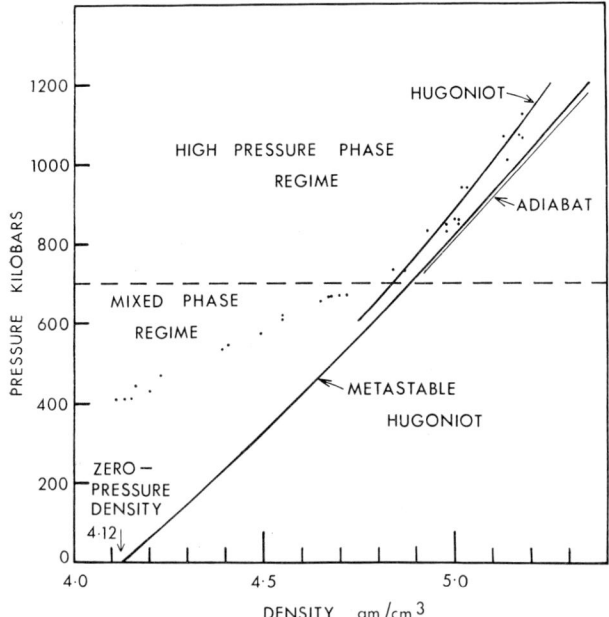

Fig. 12. Pressure-density plot of Twin Sisters dunite under shock compression. Data points from McQueen et al. (1967). High pressure regime represents Hugoniot of material which has been completely converted to high pressure phase with zero pressure density obtained by extrapolation using finite strain theory and empirical ϕ–ρ relationship. Mixed phase regime represents state of incomplete conversion to high pressure phase. Metastable Hugoniot and adiabat for high pressure phase are centred at state $P = 0$ and $T = 25$ °C. After Ahrens et al. (1969).

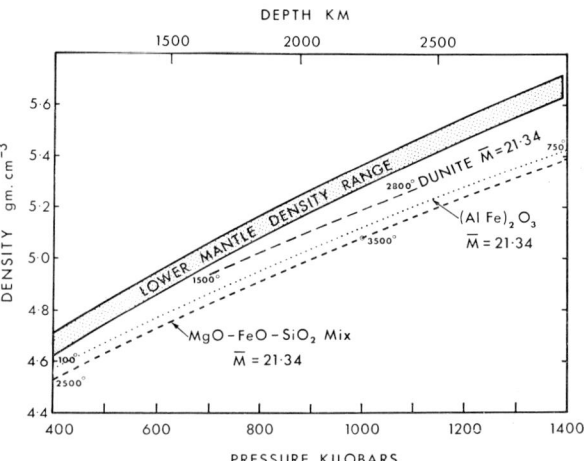

Fig. 13. Comparison of density distribution in lower mantle (consistent with pyrolite upper mantle) with densities of some model mantle materials derived principally from shock wave data. Density–pressure relationships for dunite and $(Al,Fe)_2O_3$ are along the respective Hugoniots (McQueen et al., 1967) and the estimated temperatures (Ahrens, personal communication) are shown. In the case of the MgO–FeO–SiO_2 mix the Hugoniot densities have been corrected to allow for the probable higher temperatures in the lower mantle. The indicated temperatures are notional.

positions lying between those of pyrolite and eclogite (i.e. basalt) the corresponding lower mantle density distributions can be obtained by interpolation between the distributions for the pyrolite and eclogite models. In the discussion of the pyrolite composition (section 7.1) it was pointed out that there was some flexibility in the ratio of peridotite to basalt assumed but that this was likely to lie between 4 and 2. Figure 13 shows the density distribution range for the lower mantle which follows from the assumption that the upper mantle pyrolite composition lies between these limits. The Mg/(Mg+Fe)* ratio of pyrolite is 88.8 (atomic) and the mean atomic weight for the pyrolite composition given in table 14 is 21.34.

If the lower mantle is assumed to be approximately homogeneous** following Birch's (1952) conclusion, then the density distribution over the pressure range

* Fe = Fe+Ni+Mn+Co.
** i.e. as a first approximation the velocity abnormalities determined by Johnson (1969) are ignored.

400–1200 kb can be extrapolated back to zero pressure in order to obtain the equivalent zero-pressure density and elastic ratio of the lower mantle mineral assemblage. Clark and Ringwood (1964) fitted the Birch–Murnaghan equation (below) based on the third order theory of finite strain to the density distributions in the lower mantle corresponding to pyrolite and to eclogite upper mantles:

$$P = 3K_0 f(1+2f)^{\frac{5}{2}}(1-2\xi f),$$

where f = strain and ξ is a dimensionless parameter proportional to the coefficient of the third order terms. Their results are shown in table 17.

The values of ρ_0 were found to be insensitive to the presence of superadiabatic temperature gradients in the lower mantle of the order of 1–2 °C/km.

The calculated zero-pressure–zero-temperature density of the lower mantle was found to be about 5% greater than that of a mixture of oxides (SiO_2 as stishovite) with the pyrolite composition. It was suggested that this discrepancy might be caused

(a) by the occurrence of transformations in the transition zone leading to the formation of phases which were significantly denser than the isochemical mixed oxides (e.g. $MgSiO_3$ perovskite), or

TABLE 17

Zero-pressure parameters* for mantle models after CLARK and RINGWOOD (1964)

	Pyrolite model		Eclogite model	
	1600 °C	0 °C	1600 °C	0 °C
Mean density of upper mantle	3.40		3.63	
ρ_0 (lower mantle) g/cm^3	4.11	4.25 ± 0.05	3.90	4.05 ± 0.05
K_0 (lower mantle) kb	2300	2680 ± 160	2130	2550 ± 150
ϕ_0 (lower mantle) (km/s)^2	56.0	63 ± 3	54.5	63 ± 3
ξ (lower mantle)	0.22		0.20	

* The uncertainties quoted apply only to the transformation of elastic properties from 1600 °C to 0 °C ($P = 0$).

(b) by an increase in the Fe/(Fe+Mg) ratio in the lower mantle from 0.1 to about 0.2, or

(c) by a combination of these factors.

The recent studies of velocity distribution in the lower mantle by JOHNSON (1969) discussed previously suggest the occurrence of further phase transformations in the lower mantle, and thus contradict the assumption of homogeneity in this region adopted by Bullen, Birch and Clark and Ringwood. Because of the relatively small magnitude of these velocity changes, the contradiction is not severe, and the homogeneity assumption remains useful as a first approximation to the properties of this region. Nevertheless the presence of these small inhomogeneous regions will have a significant effect upon estimates of zero-pressure parameters for the lower mantle to the extent that these are obtained by an extrapolation procedure. The inhomogeneities will cause the values of ρ_0 and ϕ_0 in table 17 to be systematically underestimated. This strengthens the conclusion that the zero-pressure density of the lower mantle is significantly higher than mixed oxides of pyrolite composition.

BIRCH (1952, 1964) has estimated the zero-pressure density and elastic ratio of the lower mantle using the assumptions of homogeneity, adiabaticity and second order finite strain theory. The density distributions used by Birch in the upper mantle were somewhat arbitrary – one of these (solution II) is close to the pyrolite model, the other (solution I) is about half way between the pyrolite and eclogite models of Clark and Ringwood. For solution II Birch estimates a zero-pressure density for the lower mantle which is about 0.15 g/cm^3 smaller than the value obtained by Clark and Ringwood. Birch suggests that the lower mantle may contain about 10 weight % of iron oxide, which is slightly higher than that of pyrolite (table 14).

The invaluable shock wave data on the Hugoniot equations of state of a number of rocks and minerals of geophysical interest obtained by McQUEEN and MARSH (1966) and McQUEEN et al. (1967) have provided the raw material for many interpretive papers dealing with the earth's lower mantle (BIRCH, 1961b, 1964; McQUEEN et al., 1964, 1966, 1967; TAKEUCHI and KANAMORI, 1966; WANG, 1967, 1968, 1969; D. L. ANDERSON and KANAMORI, 1968; KNOPOFF and SHAPIRO, 1969; SHAPIRO and KNOPOFF, 1969; AHRENS et al., 1969). In view of the wide use of this data, it is wise to pay some attention to the assumptions and uncertainties involved.

Experimental data points obtained by McQueen et al. for the Twin Sisters dunite are shown in fig. 12. The area below 700 kb is interpreted as a mixed phase region in which phase transformations have not gone to completion. Above 800 kb, the data points are interpreted as referring to a single high pressure phase. The transformation pressure does not relate to thermodynamic equilibrium conditions. The individual P–ρ data points are derived from experimental measurements via the Rankine–Hugoniot relationships and can be determined quite accurately under favourable conditions. Nevertheless in this case there is a substantial amount of scatter due to experimental uncertainties in the primary data to which a curve (the raw Hugoniot of fig. 12) must be fitted. Uncertainties in the location of this curve may also be caused by subjective judgements of the position of the boundary between the mixed phase and high pressure phase regimes, and by the assumption of homogeneity in the high pressure phase regime. Phase transformations accompanied by small density changes are very difficult to detect but they have a strong influence on the slope of the Hugoniot. In the case of the Twin Sisters dunite, a reasonably large number of data points is available so that the uncertainties are minimized and the location of the raw Hugoniot is reasonably well established. This, however, is not the case for several of the other rocks and minerals studied.

Temperature increases along the Hugoniot more rapidly than the adiabatic gradient and its estimation requires assumptions concerning the volume dependence of Gruneisen's parameter. The limitations of these assumptions have been discussed in some detail by KNOPOFF and SHAPIRO (1969) and O. L. ANDERSON and LIEBERMANN (1970). For some geophysical purposes, it is useful to transform the raw Hugoniot data to the metastable Hugoniot – this is the locus of P, ρ, T points which the high pressure phase would follow if it were possible to use it as metastable specimen material for the shock experiments. Other transformations of importance are conversion to adiabats and isotherms (fig. 12). These transformations require several assumptions concerning the nature of the equation of state, the transformation energy associated with phase changes, and the magnitude of the correction required for finite strength of the specimen. These assumptions lead to corrections which can be carried out in different ways. Accordingly it is not surprising that significant differences often arise between the results of different investigators.

In the case of the Twin Sisters dunite, for which fairly good data are available, the $P-\rho$ relationships appear to be reasonably well determined and the data on this dunite and also upon the hortonalite dunite and fayalite have been interpreted to imply that these olivine-rocks are shocked into a high pressure phase which has a zero-pressure density very similar to the density of the isochemical mixed oxides (MgO, FeO and SiO_2 stishovite) (WANG, 1967, 1968; D. L. ANDERSON and KANAMORI, 1968; AHRENS et al., 1969). (On the other hand, the estimates of ϕ, which involve the derivative $(\partial P/\partial \rho)_S$ from the primary data by different workers differ substantially because of the different approaches. This aspect is considered in the next subsection.)

The mean atomic weight for the Twin Sisters dunite is 20.94 compared to 21.34 for pyrolite. Following WANG (1968) the $P-\rho$ data for the hortonolite dunite ($\overline{M} = 24.3$) and fayalite ($\overline{M} = 29.1$) can be used to interpolate the $P-\rho$ relations for dunites of intermediate compositions. In fig. 13 the $P-\rho$ relationship for a dunite of mean atomic weight equal to pyrolite (21.34) is shown. The correction to the Twin Sisters data is small and does not introduce a significant error. These $P-\rho$ values lie along the raw Hugoniot and the temperatures have been estimated by AHRENS (personal communication).

It is seen from fig. 13 that the $P-\rho$ curve for the high pressure phase of dunite ($\overline{M} = 21.34$) lies significantly below that of the density appropriate for a pyrolite mantle. Since the temperatures in the lower mantle are probably substantially higher than the temperatures along the Hugoniot (fig. 13), a temperature correction would only increase the discrepancy. This may be taken as strong evidence that the density of the lower mantle is significantly higher than would be expected for pyrolite ($\overline{M} = 21.34$) existing in a phase assemblage with a density equal to the isochemical mixed oxides. Corresponding arguments along the same lines have been made by BIRCH (1964) WANG (1967, 1969) and D. L. ANDERSON (1967a, 1968).

A similar picture emerges if other model systems are used to represent the lower mantle. RINGWOOD (1969a) compared the density of an $MgO-FeO-SiO_2$ (stishovite) mixture (mean atomic weight 21.34) of a composition appropriate to the lower mantle with the density of a pyrolite lower mantle (fig. 13). The density of the model mixture obtained by interpolation from shock wave data on MgO, SiO_2 and Fe_2SiO_4, also lies substantially below that of the lower mantle. A third model was based upon $Al_2O_3-Fe_2O_3$ corundum type solid solutions with mean atomic weight 21.34. The $P-\rho$ relationship for Al_2O_3 was obtained from shock data and for Fe_2O_3 was calculated using the ultrasonic compressibility data of ANDERSON et al. (1968). In this case also, allowance for the probable temperature difference between the models and the lower mantle would only serve to increase the discrepancy.

The conclusion that the density of the lower mantle (appropriate to a pyrolite upper mantle) is significantly higher than the density of mixed isochemical oxides of pyrolite composition is thus strongly supported both by methods based on the extrapolation of lower mantle densities to low pressure and by direct comparisons with a variety of shock data. In both cases, the principal sources of uncertainty (possibility of small phase changes in the lower mantle in the first case, temperature in the second case) are in a direction so that corrections, if applied, would only increase the magnitude of this difference.

7.6. *Elasticity of the lower mantle*

We have already noted that the inferred higher density of the lower mantle as compared with the corresponding density of isochemical mixed oxides of pyrolite composition could be explained either by further phase transformations leading to denser structures and/or by an increase in iron content. The latest seismic evidence for further phase transformations occurring below 700 km (figs. 1 and 11) favours the first alternative. However this evidence cannot be considered decisive, until the new velocity structures in the lower mantle are confirmed and the differences between the JOHNSON (1969) and ARCHAMBEAU et al. (1969) models between 800 and 1200 km are resolved. There is also the assumption that the phase transformations near 650 km result in a mineral assemblage with a density similar to the density of isochemical mixed oxides. Although this assumption is strongly supported by static high pressure investigations on analogue systems, it is not finally proven. Hence, an explanation of the density of the lower mantle in terms of an increase in iron content with depth cannot be entirely excluded on these grounds.

In principle, the property which is likely to be of most use in distinguishing between the two alternatives is the elastic ratio ϕ. A proportional change in density is related to the corresponding proportional change in elastic ratio by

$$\frac{\Delta \phi}{\phi} = n \frac{\Delta \rho}{\rho}.$$

For an increase in density caused by a phase change, the seismic equation of state (D. L. ANDERSON, 1967b) gives an *n* value of $+3$. For an increase of density caused by normal thermal contraction, CLARK and RINGWOOD (1964) estimated *n* to be approximately $+2$. However for an increase in density resulting in an increase of iron content, *n* is negative, with a value of approximately -1. The fact that both phase transformations to denser states and thermal contraction change ϕ in the opposite direction to the change caused by an increase in iron content provides an important constraint on proposed solutions. It is possible to compare ϕ values for the mantle obtained directly from the observed seismic velocities ($\phi = V_P^2 - \frac{4}{3} V_S^2$) with ϕ values derived from shock wave experiments on rocks under pressures similar to those in the lower mantle. However the procedure must be used with caution since, as noted in section 7.3 there are several uncertainties and assumptions involved in the derivation of ϕ from shock wave data and the results of different authors working on the same primary data differ substantially (fig. 14).

WANG (1968) has calculated the hydrodynamical sound velocity ($C = \sqrt{\phi}$) and density for the high pressure phase of the Twin Sisters dunite, along an adiabat centered at 1200 °C, atmospheric pressure. This experimentally derived adiabatic C–ρ plot is then compared with a C–ρ plot for the lower mantle assuming homogeneity. (Density distribution used was the Birch model I). The two curves are nearly parallel, and Wang concludes that an increase of iron content in the experimental dunite would bring them into coincidence. He estimates the FeO content of the lower mantle as about $14 \pm 3\%$, dependent upon the temperature distribution in this region. However, this value is obtained by matching the Birch model I density distribution in the

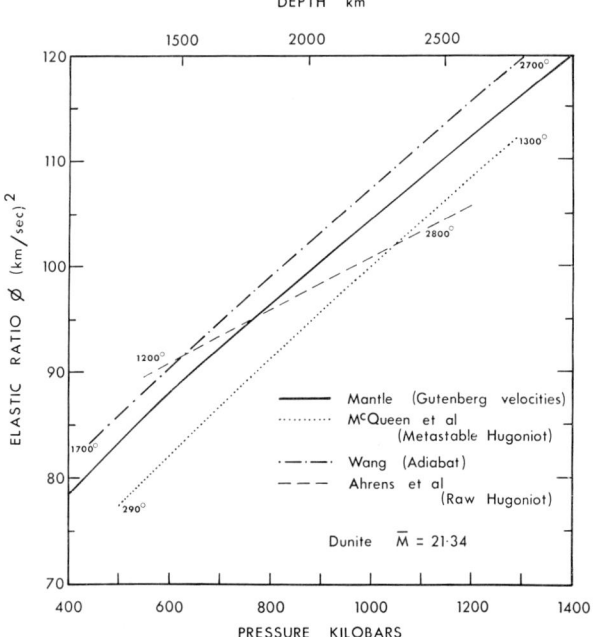

Fig. 14. Comparison of ϕ–P relationship for lower mantle (obtained from Gutenberg velocity distributions) with ϕ–P relationships for Twin Sisters dunite (corrected to $\bar{M} = 21.34$) derived from shock wave data by McQUEEN et al. (1967), WANG (1968) and AHRENS et al. (1969). The temperatures calculated by these authors are indicated.

lower mantle. In order to match the pyrolite density distribution (fig. 13) using Wang's methods, an iron oxide content in the lower mantle of about $18 \pm 3\%$ would be required. Thus Wang concludes that the lower mantle may be composed of the high pressure modification of a dunite, possessing a density equal to that of the isochemical mixed oxides, but possessing a higher iron content than the Twin Sisters Dunite (and pyrolite). BIRCH (1964) reaches a similar conclusion.

However, there may be a certain ambiguity about matching shock C–ρ curves with the C–ρ curve for the mantle and I prefer to make the comparison by plotting density and velocity (or ϕ) separately against pressure, rather than against each other. The justification for this procedure is that in the lower mantle, the proportional differences in pressure for different models (pyrolite and eclogite) are smaller than the corresponding differences in density at similar depths. (CLARK and RINGWOOD, 1964. Thus for this range of models, pressure in the mantle can be treated approximately as an independent variable, whereas density is much more dependent upon the particular mantle model. The P–ρ plots were shown in fig. 13, and the P–ϕ plots are given in fig. 14*. The experimental ϕ values used are for the Twin Sisters dunite, corrected to a mean atomic weight of 21.34. The ϕ values for the mantle are those tabulated by CLARK and RINGWOOD (1964) and are based upon the Gutenberg velocities.

It is seen that substantial differences exist in positions and slopes of the shock ϕ–P curves given by different authors. These are not principally due to the fact that we are using raw Hugoniot, metastable Hugoniot and adiabat data characterised by different temperature distributions. They are primarily the result of different methods of treating the raw data used by the different authors and reflect some of the uncertainties implicit in the methods. Clearly the determination of ϕ from shock data is a much more uncertain process than the determination of ρ. The temperatures estimated by the various authors to apply along the respective ϕ–ρ curves are also shown in fig. 14.

We see in fig. 14 that the ϕ values calculated from shock data are generally close to those of the mantle (mostly within 5%). Taken at face value this might be held to support Wang's interpretation. Closer inspection, however, reveals some difficulties. Referring to fig. 14, if we increase the iron content of the dunite sufficiently to bring its density–pressure relationship (fig. 13) into agreement with the lower mantle (Wang's hypothesis) we require a density increase of 0.13 g/cm^3 corresponding to a mean atomic weight of 22.0, equivalent to 21% fayalite in the dunite or 20% FeO. Increasing the iron content results in a decrease of ϕ. Making the appropriate correction, the calculated values of ϕ for a dunite $\overline{M} = 22.0$ according to Wang's calculation are very close, but nevertheless slightly below ($\approx 1\%$) the mantle ϕ values. The disagreement is minor, and to this point Wang's hypothesis appears sound. However, the ϕ values calculated by MCQUEEN et al. (1967) for the high pressure phase of the Twin Sisters dunite (adjusted to $\overline{M} = 21.34$) fall substantially below the mantle ϕ curve and likewise those calculated by Ahrens et al. (fig. 14) at pressures greater than 800 kb. If in addition the ϕ–P curves of these authors are corrected downwards by about -4 (km/s)^2 to allow for a mean atomic weight of 22.0, they become clearly discordant with the mantle P–ϕ distribution. Thus we may conclude that to this stage, the treatments of shock wave data by MCQUEEN et al. (1957) and AHRENS et al. (1969) (above 800 kb) do not favour an increase of iron content with depth. On the other hand Wang's calculated ϕ–P relationship is consistent with a modest increase of iron content.

However, the differences in temperature distributions between the mantle and the shock wave ϕ–P curves have not been taken into account. The temperatures along Wang's adiabat may well be about 1000 °C smaller than the temperatures at equivalent depths in the mantle. BIRCH (1952; table 8) has estimated the effect of compression on thermal expansion using finite strain theory. Applying this correction to an estimated high-temperature thermal expansion-coefficient at zero pressure, an average thermal expansion-coefficient of $20 \times 10^{-6}/°C$ for the lower mantle appears reasonable*.

---

* I am grateful to Dr. Wang for a detailed tabulation of his calculated densities and pressures along the adiabat of the high pressure phase of the Twin Sisters dunite.

* BIRCH (1952; table 16) estimates an average thermal expansion coefficient of $15 \times 10^{-6}/°C$ for the lower mantle. This appears to be on the low side owing to the use of experimental measurements of thermal expansion coefficients obtained in the vicinity of 1000 °C. The thermal expansion coefficient α_0 (zero-pressure) increases with temperature, and the value of α_0 at temperatures appropriate for the lower mantle (2000–4000 °C) would probably be significantly higher than estimated by Birch.

Thus a temperature difference of 1000 °C would correspond to a 2% difference in density. The corresponding effect on ϕ as discussed earlier in this section amounts to about 4%. Thus Wang's experimental ϕ values need to be reduced by about 4 (km/s)^2 if the mantle temperatures are as suggested. Wang's ϕ values then fall significantly below the mantle distribution, whilst those of MCQUEEN et al. (1967) and AHRENS et al. (1969) are well below.

Two other effects remain to be considered. Referring to fig. 13, we see that if Wang's temperatures are 1000 °C below the mantle temperatures, an additional increase of mean atomic weight of the dunite is necessary in order to obtain agreement between the density distribution in the mantle, and the shock-wave density distribution. Thus the mean atomic weight of the dunite would need to be increased to 22.4 equivalent to 26% fayalite molecule in the dunite or 24% FeO. This will cause a further reduction of ϕ by about 2 (km/s)^2 thus further increasing the discrepancy. Finally, recent revisions of the P and S velocity distribution in the mantle suggest that the ϕ values in the deep mantle may be about 2 (km/s)^2 higher than shown in fig. 14 (D. L. ANDERSON and JULIAN, 1969).

The above discussion suggests the following conclusions:

1. The ϕ values for the high pressure phase of dunite obtained from shock data by MCQUEEN et al. (1967) and AHRENS et al. (1969), when compared to the ϕ distribution in the mantle, appear to be inconsistent with the increase in iron content of the lower mantle needed to match the $P-\rho$ curves of shocked dunite and lower mantle (fig. 13).

2. On the other hand the ϕ values obtained from shock data by WANG (1968) are consistent with his conclusion that a small increase of iron occurs in the lower mantle, provided that the lower mantle is relatively cool (≈ 2700 °C at the core–mantle boundary). If however the temperatures throughout the lower mantle are about 1000 °C higher than along Wang's adiabat, the ϕ distribution appears to be inconsistent with an increase in iron content, as in the cases above.

3. The discrepancy in density between a mineral assemblage with $\overline{M} = 21.34$ and possessing the density of an isochemical oxide mixture with the density in the lower mantle (fig. 13) can be explained if it is assumed that the phases in the deep mantle have undergone additional transformations to states which are a few percent denser than the isochemical mixed oxides (at zero pressure). If dunite ($\overline{M} = 21.34$) were transformed into such a mineral assemblage, a corresponding increase in ϕ would result, and a self-consistent explanation of the discrepancies between shock wave and mantle ϕ values may emerge for the treatments of MCQUEEN et al. (1967) and AHRENS et al. (1969) and for the high temperature case of WANG (1968).

Conclusion 3 above raises the question of why the dunites were not shocked into the ultra-dense state if the mantle within the same pressure range is in this state as inferred. A possible explanation is to be found in fig. 12, where it is seen that complete transformation of the Twin Sisters dunite into a high pressure phase assemblage with a density similar to isochemical mixed oxides does not occur until 700–800 kb, whereas in the earth, and from static high pressure work, this transformation occurs at 200–300 kb. Evidently a large non-equilibrium overpressure is required to drive the transformation under shock conditions. It is possible that the required overpressure for the later transformation into the ultradense state was not attained in the shock experiments, which ranged up to 1100 kb. (However, in the case of pure forsterite MCQUEEN and MARSH (1966) obtained evidence of transformation into a state which was substantially denser than the mixed oxides.)

The differences which exist in the reduction of shock wave data by different investigators and the inherent uncertainties about some of the assumptions involved introduce corresponding uncertainties when the data are used in an attempt to resolve fine details of mantle constitution. This applies particularly to the discussion of ϕ values above. Nevertheless, taking an overall view of available data, the author believes that the shock wave arguments generally favour the hypothesis that the mineral assemblage in the lower mantle has transformed to a state somewhat denser than the isochemical mixed oxides in preference to the alternative hypothesis that the lower mantle consists of a mineral assemblage with the density of isochemical mixed oxides, but with a higher iron content than the upper mantle. This conclusion, however, must be regarded as tentative, and may need revision, particularly if temperatures in the lower mantle prove to be much lower than now appears probable.

8. Conclusions

The devotion of much of the previous section to consideration of whether the lower mantle has a somewhat higher iron content than the upper mantle, or whether the mineral assemblage present is slightly denser than isochemical mixed oxides is an indication of the extent of progress in knowledge of the constitution of the mantle during recent years. The differences between these two hypotheses, in terms of providing an explanation of the physical properties of the lower mantle, are rather small and point towards future developments as being concerned more with questions of fine structure, than with the major problems of constitution which were being debated 10 to 20 years ago. Outstanding problems of this period, such as whether major phase transformations occurred in the mantle and if so, what was their role and to what extent might they explain the gross physical properties of the mantle, are now settled. During the next ten years we may confidently expect that answers to many problems of fine structure will be found as seismic velocity depth distributions are refined using arrays, as our knowledge of elastic properties of minerals, their $P–T$ dependence, and of their systematic relationships to crystal structure expands, and as further improvements in static and dynamic high pressure experimental techniques are made. Although such debates as whether iron content increases with depth or remains approximately constant involve interpretation of second order physical properties, answers to these questions may be of profound genetic significance and are bound to influence our ideas on the origin and early evolution of the earth.

In this paper we have reviewed at length recent progress in the study of high pressure phase transformations in silicates and analogue compounds, and applied the results to investigate the mineralogy of the mantle and the extent to which the major physical properties of the mantle may be explained in terms of mineralogy. The following conclusions may be drawn:

1. The occurrence of major phase transformations in the mantle is no longer in doubt. BIRCH's (1952) hypothesis that the anomalous elastic properties between 400 and 900 km are predominantly caused by phase transformations is confirmed.

2. Directly discovered and indirectly inferred phase transformations in a mantle of pyrolite composition quantitatively explain the occurrences, depths, and velocity-changes of the two major seismic discontinuities near 400 and 650 km.

3. Experimental investigations show that the regions between 150 and 350 km and between 450 and 600 km probably do not contain important phase changes. This is consistent with recent determinations of seismic velocity-gradients in these regions which are inferred to be homogeneous.

4. Considering (2) and (3) above, we find that the principal features of the seismic velocity distribution from 150 to 800 km are explained by directly discovered and indirectly inferred phase transformations occurring in a mantle of pyrolite composition. Furthermore these provide explanations of the seismic velocity and density distributions throughout the lower mantle which are correct to within about five percent.

5. The five percent discrepancy in density could be resolved either by the occurrence of additional small phase transformations below 800 km or by an increase in iron content with depth or by some combination of these effects.

6. Seismic evidence suggests the occurrence of further small phase transformations below 800 km of a magnitude which would resolve the density discrepancy. The presence of these phase transformations in the lower mantle is generally favoured by an interpretation of shock wave velocity data. However, in view of possible uncertainties in the seismic and shock wave data, this conclusion cannot yet be regarded as firmly established and the possibility of a small increase in the iron content of the lower mantle may warrant reconsideration as further shock and seismic data become available.

7. Systematic crystal chemical considerations combined with high pressure investigation suggest that if the occurrence of further phase transformations in the deep mantle is finally required by seismic and shock wave data, the most probable structures of minerals in this region would be of the perovskite and calcium ferrite types.

8. A satisfying and widely consistent interpretation of the physical constitution of the mantle in terms of its mineralogical constitution has emerged.

Acknowledgements

Much of the work described in this paper has been carried out in collaboration with Mr Alan Major of the

Australian National University and with Dr. A. F. Reid of the Division of Mineral Chemistry, CSIRO. The many valuable contributions from these colleagues are gratefully acknowledged.

References

AHRENS, L. H. (1952) The use of ionization potentials I, Geochim. Cosmochim. Acta **2**, 155–169.

AHRENS, T. J. and Y. SYONO (1967) Calculated mineral reactions in the Earth's mantle, J. Geophys. Res. **72**, 4181–4188.

AHRENS, T. J., D. L. ANDERSON and A. E. RINGWOOD (1969) Equations of state and crystal structures of high pressure phases of shocked silicates and oxides. Rev. Geophys., in press.

AKIMOTO, S. and H. FUJISAWA (1968) Olivine–spinel solid solution equilibria in the system Mg_2SiO_4–Fe_2SiO_4, J. Geophys. Res. **73**, 1467–1479.

AKIMOTO, S. and Y. IDA (1966) High pressure synthesis of Mg_2SiO_4 spinel, Earth Planet. Sci. Letters **1**, 358–359.

AKIMOTO, S. and Y. SYONO (1967) High pressure decomposition of some titanate spinels, J. Chem. Physics **47**, 1813–1817.

AKIMOTO, S. and Y. SYONO (1968) The coesite–stishovite transition, Tech. Rept. Inst. Solid State Physics Tokyo Univ. Ser. A no. 327.

ANDERSON, D. L. (1964) Densities of the mantle and core (abstract) Trans. Am. Geophys. Union **45**, 101.

ANDERSON, D. L. (1967a) Phase changes in the upper mantle, Science **157**, 1165–1173.

ANDERSON, D. L. (1967b) A seismic equation of state, Geophys, J. **13**, 9–30.

ANDERSON, D. L. (1968) Chemical inhomogeneity of the mantle. Earth Planet. Sci. Letters **5**, 89–94.

ANDERSON, D. L. and B. R. JULIAN (1969) Shear velocities and elastic parameters of the mantle, in press.

ANDERSON, D. L. and H. KANAMORI (1968) Shock wave equations of state for rocks and minerals, J. Geophys. Res. **73**, 6477–6502.

ANDERSON, D. L. and N. TOKSOZ (1963) Surface waves on a spherical earth I, Upper mantle structure from Love waves, J. Geophys. Res. **68**, 3483–3500.

ANDERSON O. L. and R. C. LIEBERMANN (1970) Equations for the elastic constants and their pressure derivates for three cubic lattices and some geophysical applications, Phys. Earth Planet. Interiors **3**, 61–86.

ANDERSON, O. L. and J. E. NAFE (1965) The bulk modulus–volume relationship for oxide compounds and related geophysical problems, J. Geophys. Res. **70**, 3951–3962.

ANDERSON, O. L., E. SCHREIBER, R. C. LIEBERMANN and N. SOGA (1968) Some elastic constant data on minerals relevant to geophysics, Rev. Geophys. **6**, 491–524.

ARCHAMBEAU, C. B., E. A. FLINN and D. G. LAMBERT (1969) Fine structure of the upper mantle, in press.

BAYER, G. and W. HOFFMAN (1966) Complex alkali titanium oxides $A_x(B_yTi_{8-y})O_{16}$ of the α-MnO_2 structure-type, Am. Mineralogist **51**, 511–516.

BERNAL, J. D. (1936) Discussion, Observatory **59**, 268.

BINNS, R. A. (1970) $(Mg,Fe)_2SiO_4$ spinel in a meteorite, Phys. Earth Planet. Interiors **3**, 156–160.

BINNS, R. A., R. J. DAVIS and S. B. J. REED (1969) Ringwoodite, natural $(MgFe)SiO_4$ spinel in the Tenham meteorite, Nature **221**, 943–944.

BIRCH, F. (1939) The variation of seismic velocities within a simplified earth model in accordance with the theory of finite strain, Bull. Seismol. Soc. Am. **29**, 463–479.

BIRCH, F. (1952) Elasticity and constitution of the Earth's interior, J. Geophys. Res. **57**, 227–286.

BIRCH, F. (1961a) The velocity of compressional waves in rocks to 10 kilobars 2, J. Geophys. Res. **66**, 2199–2224.

BIRCH, F. (1961b) Composition of the earth's mantle, Geophys. J. **4**, 295–311.

BIRCH, F. (1964) Density and composition of mantle and core, J. Geophys. Res. **69**, 4377–4388.

BIRCH, F. and P. LE COMTE (1960) Temperature–pressure plane for albite composition, Am. J. Sci. **258**, 209–217.

BOLT, B. (1969) PdP and PKiKP waves and diffracted PcP waves, preprint.

BOLT, B. A., M. O'NEILL and A. QAMAR (1968) Seismic waves near $110°$: is structure in core or upper mantle responsible? Geophys. J. **16**, 475–487.

BORN, M. and W. HEISENBERG (1924) Über den Einfluss der Deformierarkeit der Ionen auf optische und chemische Konstanten I, Z. Phys. **23**, 388–410.

BOYD, F. (1964) Geological aspects of high-pressure research, Science **145**, 13–20.

BOYD, F. and J. L. ENGLAND (1961) Melting of silicates at high pressures, Carnegie Inst. Wash. Yearbook **60**, 120.

BULLEN, K. E. (1936) The variation of density and the ellipticities of strata of equal density within the earth, Monthly Notices Roy. Astron. Soc. Geophys. Suppl. **3**, 395–401.

BUNDY, F. P. (1963) Direct conversion of graphite to diamond in static pressure apparatus, J. Chem. Physics **38**, 631–643.

BYSTRÖM, A. and A. M. BYSTRÖM (1950) The crystal structure of hollandite, the related manganese oxide minerals, and α-MnO_2, Acta Cryst. **3**, 146–154.

CLARK, S. P. (1961) A redetermination of equilibrium relations between kyanite and sillimanite, Am. J. Sci. **259**, 641–650.

CLARK, S. P. and A. E. RINGWOOD (1964) Density distribution and constitution of the mantle, Rev. Geophys. **2**, 35–88.

CLARK, S. P., F. SCHAIRER and J. DE NEUFVILLE (1962) Phase relations in the system $CaMgSi_2O_6$–$CaAl_2SiO_6$–SiO_2 at low and high pressure, Carnegie Inst. Wash. Yearbook **61**, 61.

DACHILLE, F. and R. ROY (1960) High pressure studies of the system Mg_2GeO_4–Mg_2SiO_4 with special reference to the olivine–spinel transition, Am. J. Sci. **258**, 225–246.

DRICKHAMER, H. (1963) Electronic structure at high pressure, in: W. Paul and D. Warschauer, eds. *Solids under pressure*, (McGraw-Hill, New York).

EVERNDEN, J. (1958) Finite strain theory and the earth's interior, Geophys. J. **1**, 1–8.

GELLER, S. (1967) Crystal chemistry of the garnets, Z. Krist. **125**, 1–47.

GELLER, S. (1969) Alternative explanation of the garnet occurrence in a meteorite, Science **163**, 829–830.

GOLDSCHMIDT, V. M. (1931) Zur Kristallchemie des Germaniums, Nachr. Akad. Wiss. Göttingen, Math. Physik. Kl. **1** no. 2, 184–190.

GREEN, D. H. (1970) A review of experimental evidence on the origin of basaltic and nephelinitic magmas, Phys. Earth Planet. Interiors **3**, 219–233.

GREEN, D. H. and A. E. RINGWOOD (1963) Mineral assemblages in a model mantle composition, J. Geophys. Res. **68**, 937–945.

GREEN, D. H. and A. E. RINGWOOD (1967a) The genesis of basaltic magmas, Contrib. Mineral. Petrol. **15**, 103–190.

GREEN, D. H. and A. E. RINGWOOD (1967b) An experimental investigation of the gabbro to eclogite transformation and its petrological applications, Geochim. Cosmochim. Acta **31**, 767–833.

GREEN, D. A. and A. E. RINGWOOD (1967c) The stability fields of aluminous pyroxene peridotite and garnet peridotite and their relevance in upper mantle structure, Earth Planet. Sci. Letters **3**, 151–160.

GREEN, D. H., J. W. MORGAN and K. S. HEIER (1968) Thorium, uranium and potassium abundances in peridotite inclusions and their host basalts, Earth Planet. Sci. Letters **4**, 155–166.

GRIGGS, D. (1954) The Earth's mantle: discussion, Trans. Am. Geophys. Union **35**, 93–96.

GUTENBERG, B. (1958) Velocity of seismic waves in the earth's mantle, Trans. Am. Geophys. Union **39**, 486–489.

GUTENBERG, B. (1959) *Physics of the Earth's interior*, 1, J. V. Mieghem ed., Intern. Geophys. Ser. (Academic Press, New York) 240 pp.

HALES, A. L., J. CLEARY, H. DOYLE, R. GREEN and J. ROBERTS (1968) P wave station anomalies and the structure of the upper mantle, J. Geophys. Res. **73**, 3885–3896.

JEFFREYS, H. (1937) On the materials and density of the earth's crust, Monthly Notices Roy. Astron. Soc. Geophys. Suppl. **4**, 50–61.

JEFFREYS, H. (1939) The times of P, S and SKS and the velocities of P and S, Monthly Notices Roy. Astron. Soc. Geophys Suppl. **4**, 498–533.

JOHNSON, L. (1967) Array measurements of P-velocities in the upper mantle, J. Geophys. Res. **72**, 6309–6325.

JOHNSON, L. (1969) Array measurements of P-velocities in the lower mantle, J. Geophys. Res.

JULIAN, B., and D. L. ANDERSON (1968) Travel times, apparent velocities and amplitudes of body waves, Bull. Seismol. Soc. Am. **58**, 339–366.

KANAMORI, H. (1967) Upper mantle structure from apparent velocities of P-waves recorded at Wakayama micro-earthquake observatory, Bull. Earthquake Res. Inst. Tokyo Univ. **45**, 657–678.

KAWAI, N. (1966) A static high pressure apparatus with tapering multi-pistons forming a sphere I, Proc. Japan Acad. **42**, 385–388.

KLEEMAN, J. D. and J. A. COOPER (1970) Geochemical evidence for the origin of some ultramafic inclusions from Victorian Basanites, Phys. Earth Planet. Interiors **3**, 300–306.

KLEEMAN, J. D., D. H. GREEN and J. F. LOVERING (1969) Uranium distribution in ultramafic inclusions from Victorian basalts, Earth Planet. Sci. Letters **5**, 449–458.

KNOPOFF, L. and J. N. SHAPIRO (1969) Comments on the interrelationships between Grüneisen's parameter and shock and isothermal equations of state, J. Geophys. Res. **74**, 1439–1450.

KOVACH, R. L. and D. L. ANDERSON (1964) Higher mode surface waves and their bearing on the structure of the earth's mantle, Bull. Seismol. Soc. Am. **54**, 161–182.

KUME, S., T. MATSUMOTO and M. KOIZUMI (1966) Dense form of germanium orthoclase ($KAlGe_3O_8$), J. Geophys. Res. **71**, 4999–5001.

LIEBETZ, J. and C. J. M. ROOYMANS (1965) Die ilmenit–perowskit Phasenumwandlung von $CdTiO_3$ unter hohem Druck, Z. Phys. Chem. Frankfurt **44**, 242–249.

LEGGO, P. J. and R. HUTCHINSON (1968) A Rb-Sr study of ultrabasic xenoliths and their basaltic host rocks from the Massif Central, France, Earth Planet. Sci. Letters **5**, 71–75.

MAGNITSKY, V. A. and V. A. KALININ (1959) The properties of the earth's mantle and the physical properties of the transition layer, Bull. Acad. Sci. U.S.S.R. Geophys. Ser. (English Transl.) 1-6, 49–54.

MAREZIO, M., J. P. REMEIKA and A. JARARAMAN (1966) High pressure decomposition of synthetic garnets, J. Chem. Phys. **45**, 1821–1824.

MASON, B., J. NELEN and J. S. WHITE (1968) Olivine-garnet transformation in a meteorite, Science **160**, 66–67.

MCDONALD, G. J. F. (1956) Quartz-coesite stability relations at high temperatures and pressures, Am. J. Sci. **254**, 713–721.

MCDONALD, G. J. F. (1962) On the internal constitution of the inner planets, J. Geophys. Res. **67**, 2945–2974.

MCQUEEN, R. G. and S. P. MARSH (1966) in: S. P. Clark Jr., ed., *Handbook of physical constants*, Geol. Soc. Am. Mem. **97**.

MCQUEEN, R. G., J. N. FRITZ and S. P. MARSH (1964) On the composition of the earth's interior, J. Geophys. Res. **69**, 2947–2978.

MCQUEEN, R. G., S. P. MARSH and J. N. FRITZ (1967) Hugoniot equation of state of twelve rocks, J. Geophys. Res. **72**, 4999–5036.

MEIJERING, J. L. and C. J. M. ROOYMANS (1958) On the olivine–spinel transition in the earth's mantle, Koninkl. Ned. Akad. Wetenschap. Proc. Ser. B **61**, 333–344.

MORIMOTO, N., S. AKIMOTO, K. KOTO and M. TOKONAMI (1970) Crystal structures of high pressure modifications of Mn_2GeO_4 and Co_2SiO_4, Phys. Earth Planet. Interiors **3**, 161–165.

MÜLLER-BUSCHBAUM, H. (1966) Über Oxoscandate II zur Kenntnis des $MgSc_2O_4$, Z. Anorg. Allgem. Chem. **343**, 113–224.

NIAZI, M. and D. L. ANDERSON (1965) Upper mantle structure of western North America from apparent velocities of P waves, J. Geophys. Res. **70**, 4633–4640.

O'HARA, M. J. (1968) The bearing of phase equilibria studies in synthetic and natural systems on the origin and evolution of basic and ultrabasic rocks, Earth Sci. Rev. **4**, 69–133.

O'HARA, M. J. (1970) Upper mantle composition inferred from laboratory experiments and observations of volcanic products, Phys. Earth Planet. Interiors **3**, 234–243.

PISTORIUS, W. and G. C. KENNEDY (1960) Stability relations of grossularite and hydrogrossularite at high temperature and pressures, Am. J. Sci. **258**, 247–257.

PRESS, F. (1968) Earth models obtained by Monte Carlo inversion, J. Geophys. Res. **73**, 5223–5234.

PREWITT, C. T. and A. W. SLEIGHT (1969) Garnet-like structures of high-pressure cadmium germanate and calcium germanate, Science **163**, 386–387.

REID, A. F. and A. E. RINGWOOD (1968) High pressure $NaAlO_2$, an α-$NaFeO_2$ isomorph, Inorg. Chem. **7**, 443–445.

REID, A. F. and A. E. RINGWOOD (1969a) High pressure scandium oxide and its place in the molar volume relationships of dense structures of M_2X_3 and ABX_3 type, J. Geophys. Res., in press.

REID, A. F. and A. E. RINGWOOD (1969b) Six-coordinate silicon: High pressure strontium and barium aluminosilicates with the hollandite structure, Solid State Chem., in press.

REID, A. F. and A. E. RINGWOOD (1969c) Newly observed high pressure transformations in Mn_3O_4, $CaAl_2O_4$ and $ZrSiO_4$, Earth Planet. Sci. Letters, in press.

REID, A. F., A. D. WADSLEY and A. E. RINGWOOD (1967) High pressure $NaAlGeO_4$, a calcium ferrite isomorph and model structure for silicates at depth in the earth's mantle, Acta Cryst. **23**, 736–739.

RINGWOOD, A. E. (1956) The olivine–spinel transition in the earth's mantle, Nature **178**, 1303–1304.

RINGWOOD, A. E. (1958a) Constitution of the mantle I: Thermodynamics of the olivine–spinel transition, Geochim. Cosmochim. Acta **13**, 303–321.

RINGWOOD, A. E. (1958b) Constitution of the mantle II: Further data on the olivine–spinel transition, Geochim. Cosmochim. Acta, **15**, 18–29.

RINGWOOD, A. E. (1958c) Constitution of the mantle III: Consequences of the olivine–spinel transition, Geochim. Cosmochim. Acta **15**, 195–212.

RINGWOOD, A. E. (1960) Some aspects of the thermal evolution of the earth, Geochim. Cosmochim. Acta **20**, 241–259.

RINGWOOD, A. E. (1962a) Prediction and confirmation of olivine–spinel transition in Ni_2SiO_4, Geochim. Cosmochim. Acta **26**, 457–469.

RINGWOOD, A. E. (1962b) Mineralogical constitution of the deep mantle, J. Geophys. Res. **67**, 4005–4010.

RINGWOOD, A. E. (1962c) A model for the upper mantle, J. Geophys. Res. **67**, 857–866.

RINGWOOD, A. E. (1962d) A model for the upper mantle 2, J. Geophys. Res. **67**, 4473–4477.

RINGWOOD, A. E. (1962e) Present status of the chondritic earth model, in: C. B. Moore ed., *Researches on meteorites* (Wiley, New York) 198–216.

RINGWOOD, A. E. (1963) Olivine–spinel transformation in cobalt orthosilicate, Nature **198**, 79–80.

RINGWOOD, A. E. (1966a) The chemical composition and origin of the earth, in: P. M. Hurley ed., *Advances in Earth Science* (M.I.T. Press, Boston) 287–356.

RINGWOOD, A. E. (1966b) Mineralogy of the mantle, in: P. Hurley ed., *Advances in Earth science* (M.I.T. Press, Boston, 357–398.

RINGWOOD, A. E. (1966c) Genesis of chondritic meteorites, Rev. Geophys. **4**, 113–175.

RINGWOOD, A. E. (1967) The pyroxene-garnet transformation in the earth's mantle, Earth Planet. Sci. Letters **2**, 255–263.

RINGWOOD, A. E. (1969a) Phase transformations in the mantle, Earth Planet. Sci. Letters **5**, 401–412.

RINGWOOD, A. E. (1969b) Composition and evolution of the upper mantle, in Upper mantle committee monograph, in press.

RINGWOOD, A. E. and D. H. GREEN (1966) An experimental investigation of the gabbro-eclogite transformation and some geophysical implications, Tectonophysics **3**, 383–427.

RINGWOOD, A. E. and A. MAJOR (1966a) High pressure transformation of $FeSiO_3$ pyroxene to spinel plus stishovite, Earth Planet. Sci. Letters **1**, 135–136.

RINGWOOD, A. E. and A. MAJOR (1966b) High pressure transformations in pyroxenes, Earth Planet. Sci. Letters **1**, 351–357.

RINGWOOD, A. E. and A. MAJOR (1966c) Synthesis of Mg_2SiO_4–Fe_2SiO_4 solid solutions, Earth Planet. Sci. Letters **1**, 241–245.

RINGWOOD, A. E. and A. MAJOR (1966d) High pressure transformation in $CoSiO_3$ pyroxene and some geochemical implications, Earth Planet. Sci. Letters **1**, 209–211.

RINGWOOD, A. E. and A. MAJOR (1967a) Some high pressure transformations of geophysical interest, Earth Planet. Sci. Letters **2**, 106–110.

RINGWOOD, A. E. and A. MAJOR (1967b) High pressure reconnaissance investigation in the systems Mg_2SiO_4–MgO–H_2O, Earth Planet. Sci. Letters **2**, 130–133.

RINGWOOD, A. E. and A. MAJOR (1967c) The garnet–ilmenite transformation in Ge–Si pyrope solid solutions, Earth Planet. Sci. Letters **2**, 331–334.

RINGWOOD, A. E. and A. MAJOR (1968a) Apparatus for phase transformation studies at high pressures and temperatures, Phys. Earth Planet. Interiors **1**, 164–168.

RINGWOOD, A. E. and A. MAJOR (1968b) High pressure transformations in pyroxenes II, Earth Planet. Sci. Letters **5**, 76–78.

RINGWOOD, A. E. and A. MAJOR (1970) The system Mg_2SiO_4–Fe_2SiO_4 at high pressures and temperatures, Phys. Earth Planet. Interiors **3**, 89–108.

RINGWOOD, A. E. and A. F. REID (1968a) High pressure polymorphs of olivines: the K_2NiF_4 type, Earth Planet. Sci. Letters **5**, 67–70.

RINGWOOD, A. E. and A. F. REID (1968b) High pressure transformations of spinels I, Earth Planet. Sci. Letters **5**, 245–250.

RINGWOOD, A. E. and A. F. REID (1969c) High pressure transformations of olivines II, in press.

RINGWOOD, A. E. and M. SEABROOK (1962a) Olivine–spinel equilibria at high pressure in the system Ni_2GeO_4–Mg_2SiO_4, J. Geophys. Res. **67**, 1975–1985.

RINGWOOD, A. E. and M. SEABROOK (1962b) High pressure transition of $MgGeO_3$ from pyroxene to corundum structure, J. Geophys. Res. **67**, 1690–1691.

RINGWOOD, A. E. and M. SEABROOK (1963) High pressure phase transformations in germanate pyroxenes and related compounds, J. Geophys. Res. **68**, 4601–4609.

RINGWOOD, A. E., A. F. REID and A. D. WADSLEY (1967a) High pressure $KAlSi_3O_8$, an aluminosilicate with 6-fold coordination, Acta Cryst. **23**, 1093–1095.

RINGWOOD, A. E., A. F. REID and A. D. WADSLEY (1967b) High pressure transformations of alkali aluminosilicates and alumino-germanates, Earth Planet. Sci. Letters **3**, 38–40.

ROBERTSON, E., F. BIRCH and C. J. F. MCDONALD (1957) Experimental determination of jadeite stability relations to 25,000 bars, Am. J. Sci. **255**, 115–137.

ROBBINS, L. R. and E. M. LEVIN (1959) The system magnesium oxide–germanium dioxide, Am. J. Sci. **257**, 63–70.

ROOYMANS, C. J. M. (1967) Structural investigations on some oxides and other chalcogenides at normal and very high pressures, doctoral thesis, Univ. Amsterdam.

SCLAR, C. B. (1970) High pressure studies in the system MgO–SiO_2–H_2O, Phys. Earth Planet. Interiors **3**, 331.

SCLAR, C. B., L. C. CARRISON and O. M. STEWART (1967) High pressure synthesis of a new hydroxylated pyroxene in the system MgO–SiO_2–H_2O (abstract) Trans. Am. Geophys. Union **48**, 226.

SHAPIRO, J. N. and L. KNOPOFF (1969) Reduction of shock-wave equations of state to isothermal equations of state, J. Geophys. Res. **74**, 1435–1438.

SHIMAZU, Y. (1958) A chemical phase transition hypothesis on the origin of the C-layer within the mantle of the earth, J. Earth Sci. Nagoya Univ. **6**, 12–30.

SMITH, J. V. and P. MOORE (1969) Pyroxene–garnet transformation in Coorara Meteorite., Science, in press.

STEUBER, A. M. and V. R. MURTHY (1966) Strontium isotope and alkali element abundances in ultramafic rocks, Geochim. Cosmochim. Acta **30** 1243–1253.

STISHOV, S. M. (1962) On the internal structure of the earth, Geokhimiya 1962, **8**, 649–659.

STISHOV, S. M. (1963) Equilibrium line between coesite and the rutile-like modification of silica (in Russian), Dokl. Akad. Nauk SSSR **148** no. 5, 1186–1188.

STISHOV, S. M. and S. V. POPOVA (1961) New dense polymorphic modification of silica, Geochimiya 1961– **10**, 837–839.

SYONO, Y., H. SAWAMOTO and S. AKIMOTO (1969) Disordered ilmenite $MnSnO_3$ and its magnetic property, Solid State Comm. **1**, 113–116.

TAKEUCHI, H. and H. KANAMORI (1966) Equations of state of

matter from shock wave experiments, J. Geophys. Res. **71**, 3985–3994.

TAUBER, A., E. BANKS and H. KEDESY (1958) Synthesis of germanate garnets, Acta Cryst. **11**, 893–894.

VERHOOGEN, J. (1953) Elasticity of olivine and constitution of the earth's mantle, J. Geophys. Res. **58**, 337–346.

WADA, T. (1960) On the physical properties within the B-layer deduced from olivine model and on the possibility of polymorphic transition from olivine to spinel at the 20° discontinuity, Disaster Prevent. Res. Inst. Kyoto Univ. Bull. **37**, 1–20.

WADSLEY, A. D. (1964) in: L. Mandelcorn, ed., *Non-stoichiometric compounds* (Academic Press, New York) 108 et seq.

WADSLEY, A. D., A. F. REID and A. E. RINGWOOD (1968) The high pressure form of Mn_2GeO_4, a member of the olivine group, Acta Cryst. **B24**, 740–742.

WANG, C. (1967) Phase transitions in rocks under shock compression, Earth Planet. Sci. Letters **3**, 107–113.

WANG, C. (1968) Constitution of the lower mantle as evidenced from shock wave data for some rocks, J. Geophys. Res. **73**, 6459–6476.

WANG, C. (1969) Equation of state of periclase and some of its geophysical implications, J. Geophys. Res. **74**, 1451–1457.

WYCKOFF, R. W. G. (1964) *Crystal structures*, 1, 2 and 3, 2nd ed. (Interscience, New York).

$(Mg,Fe)_2SiO_4$ SPINEL IN A METEORITE

R. A. BINNS

Department of Geology, University of New England, Armidale, N.S.W., Australia

Purple isotropic particles replacing olivine in black veins cutting the Tenham meteorite prove to be polycrystalline aggregates of $(Mg,Fe)_2SiO_4$ spinel (ringwoodite) with properties conforming to those of synthetic equivalents. Mineralogical evidence for pressure gradients near the veins indicates that these structures and their enclosed ringwoodite particles have been formed by interaction between irregular shock waves passing through the chondrite parent body after a major cosmic impact. Ringwoodite is probably preserved by extremely rapid post-shock cooling.

Ringwoodite, the spinel or γ form of $(Mg,Fe)_2SiO_4$, has recently been described as a natural phase in the Tenham meteorite by BINNS et al. (1969). Although the occurrence is not directly relevant to the constitution of planetary interiors, it provides interesting corroboration of experimental studies dealing with the stability at high presures of a spinel polymorph of olivine.

Tenham is a white olivine–hypersthene chondrite of distinctly recrystallized appearance (cf. VAN SCHMUS and WOOD, 1967; BINNS, 1967a). Its chondritic structure is poorly preserved, and the meteorite is essentially a granoblastic assemblage of olivine (Fa_{26}), orthopyroxene (Fs_{22}), sodic plagioclase (An_{10-12}), nickel–iron alloy, and troilite. After recrystallization, however, the plagioclase has become largely converted to its glassy or amorphous equivalent *maskelynite*, a widely accepted indicator of high pressure shock (MILTON and DE CARLI, 1963; BINNS, 1967b). Unaltered plagioclase remains only in areas remote from the black vein structures that cut irregularly across the Tenham stones (see fig. 1A).

Ringwoodite occurs as purple isotropic inclusions up to 100 μ diameter in the black vein material (fig. 1B), and also as pseudomorphs after olivine in chondrite immediately adjacent (within 10–20 μ) to certain thicker veins or at the margins of chondritic inclusions within very thick breccia-like veins (fig. 1C). Electron probe microanalyses (table 1) reveal compositions essentially identical to that of unaltered olivine away from the veins. The refractive index ($n = 1.768 \pm 0.003$) and cell dimension ($a_0 = 8.113 \pm 0.003$ Å) of this purple ringwoodite agree closely with those interpolated for synthetic spinel of composition $(Mg_{0.74}Fe_{0.26})_2SiO_4$ (see fig. 2). X-ray diffraction photographs of uncrushed particles show the ringwoodite to be very finely polycrystalline. The calculated density of ringwoodite, 3.90 g/cm^3, is some 12% greater than that of its isochemical olivine polymorph (3.48 g/cm^3).

Slightly more distant from some of the veins (up to ca. 200 μ), and more particularly within chondritic fragments in the breccia-like zones, olivine has been transformed to a bluish-grey or grey isotropic material (see fig. 1C). This also gives the diffraction pattern of spinel, but with an increased cell dimension ($a_0 = 8.127 \pm 0.003$ Å) suggestive of a more iron-rich composition $(Mg_{0.66}Fe_{0.34})_2SiO_4$ (see fig. 2). The pattern has less well resolved lines than that of purple ringwoodite, which indicates compositional inhomogeneity. However electron probe data also show the overall composition of this grey material to resemble that of the original olivine in the stone (table 1). Although the purple ringwoodite may well have been compressed during inversion, there is no textural indication of shrinkage in the grey pseudomorphs after olivine. Despite the known experimental difficulty of quenching glasses with forsteritic compositions (cf. however DOBRETSOV et al., 1970), it is suggested that the latter consist of iron-enriched ringwoodite dispersed through a low density magnesian orthosilicate which, on account of the lack of optical birefringence or additional X-ray diffraction

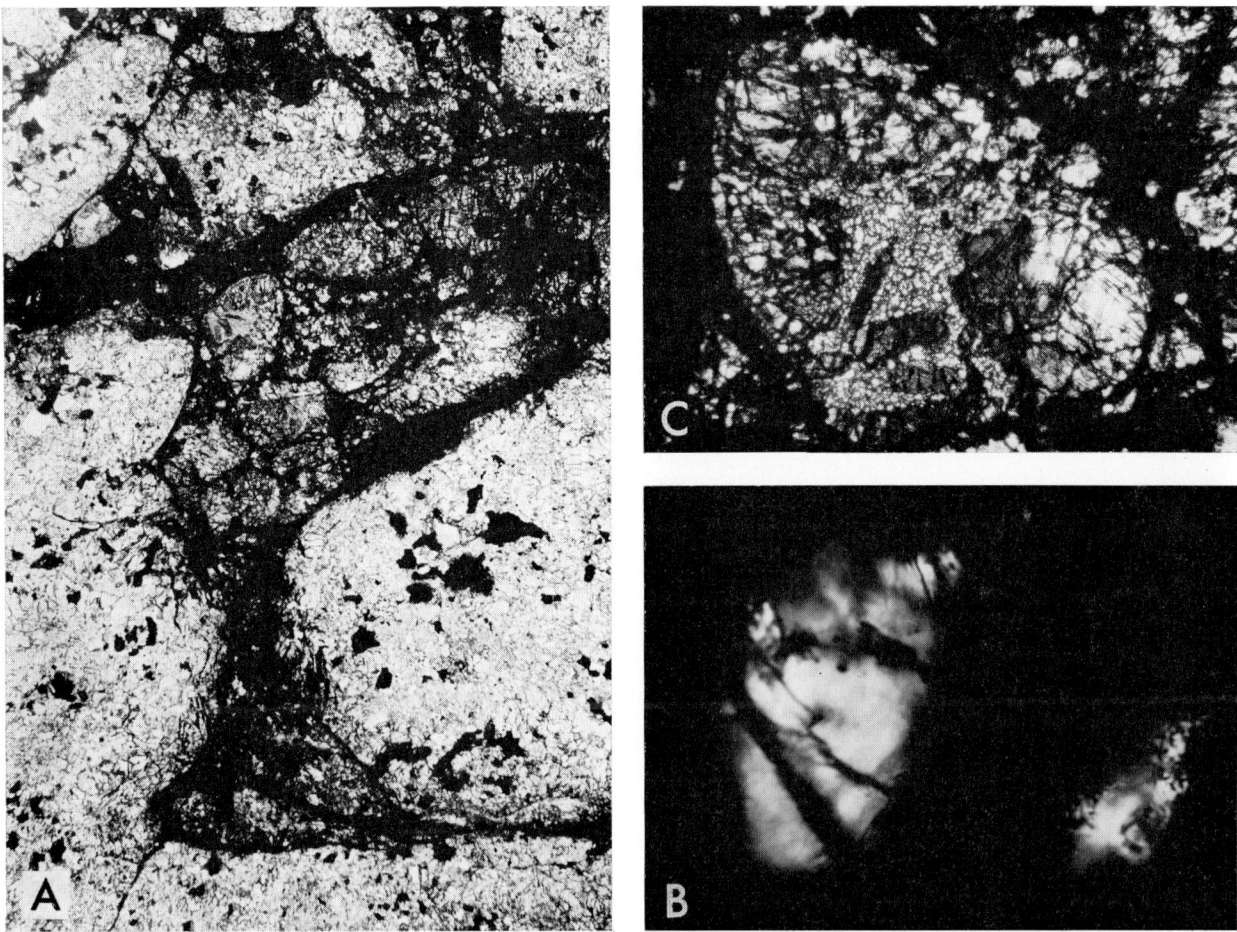

Fig. 1. A. Portion of a thick vein structure in the Tenham meteorite. The light areas represent the recrystallized chondrite of which the meteorite is largely composed, and consist of granular olivine, orthopyroxene, maskelynite, with opaque metal and troilite. ($\times 15$). B. Purple ringwoodite particles (finely polycrystalline spinel) from the black portion of a vein. The black area is partly composed of the garnet mentioned in the text ($\times 560$). C. Enlarged view of a chondritic inclusion occurring just above the centre of fig. 1A. The dark grey areas (three of which are sub-rectangular in outline) were originally olivine micro-phenocrysts but are now composed of bluish-grey or grey spinel-bearing material. Light areas are pyroxene and maskelynite. At the margins of the inclusion the grey material is zoned outwards to purple ringwoodite ($\times 70$).

TABLE 1

Electron probe microanalyses (BINNS et al., 1969) of spinel and olivine in the Tenham meteorite

	Purple ringwoodite	Grey pseudomorphs	Unaltered olivine
SiO_2	38.9	38.7	38.3
FeO	23.4	23.2	23.2
MgO	37.0	38.8	38.6
CaO	nil	nil	nil
Total	99.3	100.7	100.1

lines, must be amorphous or vitreous (due to shock, see below).

If this suggestion is correct, the grey pseudomorphs more remote from veins may be interpreted as inversion products formed at lower pressures than were the purple ringwoodite particles (see fig. 3). This is supported by other evidence of pressure gradients increasing sharply towards the veins, including the distribution of plagioclase and maskelynite, and the presence of martensite-like etch textures in kamacite (α-$Ni_{0.05}Fe_{0.95}$) particles close to the veins. The latter suggest localized inversion to the high pressure ε-Fe phase (SMITH, 1958; KNOX, 1963). Moreover the black material of which the veins are composed (deep brown, isotropic, and apparently microcrystalline in very thin sections) gives a garnet X-ray diffraction pattern ($a_0 = 11.507 \pm$

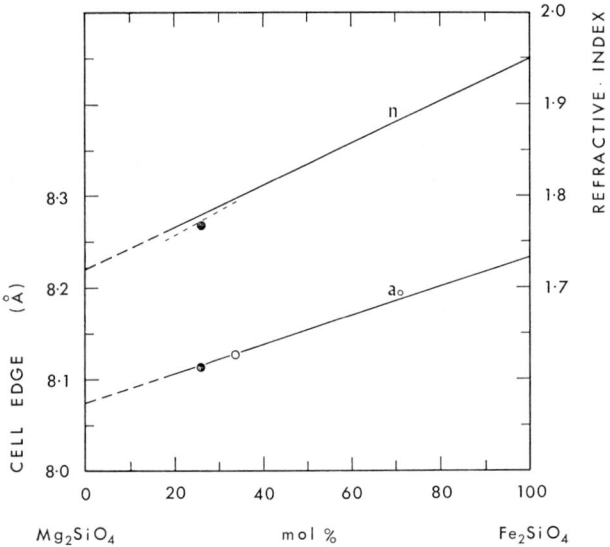

Fig. 2. Variation in refractive index (n) and cell dimension (a_0) with composition in synthetic $(Mg,Fe)_2SiO_4$ spinels (RINGWOOD and MAJOR, 1966; RINGWOOD, 1970). The solid line for refractive index is calculated from the Gladstone–Dale rule; the dashed line slightly below it is the best fit to actual measurements. Both lines are extrapolated below $Mg_{0.80}Fe_{0.20}$, where orthorhombic β-$(Mg,Fe)_2SiO_4$ rather than spinel is the high pressure polymorph (RINGWOOD, 1970). The full circles denote measured properties of purple ringwoodite ($Mg_{0.74}Fe_{0.26}$) from the Tenham meteorite. The high cell dimension of spinel from bluish-grey pseudomorphs (open circle) indicates a more iron-rich composition.

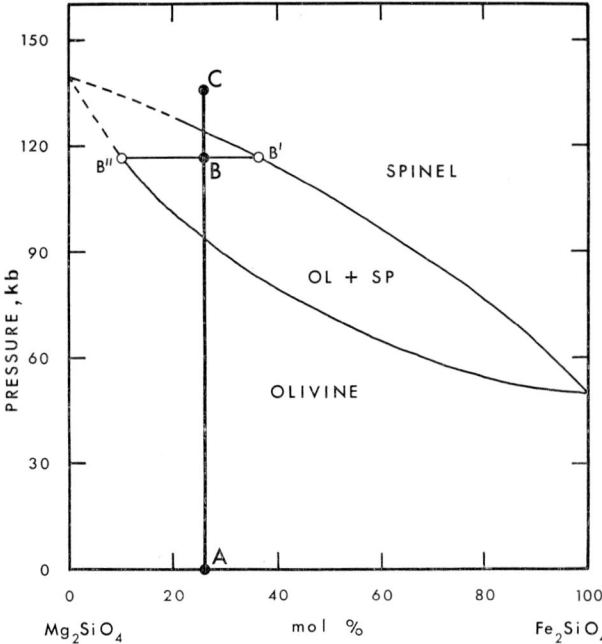

Fig. 3. Phase diagram at 1000 °C for the system Mg_2SiO_4–Fe_2SiO_4 (after AKIMOTO and FUJISAWA, 1969, but neglecting the stability of β-$(Mg,Fe)_2SiO_4$ at magnesian compositions, RINGWOOD, 1970). Under conditions of increasing pressure, olivine (A) would invert first (B) to a mixture of iron-enriched spinel (B′) and magnesian olivine (B″), then to spinel of the original composition (C). B is appropriate to the bluish-grey pseudomorphs in Tenham, except that the magnesian olivine is apparently shocked to an amorphous or glassy equivalent, and C is appropriate to the purple ringwoodite.

0.003 Å) which probably conceals the strong lines of ringwoodite. Since its cell dimension is acceptably close to the value 11.52 ± 0.01 Å predicted by RINGWOOD (1970, and personal communication) for a garnet with composition $M_3(MSi)Si_3O_{12}$, where $M = Mg_{0.78}Fe_{0.22}$, it appears highly probable that the garnet phase in the veins represents a high pressure polymorph of the orthopyroxene (virtually free of Ca and Al) which elsewhere forms an important constituent of the stone. An apparently similar garnet ($a_0 = 11.515$ Å) was recorded by MASON et al. (1968) from veins within the Coorara meteorite, also an olivine–hypersthene chondrite.

Laboratory investigations indicate that $MgSiO_3$ pyroxene will remain stable at pressures sufficient to bring about the olivine → spinel transformation in Mg_2SiO_4, but may invert to a garnet structure at still higher pressures (RINGWOOD, 1967, 1970; RINGWOOD and MAJOR, 1967). Accepting the suggested identity of the Tenham garnet, the fact that ringwoodite lies in contact with garnet within the veins, but coexists with orthopyroxene in their immediate vicinity, again implies pronounced pressure gradients towards the veins. That these were accompanied by increase in temperature is suggested by the tiny globules and threads, composed of an intimate intergrowth between troilite and either kamacite or taenite (Ni-rich γ-FeNi) or both, which are abundantly dispersed throughout the garnetiferous vein material and which, without taking account of pressure effects, suggest that temperatures in the order of 1000 °C were exceeded (cf. RAMDOHR and KULLERUD, 1962; KULLERUD, 1963). Away from the veins there is no textural evidence of homogenization or reaction between adjacent kamacite, taenite, or troilite, although some polygonization or granulation may be apparent in the general proximity of vein structures.

A close genetic relationship between the vein structures and the pressure gradients discussed above is clearly evident. Both are most satisfactorily accounted for by the additive interaction of shock wave fronts passing irregularly through a chondritic parent body as the result of a major cosmic impact, a hypothesis supported by reproduction of veined structures during explosive shock experiments on chondrite (FREDRIKSSON et al., 1963).

Unfortunately, evaluation of the actual pressures under which ringwoodite was formed in Tenham is somewhat hindered by currently imperfect understandings of the behaviour of complex chondritic mineral assemblages under shock. However a general magnitude of ca. 300 kb for the advancing shock waves is indicated by the almost ubiquitous transformation of plagioclase to maskelynite. Maskelynite-like pseudomorphous glass has been produced experimentally from bytownite (An_{80}) in gabbro shocked to 250–350 kb at an estimated transient temperature of only 200–300 °C (MILTON and DE CARLI, 1963), and recent shock experiments covering the full plagioclase compositional range indicate that maskelynite formation commences at 150 kb and is completed at 325 kb, irrespective of An content (DE CARLI et al., 1967). Somewhat lower shock wave pressures are indicated by the more restricted distribution of kamacite showing martensitic etch patterns, since the $\alpha \rightarrow \varepsilon$ iron transformation has been achieved experimentally at 130 kb under both shock and static conditions (cf. TAKAHASHI and BASSETT, 1964; LOREE et al., 1966), but this apparent inconsistency may be explained by the widely dispersed character of metal particles through the silicate constituents of the Tenham meteorite (see the light areas in fig. 1A).

Additive interaction between colliding 300 kb shock fronts, arising from irregularities in the collision process and in the shape or nature of the chondrite parent body, could explain the erratic distribution of vein structures, and produce locally the higher pressures (ca. 450 kb) necessary to bring about inversion of olivine under disequilibrium shock conditions (cf. McQUEEN et al., 1967). The transient temperatures reached in the veins, and the meaning, in relation to occurrences of ringwoodite and garnet in Tenham, of experiments indicating that shocked orthopyroxene undergoes a phase transformation at slightly lower pressures than will olivine of comparable iron content (McQUEEN et al., 1967), are not clear at present. A more detailed understanding of physical conditions implied by the mineralogical relationships in Tenham must await further experimentation on chondritic samples.

Although Hugoniot characteristics of shocked dunite clearly show a phase transition in olivine at elevated pressure (McQUEEN et al., 1967), the recovery of spinel or other high pressure products has not so far been reported. In planar shock wave experiments, post shock conditions including high residual temperatures and slow cooling rates apparently permit reversal to the low pressure products. Consequently the nature of the polymorphic transition or breakdown might be debated (cf. WANG, 1967). The unique conditions whereby very high pressures were established in the relatively narrow veined portions of Tenham evidently allowed almost instantaneous quenching by conduction following passage of the shock waves to the comparatively low residual temperature of the maskelynite-bearing bulk of the stone. Thereby ringwoodite and (?) $(Mg,Fe)SiO_3$ garnet were able to be preserved. X-ray diffraction photographs of various parts of the veins and adjacent material have provided no evidence suggesting the presence of either stishovite or of an ilmenite structure, possible alternative high pressure breakdown products of Mg–Fe olivine and pyroxene (cf. RINGWOOD, 1970). Thus the Tenham occurrences indicate that the first reactions to take place with increasing pressure are the polymorphic inversions olivine (Fa > 20) $\rightarrow$ spinel and pyroxene $\rightarrow$ garnet. Perhaps ironically, the occurrence also suggests that recovery of high pressure products may be more successful in experiments where the geometry of the shock waves is less carefully controlled.

References

AKIMOTO, S. and H. FUJISAWA (1968) J. Geophys. Res. **73**, 1467.
BINNS, R. A. (1967a) Earth Planet. Sci. Letters **2**, 23.
BINNS, R. A. (1967b) Nature **213**, 1111.
BINNS, R. A., R. J. DAVIS and S. J. B. REED (1969) Nature **221**, 943.
DE CARLI, P. S., T. J. AHRENS and J. T. ROSENBERG (1967) Abstr. 30th Ann. Meeting Am. Meteorit. Soc.
DOBRETSOV, N. L., A. A. DERIBAS and V. J. MALY (1970) Phys. Earth Planet. Interiors **3**, 346.
FREDRIKSSON, K., P. S. DE CARLI and A. AARAMÄE (1963) Shock-induced veins in chondrites, in: W. Priester, ed., *Space Research* III (North-Holland Publ. Co., Amsterdam) 974.
KNOX, R. (1963) Geochim. Cosmochim. Acta **27**, 261.
KULLERUD, G. (1963) Carnegie Inst. Wash. Yearbook **62**, 175.
LOREE, T. R., R. H. WARNES, E. G. ZUKAS and C. M. FOWLER (1966) Science **153**, 1277.
MASON, B., J. NELEN and J. S. WHITE (1968) Science **160**, 66.
McQUEEN, R. G., S. P. MARSH and J. H. FRITZ (1967) J. Geophys. Res. **72**, 4999.

MILTON, D. J. and P. S. DE CARLI (1963) Science **140**, 670.
RAMDOHR, P. and G. KULLERUD (1962) Carnegie Inst. Wash. Yearbook **61**, 163.
RINGWOOD, A. E. (1967) Earth Planet. Sci. Letters **2**, 255.
RINGWOOD, A. E. (1970) Phys. Earth Planet. Interiors **3**, 109–155.
RINGWOOD, A. E. and A. MAJOR (1966) Earth Planet. Sci. Letters **1**, 241.
RINGWOOD, A. E. and A. MAJOR (1967) Earth Planet. Sci. Letters **1**, 351.
SMITH, C. S. (1958) Trans. AIME **212**, 574.
TAKAHASHI, T. and W. A. BASSETT (1964) Science **145**, 483.
VAN SCHMUS, W. R. and J. A. WOOD (1967) Geochim. Cosmochim. Acta **31**, 747.
WANG, C. (1967) Earth Planet. Sci. Letters **3**, 107.

CRYSTAL STRUCTURES OF HIGH PRESSURE MODIFICATIONS OF Mn_2GeO_4 AND Co_2SiO_4

N. MORIMOTO, S. AKIMOTO, K. KOTO and M. TOKONAMI

*Institute of Scientific and Industrial Research, Osaka University, Osaka, Japan and
Institute for Solid State Physics, The University of Tokyo, Tokyo, Japan*

The stability field and the crystal structure of β-Mn_2GeO_4 and the crystal structure of the β-Co_2SiO_4 have been determined. The β phase of Mn_2GeO_4 is stable over a wide range of temperature at pressure range intermediate between the fields of α-Mn_2GeO_4 (olivine-type) and δ-Mn_2GeO_4 (Sr_2PbO_4-type), another high pressure polymorph.
The space group and cell dimensions of β-Mn_2GeO_4 are $Imma$ and $a = 6.025$, $b = 12.095$ and $c = 8.752$ Å. The structure was determined by using the three-dimensional data to $R = 0.061$.

Oxygen atoms are in a cubic close packing as in the spinel structure. Mn and Ge atoms are within octahedra and tetrahedra of oxygen atoms respectively. GeO_4 tetrahedra share one of their oxygen atoms resulting in a Ge_2O_7 group and an oxygen atom which is not bounded to any Ge atom. The structure of β-Co_2SiO_4 is isostructural with that of β-Mn_2GeO_4.
The structure of the β phase is compared with that of the γ phase (spinel-type) and their stability is discussed.

1. Introduction

High-pressure transformations in R_2MX_4 type compounds (WYCKOFF, 1965) are of great importance in the interpretation of the sudden increase of seismic velocity at certain depths in the Earth's mantle. The olivine-spinel transformation of R_2MX_4 compounds has hitherto been observed in Mg_2GeO_4, Fe_2SiO_4, Ni_2SiO_4 and Co_2SiO_4. RINGWOOD and MAJOR (1966) reported occurrence of a noncubic high-pressure polymorph, "distorted" or "modified" spinel, in the transformations of $(Mg_{0.85}Fe_{0.15})_2SiO_4$ and Mg_2SiO_4. They considered this high-pressure polymorph, β phase, to be produced from a true spinel when pressures were released. However, their recent extensive investigation (RINGWOOD, 1969) of the phase relations of the system Mg_2SiO_4–Fe_2SiO_4 at high pressures indicates the β phase is thermodynamically stable in its synthesis field. AKIMOTO and SATO (1968) demonstrated in their high-pressure experiments on Co_2SiO_4 that β-Co_2SiO_4, analogous to β-Mg_2SiO_4, is stable in a special field intermediate between the olivine and spinel fields.

In the study of the phase relations of Mn_2GeO_4, β-Mn_2GeO_4 was obtained. Because the β phase is of great importance in the high-pressure transformations of R_2MX_4 compounds, the stability field and crystal structure of β-Mn_2GeO_4 have been investigated. The crystal structure of β-Co_2SiO_4 has also been studied.

2. Phase relations of Mn_2GeO_4 at high pressures

Phase relations of Mn_2GeO_4 were studied over the temperature range 790° to 1240° in the pressure range 31 to 70 kb with the tetrahedral anvil type of high-pressure apparatus (AKIMOTO et al., 1965). The high-pressure polymorph, β-Mn_2GeO_4, is stable in the pressure range intermediate between the fields of the olivine-type polymorph, β-Mn_2GeO_4, and another high-pressure polymorph, δ-Mn_2GeO_4 (fig. 1).* In all experiments, α-Mn_2GeO_4 was used as the starting material. The cell dimensions, space groups and densities at the atmospheric pressure of the three polymorphs are given (table 1). A true spinel, γ-phase, has not been observed in Mn_2GeO_4; the densest polymorph, δ-Mn_2GeO_4, has the Sr_2PbO_4-type structure in which Ge atoms are in octahedra of oxygen atoms (WADSLEY et al., 1968). The increases of density from α-Mn_2GeO_4

---
* We call the polymorphs of the R_2MX_4 compounds as follows: α phase (olivine type), β phase ("modified" spinel type), γ phase (spinel type) and δ phase (strontium plumbate type). This decision was made at a conference by S. Akimoto, N. Kawai, P. Moore, N. Morimoto, A. E. Ringwood and T. Takahashi at the Department of Geophysics and Geochemistry, the Australian National University in January 1969.

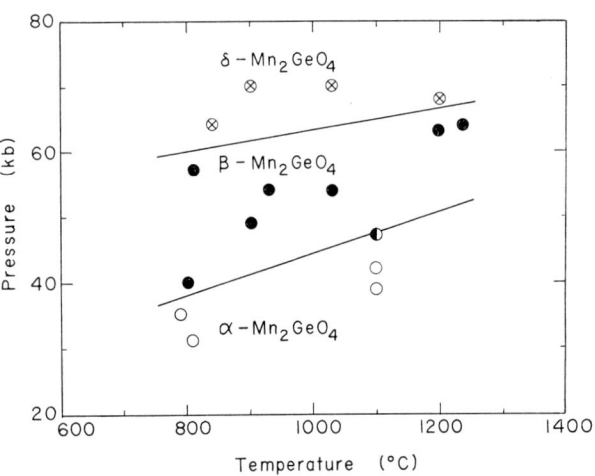

Fig. 1. Stability diagram for the high-pressure and high-temperature transformations of Mn_2GeO_4. Starting material is α-Mn_2GeO_4.

TABLE 1

Cell dimensions, space group and density of Mn_2GeO_4 polymorphs (standard deviations given between parentheses are expressed in units of the last digit stated; Radiation used, CuK_α, (1.5405 Å)

Polymorph	a(Å)	b(Å)	c(Å)	Space group	Density (gcm^{-3})
α-Mn_2GeO_4	5.061(1)	10.719(1)	6.295(1)	Pbnm	4.79
β-Mn_2GeO_4	6.025(2)	12.095(4)	8.752(2)	Imma	5.13
δ-Mn_2GeO_4	5.262(1)	9.274(1)	2.954(1)	Pbam	5.68

to β-Mn_2GeO_4 and from β-Mn_2GeO_4 to δ-Mn_2GeO_4 are 7.1% and 10.7% respectively.

3. Structure determination of the β phase

The structure of β-Mn_2GeO_4 has been determined by using single crystals, synthesized at 1240 °C and 64 kb. The crystals were first examined by Weissenberg and precession methods. The orthorhombic cell dimensions are comparable to those of β-Mg_2SiO_4 and β-Co_2SiO_4 (table 2). The calculated density is 5.15 gcm^{-3} with the cell content of 8 [Mn_2GeO_4]. The diffraction aspect is $I_{**}a$, giving the possible space groups $Imma$ and $Im2a$. A nearly spherical crystal, 0.1 mm in diameter, was used for collecting the intensity data. Three-dimensional intensities for 521 symmetrically independent reflections were measured by the automatic four-circle diffractometer, Rigaku Co., using molybdenum radiation out to a maximum diffraction angle of $2\theta = 60°$ by the ω-2θ scan method. Of these reflections, 31 were less than twice the background value and were regarded to be zero in intensity. The intensities were corrected for Lorentz and polarization factors. No absorption correction was made.

Because the cell dimensions of β-Mg_2SiO_4 and β-Co_2SiO_4 are obtained by the transformation matrix $\frac{1}{2}\bar{1}0/110/001$ from those of the corresponding true spinels (fig. 2), it is reasonable to assume that oxygen atoms occupy similar positions in both structures with a cubic close packing. If we assume symmetry centers in the structure of the β phase, the possible space group is $Imma$. The Patterson projection on (100) of β-Mn_2-

TABLE 2

Cell dimensions of the β phases for three compositions (standard deviations given between parentheses are expressed in units of the last digit stated)

Composition	a(Å)	b(Å)	c(Å)	References
Mg_2SiO_4	5.710(4)	11.45(2)	8.248(9)	Ringwood and Major (1966)
Co_2SiO_4	5.753(1)	11.522(4)	8.337(2)	Akimoto and Sato (1968)
Mn_2GeO_4	6.025(2)	12.095(4)	8.752(2)	Akimoto and Sato (1968)

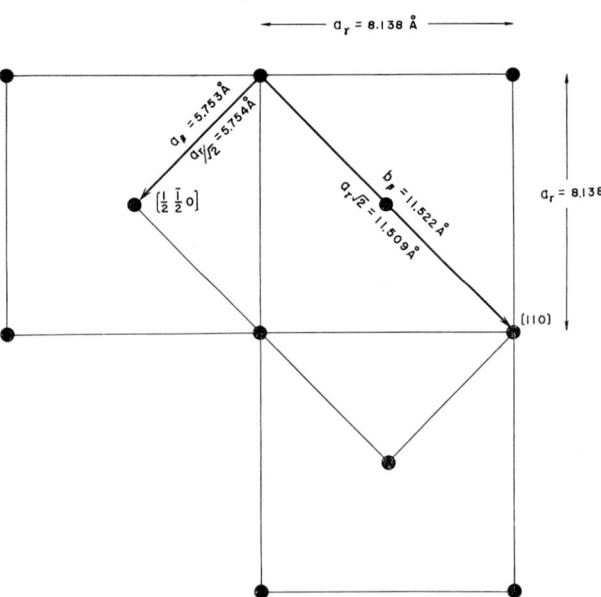

Fig. 2. Relationship between the unit cells of β-Co_2SiO_4 and γ-Co_2SiO_4.

GeO_4 enabled us to decide the positions of metal atoms, tetrahedral sites (A sites) for Ge atoms and octahedral sites (B sites) for Mn atoms. The initial parameters of metal atoms determined in this way and those of oxygen atoms obtained from the spinel structure were used for the structure factor calculations for $hk0$ and $0kl$. Then the Fourier projections on (001) and (100) were synthesized to obtain the displacements of oxygen and metal atoms from the initial positions. Further refinements were carried out by the full-matrix, least squares method (BUSING et al., 1962). Three cycles of refinement of the atomic coordinates and one scale factor were made, keeping the isotropic temperature factor $B = 1.0$ for all atoms. Three more cycles of refinement of individual isotropic temperature factors, in addition to the atomic coordinates and one scale factor, reduced the residual R to 0.061 for all 521 reflections and 0.055 for 490 observed reflections. The final atomic coordinates and individual isotropic temperature factors are given (table 3).

TABLE 3

Atomic coordinates (expressed in cell edges) and isotropic temperature factors of β-Mn_2GeO_4 (upper) and β-Co_2SiO_4 (the standard deviations given between parentheses are expressed in the unit of the last digit stated)

	Atom	x	y	z	B
R(1)	Mn(1)	0.0000(0)	0.0000(0)	0.0000(0)	0.49(6)
	Co(1)	0.0000(0)	0.0000(0)	0.0000(0)	0.35(9)
R(2)	Mn(2)	0.0000(0)	0.2500(0)	−0.0319(3)	0.43(5)
	Co(2)	0.0000(0)	0.2500(0)	−0.0283(6)	0.34(9)
R(3)	Mn(3)	0.2500(0)	0.1281(2)	0.2500(0)	0.49(3)
	Co(3)	0.2500(0)	0.1239(3)	0.2500(0)	0.38(5)
M	Ge	0.0000(0)	0.1193(1)	0.6165(2)	0.37(2)
	Si	0.0000(0)	0.1208(7)	0.6168(7)	0.25(8)
X(1)	O(1)	0.0000(0)	0.2500(0)	0.2123(13)	0.80(23)
	O(1)	0.0000(0)	0.2500(0)	0.2161(25)	0.25(30)
X(2)	O(2)	0.0000(0)	0.2500(0)	0.7191(14)	1.10(25)
	O(2)	0.0000(0)	0.2500(0)	0.7110(31)	1.21(43)
X(3)	O(3)	0.0000(0)	−0.0088(7)	0.2546(12)	0.89(18)
	O(3)	0.0000(0)	−0.0136(12)	0.2593(21)	0.05(23)
X(4)	O(4)	0.2586(9)	0.1224(8)	−0.0033(6)	0.94(9)
	O(4)	0.2627(16)	0.1242(20)	−0.0087(12)	0.52(17)

The structure of β-Co_2SiO_4 has been determined by using single crystals, synthesized at 1420 °C and 81 kb. Three-dimensional intensities for 492 reflections were obtained in the same way as β-Mn_2GeO_4. Refinement was initiated using the atomic coordinates of β-Mn_2GeO_4, with the individual isotropic temperature factors: 0.5 for Co, 0.4 for Si and 1.0 for oxygen atoms. After three cycles of least squares refinement, the residual R reduced to 0.105 for all reflections and 0.093 for 454 observed reflections. The final atomic coordinates and individual isotropic temperature factors are given (table 3).

4. Description of the structure of the β phase

The structure of β-Mn_2GeO_4 is viewed in [100] direction (fig. 3). The oxygen atoms are seen to be approximately in cubic close packing as in spinel. The Ge atoms occupy the A sites and the Mn atoms, the B sites. However, the arrangements of A and B sites in β-Mn_2GeO_4 are different from those in spinel. For comparison, the structure of spinel with R_2MX_4 composition, where the M atoms are on the A sites and the R atoms on the B sites, is shown (fig. 4). Two GeO_4 tetrahedra, which would be isolated in the spinel structure, share one of their oxygen atoms resulting in a Ge_2O_7 group and an oxygen atom not bonded to any Ge atom. Thus we must modify the spinel structure by displacing four M and four R atoms out of the eight M and sixteen R atoms in the cell in order to obtain the structure of the β phase.

It is of interest to notice that O(1) is bonded not to any Ge atom but to five Mn atoms, and that O(2) is bonded to two Ge atoms and one Mn atom, because of the formation of the Ge_2O_7 groups. O(3) and O(4) are bonded to one Ge atom and three Mn atoms. The Ge–O distances of the tetrahedra range from 1.749 to 1.818 Å with the mean value of 1.772 Å and the Mn–O distances of the octahedra from 2.133 to 2.239 Å with the mean value of 2.188 Å. The Ge_2O_7 group is schematically shown with the bond distances and angles (fig. 5).

Pauling's electrostatic balance rule fails for O(1) and O(2), the charge balance being $-1/3$ for O(1) and $+1/3$ for O(2). The negative charge on O(1), responsible for short distances of this oxygen to Mn atoms, compensates for the surplus charge of O(2), in the form of long distances of this oxygen to Ge and Mn atoms.

The similar discussion is given for the structure of β-Co_2SiO_4. The Si_2O_7 group in β-Co_2SiO_4 is shown for comparison with the Ge_2O_7 group (fig. 5).

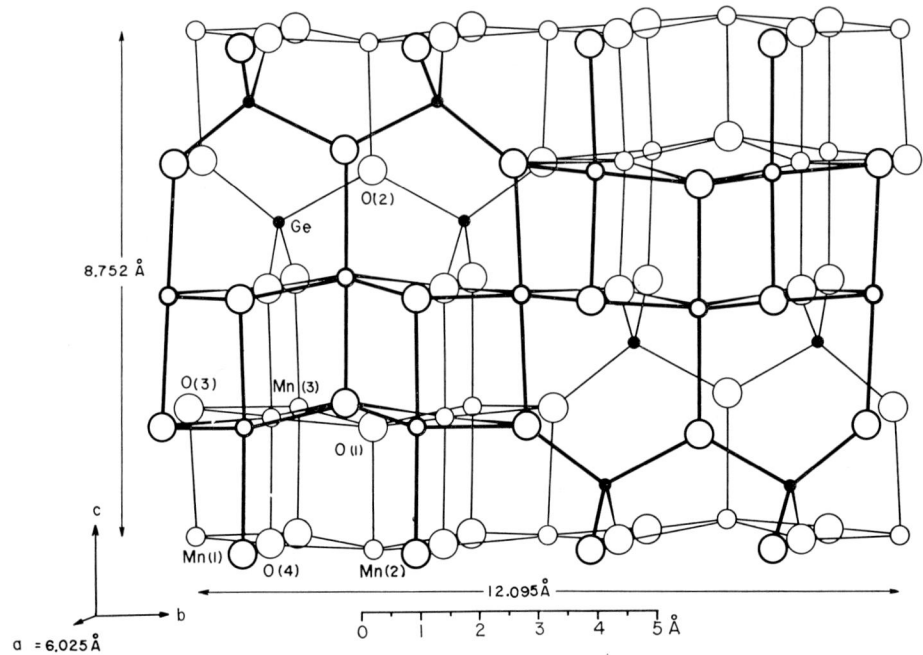

Fig. 3. Orthorhombic crystal structure of β-Mn_2GeO_4.

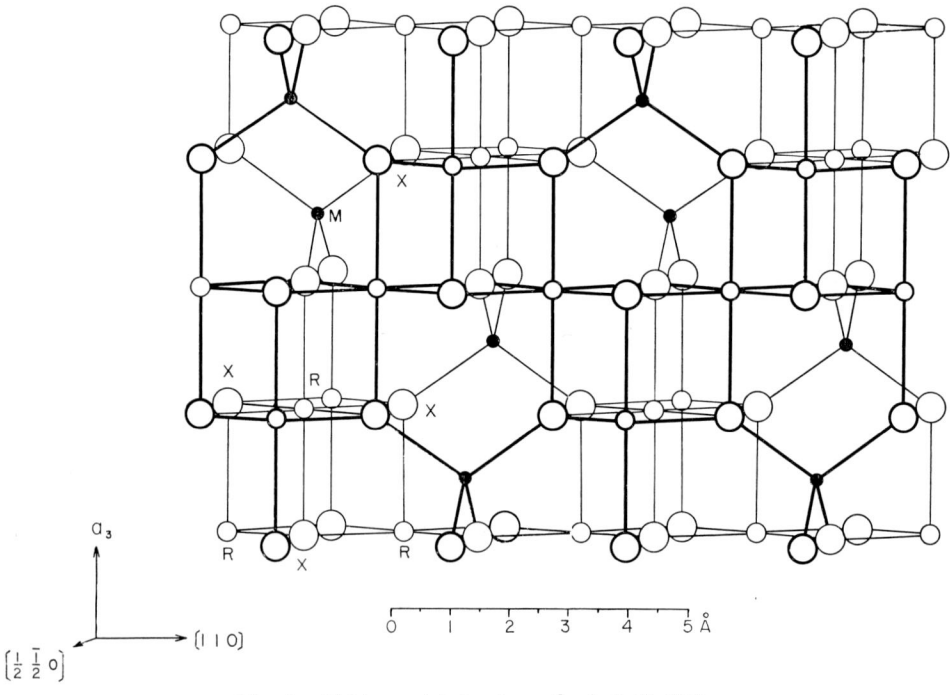

Fig. 4. Cubic crystal structure of spinel, R_2MX_4.

5. Comparison of the structures of the β and γ phases

From the lattice dimension, $a = 8.138$ Å for γ-Co_2SiO_4, we can deduce the value of $x = 0.366$ for the x coordinate of oxygen atoms, on the basis of the assumption that the Si–O distance is the same as the average distances of the nonbridging bonds in β-Co_2SiO_4, 1.63 Å. Then the corresponding Co–O dis-

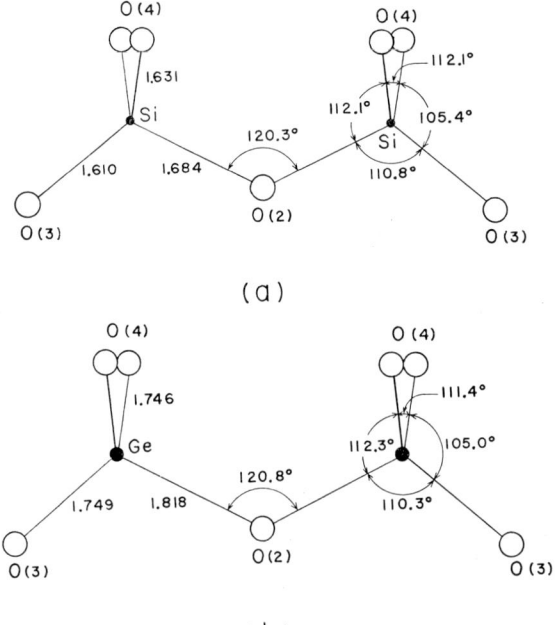

Fig. 5. Independent Si_2O_7(a) and Ge_2O_7(b) in the β phases. The bridging bonds are longer than the nonbridging bonds in both groups.

tance becomes 2.11 Å which is the same as the average Co–O distance in the β phase. For a hypothetical structure of γ-Mn_2GeO_4, we obtain the value of $x = 0.369$ on the basis of the assumption that the Ge–O and Mn–O distances are the same as the corresponding average values in the β phase (1.76 Å and 2.19 Å respectively).

In the structures of the γ phases, the shared edges are longer than the unshared edges for the A-site octahedra (table 4). However, for the structure of the β phases, the shared edges are the same as or shorter than unshared edges. Because a shortening of the shared polyhedral edges is considered to be an important stabilizing factor for the ionic crystals (PAULING, 1960; KAMB, 1968), the stability of the β phases might be explained by this shortening.

TABLE 4

Shared and unshared edges, expressed in Å, of the A-site-octahedra in the β and γ phases (the mean values are given for the β phases; the γ phase for Mn_2SiO_4 is hypothetical)

	Co_2SiO_4		Mn_2SiO_4	
	Shared edges	Unshared edges	Shared edges	Unshared edges
β phase	2.99	3.00	3.06	3.12
γ phase	3.08	2.88	3.16	3.02

References

AKIMOTO, S., H. FUJISAWA and T. KATSURA (1965) J. Geophys. Res. **70**, 1969.
AKIMOTO, S. and Y. SATO (1968) Technical Report of ISSP, Series A. No. 328.
BUSING, W. R., K. O. MARTIN and H. A. LEVY (1962) U.S. Atomic Energy Commission Report No. ORNL-TM-305.
KAMB, B. (1968) Am. Mineralogist **53**, 1439.
PAULING, L. (1960) *The Nature of the Chemical Bond*, 3rd ed. (Cornell Univ. Press).
RINGWOOD, A. E. (1969) Publication 666, Dept. of Geophys. and Geochem. (Australian National Univ., Canberra).
RINGWOOD, A. E. and A. MAJOR (1966) Earth Planet. Sci. Letters **1**, 241.
WADSLEY, A. D., A. F. REID and A. E. RINGWOOD (1968) Acta Cryst. B **24**, 740.
WYCKOFF, W. G. (1965) *Crystal Structures*, 2nd ed. (Interscience).

CRYSTAL STRUCTURE OF β-Mg$_2$SiO$_4$: CRYSTAL-CHEMICAL AND GEOPHYSICAL IMPLICATIONS

P. B. MOORE and J. V. SMITH

Department of the Geophysical Sciences, University of Chicago, Chicago, Illinois, 60637, U.S.A.

The crystal structure of β-Mg$_2$SiO$_4$ is based on cubic close packing of O atoms with Si atoms sharing pairs of adjacent tetrahedral sites yielding (Si$_2$O$_7$) groups, and with Mg atoms occupying octahedral sites. Close-packed structures are discussed in terms of topology and of the simple ionic packing model.

1. Introduction

RINGWOOD and MAJOR (1966) demonstrated the existence of a third polymorph of Mg$_2$SiO$_4$, beside the well-known olivine (α) and spinel (γ) structure types. This structure, β-Mg$_2$SiO$_4$, appears for compositions more magnesian than (Mg$_{0.8}$Fe$_{0.2}$)$_2$SiO$_4$ at pressures above 150 kb. Crystal-chemical knowledge of this new structure type is of great importance in understanding the phase transitions in the Earth's mantle since its bulk composition is believed to be sufficiently rich in Mg and Si to fall within the stability field of this phase.

We determined the crystal structure of isostructural β-(Mg$_{0.9}$Ni$_{0.1}$)$_2$SiO$_4$ from X-ray patterns prepared from a synthetic powder kindly supplied by Dr. A. E. Ringwood. The space group could not be uniquely determined from the powder pattern; however, AKIMOTO and SATO (1968) reported that Dr. N. Morimoto had found the space group *Ibmm* for single crystals of isostructural β-Mn$_2$GeO$_4$ and β-Co$_2$SiO$_4$. Using this space group two structures were proposed, of which one was more probable (MOORE and SMITH, 1969). Using single crystal data MORIMOTO et al. (1970) independently determined the structure of Mn$_2$GeO$_4$, obtaining the same topology as for the probable structure of MOORE and SMITH (1969). Since the atomic parameters obtained by MORIMOTO et al. (1970) should be considerably more accurate than those obtained by us, we shall concentrate here on the general topologic and crystal-chemical relations between the β-structure and other related structures based on close-packing of oxygen atoms.

2. Structure analysis

Powder mounts of β-(Mg$_{0.9}$Ni$_{0.1}$)$_2$SiO$_4$ were used for the structure analysis because the high-angle lines were sharper than those for the pure Mg$_2$SiO$_4$ material. (Incidentally the composition Mg$_{0.85}$Ni$_{0.15}$ reported in MOORE and SMITH (1969) is in error.) Fine ground powder was rolled in rubber cement to form a tiny sphere with minimum preferred orientation. Six film patterns using FeK$_\alpha$ and CuK$_\alpha$ radiations and a 114.6 mm diameter Buerger-type powder camera were obtained from 12-hour, 1-day and 2-day exposures. The films were corrected for shrinkage, the line positions were measured, and the intensities estimated by eye using a calibrated scale of line segments.

Table 1 reports the indexing of the 102 observed powder lines using the 358 possible independent diffraction planes for the space group *Ibmm*. Four lines were unindexable, and are believed to arise from impurities: they do not correspond to obvious impurities such as olivine and spinel. Least squares refinement of high-angle lines on the FeK$_\alpha$ pattern yielded $a = 8.248(1)$, $b = 11.444(2)$, $c = 5.6960(5)$ Å. The standard deviation listed in brackets to the same decimal significance is probably an under-estimate because of uncorrected absorption error.

The limitations on polyhedral arrangement enforced by the space group, and the relationship of the cell geometry and space group to those of spinel, led to only two sensible models for the β-structure. In the first model, the silicate tetrahedra are paired as (Si$_2$O$_7$) groups while in the second they are insular. Denoting

CRYSTAL STRUCTURE OF β-Mg_2SiO_4

TABLE 1

X-ray powder diffraction data

d_{obs}	d_{calc}	I	hkl	d_{obs}	d_{calc}	I	hkl	d_{obs}	d_{calc}	I	hkl	d_{obs}	d_{calc}	I	hkl
6.69	6.69	5	110		1.527		521	1.085	1.086	10b	482	0.9060	0.9065	10	484
5.79	5.72	2	020	1.514	1.514	10	530*		1.085		444		0.9062		772
4.67	4.69	10	101		1.511		361		1.084		712	0.9039	0.9048	10	901
3.463	3.462	5	130		1.507		323	1.071	1.070	5	741		0.9043		0,12,2
3.342	3.346	10	220	1.497	1.497	5	352		1.070		392		0.9032		813
3.206	3.207	40	211	1.479	1.479	10	451		1.069		633	0.9012	0.9010	10	046
3.053	—	5	ni		1.479		262	1.062	1.062	10	473	0.8972	0.8981	2	781
2.964	—	5	ni	1.469	1.468	10	271		1.062		581*		0.8966		394*
2.855	2.861	20	040	1.444	1.442	100	442		1.062		0,10,2	0.8905	0.8911	5	930
	2.848		002	1.427	1.431	40	080	1.056	1.055	5	235		0.8908		2,11,3*
2.761	—	2	ni		1.424		004		1.055		383		0.8908		545
2.677	2.673	40	310	1.406	1.405	10	370	1.051	1.050	5	145	0.8811	0.8815	2b	833
2.622	2.621	50	112	1.400	1.400	10	460	1.047	1.048	10	750		0.8803		246
2.556	2.550	5	022		1.397		172*		1.047		732*		0.8795		5,11,0
2.511	2.513	20	231	1.382	1.382	2	024	1.0374	1.0385	5	662	0.8744	0.8746	2	615
2.481	2.476	20	301	1.378	1.377	5	253*		1.0373		534*	0.8720	0.8720	2	156
2.437	2.442	100	141		1.375		600	1.0304	1.0322	10	1,11,0		0.8719		2,10,4
2.348	2.351	10	240	1.340	1.338	10	550*		1.0310		800*	0.8694	0.8699	5	912
	2.344		202		1.337		532		(1.0235)		(293)	0.8623	0.8627	10	941
2.273	2.272	5	321		1.337		620	1.0145	1.0147	2	820		0.8623		406
2.232	2.230	20	330	1.327	1.328	5	163		(1.0106)		(811)		0.8621		871
2.199	2.205	40	150		1.327		611*	1.0092	1.0092	20	084	0.8551	0.8549	2	635
	2.199		132*	1.313	1.312	10	433*	1.0001	1.0006	10	4,10,0	0.8499	0.8508	2	950
2.132	—	2	ni		1.310		224		1.0002		374*		0.8505		932
2.064	2.062	40	400	1.277	1.278	10	082	0.9930	0.9934	2	415		0.8499		066
2.019	2.018	100	042		1.275		044		0.9933		2,11,1	0.8476	0.8477	10	385
1.938	1.940	5	420	1.259	1.257	20	314*	0.9830	0.9832	10	752		0.8477		1,12,3
1.911	1.912	10	411		1.257		190	0.9752	0.9752	5	554*	0.8365	0.8367	15	1,13,2
	1.907		060		1.257		462		0.9746		624		0.8364		880*
1.888	1.888	10	251	1.246	1.245	10	503	0.9699	0.9704	10	1,11,2	0.8348	0.8351	20	804
1.875	1.872	10	341	1.239	1.239	10	640		0.9700		8,4,0	0.8283	0.8282	20	4,12,2
1.851	1.850	5	103		1.239		381		0.9695		8,0,2	0.8259	0.8256	30	446
1.817	1.813	2	242		1.238		602	0.9548	0.9559	2	770	0.8192	0.8191	2	655
1.758	1.767	15	161	1.201	1.200	5	334		0.9558		822		0.8190		705
	1.761		123	1.188	1.188	5	291		0.9537		0,12,0	0.8171	0.8175	5	961
	1.759		350*		1.187		273	0.9497	0.9495	2	592		0.8174		0,14,0
	1.756		332*	1.176	1.175	20	480*		0.9493		006*		0.8171		5,12,1
1.743	1.744	10	152		1.172		710	0.9392	0.9399	10	116		0.8169		176
1.726	1.731	10	260	1.174	1.172	20	404*		0.9393		583*		0.8169		923
	1.728		431*	1.163	1.161	2	570		(0.9374)		(505)	0.8034		2	
1.707	1.705	10	213	1.155	1.154	5	701	0.9341	0.9348	10	644	0.8017		10	
1.671	1.673	15b	440		1.154		651		0.9345		1,12,1	0.7984		5	
	1.670		402*		1.154		390	0.9210	0.9215	2	365	0.7927		10	
1.638	1.633	10	510	1.151	1.150	10	192		0.9212		691	0.7874		5	
1.605	1.604	5	170		1.148		424		0.9208		3,11,2	0.7840		5	
	1.603		422	1.132	1.132	2	183*		0.9204		673	0.7804		2	
1.586	1.585	10	062		1.128		105	0.9179	0.9182	5	842	0.7793		2	
	1.585		501*	1.107	1.108	10	613		(0.9167)		(4,11,1)				
1.572	1.572	2	233		1.107		125	0.9156	0.9155	5	136				
1.562	1.562	10	303		1.107		354		(0.9142)		(455)				
1.554	1.554	20	143	1.094	1.093	5	215		(0.9116)		(275)				

* principal component.

octahedrally coordinated populations by M and tetrahedral ones by T, the corresponding cell contents may be written $M_{16}O_4(T_2O_7)_4$ and $M_{20}O_{16}(TO_4)_4$. Least-squares refinements led to a discrepancy index of 0.19 for the former and 0.23 for the latter. The index was computed only for structure amplitudes derived from uniquely-indexed lines. This small difference between the discrepancy indices arises principally from the identity of the positions of the oxygen atoms and of some of the cations in the two arrangements. Careful examination of the weak, low-angle reflections showed that the first model was definitely superior to the second one.

Refinement of the powder data was tedious and uncertain because of the overlap of many reflections as shown in table 1. The uncertainty was particularly serious in the region where α_1 lines overlapped with α_2 lines produced by other reflections. This uncertainty was partly resolved by using FeK_α radiation whose greater dispersion permitted resolution of some merged lines on the CuK_α pattern. Between each set of refinements, intensities of overlapped reflections were proportioned from the observed single intensity on the basis of calculated intensities. It was not possible to obtain accurate independent B displacement factors for all the atoms. Consequently the oxygen atoms were assigned B values greater than those of the cations by 0.25, and a single overall temperature factor was refined yielding $B = 0.31$ for cations and 0.56 for oxygen atoms. These values are similar to those for kyanite (BURNHAM, 1963: 0.25–0.33 for cations and 0.33–0.57 for oxygens), indicating no significant error from neglect of an absorption correction. No significant change of positional parameters resulted when the B values were changed arbitrarily by amounts up to 0.5. The interatomic distances are consistent with assignment of all the Si atoms to the T sites. For purposes of calculation the Mg and Ni atoms were randomly distributed among the three types of M sites. Half-ionized scattering factors were used. No correction was made for anomalous scattering.

The final observed and calculated structure amplitudes are given in table 2; the latter amplitudes for overlapping reflections were proportioned from the observed intensity according to the squared calculated amplitudes. Most unobserved lines have low values of the calculated intensity. It should be noted that because the β-structure is essentially a superstructure with a pseudo-cell, there are few strong reflections and many weak ones. The reflections 240 and 202 were omitted from the final refinement because of the possibility that the observed intensity really arises from an impurity. The weights were assigned empirically using estimates of errors in intensity measurement and uncertainties of overlapping reflections. To save space they are omitted: little change occurred in the atomic coordinates when the weights were changed within reasonable limits.

Table 3 shows the final atomic coordinates with the standard errors from the least-squares refinement. Table 4 contains the principal interatomic distances. The estimated standard errors are around 0.02 Å; however, this value is too low because of the false impression given by the overlapping reflections, and possibly because of strong correlations resulting from the pseudo-structure.

3. Atomic structure of $M_4O(T_2O_7)$

Fig. 1 shows the structure of the β-form projected down the c-axis. Oxygen atoms occur both at heights 0.25 and 0.75 exactly defined by the mirror planes perpendicular to c, and at heights near 0.0 and 0.5. Pairs of silicon atoms either at 0.25 or 0.75 form Si_2O_7 doublets constrained by point symmetry mm at the shared oxygen. The M1 and M2 cations also lie at 0.25 or 0.75, and each one is bonded to two oxygen atoms at the same height and two pairs of oxygen atoms near 0.0 and 0.5 as first neighbors forming the

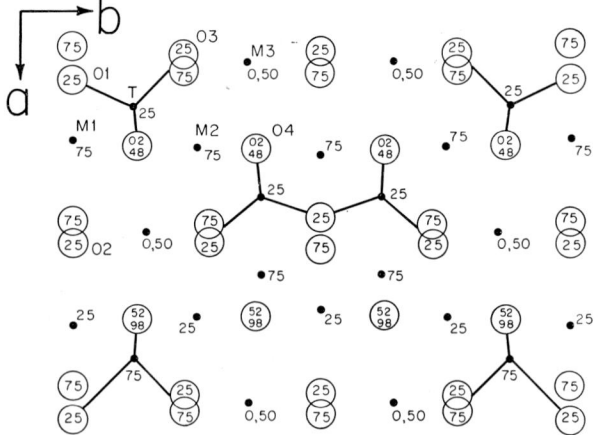

Fig. 1. Projection of the β-structure down the c-axis. The heights of the atoms are shown as percentages of the c-repeat. Oxygen atoms are shown by open circles, and cations by small solid circles. The Si–O bonds within one unit cell are shown. Tetrahedra at the corners also form Si_2O_7 groups.

CRYSTAL STRUCTURE OF β-Mg$_2$SiO$_4$

TABLE 2

Observed and calculated structure amplitudes

h	k	l	F_o	F_c	h	k	l	F_o	F_c	h	k	l	F_o	F_c	h	k	l	F_o	F_c	h	k	l	F_o	F_c
1	1	0	5	−5	5	1	2	a	−15	1	8	3	20	−14	5	5	4	26	23	7	6	3	a	2
0	2	0	6	5	3	7	0	56	−56	1	0	5	7	−4	6	2	4	7	−6	1	8	5	a	−3
1	0	1	9	7	4	6	0	6	6	7	3	0	a	−8	1	10	3	a	−5	7	3	4	a	−6
2	0	0	a	−3	1	7	2	30	−33	6	6	0	a	3	3	11	0	a	20	2	12	2	a	4
1	2	1	a	4	1	1	4	a	−3	1	10	1	a	5	1	6	5	2	2	8	3	3	10	11
1	3	0	10	11	4	1	3	a	15	6	1	3	23	−18	8	4	0	35	−37	2	4	6	6	−6
2	2	0	12	−13	5	4	1	a	10	1	2	5	4	3	8	0	2	3	3	5	11	0	21	23
2	1	1	28	−29	0	2	4	17	3	3	5	4	26	20	1	11	2	31	32	6	10	0	a	−11
0	4	0	41	−38	2	5	3	24	−17	2	10	0	a	−12	4	3	5	a	−15	6	6	4	a	2
0	0	2	37	−35	6	0	0	1	0	2	6	4	a	3	8	2	2	7	7	6	1	5	16	12
3	1	0	50	−58	3	4	3	a	−3	2	1	5	26	−14	7	7	0	27	−22	1	13	0	a	3
1	1	2	41	−46	1	8	1	a	5	4	8	2	27	27	0	12	0	22	−19	3	3	6	a	12
0	2	2	14	−10	2	8	0	a	−6	4	4	4	22	14	0	0	6	37	−21	1	5	6	14	12
2	3	1	23	20	2	0	4	a	2	7	1	2	25	16	5	9	2	14	−9	2	10	4	12	−9
3	0	1	37	43	5	5	0	45	32	5	7	2	a	−13	7	4	3	a	3	9	1	2	23	−21
1	4	1	54	67	5	3	2	7	5	5	1	4	a	−14	4	10	2	a	−12	4	12	0	a	8
2	4	0	12*	4	6	2	0	13	−9	7	4	1	27	−21	1	9	4	a	7	8	6	2	a	10
2	0	2	23*	4	6	1	1	26	19	6	3	3	17	14	4	9	3	a	7	7	9	0	a	−8
3	2	1	11	−15	1	6	3	3	−2	3	9	2	3	2	1	1	6	19	−23	9	4	1	30	31
3	3	0	33	35	1	3	4	a	8	1	7	4	a	−12	5	8	3	32	−37	4	0	6	9	16
1	5	0	9	14	4	3	3	26	20	4	9	1	a	−8	5	0	5	20	23	8	7	1	15	−20
1	3	2	39	46	2	2	4	10	−8	4	7	3	14	14	0	2	6	a	−5	6	3	5	19	−9
2	2	2	a	4	0	8	2	30	−22	5	8	1	35	34	6	8	2	a	12	4	2	6	a	−2
4	0	0	88*	91	0	4	4	29	−21	0	10	2	16	−14	6	4	4	19	−22	4	7	5	a	−9
0	4	2	92*	−92	6	3	1	a	−16	2	3	5	13	14	1	12	1	28	32	2	13	1	a	10
4	1	1	23	−19	3	7	2	a	−2	3	8	3	34	37	2	12	0	a	−10	9	5	0	13	10
0	6	0	27	20	3	1	4	42	−38	3	0	5	a	−18	5	10	1	a	−4	9	3	2	12	9
2	5	1	20	20	4	6	2	17	16	1	4	5	30	45	8	5	1	a	−9	0	6	6	1	1
3	4	1	21	−13	1	9	0	9	8	7	5	0	10	−9	2	0	6	a	−18	3	8	5	20	−15
1	0	3	22	−24	4	7	1	a	−10	7	3	2	51	−46	5	2	5	a	−4	1	12	3	28	−21
2	4	2	12	−3	5	0	3	46	−53	5	6	3	a	3	3	10	3	a	4	7	5	4	a	−7
1	6	1	3	2	6	4	0	26	−25	3	10	1	a	−5	3	6	5	5	−7	8	5	3	a	7
1	2	3	4	−4	6	0	2	16	14	6	6	2	12	10	6	9	1	12	15	5	10	3	a	3
3	5	0	34	31	3	8	1	34	−32	5	3	4	32	28	6	7	3	9	−12	5	6	5	a	−2
3	3	2	27	24	2	8	2	a	16	3	2	5	a	−7	3	11	2	15	19	5	11	2	a	8
1	5	2	22	21	5	6	1	a	−3	6	7	1	a	13	8	4	2	27	−26	6	10	2	a	−5
2	6	0	5	5	2	4	4	a	−14	8	0	0	76	67	1	3	6	19	23	6	9	3	a	−14
4	3	1	27	−28	5	2	3	a	6	1	11	0	11	5	4	11	1	a	−16	3	13	0	a	10
2	1	3	26	21	5	5	2	a	−7	2	10	2	a	−9	4	5	5	a	−12	8	8	0	48	49
4	4	0	18	15	6	2	2	a	3	2	9	3	19	10	9	1	0	a	2	1	13	2	2	2
4	0	2	47	40	3	6	3	a	9	8	2	0	11	−6	2	2	6	a	2	8	0	4	41	49
5	1	0	25	−20	3	3	4	33	24	8	1	1	21	−20	2	7	5	a	−14	1	11	4	a	4
4	2	2	4	−3	1	5	4	a	9	0	8	4	88	81	8	6	0	a	1	3	5	6	a	4
1	7	0	31	−18	4	5	3	a	15	5	9	0	a	−2	4	8	4	36	34	4	11	3	a	14
5	0	1	51	51	2	9	1	16	−12	7	0	3	a	5	7	7	2	20	−15	2	6	6	a	7
0	6	2	1	2	2	7	3	22	18	6	5	3	a	8	7	1	4	a	−15	2	9	5	a	−8
2	3	3	6	−17	4	8	0	54	50	3	7	4	49	−40	9	0	1	3	1	4	12	2	48	−56
3	0	3	44	52	7	1	0	23	−21	4	10	0	6	−4	8	1	3	24	19	7	9	2	a	−7
1	4	3	46	−38	4	0	4	54	50	4	6	4	a	5	0	12	2	50	−37	8	2	4	a	−8
5	2	1	a	−8	5	7	0	18	−26	4	1	5	10	−9	0	4	6	42	−38	4	4	6	45	−57
5	3	0	33	37	7	0	1	13	4	2	11	1	14	12	5	7	4	a	−20	9	0	3	a	−7
3	6	1	14	−11	6	5	1	9	−8	6	8	0	a	−2	7	8	1	1	1	10	0	0	a	−10
3	2	3	11	10	3	9	0	13	−21	2	5	5	a	13	3	9	4	15	−16	5	9	4	a	−2
3	5	2	19	7	1	9	2	30	−25	6	0	4	a	4	3	1	6	a	1	5	1	6	a	−9
4	5	1	23	−19	4	2	4	9	−2	3	4	5	a	−15	9	2	1	a	−7	6	11	1	a	−12
2	6	2	21	13	0	10	0	a	−4	7	6	1	a	−3	0	10	4	a	−3	7	8	3	a	5
2	7	1	32	−23	5	4	3	a	2	7	2	3	a	4	9	3	0	a	14					
4	4	2	99	−116	0	6	4	a	14	7	5	2	36	33	5	4	5	20	13					
0	8	0	89*	108	6	4	2	a	10	8	3	1	a	−14	2	11	3	12	−11					
0	0	4	89*	114	7	2	1	9	−5	2	8	4	a	−2	3	12	1	a	−8					

a absent.   * omitted in final refinement.

Table 3
Atomic coordinates

Atom type	x	y	z	B
T	0.1316(12)	0.1268(8)	0.25	0.31
M1	0.2271(13)	0	0.75	0.31
M2	0.25	0.25	0.75	0.31
M3	0	0.3675(7)	0	0.31
O1	0.0496(38)	0	0.25	0.56
O2	0.5298(37)	0	0.25	0.56
O3	0.9779(25)	0.2353(18)	0.25	0.56
O4	0.2550(19)	0.3757(12)	0.0245(26)	0.56

Standard deviation shown in brackets to same decimal level.

octahedron. The point symmetry of M1 is mm and of M2 is $2/m$. The M3 cations lie at 0.0 or 0.5 and have point symmetry 2.

Not obvious in fig. 1 is the close-packing of the oxygen atoms. This will be more apparent in fig. 2 where the β-structure is compared with spinel and with the alternative arrangement for symmetry Ibmm. In the β-structure the layers of close-packed oxygens occupy the crystallographic planes (240), (2$\bar{4}$0), (202) and (20$\bar{2}$). The oxygens have the topology of the well-known cubic close-packed arrangement, as may be seen from fig. 1. Starting at the extreme bottom left with the pair of atoms of type O1 at heights 0.25 and 0.75 and traversing perpendicular to the planes (240), it is necessary to reach the third (240) plane in order to superimpose the atoms directly on the starting plane: the atoms of type O1 at the center of the diagram then superimpose on the initial pair of atoms. The (240) planes correspond to the conventional ...ABC... arrangement of cubic close packing. A similar analysis holds for the (202) planes.

The relations between the polyhedra may be obtained by comparing fig. 1 with fig. 2b. The Si atoms share an O1 oxygen and use two O3 and four O4 atoms to form (Si$_2$O$_7$) doublets. No edges or faces of the (Si$_2$O$_7$) groups are shared with MO$_6$ octahedra. The M3 octahedra form double columns parallel to c by sharing O2–O3 edges lying parallel to b and O2–O2′ edges parallel to c. Each M3 octahedron also shares an O2–O4 edge with each of two M1 octahedra and an O3–O4 edge with each of two M2 octahedra. Each M1 octahedron shares an O2–O4 edge with each of four M3 octahedra and an O4–O4′ edge with each of two M2 octahedra. The M2 octahedron shares on O4–O4′ edge with each of two M1 octahedra and an O3–O4 edge with each of four M3 octahedra.

If one assumes that all bonds in a polyhedron have equal electrostatic strength, O1 is oversaturated since it is bonded to two Si atoms and one M1 yielding 7/3 electrostatic units. The O2 atom has a compensating undersaturation since it is bonded to five M atoms with a total of 5/3 charge units. The O3 and O4 atoms are exactly saturated since they are each bonded to one Si and three M atoms.

Although the individual distances may have errors which make comparisons statistically insignificant at the 0.1 Å level, mean distances averaged over a polyhedron should be more meaningful especially as many positional parameters are restricted by symmetry.

The mean T–O distance of 1.64 Å is compatible with the mean distance of 1.634 ± 0.003 Å for Si–O in forsterite (BIRLE et al., 1968) and with the mean distances of 1.623 ± 0.003 and 1.633 ± 0.003 Å for Si–O in kyanite (BURNHAM, 1963), thus confirming the crystal-chemical likelihood that the T site is populated prin-

Table 4
Interatomic distances (Å)

T–O1	1.60	O1–O3	2.76	2M2–O3	1.89	4O3–O4	2.73
T–O3	1.77	2O1–O4	2.50	4M2–O4	2.13	4O3–O4′	2.95ς
2T–O4	1.59	2O3–O4	2.85		*2.05*	2O4–O4′	2.88
	1.64	O4–O4′	2.57			2O4–O4″	3.13†
M1–O1	2.28	4O1–O4	3.22	2M3–O2	2.09	O2–O2′	2.89**
M1–O2	2.00	4O2–O4	2.81*	2M3–O3	2.09	O3–O3′	2.87
4M1–O4	2.12	2O4–O4′	2.84	2M3–O4	2.11	2O2–O3	3.03
	2.13	2O4–O4″	3.13		*2.10*	2O2–O4	2.81*
						2O2–O4′	3.03
						2O3–O4	2.95
						2O3–O4′	3.07

* † ς ** shared edges.

CRYSTAL STRUCTURE OF β-Mg$_2$SiO$_4$

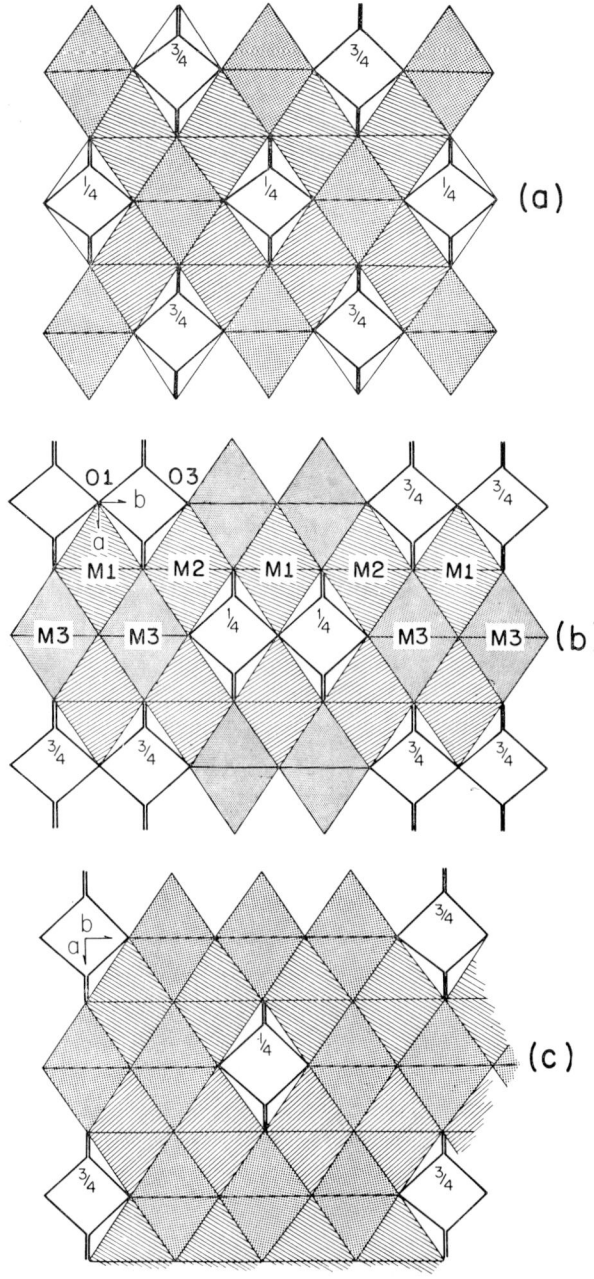

Fig. 2. Comparison of the crystal structures of a) spinel, b) β-structure and c) hypothetical structure as seen in a c-axis projection. The spinel structure is projected down an isometric (110) axis with another (110) axis horizontal and a (100) axis towards the reader. The Si tetrahedra are represented by four spokes for the Si–O bonds. The height of one tetrahedron out of each adjacent pair is shown by the fraction $\frac{1}{4}$ or $\frac{3}{4}$: the other tetrahedron is displaced by $\frac{1}{2}$. Columns of octahedra formed by edge-sharing of octahedra centered at heights 0 and $\frac{1}{2}$ are stippled. Single octahedra at $\frac{1}{4}$ are ruled NE–SW while those at $\frac{3}{4}$ are ruled NW–SE. In the β-structure, the M1, M2, M3, O1 and O3 atoms are depicted schematically.

cipally, if not entirely, by Si. Tetrahedrally-coordinated Mg would yield a mean distance near 1.9 Å as for akermanite (LOUISNATHAN, 1969).

The mean M–O distances of the three unique octahedra (M1 = 2.13; M2 = 2.05; M3 = 2.10 Å) vary more than the indicated experimental error. Since the distance correlates with the number of shared edges (2.05–10; 2.10–7; 2.13–6), it is tempting to explain the variation by cation repulsion. In forsterite (BIRLE et al., 1968) the situation is similar (2.10–6; 2.13–3), but not in the AlO$_6$ groups of kyanite (BURNHAM, 1963) – (1.918–4; 1.916–4; 1.906–4; 1.897–3).

4. Geometric relations between the β-structure and other close-packed structures

The β-structure is based on cubic close-packing of oxygen atoms and is therefore topologically related to a host of other structures. Here we consider only the closest relatives.

Fig. 2 compares the β-structure with those for spinel and a hypothetical structure with the same space group and cell dimensions of the β-structure (MOORE and SMITH, 1969). Fig. 3 shows the structures of manganostibite, staurolite and kyanite. All the structures are idealized so that the oxygens occupy the ideal positions of cubic close-packing while the cations occupy the exact centers of the tetrahedral and octahedral voids. The oxygen atoms lie at the intersections of the lines outlining the polyhedra. Tetrahedral cations are indicated by the T–O bonds. Octahedral cations occupy the centers of the octahedra shown in projection down a [110] axis. Oxygen atoms lie at four equally-spaced levels approximately 1.4 Å apart yielding a repeat distance of 5.6 Å perpendicular to the layers. Stippled areas represent columns of octahedra sharing horizontal edges at heights $z = \frac{1}{4}$ and $\frac{3}{4}$ with their M cation lying at $z = 0$ and $\frac{1}{2}$. Octahedra with rules running NW–SE have M cations lying at $z = \frac{3}{4}$; since there is no cation at $z = \frac{1}{4}$, columns are not formed. Octahedra with rules running NE–SW have M cations only at $z = \frac{1}{4}$.

The axial orientations have been chosen to provide a simple comparison between all the structures as explained in the figure legend.

Spinel (fig. 2a) consists of a framework of edge-shared octahedra containing single tetrahedra sharing corners with the octahedra. In projection, the framework appears to consist of a braiding of vertical sheets,

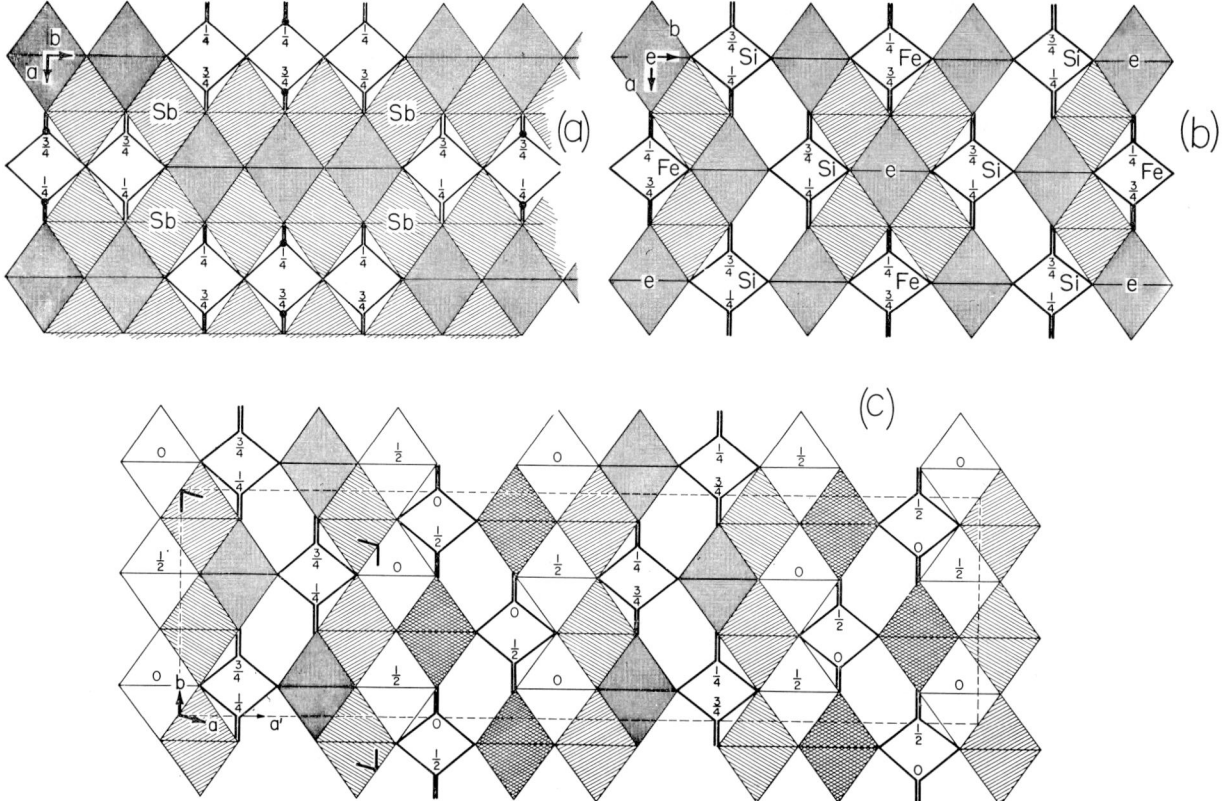

Fig. 3. Comparison of the crystal structures of a) manganostibite, b) staurolite and c) kyanite. See fig. 2 for explanation of the drawing conventions. For manganostibite, the positions of the Sb atoms are shown; other octahedral sites are occupied by Mn: the As atoms lie at the tetrahedral nodes marked by solid circles; other sites are occupied by Mn. For staurolite, tetrahedral sites are marked by the occupants Fe or Si. Octahedral sites marked by e are only half-occupied in staurolite from a metamorphic rock, but might be fully occupied in a synthetic Mg-staurolite. The kyanite structure is projected down c^* yielding a nearorthogonal cell with $a' = 4a+b$ and $b' = b$. The outlines of the true triclinic cell and of the super-cell are shown. Cross-hatching of octahedra means occupancy at both $\frac{1}{4}$ and $\frac{3}{4}$.

each composed of columns of octahedra cross-linked by non-column octahedra (the choice of stippling and hatching in fig. 2a shows this clearly). However the apparent sheets lie parallel to isometric (111) and are repeated by the symmetry such that they integrate into a framework. Although the tetrahedra appear in projection to share edges, they do not since tetrahedra adjacent in projection differ in height by $\frac{1}{2}c$.

The β-structure contains double columns of M3 octahedra (stippled) cross-linked by M1 and M2 octahedra to form a continuous framework. The tetrahedra are linked in pairs. The cross-linking octahedra are topologically identical to those of spinel, and the only difference between the β-structure and spinel is in the intervening octahedral columns and tetrahedra. In spinel both the octahedral columns and silicate groups are single whereas in the β-structure both are double.

Obviously an infinite family of structures with symmetry $Ibmm$ can be envisaged in which any integral number of linked tetrahedra can alternate with any number of octahedral columns condensed into a ribbon (or wall). Thus the hypothetical structure of fig. 2b has single tetrahedra alternating with a triple ribbon. Manganostibite (MOORE, 1968) has triple tetrahedra alternating with a triple wall. The general formula for the homologous series is obtained as follows: let the octahedral wall be n octahedra wide and let there be m tetrahedra in each condensed unit. The cross-linking octahedra must number $\frac{1}{2}(n+m)$. The total number of oxygen atoms is $2(n+m)$. Hence the formula is $\frac{1}{2}(3n+m)$M, mT and $2(n+m)$O. Since the silicate cluster would have composition $(T_m O_{3m+1})$, the overall formula can be written as $M_{(\frac{3}{2}n+\frac{1}{2}m)}O_{(2n-m-1)}(T_mO_{3m+1})$. For the sub-series with $n = m$, the overall formula can be written as $M_{2n}O_{n-1}(T_nO_{3n+1})$ as given by MOORE and SMITH (1969). The numbers $n = 1$, 2 and 3 yield the formulae for spinel, β-structure and manganostibite respectively.

If more than one type of chemical atom occupies the M or the T sites, cation ordering is possible. Each member of the homologous series can be subdivided into various structures with cations ordered into different positions. Manganostibite is an example since the M sites are regularly occupied by one Sb to each five Mn and the T sites by one As for each two Mn yielding the ideal formula $(Mn^{2+}_5Sb^{5+})O_2(Mn^{2+}_2As^{+5}O_{10})$.

MOORE and SMITH (1969) discussed other structures formed from cubic close-packing of oxygens in which the percentage of occupied voids changes drastically. Thus there is a series in which the tetrahedral sites of spinel are emptied and the released octahedral sites occupied, leading to formulae $M_{16}T_8O_{32}$, $M_{20}T_4O_{32}$, $M_{24}O_{32}$, $M_{32}O_{32}$. The second formula is for the hypothetical structure of fig. 2c. The $M_{24}O_{32}$ (or more simply M_3O_4) composition corresponds to another hypothetical structure with $a \sim 8.2$, $b \sim 2.85$, $c \sim 5.7$ Å, $z = 2$, $Pcmm$ (fig. 4a). The $M_{32}O_{32}$ (or MO) composition obtained by filling remaining octahedral voids yields rock salt whose isometric cell can be expressed by the orthorhombic sub-group $a \sim 4.1$, $b \sim 2.85$, $c \sim 5.7$ Å, $z = 4$, $Immm$ (fig. 4b).

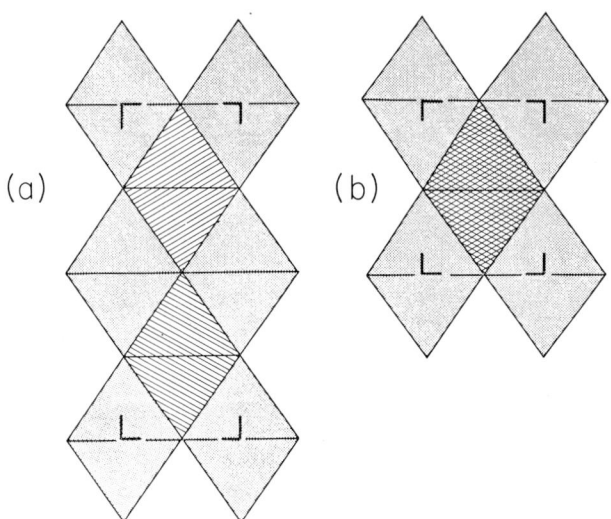

Fig. 4. Comparison of crystal structures of a) hypothetical structure for M_3O_4, and b) rock salt.

Staurolite and kyanite are examples of structures in which a greater percentage of voids is unoccupied than in spinel. Fig. 3b shows that some of the possible octahedra are missing. In addition, there are complex changes of height which differentiate the structures from those previously described. In the projection of staurolite down c there are single columns of octahedra cross-linked by single octahedra to yield braided sheets extending infinitely in the (010) plane but *not* linked by octahedra to adjacent sheets. Hence there is no framework of octahedra as in spinel and the β-structure. The octahedral sheets are cross-linked entirely by tetrahedra, and the absence of an octahedral framework results from unoccupied octahedral voids. Although at a casual glance there is a sheet of octahedral columns and cross-linking tetrahedra identical to that of spinel, closer examination shows that alternate tetrahedra have moved by $\frac{1}{2}c$ thereby doubling the horizontal repeat. This alternation breaks up the chains of octahedra which in spinel lie at the same height parallel to the horizontal axis. In staurolite the equivalent of these chains (the second row down of octahedra) consists of pairs of octahedra alternatively at heights $\frac{1}{4}$ and $\frac{3}{4}$, separated by voids.

The staurolite structure has ordered arrangements of cations which make the structure even more complex. Two-thirds of the tetrahedral voids are occupied by Si and one-third by Fe (SMITH, 1968). In natural staurolite, typical of metamorphic rocks, alternate octahedra in the stippled areas marked e in fig. 3b are unoccupied. In synthetic Mg-staurolite, these voids might be occupied since the lower valence of Mg would permit more octahedral cations per unit cell. However the oxide ratio and the number of hydroxyl groups replacing oxygens are uncertain in synthetic Mg-staurolites (SCHREYER and SEIFERT, 1969). Topologically the formula of natural staurolites is close to $M_{18}T_{12}O_{48}$ with a chemical formula near $(Al, Mg)_{18}(Fe, Al)_4(Si Al)_8$-$O_{48}H_{\sim 4}$. SMITH (1968) showed that there is weak occupancy of some octahedral sites which share faces with tetrahedra: simultaneous occupancy is unlikely on account of severe repulsion.

The kyanite structure has a small triclinic unit cell (fig. 3c) but in order to show the analogy to the other structures it is necessary to project down c^* thereby producing a large near-orthogonal super-cell with $a' = 4a+b$ and $b' = b$. The most obvious relation to staurolite is the zigzag rows of Si tetrahedra interspersed with voids. The braided sheets of octahedra in staurolite are replaced by a complex corrugated layer, varying in width between one octahedron and three octahedra. The octahedra nearest to the silicate tetrahedra are identical in topology to those of staurolite but the next zigzag row of octahedra is different. Instead of a band of alternate tetrahedra and octahedral columns, there is an alternation of octahedra at heights 0 and $\frac{1}{2}$. These in turn are bonded to a band of tetrahedra and octahedral columns displaced by $\frac{1}{4}c$ with respect to those in staurolite. In order to distinguish these octahedral columns from those at heights 0 and $\frac{1}{2}$, cross-hatching is used. The change-of-step of $\frac{1}{4}c$ results in the necessity for $a' = 4a+b$ in order that the pseudo-orthogonal cell repeats in the a' direction.

5. Close-packed structures in relation to pressure and temperature: geophysical implications

Close-packed structures can be studied with several aims: a mathematician will examine the geometrical possibilities and consequent restrictions on the allowable patterns; a physical chemist will look for the underlying principles which determine the limits of chemical substitution in each structure type – these limits will vary with the temperature-pressure-time situation; a geophysicist will attempt to interpret incomplete observations on the physical properties of planets in terms of likely element abundances existing as specific atomic packings. The subject is much too vast and is much too uncertain for a systematic, definitive description. Here we shall merely attempt to explore some ideas, beginning with simple geometrical arguments which can be fully defined and ending with plausible chemical arguments. In the next section these general arguments will be applied to the olivine, spinel and β-forms of Mg_2SiO_4.

The gravitational potential of spherical bodies increases with depth providing the driving force for concentration of denser material. The density of matter can increase in two ways: a purely geometric effect in which void space is minimized, and a chemical effect in which the inherent atomic density is determined by the total mass divided by a volume determined from the interaction of the atomic constituents. The occurrence of a particular geometry and chemistry ultimately resides in the quantum-mechanical forces, commonly approximated by a simple ionic model. The simplest ionic model utilizes hard, incompressible spheres and the concept of local electrostatic neutrality in which the charge on the anion is balanced by the valence bonds from the first neighbors. The next simplest utilizes compressible spheres which distort so as to minimize cation-cation repulsion. A more fundamental approach minimizes the electrostatic potential. Unfortunately calculations of the electrostatic potential depend critically on the details of the atomic positions, which of course are unknown for hypothetical structures, and which require very accurate measurement even for experimentally-observed materials. Here we shall perforce be restricted to the second approximation which is typically expressed by the geometrical approach of PAULING (1929) in which cation-cation repulsion results in a shortening of shared edges. KAMB (1968) explained the packing relationships of the olivine and spinel forms of Mg_2SiO_4 by three subtle effects which principally depend on cation-cation repulsion (see next section). We believe that it will be necessary even for the ionic model to consider the Madelung potential (which is calculated over the entire structure rather than considering just first and second neighbors), and that the simpler approaches are too limiting.

Nevertheless for an initial geometrical analysis of close-packed structures, we must utilize simple concepts such as shared polyhedral elements. Geometrically cation-cation repulsion increases as corners, edges and finally faces are shared.

1) Since anions are larger than cations, the former tend towards a close-packed arrangement while the latter fit into the voids (PAULING, 1929). The densest arrangement of anions in two dimensions uses the nodes of a tessellation of equilateral triangles. In three dimensions, each triplet of adjacent layers may be arranged in two ways, the familiar cubic and hexagonal close packing. Although there is an unrestricted number of ways of stacking additional layers, we shall consider here only the two simplest *regular* sequences. Nevertheless it would be unwise to rule out the existence deep in the Earth of structures based on a more complex sequence of layers, such as the double h.c.p. arrangement.

2) Between the anions are voids of two sizes; those between four anions have tetrahedral point symmetry whereas those between six anions have octahedral sym-

metry. When the anion layers are undistorted, there is no difference between the c.c.p. and h.c.p. arrangements concerning the ratio of void volume to anion volume. Perfect geometrical fitting of cations in the voids occurs when $r_M/r_O = 0.41\ldots$ and $r_T/r_O = 0.22$, where r is the radius and O, M and T denote the anions and octahedral and tetrahedral cations. Geometrically it is convenient to envisage an imaginary polyhedron whose apices are specified by the first-neighbor anions of the central cation.

3) Sharing of polyhedral elements depends on two factors a) the specific distribution of cations b) the anion arrangement.

a) In h.c.p. for stoichiometries M:O $\leq$ 1:2, cation arrangements can occur in which no populated M polyhedra share faces. When M:O is greater than 1:2 some faces must be shared.

b) In h.c.p. for stoichiometry $\leq$1:2, M cations can be arranged so that trigonal triplets of populated octahedra are formed by sharing edges: one edge from each octahedron forms an open equilateral triangle which can accommodate a face of an SiO_4 tetrahedron. In this arrangement, which is found in olivine, only edges are shared between the populated polyhedra.

c) In c.c.p. no faces are shared between populated octahedra even when all octahedral sites are filled i.e. for M:O = 1:1.

d) In c.c.p. a tetrahedron can only share an edge with an octahedron when it actually shares a face.

e) In c.c.p. when M:O < 1:1, arrangements can be found in which populated tetrahedra share only corners with populated octahedra.

f) In c.c.p. when T:O > 1:1, populated tetrahedra must share edges with other tetrahedra.

4) Turning now to chemical bonding, Pauling's arguments on cation-cation repulsion can be used to predict the order of stability of shared polyhedral components. The electrostatic force is given by $Z_1 Z_2/r^2$ where Z_1 and Z_2 are the charges of the two cations and r is the inter-cation distance. The higher the charge of a cation, the smaller its radius and the greater its tendency to go into a T rather than an M site. This argument applies strictly only to iso-nuclear cations such as Fe^{2+} and Fe^{3+}. Taking the entire Periodic Table it is not strictly true since by changing the nucleus, ions of lower charge can be smaller than ions of higher charge. For the rock-forming elements entering tetrahedral and octahedral sites in close-packed oxygen atoms it should be *generally* true that the order of stability is no sharing > M–M shared corners > M–T shared corners > T–T shared corners $\gg$ M–M shared edges > M–T shared edges $\gg$ M–M shared faces $\gg$ T–T shared edges > M–T shared faces $\gg$ T–T shared faces.

5) No simple rules can be established to express the concept of local charge balance since the valences of the occupants of the octahedral, or of the tetrahedral sites, are not constant. Thus in spinels both divalent and tetravalent ions have been found to occupy the tetrahedral site. Nevertheless it is extremely important to consider the extent to which there is local neutralization of charge when considering a particular chemical composition and several possible structures.

6) Increase of pressure leads to denser packed geometries as a result of the favorable PV term in the Gibbs function. Such diminution of volume leads to increasing cation-cation repulsion unless the cations can redistribute into a new pattern. Considering just nearest-cation repulsions, a pressure-induced polymorphic change should involve movement up the sequence given under 4). For example, there might be a reduction of the number of shared edges, or there might be a change from M–T to M–M sharing with constancy of number. Considering repulsion among *all* cations, the overall repulsion could be reduced by retention of the number and type of shared elements, but with a more spatially-isotropic distribution. Thus local concentrations of shared edges at low pressure should be smoothed out by the polymorphic change.

7) On Earth, increase of gravitationally-induced pressure is normally accompanied by increase of temperature. This temperature increase favors an increase of entropy as a result of the TS term in the Gibbs function. STRENS (1967) summarized relations between symmetry, entropy and volume in polymorphism. The entropy in close-packed structures might increase because of positional disorder of cations between the sites (such as occurs in $MgAl_2O_4$ and other spinels – e.g. HAFNER and LAVES, 1961), or because of a change in the scheme of lattice vibrations typically involving transformation to a higher symmetry.

8) Change of valence of a cation produces conflicting factors on the density. Increase of valence causes a decrease of atomic radius thereby tending towards a smaller volume *unless* the coordination number chan-

ges. However for a given number of oxygen atoms the number of cations is reduced. Each situation requires specific examination but in general there is a tendency for decrease of valence to favor higher density.

The above emphasis on geometrical close-packing is appropriate to conditions on Earth, since the dominant anion (oxygen) when close-packed provides voids suitable for the common cations. Under surface conditions the following cations are known to enter close-packed oxygen structures: B^{3+}, Mg^{2+}, Al^{3+}, Si^{4+}, P^{5+}, S^{6+}, Ti^{4+}, Cr^{3+}, Mn^{2+}, Mn^{3+}, Mn^{4+}, Fe^{2+}, Fe^{3+}, Co^{2+}, Co^{3+}, Ni^{2+}, Cu^{2+}, Ga^{3+}, Zn^{2+}, Ge^{4+}, As^{5+}, V^{5+}. The anions OH^- and F^- replace O^{2-}. The relatively rare chalcogenides are not considered here because the bonding is angularly directed.

In general the close-packed oxide structures found either in nature or by laboratory synthesis obey the rules predicted for simple ionic bonding: e.g. brucite, spinel, olivine, β-Mg_2SiO_4 and rutile share polyhedral components high in the above list under (4). A well-known exception is the corundum structure occurring for Al_2O_3, Fe_2O_3, $FeTiO_3$, etc. Pairs of octahedra share faces as required mathematically in h.c.p. for M:O > 1:2. Several structures can be devised in c.c.p. for this M:O ratio in which no polyhedral faces are shared: e.g. a rock salt structure with some vacant sites. Perhaps all such possible structures have a poor local charge balance thus countering the advantage of shared polyhedral elements. At constant temperature, the olivine, β- and spinel forms of Co_2SiO_4 follow a sequence with increasing pressure (AKIMOTO and SATO, 1968) consistent with the order of stability predicted in 4) as discussed in detail in the next section.

An example of a polymorphic transition with no change of number and type of shared elements but in which the anisotropy of shared elements is reduced by pressure is the rutile → α-PbO_2 transformation. Both have h.c.p. oxygens and both have octahedra which share two edges on average. However in rutile the shared edges form linear octahedral chains whereas in α-PbO_2 the octahedral chains are kinked into a zigzag arrangement. This structural transformation actually occurs under pressure for the TiO_2 composition, and may occur for other important compositions such as SiO_2.

The two arrangements which have complete octahedral occupancy in close-packed systems – rock salt, c.c.p. and niccolite, h.c.p. – offer an interesting contrast. All twelve octahedral edges are shared in rock salt, but in niccolite two faces are shared in addition to six edges. Ionic compounds should prefer the rock salt to the niccolite structure because of the difference in sharing of polyhedral elements. Indeed many sulfides and sulpho-salts have either the rock salt structures e.g. galena or a superstructure e.g. miargyrite.

To conclude this section, it appears from qualitative examination of known structures, that c.c.p. arrangements allow for more diversity in M:T ratios than h.c.p. arrangements while at the same time minimizing cation repulsions. Systematic topologic analysis of all simple arrangements is desirable in light of the foregoing arguments.

6. Crystal chemical relations between olivine, spinel and the β-Mg_2SiO_4 structure

KAMB (1968) discussed in detail the crystal-chemical basis for the relative stabilities of the olivine and spinel forms of Mg_2SiO_4. Recognizing the desirability of complete calculations of crystal energy by summation of electrostatic forces over the entire crystal, he urged the value of simple considerations based on Pauling's rules. Kamb hypothesized that the as yet unobserved spinel form of Mg_2SiO_4, assumed to exist at high pressure, would be a normal spinel with Mg in M sites and Si in T sites, and carried out calculations on this basis. Since the oxygen atoms of both olivine and spinel are electrostatically neutralized by their first neighbors, he compared the edge-sharing properties of the structures.

Kamb pointed out that in order for the shared edges of spinel to be shorter than the unshared edges, the T ions must be inherently larger than the M ions (the ionic radius normalized to the same coordination number is used for the comparison: a 7% reduction of radius is assumed to occur from octahedral to tetrahedral coordination). This explains the seemingly unusual occurrence of Mg^{2+} in 4-coordination and Al^{3+} in 6-coordination for Al_2MgO_4 spinel. Since the Mg–O bond is more compressible than the Si–O bond, pressure should reduce the unfavorable difference in length between shared and unshared edges in Mg_2SiO_4 normal spinel thereby helping to stabilize it with respect to olivine. However the effect seemed too small and Kamb ascribes at least some of the pressure stabilization of spinel over olivine to the $P\Delta V$ contribution to the Gibbs function which favors the denser spinel structure.

Kamb pointed out that at first sight the olivine structure should be less stable than the spinel structure since although both have six shared edges per formula unit, the former has 3Mg–Si and 3Mg–Mg whereas the latter has 6Mg–Mg shared edges. However, he demonstrated that this unfavorable cation-cation repulsion effect is compensated by several features of the olivine structure permitted by the lower symmetry and the specific arrangement of cations. In particular the smaller size of the Si atom in relation to the Mg atom permits the oxygen atoms to adjust themselves so that the shared edges are shortened. The geometry of the olivine structure allows the cations to withdraw from each other in contrast to spinel which has severe symmetry restrictions. Kamb ascribes the 11% increase in density from olivine to the hypothetical spinel form of Mg_2SiO_4 to three effects in olivine a) repulsion of Mg atoms across a shared edge thereby lengthening c, b) anisotropic distortion of the SiO_4 tetrahedron by edge sharing thereby lengthening a and c) a general withdrawal of Mg atoms thereby lengthening b.

The fundamental implication of Kamb's arguments is that cation repulsion can only be minimized when the symmetry restrictions and inherent ionic radii are favorable. Hence each structure type must be examined in detail.

Since 4 oxygen atoms of the β-Mg_2SiO_4 structure are oversaturated (7/3 charge units) and 4 are undersaturated (5/3 charge units), this aspect of the structure is inferior to both olivine and spinel forms. The β-structure has $6\frac{1}{2}$ shared M–M edges in comparison to the 3M–M and 3M–T edges of olivine and the 6M–M edges of spinel. As described earlier, there is considerable experimental uncertainty concerning the individual interatomic distances of the β-structure. The range of O–O distances in the octahedra, 2.8–3.2 Å, is only half the range 2.6–3.4 in forsterite. Moreover there is no systematic pattern of shortening of shared edges in the β-structure.

Obviously detailed analysis must await more accurate single-crystal data, but one might hazard the suggestion that the intermediate density at low pressure of β-Mg_2SiO_4 in relation to the olivine and spinel forms results partly from the local charge balance and partly from cation-cation repulsions which can be interpreted on the basis of the number and type of shared edges, and on the symmetry restrictions of geometrical distortions.

The possibility of an inverse spinel $(Mg_{0.5}Si_{0.5})_2^{VI}$-$Mg^{IV}O_4$ is worth considering since it has tetrahedral ions inherently larger than the octahedral ones. The octahedral ions might be spatially disordered or might be regularly arranged in a superstructure. Whatever the arrangement there is the disadvantage over the normal spinel that 4-valent Si ions would be sharing octahedral edges with either or both of Si and Mg ions. Qualitatively an inverse or fully-disordered spinel seems less likely to exist than a normal Mg_2SiO_4 spinel. Measurements are in progress to attempt to determine the site occupancies of Mg-rich spinels produced by Ringwood in the $(Mg, Fe)_2SiO_4$ system at high pressure.

Acknowledgements

We thank Scott Baird and S. John Louisnathan for taking X-ray photographs and making calculations. Prof. N. Morimoto kindly discussed his results. The work was supported by NASA grant NAS 9-8086.

References

AKIMOTO, S. and Y. SATO (1968) Tech. Rept. Inst. Solid State Physics (Univ. Tokyo) Ser. A, No. 328.
BIRLE, J. D., G. V. GIBBS, P. B. MOORE and J. V. SMITH (1968) Am. Mineralogist **53**, 807.
BURNHAM, C. W. (1963) Z. Krist. **118**, 337.
HAFNER, S. and F. LAVES (1961) Z. Krist. **115**, 321.
KAMB, B. (1968) Am. Mineralogist **53**, 1439.
LOUISNATHAN, S. J. (1969) Ph.D. thesis (Univ. Chicago).
MOORE, P. B. (1968) Arkiv. Mineral. Geol. **4**, 449.
MOORE, P. B. and J. V. SMITH (1969) Nature **221**, 653.
MORIMOTO, N., S. AKIMOTO, K. KOTO and M. TOKONAMI (1970) Phys. Earth Planet. Interiors **3**, 161–165.
PAULING, L. (1929) J. Am. Chem. Soc. **51**, 1010.
RINGWOOD, E. A. and A. MAJOR (1966) Earth Planet. Sci. Letters **1**, 241.
SCHREYER, W. and F. SEIFERT (1969) Am. J. Sci. **267A**, 407.
SMITH, J. V. (1968) Am. Mineralogist **53**, 1139.
STRENS, R. G. J. (1967) Mineral. Mag. **36**, 565.

INTERPRETATIONS OF THE LOW-VELOCITY ZONE

FRANCIS BIRCH

Hoffman Laboratory, Harvard University, Cambridge, Mass., U.S.A.

The physical interpretation of the low-velocity zone of the upper mantle depends critically upon details of the velocity variation which are still uncertain. As examples, Gutenberg's solution is roughly consistent with a homogeneous layer of peridotite affected only by temperature and pressure, while Anderson's is not. Partial melting, relaxation of elasticity and compositional changes are among the suggested explanations. New evidence derived from the free oscillations by Press indicates that the low-velocity layer may also be a high-density layer; this might be eclogite or iron-rich peridotite.

The concept of a "low-velocity" layer in the upper mantle was introduced by Gutenberg in 1926 to explain the variation of amplitudes with distance for California earthquakes. In subsequent studies, low-velocity channels have frequently been postulated, especially for the sub-oceanic mantle. A full discussion of the seismic work is given by LEHMANN (1967). In the following, I should like to examine the question of the physical interpretation of low-velocity layers, taking several published examples as representative, and considering only the sub-oceanic mantle.

Where low-velocity channels exist, the normal methods of dealing with the refraction of seismic waves break down, and the distribution of velocity with depth becomes indeterminate. Much ingenuity and intuition have gone into the construction of the published velocity-depth functions. It is difficult to exclude the possibility that some degree of interpretation is already present in the various solutions, and thus we are likely to reproduce the original bias, especially since the published distributions differ in just the details which are critical for physical interpretation.

Current explanations may be classified under two main headings: (1) uniform composition for the upper mantle; (2) non-uniform composition. If the composition is assumed to be uniform, then the low-velocity zone may be ascribed to temperature and pressure acting together on a medium of uniform mineralogy, with or without relaxation of elasticity, to partial melting, with a liquid phase dispersed along grain boundaries or in "pockets", or to phase changes conditioned by pressure and temperature. If variation of composition is included, then intrinsic variations of velocity are superimposed on the effects mentioned above.

As some variety of peridotite is widely regarded as the most probable composition, we may begin with the limiting case of a uniform layer of olivine (BIRCH, 1969) of the composition most frequently encountered (about Fa_{10}), for which a considerable amount of information is available. As representative of the seismic distributions, we take those of GUTENBERG (1959) and of ANDERSON (1967, CIT 11A). Gutenberg's distributions have the characteristics to be expected for a uniform layer affected only by temperature and pressure in "normal" respects: the depth to the minimum for V_P is less than that to the minimum for V_S, and the width of the region of low V_P is less than that of low V_S. With the pressure and temperature coefficients for forsterite given by SCHREIBER and ANDERSON (1967) and SOGA et al. (1966) the required temperatures for V_P and V_S are reasonably consistent with one another, and only slightly higher than estimates based on petrological arguments, a difference which might be eliminated by the selection of a suitable peridotite in place of dunite. The temperatures required to give Gutenberg's velocities, with these assumptions, are shown in fig. 1. The corresponding density curve will also have a shallow minimum near that of V_P.

In contrast Anderson's CIT 11A distributions show the two low-velocity zones (for V_P and V_S) in the same

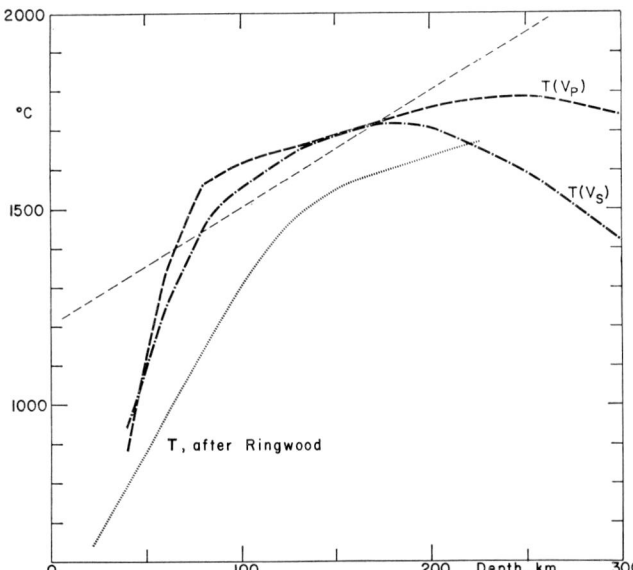

Fig. 1. Temperatures for an olivine layer computed from Gutenberg's (1959) velocities; the broken line is the dry pyrolite solidus, the dotted line, oceanic temperatures. After Ringwood (1966).

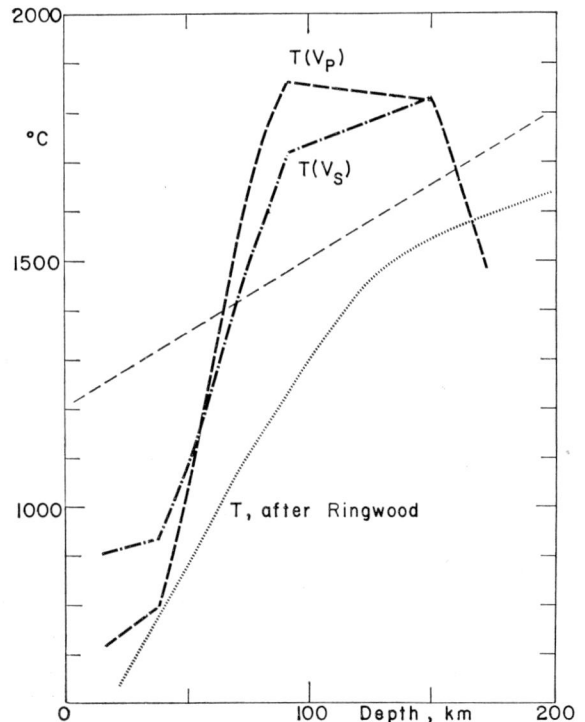

Fig. 2. Temperatures for an olivine layer computed from Anderson's (1967) velocities; other curves as in fig. 1.

range of depth, with fairly sharply marked beginnings of negative velocity gradients at 36 km, and of positive gradients near 151 km. The interpretation in terms of pressure and temperature alone with the coefficients for forsterite gives roughly consistent temperatures for the two velocities (fig. 2), but with the implausible features of very low temperature gradients between 16 and 36 km, and strongly negative temperature gradients below 151 km; an interpretation in terms of simple pressure and temperature effects is incompatible with the CIT 11A velocities. Instead, we may postulate either partial melting for the interval 36 to 151 km, which is the interpretation preferred by Spetzler and Anderson (1968), or a change of composition to a material of intrinsically lower velocity (iron-rich olivine or eclogite) or possibly a relaxation of elasticity associated with a particular range of temperature. It is necessary to explain not only the decrease of velocity but also the rise at the base of the layer, and this consideration raises doubts concerning the adequacy of relaxation as the sole cause. Behavior in the 16 to 36 km interval probably requires compositional change of some sort; it seems unlikely that the temperature gradient here can be low enough to allow the increase shown in the CIT 11A velocities for any uniform composition.

The hypothesis of partial melting has several attractions. It may be supposed that the temperature follows the solidus in the low-velocity layer, with the amount of liquid rising to some maximum, then decreasing, to disappear at the base of the layer where the temperature falls below the solidus. The two velocities would be affected at the same depths, though not necessarily in the same proportion. There are indications that the effects, for a given amount of melt, may be considerably larger than predicted by simple theories (Spetzler and Anderson, 1968); qualitatively, there is agreement to the extent that both velocities and attenuation would be affected at the same places. These questions have been discussed in several recent papers (Walsh, 1968; Gordon and Davis, 1968; Vaisnys, 1968).

An important new element, greatly influencing the interpretation, has been introduced by Press (1970), with the determinations of density required to satisfy the high-order modes of the free oscillations. Of 51 "acceptable" solutions obtained by Press, 30 show densities greater than or equal to 3.6 g/cm³ somewhere between 70 and 370 km, and all show densities greater than 3.5 g/cm³ over substantial intervals. The mean density for the outermost few hundred kilometers seems to be determined by these calculations to within a few percent. Densities greater than 3.5 g/cm³ have appeared in various solutions (BIRCH, 1961a; ANDERSON and TOKSÖZ, 1963; PEKERIS, 1967), but have usually appeared to result from some physical hypothesis which might not be valid; other solutions of the problem of free vibration have seemed consistent with lower densities (e.g. LANDISMAN et al., 1965; BULLEN and HADDON, 1967). With the continual revision of travel times and periods, successive calculations may not be strictly comparable. The Monte Carlo method employed by Press has the advantage of dispensing with the customary assumptions concerning relations among velocities and densities, and thus features common to the acceptable solutions must be given much weight. It is noteworthy that in this group of solutions V_S, which is allowed to vary, always increases with depth at the "discontinuities" (371–421 km and 621–721 km), and except for 3 solutions where small negative gradients occur between 421 and 621 km, shows positive gradients everywhere below 371 km. On the other hand, density decreases with depth across the 371–421 km, discontinuity in 14 of the 51 solutions. This leaves a large majority conforming with our preference for correlated variation of these quantities, and if all these solutions satisfy the seismic data within reasonable limits, it is evidently possible to select examples which also satisfy our prejudices regarding physical consistency. The considerable variation in detail among these solutions, despite the large input of observational data, shows the importance of well-founded general assumptions regarding physical behavior for an eventual "unique" solution.

If the high densities in the upper mantle prove to be indispensable, current models of chemical and mineralogical composition will require revision. The large amount of experimental work on systems having the composition of "pyrolite" (GREEN and RINGWOOD, 1967, 1970; RINGWOOD, 1970) has not produced phases with the required high density. The Press models of the upper mantle resemble the "eclogite model" of CLARK and RINGWOOD (1964), except for the "low density layer" which in the Press models underlies the layer of high density. An interpretation of the low-velocity region in terms of iron-rich olivine has been discussed (BIRCH, 1969), and this would also lead to high densities. But whether eclogite or chrysolite, it seems likely that the iron content must be higher than is usually supposed.

By comparison with dunite, the experimental material on the elastic properties of eclogites is meager. As Clark and Ringwood point out, most of the eclogites which have been studied are of relatively low density, either because of alteration or because of low proportions of garnet. Their discussion is based on an assumed average density of 3.55 g/cm³. Several eclogites from Norway, collected and described by SCHMITT (1964), have densities close to this, and their velocities have been determined by SIMMONS (1964, and unpublished). The available data are given in table 1. The 10 kb

TABLE 1

Velocities in eclogites, room temperature and 10 kb*

Origin	Density (g/cm³)	V_P (km/s)	V_S (km/s)	Ref.
Norway, Eiksundal complex				
1552	3.453	8.21		2
	3.577		4.60	1
1553	3.559	8.35		2
	3.578		4.66	1
California				
Healdsburg	3.441	8.01		3
	3.444		4.58	1
Norway				
Sondmøre	3.376	7.69		3
Oahu				
Salt Lake Crater	3.39	7.94		4
Africa				
Kimberley	3.376	7.87		3
Tanganyika	3.328	7.71		3
Oahu				
"garnet peridotite"	3.23	7.85		4

* All values are means for three samples, except those of ref. 4.
1. SIMMONS (1964).
2. SIMMONS (unpublished).
3. BIRCH (1960).
4. MANGHNANI and WOOLLARD (1968).

velocities in eclogite of density 3.55 g/cm³ are about 8.3 and 4.6 km/s at ordinary temperatures, somewhat lower than those of olivine (Fa_8). The velocities are approximately proportional to the densities, with $\Delta V/\Delta \rho$ about 3 (km/s)/(g/cm³), as for other rocks with nearly constant mean atomic weight, and these points define a line roughly consistent with the measurements on basic rocks (BIRCH, 1960, 1961b). The normal temperature coefficients are probably somewhat smaller for eclogite than for olivine (BIRCH, 1969, table 3); this is qualitatively consistent with the smaller thermal expansion of eclogite. Thus, while an eclogite layer would account for the high density, the low velocities and high attenuation still require partial melting or elastic relaxation. The average amount of melt must, however, not exceed a few percent if the high density is to be retained.

In place of eclogite, which is essentially "garnet pyroxenite", it may be preferable to try to satisfy the requirements with "garnet peridotite" or "garnet pyrolite", but in either case, the need for high density will lead to much higher proportions of dense garnet than are usually associated with these terms (e.g. fig. 3 of GREEN and RINGWOOD, 1967). The eclogites with high densities such as Schmitt's samples are those with large fractions of iron-rich garnets. Thus the upper mantle may, as a whole, be richer in iron, lime and alumina than the currently favored "pyrolite" or "garnet pyrolite" as defined by Green and Ringwood, and closer to the eclogitic composition.

Acknowledgments

I am indebted to Prof. Frank Press for the opportunity of seeing his calculations in advance of the Canberra meeting, and for permission to comment on them in this publication, and to the Australian National University for a generous travel grant which made it possible to participate in this conference.

Added in proof: For additional data on eclogites, see BAJUK, E. I., M. P. VOLAROVICH, K. KLÍMA, Z. PROS, and J. VANĚK (1967), Studia Geophys. Geodaet. Ceskoslov. Akad. Ved. **11**, 271.

References

ANDERSON, D. L. (1967) in: T. F. Gaskell, ed., *The Earth's mantle* (Academic Press, New York).
ANDERSON, D. L. and M. N. TOKSÖZ (1963) J. Geophys. Res. **68**, 3483.
BIRCH, F. (1960) J. Geophys. Res. **65**, 1083.
BIRCH, F. (1961a) Geophys. J. **4**, 295.
BIRCH, F. (1961b) J. Geophys. Res. **66**, 2199.
BIRCH, F. (1969) in: V. V. Beloussov and P. J. Hart, eds., *The Earth's crust and upper mantle*, Geophys. Monograph **13** (Am. Geophys. Union, Washington, D.C.).
BULLEN, K. and R. A. W. HADDON (1967) Proc. Nat. Acad. Sci. U.S. **58**, 846.
CLARK, JR. S. P. and A. E. RINGWOOD (1964) Rev. Geophys. **2**, 35.
GORDON, R. B. and L. A. DAVIS (1968) J. Geophys. Res. **73**, 3917.
GREEN, D. H. and A. E. RINGWOOD (1967) Earth Planet. Sci. Letters **3**, 151.
GREEN, D. H. and A. E. RINGWOOD (1970) Phys. Earth Planet. Interiors **3**, 359.
GUTENBERG, B. (1926) Z. Phys. **27**, 11.
GUTENBERG, B. (1959) Geophys. J. **2**, 348.
LANDISMAN, M., Y. SATÔ and J. NAFE (1965) Geophys. J. **9**, 439.
LEHMANN, I. (1967) in: T. F. Gaskell, ed., *The Earth's mantle* (Academic Press, New York).
MANGHNANI, M. H. and G. P. WOOLLARD (1968) in: L. Knopoff, C. L. Drake and P. J. Hart, eds., *The crust and upper mantle of the Pacific area*, Geophys. Monograph **12** (American Geophys. Union, Washington, D.C.).
PEKERIS, C. L. (1966) Geophys. J. **11**, 85.
PRESS, F. (1970) Phys. Earth Planet. Interiors **3**, 3.
RINGWOOD, A. E. (1966) in: P. M. Hurley, ed., *Advances in Earth science* (M.I.T. Press, Cambridge, Mass.).
RINGWOOD, A. E. (1970) Phys. Earth Planet. Interiors **3**, 109.
SCHMITT, H. H. (1964) Dissertation, Harvard University, Cambridge, Mass.
SCHREIBER, E. and O. L. ANDERSON (1967) J. Geophys. Res. **72**, 762, 3751.
SIMMONS, G. (1964) J. Geophys. Res. **69**, 1123.
SOGA, N., E. SCHREIBER and O. L. ANDERSON (1966) J. Geophys. Res. **71**, 5315.
SPETZLER, H. and D. L. ANDERSON (1968) J. Geophys. Res. **73**, 6051.
VAISNYS, J. R. (1968) J. Geophys. Res. **73**, 7675.
WALSH, J. B. (1968) J. Geophys. Res. **73**, 2209.

SPLIT SPHERE HIGH PRESSURE VESSEL AND PHASE EQUILIBRIUM RELATION IN THE SYSTEM Mg_2SiO_4-Fe_2SiO_4

NAOTO KAWAI, SHOICHI ENDHO and KEISUKE ITHO

Department of Material Physics, Faculty of Engineering Science, Osaka University, Toyonaka, Osaka, Japan

In part 1 of this paper a split sphere multipiston assemblage is introduced with which it is possible to produce ultra high pressure and temperature for synthetic experiments.

The second part of this paper reports a study on high pressure and temperature stability relation in Mg-rich mineral olivine, i.e. the most occurring mineral in the earth's mantle. Then, a phase diagram in the system Mg_2SiO_4 and Fe_2SiO_4 is shown together with some technical remarks.

1. Split sphere high pressure apparatus

Recently the present authors have designed a special apparatus in order to obtain ultra high pressure, higher than 0.5 Mega bar under which it is possible to carry out a simultaneous heating of a relatively ample test specimen.

The central part of this equipment is composed of eight wedge-shaped multipistons forming a solid spherical anvil with diameter 200 mm and a small central octahedral chamber for our specimen to be compressed. This spherical anvil is covered with a pair of thick semispherical shells made of rubber. Then it is soaked in a reservoir containing a compressed liquid. Since each wedge-shaped piston in the spherical assemblage has a narrow inner (front) surface in comparison to its larger outer (bottom) surface, the applied liquid pressure can be greatly magnified and directed towards center of the sphere.

In order that the pressure vessels may resist such a high pressure and temperature as they themselves can produce, not only the anvil materials ought to be correspondingly strong and hard but also the array of the multipistons should be in a highly geometrical symmetry around a point at which the high pressure in question is to be generated. Stress distribution to occur in such a symmetrical array would be so uniform that one can expect a much higher yielding pressure for the vessel, as compared to those in the other asymmetrical arrays.

The multipistons also should possess a large pressure magnifying ratio $n = s/S$ where s and S represent respectively the area of the inner (front) and that of the outer (bottom) surface of each piston.

It is also required that the multipiston assemblage can shrink as effectively as possible. This shrinkage accompanies a large volume reduction of the sample chamber ΔV from its originally rather small volume, V. In such cases the ratio $m = \Delta V/V$ becomes so large that easy production of an extremely high pressure can be expected.

Needless to say that the inward movements of all multipistons should be strictly synchronous. Neither retardation nor precedence of one piston from the rest should, therefore, be allowed at all.

Besides the above-mentioned intrinsic characteristics in an ideal vessel, the apparatus in general is desired to be manufactured as easily and operated as simply as possible.

Fulfilling, therefore, the several above-mentioned requirements, a sphere made of tungsten carbide containing 3% Co binder was first prepared. This sphere was then enclosed in a spherical shell made of hardened steel. By the mutually perpendicular three planes, each of which is passing through the center of the sphere, the entire body was split up to form the above-mentioned eight wedge-shaped pistons. Each piston thus made has a curved bottom and three inclining side planes which meet at the top as shown in fig. 1. The top of each piston was so truncated that there would remain an octahedral hollow space in the center of the spherical anvil comprising all pistons.

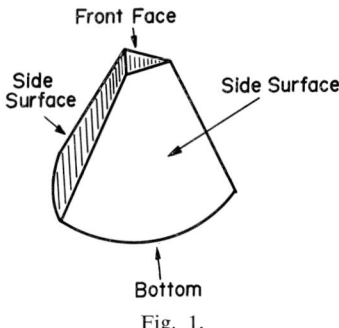

Fig. 1.

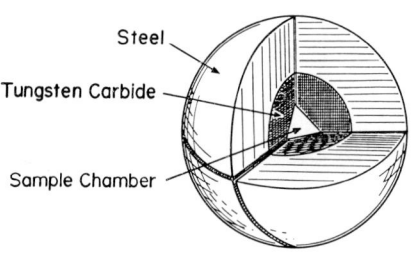

Fig. 2.

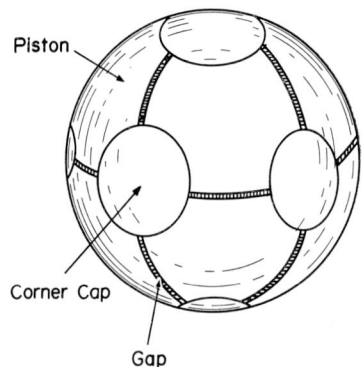

Fig. 3.

Into this central hollow space an octahedral pyrophillite pressure transmitting media was inserted.

Usually a cylindrical heater made of graphite tube was buried in the center of the octahedron. The test specimen or the material to be synthesized was placed in the middle of the graphite tube. To heat the specimen up to a desired temperature, an electric current was driven through the graphite tube. The front surfaces of the two opposing multipistons were employed as the thresholds of the electric current. With the above-mentioned front surfaces two terminals of our graphite heater were made in good electric contact.

On the inclining side surfaces of each piston a number of soft insulator sheets were placed and glued, so that the latter were sandwiched in between the neighbouring pistons. With these sheets it was possible to keep all gaps between the pistons equal in the entire assemblage. The above insulator sheets played the following three important roles in the course of the pressure production. First, these sheets separated one piston electrically from the others. So they played an important role as an insulator. Secondly, these sheets were easily compressed and stretched to cover broadly the side surfaces of each piston. One piston, therefore, was surrounded by three adjacent pistons and supported firmly at its shoulder by these stretched and deformed materials. The situation mentioned above made occurring shearing stresses in the central part of the pistons progressively decreased with increasing applied pressure. Finally, these stretched spacer (insulator) sandwiched between the neighbouring pistons played a role as an effective packing or a special seal of the elevated pressure preventing its extrusion out of the sphere.

In fig. 2 it was shown how eight pistons could be arrayed. A place where the four pistons meet on the spheres outer surface was flattened out. A corner cap previously prepared was placed on this flat face (see fig. 3). Altogether six corner caps thus placed were very important in keeping the correct geometrical shape of the array, since any free rotation or shift of one piston relative to the rest could be prevented by the help of these corner caps during the shrinkage of the assemblage.

Fig. 4 shows the heating tube placed inside of the pyrophillite octahedron.

Calibration of pressure produced within the octahedral media was undertaken by the use of a number of phase changes in several materials already studied and reported.

Small specimens of Bi, Ba, Fe, Pb, Ge, GaAs, and Si were prepared and were placed in the octahedral specimen media to be pressurized. Observing the abrupt resistance changes of these test specimens in term and measuring simultaneously the liquid pressures in the reservoir, a pressure standard curve was determined.

Besides the above-mentioned transition points, resistance maximum in CdS to occur under 465 kb at room temperature was also employed as another fixed point with which our standard was completed.

According to the repeated experiments continued so

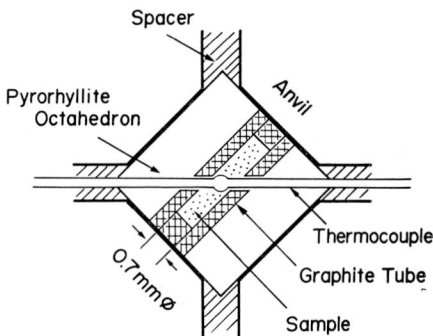

Fig. 4. Details of the present furnace assembly.

far the present apparatus is able to produce a pressure greater than 450 kb at room temperature and slightly higher than 300 kb at 1200 °C.

2. Phase equilibrium relation in the system Mg_2SiO_4-Fe_2SiO_4

By using the high pressure equipment we introduced in the previous section, stability relation in Mg-rich mineral olivine was investigated with a natural crystal taken from the Red sea volcanic area. Chemical analysis on this natural crystal showed that it can be represented by a formula $(Mg_{0.93}, Fe_{0.07})_2SiO_4$. The mineral was pressurized in the vessel and heat-treated at 1000 °C for a sufficient duration. The converted dense forms of this mineral were taken out of the vessel. Then, they were identified under the microscope and examined by the use of the X-ray analysis.

Results of several experimental runs at 1000 °C under various high pressures are shown in table 1. Following descriptions such as those given beneath are valid in the stability relation of this Mg-rich silicate:

1) At 1000 °C this Mg-rich olivine remains as a stable phase in a pressure range from 0 up to 135 kb. Under pressure higher than 135 kb the olivine breaks up into two mineral phases, one being spinel and the other olivine phase respectively. This region of bimineral assemblage olivine+spinel was predicted and confirmed by RINGWOOD (1962, 1963), RINGWOOD et al. (1963) and AKIMOTO et al. (1965, 1966, 1968). At 1000 °C this bimineral region was found to retain itself until the pressure would be elevated above 145 kb.

2) In a pressure range from 145 kb up to 155 kb and at temperature 1000 °C there appears another bimineral assemblage olivine+the so-called distorted spinel.

3) In a range from 155 kb up to 195 kb and at temperature 1000 °C only distorted spinel remains as a stable phase.

4) In a pressure range from 195 kb up to approximately 225 kb there appears again another bimineral assemblage. In this range both the distorted spinel and the spinel phase of the silicate become stable.

5) At 1000 °C and under pressure higher than 225 kb the crystal is stable only in a form of the spinel phase.

The results obtained in our experiments were compared with the experimental data on $(Mg, Fe)_2SiO_4$ already studied and reported by RINGWOOD (1962, 1963), RINGWOOD et al. (1963) and AKIMOTO (1965, 1966, 1968). As these results show it was found that the polymorphism in this silicate mineral might be well explained by such a phase diagram as shown in fig. 5.

Needless to say, polymorphic transitions in this silicate mineral, especially those in Mg-rich olivine are of special interest and importance in geophysics, since it has been a legitimate but long buried concept since BERNAL (1936) that the seismic discontinuities in the earth mantle beginning at 370 km below the sea level can reasonably be assumed to be the mineral discontinuities in the mantle arising mainly from the polymorphic transitions.

If, for instance, the mantle were assumed to be composed of such a Mg-rich silicate as that with composition $(Mg_{0.93}, Fe_{0.07})_2SiO_4$, then the temperature at 370 km below the earth's surface is to be determined.

TABLE 1

Run no.	Pressure (kb)	Temperature (°C)	Time (min)	Results
52	190	1000	10	olivine+distorted spinel
53	210	1000	20	olivine+distorted spinel +spinel
64	155	800	20	olivine+spinel
65	145	800	20	olivine+spinel
66	134	780	20	olivine+spinel
67	120	740–800	18	olivine
72	125	800	10	olivine+(trace)spinel
73	134	900–1000	4.5	olivine
74	134	900	4	olivine+spinel
76	115	700	30	olivine
77	120	700	28.5	olivine+spinel
78	130	900	9.5	olivine
79	138	945	2.5	olivine+spinel
80	138	1000	4	olivine
81	141	1000	5.5	olivine
82	123	775	17	olivine+(trace)spinel
83	144	1000	5	olivine+spinel

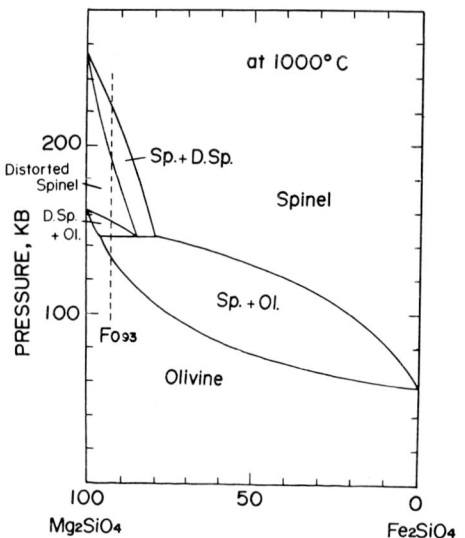

Fig. 5. Possible phase diagram in the Mg_2SiO_4–Fe_2SiO_4 system at 1000 °C.

The temperature there was found to be 800 °C if Bridgman's volume pressure scale were taken as standard. Whereas it turned out to be 1200 °C if NaCl pressure scale were taken into account.

Our interpretation regarding the structure of the earth's mantle is based upon the results of our high pressure experiments in which the pressure standard was so established that the transition pressures in several standard materials were really as correct as BALCHAN and DRICKAMER (1961), MINOMURA and DRICKAMER (1962), BUNDY (1967) and other high pressure researchers reported several years ago. However, recent high pressure experiments are revealing the fact that there still remains a number of questions to be solved as to the true values of these occurring pressures of the transitions, in particular when they are higher than 100 kb. For example transition of Pb was assumed by BLACHAN and DRICKAMER (1961) to occur at 160 kb, whereas by BASSET et al. (1969) to occur at 120 kb. Thus, the inconsistencies regarding the pressure scale are becoming more and more serious as well as conspicuous. In conclusion the authors would like to confess that too much still remains to be solved to really clarify the physical aspect and chemical constituent of the earth's mantle in the present stage of experiments.

References

AKIMOTO, S. and H. FUJISAWA (1966) Earth Planet. Sci. Letters **1**, 241.
AKIMOTO, S. and H. FUJISAWA (1968) J. Geophys. Res. **73**, 1467.
AKIMOTO, S., H. FUJISAWA and T. KATSURA (1965) J. Geophys. Res. **70**, 1969.
BALCHAN, A. S. and H. C. DRICKAMER (1961) Rev. Sci. Instr. **32**, 308.
BASSET, A. et al. (1969) to be published.
BERNAL, J. D. (1936) Observatory **59**, 268.
BUNDY, F. P. (1967) J. Appl. Phys. **38**, 157.
MINOMURA, S. and H. G. DRICKAMER (1962) J. Phys. Chem. Solids **23**, 451.
RINGWOOD, A. E. (1962) Geochim. Cosmochim. Acta **26**, 457.
RINGWOOD, A. E. (1963) Nature **198**, 79.
RINGWOOD, A. E. and M. SEABROOK (1963) J. Geophys. Res. **68**, 4606.

HIGH-PRESSURE DECOMPOSITION OF THE SYSTEM $FeSiO_3$-$MgSiO_3$

SYUN-ITI AKIMOTO and YASUHIKO SYONO

Institute for Solid State Physics, University of Tokyo, Minato-ku, Tokyo, Japan

Stability relations of (Mg, Fe)SiO_3 solid solutions ranging in composition from pure $FeSiO_3$ to ($Mg_{0.3}$ $Fe_{0.7}$)SiO_3 have been studied at high temperatures in the pressure range to 96 kb. High-pressure decomposition of (Mg, Fe)SiO_3 clinopyroxene into (Mg, Fe)$_2SiO_4$ spinel solid solution and stishovite was established. The decomposition curve for $FeSiO_3$, clinoferrosilite, was found to coincide with the coesite–stishovite transformation curve.

The pyroxenes are accepted to be the most abundant class of minerals next to the olivines in the upper mantle. Reconnaissance study of the high-pressure transformation of A^{2+} $B^{4+}O_3$ type silicates and germanates with pyroxene structure has already been reported by Ringwood et al. (RINGWOOD, 1967; RINGWOOD and MAJOR, 1966a, b, c, 1967, 1968; RINGWOOD and SEABROOK, 1963). They successfully discovered various types of transformations such as the transformation to ilmenite, to garnet and to perovskite, and the decomposition into spinel and rutile-type structure. Construction of the reliable transformation diagram of the $MgSiO_3$–$FeSiO_3$ system over the wide range of pressure and temperature is, however, indispensable for the quantitative analysis of the high-gradient zone of seismic wave velocities in the mantle. In the present study, some preliminary experimental results upon phase equilibria at high pressures in the iron-rich side of the pyroxene system $MgSiO_3$–$FeSiO_3$ are reported.

High-pressure and high-temperature experiments were carried out by means of a tetrahedral press using pyrophyllite as pressure transmitting material. Two different sizes of cemented tungsten carbide anvils with 9 mm and 7 mm edge length were used. Pressure values were calibrated at room temperature on the basis of the pressure standard proposed by JEFFERY et al. (1966).

Powder samples were directly embedded in the graphite tube which was used as a heating element as well. Temperatures were measured with a Pt/Pt–13% Rh thermocouple without any corrections for the effect of pressure on the emf of the thermocouple.

Starting materials used in the present study are in general orthopyroxenes of known composition which were synthesized at 40 kb and at 1200 °C. Equimolar mixtures of fayalite and amorphous silica were also used as starting materials for the study of high-pressure transformation in ferrosilite. The experimental method used in determining the stability diagram of (Mg, Fe)SiO_3 at high-pressures and high-temperatures was the usual quenching method. Samples were subjected to desired pressures at desired temperatures for a predetermined interval of time and quenched isobarically by turning off the heating power. After relaxation of the pressure and recovery of the samples, the phases present were examined by optical and X-ray diffraction technique.

Equilibrium data of ferrosilite at various pressure–temperature–time conditions are summarized in table 1 and illustrated in fig. 1. A previous report by RINGWOOD and MAJOR (1966a) that clinoferrosilite decomposes into Fe_2SiO_4 spinel plus stishovite was confirmed over a wide temperature range. The present data also indicate that the boundary curve for this decomposition of clinoferrosilite is well represented by the boundary curve for the coesite–stishovite transition which was determined as P (kb) $= 69 + 0.024\,T\,(°C)$ in our recent study (AKIMOTO and SYONO, 1969) and reproduced by the solid straight line in fig. 1. The olivine–spinel transformation in Fe_2SiO_4 is known to take

TABLE 1

Results of runs on the high-pressure and high-temperature phase transformations in FeSiO$_3$

Run no.	Pressure (kb)	Temperature (°C)	Time (min)	Phases present
2*	83	790 ± 5	30	clinoferrosilite
3	89	790 ± 5	30	spinel+stishovite +clinoferrosilite (small amount)
1*	94	790 ± 5	30	spinel+stishovite
5	88	1000 ± 5	20	clinoferrosilite
4	94	1000 ± 5	20	spinel+stishovite
7	98	1150 ± 10	15	spinel+stishovite
6	94	1210 ± 10	10	clinoferrosilite

* Starting materials used in the runs with asterisk mark are equimolar mixtures of fayalite and amorphous silica. Unless otherwise designated, starting materials are orthoferrosilite.

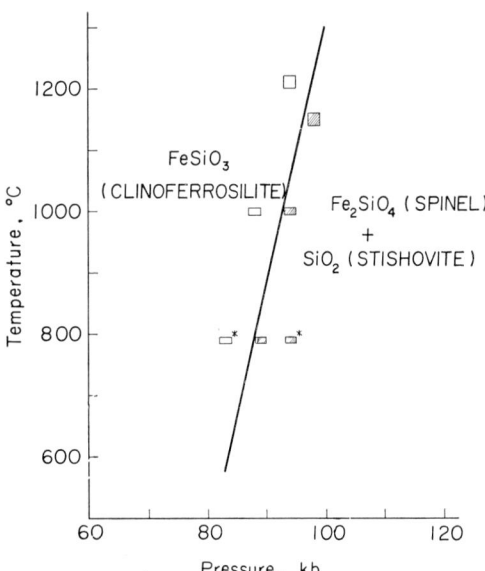

Fig. 1. Stability diagram for the high-pressure and high-temperature decomposition of FeSiO$_3$. * indicates the runs using equimolar mixtures of fayalite and amorphous silica as the starting materials. Unless otherwise designated, the starting material is orthoferrosilite.

place at pressures of about 45 kb to 55 kb in the temperature range of 800 °C to 1200 °C (AKIMOTO et al., 1967). These pressure values are much lower than the transition pressure of coesite to stishovite at the corresponding temperatures. Consequently, the formation of the stishovite plays a decisive role in realizing the reaction, FeSiO$_3$ → Fe$_2$SiO$_4$ (spinel)+SiO$_2$ (stishovite). Substantial agreement of the boundary curve for the decomposition of clinoferrosilite with the coesite–stishovite transition curve may be caused by these reasons. It is similarly suggested from the present study that the decomposition curve of CoSiO$_3$ clinopyroxene to Co$_2$SiO$_4$ spinel plus stishovite coincides with the coesite–stishovite transition curve.

Experimental results for the decomposition of the FeSiO$_3$–MgSiO$_3$ system over the pressure range to 96 kb at 800 °C are shown in table 2 and fig. 2. It is seen that at this temperature the starting orthopyroxene solid solutions transform to clinopyroxene solid solutions at high pressures and further decomposes to stishovite plus spinel solid solutions of the Fe$_2$SiO$_4$–Mg$_2$SiO$_4$ system at higher pressures. This agrees well with the previous report by RINGWOOD and MAJOR (1966c, 1968).

The compositions of spinel s.s. (solid solutions) in the three phase region (spinel s.s. + stishovite + clinopyroxene s.s.) were determined from their lattice parameters based on the lattice parameter–composition relationship for (Fe, Mg)$_2$SiO$_4$ spinel s.s. previously published by us (AKIMOTO and FUJISAWA, 1968). This information made it possible to estimate the boundary curve in fig. 2 separating the two phase region (spinel

TABLE 2

Results of runs on the high-pressure and high-temperature phase transformations in the system FeSiO$_3$–MgSiO$_3$

Run no.	Composition*	Pressure (kb)	Temperature (°C)	Time (min)	Phases present**	Lattice parameter of spinel s.s. (Å)
1004	10M90F	86	805 ± 5	30	cpx	
1001	idem	90	800 ± 5	30	sp+st+cpx	8.227 ± 0.001
1003	idem	96	800 ± 5	30	sp+st+cpx	8.222 ± 0.002
2001	20M80F	90	800 ± 5	30	cpx	
2002	idem	93	810 ± 5	30	sp+st+cpx	8.222 ± 0.001
3002	30M70F	96	810 ± 5	30	cpx+sp+st	8.220 ± 0.002

* M = MgSiO$_3$; F = FeSiO$_3$.
** cpx = clinopyroxene solid solutions; sp = spinel solid solutions; st = stishovite.

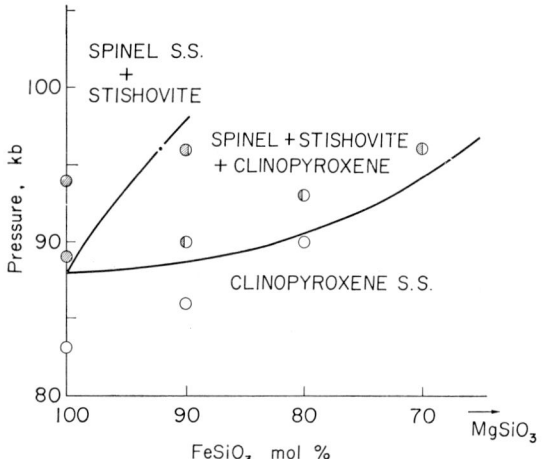

Fig. 2. Phase diagram for the high-pressure and high-temperature decomposition of the system $FeSiO_3$–$MgSiO_3$ at 800 ± 10 °C. The dot indicates the boundary estimated from the spinel composition obtained from the lattice parameter.

s.s.+stishovite) from the three phase region (spinel s.s.+stishovite+clinopyroxene s.s.). Since the boundary curves determined in this study are only limited to the narrow composition range of the iron-rich side of the $FeSiO_3$–$MgSiO_3$ system, it is still difficult to elucidate the general transformation behavior of the whole composition range of the system. It is clear, however, that the three phase region ranges over some twenty or thirty kilobars in the intermediate composition of the $FeSiO_3$–$MgSiO_3$ system. Thus, the decomposition of the (Mg, Fe)SiO_3 clinopyroxene may probably contribute to the formation of the transition zone in the mantle. The present results also support the previous conclusion by RINGWOOD and MAJOR (1966c) that (Mg, Fe)SiO_3 clinopyroxenes remain stable at pressures which are substantially higher than required to cause olivines with similar Mg/Fe ratios to transform to spinels.

References

AKIMOTO, S. and H. FUJISAWA (1968) J. Geophys. Res. **73**, 1467.
AKIMOTO, S., E. KOMADA and I. KUSHIRO (1967) J. Geophys. Res. **72**, 679.
AKIMOTO, S. and Y. SYONO (1969) J. Geophys. Res. **74**, 1653.
JEFFERY, R. N., J. D. BARNETT, H. B. VANFLEET and H. T. HALL (1966) J. Appl. Phys. **37**, 3172.
RINGWOOD, A. E. (1967) Earth Planet. Sci. Letters **2**, 255.
RINGWOOD, A. E. and A. MAJOR (1966a) Earth Planet. Sci. Letters **1**, 135.
RINGWOOD, A. E. and A. MAJOR (1966b) Earth Planet. Sci. Letters **1**, 209.
RINGWOOD, A. E. and A. MAJOR (1966c) Earth Planet. Sci. Letters **1**, 351.
RINGWOOD, A. E. and A. MAJOR (1967) Earth Planet Sci. Letters **2**, 106.
RINGWOOD, A. E. and A. MAJOR (1968) Earth Planet Sci. Letters **5**, 76.
RINGWOOD, A. E. and M. SEABROOK (1963) J. Geophys. Res. **68**, 4601.

HIGH-PRESSURE SYNTHESIS OF A "MODIFIED" SPINEL AND SOME GEOPHYSICAL IMPLICATIONS

SYUN-ITI AKIMOTO

Institute for Solid State Physics, University of Tokyo, Minato-ku, Tokyo, Japan

Stability relations of Mn_2GeO_4 have been studied over the temperature range 790 to 1240 °C in the pressure range 31 to 70 kb. β-Mn_2GeO_4 with a "modified" spinel structure was observed in a large synthesis field between the fields of the olivine polymorph (α-Mn_2GeO_4) and another high-pressure polymorph (δ-Mn_2GeO_4) with Sr_2PbO_4-type structure. Density increases associated with the α-β transformation and the β-δ transformation were estimated to be 7.1% and 10.7%. The mean volume thermal expansivity for β-Mn_2GeO_4 has been determined to be $(30 \pm 1) \times 10^{-6}$ °C^{-1} in a temperature range of room temperature to 460 °C and at ambient pressure.

A high-pressure transformation diagram in $(Mg, Co)_2SiO_4$ solid solutions ranging in composition from pure Co_2SiO_4 to $(Mg_{0.5}Co_{0.5})_2SiO_4$ has been studied over the pressure range 61 to 94 kb at 800, 1000 and 1200 °C. The phase relationships of the system were found to be strongly influenced by temperature; at 800 °C any indication of the phases other than olivine and spinel solid solution was not found, but at 1200 °C the formation of the "modified" spinel solid solutions was distinguishable in a pressure range between the fields of the olivine and the true spinel solid solutions.

It is plausible from these experimental results to conclude that the "modified" spinel solid solutions synthesized by Ringwood and Major for compositions more magnesian than $(Mg_{0.85}Fe_{0.15})_2SiO_4$ in the Mg_2SiO_4–Fe_2SiO_4 system may represent a thermodynamically stable phase and not a metastable phase formed during quenching. If the mantle olivine transformed stepwise to the true spinel phase through the "modified" spinel phase, the previous estimates of the constitution and width of the transition zone of the mantle should be reexamined.

1. Introduction

It has recently been accepted that high-pressure transformations in $(Mg, Fe)_2SiO_4$ are responsible for the formation of the high-gradient zone of seismic-wave velocities at a depth of about 350 to 1000 km in the Earth's mantle. Successful interpretation of the sharp velocity increase around 370 km has been given by ANDERSON (1967) and FUJISAWA (1968) on the basis of the simple olivine-spinel equilibrium diagram which was established for the $(Mg, Fe)_2SiO_4$ solid solutions ranging in composition from pure Fe_2SiO_4 to $(Mg_{0.8}Fe_{0.2})_2SiO_4$ (AKIMOTO and FUJISAWA, 1966 and 1968). However, some complexity in the transformation in more magnesian olivines than $(Mg_{0.85}Fe_{0.15})_2SiO_4$ was pointed out by RINGWOOD and MAJOR (1966). They synthesized another non-cubic high-pressure phase (β phase*) at 900° to 1000 °C and at pressures just above the stable range of olivine. An X-ray powder diffraction pattern of this β phase showed a marked resemblance to spinel but contained many extra lines. RINGWOOD and MAJOR (1966) suggested that the β phase may be a "distorted" or "modified" spinel, and that the distortion occurred during quenching of an original true spinel. The crystal structure of the β phase, however, has been remained unknown on account of the lack of a suitable single crystal for the structure analysis.

In these situations, it is of great importance to determine what is the crystal structure of the β phase and to examine whether or not the β phase is a metastable phase formed during quenching from high-pressure and high-temperature conditions. We have recently found the occurrence of a new high-pressure polymorph of Co_2SiO_4 which is the isotype of the β phase of Mg_2SiO_4. The stability relationships between this new polymorph and olivine and spinel polymorphs of Co_2SiO_4 at high-pressures and high-temperatures have been reported in our recent paper (AKIMOTO and SATO, 1968). Further, successful synthesis of a single crystal of the β phase of Co_2SiO_4 made it possible to determine

---

* In the present paper, the terms, α, β, γ and δ phase in R_2MX_4 type compounds are used for olivine, "modified" spinel, true spinel, and Sr_2PbO_4-type polymorph respectively.

the crystal structure of the β phase. Structure analysis carried out by MORIMOTO et al. (1969) revealed that the β phase of Co_2SiO_4 could be assigned an orthorhombic structure with space group of *Imma*, and justified that the β phase is termed "modified" spinel. Detailed report is given elsewhere in this volume (MORIMOTO et al., 1970).

In the present paper, other examples of the high-pressure synthesis of the "modified" spinel are given in Mn_2GeO_4 and in $(Mg, Co)_2SiO_4$ solid solutions.

2. High-pressure transformations of Mn_2GeO_4

Preliminary reports of the phase relations of Mn_2GeO_4 have already been given by a few investigators. RINGWOOD and SEABROOK (1963) noted that olivine polymorph of Mn_2GeO_4 decomposed into $MnGeO_3$ ilmenite plus an additional phase (probably MnO) at about 90 kb and at 700 °C. Recently, above 90 kb and at 900 °C WADSLEY et al. (1968) synthesized a new high-pressure polymorph of Mn_2GeO_4 (δ phase) which was isomorphous with Sr_2PbO_4. In the present study we found that a third high-pressure polymorph of Mn_2GeO_4 was synthesizable in the pressure range intermediate between the olivine and Sr_2PbO_4-type field.

High-pressure and high-temperature experiments were done using the tetrahedral-anvil type of high-pressure apparatus. Two different sizes of cemented tungsten carbide anvils with 15 mm and 20 mm edge length were used, depending upon maximum pressure desired. Pressure values were calibrated at room temperature on the basis of the pressure scale proposed by JEFFERY et al. (1966).

Starting materials used in the ordinary runs are the olivine polymorph of Mn_2GeO_4 (α phase) which was prepared by sintering a stoichiometric mixture of MnO and GeO_2 sealed in an evacuated silica tube at 1000 °C for 40 hours. The unit cell dimensions of the Mn_2GeO_4 olivine are $a = 5.061 \pm 0.001$ Å, $b = 10.719 \pm 0.001$ Å and $c = 6.295 \pm 0.001$ Å. In order to determine the transition diagram of Mn_2GeO_4 accurately and to settle the question whether or not the newly found high-pressure polymorph of Mn_2GeO_4 (β phase) is a real stable phase in its synthesis field, a number of reverse runs, in which the β phase and the δ phase of Mn_2GeO_4 were used as starting materials, were also made.

Powder samples of starting materials were directly embedded in the tubular graphite furnace which was placed diagonally with the axis of the cylinder between opposite edges of the baked pyrophyllite tetrahedron. Temperatures were measured in the central part of the samples with a Pt/Pt-13% Rh thermocouple without any correction for the effect of pressure on the emf of the thermocouple.

The experimental method used in determining the stability diagram of Mn_2GeO_4 at high-pressures and high-temperatures was the usual quenching method. After pressure was applied to the sample, the temperature was brought to the desired value and held for the desired interval of time. Then the sample was quenched under the working pressure by turning off the heating power. After relaxation of pressure, the phases present in the central part of the quenched samples were examined by the X-ray diffraction technique.

Results of runs on the high-pressure and high-temperature phase transformations in Mn_2GeO_4 are summarized in table 1. The high-pressure synthesis of the δ phase of Mn_2GeO_4 was confirmed in the present study over the wide temperature range. Unit cell dimensions and the calculated density of the δ-Mn_2GeO_4; $a = 5.262 \pm 0.001$ Å, $b = 9.274 \pm 0.001$ Å, $c = 2.954 \pm 0.001$ Å and $\rho_\delta = 5.676$ g/cm^3, are in good agreement with the values reported by WADSLEY et al. (1968). However, a new high-pressure polymorph of Mn_2GeO_4 with greyish-blue colour was observed in the pressure range intermediate between the α and δ phase field. X-ray diffraction data of the newly found high-pressure polymorph of Mn_2GeO_4 were found to be successfully indexed by the β-Co_2SiO_4 structure. Structure analysis carried out bij MORIMOTO et al. (1969 and 1970), using single crystals of Mn_2GeO_4 synthesized at 64 kb and at 1240 °C, established that the new high-pressure polymorph of Mn_2GeO_4 was assigned the "modified" spinel structure with space group of *Imma*. Unit cell dimensions and the calculated density of this β-Mn_2GeO_4 were determined to be $a = 6.025 \pm 0.002$ Å, $b = 12.095 \pm 0.004$ Å, $c = 8.752 \pm 0.002$ Å and $\rho_\beta = 5.13$ g/cm.

A true spinel phase could not be synthesized in the Mn_2GeO_4 sample at the pressure-temperature condition studied in this work. This is a marked difference from the high-pressure transformations in Co_2SiO_4. Further, any evidence suggesting a decomposition of

TABLE 1

Results of runs on the high-pressure and high-temperature phase transformations in Mn_2GeO_4

Run no.	Pressure*, kb	Temperature, °C	Time, min	Phases present
		Starting material, α-Mn_2GeO_4 (Olivine type)		
6	31	810 ± 5**	50	α phase
7	35	790 ± 10	50	α phase
9	39	1100 ± 20	18	α phase
8	40	800 ± 10	50	β phase
10	42	1100 ± 20	15	α phase
11	47	1100 ± 20	15	β phase + α phase
15	49	900 ± 5	40	β phase
4	54	930 ± 5	25	β phase
13	54	1030 ± 20	30	β phase
12	56	810 ± 5	31	β phase
1	63	1200 ± 20	20	β phase
3	64	840 ± 5	45	δ phase
2	64	1240 ± 20	20	β phase
5	68	1200 ± 20	20	δ phase
16	70	900 ± 5	40	δ phase
14	70	1030 ± 10	30	δ phase
		Starting material, β-Mn_2GeO_4 (Modified spinel type)		
21	41	1030 ± 15	25	α phase
22	66	1100 ± 10	20	δ phase
		Starting material, δ-Mn_2GeO_4 (Sr_2PbO_4 type)		
31	47	1130 ± 20	15	α phase
32	57	940 ± 5	30	β phase
33	60	930 ± 5	35	δ phase

* Precision of pressure control is about ±1 kb.
** Precision of temperature control.

Mn_2GeO_4 into $MnGeO_3$ ilmenite and MnO was not observed at pressures below 70 kb.

Experimental runs using δ-Mn_2GeO_4 as starting material revealed that the β-Mn_2GeO_4 could be formed not only from olivine but also from the Sr_2PbO_4-type polymorph. Further, complete conversion from β-Mn_2GeO_4 to δ phase or α phase was confirmed by the reverse reactions using β-Mn_2GeO_4 as starting material. These mutual transformation behaviors among the α, β, and δ phases of Mn_2GeO_4 may suggest that the β-Mn_2GeO_4 synthesized in this work is a real stable phase and not a metastable phase formed from a true spinel during quenching.

A possible stability field of Mn_2GeO_4 is shown in fig. 1, where the α–β and β–δ transformation curve at the temperature range from about 800 °C to 1200 °C is tentatively fitted by the linear relation $P(kb) = 13 + 0.031\ T(°C)$ and $P(kb) = 39 + 0.022\ T(°C)$ respectively.

In the course of high-pressure transformation of Mn_2GeO_4, the density increase was calculated to be

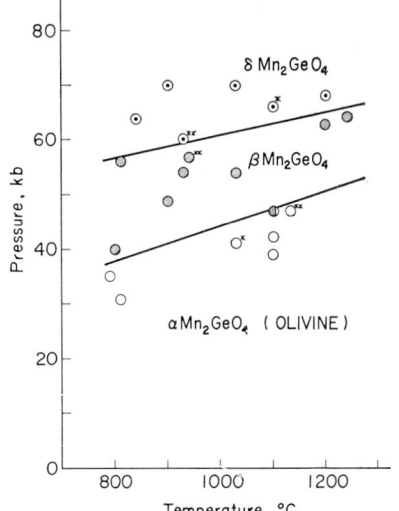

Fig. 1. Stability diagram for the high-pressure and high-temperature transformation of Mn_2GeO_4. * indicates the runs using β-Mn_2GeO_4 as the starting materials. ** indicates the runs using δ-Mn_2GeO_4 as the starting materials. Unless otherwise designated, the starting material is the olivine polymorph of Mn_2GeO_4 (α-Mn_2GeO_4).

7.1% for the α–β transformation and 10.7% for the β–δ transformation, using the lattice dimensions of the quenched α, β and δ phase. The total density increase from α phase to δ phase amounts to 18.4%.

The thermal expansion of the α, β and δ phase of the Mn_2GeO_4 sample was determined in a temperature range of room temperature to 460 °C and at ambient pressure by means of an X-ray diffractometer equipped with a heating device. Six to seven strong diffraction lines were used in determining the lattice dimensions of each phase at high temperatures. No indication of the phase transformation was observed in the β and δ phase of Mn_2GeO_4 after 60 minutes at 460 °C. The mean thermal expansivity, $\Delta V/(V_0 \times \Delta T)$, was determined to be $(36 \pm 1) \times 10^{-6}\,°C^{-1}$ for α-Mn_2GeO_4, $(30 \pm 1) \times 10^{-6}\,°C^{-1}$ for β-Mn_2GeO_4 and $(39 \pm 3) \times 10^{-6}\,°C^{-1}$ for δ-Mn_2GeO_4 respectively. It is interesting to note that the thermal expansivity of the δ-Mn_2GeO_4 with the highest density shows the largest value among the three polymorphs of Mn_2GeO_4.

3. High-pressure transformation in the system Mg_2SiO_4–Co_2SiO_4

High-pressure transformation mode in Co_2SiO_4 olivine is known to be so temperature-dependent that the occurrence of the β phase is restricted in a relatively high temperature range (AKIMOTO and SATO, 1968). This motivated us to investigate the transformation diagram of the Mg_2SiO_4–Co_2SiO_4 system over the wide range of temperature and pressure. A knowledge obtained from such a study may help towards the better understanding of the high-pressure transformation behavior of pure Mg_2SiO_4, which is still in controversy.

Microcrystalline olivines prepared at the interval of 12.5 mole percent in the compositional range from pure Co_2SiO_4 to $(Mg_{0.5}Co_{0.5})_2SiO_4$ were used as starting compound in the present high-pressure experiments. Olivine solid solution of $(Mg_{0.05}Co_{0.95})_2SiO_4$ was also used as starting material with a view to investigate the detailed transformation behavior of the extremely cobalt-rich side of the system. In the synthesis of the olivine solid solutions intimate powder mixtures of forsterite and Co_2SiO_4 olivine, which had been prepared in advance, were made to react in air at temperatures ranging from 1300 °C to 1400 °C for more than 8 hours. Formation of uniform olivine solid solutions was justified from well-defined X-ray diffraction peaks.

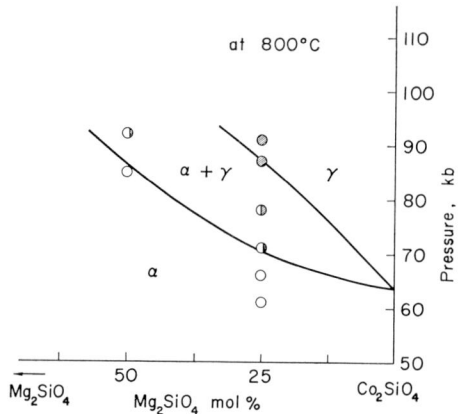

Fig. 2. Phase diagram for the high-pressure transformation of the system Mg_2SiO_4–Co_2SiO_4 at 800 °C.

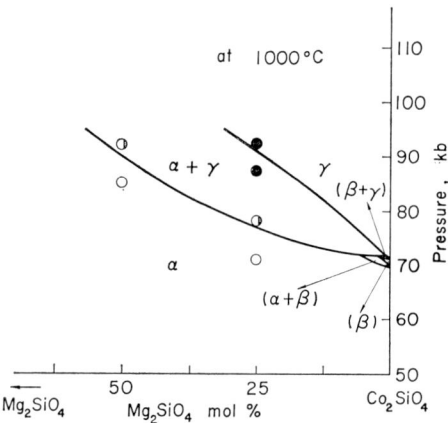

Fig. 3. Phase diagram for the high-pressure transformation of the system Mg_2SiO_4–Co_2SiO_4 at 1000 °C.

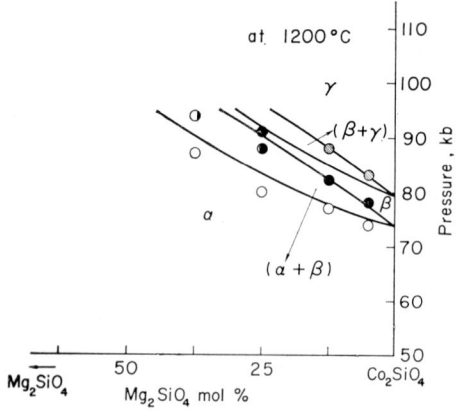

Fig. 4. Phase diagram for the high-pressure transformation of the system Mg_2SiO_4–Co_2SiO_4 at 1200 °C.

TABLE 2

Results of high-pressure and high-temperature experiments in the system Mg_2SiO_4–Co_2SiO_4

Run no.	Composition*, mole %	Pressure**, kb	Time, min	Phases present***	Lattice parameter of spinel s.s. (γ phase), Å
\multicolumn{6}{c}{Run temperature, 800 ± 10 °C}					

Run no.	Composition*, mole %	Pressure**, kb	Time, min	Phases present***	Lattice parameter of spinel s.s. (γ phase), Å
		Run temperature, 800 ± 10 °C			
2501	25M75C	61	60	α	
2502	idem	66	60	α	
2503	idem	71	60	$\alpha+\gamma$	8.134 ± 0.001
2505	idem	78	60	$\alpha+\gamma$	8.128 ± 0.001
2507	idem	87	60	γ	8.122 ± 0.001
2509	idem	91	60	γ	8.122 ± 0.001
5001	50M50C	85	60	α	
5003	idem	92	60	$\alpha+\gamma$	
		Run temperature, 1000 ± 10 °C			
2504	25M75C	71	40	α	
2506	idem	78	40	$\alpha+\gamma$	8.133 ± 0.002
2508	idem	87	40	$\gamma+\alpha$	8.124 ± 0.001
2510	idem	92	40	γ	8.122 ± 0.001
5002	50M50C	85	40	α	
5004	idem	92	40	$\alpha+\gamma$	
		Run temperature, 1200 ± 20 °C			
0503	5M95C	74	10	α	
0501	idem	78	10	β	
0502	idem	83	10	γ	8.136 ± 0.001
1251	12.5M87.5C	77	15	α	
1253	idem	82	15	$\beta+(\alpha)$	
1252	idem	88	15	γ	8.130 ± 0.001
2512	25M75C	80	20	α	
2514	idem	88	20	$\beta+(\alpha)$	
2513	idem	91	20	β	
3752	37.5M62.5C	87	20	α	
3751	idem	94	20	$\alpha+\beta$	

* M: Mg_2SiO_4, C: Co_2SiO_4;
** Precision of pressure control is about ±1 kb;
*** α: olivine solid solution, β: "modified" spinel solid solution, γ: true spinel solid solution.

High-pressure and high-temperature techniques used in determining the equilibrium diagram of the Mg_2SiO_4–Co_2SiO_4 system are essentially the same as described in the section of the high-pressure transformations in Mn_2GeO_4. Cemented tungsten carbide anvils with 9 mm edge length were used for the runs above 80 kb. Three discrete values of temperatures, 800°, 1000° and 1200 °C, were adopted as run temperatures.

Results of experimental runs are summarized in table 2. Diagrams showing the phases present and their stability field are given in figs. 2, 3 and 4 corresponding to 800°, 1000° and 1200 °C. Phases other than olivine and spinel solid solution were not observed in the run products at 800 °C (fig. 2). The equilibrium diagram at 1000 °C (fig. 3), in which a small stability field of the β phase was shown in a limited region close to Co_2SiO_4, was prepared in taking the transformation mode of pure Co_2SiO_4 into consideration (AKIMOTO and SATO, 1968). At 1200 °C, a remarkable expansion of the β phase region was established by the successful synthesis of the "modified" spinel solid solutions (fig. 4). In these figures, the olivine, spinel and "modified" spinel solvus curves were estimated from the relative content of each phase which was determined approximately from both the microscopic observation and the intensity ratio of the X-ray diffraction chart. Lattice parameters of the homogeneous spinel solid solutions synthesized in the present work were also given in table 2. Linear decrease of the lattice parameters with increasing amount of Mg_2SiO_4 was observed to $(Mg_{0.25}Co_{0.75})_2SiO_4$. This information also served for estimating the compositions of spinel solid solutions in the run products within the olivine-spinel transformation interval. Provisional estimate by linear extrapolation of the lattice parameter of the pure Mg_2SiO_4 spinel is 8.08 Å. This value agrees reasonably with the value estimated from the previous study of the Mg_2SiO_4–Fe_2SiO_4 system (RINGWOOD and MAJOR, 1966; AKIMOTO and FUJISAWA, 1968).

4. Discussion

The preceding experimental results demonstrate the high-pressure synthesis of the "modified" spinel phase in Mn_2GeO_4 and Mg_2SiO_4–Co_2SiO_4 solid solutions. The most notable feature of the high-pressure transformation in Mn_2GeO_4 is that there exists a large synthesis field of the β phase and that there is no indication of the occurrence of the γ phase. Reversibility of the phase transformations was also confirmed through the runs using the different polymorphs of Mn_2GeO_4 as starting material. These stability relations of Mn_2GeO_4 may strongly support the previous suggestion, obtained from the high-pressure and high-temperature study of Co_2SiO_4, that the "modified" spinel phase is a thermodynamically stable phase at the specified field of pressure and temperature (AKIMOTO and SATO, 1968). However, recent structure analysis of the β-Co_2SiO_4 and β-Mn_2GeO_4 revealed that from the crystalchemical viewpoint the crystal structure of the "modified" spinel seemed to be unusual compared with the structure of the true spinel on account of the existence of Si_2O_7 or Ge_2O_7 groups leaving oxygen atoms unbonded to any Si or Ge atoms (MORIMOTO et al., 1969 and 1970). Thus the newly-added evidences for the stability of the β phase may still be insufficient to exclude the possibility that the β phase is a metastable quench product. Definite conclusion on the stability of the "modified" spinel structure will be obtained from a high-pressure and high-temperature X-ray diffraction study.

Since the present data on the high-pressure transformation of $(Mg, Co)_2SiO_4$ olivine are only limited to the composition range of the cobalt-rich side, it is still difficult to estimate the high-pressure transformation mode of pure forsterite. However, the present work showed that the phase relationships of the Mg_2SiO_4–Co_2SiO_4 system were strongly influenced by temperature. It is highly probable that pure forsterite transforms directly from olivine to true spinel at low temperatures around 800 °C, but at high temperatures above about 1000 °C it transforms from olivine to true spinel through the intermediate phase of the "modified" spinel. At the present stage of knowledge it is plausible to assume that the β phase solid solutions synthesized by RINGWOOD and MAJOR (1966) for compositions more magnesian than $(Mg_{0.85}Fe_{0.15})_2SiO_4$ in the Mg_2SiO_4–Fe_2SiO_4 system may possess a real stability field at high-pressures and high-temperatures.

If the mantle olivine transformed stepwise to the γ phase through the β phase, the previous estimates (ANDERSON, 1967; AKIMOTO and FUJISAWA, 1968; FUJISAWA, 1968) of the constitution and width of the transition zone of the mantle should be reexamined. Since all these estimates were based on the simple olivine–true spinel equilibrium diagram in the Mg_2SiO_4–Fe_2SiO_4 system, the transition width from olivine to true spinel would have been estimated at a considerably lower value. From the recent study of the high-pressure polymorphs of Co_2SiO_4 (AKIMOTO and SATO, 1968) the density increase associated with the olivine–"modified" spinel transformation is estimated to be about 80% of that for the olivine–true spinel transformation. This suggests that the sharp increase of the seismic wave velocities would be expected even if the "modified" spinel is the first high-pressure polymorph of the mantle olivine. Direct determination of the elastic wave velocities of both the "modified" spinel and the true spinel by ultrasonic method may hold the key for the more quantitative analysis of the transition zone in the Earth's mantle. The more detailed investigation of the high-pressure transformation diagram of the magnesium-rich side of the Mg_2SiO_4–Fe_2SiO_4 system at temperatures higher than 1000 °C is also indispensable for such analysis.

Acknowledgements

The assistance and collaboration of Mr. Y. Ida and Miss Y. Sato in much of the experimental work described here are gratefully acknowledged. I am also indebted to Dr. Y. Syono for his helpful discussion.

References

AKIMOTO, S. and H. FUJISAWA (1966) Earth Planet. Sci. Letters **1**, 237.
AKIMOTO, S. and H. FUJISAWA (1968) J. Geophys. Res. **73**, 1467.
AKIMOTO, S. and Y. SATO (1968) Phys. Earth Planet. Interiors **1**, 498.
ANDERSON, D. L. (1967) Science **157**, 1165.
FUJISAWA, H. (1968) J. Geophys. Res. **73**, 3281.

JEFFERY, R. N., J. D. BARNETT, H. B. VANFLEET and H. T. HALL (1966) J. Appl. Phys. **37**, 3172.

MORIMOTO, N., S. AKIMOTO, K. KOTO and M. TOKONAMI (1969) Science **165**, 586.

MORIMOTO, N., S. AKIMOTO, K. KOTO and M. TOKONAMI (1970) Phys. Earth Planet. Interiors **3**, 161.

RINGWOOD, A. E. and A. MAJOR (1966) Earth Planet. Sci. Letters **1**, 241.

RINGWOOD, A. E. and M. SEABROOK (1963) J. Geophys. Res. **68**, 4601.

WADSLEY, A. D., A. F. REID and A. E. RINGWOOD (1968) Acta Cryst. **B24**, 740.

LATTICE ENERGIES, PHASE TRANSFORMATIONS AND VOLATILES IN THE MANTLE

W. S. FYFE

Manchester University

While an exact calculation of the conditions of a phase transition is yet impossible in most chemically complex materials, it may be possible to show that a given transition is at least likely at some position in the mantle. The necessary conditions for a transition are that the volume change be negative and that the free energy barrier does not exceed 36 kcal for each cm³ of negative volume change. It is this rather large energy change ($\Delta P.\Delta V$) involved over the range of mantle pressures that allows any reasonable prediction. With simple substances with relatively small lattice energies predictions can be made with some degree of certainty; e.g. AB, AB$_2$ compounds such as MgO, SiO$_2$ etc. It is shown that none of the common types of hydrates minerals, amphiboles and micas, can carry water into the deep mantle. If water is trapped, it is more likely to be found in phases such as hydroxy silicates where (OH)$_n$ species replace Si–O groups and are partially stabilized by dilution in solid solutions.

1. Introduction

One of the ultimate and most difficult problems of Earth Science is to deduce the chemical composition and structural states of matter in the inaccessible portions of the earth. It is almost certain that our present views on the composition of the earth (i.e. the mantle and core), so largely influenced by the composition of an unbalanced meteorite sample, will require modification in future years. There are still innumerable problems in rationalizing views on the change in composition with depth with changes to be expected from thermodynamic equilibrium in a gravitational field. This latter problem has received scant attention (e.g. BREWER, 1951) and some calculations on the MgO–FeO–SiO$_2$ system (KERN and WEISBROD, 1967) seem hardly in accord with present proposals concerning the composition of the mantle. Perhaps the earth is still a long way from being in gravitational equilibrium or perhaps our knowledge of phases and their compressibilities at high pressures still leaves something to be desired.

If a given chemical composition is to be in stable equilibrium at a given position in the earths gravitational field it is necessary that the phases formed be of appropriate density. While such perfect equilibrium states may not be achieved, we are continually searching for phase changes and phase chemistry which are feasible and any approach in this test of feasibility can guide experimental approaches. The problem is difficult and I would agree with KAMB's (1968) remark: "In proposing structural explanations for physical properties, one must guard against facile a posteriori rationalizations that are arbitrarily contrived to fit particular facts, and that have no general validity or significance".

If we consider the mantle with its rather small thermal gradient it is clear that in a general way pressure is a more significant thermodynamic variable than temperature. For most phase changes (A→B), entropy and volume changes are sympathetic in sign and to some extent in magnitude. Thus when plotted on P–T coordinates, slopes of phase changes tend to be similar. On average, for a ΔV of -1 cm³/mol, ΔS is near -0.5 cal/mol °C. Given the mantle pressure range of 1.5×10^6 b and temperature range of about 5000 °C, over this range, changes in ΔG of reaction for the above ΔV and ΔS averages will be

$$\left(\frac{\partial \Delta G}{\partial P}\right)_T = \Delta V \approx -36 \text{ kcal},$$

$$\left(\frac{\partial \Delta G}{\partial T}\right)_P = -\Delta S \approx +2.5 \text{ kcal}.$$

The $P\Delta V$ term is comparable to chemical bonding energies and indicates that structures most unlikely at the surface may be easily achieved in the deep mantle. For example the ΔV of the graphite–diamond transition is about -2 cm^3/mol and if diamond was to be impossible in the mantle, the bond energy of graphite would need to be 80 kcal greater than diamond. Even if diamond had never been seen, data on the energies and lengths of carbon bonds as seen in organic molecules would certainly lead us to expect diamond at quite moderate pressures. In fact quantum mechanical calculations would be adequate in this case.

The problem of predicting the possibility of a phase change over the mantle P–T range thus resolves itself into two questions (if we assume that T and S are rather unimportant to a first approximation). First, can the volume change be guessed? Second, can the bond energies or lattice energies be guessed with any degree of significance? It is normally the latter question that involves difficulty; but less so if we do not require an accurate transition pressure. For a long time to come experimental methods will be required, but at least some guidance for experiment seems possible. The limited success of Fersman's approach to bond energies (see MASON, 1966, p. 85) from which it follows that

$$\sum \Delta H_f \text{ oxides} \approx \sum \Delta H_f \text{ compound}$$

emphasizes the difficulty of finding energy differences between structural states as these are often small differences between very large values (e.g. compare the lattice energies of quartz and cristobalite).

2. Changes in coordination

Phase changes induced by high pressures most frequently involve changes in coordination numbers. There are important exceptions (e.g. the olivine–spinel transition discussed by KAMB, 1968). Most of the discussion that follows is involved with examples where changes in coordination occur.

A typical simple example is illustrated by KCl. This is known to change from the NaCl structure (6-coordination) to the CsCl structure (8-coordination) at moderate pressures. The ratio of the radii of K$^+$ and Cl$^-$ (0.734) is almost exactly on the theoretical crossover from 6- to 8-coordination. Without any change in bond lengths ΔV would be about -8 cm^3/mol. For the transition to occur in the mantle pressure range, the ΔH of transition must not exceed 280 kcal/mol, and as this figure is greater than the lattice energy of the 6-coordinated form (163 kcal) the transition must be considered feasible.

If we consider the simple Born lattice energy equation:

$$U = \frac{NAe^2 z_1 z_2}{R}\left(1 - \frac{1}{n}\right),$$

and assume that both R and n are similar, then the lattice energies of the two forms are given by

$$\frac{U_1}{U_2} = \frac{A_1}{A_2},$$

where U is the lattice energy and A the Madelung constant for the structure concerned. In this case, on account of the small value of the U's, a reasonable prediction can be made of the energy barrier (see FYFE et al., 1958).

Another such simple case is shown by the rather high pressure transition of ZnO from the four coordinated sphalerite structure to the halite structure. Again, assuming no change in bond lengths, ΔV is about -2.5 cm^3/mol. In this case a rather good lattice energy for the NaCl structure can be obtained from analogous transition metal oxides allowing for crystal-field terms (FYFE, 1963). The transition is feasible and the transition pressure can be estimated with significance.

An interesting case is provided by magnesium oxide. This oxide is certainly a possible mantle phase, particularly where silicates are replaced by oxides including stishovite. It has been suggested that MgO may change from the halite to the CsCl structure. Simple spherical ions such as Mg^{2+} and O^{2-} are rather incompressible. In this case, unless large repulsive oxygen overlap is to be introduced, the Mg–O bond length in periclase must lengthen in an 8 coordinate structure. If the volume of the latter is calculated assuming that O–O penetration does not occur, it will be larger than in the periclase structure. If the volume is made smaller, ΔV will still be small and a large barrier introduced. Thus, from present data, MgO in the CsCl structure does not appear easily accessible at least not till such pressures are achieved that the ionic model ceases to be relevant.

Magnesium does occur in 8-coordination in some silicate structures, for example pyrope. But in this case, oxygen atoms linked to magnesium are also linked to either aluminium or silicon with quite covalent bonds. These linkages all reduce the ionic charge on oxygen and can remove the spherical symmetry of the charge cloud. Such influences are unlikely with the simple oxide.

That the stishovite transition in SiO_2 is feasible is also easily shown but in this case, on account of the large lattice energies involved, calculation of the barrier with significant accuracy is difficult. First, SiO_2 in the rutile structure, even with the same cell volume as TiO_2, would have a smaller volume than quartz, by about 4 cm^3/mol. The lattice energies of rutile and quartz differ by about 240 kcal (experimental values) so that given the true volume of stishovite this barrier can be overcome and the real barrier must be less. A reasonable guess of both cell volume and barrier can be derived by consideration of bond distances and energies involved in the formation of TiF_6^{2-} and SiF_6^{2-} in aqueous solution (FYFE, 1963). We may note, that once pressures are high enough that SiO_2 in the rutile structure is favoured, we may then expect a large array of silicates with 6-coordinated silicon to also be in the realms of possibility; the heats of formation of most compounds from the oxides in the appropriate structural state being rather small (cf. $FeO + SiO_2 \rightarrow FeSiO_3$ and $FeO + TiO_2 \rightarrow FeTiO_3$). Thus phases such as $KAlSi_3O_8$ (orthoclase) in a dense structure should be possible in the stishovite stability field as suggested by KUME et al. (1966).

The high-spin–low-spin transition in FeO and hence in all iron oxygen compounds has been discussed by FYFE (1960) and more recently by STRENS (1969). This transition, which seems feasible in the mantle, has considerable bearing on iron-magnesium geochemistry in the deep mantle and on the optical properties of this region. While it must lead to denser iron compounds (the diamagnetic ferrous ion being small) it may also lead to iron compounds with rather different compressibilities than those with which we are familiar. The transition in octahedrally coordinated iron oxygen compounds has not been observed. The calculations of the transition region in the mantle both by Fyfe and Strens seem inadequate as neither have taken into account all the factors involved, but at least Strens calculations show it to be feasible (more detailed work is in progress by Fyfe and McLellan).

3. Phases containing volatiles

While phase changes in rather simple compounds can often be rationalized and predicted, as soon as we come to consider multi-atom compounds, problems increase. In the lower crust and upper mantle, most water is contained in rather complex structures such as micas and amphiboles while in some regions, serpentine minerals may be significant. With our present knowledge, carbon dioxide is mainly found in rather simple carbonates. RUBEY (1951) in his classic essay on sea water discussed the evolution of the hydrosphere, and showed convincingly that water and other volatiles have been slowly added to the surface from the mantle. It is of some importance to understand how volatiles may be trapped in the deep mantle for this has bearing on many geochemical problems including evolution of the oceans, atmosphere and life.

It is a striking fact, that most of the complex hydrate minerals with which we are familiar cannot persist far into the mantle. First we may consider the serpentine minerals. BOWEN and TUTTLE (1949) showed that at low water pressure (1–2 kb) serpentine breaks down to less hydrous phases at temperatures near 500 °C. Let us consider the simplified reaction

$$\underset{\text{serpentine}}{Mg_3Si_2O_5(OH)_4} \rightarrow \underset{\text{enstatite}}{MgSiO_3} + \underset{\text{forsterite}}{Mg_2SiO_4} + 2H_2O.$$

The ΔV_0 of this reaction is about 5 cm^3/mol. As breakdown occurs in the region 5–600 °C, we would anticipate maximum thermal stability where ΔV tends to zero. From data in SHARP's (1962) tables we would expect this inflection to occur near 30 kb. KITAHARA et al. (1966) have observed such inflection in the vapour pressure curve. The general form of the phase diagram is shown in fig. 1.

The form of vapour pressure curve of fig. 1 is not normally observed in low pressure studies, but once the mantle pressure range is considered, most curves will follow such a pattern. Two types of mineral are of particular interest in deep water retention; the micas (in particular phlogopite) and amphiboles (in particular hornblendes). Not much data is available for hornblendes as yet, but tremolite provides a model.

For the reaction

$$\underset{\text{tremolite}}{Ca_2Mg_5Si_8O_{22}(OH)_2} \rightarrow 2\,\underset{\text{diopside}}{CaMgSi_2O_6} + 3\,\underset{\text{enstatite}}{MgSiO_3} +$$
$$+ \underset{\text{quartz}}{SiO_2} + H_2O$$
$$\Delta G_0 = 18 \text{ kcal} \quad \Delta V_0 = -5.5 \text{ cm}^3.$$

From the available data (CLARK, 1966) it would be expected that tremolite would have a maximum thermal stability at about 1000 °C, and after about 10 kb the

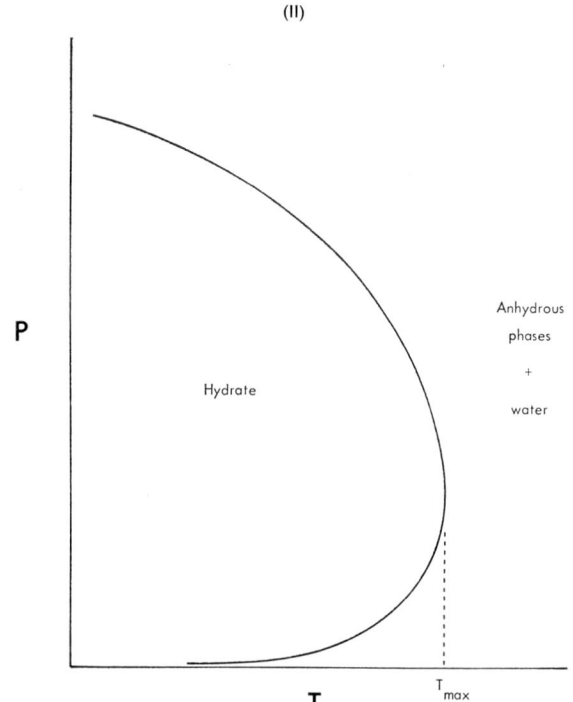

Fig. 1. General form of the stability field of most hydrates. T_{max} occurs near 600 °C for serpentine, 1000 °C for amphiboles, and rather higher for phlogopite.

vapour pressure curve will become negative and reach about 100 kb at room temperature.

We would anticipate a similar curve for the transition

amphibolite → eclogite + vapour.

Previously, ESSENE and FYFE (1967) suggested that this curve could have positive slope on account of the large negative ΔV of the reaction

amphibolite → eclogite + water,

and the small amount of water. Essene and Fyfe suggested that ΔS could also be negative. It seems however that this is inconsistent with all data on the reaction and with the anomalously low entropy of amphiboles as compared with pyroxenes. In fact, the limits of hornblende may be similar to that indicated for tremolite. It seems that YODER and TILLEY's (1962) statement that eclogite cannot exist stably in the crust with excess water is probably correct (this point will be discussed separately by the writer in a paper in press).

Of all micas studied, phlogopite has the greatest thermal stability. The reaction

$$2 \, KMg_3(AlSi_3O_{10})(OH)_2 \underset{\text{phlogopite}}{} \to 6 \, MgSiO_3 \underset{\text{enstatite}}{} + K AlSi_2O_6 \underset{\text{leucite}}{} +$$
$$+ \, KAlSiO_4 \underset{\text{kalsilite}}{} + 2 \, H_2O$$

occurs in quartz free systems and the vapour pressure reaches 5 kb only over 1000 °C. The ΔV_0 of the reaction is 73.5 cm^3 on account of the low densities of leucite and kalsilite. There is no way of changing the slope of the vapour pressure curve unless it is changed by melting of phlogopite or unless phase changes occur in the K-Al silicates. In this connection the possible phase change of $KAlSi_3O_8$ to a dense phase suggested by KUME et al. (1966) is of considerable interest. If this occurs in the region of stishovite stability we may then have the reaction

$$KMg_3(AlSi_3O_{10})(OH)_2 \to \text{dense } KAlSi_3O_8 + \\ + \, 3 \, MgO + H_2O.$$

If the molar volume of dense $KAlSi_3O_8$ is placed at the same value as the germanate $KAlGe_3O_8$ (and it must be less), the reaction now has a large negative ΔV and the reaction must reach a thermal maximum. Other changes in leucite and kalsilite may occur and cause reversal even sooner.

It is of considerable interest to note that hydrates which can show surprising thermal stability at high pressures are simple ones. Thus

$$Mg(OH)_2 \underset{\text{brucite}}{} \to MgO \underset{\text{periclase}}{} + H_2O, \quad \Delta V_0 = +4.6 \, cm^3,$$
$$2 \, AlOOH \underset{\text{diaspore}}{} \to Al_2O_3 \underset{\text{corundum}}{} + H_2O, \quad \Delta V_0 = +8.05 \, cm^3.$$

To reverse the sign of these reactions (it has already been stated that it is unlikely that a major phase change can occur with MgO) water will have to be compressed to very large densities, about two in the case of the diaspore reaction. For example, at 1000 °C the entropy of this reaction and the ΔV will probably still be positive at 100 kb (SHARP, 1962). KENNEDY (1959) has studied diaspore stability to 50 kb and shows a linear relation of $\log P$ with T. But even in this case the geothermal gradient is probably too steep to allow persistence to great depths.

In summary, it seems that few known hydrated silicates are likely to remain stable below a few hundred kilometers in the mantle of the earth. Phlogopite stability will exceed amphibole stability unless unknown phase changes in K-Al silicates occur at rather low pressures. This conclusion poses the question: are terrestrial processes concerned with water restricted only to the upper mantle or is water still present at much deeper levels?

4. Hydro-silicates and related compounds

To increase the stability of hydrates at high pressures, much denser hydroxy compounds are required, or phases must form which will readily mix with common phases in the mantle leading to lower chemical potentials. In this connection phases which may be of interest are hydroxy silicates such as the hydrogarnets. In these phases $(OH)^{4-}$ replaces SiO^{4-}, the best known case being the hydrogarnets. YODER (1950) showed that these can have considerable thermal stability.

Volume data for the grossularite–hydrogrossular series (YODER, 1950) indicate that along the series

$$Ca_3Al_2(SiO_4)_3 \rightarrow Ca_3Al_2(OH)_{12}$$

replacement of one SiO_4 by $(OH)_4$ leads to a ΔV of 8 cm^3. The structure of the $(OH)_4$ group has been studied by neutron diffraction and indicates that the structure contains unsymmetrical hydrogen bonds and that the O–O distances of the $(OH)_4$ group are much longer (3.17–3.39 Å) compared to the O–O distances in the normal SiO_4 group (2.7 Å). Nevertheless, even with this large O_4 group some processes of interest appear almost possible.

Thus a reaction such as

$$\underset{\text{brucite}}{2\ Mg(OH)_2} \rightarrow \underset{\text{olivine}}{Mg_2(OH)_4}$$

would have a ΔV of only $+2$ cm^3 (we should perhaps note that brucite may well change to a rutile structure like MgF_2 at high pressures).

It seems to this author that the $(OH)_4$ grouping might show abnormal compressibility. If the hydrogen bonds were to become symmetrical, considerable shortening of O–O distances would follow. Such shortening is observed in the F–H–F distances in the bifluoride ion HF_2^- as compared with polymerized HF. A symmetrical O–H–O bridge might have an O–O distance of as little as 2.5 Å. In this case a hydrogarnet might well be denser than a normal garnet. Overlap calculations (FYFE, 1953) show that changes from unsymmetrical to symmetrical hydrogen bonds are probable as bond distances shorten. The state of our knowledge on the structures of hydroxy compounds at high pressures is so meagre that one hesitates to predict, but it appears that more work on hydroxysilicates is needed. Unless denser compounds can be produced, it seems unlikely that much water will occur in deep mantle phases. The situation with carbonates is equally tantalizing. Are there spinels such as Mg_2CO_4 based on CO_4 groups? Reactions such as

$$MgCO_3 + MgO \rightarrow Mg_2CO_4 \text{ (spinel)}$$

would certainly have negative volume changes.

References

BOWEN, N. L. and O. F. TUTTLE (1949) Bull. Geol. Soc. Am. **60**, 439.
BREWER, L. H. (1951) J. Geol. **59**, 490.
CLARK, S. P. (1966) *Handbook of physical constants*, Geol. Soc. Am. Mem. **97**.
ESSENE, E. J. and W. S. FYFE (1967) Contrib. Mineral. Petrol. **15**, 1.
FYFE, W. S. (1953) J. Chem. Phys. **21**, 2.
FYFE, W. S. (1960) Geochim. Cosmochim. Acta **19**, 141.
FYFE, W. S., F. J. TURNER and J. VERHOOGEN (1958) *Metamorphic reactions and metamorphic facies*, Geol. Soc. Am. Mem. **73**.
KAMB, B. (1968) Am. Mineralogist **53**, 1439.
KENNEDY, G. C. (1959) Am. J. Sci. **257**, 563.
KERN, R. and A. WEISBROD (1967) *Thermodynamics for Geologists* (Freeman, Cooper, and Co., San Francisco).
KITAHARA, S., S. TAKEMOUCHI and G. C. KENNEDY (1966) Am. J. Sci. **264**, 233.
KUME, S., T. MATSUMOTO and M. KOIZUMI (1966) J. Geophys. Res. **71**, 4999.
MASON, B. (1966) *Principles of geochemistry* (Wiley, New York).
RUBEY, W. W. (1951) Bull. Geol. Soc. Am. **62**, 1111.
STRENS, R. G. J. (1967) Proc. Newcastle Conf. 1967 Appl. Mod. Phys. Earth Planet. Interiors, in press.
YODER, H. S. (1950) J. Geol. **58**, 221.
YODER, H. S. and C. E. TILLEY (1962) J. Petrol. **3**, 342.

THE PHASE BEHAVIOR OF SIMPLE COMPOUNDS

JOHN C. JAMIESON

Department of the Geophysical Sciences, University of Chicago, Chicago, Illinois 60637, U.S.A.

The phase behavior of simple compounds is not necessarily simple. Several examples are given in which either common approximations fail, or "expected" phases do not appear. These include TiO_2, ZnO and Ag-halides. In addition, an example is given of development of an inhomogeneous stress system in a situation where homogeneity is usually assumed.

1. Introduction

It is of geophysical interest to study the polymorphic behavior of materials, not only to ascertain density changes for equations of state but for more subtle reasons, i.e. to determine the actual solid state physical properties of potential earth-forming materials. The detailed crystal structure determines such physical parameters as the electrical conductivity, magnetic properties, anisotropic elastic constants and the crystal field effect on transition elements with its concommittant effect on radiative transfer. Chemical effects which may be ascribed to detailed structures include solid solubility ranges and reaction kinetics. To point out that simple compounds do not necessarily have simple phase behavior a few specific examples follow:

2. Titanium dioxide

The thermodynamically rigorous condition for one material of fixed composition (compound or element) to exist in two phases in equilibrium at a constant hydrostatic pressure P and temperature T is

$$-\Delta G(P_0, T) = \int_{P_0}^{P} \Delta V(P, T)\, dP, \qquad (1)$$

in which ΔG is the difference of their Gibb's free energies and ΔV the same difference of their molal volumes. If $\Delta G(P_0, T)$ is measured at $P_0 = 1$ bar and as a function of temperature say by calorimetric means, and the volume behavior of both phases is known over the region of (P, T) space of interest, then the integral may be solved in terms of its end-point P and a phase diagram developed. In order to make rough estimates of phase behavior when volume behavior are unknown, frequently ΔV is set equal to a constant, and then eq. (1) becomes simply

$$-\Delta G(P_0, T) = \Delta V(P_0, T)(P - P_0), \qquad (2)$$

from which an equilibrium pressure can be obtained. One's intuitive feeling is that errors made in this way are percentage errors, possibly large, but not orders of magnitude. Over large pressure ranges such as 100 kb plus there is still very little compressibility data on *both* phases of any substance exhibiting a polymorphic transition. JAMIESON and OLINGER (1968) have presented such data for TiO_2 in its rutile and α-PbO_2 forms to almost 200 kb using X-ray diffraction. There they found ΔV to be highly non-linear with a minimum at *circa* 140 kb. In terms of an experimental error analysis, the equilibrium pressure for the transition rutile to α-PbO_2 form could be as low as 60 kb or as high as no upper bound at all, i.e. it wouldn't occur at the experimental temperatures (room)! In the absence of actual compression data for all phases under consideration one should be suspicious of the validity of other calculations of this type where a small ΔV and very high pressures are involved. For a large ΔV and small energy differences the approximation may suffice.

3. Zinc oxide

The ambient form of ZnO is in the hexagonal wurtzite structure. BATES *et al.* (1962) discovered that a NaCl type modification could be quenched from pressures above 100 kb and temperatures above 100 °C. These conversions could only be obtained in the presence of ammonium chloride. This study has the features of many phase studies in geophysical materials. A phase line is developed by exposing the substance to given (P, T) conditions, and a post-mortem analysis is made to see if a reaction has occured. Almost invariably there is a (lower) cut-off temperature below which no apparent reaction has occured, and the very plausible explanation is given that the kinetics of the reaction are simply too slow for any products to appear in reasonable experimental times. In a survey of divalent oxides JAMIESON et al. (1964) have found this kinetic explanation not to hold for ZnO. We have found the wurtzite–NaCl transition to be *reversible* at room temperature using *in situ* X-ray diffraction. Several runs were made both with and without internal NaCl standards. The runs containing standards gave the pressure of transition in the neighborhood of 90 kb. Quite clear patterns could be indexed as the NaCl structure with all expected lines appearing with $h^2+k^2+l^2$ up to 24. The 90 kb transition is in excellent agreement with the extrapolation to room temperature of the data of BATES et al. (1962). Actually EDWARDS et al. (1959) had earlier optically observed a transition in ZnO at 100000 at which they ascribed to the appearance of the zincblende form. We believe this is actually a transition to the NaCl type as their optical study could not be diagnostic as to structure type.

In the absence of further X-ray studies over a region of (P, T) space it is only speculative to attempt to explain why a structural transition should be rapid enough to be reversible at room temperature, yet sluggish enough to be quenched from higher temperatures but we do suggest two possibilities. One is that the NH_4Cl used as flux may enter into the reaction to stabilize the NaCl structure, or that the elevated temperatures may lead to grain growth of the NaCl phase, and that such larger grains are less likely to undergo a reverse transition to the zincite type. In any event the primary moral to be drawn is that the lack of an observed transition in quench studies is no proof that the material is not in a different structure at high pressures.

An even more complicated situation occurs in rutile types MnF_2 and ZnF_2 in which a new phase (α) appears in quench studies. This phase is not the phase actually present at elevated pressures but forms on release of pressure. This has been discussed extensively by DANDEKAR and JAMIESON (1969) elsewhere. For MnF_2 at room temperature with increasing pressure we find: rutile → tetragonal → orthorhombic; while on pressure relief: orthorhombic → tetragonal → (α). The overall volume change approaches 25%. Needless to say this situation could not have been resolved without *in situ* X-ray diffraction initially performed by VERESHCHAGIN et al. (1966) and KABALKINA et al. (1968).

4. Silver halides

Sodium, potassium and rubidium halides appearing in the NaCl structure at ambient conditions are known to transform to the CsCl structure at elevated pressures. This is a very reasonable occurance since the CsCl type has a higher cation-anion coordination. A frequent suggestion for the constitution of the middle and lower mantle is that it is a mixture of oxides. Since the majority of these oxides exist in NaCl structures at one bar, the question arises as to whether or not they will be in a CsCl structure at depth in the earth. We have no direct evidence on this point but we would like to point out that there are substances in which the CsCl structure does not appear as a high pressure form, at least in the range of pressure and temperature studied.

SCHOCK and JAMIESON (1969) have recently studied the high pressure forms of AgCl, AgBr and AgI. The first two transform (reversibly) from their normal NaCl structure to the cinnabar type near 90 kb. This is a distortion of the NaCl type which permits covalent Ag bonding with each Ag having two linear nearest neighbors. The latter, AgI, adapts the NaCl structure above 4 kb, then at 90 kb it transforms to neither the cinnabar structure nor the CsCl. A tentative tetragonal indexing has been offered for this form. Electrical conductivity measurements suggest that it too is more covalent than the NaCl type. The volumes of these high pressure forms are the order of 10% less than their NaCl counterparts. How these would compare with volumes of hypothetical CsCl structures is unknown. But it doesn't seem impossible that divalent oxides at high pressures (and temperatures) might adopt the above or other distortions of the NaCl (or even the CsCl) structure type. This could make interpretation of the earth's structures and coordination numbers quite inaccurate.

5. Pressure intensification

Currently all static ultra high pressure phase studies are made using mechanical mixtures of solids. Either the sample is embedded in a matrix such as pyrophyllite, etc. or admixed with some internal standard to determine pressures. In the final analysis, when a material transforms it is a mechanical mixture of its own two phases. It has been earlier suggested by BOBROWSKY (1963) and later by CORLL (1967) and CORLL and WARREN (1965) that "pressures" would not be uniform in such mixtures. This has been found the case in mixtures of Nb and NaCl by JAMIESON and OLINGER (1968) using *in situ* X-ray diffraction. In this study the compression of Nb at a constant compression of NaCl was found to be a function of the (volume) amount of Nb present. This would seem to have petrological significance when a pressure is assigned to a given rock unit using "indicator" minerals as these minerals may have formed at a pressure different from other areas of the same rock.

Acknowledgements

This research has been supported by NSF Grant NSF GA 1270 and ARPA grant SD-89-Research and by the Petroleum Research Fund of the American Chemical Society. Grateful acknowledgment is made to the donors of that fund.

References

BATES, C. H., W. B. WHITE and R. ROY (1962) Science **137**, 993.

BOBROWSKY, A. (1963) Pressure alterations through use of solid containers or jackets in high pressure environment, in: A. A. Giardini and E. C. Lloyd, eds., *High Pressure Measurement* (Butterworth, Washington D.C.).

CORLL, J. A. (1967) J. Appl. Phys. **38**, 2708.

CORLL, J. A. and W. E. WARREN (1965) J. Appl. Phys. **35**, 3655.

DANDEKAR, D. P. and J. C. JAMIESON (1969) Some high pressure phases of RX_2 fluorides, in: D. B. McWhan, ed., *High pressure symposium volume*, Am. Cryst. Assoc. 1969 meeting.

EDWARDS, A. L., T. E. SLYKHOUSE and H. G. DRICKAMER (1959) Phys. Chem. Solids **11**, 140.

JAMIESON, J. C., Q. C. JOHNSON and D. K. SMITH (1964), Unpublished research.

JAMIESON, J. C. and B. OLINGER (1968a) Science **161**, 893.

JAMIESON, J. C. and B. OLINGER (1968b) Pressure inhomogeneity, a possible source of error in using internal standards for pressure gages, in: *Symposium on the accurate characterization of the high pressure environment*, Oct. 14–18, 1968, Natl. Bur. Std. (U.S.).

KABALKINA, S. S., L. P. VERESHCHAGIN and L. M. LITYAGINA (1968) Soviet Phys. "Doklady" (English Transl.) **12**, 946.

SCHOCK, R. N. and J. C. JAMIESON (1969) Phys. Chem. Solids, to be published.

VERESHCHAGIN, L. P., S. S. KABALKINA and A. A. KOTILEVETS (1966) Soviet Phys. JETP (English Transl.) **22**, 1181.

CRYSTAL CHEMISTRY OF HIGH PRESSURE POLYMORPHS OF ABO_3, AB_2O_4 AND AB_4O_8 COMPOUNDS, AND THEIR POSSIBLE IDENTIFICATIONS WITH THE PHASES OCCURRING IN THE MANTLE

A. F. REID

Division of Mineral Chemistry, CSIRO, Australia

The volumes of specific coordination polyhedra bear a direct proportionality to the unit cell volume of the structure which contains them. Averaged metal-oxygen bond lengths, or sums of ionic radii, or other related structural parameters, can be plotted versus the molar volumes of an isomorphous series of compounds to give a smooth curve. Such curves can then be used to extrapolate to the molar volumes of silicates or other minerals in any of a number of postulated high pressure forms.

This concept has been used to unify a number of current observations on high pressure forms of ABO_3, AB_2O_4 and AB_4O_8 compounds, including the high density hollandite polymorphs of the feldspars of K, Sr and Ba, and to predict the molar volumes of the as yet unobserved high pressure polymorphs of a number of silicate minerals believed to be of importance in the mantle.

STABILITY OF MANTLE MINERALS FROM LATTICE CALCULATIONS AND SHOCK WAVE DATA*

EDWARD S. GAFFNEY and THOMAS J. AHRENS

Seismological Laboratory, California Institute of Technology
Passadena, California, U.S.A.

Shock wave and static high pressure data for mantle minerals have indicated that at high pressures a series of denser polymorphs form whose crystal structures can at present only be inferred from calculated densities and crystal chemical arguments. In order to determine the admissibility of some of these proposed structures theoretical Madelung lattice energies are calculated for several oxides (FeO, Al_2O_3, Cr_2O_3, Fe_2O_3, SiO_2, TiO_2) spinels (Al_2MgO_4, Mg_2SiO_4, Fe_2SiO_4, Ni_2SiO_4, $FeCr_2O_4$, Fe_2TiO_4, Fe_3O_4) and perovskites ($CaTiO_3$, $SrTiO_3$, $MgSiO_3$, $Fe_2^{3+}O_3$, $Fe^{2+}Fe^{4+}O_3$). Comparison of calculated enthalpies of formation with measured values yield approximate values for the effects of covalency on enthalpies of formation for $Al-O_6$, $Ti-O_6$, $Si-O_4$, $Si-O_6$, $Fe^{3+}-O_6$, $Cr^{3+}-O_6$, $Fe^{3+}-O_4$ and $Fe^{2+}-O_4$. This effect is seen to be very similar for the same ion pair in the same coordination but in different compounds. The calculations indicate that enstatite ($MgSiO_3$) can not enter a perovskite with a density greater than about 3.9 g/cm³ and that the high pressure phase of Fe_2O_3 can be a perovskite only if the Fe^{3+} disproportionates into Fe^{2+} and Fe^{4+} and the 3d electrons in the latter are spin paired.

1. Introduction

Recent seismological studies (JOHNSON (1968), ARCHAMBEAU et al. (1969)) have shown that the marked increase in elastic velocity, long known to occur in the C-region of the earth at a depth between 200 and 900 km arises from at least two distinct zones within this range which are approximately 50 km thick. The seismological data indicate the velocity increases sharply at about 375 km in the shallow zone, and again at about 700 km in the deeper zone. The results of both static high pressure studies (RINGWOOD (1970), AKIMOTO amd FUJISAWA (1968), SCLAR (1964)) and thermochemical studies (AHRENS and SYONO (1967) and ANDERSON (1967)) have shown that the probable mantle minerals, olivine and pyroxene, transform near 100 kb to denser structures according to the reactions:

$(Mg, Fe)_2SiO_4$ (olivine) → $(Mg, Fe)_2SiO_4$ (spinel), (I)

$2(Mg, Fe)SiO_3$ (pyroxene) → $(Mg, Fe)_2SiO_4$ (spinel) + SiO_2 (stishovite). (II)

Because both reactions (I) and (II) involve large increases in elastic moduli and hence elastic velocities,

ANDERSON (1967) has suggested that these take place in the upper transition zone. In the case of reaction (I), the bulk modulus increases from about 1.2 Mb to about 2.1 Mb. Although there is a large increase in mean bulk modulus and density ($\approx 10\%$) in both reactions (I) and (II), the coordination of Mg^{++} or Fe^{++} and of Si^{+4} is octahedral and tetrahedral, respectively, in both olivine and spinel. In reaction (II) one half of the silicon ions go from tetrahedral to octahedral coordination (in stishovite) with oxygen ions. Reactions (I) and (II) thus represent relatively large increases in density without large accompanying changes in ion coordination.

In contrast to the upper transition zone, the lower 750 km or "post-spinel" transition zone (ANDERSON, 1967), presumably involves a transition of all the Si^{+4} to octahedral coordination and perhaps of the divalent metals to 8 or higher coordination with oxygen. The available shock-wave Hugoniot data (McQUEEN et al. (1967), also quoted in BIRCH (1966)) for some of the likely mantle minerals and some of their structural analogs display strong evidence of transition to the so-called post-spinel phases. These shock-wave data have been analyzed by McQUEEN et al. (1967), WANG (1968), ANDERSON and KANAMORI (1968), and AHRENS et al. (1969) in order to obtain the density and equation of state parameters of the shock-induced high pressure

---

* Contribution No. 1625, Division of Geological Sciences, California Institute of Technology.

phases. Ahrens *et al.* have suggested that, at least in the shock-wave case, the so-called "post-spinel" transformations might correspond to the following reactions

(Mg, Fe)$_2$SiO$_4$ (olivine) → (Mg, Fe)$_2$SiO$_4$ (strontium plumbate or potassium nickel fluoride structure), (IIIa)

Al$_2$MgO$_4$ (spinel) → Al$_2$MgO$_4$ (calcium ferrite structure or calcium titanite structure), (IIIb)

(Mg, Fe)SiO$_3$ (pyroxene) → (Mg, Fe)SiO$_3$ (ilmenite or prerovskite strucutre), (IV)

Fe$_2$O$_3$ (hematite) → Fe$_2$O$_3$ (perovskite or β-rare earth structure). (V)

In the analysis of the shock wave data for the high pressure phases, the zero-pressure bulk modulus and density were calculated using ANDERSON's (1967) seismic equation of state. The probable structures of the high pressure phases were inferred on the basis of the calculated densities, crystal-chemical arguments, and the results of static high pressure experiments on analog compounds. In all of the proposed high-pressure phases, silicon is in sixfold coordination and the other cations are in six, eight, or twelvefold coordination with oxygen.

In order to determine the admissibility of one or more of the proposed high-pressure structures, and to evaluate the heats of formation and types of bonding involved in these polymorphs, we have calculated theoretical lattice energies for some of the pertinent simple oxides, spinels, and perovskites. These polymorphs are presumably present in the mantle. A modified Born-type calculation for ionic bonding in the crystals is used with the available data for bulk moduli and density (lattice parameters) in calculating theoretical enthalpies of formation.

2. Theory

2.1. Born-Haber cycle

The Born-Haber cycle may be used to calculate the heat of formation of an essentially ionic crystal if the lattice energy is known, viz.,

M(std.st.) + X(std.st) → M$^+$(ideal gas) + X$^-$(ideal gas), (1)

M$^+$(ideal gas) + X$^-$(ideal gas) → M$^+$X$^-$(crystal). (2)

The enthalpy of formation of M$^+$ (ideal gas) which consists of vaporization and ionization enthalpies is obtained from standard thermochemical tables (e.g. ROSSINI *et al.* (1952)). The enthalpy of formation of the anion, such as that of O$^=$, the principal anion of interest to the study of the mantle, must be calculated theoretically (GAFFNEY and AHRENS (1969)). The enthalpy change associated with eq. (2) is just equal to the lattice energy W_L (discussed below) plus NC_pT where N is the number of moles of ionic gas per mole of solid. Assuming that the ionic gas is an ideal gas, $C_p = \frac{5}{2}R$, where T and R have their usual meanings.

2.2. Calculation of the lattice energy

The lattice energy W_L is the energy change of (2) above. The Born-Mayer form of the potential is

$$U_j = z\lambda \exp(-R/\rho) - \frac{\alpha_R}{R},$$

where z is the number of nearest neighbors, λ and ρ are repulsive force constants, R is some scale length (we use the cube root of the molecular volume), and α_R is the Madelung constant for the same scale length. Using the equilibrium lattice dimensions we can eliminate $z\lambda$ and summing over the lattice we get the lattice energy

$$W_L = -\frac{N_A \alpha_R q^2}{R_0}\left(1 - \frac{\rho}{R_0}\right). \quad (3)$$

where N_A is Avagadro's number. The parameter ρ is evaluated using the relation with the bulk modulus, K_T:

$$\frac{R_0}{\rho} = \frac{9R_0 V K_T}{\alpha_R q^2} + 2. \quad (4)$$

2.3. Other forces

In the above derivation of the lattice energy W_L we have considered a "purely ionic" crystal with only two kinds of forces, coulombic and repulsive. However, there are other forces which may contribute to the lattice energy. Among these are van der Waals forces, covalent bonds and dipole and higher order multipole forces. In addition, there is zero-point and vibrational energy in the lattice. These last two and the van der Waals terms are fairly small, less than about 10 kcal/mole combined (GAFFNEY and AHRENS (1969)) and their omission is somewhat compensated for since the repulsive parameter ρ is obtained from empirical data. The largest contribution to non-ionic lattice energy is due to covalent bonds. For crystals whose enthalpy of formation (ΔH_f°) is known, the difference between calculations based on ionic theory and the measured (known) value of ΔH_f° give an apparent value for the covalent enthalpy in the lattice. There will, however, be some error in taking the actual enthalpy and the calculated ionic enthalpy equal to the covalent bond energy. This is because the lattice parameters and bulk moduli used to calculate the ionic enthalpy do themselves reflect the actual potentials within the crystal and not just the ionic portion of the potential. Also the "resonance" between ionic and covalent bonding arrangements will contribute to the lattice energy. However, in general the difference between the calculated ionic enthalpy and the actual enthalpy should be a good index of the relative proportion of covalent bonding involved.

In some of the crystal structures considered, notably rutile, α-quartz and corundum, non-radially-symmetric electric fields are known to be present at some of the lattice sites. In such cases the charge distribution associated with the ion occupying that site will be deformed into a dipole or higher order multipole. As a result interactions other than monopole interactions should be included in calculating the ionic lattice energy. We have taken such interaction into account only for SiO_2 (stishovite), TiO_2 (rutile), and Al_2O_3 (corundum). In the first, the permanent dipole effect can be estimated to be about 62 kcal/mole (by analogy with KINGSBURY's (1968) calculation of this same effect in rutile). For rutile it is 51 kcal/mole (KINGSBURY (1968)) and for Al_2O_3 multipole interactions account for about 25 kcal/mole (HAFNER and RAYMOND (1968)).

3. Results

Equations (3) and (4) were used to calculate the lat-

TABLE 1

Data for calculation of lattice energies

Compound	Structure	V(Å^3)	R_0(Å)	$\alpha_R^{(1)}$	K_T(Mb)	$q^2(e^2)$
FeO	halite	20.197	2.723	2.2018	1.42[2]	2
SiO_2	α-quartz	37.672	3.352	9.168[3]	0.374[4]	4
SiO_2	rutile	23.269	2.855	7.7219	3.627[5]	4
TiO_2	rutile	31.225	3.149	7.7191[6]	2.125[7]	4
Al_2O_3	corundum	42.466	3.489	45.7726	2.505[4]	1
Cr_2O_3	corundum	48.30	3.64	45.282	2.237	1
$Fe_2^{3+}O_3$	corundum	50.268	3.691	45.679	2.027[4]	1
$Fe_2^{3+}O_3$	perovskite	45.716[5]	3.576	44.5549	3.814[5]	1
$Fe^{2+}Fe^{4+}O_3$	perovskite	45.716[5]	3.576	12.3775	3.814[5]	4
$MgSiO_3$ - (a)	perovskite	39.225[5]	3.398	12.3775	4.188[5]	4
$MgSiO_3$ - (b)	perovskite	40.957	3.4	12.3775	3.49[5]	4
$MgSiO_3$ - (c)	perovskite	44.36	3.54	12.3775	2.6[5]	4
$SrTiO_3$	perovksite	59.558	3.905	12.3775	1.787[8]	4
$CaTiO_3$	perovskite	55.8325	3.822	12.3775	1.633[9]	4
Al_2MgO_4	spinel	65.939	4.040	67.535	1.95[10]	1
Mg_2SiO_4	spinel	65.817	4.038	71.99	2.02[9]	1
Fe_2SiO_4	spinel	69.782	4.117	72.225	2.12[2]	1
Ni_2SiO_4	spinel	65.0376	4.0215	72.1 (est.)	2.11[11]	1
$Fe_2Cr_2O_4$	spinel	73.455	4.188	64.30	1.87	1
Fe_2TiO_4	spinel	76.766	4.25	68.25	1.76	1
Fe_3O_4	spinel	73.982	4.198	65.475	1.872	1

[1] WADDINGTON, J. C. (1959) Advan. Inorg. Chem. Radiochem. **1**, 157.
[2] MAO, H. (1967) Ph.D. thesis, Univ. Rochester, N.Y.
[3] J. SHIMIN (1966) Konstanta Madelunga dlia α-kvartsa, Lietuvos Fiz. Rink., **VI (3)**, 383.
[4] ANDERSON, O. L., E. SCHREIBER, R. C. LIEBERMANN and N. SOGA (1968) Rev. Geophys. **6**, 491.
[5] Estimated from Hugoniot data, AHRENS et al. (1969).
[6] KINGSBURY (1968).
[7] Average value from G. SIMMONS (1965) J. Grad. Res. Center **34**, 1.
[8] BELL R. O. and G. RUPPRECHT (1963) Phys. Rev. **129**, 90.
[9] Estimated from ANDERSON's (1967) seismic equation of state.
[10] LEWIS, M. F. (1966) J. Acoust. Soc. Am. Letters **40 (3)**, 728.
[11] MAO, H., T. TAKAHASHI and W. A. BASSETT (1970) Phys. Earth Planet. Interiors **3**, 51

tice energy of several mantle minerals using data given in table 1. The cube root of the molecular volume is used as the scale length R. For the compounds mentioned above (e.g. stishovite and corundum) we have estimated multipole contributions to W_L. The heats of formation have been calculated by the Born–Haber cycle and are shown, with the other energies in the cycle, in table 2.

4. Discussion

Several of the compounds shown in table 2 have known heats of formation. These serve as a check on the validity of our calculations: a value of ΔH_f° that is more than the observed value is in most cases explained by an appreciable covalent contribution to lattice energy. If on the other hand a value of ΔH_f° is calculated to be considerably less than that which is thermochemically measured we must conclude that substantial covalent and/or strong dipole or higher multipole interaction takes place in the mineral, and the simple ionic model is inappropriate. A positive contribution to the lattice energy can arise only from repulsive forces all of which have been included empirically regardless of their mathematical form. (Failure to include all attractive forces will have a small effect on calculation of ρ/R which could presumably give ΔH_f°'s slightly less than the observed, e.g. in Cr_2O_3.)

For minerals with known heats of formation (e.g. FeO, $MgAl_2O_4$) we find that the calculated ΔH_f° is almost always greater than the observed value. With the exception of α-quartz discrepancies are from 6 to 280 kcal/mole, and lie mostly between about 50 and 250 kcal/mole. These greater values arise from an omission of covalent bond energies. Also there are small contributions from multipole forces in the cases for which they have not been included. We conclude

TABLE 2

Born-Haber cycle energies (kcal/mole)

Compound	Structure	W_L[1]	Multiple terms	Cations[2] ionization	Anions[3] ionization	Crystal field	Heat of formation calculated[4]	Heat of formation observed[5]
FeO	halite	−877		651	193	−13	−46	−54
SiO_2	α-quartz	−2182		2469	386	—	+670	−217
	rutile	−2880	−62[6]	2469	386	—	−101	−206
TiO_2	rutile	−2560	−51[7]	2224	386	—	−1	−226
Al_2O_3	corundum	−3513	−25[8]	2615	579	—	−344	−399
Cr_2O_3	corundum	−3366		2620	579	−120	−287	−273
$Fe_2^{3+}O_3$	corundum	−3325		2708	579	—	−45	−197
$Fe_2^{3+}O_3$	perovskite	−3587		2708	579	—	−307	>−197[9]
$Fe^{2+}Fe^{4+}O_3$	perovskite	−3931		3318[9]	579	≈−137[9]	−181	>−197[9]
$MgSiO_3$ - a	perovskite	−4086		3031	579	—	−476	>−370[9]
$MgSiO_3$ - b	perovskite	−3958		3031	579	—	−348	>−370[9]
$MgSiO_3$ - c	perovskite	−3755		3031	579	—	−145	−370[9]
$SrTiO_3$	perovskite	−3413		2646	579	—	−189	−397[10]
$CaTiO_3$	perovskite	−3397		2687	579	—	−130	−397[10]
Al_2MgO_4	spinel	−4447		3177	772	—	−507	−553
Mg_2SiO_4	spinel	−4714		3593	772	—	−349	−512
Ni_2SiO_4	spinel	−4761		3869	772	−58	−176	−328[10]
Fe_2SiO_4	spinel	−4724		3771	772	−23	−204	−350
$FeCr_2O_4$	spinel	−4171		3271	772	−131	−259	−342[2]
Fe_2TiO_4	spinel	−4325		3526	772	−19	−46	−356
Fe_3O_4	spinel	−4228		3359	772	−11	−108	−267

[1] Calculated from eq. (3) in the text.
[2] ROSSINI et al. (1952) Nat. Bur. Std. Bull. **500** except as otherwise noted.
[3] GAFFNEY and AHRENS (1969).
[4] From equations (3) and (4).
[5] ROBIE and WALDBAUM (1968) U.S. Geol. Surv. Bull. **1258**, except as otherwise noted.
[6] Scaled from data of KINGSBURY (1968) for TiO_2 according to $r_{TiO_2^2}/r_{SiO_2^2}$.
[7] KINGSBURY (1968).
[8] HAFNER and RAYMOND (1968).
[9] See text.
[10] TAYLOR and SCHMALZREID (1964) J. Phys. Chem. **68**, 2444, and AKIMOTO, FUJISAWA and KATSURA (1965).

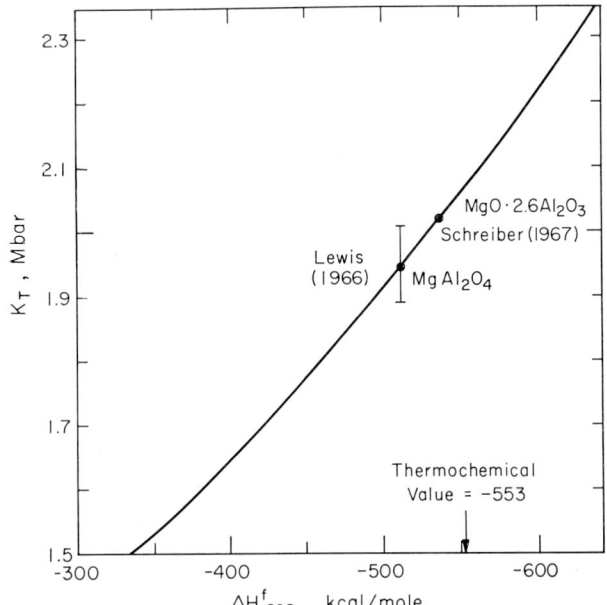

Fig. 1. Variation of enthalpy of formation (ΔH_f°) of Al_2MgO_4 (spinel) with K_T.

that if the structure is known our estimates will not be extremely low even if non-ionic bonding is important.

The enthalpy calculated for α-quartz is in much poorer agreement with the observed heat of formation than any of the above compounds. This possibly arises from its low bulk modulus (which may in itself result from covalency). If however the 160 kcal/mole covalent contribution of Si–O bond determined from the silicate spinels is valid for tectosilicates then very little of the almost 900 kcal/mole discrepancy in quartz can be attributed to covalency. We conclude that the lattice energy calculated as we have done it is not valid for oxides in fourfold coordination which are as compressible as quartz.

Table 3 lists the differences between calculated and measured enthalpies of formation for several of the compounds. There are five cases in which a particular coordination is represented by more than one compound: $Al–O_6$, $Ti–O_6 Fe^{2+}–O_4$, $Fe^{3+}–O_6$ and $Si–O_4$. For both corundum and spinel ($MgAl_2O_4$) the apparent enthalpy of covalency is about 50 kcal/mole, for three silicate spinels it is about 160 kcal/mole and for rutile and two titanites it is about 235 kcal/mole. The second case shows that for different compounds in the same structure the same ion pairs have nearly identical enthalpies of covalency. The other cases show us that this holds even for different structures if the coordination is the same. However, comparison of stishovite and the silicate spinels shows that this is now true if there is a coordination change. Therefore the following list of enthalpies of covalency can be inferred for future use:

$Al–O_6$ -25 ± 5 kcal/mole;
$Ti–O_6$ -217 ± 10 kcal/mole ($CaTiO_3$ omitted because of an unreliable bulk modulus);
$Si–O_4$ -154 ± 9 kcal/mole;
$Si–O_6$ ≈ -105 kcal/mole);

TABLE 3

Apparent enthalpies of covalency

Compound	Structure	Enthalpy of formation (kcal/mole) observed calculated		Apparent enthalpy of covalency (kcal/mole)	Predominant covalent bond
Al_2O_3	corundum	−399	−344	−55	$Al–O_6$
Al_2MgO_4	spinel	−553	−507	−46	$Al–O_6$
Mg_2SiO_4	spinel	−512	−349	−163	$Si–O_4$
Ni_2SiO_4	spinel	−328	−176	−152	$Si–O_4$
Fe_2SiO_4	spinel	−350	−204	−146	$Si–O_4$
SiO_2	rutile	−206	−101	−105	$Si–O_6$
Fe_2O_3	corundum	−197	−45	−80*	$Fe^{+3}–O_6$
TiO_2	rutile	−226	−1	−225	$Ti–O_6$
$SrTiO_3$	perovskite	−397	−189	−208	$Ti–O_6$
$CaTiO_3$	perovskite	−397	−130	−267	$Ti–O_6$
Cr_2O_3	corundum	−273	−287	+14	$Cr–O_6$
$FeCr_2O_4$	spinel	−342	−259	−83	$Fe^{2+}–O_4$
Fe_2TiO_4	spinel	−356	−46	−310	$Ti–O_6, Fe^{2+}–O_4$
Fe_3O_4	spinel	−267	−108	−159	$Fe^{3+}–O_6, Fe^{3+}–O_4$

* See text.

Fe^{3+}–O_6 ≈ –80 kcal/mole (see further discussion below for Fe_2O_3).

Fe^{2+}–O_4 –88 ± 5 kcal/mole

Fe^{3+}–O_6 ≈ –110 kcal/mole

(The quoted uncertainties represent the total spread between values calculated for different compounds.)

The relative covalent energy of stishovite and the silicate spinels is as one would expect. In stishovite each Si^{+4} is bonded to six $O^=$ ions at a distance of ≈ 1.77 Å, whereas in the spinels each Si is bonded to only four $O^=$ at ≈ 1.62 Å. The lower coordination will favor covalent bonding more than the higher. Similarly shorter bonds may also facor covalency. This relation between covalency and coordination also holds for the two iron ions, Fe^{2+} and Fe^{3+}.

The energies for hematite indicate a rather large covalent contribution of about 150 kcal/mole. However, a closer examination of the isostructural Al_2O_3 indicates that such an estimate is much too high. The multipole term in general is due primarily to dipole effects with a smaller effect due to quadrapoles. However, in Al_2O_3 the dipole terms are negligible and the quadrapole terms dominate (HAFNER and RAYMOND (1968)). This is not required by the general corundum lattice but only by the specific one for Al_2O_3. We should expect therefore that for Fe_2O_3 as for most oxides the dipole terms would be larger than the 25 kcal/mole in Al_2O_3. This will decrease the covalent contribution (probably to less than 100 kcal/mole).

For the compounds whose heat of formation and structure is unknown we can use the arguments given at the beginning of this section to evaluate the correctness of the proposed structure. If our calculated heat of formation is much less than that of a stable phase we conclude that the structure is not correct in some respect.

AHRENS et al. (1969) have proposed several possible shock-induced high-pressure structures for $MgSiO_3$ and Fe_2O_3. The high-pressure equation of state as well as zero-pressure pressure-density of $MgSiO_3$ (≈ 4.25 g/cm³) is poorly known. This severely limits the accuracy of our calculation.

For $MgSiO_3$ the high-pressure phase proposed for the shocked state was either a perovskite structure or an ilmenite structure. The latter was favored because it gives a density which is closer to that inferred from the shock data. Our calculation for $MgSiO_3$ (perovskite) for a density of 4.25 g/cm³ gives a heat of formation about 100 kcal/mole less than that of the natural phase, enstatite. In addition, Si^{+4} in six-fold coordination with oxygen should contribute about –150 kcal/mole to heat of formation (cf. stishovite) making $MgSiO_3$ (perovskite) much more stable than $MgSiO_3$ (enstatite). We know that this can not be true, so we conclude that either this proposed structure or the density is incorrect. A density of 4.07 g/cm³ gives a heat of formation about 20 kcal/mole less than that of enstatite even after including the covalent effects (see fig. 2). A density of ≈ 3.95 g/cm³ gives an enthalpy of formation which would be consistent with the perovskite structure. The reported shock data for enstatite (MCQUEEN and MARSH (1966)) are not sufficiently definitive to exclude this value. Unfortunately no Madelung constant is available for ilmenite so we can not check that structure.

AHRENS et al. (1969) also proposed that Fe_2O_3, which has the corundum structure at low pressure, goes into a perovskite structure at high pressures. We have investigated two cases: the first in which the iron remains trivalent (forming a 3–3 perovskite) and the second in which an electron is transferred from one iron ion to the other yielding one divalent and one tetravalent ion for each pair of Fe^{3+} (forming a 2–4 perovskite (REID and RINGWOOD, 1969)). The latter case leads to some major difficulties which will be discussed after considering the first, simpler, case.

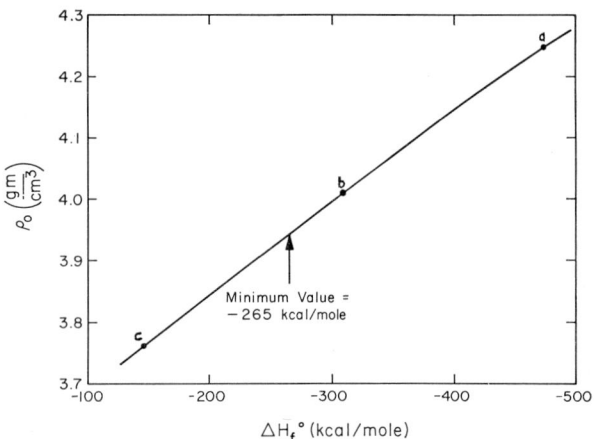

Fig. 2. Relation between calculated ionic enthalpy of formation and density for $MgSiO_3$ (perovskite). Arrow indicates minimum permissible value for this high pressure phase. Perovskites (a) and (b) are not stable.

For the 3–3 perovskite we calculate a heat of formation of -307 kcal/mole or 110 kcal/mole less than that of hematite. This energy is probably even lower when we allow for the covalence of the octahedrally coordinated Fe^{3+} and for multipole terms. We therefore conclude that a 3–3 perovskite is not a reasonable high-pressure phase of hematite.

Evaluation of the heat of formation of Fe_2O_3 as a 2–4 perovskite presents some problems. The heat of formation of Fe^{4+} is given by Allen in *Astrophysical Quantities* (1963), but the precision is poor and may be as uncertain as $+200$ kcal/mole. The effect of crystal fields on the heat of formation is also difficult to determine. There are no measured values for this quantity. We have estimated it as follows: Fe^{4+} has four 3d electrons and is isoelectronic with Mn^{3+} and Cr^{2+}. The crystal field splitting of the energy levels in Mn^{3+} is about 6000 cm^{-1} greater than in Cr^{2+}. Other pairs of isoelectronic ions (V^{4+}–Ti^{3+}, Cr^{3+}–V^{2+}, Fe^{3+}–Mn^{2+}, Co^{3+}–Fe^{2+}) behave similarly with the higher charge ion having a splitting of ≈ 6000 cm^{-1} more than the lower charges ion. We therefore assume that the splitting for Fe^{4+} is about 6000 cm^{-1} greater than for Mn^{3+}. This leads to a crystal field energy of about -46 kcal/mole if the electrons are not paired in the 3d orbitals or about -139 kcal/mole if they are paired. It seems probable that the splitting of the energy levels is sufficient to induce such pairing. If such pairing does not occur we must be cautious because Jahn–Teller distortion can be very large in 3d^4 ions and this will destabilize them. We have therefore considered only the spin-paired case because it is both probable and more tractible. In this case the enthalpy of formation is calculated to be -181 kcal/mole, only 16 kcal/mole above that of hematite.

Contributions due to covalency are probably quite small for this structure. Fe^{4+} has only four valence electrons for six bonds so it should be nearly ionic. For Fe^{2+} in twelve coordination bonding will likewise be nearly ionic. (The long bond lengths (≈ 2.5 Å) accompanying this high coordination will make crystal field effects on Fe^{2+} negligible.) It seems therefore subject to our estimations about the energetics of Fe^{4+}, that the heat of formation of $Fe^{2+}Fe^{4+}O_3$ (perovskite) is slightly larger than that of hematite and that it therefore is an admissible high-pressure structure.

5. Conclusions

The above calculations lead to the following conclusions:

1. In most cases lattice energy calculations suitably corrected for permanent multipole energies will give a good estimate of the covalency of a compound.
2. This is not true for oxides with very high compressibility such as quartz.
3. For nearly ionic bonds the energy due to covalency of a particular bond is nearly constant in minerals with the same coordination (e.g. Fe_2SiO_4–Mg_2SiO_4; Al_2O_3–Al_2MgO_4).
4. Covalency in a particular bond appears to decrease as coordination increases (e.g. Mg_2SiO_4 (spinel)–SiO_2 (stishovite)).
5. If enstatite converts to the perovskite structure at high pressure, it should have an equivalent zero-pressure density of ≈ 3.9 g/cm^3.
6. Hematite does not form a perovskite at high pressures unless the Fe^{3+} disproportionates into Fe^{2+} and Fe^{4+}. If the d electrons in the latter would be spin paired under the conditions of its formation, then a 2–4 perovskite structure appears compatible with the properties of the high pressure phase inferred from the shock data.

Although it is not possible to calculate a theoretical enthalpy of formation with sufficient accuracy to predict solid–solid transition pressures, this type of calculation permits bounds to be placed on the density and bulk modulus of proposed high-pressure structures.

Acknowledgements

This research was supported by the National Science Foundation, Grant GA-1650. We have profited from discussing this work in various stages with A. E. Ringwood, A. L. Reid and B. Kamb, and from written communication with M. Tosi.

References

AHRENS, T. J., D. L. ANDERSON and A. E. RINGWOOD (1969) Rev. Geophys., in press.
AHRENS, T. J. and Y. SYONO (1967) J. Geophys. Res. 72, 4181.
AKIMOTO, S. and H. FUJISAWA (1968) J. Geophys. Res. 73, 1467.
ALLEN, C. W. (1963) *Astrophysical quantities*, 2nd ed. (Athlone Press, London).
ANDERSON, D. L. (1967) Science 157, 1165.
ANDERSON, D. L. and H. KANAMORI (1968) J. Geophys. Res. 73, 6477.

ARCHAMBEAU, C. B., E. A. FLINN and D. G. LAMBERT (1969), in press.

BIRCH, F. (1966) Compressibility; elastic constants, in: S. P. Clark, ed., *Handbook of physical constants*, Geol. Soc. Am. Mem. **97**.

GAFFNEY, E. S. and T. J. AHRENS (1969) J. Chem. Phys., in press.

HAFNER, S. and M. RAYMOND (1968) J. Chem. Phys. **49**, 3570.

JOHNSON, L. (1969), to be published.

KINGSBURY, P. I. (1968) Acta Cryst. **A24**, 578.

KITTEL, C. (1966) *Introduction to solid state physics*, 3rd ed. (Wiley and Sons, New York).

MCQUEEN, R. G., S. P. MARSH and J. N. FRITZ (1967) J. Geophys. Res. **72**, 4999.

REID, A. F. and A. E. RINGWOOD (1969), to be published.

RINGWOOD, A. E. (1970), Phys. Earth Planet. Interiors **3**, 109.

ROSSINI, R. D., D. D. WAGMAN, W. H. EVANS, S. LEVINE and I. JAFFE (1952) Nat. Bur. Std. (U.S.) Circ. **500**.

SCLAR, C. B., L. C. CARRISON and C. M. SCHWARTZ (1964) J. Geophys. Res. **69**, 325.

WANG, C. (1968) J. Geophys. Res. **73**, 6459.

CAN MANTLE MINERALS HAVE THE NiAs STRUCTURE?

CHI-YUEN WANG

Department of Geology and Geophysics, University of California, Berkeley, California, U.S.A.

From the point of view of crystal chemistry, the nature of the chemical bonds in the NiAs structure is examined in an attempt to offer some criteria for the possible correctness of the prediction that this structure may be an important structure for mantle minerals. It is concluded that the nontransition-metal minerals will not have the NiAs structure; on the other hand, the possibility cannot be excluded that iron oxides may assume this structure at great depth. A number of geophysically interesting implications may arise if iron oxides have the NiAs structure in the lower mantle. A definite answer for the geophysical significance for the NiAs structure cannot be given at the present, however, but must be remained as a task for future experimental investigation.

1. Introduction

In several recent studies of the Hugoniot data for some rocks shocked to about 1 megabar (McQueen et al., 1967; Anderson and Kanamori, 1968; Wang, 1968a), the physical parameters which characterize the dense forms of these rocks were deduced. The zero-pressure densities for these dense forms were found to be much greater than the original densities. The density increases were interpreted in the light of known phase changes in some rock-forming minerals (Wang, 1968a). The density of the shocked Twin Sisters dunite (about 4.1 g/cm^3), however, was found to be much greater than what would be expected from the olivine → spinel transition. Thus an unknown phase change must have occurred in dunite under shock compression. The geophysical significance of this unknown phase transition can hardly be over-emphasized. Studies of the equation of state of shocked rocks suggested that the lower mantle may have a "dunitic" composition (McQueen et al., 1967; Wang, 1968a). Furthermore, successful applications of the olivine → spinel transition were made to interpret a high gradient in the seismic velocity profile at the depth of about 400 km (Anderson, 1967; Akimoto and Fujisawa, 1968). The other high gradient in the seismic velocity profile at the depth of about 600–700 km could conceivably be interpreted by a further transition of the spinel phase to a denser structure.

Several suggestions have been made in regard to the phase change of silicate minerals near the bottom of the C layer. One is that the silicates break down to aggregates of closely packed, simple oxides. Shock-wave data for some minerals showed, however, some puzzling evidence. It was shown (Wang, 1968b) that if spinel ($MgAl_2O_4$) and forsterite are assumed to break down to mixtures of closely packed, simple oxides, a systematic error of about 4% must be assigned to the Hugoniot determinations for the dense phases of these two minerals. Another suggestion is that olivine may break down to an aggregate of minerals with the ilmenite and the sodium chloride structures. This suggestion has been substantiated considerably by the works of Ringwood (1966, for example) and by the recent success of Akimoto and Syono (1967) in transforming some titanate spinels to aggregates of minerals with the ilmenite and the sodium chloride structures. It is known, however, that the titanate spinels have the inverse spinel structure, so that the titanate ions are six coordinated. On the other hand, the silicate spinels are most likely to have the normal spinel structure; thus the silicon ions are four coordinated. It would seem, therefore, that the phase transition in the silicate spinels may not be simply correlated with that in the titanate spinels. Other structures, such as the strontium plumbate structure, were proposed for the dense phase by Ringwood (1968), but none has yet been synthesized for the common rock-forming silicates.

Recently, ALBERS and ROOYMANS (1965) suggested that some mantle minerals may have the NiAs structure. Their suggestions were based upon the following evidence: In the series of compounds $FeCr_2X_4$, with X respectively O, S, Se and Te, structures of the spinel-type (when X is O or S) and the NiAs-type (when X is Se or Te) are known to occur. ALBERS and ROOYMANS (1965) succeeded in transforming the $FeCr_2S_4$ spinel to the NiAs structure at high pressure and temperature, with a density increase of about 9%. Similar phase transitions for the $CoCr_2S_4$ spinel and the $MnCr_2S_4$ spinel were recently brought about by BOUCHARD (1967) and TRESSLER and STUBICAN (1968). ALBERS and ROOYMANS (1965) suggested that the corresponding oxyspinels may also be transformed to the NiAs structure under high pressure and temperature; AHRENS and SYONO (1967) extended this suggestion to the spinels $MgAl_2O_4$, Mg_2SiO_4, and Fe_3O_4. The geophysical significance of this possibility lies in the implication that, as will be clear from the following discussion, the transition metal compounds may become immiscible with the nontransition-metal compounds at great depth in the earth's mantle, in direct contrast with observation in the common rock-forming minerals.

The present study is an attempt to examine, from the point of view of crystal chemistry, the likelihood for the NiAs structure to be an important structure for mantle minerals. Although necessarily qualitative and sometimes intuitive, this approach has been known to be useful in offering some criteria for the probable correctness of predicted structures. The possible value of this study is that it may provide some positive or negative justification for the pursuance of certain experimental work.

2. Geometrical configuration

To facilitate our discussion, the geometrical configuration of the NiAs structure is briefly given below. A comprehensive review of the compounds with this structure was given by KJEKSHUS and PEARSON (1964). For convenience the term "ion" used below is to be interpreted as meaning that the bonds are largely ionic, even though the chemical bonds may have considerable covalent character.

Let the general composition of the materials with the NiAs structure be represented by M_mX_n, where M is a metal ion and X is an ion of a B group element. The X ions are in approximately close-packed layers which in turn are stacked in the ABAB...sequence. For m smaller than or equal to n in the composition, all the M ions in this structure occupy the octahedral interstices. Now, the octahedral interstices between the layers of the X ions in such arrangement are stacked along the c-axis of the structure and each of them shares two of its faces with two other octahedral interstices, one directly above and one below. This feature may be contrasted with that in the NaCl structure in which the layers of the X ions are stacked in the ABCABC... sequence where none of octahedral interstices shares faces with others.

For compounds with the composition MX, all the octahedral interstices are occupied. In this case, each X ion is surrounded by six M ions at the corners of a trigonal prism. When m is smaller than n in M_mX_n, vacant octahedral interstices appear. A very characteristic feature of the vacant octahedral interstices is that they appear in *every other* layer of the M ions. The ideal NiAs structure has hexagonal symmetry; frequently, however, the structure is distorted for one reason or another, and the resultant structure has a lower symmetry.

The axial ratio (c/a) of the NiAs structure is found to vary from 1.22 to 1.97 for different compounds. The ideal axial ratio for the hexagonally close-packed spheres is $c/a = 1.633$. Structures with axial ratios below the ideal ratio are probably stabilized by the cation-cation bonding through overlapping of their d orbitals along the c-axis (KJEKSHUS and PEARSON, 1964) or by similar interaction between the cations in the octahedral interstices and those in the bipyramidal interstices (for m greater than n in the compound M_mX_n, GOODENOUGH, 1963). These compounds usually have some metallic character as evidenced from their metallic conductivity. Though of considerable interest to the chemist, the cation-cation bonding is not important for the mantle minerals and is therefore omitted in the following discussion of the chemical bond in the NiAs structure.

3. Chemical bonds

KJEKSHUS and PEARSON (1964) suggested that compounds may adopt the nickel arsenide structure (i) because of the directional character of the chemical bonds formed by the component atoms (axial ratio close to 1.63), and (ii) because relative ionic sizes will not allow the compound to take the rocksalt structure even though the ionicity is relatively high (axial ratio higher than 1.63). These suggestions are obviously too general to be of any use for the present discussion of the NiAs structure as a possible form for mantle minerals. Unfortunately, the bonding properties required to stabilize the NiAs structure are not well understood. It is thus necessary to draw rather broad inferences from various existing evidences. In the following sections, I will first present briefly some evidence to show that the NiAs structure is electrostatically unstable as compared to some other structures and thus to indicate that appreciable covalent character must exist in the chemical bonds to stabilize the structure. Secondly, the question of why only transition-metal ions are found in the NiAs structure is analyzed; the answer to this question is found to suggest that it is not likely that nontransition-metal compounds can have this structure even under high pressure and temperature. Thirdly, the orbital configuration for the X ions in the NiAs structure is discussed in an effort to examine whether it is at all possible for the oxygen ions to have similar orbitals.

4. Electrostatic considerations

It is easy to show by using BRAGG's (1937) rule of lines of force that in compounds with composition similar to the spinels (M_3X_4 or $M_1N_2X_4$), the NiAs structure cannot be electrostatically stable. Lines of force start from cations in numbers proportional to their valence, and end on anions. These lines of force for each cation were divided equally among the bonds to the corners of its coordinated polyhedron; the rule then states that each anion receives from its immediate neighboring cations enough lines of force to satisfy its valence. Thus, for example, an oxygen ion O^{--} may be satisfied by various combinations of octahedrally coordinated metal ions such as six Mg^{2+} (periclase), four Al^{3+} (corundum), three Si^{4+} (stishovite), two Ti^{4+} and two Fe^{2+} (ilmenite), etc. For compounds with composition M_3X_4 or $M_1N_2X_4$ in the NiAs structure the X ions cannot be coordinated with equal numbers of cations throughout the crystal structure. The X ions may be coordinated alternately with five and four cations, but in this case no simple composition may satisfy Bragg's rule.

Another simple argument for the same problem may be given for compounds with composition MX in the NiAs structure. An anion in an ideally close-packed hexagonal array ($c/a = 1.633$) occupies the same space as that it occupies in a cubic close-packed array. If all the octahedral interstices in these arrays are occupied by metal ions, the cation-cation distance along the c-axis of the hexagonal close-packed array is considerably shorter than the cation-cation distance in the cubic-packed array. Cation-cation repulsion along the c-axis of the hexagonal array tends to increase the axial ratio and to result in an inefficient use of space. Thus if electrostatic forces are the only interaction among the ions, the NaCl structure is more stable than the NiAs structure. ZEMANN (1958) computed the Madelung constants for the NiAs structure of various axial ratios. The maximum of these constants is found to be smaller than the Madelung constant for the NaCl structure.

The above arguments are substantiated by a diagram of MOOSER and PEARSON (1959) for simple transition-metal compounds. In this diagram, the average principal quantum number (n) for these compounds is plotted versus their differences in electronegativity (ΔX). It was found that there was a surprisingly sharp separation of compounds with the NiAs structure on one side and the NaCl structure on the other. Given n, the NiAs structures invariably lie on the side of lower ΔX. The sharpness of the separation between the two structures probably suggests that the factor which stabilizes the NiAs structure for crystals with lower ionicity is due to a simple mechanism.

5. Why transition metals?

MOOSER and PEARSON (1958) also plotted n and ΔX for the nontransition-metal compounds. It was found that the same boundary which divides the diagram for the transition-metal compounds also divides the nontransiton-metal compounds into two regions. However, here the region with the lower ΔX is occupied by compounds with the tetrahedrally coordinated sphalerite and wurtzite structures, instead of the NiAs structure. The similarity in the locations of the two boundaries on the diagrams suggests that the mechanisms which control the division of the structures on the two diagrams are closely related. On the other hand, the NiAs structure suggests that there is a strong tendency for the transition-metal ions to coordinate with six anions instead of four when the chemical bonds have appreciable covalent character.

When a transition-metal ion is surrounded by several anions at the corners of a regular polyhedron the five d orbitals no longer have the same energy but split into groups. For example, in an octahedral arrangement of the anions, the d orbitals split into a lower triplet and an upper doublet. The physical basis for this splitting of the energy levels is entirely electronic and arises from the fact that the d shell is not spherically symmetric. Splitting of the d shell ends in a lowering of the energy of the system with reference to that in which the cation has a spherically symmetric configuration. It may be shown (PAULING, 1961; ORGEL, 1966) that two of the d orbitals tend to hybridize with the s and p orbitals to form six stable orbitals, d^2sp^3, pointing towards the corners of a regular octahedron. Using the spectroscopic data for the metal ions surrounded by oxygen ions, MCCLURE (1957) found that in most cases the extra stabilization due to d shell splitting is greater in an octahedral coordination than in a tetrahedral coordination. Although this "site preference" energy is in general small, it points to the same direction that stabilization is gained for most of the transition-metal ions to assume the octahedral rather than the tetrahedral sites. For the nontransition-metal compounds, the tendency for the metal ions to assume the octahedral interstices is not obvious and their structures may directly reflect the orbital configurations for the anion.

Without better evidence to the contrary, it seems reasonable to suggest that the NiAs-type structure is not a likely structure for compounds of magnesium or silicon.

6. Bonding orbitals for the anions

In the previous section it is suggested that the formation of the NiAs structure is closely associated with a strong tendency for the transition-metal ions to coordinate with six anions at the corner of a regular octahedron when the chemical bonding has appreciable covalent character. However, the tendency for the transition-metal ions to assume the octahedral interstices cannot be the sole factor to stabilize the NiAs structure as against the NaCl structure. The stabilization of the NiAs structure must be related to the characteristics of the anions.

The familiar p^3 orbitals and the sp^3d^2 orbitals may be excluded for the anions on the ground that they favor the NaCl structure rather than the NiAs structure. Theoretical considerations suggest that several kinds of hybrid orbitals involving the d orbital, such as spd^4 and pd^5 have a trigonal-prismatic configuration. Although such orbital configurations have been suggested for anions in certain crystals, they are not particularly interesting for geophysical considerations because it is most likely that oxygen ions cannot use d-orbitals.

PAULING (1961) suggested that it is not necessary for the central anion to have six orbitals to form covalent bonds with six coordinated M ions. Instead, it is enough for the central X ion to have the sp^3 hybrid orbitals; the distorted tetrahedral orbitals "resonate" at the six positions by "pivoting around the central ion". This theory explains why trigonal-prismatic sites are preferred to octahedral sites by ions which form quadricovalent bonds: In an octahedral site the angle between at least two of the tetrahedral orbitals of the X ion must be distorted from the original 109° to 180° while in the trigonal-prismatic site the largest distortion is from 109° to 130° and this angle can be reduced by adjusting the position of the anion a little away from the center of the trigonal prism. It may be worthwhile to note that, assuming the current theory on superexchange interact on among ions (for example, GOODENOUGH, 1963; HALPERN, 1965), Pauling's resonating quadricovalent bond may be used to explain the general magnetic properties of minerals with the NiAs structure.

An important evidence for the argument that it is not impossible for the transition metal oxides to assume the NiAs structure comes from the fact that ions of some elements in the same row in the periodic table as oxygen are known to coordinate with six transition-metal ions at the corner of a trigonal prism, for example, N^{--} and C^{--} in the compounds MoN, WN, MoC, WC, which have the WC structure and N^{--} in the compound NbN which has the TiP structure. Both the WC and the TiP structures closely resemble the NiAs structure.

Assuming Pauling's suggestion of the resonating quadricovalent bonds, we may imagine that there exist for the transition-metal compounds two competing tendencies: electrostatic forces among the ions tend to arrange the ions in the NaCl structure in order to lower the Coulomb energy of the system; the covalent bonds, however, tend to arrange the ions in the NiAs structure in order to lower the distortion energy in the covalent bonds. The choice between these two crystal structures is determined by the "covalency" of the bonds.

At high pressure, covalency of the chemical bonds in the transition metal compounds increases because of greater interaction between the ligand field with the electrons in the d shell which tends to spread the d-orbitals of the metal ions and make them more available for the orbitals of the anions. Possible change of number of unpaired electrons in the iron ions (FYFE, 1960; STERN, 1966) also tends to increase the covalency in the bond (ORGEL, 1966). Since available information on the pressure and temperature effects on the electronic properties of the ions does not allow me to make quantitative estimate of the change of covalency with depth, the question as to whether iron oxides may assume the NiAs structure in the deep mantle has to be answered by future experimental studies.

7. Concluding remarks

The NiAs structure is not a likely structure for the nontransition-metal compounds in the mantle. However, the possibility exists that iron oxide may assume this structure at some great depths. If so, the easy replacement of magnesium by iron in the common rock-forming minerals will no longer be observed at such depths. A number of geophysically interesting implications may arise: 1) depletion of iron from the mantle due to gravitational separation will become more effective at such depths, and 2) solution of liquid iron in the solid mantle may not proceed by the process of iron replacement of magnesium or silicon in the mantle minerals. Future experimental investigation on phase transition in iron oxides is needed to provide a definite answer on the geophysical significance of the NiAs structure.

Acknowledgments

I wish to thank Professor John Verhoogen and Hans-Rudolf Wenk for discussion and criticism. Professor Richard E. Powell's comments are much appreciated. I also wish to thank Dr. C. J. M. Rooymans for suggesting several references on the NiAs structure. This research is supported in part by the U.S. National Science Foundation under grant NSF GA-1683. A grant from the Organizing Committee made it possible for me to participate in this Symposium.

References

AHRENS, T. J. and Y. SYONO (1967) J. Geophys. Res. **72**, 4181.
AKIMOTO, S. and H. FUJISAWA (1968) J. Geophys. Res. **73**, 1467.
AKIMOTO, S. and Y. SYONO (1967) J. Chem. Phys. **47**, 1813.
ALBERS, W. and C. J. M. ROOYMANS (1965) Solid State Communications **3**, 417.
ANDERSON, D. L. (1967) Science **157**, 1165.
ANDERSON, D. L. and H. KANAMORI (1968) J. Geophys. Res. **73**, 6477.
BOUCHARD, R. J. (1967) Mat. Res. Bull. **2**, 459.
BRAGG, W. L. (1937) Cornell University Press.
FYFE, W. S. (1960) Geochim. Cosmochim. Acta **19**, 141.
GOUDENOUGH, J. B. (1963) *Magnetism and the Chemical Bond* (Interscience Publishers).
KJEKSHUS, A. and W. B. PEARSON (1964) in: H. Reiss, ed., *Progress in Solid State Chemistry* **1** (Pergamon Press).
MCCLURE, D. S. (1957) J. Phys. Chem. Solids **3**, 311.
MCQUEEN, R. G., S. P. MARSH and J. N. FRITZ (1967) J. Geophys. Res. **72**, 4999.
MOOSER, E. and W. B. PEARSON (1959) Acta Cryst. **12**, 1015.
ORGEL, L. E. (1966) *An Introduction to Transition-Metal Chemistry: Ligand Field Theory* (Methuen).
PAULING, L. (1960) *The Nature of the Chemical Bond*, 3rd edition (Cornell Univ. Press).
RINGWOOD, A. E. (1966) in: P. M. Hurley, ed., *Advances in the Earth Sciences* (M.I.T. Press).
RINGWOOD, A. E. (1968) Publ. 666, Dept. of Geophysics and Geochemistry, Australian National University, Canberra.
STERN, R. G. J. (1966) Chem. Commun. 777.
TRESSLER, R. E. and V. S. STUBICAN (1968) J. Am. Ceram. Soc. **51**, 391.
WANG, C.-Y. (1968) Nature **218**, 560.
WANG, C.-Y. (1968) J. Geophys. Res. **73**, 6459.
ZEMANN (1958) Acta Cryst. **11**, 55.

MAGMAS, XENOLITHS AND EXPERIMENTAL PETROLOGY

A REVIEW OF EXPERIMENTAL EVIDENCE ON THE ORIGIN OF BASALTIC AND NEPHELINITIC MAGMAS

D. H. GREEN

Department of Geophysics and Geochemistry, Australian National University, Canberra, Australia

Geological and petrological arguments are used to select those basaltic compositions which have been directly derived from the upper mantle with minimum opportunity for crystal fractionation or for contamination within the crust. The experimental study of the crystallization of these magmas at high pressures, under both anhydrous and hydrous conditions provides evidence of genetic links between magma types. These studies also provide data necessary to evaluate a major constraint on "primary" magmas, i.e. that magmas derived by direct partial melting of the upper mantle must have olivine and enstatite among their liquidus phases at the P, T conditions of magma segregation. The characteristic magma derived from either direct partial melting (~20–25%) of pyrolite or fractional crystallization of olivine-rich tholeiite at depths of 15–35 km is high alumina olivine tholeiite. At 35–70 km partial melting of anhydrous pyrolite or fractional crystallization of olivine-rich tholeiite or tholeiitic picrite produces a series of liquids from olivine-rich tholeiite through olivine basalt and alkali olivine basalt to basanite (~25% Ol, ~5% Ne). With availability of water in the pyrolite source or during crystal fractionation, the pyrolite solidus is depressed. This produces changes in the subsolidus mineralogy and in the nature of liquids formed by low degrees of partial melting or by extensive crystal fractionation of less undersaturated magmas. Olivine nephelinites and basanites may be generated by small degrees of melting of water-bearing (0.1–0.2% H_2O) pyrolite at ~50–70 km and olivine melilite nephelinites, olivine-rich nephelinites and olivine-rich basanites form in a similar way, as hydrous magmas at 70–100 km depth.

1. Introduction

The processes of generation of basaltic magmas within the earth's mantle involve crystal-liquid equilibria in chemically complex systems at pressures at least up to 30 kb. The availability of solid media high pressure apparatus (BOYD and ENGLAND, 1960) and the application of the electron probe microanalyser to the chemical analysis of complex phases synthesized at high pressure, have resulted in important advances in the study of basalt genesis in recent years. With the ability to quantitatively study the behaviour of natural, complex basalt compositions at high pressures, the need to attempt extrapolations from simple 3 and 4 component systems to the behaviour of natural basalts has largely disappeared, although the intrinsic value of studies of simple systems has not diminished. However, working with natural basalts involves a large element of choice and a dependence on geological and petrological arguments to select those basaltic compositions which are of direct mantle derivation and thus readily justify, on petrogenetic grounds, the study of their crystallization behaviour at high pressures.

In studies of basalts where the chemical composition of a magma is the principal concern, it is convenient to adopt a normative (i.e. indirect chemical) classification rather than a modal (mineralogical) classification. The nomenclature used is as follows:

Tholeiite: basalt with normative hypersthene;

Quartz tholeiite: basalt with normative hypersthene and quartz;

Olivine tholeiite: basalt with normative hypersthene and olivine, hypersthene >3%;

Olivine basalt: with normative olivine and with 0–3% normative hypersthene; no normative nepheline;

Alkali olivine basalt: with normative olivine and nepheline; nepheline <5%;

Basanite: basalt with normative olivine, nepheline and albite and with nepheline >5%, albite >2%;

Olivine nephelinite: basalt-like composition with major normative olivine and nepheline; albite <2%, normative orthoclase and/or leucite but no normative larnite;

Olivine melilite nephelinite: basalt-like composition with normative olivine, nepheline, leucite and larnite. The principal variation between the above magma types is the degree of silica saturation. Most workers recognize continuity and transition in chemical composition between the various basalts. Silica content decreases from 47–49% in the olivine tholeiites to <38% in olivine melilite nephelinite, whereas alkalis ($Na_2O + K_2O$) increase and CaO also increases, particularly in the nephelinitic compositions. A further variation in basalt compositions is apparent in the recognition of distinctive high-alumina basalts (TILLEY, 1950; KUNO, 1960). These are aphyric basalts with normative olivine and either normative hypersthene or low normative nepheline contents but with Al_2O_3 contents distinctly higher (at 16–20% Al_2O_3) than "normal" olivine tholeiites or alkali olivine basalts of similar SiO_2 and $Na_2O + K_2O$ contents (KUNO, 1960).

The following review does not attempt to be comprehensive and in particular, recent papers by KUSHIRO (1968), ITO and KENNEDY (1968), O'HARA and YODER (1967) contain relevant experimental data, and illustrate other approaches to the problems of basalt genesis. The paper is mainly concerned with summarizing data and presenting inferences from those experimental studies with which the author has been associated and also with rebuttal of recent criticism of earlier papers.

2. Constraints on magma composition in the upper mantle

The geophysical and petrological arguments for a peridotitic composition for the upper mantle are considered to be compelling. It is assumed that the average mantle composition, in regions unaffected by earlier extraction of basaltic or other liquids, is that of a peridotite capable of providing 20–30% of basaltic liquids, or 40% of picritic liquids, by partial melting leaving residual olivine and enstatite. This is the rationale of the "pyrolite" model (RINGWOOD, 1962; GREEN and RINGWOOD, 1963; and RINGWOOD, 1966). It is emphasized that the concept of pyrolite is a general one but the actual calculation of a specific "pyrolite" composition will change with improved knowledge of liquid-residue relationships in the upper mantle. Thus the "pyrolite" composition calculated by GREEN and RINGWOOD (1963) utilized NOCKOLDS' (1954) average basalt and highly refractory dunite, whereas RINGWOOD's (1966) calculation taking into account the studies in basalt crystallization at high pressures, utilized a Hawaiian tholeiite as the basaltic liquid and an enstatite-peridotite as the residue. The particular proportions of basalt and residual peridotite were chosen to yield compositions similar in major elements to a class of peridotites (high temperature peridotites, lherzolite nodules in basalts and in kimberlites) which are of mantle derivation and cannot be less refractory than the parental material for basalts. The deduced pyrolite composition was also compared with the mantle composition inferred from the chondritic earth model.

The acceptance of a pyrolite composition for the upper mantle imposes a major constraint on basalt compositions in that liquids derived by partial melting of the mantle must be in equilibrium, at their depth of magma segregation, with residual minerals of pyrolite. The choice of an eclogite composition for the upper mantle would impose very different constraints on partial melting but these can be shown to be inconsistent with natural basalts and their high pressure crystallization products. The subsolidus mineralogy of pyrolite has been investigated experimentally (GREEN and RINGWOOD, 1967b, and this volume) and some data also obtained on partial melting of pyrolite. From these data we can infer that at very low degrees (1–3%) of melting, magmas would be in equilibrium with olivine, enstatite, clinopyroxene and plagioclase at low pressures; olivine, aluminous enstatite, aluminous clinopyroxene ± spinel* at intermediate pressures; and olivine, enstatite, clinopyroxene and garnet at high pressures. With increasing degree of melting, minerals enter the melt in proportions and sequence dependent on the P, T conditions, until at high degrees (>15–20%) of melting, only olivine and enstatite remain as residual phases. A self-consistent model for basalt genesis from a pyrolite mantle thus requires that basaltic liquids which are unfractionated products of direct partial melting should include olivine and enstatite as liquidus phases at pressures and temperatures matching those of their origin. Magmas which segregate from residual peridotite and begin to fractionate by removal of liquidus crystals at the depth of magma segregation may yield liquids which do not have olivine and orthopyroxene as liquidus phases, provided that there are reac-

---

* The presence or absence of spinel at the solidus depends sensitively upon the bulk composition and on the solidus temperature, i.e. whether the solidus is depressed by the presence of water or not.

tion relationships* between early precipitated phases and basaltic liquids. It may be noted that because of the complex solid solutions involved, reaction relationships may exist between early precipitated phases and basaltic liquids which are not matched by incongruent melting relationships in the simple, end-member system.

The previous discussion illustrates the constraints imposed by self-consistency arguments in the interdependent problems of upper mantle composition and basalt magma genesis. However, more positive evidence of mantle liquid composition is provided by selecting those basalts which contain xenoliths or "xenocrysts" demonstrably of high pressure origin, with density greater than that of the magma, and thus precluding the possibility of crystal fractionation of the host magma at depths less than those at which the xenolithic material was picked up. The paper by GREEN and HIBBERSON (1970) illustrates how very specific conditions of origin of some basalts can be deduced from their xenolith and xenocryst content. By selection of such basalts, it is possible to eliminate chemical variation due to low pressure fractionation and to establish a range or trend of chemical variation among basalts imposed by processes acting within the mantle (BULTITUDE and GREEN, 1968; GREEN, 1969). Basalts selected in this way are characteristically rich in olivine, have 100 $Mg/(Mg+Fe^{++})$ atomic ratios of 65–75 and range in normative composition from olivine basalt to olivine melilite nephelinite and kimberlite. As olivine tholeiite magmas do not contain lherzolite xenoliths, the choice of an olivine tholeiite as relevant for high pressure experimental study was made on the basis of the extensive studies of the 1959 Kilauea Iki eruption. The composition chosen for experimental study (GREEN and RINGWOOD, 1967a) was close to MACDONALD and KATSURA's (1961) estimate of a "parental magma" composition of the Kilauea Iki lava lake. It was considered to be a possible magmatic liquid composition at depths of 40–60 km. The reasons for the selection of other compositions listed in table 1 have been published previously. Compositions 6 and 9 are lherzolite-bearing natural basalts on which the detailed high pressure crystallization experiments have not as yet been reported.

3. Crystal fractionation at high pressures in anhydrous magmas – experimental data and implications

The experimental crystallization of the compositions in table 1 shows that the nature of the liquidus phases changes with pressure, olivine giving way with increasing pressure to orthopyroxene and/or clinopyroxene and then to garnet or garnet + clinopyroxene. The compositions of liquidus and near-liquidus phases were determined by electron probe analysis. With these data and with estimates of the degree of crystallization, the directions of fractionation of the basaltic liquids have been calculated for various depths in the crust and mantle.

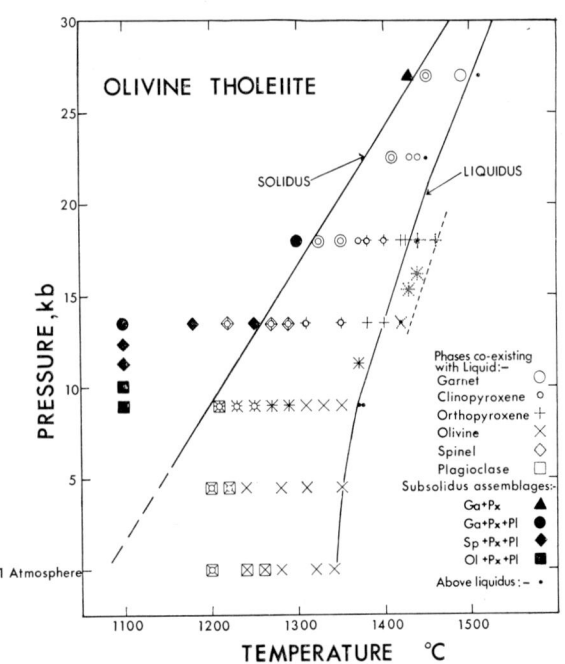

Fig. 1. Experimental crystallization of olivine tholeiite (table 1) at various pressures (GREEN and RINGWOOD, 1967a). The data for the olivine-enriched tholeiite (table 2) are shown as dotted symbols and the approximate liquidus shown by a dashed line. This composition is that of the olivine tholeiite to which has been added 5% olivine (Fo_{20}).

The experimental data published for the olivine tholeiite composition are reproduced in fig. 1 (GREEN and RINGWOOD, 1967a, fig. 4). Olivine is the liquidus phase at low pressure (<10 kb), olivine and orthopyroxene

---

* Such reaction relationships include
Olivine + liquid$_A$ → clinopyroxene$_{ss}$ + liquid$_B$
Orthopyroxene + liquid$_A$ { → clinopyroxene$_{ss}$ + liquid$_B$
or
→ clinopyroxene$_{ss}$ + garnet + liquid$_C$

TABLE 1

Compositions of basaltic glasses used for experimental studies of crystallization at high pressures

	1 Olivine Tholeiite (Green and Ringwood, 1967a)	2 Olivine Basalt	3 Picrite (Green and Ringwood, 1967a)	4 Auckland Id. Olivine Basalt (Green and Hibberson, 1970)	5 Alkali Basalt (Green and Ringwood, 1967a)	6 Olivine-rich basanite (Mt. Leura, Victoria) (Green, unpublished)	7 Olivine nephelinite (Bultitude and Green, 1968)	8 Picritic nephelinite (Bultitude and Green, 1968)	9 Olivine nephelinite (Scottsdale, Tasmania) (Green, unpublished)	10 Pyrolite (Ringwood, 1966)	
SiO_2	46.95	47.05	45.51	46.55	45.39	44.63	44.3	43.7	39.31	45.16	SiO_2
TiO_2	2.02	2.31	1.93	3.18	2.52	2.92	1.5	1.3	3.87	0.71	TiO_2
Al_2O_3	13.10	14.17	12.44	12.70	14.69	11.67	14.2	12.3	9.45	3.54	Al_2O_3
Fe_2O_3	1.02	0.42	0.92	2.98	1.87	2.95	0.5	1.1	5.07	0.46	Fe_2O_3
FeO	10.07	10.64	8.67	9.72	12.42	9.39	9.7	10.8	10.69	8.04	FeO
MnO	0.15	0.16	0.15	0.17	0.18	0.15	0.2	0.2	0.20	0.14	MnO
MgO	14.55	12.73	18.79	10.63	10.37	13.85	13.3	17.0	13.90	37.47	MgO
CaO	10.16	9.87	9.67	8.66	9.14	7.68	11.2	9.7	11.20	3.08	CaO
Na_2O	1.73	2.21	1.64	2.95	2.62	3.65	3.6	2.7	2.98	0.57	Na_2O
K_2O	0.08	0.44	0.08	0.95	0.78	2.00	1.0	0.8	1.53	0.13	K_2O
P_2O_5	0.21	—	0.20	0.60	0.02	1.03	0.5	0.4	2.30	0.06	P_2O_5
Cr_2O_3										0.43	Cr_2O_3
$\frac{100\,Mg}{Mg+Fe^{++}}$	72.0	68.1	79.4	66.2	59.8	72.5	71.0	73.8	70.0	90.0	$\frac{100\,Mg}{Mg+Fe^{++}}$
CIPW Norms											
Lc	—	—	—	—	—	—	—	—	0.3	—	Lc
Or	0.6	2.7	0.5	5.6	4.5	11.7	6.1	4.5	8.3	1.1	Or
Ab	14.7	18.9	13.9	24.8	18.0	12.0	2.0	3.4	—	4.7	Ab
Ne	—	—	—	—	2.2	10.2	15.3	10.6	13.4	—	Ne
An	27.6	27.3	26.3	18.9	26.2	9.8	19.4	19.2	8.4	6.6	An
Di	17.0	17.6	16.5	17.7	15.7	18.7	26.4	21.7	29.1	6.7	Di
Hy	12.3	1.3	2.8	2.7	—	—	—	—	—	15.4	Hy
Ol	21.9	27.2	34.6	18.8	25.8	25.8	25.9	35.7	22.3	62.7	Ol
Ilm	3.8	4.4	3.7	6.1	4.8	5.6	2.9	2.4	6.4	1.4	Ilm
Mt	1.4	0.6	1.3	4.4	2.9	4.3	0.7	1.6	7.4	0.6	Mt
Ap	0.5	—	0.4	1.3	—	2.0	1.3	1.0	4.5	0.1	Ap
Chr										0.7	Chr

TABLE 2

Experimental crystallization at high pressures of an olivine tholeiite composition prepared from olivine tholeiite (table 1 column 1) and olivine (Fo$_{90}$) mixed in proportions 95:5 = olivine tholeiite: olivine. Also experimental runs on olivine tholeiite (table 1, column 1) in graphite rather than Pt capsules. All runs in graphite capsules under "dry" conditions.

Pressure (Kb)	Temperature (°C)	Time (mins)	Products	Comments
Olivine tholeiite + olivine (Fo$_{90}$)				
13.5	1420	30	Ol+glass	Many small olivine crystals
15.3	1430	30	Ol+Opx+glass	Many small olivine and fewer but much larger orthopyroxene crystals
16.2	1440	30	Opx+Ol+glass	Rare small olivine and more common, much larger orthopyroxene
18.0	1460	30	Opx+glass+quench	Rare orthopyroxene
18.0	1440	30	Opx+glass+quench	More common orthopyroxene, possible clinopyroxene
18.0	1420	30	Opx+Cpx+quench	Cpx > Opx
Olivine tholeiite				
9	1360	30	Ol+glass	Many small olivine crystals
13.5	1380	30	Opx+glass	Large orthopyroxene crystals
13.5	1360	30	Opx+Cpx+glass	Cpx > Opx

occur together on the liquidus at approximately 11.3 kb and orthopyroxene is the liquidus phase at 13–18 kb, followed by clinopyroxene and garnet at higher pressures. Although fig. 1 includes only runs in Pt capsules, the problem of Fe-loss and sample composition change was previously evaluated and shown to be of minor significance. This has been further checked by runs in graphite capsules at 9 kb and 13.5 kb confirming the role of olivine at 9 kb and orthopyroxene at 13.5 kb as liquidus phases (table 2). The data on the olivine tholeiite composition suggest that crystal fractionation of olivine tholeiitic magmas at 13–18 kb will be dominated by orthopyroxene separation. Analyses of the orthopyroxenes crystallized and estimation of the degree of crystallization enabled the calculation of derivative liquids – the fractionation trend was shown to lead very directly from hypersthene-normative olivine tholeiite through olivine basalt to alkali olivine basalt compositions. The effectiveness of orthopyroxene crystallization throughout this spectrum of compositions was confirmed by detailed experiments on the olivine basalt and alkali olivine basalt compositions (table 1) (GREEN and RINGWOOD, 1967a). Although these compositions had orthopyroxene as a liquidus phase at 13–18 kb, they were undersaturated in olivine and thus these three experimentally-studied compositions could not be direct partial melting products of a pyrolite mantle at 13–18 kb (although the olivine tholeiite and possibly the alkali basalt could be such partial melting products at 11.3 kb – see GREEN and RINGWOOD, 1967a). On the other hand, the picrite composition (table 1), which is closely similar to the olivine basalt composition except for substantially greater olivine content, has olivine as the liquidus phase at 13.5 kb and 18 kb and orthopyroxene appears some 20–30 °C below the liquidus. Thus, bearing in mind the similarities of the olivine basalt and picrite compositions and their relationships to the olivine tholeiite and alkali olivine basalt compositions, it was inferred (GREEN and RINGWOOD, 1967a, page 166) that liquids derived by partial melting of the pyrolite source at 13–18 kb should be slightly richer in olivine than the chosen basalt series, but not so rich as the picrite composition. In spite of the earlier discussion, O'HARA (1968) has strongly criticized the hypotheses of crystal fractionation and of partial melting presented by Green and Ringwood, on the grounds that the chosen basalts were not possible partial melting products from a mantle peridotite. O'Hara's paper implies that it is not possible to deduce or extrapolate data from specific basalt compositions to slightly dissimilar compositions without the device of reduction of complex systems to 3 or 4 component "model" systems, which then can be understood in terms of the simple system. To further demonstrate the correctness of the previous deductions, crystallization experiments have been carried out on a new glass prepared from a mix of olivine tholeiite (table 1) and olivine (Fo$_{90}$) in proportions 95:5 (yielding a composition with 25.8 % normative olivine, 11.7% hypersthene, 16.2% diopside etc.). The results of these experimental runs are listed in table 2 – olivine precedes orthopyroxene at 13.5 kb, the two phases occur together on the liquidus at 15–16 kb (cf. the 11.3 kb run on the original tholeiite) and orthopyroxene is the liquidus phase at 18 kb. A composition of olivine tholeiite and olivine mixed in 92:8 proportions should yield olivine and orthopyroxene on the liquidus at 18 kb and would contain 28.2% normative olivine, 11.3% hypersthene etc. The previous analytical data on the olivine and orthopyroxene from the olivine tholeiite show 100 Mg/(Mg+Fe^{++}) ratios consistent with the values attributed to pyrolite.

Although the above discussion should clarify questions on the nature of liquids in equilibrium with residual olivine and enstatite at 9–18 kb, there remains a major divergence of opinion on the nature of crystal fractionation in batches of magma separated from residual peridotite at 13–18 kb or derived from deeper levels and partially crystallizing in the 13–18 kb pressure range. GREEN and RINGWOOD (1967a) considered that crystal fractionation in this region would be dominated by orthopyroxene in olivine tholeiite and olivine basalt magmas and by orthopyroxene and subcalcic clinopyroxene in alkali olivine basalt magmas. The role of olivine was incorrectly ignored in the sections on crystal fractionation (fig. 10, GREEN and RINGWOOD, 1967a). On the other hand, O'HARA (1968, p. 92–93) asserts that olivine is the major phase precipitating from possible magmas in this depth region, with lesser orthopyroxene and clinopyroxene. O'HARA (1968 p. 93) considers that the spinel-lherzolite inclusions in nepheline-normative magmas are cognate accumulates in relation to their host magmas, and that their precipitation is an essential part of the process by which some hypersthene normative magmas become nepheline normative. The basis for inferring a cognate origin for spinel-lherzolite xenoliths was given in O'HARA and MERCY (1963, p. 283–286) and expanded in O'HARA (1968b) but does not withstand critical examination of the assumptions involved nor consideration of additional data on lherzolite assemblages and their conditions of equilibration (GREEN and RINGWOOD, 1967a, 181–184). In addition, later work involving isotopic and trace element studies (LEGGO and HUTCHISON, 1969; GREEN et al., 1968; KLEEMAN et al., 1969; COOPER and GREEN, 1969; KLEEMAN and COOPER, 1970) has demonstrated that typical lherzolite inclusions are of accidental origin and represent mantle fragments of very much greater age than their host magmas. Some lherzolites are still capable of yielding basaltic magmas by partial melting i.e. they are not residues or accumulates from much earlier basaltic magma episodes (KLEEMAN et al., 1969).

If a magma segregates from residual olivine and enstatite and then begins to cool and fractionally crystallize then olivine will indeed be a liquidus phase at the depth of segregation or at shallower depths. However, this does not imply nor require that olivine is volumetrically a *major* phase in the precipitate material. The basalt is a complex, multi-component system, in which the cooling liquid will continuously react with precipitated olivine (if it is not removed from the liquid), converting it to more Fe-rich olivine. The *amount* of olivine present in the total precipitate, may, however, increase rapidly or imperceptibly, or it may decrease – the latter effect would require a reaction relationship such as has been observed at 9 kb (m olivine$_1$ + liquid$_A \rightleftharpoons n$ olivine$_2$ + clinopyroxene + liquid$_B$ ($m > n$) GREEN and RINGWOOD, 1967a, pp. 128, 129, 143)) but for which there is no unequivocal evidence at 13–18 kb. The published experimental data on the various compositions and the data included herein on the olivine-enriched tholeiite show clearly that olivine will be a minor phase in the range of accumulates from basaltic magmas fractionally crystallizing near their depth of segregation at 13–18 kb – these precipitates would range from olivine-poor orthopyroxenite to olivine-poor pyroxenites. If a tholeiitic magma segregates from residual material at 18 kb, moves rapidly to 13 kb and then begins to crystallize, then the initial precipitate will be olivine, and orthopyroxene would only appear after approximately 5% crystallization. Further precipitation would be dominated by orthopyroxene or, at lower temperatures, clinopyroxene. These precipitates from basaltic magmas must have lower 100 Mg/(Mg+Fe^{++}) ratios than the corresponding phases in the residual peridotite*. The iron-enrich-

---

* "Xenocrystal" and "xenolithic" material matching the anticipated precipitates from basaltic magmas includes the Salt Lake Crater pyroxenites and garnet pyroxenites (GREEN, 1966; JACKSON and WRIGHT, this volume) and other similar examples (see paper by IRVING and GREEN, 1970). The processes of crystal accumulation are themselves complex in that the bulk composition of an accumulate will depend on the presence or absence

ment trend in basalt fractionation is a fundamental constraint on the amount of fractional crystallization, particularly of olivine, possible in attempting to relate one basalt magma type to another and all basalts to an ultimate equilibrium with mantle peridotite (with olivine ~ Fo_{90}). This forms a principal objection to O'HARA's (1968, p. 118) requirement of very high-grees of crystallization (40% olivine separation or 50% eclogite+40% olivine separation from picritic parents for tholeiitic and alkali olivine basalt magmas respectively) since microprobe data on a variety of compositions have shown that olivine, clinopyroxene and orthopyroxene separation strongly enrich (and garnet slightly enriches) liquids in Fe relative to Mg. O'Hara's models and arguments utilize projections and simple system analogies in which Fe and Mg are treated as equivalent. These models must finally be translated into actual magma and mineral compositions and the generalized, qualitative "trends" be evaluated in simple, multicomponent calculations. The evidence from natural lherzolite-bearing alkali basalts, basanites, olivine nephelinites and olivine melilite nephelinites shows that this whole spectrum of compositions of mantle derivation is remarkably magnesian, posing some difficulty in invoking even the relatively small degrees of crystallization required by the GREEN and RINGWOOD (1967a) crystal fractionation models and favouring direct partial melting hypotheses.

The implications of the crystallization studies under dry conditions are summarized in the upper part of fig. 2, amended from fig. 10 (GREEN and RINGWOOD, 1967a) in recording the minor role of olivine in crystal fractionation at 13–18 kb. The diagram illustrates the spectrum of magma compositions and the nature of accumulate minerals which may be derived by crystallization of a parental olivine tholeiite magma at 30–70 km depth. The maximum degree of undersaturation developed by dry fractionation appears to be a basanitic liquid with ~5% normative nepheline, further fractionation being controlled by garnet and clinopyroxene with little change in the degree of under-

of entrapped liquid. Assuming elimination of entrapped liquid in very slow accumulation (WAGER and WADSWORTH, 1960) the anticipated accumulates include dunites (<Fo_{89}) and olivine-poor pyroxenites but not spinel lherzolites (Fo_{88}–Fo_{92}). CARTER (1966) reports preliminary data on xenolithic materials in basalt which reflect very well the divergence in Mg/(Mg+Fe^{++}) ratios between residual and precipitate material in basalt genesis.

saturation. No mechanism for derivation of more nepheline-rich basanites, olivine nephelinites or olivine melilite nephelinites was apparent from the dry melting studies and detailed studies of the dry melting behaviour of the olivine-rich basanite, olivine nephelinite and picritic nephelinite compositions (table 1) revealed only olivine, clinopyroxene and garnet as liquidus phases up to 40 kb with calculated fractionation trends athwart the natural compositional variation and with no suggestion that such liquids could form in equilibrium with residual olivine, enstatite, ± clinopyroxene ± garnet from a pyrolite source.

4. Crystal fractionation at high pressures in the presence of water - experimental data and implications

BULTITUDE and GREEN (1968) reported a reconnaissance study of the effects of water on the crystallization of olivine nephelinite and picritic nephelinite compositions (7, 8, table 1). At that time, a suitable welding unit and technique for sealing known quantities of water within the small solid media apparatus capsules was not available to the authors. Simple empirical methods, using graphite capsules or "open" (crimped, but not welded) Pt capsules and pressure media (talc) which dehydrate during a run, were found to depress the dry liquidi of basalts by 100° ± 20 °C in the pressure range 13–30 kb. The use of "open" capsules in the presence of dehydrating pressure media is not new (YODER and TILLEY, 1962; GREEN and RINGWOOD, 1964c; O'HARA and YODER, 1967). Using the same techniques but also adding water (5–25%) to the sample, the basalt liquidi were lowered 200–300 °C in the 13–30 kb pressure range. The most notable effect of lowering of the liquidi by the presence of water is the enlargement of the field of crystallization of orthopyroxene at the expense of clinopyroxene and possibly of olivine. BULTITUDE and GREEN (1968) found evidence that orthopyroxene plays a dominant role in the fractional crystallization of olivine nephelinite and picritic nephelinite compositions at 18–27 kb at temperatures of 1150 °C–1300 °C. The possibility of sample composition change by Fe loss and by migration of components, via an aqueous phase, either into or out of the sample container was recognized (op. cit. p. 328) and attempts were made to eliminate this factor by using both Pt and graphite containers, by use of boron-nitride spacers rather than ceramic or pyrophyllite and by bulk analyses for Si,

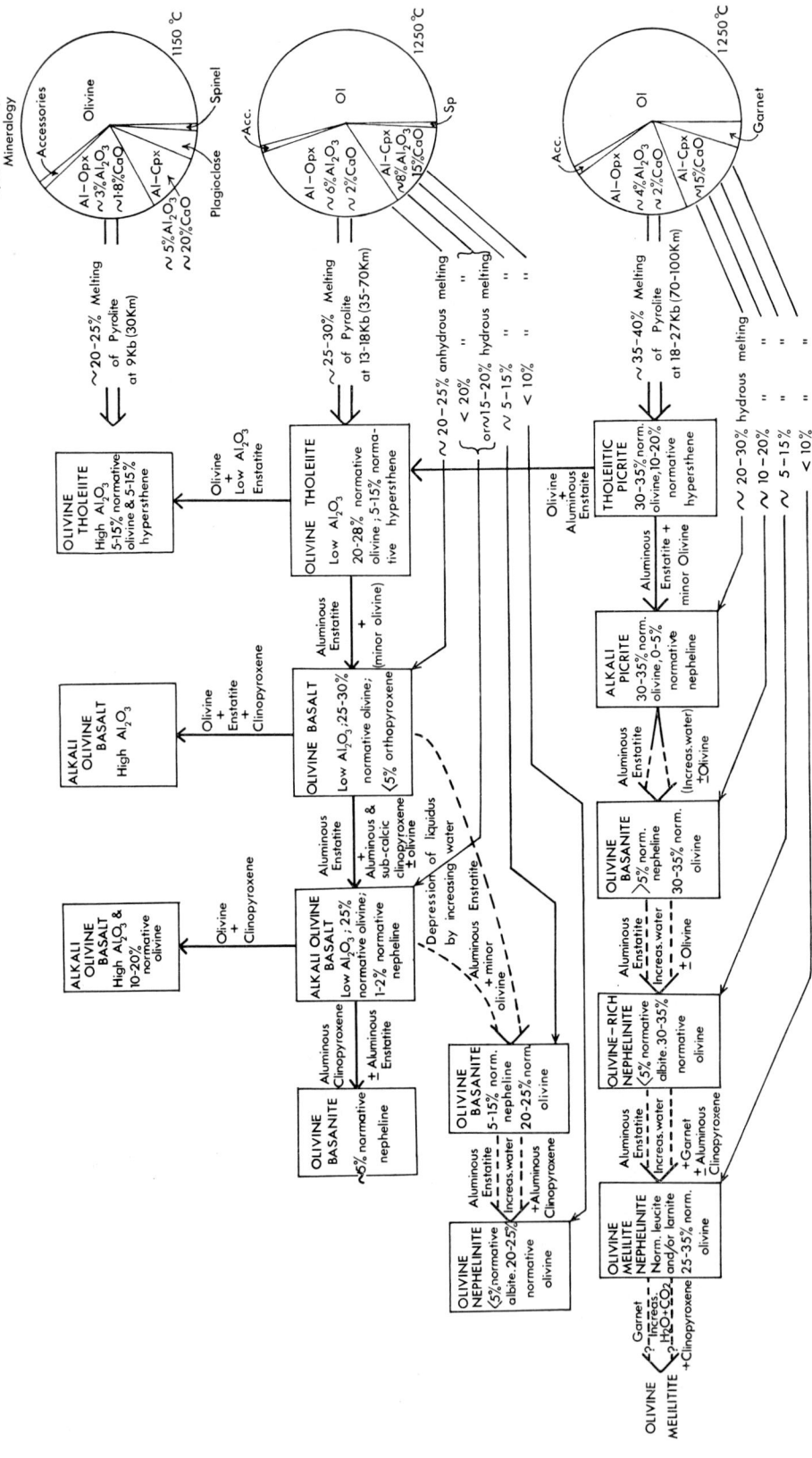

Fig. 2. Diagrammatic synthesis of relationships between basalt magma types. The rectangular boxes denote magmas related by crystal fractionation processes operating at various depth or pressure regimes. They are also derived by possible partial melting processes from a pyrolite source. The approximate mineralogy of the pyrolite source at 1150 °C, 9 kb; 1250 °C, 15 kb and 1250 °C, 27 kb is shown. This is not necessarily the immediate subsolidus mineralogy involved in the melting processes – both the mineralogy and the temperature of the solidus will be dependent on P_{H_2O}, P_{Load} relationships.

Al, Fe, Mg and Ca of important runs using the electron microprobe (5–10μ beam size). Orthopyroxene was produced under appropriate P, T conditions in all the variations of technique employed and the microprobe analyses showed that any sample compositional change could only have been of minor or of very selective character (Na_2O was not determined). KUSHIRO (1969) considered that the stated precautions against composition change did not eliminate the possibility of selective leaching of alkalis and this is indeed a general criticism of the technique of running in unsealed capsules. O'HARA (1968, p. 104) assumed that the sequence of runs in order of decreasing temperature at any one pressure should also be a sequence of decreasing Mg/Fe ratio in olivines and orthopyroxenes and ignored the effects of varying water contents (BULTITUDE and GREEN, p. 329); on this basis, O'Hara considered that compositional changes had occurred. At 22.5 kb and between 1200–1300 °C, the reported sequence of crystallization of the picritic nephelinite is liquid → olivine + liquid → olivine + orthopyroxene + liquid → olivine + orthopyroxene + garnet + liquid → olivine + orthopyroxene + garnet + clinopyroxene + liquid*. Amphibole reported in runs above 18 kb is probably entirely quench. The above sequence is that of increasing degree of crystallization and decreasing $Mg/(Mg+Fe^{++})$ ratio of analyzed phases (comparing runs in Pt or graphite capsules independently) but is not the sequence of decreasing temperature. No crystal-liquid reaction relationships are required by the data although the relative proportions of phases suggests that there may be a partial reaction of orthopyroxene or orthopyroxene + olivine with liquid to precipitate garnet and clinopyroxene with increasing degree of crystallization. O'HARA's (1968) rejection of the reconnaissance study of the role of water in these highly undersaturated compositions is not substantiated by the arguments he used or the data originally presented, but may rest on a simple belief in the inadequacy of the precautions taken to eliminate minor compositional change.

A study of the crystallization of the olivine-rich basanite composition (table 1) under dry conditions yielded olivine as the liquidus phase up to 22.5 kb but clinopyroxene was on the liquidus at 1470 °C, 27 kb, (GREEN 1969 and unpublished data). Depression of the liquidus by permitting access of water from the talc pressure medium or by adding small drops of water to unsealed Pt and graphite capsules yielded orthopyroxene as a major near-liquidus phase at 18–27 kb in the temperature range 1200–1300 °C. Later experiments on the same composition have utilized welded Pt capsules containing known quantities of water. These techniques prevent the possibility of change in sample composition by selective leaching of alkalis, migration of silica etc. and the effect of Fe-loss has been shown to be of minor significance (cf. GREEN and RINGWOOD, 1967a, c). The major role of orthopyroxene as a near-liquidus phase has been confirmed at 22–27 kb using the sealed capsules (see also GREEN and HIBBERSON, 1970). The liquidus temperature at 27 kb was ~1340 °C with 2% H_2O and ~1270 °C with 4.5% H_2O. The sequence of appearance of phases is olivine (minor) joined by orthopyroxene or by orthopyroxene + clinopyroxene. At lower temperatures orthopyroxene reacts with liquid to yield clinopyroxene and garnet and the assemblage is olivine (minor) + clinopyroxene + garnet + liquid. Microprobe analyses of the phases have been carried out and will allow the calculation of fractionation trends. Similar experiments on the picritic nephelinite and olivine nephelinite (table 1, column 7) using sealed capsules are in progress but the presence of orthopyroxene in the melting interval has not as yet been confirmed by electron microprobe data. Nevertheless the data obtained on the olivine-rich basanite and olivine basalt compositions (table 1, column 4) (GREEN and HIBBERSON, 1970) demonstrate that selective leaching and compositional change were not responsible for the appearance of orthopyroxene in these compositions and seem unlikely to be of significance in the earlier runs using the same techniques.

The data demonstrate that, provided water is available in the source regions of basaltic magmas, fractional crystallization by the separation of orthopyroxene as the major crystallizing phase, extends over the compositional range from olivine tholeiite to olivine rich basanite (<12% Ab, >10% Ne) and most probably to olivine nephelinite. In olivine-rich compositions, olivine is a co-precipitating phase and either clinopyroxene or garnet is the third phase to appear depending on P, T conditions and bulk magma compo-

---

* BULTITUDE and GREEN (1968, table 3, column 3) inadvertently omitted "olivine" as a coexisting phase with orthopyroxene and garnet at 22.5 kb, 1270 °C.

sition. The effect of fractional crystallization by separation of major orthopyroxene is to rather rapidly deplete the residual liquid in SiO_2, while enriching it in CaO and alkalies. The chemical variation accords well with the natural trend in mantle-derived basalts from alkali olivine basalt to olivine melilite nephelinites (BULTITUDE and GREEN, 1968). The more undersaturated basanite and olivine nephelinite compositions appear to be attainable only at deeper levels (>18 kb or 70 km) from more picritic parent magmas. The most undersaturated product of this fractionation pattern appears to be olivine melilite nephelinite derived by separation of aluminous orthopyroxene + pyrope-almandine garnet ± minor olivine from picritic nephelinite magma at ∼27 kb. The inferred processes and products of basalt fractionation at various pressures under both "wet" and dry conditions are summarized in fig. 2 (cf. GREEN and RINGWOOD, 1967a, fig. 10; GREEN, 1969, fig. 3). A major deduction from experiments with hydrous and anhydrous natural basalts, (although such experiments are still largely in the exploratory stage) is that basanites (with >10% nepheline) olivine nephelinites and olivine melilite nephelinites are hydrous magmas containing at least 2–5% H_2O at their depth of origin in the mantle. Less undersaturated magmas may also be hydrous, e.g. the Auckland Id. olivine basalt (table 1) contained ∼2% H_2O near its depth of origin (GREEN and HIBBERSON, 1970). These hydrous magmas would form at temperatures of 100–200 °C below their anhydrous solidi and in their movement to the surface would be expected to separate a vapour phase at some depth; to precipitate amphibole and phlogopite if crystallized at some depth without escape of the vapour phase and to have characteristically explosive eruptive character – these features may be observed in the extremely undersaturated magma suite. The source of water for magma genesis of this type must lie in the upper mantle and the interrelationship of this aspect of magma genesis and chemical fractionation of the upper mantle is discussed in a later section.

5. Partial melting in the mantle

A specific model composition for the source regions of basalt, the "pyrolite" composition of RINGWOOD (1966), is used in the following discussion (table 1, column 10) although it recognized that some inhomogeneity must be expected. The stability fields of plagioclase pyrolite, pyroxene pyrolite, spinel + pyroxene pyrolite and garnet pyrolite have been experimentally determined (GREEN and RINGWOOD, 1967b, 1970), and preliminary data also obtained on the stability amphibole in pyrolite composition at high pressures. In considering partial melting of pyrolite, we are concerned with the nature of the phases on the solidus, the extent of depression (if any) of the solidus by fluid pressures in the region and the sequence and proportions of phases entering the liquid at temperatures above the solidus. The microprobe analyses of the liquidus and near-liquidus phases of the basalts and of the pyrolite near-solidus assemblages are very similar and it is possible to apply the data from the experimentally amenable crystallization studies of the basalts to evaluate processes of partial melting at various depths. This has been done for anhydrous pyrolite assemblages by GREEN and RINGWOOD (1967a) and for hydrous conditions by BULTITUDE and GREEN (1968) and GREEN (1969). The inferred relationships are illustrated in fig. 2.

Tholeiitic magmas are derived by rather large degrees of partial melting of the pyrolite source. At depths of 15–35 km, magmas developed from an anhydrous pyrolite source involve melting of olivine + enstatite + clinopyroxene + plagioclase + spinel (Cr_2O_3-rich). With 20–25% melting, leaving residual olivine, low-Al enstatite and rare clinopyroxene, the liquids are tholeiitic but typically with high Al_2O_3 contents (15–17% Al_2O_3) and with normative olivine contents of 5–10% (15–25 km) or 10–15% (25–35 km). At depths of 35–70 km, the mineralogy of the pyrolite source would be olivine, aluminous enstatite, aluminous clinopyroxene and possibly rare aluminous spinel. Tholeiitic magmas result if the degree of melting is *sufficient to eliminate clinopyroxene* from the residual phases i.e. 25–35% melting. These tholeiitic magmas contain 15–28% normative olivine, are of low-alumina type (11–13% Al_2O_3), and if they fractionate during movement towards the earths surface, will trend towards high-Al tholeiites at 20–30 km or directly towards quartz-tholeiites at less than 10 km. At 70–100 km depth, melting at the *anhydrous* pyrolite solidus again involves olivine and aluminous pyroxenes, liquids are richer in olivine than those at lower pressures and tholeiitic picrites (30–35% normative olivine) are produced by very high degrees of partial melting (35–40%). At shallower levels tholeiitic picrites may fractionate and derivative liquids move along the tholeiite → alkali olivine basalt trend at 35–70 km; tholeiite → high Al basalt trend at 20–30 km or tholeiite → quartz tholeiite trend at <10 km.

With lower degrees of partial melting, liquids must be in equilibrium not only with residual olivine and enstatite, but also with clinopyroxene and, at the beginning of melting, with plagioclase (low pressure), possibly spinel (intermediate pressures) or with garnet (high pressures, >100 km depth). At depths ~35 km, low degrees of partial melting (<20%) would produce liquids of high alumina character, with ~47–50% SiO_2, and probably with small normative hypersthene or small normative nepheline contents, and with normative olivine (10–15%) and high normative plagioclase. Some magmas classed as hawaiites but containing xenoliths and xenocrysts of high pressure origin may originate in this manner (KUNO, 1964). At depths of 35–70 km under dry conditions, liquids coexisting with olivine, aluminous enstatite and sub-calcic, aluminous clinopyroxene range from basanite with ~5% nepheline to alkali olivine basalts and these magma types would be produced by *up to 20%* melting of the pyrolite source. Melting to 20–25% of the source pyrolite would eliminate sub-calcic clinopyroxene from the residue and produce olivine basalts. At depths of 70–100 km, lower degrees of partial melting (<30%) of anhydrous pyrolite would probably produce alkali picrites in equilibrium with olivine, aluminous enstatite and aluminous, sub-calcic clinopyroxene.

If small quantities of water (0.1–0.2%) are present in the source pyrolite then the subsolidus mineralogy will differ from that in fig. 2 in containing 5–15% hornblende or ~3% phlogopite. The stability of amphibole is limited to relatively shallow depths in the mantle <100–120 km (GREEN and RINGWOOD, 1967c, p. 806; LAMBERT and WYLLIE, 1968) and at deeper levels, amphibole may be replaced by phlogopite and/or a fluid phase. The presence of water in hydrous minerals or in a fluid phase (which may contain a large CO_2 content, ROEDDER, 1965) will cause lowering of the pyrolite solidus and important changes in the mineralogy undergoing partial melting. Amphibole and phlogopite are inferred to melt incongruently to olivine, pyroxenes ± garnet and a liquid phase because of the high (≥10%) solubility of water in basalt melts at these pressures and temperatures. Lowering of the pyrolite solidus to ~1200 °C means that garnet may occur on the solidus at depths >80 km and also the composition of clinopyroxene at the solidus will be much more Ca-rich (16–18% CaO) and lower in alumina than the clinopyroxene at the anhydrous solidus (~10% CaO) (GREEN and RINGWOOD, 1967a, 1967b; GREEN and HIBBERSON, 1970). Because of the temperature-sensitive nature of the pyroxene solid solution, the proportions of enstatite and clinopyroxene on the solidus will also differ for wet melting.

The experimental confirmation of the role of enstatite on the olivine basanite liquidus and the reconnaissance studies showing a similar role for enstatite and for garnet in the olivine nephelinite and picritic nephelinite compositions suggest that these highly undersaturated liquids may also be partial melting products of pyrolite under wet melting conditions. It is inferred that small percentages (<10%) of liquid formed at 200–300 °C below the anhydrous solidus at 80–120 km will be in equilibrium with olivine, orthopyroxene, garnet and minor clinopyroxene. Such liquids will be of olivine melilite nephelinite character. With increasing degree of melting, all clinopyroxene, much garnet and orthopyroxene and minor olivine enter the melt changing it to olivine nephelinite composition with residual olivine, enstatite ± minor garnet. It is emphasized that the melting process involves a continuous re-equilibration process between changing basaltic liquid and changing solid solutions and may also involve discontinuous melting relations such as m clinopyroxene$_1$ + liquid$_A$ → n orthopyroxene + p clinopyroxene$_2$ + liquid$_B$; or garnet + liquid$_A$ → orthopyroxene + liquid$_B$.

At depths <80 km, garnet is probably not involved in the wet melting of pyrolite and liquids produced with low degrees of partial melting (<10%) at 200 °C below the anhydrous pyrolite solidus will be olivine nephelinites, ranging with increasing degrees of partial melting, through olivine-rich basanites to alkali olivine basalts (fig. 2). The sequence of magmas derived by wet melting at 35–70 km will be lower in olivine than the sequence from olivine melilite nephelinite to alkali picrite at >80 km. In the movement of such liquids towards the surface, any crystallization which occurs at shallower depths will involve separation of olivine. Clinopyroxene or amphibole may either join or replace olivine as the liquidus phase as such crystallization proceeds; replacement of olivine will occur if, as is likely, reaction relationships exist between these phases and olivine + liquid in these particular undersaturated compositions.

Fractional crystallization and the movement of the hydrous undersaturated magmas towards the surface are two processes which will tend to develop supersaturation of the liquid in the volatile components (possibly H_2O+CO_2 mainly). A condition will commonly be reached at which separation of a fluid phase occurs – this will have important implications to the transport properties of the magma, the nature of its interaction with the wall-rock environment, the possibility of concentration of elements in the fluid phase and of precipitation of solid phases from the separated fluid phase. At very shallow depths, rapid expansion of a separated fluid phase may initiate explosive eruption. It is suggested that the correlation of magma type with type of xenolithic inclusion [i.e. low P assemblage or high P assemblage (WHITE, 1966; MACGREGOR, 1968; JACKSON and WRIGHT, 1970)] is not indicative of cognate relationship between magma and xenolith nor directly indicative of the depth at which magma segregates from its source material, but is a direct consequence of the water or water + CO_2 ("volatile") content of the magma and the depth at which vesiculation occurs in a magma. Thus it is inferred that olivine nephelinite or alnoitic magmas containing mineral assemblages indicative of 15–18 kb pressure, contained higher water contents and vesiculated at deeper levels than alkali olivine basalts or hawaiites containing xenoliths or xenocrysts indicative of pressures <10 kb.

6. Wall-rock reaction and selective element enrichment

Trace element concentration and ratios, while not appreciably affecting the modal or normative mineralogy nor the petrographic classification of a basalt, nevertheless provide important information and constraints on possible inter-relationships between various basalt magma types. In their application of hypotheses of olivine tholeiite and alkali olivine basalt inter-relationships, which were based on major element contents, GREEN and RINGWOOD (1967a, 167–169) considered that the known eruptive sequence and petrology of Hawaiian volcanoes could be interpreted in terms of crystal fractionation of parental olivine tholeiite magmas. The degree of crystallization (dominantly of aluminous enstatite) of the olivine tholeiite was calculated to be ~30%. However examination of the abundances of some minor elements (K, Ti, P) and trace elements (Rb, Sr, Cs, Ba, U, Th, Zr, Hf and lighter rare earths) in the Hawaiian alkali olivine basalt and tholeiitic series showed enrichment factors in the alkali olivine basalt which were much greater than those predicted from the inferred crystal fractionation. A similar conclusion applies to basalts in general, i.e. these elements ("incompatible elements" of GREEN and RINGWOOD, 1967a) are frequently much more abundant in the undersaturated magmas than predicted by the simple crystal fractionation relationships or differences in degree of partial melting outlined in the previous sections. It was suggested that this group of elements may be highly enriched in a fractionating magma by a process of "wall rock reaction" in which cooling and crystallization of a magma involved complementary processes of reaction and extraction of the lowest melting fraction from the wall-rock. Wall-rock reaction was envisaged as a highly selective contamination of a magma by its wall-rock environment. The incompatible elements were considered to be present in the pyrolite source mainly in accessory minerals such as phlogopite, apatite, ilmenite and, by incongruent melting of these phases or by their entry into the lowest melting liquids, the incompatible elements could be highly enriched in such liquids.

GAST (1968) has given a detailed analysis of the behaviour of trace elements during partial melting with particular reference to two subclasses of basaltic magmas – the oceanic-ridge or abyssal tholeiites and, the alkaline basalts of central volcanoes of oceanic regions. Hawaiian tholeiitic rocks are not adequately considered in Gast's paper but are treated in more detail by HUBBARD (1969) and the latter paper brings out more clearly the inter-relationships of trace-element and major element geochemistry and geographic features. GAST (1968) concurs with GREEN and RINGWOOD (1967a) that fractional crystallization is inadequate to explain the differences in trace element abundances but presents arguments against the wall-rock reaction process on the grounds of inadequacy of the mechanism. GREEN and RINGWOOD (1967a) attempted to work within the framework of a single parental mantle composition and did not pursue the incompatibility between their pyrolite composition (0.71% TiO_2, 0.13% K_2O) based on Hawaiian tholeiite (RINGWOOD, 1966) and the parental mantle (0.3–0.4% TiO_2, 0.03–0.05% K_2O) inferred from the oceanic ridge tholeiite (GREEN and RINGWOOD, 1967a, p. 171). GAST (1968) elaborated the minor and trace element data clearly distinguishing the oceanic-ridge tholeiites from alkali olivine basalts and attributed these differences to different source compositions (page 1077) and differences in degrees of partial melting; the alkali olivine basalts being derived with a small degree of melting (3–7%) from a more primitive, trace element-rich source (e.g. pyrolite of table 1) and the oceanic ridge tholeiites (15–30% melting) being derived from a partially depleted source, already chemically modified by previous partial melting of small extent. The weight of trace element data, particularly rare earths, and isotopic data, is considered to favour Gast's interpretation requiring variable source composition as well as variation in conditions of melting. GREEN (1968, p. 848–850) briefly reviewed the minor element abundances in a variety of basalts and inferred mantle inhomogeneity in accessory minerals such as phlogopite, apatite etc. on both local and regional scale. Recent data on lherzolite inclusions in basalts supports these conclusions on mantle inhomogeneity (GREEN et al., 1968; LEGGO and HUTCHISON, 1968; KLEEMAN et al., 1969) and suggests that in the lherzolite inclusions we are seeing examples of selectively depleted parental mantle material.

The recognition of the important role of water in the genesis of the undersaturated basalts has implications for processes of wall-rock reaction and trace element fractionation. The presence of a limited quantity of water (e.g. 0.1–0.2%) in the source region will cause melting well below the anhydrous solidus but there will be a considerable temperature range through which the degree of melting will remain small (<5%), liquids being basanite, olivine nephelinite or olivine melilite nephelinite. Such liquids will, following GAST (1968), have enriched and fractionated "incompatible" element contents as a direct consequence of partial melting. Liquids developed with higher degrees of partial melting, particularly if sufficient to eliminate clinopyroxene or garnet from the residual phases, will be of alkali olivine basalt*, olivine basalt or olivine tholeiite type and would be expected to show decreasing abundances and fractionation in incompatible elements.

---

* The conclusion of GREEN and RINGWOOD (1967a, p. 166) on dry partial melting to produce alkali olivine basalt magma should be clarified to state that liquids produced by *up to* 20% melting will

If hydrous minerals such as phlogopite, amphibole etc. are stable in the upper mantle then there exist conditions with $P_{H_2O}^{\text{wall-rock}} \leq P_{\text{fluid}}^{\text{wall-rock}} \leq P_{\text{Load}}$. If a body of magma, undersaturated in water, moves through or into a region of mantle containing hydrous phases then $P_{H_2O}^{\text{magma}}$ may be less than $P_{H_2O}^{\text{wall-rock}}$ and the hydrous phases may be rendered unstable in the new conditions. Breakdown of the hydrous phases produces water and other volatiles moving into the magma and modifying it. The presence of a separate fluid phase (possibly $CO_2 + H_2O$) may permit selective migration of elements from wall rock to magma – those elements formerly substituting in the amphibole or phlogopite but not compatible with pyroxenes and olivine may migrate in this way**. Thus the role of water (and CO_2 etc.) may be an important one in allowing selective migration of elements without significant change in major element abundances.

The crystal fractionation of a magma body from a high temperature anhydrous liquid to a lower temperature hydrous magma is not a closed system process but involves chemical interaction of the magma with its wall rock environment. The original concept of wall rock reaction (GREEN and RINGWOOD, 1967a) is expanded to include movement of a fluid phase containing trace element concentrations, or trace element diffusion *through* a fluid phase from wall rock to magma. Important evidence for wall-rock reaction processes operating at shallow depths may be inferred from isotopic studies showing isotopic changes in the late stage liquids of "closed" magma chambers. Such evidence may be more difficult to find for mantle processes where smaller isotopic differences would be expected.

Variations in degrees of partial melting of a pyrolite source from <5% to ~15–20% will produce liquids of alkaline type with highly to moderately fractionated trace element abundances (GAST, 1968). Olivine basalts and tholeiites of Hawaiian type would be produced from the same source by ≥20–30% melting but the oceanic ridge tholeiites and some Hawaiian tholeiites are derived from a source region already depleted in

---

be alkali olivine basalts at 10–20 kb. Alkali olivine basalts (~5% nepheline) formed by small degrees of melting under dry conditions should thus show fractionated and enriched trace-element abundances.

** FREY (1970) has described evidence for a "wall-rock reaction" type of process leading to selective movement of lighter REE from a metamorphic aureole (amphibole → pyroxenes) into high-temperature peridotite in which more magnesian amphibole is stable.

some elements. The high-alumina olivine tholeiites of the oceanic ridges require approximately 20% melting and magma segregation at shallower depths than alkali olivine basalt or Hawaiian tholeiite.

7. Conclusions

The problems of chemical and mineralogical composition of the upper mantle and the petrogenesis of basaltic magmas are interdependent. The acceptance of a peridotitic upper mantle composition, in which olivine (55–70%) and enstatite (15–30%) are major phases, imposes a major constraint on the compositions of liquids which may be derived by partial melting of the mantle. A most productive method of investigation of these interdependent problems has been the determination of the nature and composition of liquidus and near-liquidus phases of natural basalts under various P, T conditions and the parallel investigation of the mineralogy of pyrolite, the potential mantle source rock, under the same P, T conditions.

The composition of a basaltic magma within the mantle is determined initially by the partial melting and magma segregation process i.e. for a given mantle composition the liquid composition is determined by the P, T, P_{H_2O} conditions at which the magma segregates from residual crystals. These parameters will also control the degree of partial melting and the nature and proportions of residual crystalline phases. Following segregation, a batch of magma may have opportunity for cooling and crystal fractionation at various depths leading to further diversification of magma compositions or in some cases to obliteration of chemical characteristics which would identify the primary source conditions of a magma. Crystal fractionation of a magma near its depth of origin or at shallower depths cannot always be considered as a closed system process but involves reaction with its wall-rock environment. Wall-rock reaction causes highly selective contamination of magmas with "incompatible elements" by processes of extraction of small percentages of the low melting fraction or by element migration in or through a fluid phase [possibly ($H_2O + CO_2$) rich].

Tholeiitic magmas are derived by rather large degrees of melting of the peridotitic source rock. At depths of 15–35 km, magmas developed by 20–25% melting leaving residual olivine, low-Al enstatite and minor clinopyroxene, are tholeiitic with distinctively high Al_2O_3 contents (15–17% Al_2O_3) and low normative olivine (5–10% at 15–25 km or 10–15% olivine at 25–35 km). In this pressure range, magmas resembling high-Al_2O_3 hawaiites may be produced with smaller degrees of melting or by partial crystallization at 15–35 km of Mg-rich olivine basalts.

At depths of 35–70 km, partial melting of pyrolite under dry conditions is inferred to produce magmas ranging from basanite ($\sim 5\%$ Ne, 25% Ol) with small degrees of melting through alkali olivine basalts, olivine basalts (25–30% Ol) to olivine tholeiites (10–15% Hy, 15–30% Ol). Magmas are nepheline normative if clinopyroxene remains among the residual phase, i.e. up to $\sim 20\%$ melting of pyrolite; and olivine basalt and olivine tholeiite ($Al_2O_3 \sim$ 12–13%) leave residual olivine and aluminous enstatite only. As an alternative to direct partial melting of pyrolite, the higher temperature olivine tholeiite may yield lower temperature alkali olivine basalt or possibly basanite by crystal fractionation at 35–70 km. Because of possible reaction relationships between precipitated phases and the basaltic liquid, more extreme fractionation will produce liquids unlike those produced by small degrees of melting of the original pyrolite source. Data are insufficient to fully evaluate this aspect.

At depths of 70–100 km, dry melting of pyrolite probably produces liquids ranging from alkali picrite to tholeiitic picrite (30–35% olivine) leaving residual olivine, aluminous enstatite and possibly aluminous clinopyroxene.

There appears to be no satisfactory process of generating the extremely undersaturated basanite, olivine nephelinite or olivine melilite nephelinite magmas by either partial melting of anhydrous pyrolite or crystal fractionation of less undersaturated basaltic magmas under dry conditions. Suggestions of crystallization of very large amounts of "eclogite" (garnet + clinopyroxene) and olivine from picritic parent magma (O'HARA, 1968) are as yet inadequately defined but imply marked Fe-enrichment (relative to Mg) in the undersaturated magmas – this is not observed in the natural magmas. This suggested mechanism also implies that the highly undersaturated magmas cannot form as direct partial melting products from the mantle.

The data obtained on the role of water in the fractional crystallization of basaltic magma at high pressure lead to the inference, that in the presence of water, there is a continuous fractionation or partial melting sequence from essentially anhydrous olivine tholeiite (20–30% melting) to olivine nephelinite (2–5% water, <5% melting) at 50–70 km depth and from tholeiitic picrite, through olivine-rich basanite to olivine-melilite nephelinite (<5% melting) at 70–100 km depth. The hydrous olivine nephelinite, olivine-rich basanite, and olivine melilite nephelinite magmas form at temperatures of 100–250 °C below the dry pyrolite solidus – this has the effect of bringing partial melting into the garnet pyrolite stability field so that early formed liquids at 30 kb, the olivine melilite nephelinites, are in equilibrium with residual olivine, orthopyroxene and garnet. The presence of 0.1–0.2% H_2O (and possibly similar or larger CO_2 contents) would play an essential role in producing small amounts of very undersaturated liquids at temperatures well below the anhydrous pyrolite solidus and at depths near the low velocity zone.

Acknowledgments

The author is grateful to A. E. Ringwood and A. J. Irving for critically reading the manuscript. The invaluable technical assistance of W. Hibberson in carrying out experimental high pressure runs and of E. H. Pedersen in preparing polished mounts is gratefully acknowledged.

References

BOYD, F. R. and J. L. ENGLAND (1960) J. Geophys. Res. **65**, 741.
BULTITUDE, R. J. and D. H. GREEN (1968) Earth Planet. Sci. Letters **3**, 325.
CARTER, J. L. (1966) Ann. Rept. South-West Centre Adv. Studies, Dallas, Texas 65–66, 11.
COOPER, J. A. and D. H. GREEN (1969) Earth Planet. Sci. Letters **6**, 69.
FREY, F. A. (1970) Phys. Earth Planet. Interiors **3**, 323.
GAST, P. W. (1968) Geochim. Cosmochim. Acta **32**, 1057.
GREEN, D. H. (1966) Earth Planet. Sci. Letters **1**, 414.
GREEN, D. H. (1968) in: H. H. Hess and A. Poldervaart, eds., *Basalts: the Poldervaart treatise on rocks of basaltic composition*, Vol II (Wiley-Interscience) 835.
GREEN, D. H. (1969) Upper Mantle Symposium, Prague 1968; Tectonophysics **7**, 409.
GREEN, D. H. and W. HIBBERSON (1970) Phys. Earth Planet. Interiors **3**, 247.
GREEN, D. H., J. W. MORGAN and K. S. HEIER (1968) Earth Planet. Sci. Letters **4**, 155.
GREEN, D. H. and A. E. RINGWOOD (1963) J. Geophys. Res. **68**, 937.
GREEN, D. H. and A. E. RINGWOOD (1964) Nature **201**, 1276.
GREEN, D. H. and A. E. RINGWOOD (1967a) Contr. Mineral. Petrol. **15**, 103.
GREEN, D. H. and A. E. RINGWOOD (1967b) Earth Planet. Sci. Letters **3**, 151.
GREEN, D. H. and A. E. RINGWOOD (1967c) Geochim. Cosmochim. Acta **31**, 767.
GREEN, D. H. and A. E. RINGWOOD (1970) Physics Earth Planet. Interiors **3**, 359.
HUBBARD, N. J. (1969) Earth Planet. Sci. Letters **5**, 346.
IRVING, A. J. and D. H. GREEN (1970) Phys. Earth Planet. Interiors **3**, 385.
ITO, K. and G. C. KENNEDY (1968) Contr. Mineral. Petrol. **19**, 177.
JACKSON, E. D. and T. L. WRIGHT (1970) in press.
KLEEMAN J. D., D. H. GREEN and J. F. LOVERING (1969) Earth Planet. Sci. Letters **5**, 449.
KLEEMAN, J. D. and J. A. COOPER (1970) Phys. Earth Planet. Interiors **3**, 302.
KUNO, H. (1960) J. Petrol. **1**, 121.
KUNO, H. (1964) in: *Advancing Frontiers in Geology and Geophysics* (Osmania Univ. Press), Hyderabad 205.
KUSHIRO, I. (1968) J. Geophys. Res. **73**, 619.
KUSHIRO, I. (1969) Discussion to papers at Upper Mantle Symposium, Prague, 1968; Tectonophysics **7**, 423.
LEGGO, P. J. and R. HUTCHISON (1968) Earth Planet. Sci. Letters **5**, 71.
MACDONALD, G. A. and J. KATSURA (1961) Pacific Sci. **15**, 358.
MACGREGOR, I. D. (1968) J. Geophys. Res. **73**, 3737.
NOCKOLDS, S. R. (1954) Geol. Soc. Amer. Bull. **65**, 1007.
O'HARA, M. J. (1968) Earth Sci. Rev. **4**, 69.
O'HARA, M. J. (1968b) in: P. J. Wyllie, ed., *Ultramafic Rocks* (Wiley, N.Y.) 393.
O'HARA, M. J. and E. J. MERCY (1963) Trans. Roy. Soc. Edinburgh **45**, 251.
O'HARA, M. J. and H. S. YODER (1967) Scottish J. Geol. **3**, 1.
RINGWOOD, A. E. (1962) J. Geophys. Res. **67**, 857, 4473.
RINGWOOD, A. E. (1966) in: P. M. Hurley, ed., *Advances in Earth Science* (M.I.T. Press) 287.
TILLEY, C. E. (1950) Quart. J. Geol. Soc. London **106**, 37.
ROEDDER, E. (1965) Amer. Mineralogist **50**, 1746.
WAGER, L. R. and W. A. WADSWORTH (1960) J. Petrol. **1**, 73.
WHITE, R. W. (1966) Contr. Min. Petrol. **12**, 245.
YODER, H. S. and C. E. TILLEY (1962) J. Petrol. **3**, 342.

UPPER MANTLE COMPOSITION INFERRED FROM LABORATORY EXPERIMENTS AND OBSERVATION OF VOLCANIC PRODUCTS

M. J. O'HARA

Grant Institute of Geology, University of Edinburgh, UK

The significance to geophysicists and geochemists of recent developments in petrological thought on magma origins and partial melting of the upper mantle are reviewed and new arguments favouring a particular upper mantle composition are presented. A garnet-lherzolite-in-kimberlite model of upper mantle composition is satisfactory. Erupted lavas are not primary but have left behind substantial volumes of igneous eclogite accumulates in the upper mantle. Geochemical calculations which ignore the effect of this fractionation on the reaction relationships encountered in partial melting yield invalid conclusions. The depth of magma generation in the upper mantle may be much greater than 100 km. The present state of hypotheses of andesite generation suggest that the availability of water in the upper mantle may have controlled the early development of the continents but too many assumptions are involved to justify any firm inferences on this basis about differences in mantle composition under continents and oceans.

1. Introduction

Current debates among petrologists centre about four questions:
1) What is the composition of the primitive upper mantle from which the common basalt magmas are derived? The answer is vital to geophysicists because this composition, separately or together with the local geotherm, controls the phase constitution of the upper mantle, and hence the density, the seismic velocity and their temperature and pressure derivatives in the upper mantle. Small differences in major element chemistry are *critical* in determining whether the residua after basalt extraction retain garnet and are, therefore, of relatively high density. The answer is equally important to geochemists because the same factors control the chemistry of the partial melting process, determining which crystalline phases are involved with liquid in the minor and trace element partition process. The trace element contents of liquids produced in equilibrium with garnet- as opposed to clinopyroxene-bearing residua will be different. Even if the two source materials have identical trace element contents and nearly identical major element chemistry they may nevertheless be *critically* different in their partial melting behaviour. (Sodium, and the Na/K ratio of early partial melt liquids afford an obvious example, because Na is retained in clinopyroxene, but not in garnet.)
2) Are the liquids produced by partial melting within the upper mantle erupted at the surface without further change of composition (the primary magma concept), or are they modified by closed or open system processes during transport to the surface (the evolved magma concept; see O'HARA, 1965, 1968a pp. 96–97, for further details of the reasons for, and consequence of, abandoning the primary magma concept)? The answer is vital to geophysicists and geochemists because:
i) Adoption of the primary magma concept implies an upper mantle composed essentially of primitive and residual materials, whereas the evolved magma concept implies an upper mantle containing in certain zones substantial amounts of igneous precipitates whose density and seismic properties might be very different from those of primitive or residual material. The evolved magma concept in one form predicts the existence of a substantial proportion of eclogite in the upper mantle.
ii) If correct, the primary magma concept would justify the use of observed lava chemistry and of observed crystal-liquid partition coefficients for the direct calculation or assessment of source mantle compositions. The evolved magma concept however, implies that geochemists who use the primary magma concept will discard as inadequate the true upper mantle composition, and will adopt as satisfactory a range of upper

mantle compositions which are excessively rich in amphibole or phlogopite in order to balance their equations.

The third question arises directly from the last:
3) If magmas do evolve before or during their ascent to the surface, is the principal mechanism that of a) closed system eclogite fractionation which predicts the formation of possibly large volumes of high density eclogite accumulates within the upper mantle (O'HARA 1968a, pp. 94–95 and fig. 7), or, b) closed system enstatite fractionation, accompanied by extensive open system wall rock reaction processes, which does not predict the formation of any accumulates in the upper mantle with densities appreciably higher than that of peridotite (GREEN and RINGWOOD, 1967a)?

The importance of this question to geophysicists lies in the possibility of generating zones rich in material of density higher than common peridotite within the mantle. The importance to geochemists arises because eclogite fractionation provides a means of greatly enriching the concentrations of trace and minor elements in residual liquids without drastic changes in major element chemistry (a process necessary to explain the range of concentration of these elements in magmas of tholeiitic (sub-alkaline) composition) while simultaneously providing a secondary source rock (eclogite) capable of yielding magmas very poor in these same elements in a further cycle of partial melting (O'HARA, 1968a, pp. 125–126).

It is worth noting that eclogite cumulus will also be relatively poor in nickel, and tend to have high Cr/Ni ratios. These characteristics would leave their mark on any liquids produced in such a second cycle of melting and none of the ocean floor tholeiites known to the author have the geochemical imprint required to make this hypothesis preferable to an origin from a peridotite source.

4) The principal component of the oceanic crust appears to be derived from basalt magma, whereas the principal component of the continental crust may have been andesite magma (TAYLOR and WHITE, 1965). What are the origins of andesite? The answer to this question affects not only the interpretation of geophysical properties under active orogenic belts, but also the whole consideration of geophysical and geochemical differences between the sub-oceanic and sub-continental mantle.

2. Discussion

The three principal positions, advocated with respect to the first question, hold that the upper mantle is composed of either:
a) garnet-lherzolite (a peridotite containing olivine, orthopyroxene, clinopyroxene and pyrope-rich garnet) similar in composition to the garnet-lherzolite nodules in kimberlite (GLIK) which yield garnet-harzburgite (olivine-orthopyroxene-garnet) residua on partial melting at *most* pressures encountered in the upper mantle (O'HARA and YODER, 1963, 1967; O'HARA, 1965; ITO and KENNEDY, 1967; O'HARA, 1968a, pp. 105–113). This model is referred to in the following discussion as GLIK/GHR. Or, b) pyrolite (a peridotite (lherzolite) containing olivine orthopyroxene and clinopyroxene, in which Al_2O_3 will not appear as an extra phase such as garnet or spinel, at pressures and temperatures within the depth zone of magma generation), a theoretical composition which must by its nature yield lherzolite residua on partial melting (GREEN and RINGWOOD's hypothesis; development summarised by O'HARA, 1968a, pp. 105–112). This model is referred to hereafter as PL/LR, pyrolite III of GREEN and RINGWOOD (1967b) conforming to this requirement.

Between these two extreme positions it is necessary to recognise a further model based on the assumption that spinel-lherzolite (peridotite) nodules in basalt represent xenoliths of the source mantle composition (e.g. ROSS et al., 1954; WILSHIRE and BINNS, 1961). Available analyses suggest that such rocks will crystallise spinel or garnet as a separate phase over much of the pressure range of interest, but will yield lherzolite, not garnet harzburgite residua on partial melting. This model is, therefore, distinguished as PNIB/LR in the following discussion (while recognising that there may be found some provinces where the nodules in basalts may have compositions akin to the GLIK/GHR model).

Yet a fourth model is necessitated by the formulation of the first pyrolite model of upper mantle composition by GREEN and RINGWOOD (1963) (pyrolite I of GREEN and RINGWOOD, 1967b). This composition crystallises as orthopyroxene-poor garnet or spinel-peridotite under all relevant upper mantle conditions, but is predicted to yield dunite residua (olivine only) at a very early stage of partial melting (O'HARA, 1968a, fig. 9, and p. 111). This model is distinguished hereafter as PGL/DR.

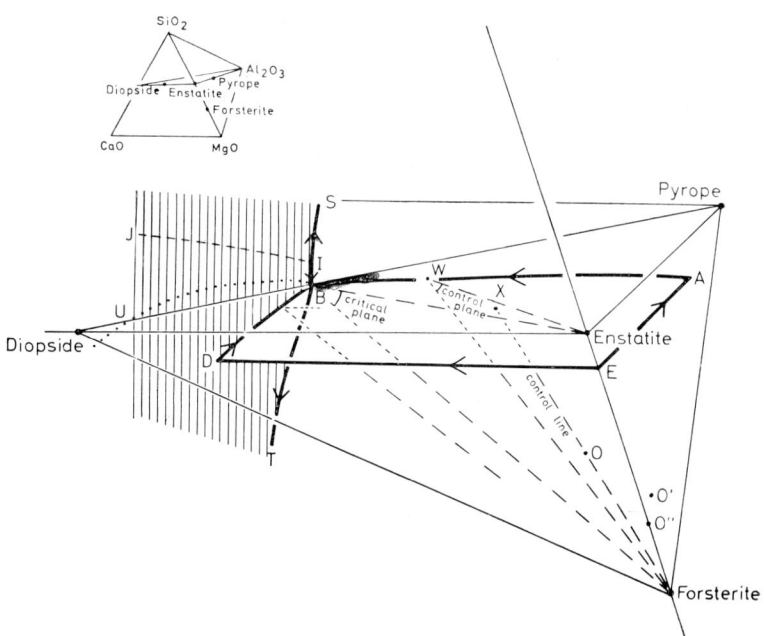

Fig. 1. Inset. Location of the larger figure within the tetrahedron CaO–MgO–Al$_2$O$_3$–SiO$_2$. Main figure. Perspective view of the two surfaces (loci) of liquids which are important in basic magma genesis at high pressures. BAED is the locus of liquids in equilibrium with olivine and orthopyroxene (the harzburgite surface) while the shaded surface extending to the left of the lines S–B–T is the locus of liquids in equilibrium with garnet and clinopyroxene (the eclogite surface) which has been discussed by O'Hara and Yoder (1967). I–J is its intersection with the diopside–enstatite–pyrope plane; B–U is the trace of residual liquids which can be obtained by closed system fractionation of eclogite from a liquid B formed by partial melting of a peridotite such as O. Further description in the text. Solid solutions are ignored for simplicity.

All four models discussed above contain so much potential olivine and orthopyroxene in their mineralogy that under many, if not all, conditions of partial melting in the mantle, olivine will be the last crystalline phase to disappear during advancing partial melting, and orthopyroxene the last but one to disappear in this process. It follows, therefore, that any understanding of the partial melting process demands a knowledge of the compositions of liquids which can form in equilibrium with olivine and orthopyroxene (i.e. by partial melting of olivine and orthopyroxene-bearing crystalline materials).

Fig. 1 is a perspective sketch of the surface which is the locus of liquids in equilibrium (at various temperatures) with olivine and orthopyroxene at one fixed pressure (ca. 30–40 kb) in the system CaO–MgO–Al$_2$O$_3$–SiO$_2$. This surface is bounded within the system by two curved lines B-W-A and B-D. B-D is the curvilinear locus of liquids in equilibrium with clinopyroxene (diopside) in addition to olivine and orthopyroxene, and runs from B within the tetrahedron to a point in the front face (CaO–MgO–SiO$_2$) C, which is the liquid coexisting with diopside, orthopyroxene and olivine in the Al$_2$O$_3$-free system. B-W-A is the curvilinear locus of liquids in equilibrium with garnet in addition to olivine and orthopyroxene. It runs from B to a point in the side face (MgO–SiO$_2$–Al$_2$O$_3$) of the tetrahedron A, which is the liquid in equilibrium with pure pyrope garnet, olivine and orthopyroxene in the CaO-free system.

In general randomly chosen materials of peridotite-like compositions such as O (fig. 1), crystallise in the solid state to a mixture of four crystalline phases, olivine, orthopyroxene, diopside and garnet. Melting begins at a particular temperature with the appearance of a liquid in equilibrium with all four crystalline phases, i.e. the unique liquid of composition B in fig. 1. As melting proceeds one or other of the crystalline phases will be totally consumed, while the liquid composition remains at B. There is a *critical plane*, joining olivine, orthopyroxene and liquid B, which divides:

i) those compositions such as O, to the far, or high Al$_2$O$_3$/CaO side, which lose diopside at this stage of partial melting, and lie within a sub-tetrahedron joining olivine, orthopyroxene and garnet (the residual crystalline phases) and the liquid B, from

ii) compositions to the near, or low Al_2O_3/CaO side, which lose garnet at this stage of partial melting and lie within a sub-tetrahedron joining olivine, orthopyroxene and diopside (the residual crystalline phases) and the liquid B.

The residua derived at this stage from partial melting of composition O have a composition O' rich in olivine and orthopyroxene, poor in garnet, i.e. the residuum is garnet harzburgite in character. B–O–O' are co-linear.

If the temperature rises further, the liquid composition migrates along B–W while the residuum composition migrates from O' to O'', garnet being reduced in amount, the lines joining residua and liquid always passing through O. When the liquid reaches W, in the control plane passing from olivine and enstatite through O, the residuum contains olivine (forsterite) and enstatite only (i.e. is harzburgite).

Any further rise in temperature leads to development of a liquid lying in the control plane, along its *intersection* with the surface BAED, i.e. along W–X, the residuum changing from O'' towards forsterite as enstatite is preferentially melted. When the liquid reaches X in composition, in the control line passing from forsterite through O, only dunite is left as a residuum. Any further rise in temperature leads to a reduction of the amount of forsterite and development of the liquid composition along the line X–O.

The locus of liquids which can be developed by partial melting at high pressure is, therefore, B–W–X–O.

Olivine (forsterite) crystals are always present during this process. It is convenient to represent the relationships by projecting them from forsterite into the plane diopside-enstatite-pyrope of fig. 1. This has been done in fig. 2, where projection of an alternative starting material N, yielding a locus of possible liquids B–Y–Z–N and various lherzolite residua is also shown.

Increasing pressure causes the whole of the surface BAED in fig. 1 to move downwards towards forsterite and away from SiO_2. Consequently, the primary liquids produced in partial melting become richer in olivine components as the pressure increases (O'HARA, 1968a, p. 82–83). The use of an olivine projection, as in fig. 2, conceals this change, which is convenient in the present instance because during their ascent to the surface these primary high pressure liquids cool in a

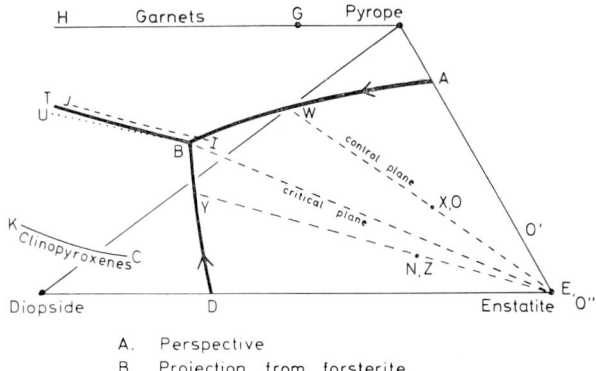

A. Perspective
B. Projection from forsterite

Figs. 2–6. Projections from olivine (forsterite) into the plane diopside–enstatite–alumina. These projections are constructed as explained in O'HARA (1968a, fig. 4) and are described in the text.

reduced pressure environment where the olivine primary phase volume has expanded, i.e. the surface BAED of fig. 1 has moved upwards in the figure. Under such circumstances substantial amounts of olivine may be lost from the ascending magma (O'HARA, 1968a, p. 95–97 and fig. 7; O'HARA, 1968b). But the resultant derived liquid appears in the same position within the olivine projection (fig. 2).

The composition of liquid B, however, does change even within the olivine projection because of the effect which increasing pressure has upon the extent of the primary phase volumes of clinopyroxene, plagioclase, spinel and garnet (O'HARA, 1968a, pp. 83–85, and figs. 3 and 4).

The locus of B from low pressure B_0 through intermediate pressure B_2 to high pressures (~30 kb) at B_3 is shown in fig. 3, as is a hypothetical upper mantle composition O. It will be apparent from the previous discussion of figs. 1 and 2 that liquids along the locus B_3–W_3–X, O can be generated by partial melting of this composition at the appropriate pressure, or B_2–W_2–X, O at lower pressures, and hence that liquids projecting anywhere inside the stippled area, or along the line W_2–W_3–W_0–X, O can be generated at suitable pressures as primary magmas. Erupted liquids which differ from these primary magmas only in having crystallised olivine will project in the same positions as shown in fig. 3. If more extensive cooling has occurred during ascent, fractional crystallisation of other phases may have occurred. All such fractionations tend to produce residual liquids displaced from the shaded locus in fig. 3 away from the enstatite corner of the

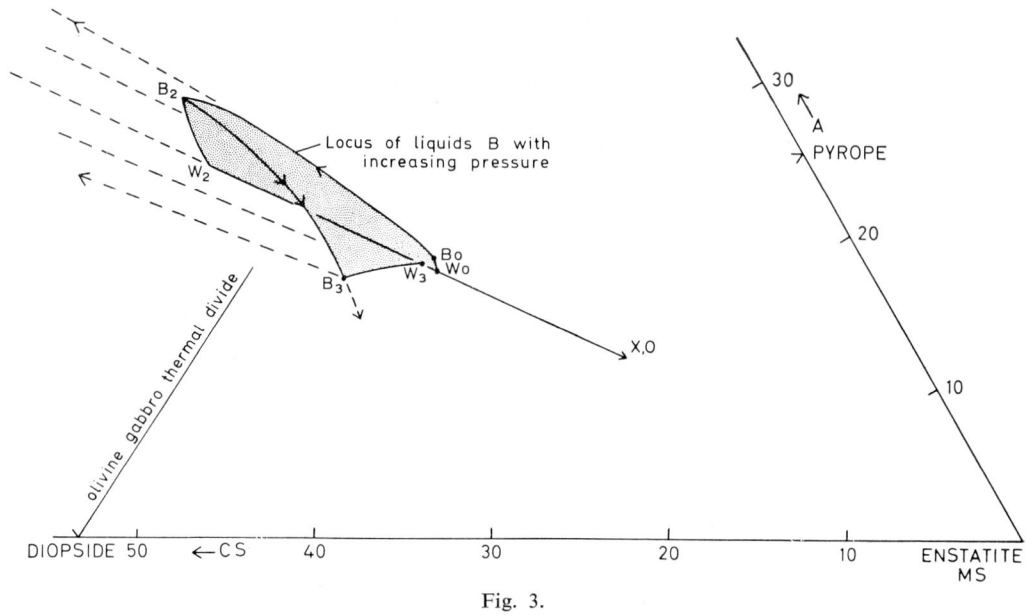

Fig. 3.

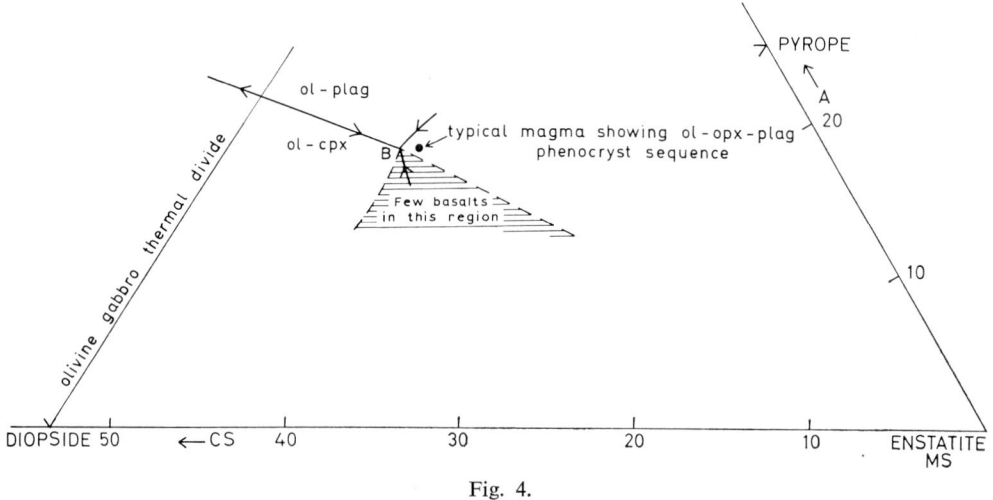

Fig. 4.

projection, as indicated by radiating broken lines.

Fig. 3 then is a typical locus of magmas derivable from a particular source mantle.

Fig. 4 shows an olivine projection derived from experimental data on natural rock compositions (O'HARA 1968a, fig. 4a), of the liquid B at atmospheric pressure, parts of the two curves analogous to BA, BD in fig. 1, and the curve separating the fields of olivine+plagioclase and olivine+clinopyroxene (diopside) crystallisation which runs through the olivine-gabbro thermal divide into nepheline-normative basalt compositions. Many natural basalts project close to this curve and show the nearly simultaneous appearance of olivine, plagioclase and clinopyroxene phenocrysts on cooling. Basalts in the shaded region would show the following sequence of phenocrysts: olivine joined by clinopyroxene and enstatite before plagioclase. Such rocks are rare.

A small filled circle shows a representative composition for a magma which would show the sequence of precipitation of phases: olivine joined by enstatite joined by plagioclase which is typical of many large layered intrusions and some lavas.

It is desirable that any proposed upper mantle com-

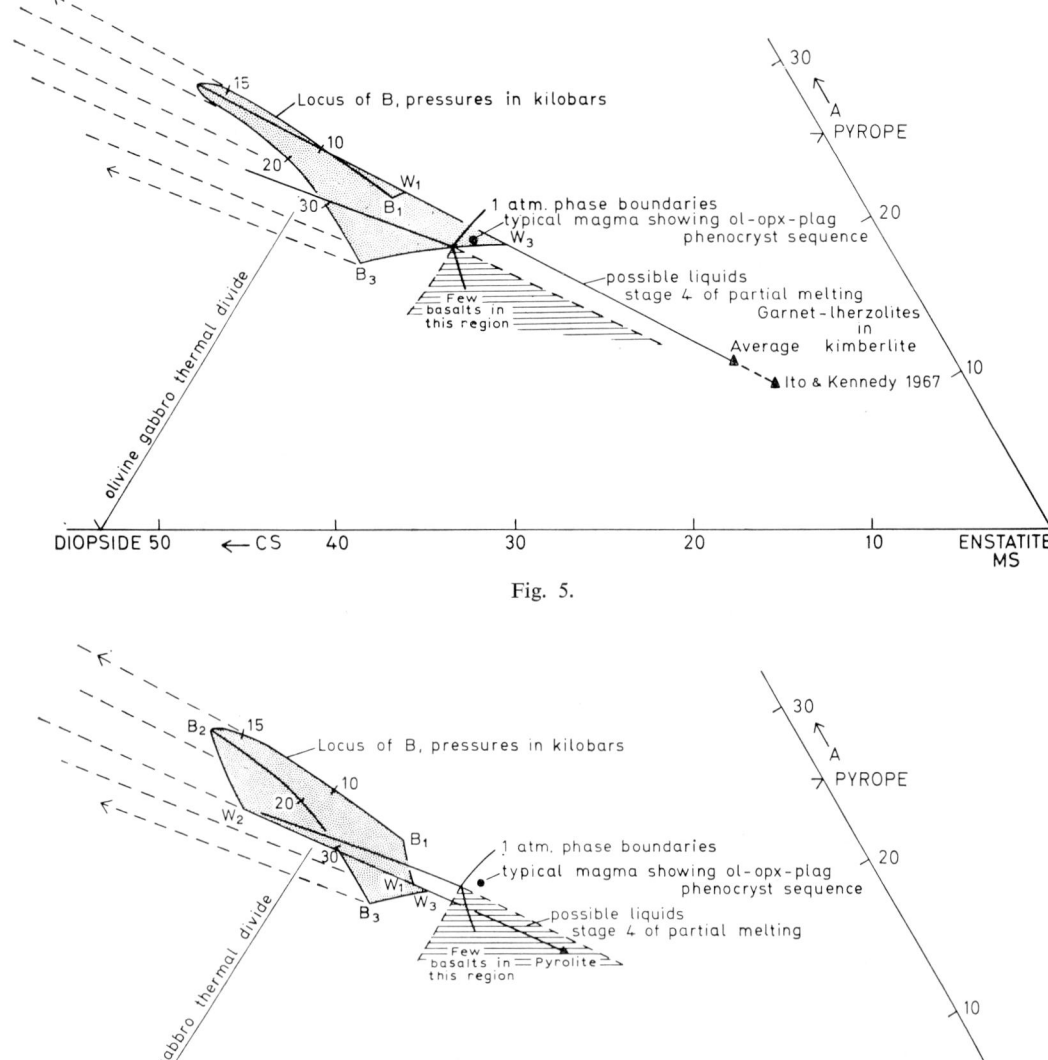

Fig. 5.

Fig. 6.

position should yield a locus of possible partial melt magma compositions, of the type displayed in fig. 3, which covers those regions in fig. 4 which are populated by observed magmas and does not in general cover those regions which are sparsely populated.

Fig. 5 combines figs. 3 and 4 for the specific case of the GLIK/GHR model, assuming a lowest possible pressure for partial melting of ca. 5 kb (B_1–W_1). This model predicts the occurrence of the commonly observed magma types and their observed low pressure crystallisation sequences, and does not predict the formation of magmas in the shaded region except at pressures greater than ca. 40 kb (B_3–W_3).

Fig. 6 combines figs. 3 and 4 for the specific case of the PL/LR model with similar assumptions. This model predicts the generation of magmas in the shaded region whenever partial melting is well advanced and it is unable to account for magmas showing the olivine joined by orthopyroxene joined by plagioclase phenocryst sequence (fig. 6).

The PNIB/LR model yields a diagram similar to fig. 6 while the PGL/DR model yields a diagram similar to fig. 5, but with the "tail" greatly shortened.

Neither the second pyrolite (PL/LR) nor the peridotite nodule in basalt (PNIB/LR) models account satisfactorily for observed magma compositions. They also predict that garnet harzburgite residua cannot form during partial melting of the mantle. WILLIAMS (1932) has noted that garnet harzburgite is relatively common but lherzolite is rare among the nodules accompanying garnet-lherzolites in kimberlite. The garnet-lherzolite (GLIK/GHR) model not only predicts the formation of garnet harzburgite residua of the type observed in kimberlite during partial melting processes, but it has been shown elsewhere (O'HARA, 1968a, pp. 89–90, and captions to figs. 4 and 5) that it yields liquids which are capable of precipitating peridotite nodules of the type found in basalts during their ascent to the surface. The converse is not true, and the GLIK/GHR model is, therefore, preferable to the PNIB/LR and PL/LR models both on the grounds of its liquid products and the nature of the residua and accumulates produced. It is moreover, based upon the composition of a natural rock found in an environment which indicates rapid transport from great depth, and crystallised in a mineral facies known to be characteristic of high pressures.

The PGL/DR model predicts that dunite residua will be encountered relatively early in the partial melting history at high pressures, in conflict with the observation that dunite is very scarce among nodules in kimberlite. Moreover, the harzburgite and garnet-harzburgite residua formed during partial melting of this composition cannot develop among the residua the relatively high ratio of orthopyroxene to olivine which is characteristic of nodules in kimberlite (O'HARA, 1968a, fig. 11). This model is therefore, less satisfactory than the GLIK/GHR model.

Two critical chemical factors to bear in mind when considering or comparing peridotite compositions as potential upper mantle compositions are, thus, seen to be:

i) the CaO/Al_2O_3 ratio which, at any given values of Na_2O/CaO and $Cr_2O_3:Fe_2O_3:Al_2O_3$ ratios, is the principal factor determining the ranges of magma type and residual peridotite mineralogy which can be created and controlling whether or not one peridotite could be a residuum from partial melting of another.

ii) the ratios of potential pyroxenes and garnet constituents, after projection from olivine, which determines how early in the partial melting sequence fully residual, i.e. dunite, mantle will be created.

No very high pressure paragenesis (garnet form of eclogite or orthopyroxene, high pressure forms of olivine, ilmenite-structure pyroxenes* have been identified in kimberlite, and the *relatively* high concentrations of Al_2O_3 in the pyroxenes of the peridotites indicates that little of the material so far investigated comes from pressures greater than ca. 50 kb (O'HARA, 1967). The assemblage of xenoliths in South African kimberlite (sparse eclogite, abundant garnet lherzolite, scarce garnet-harzburgite and harzburgite, very rare dunite) suggest that, contrary to some views, the top 150 km of the subcontinental mantle, under Southern Africa at least, has not been rendered residual in character by extensive magma extraction. Coupled with evidence of a marked culmination in the density profile between 100 and 300 km depth, and of a fall in density to levels associated with magnesian, garnet-free peridotites at slightly greater depths (PRESS, 1968), this evidence suggests that it may be too early to reject the idea that residual character in the upper mantle spreads from the bottom upwards.

The second question posed in the introduction is one which has not been faced consistently by experimental petrologists and geochemists in the past five years in the design of their experiments or studies. Experimental studies which take as their sole starting materials the compositions of actual erupted magmas, or upper mantle compositions calculated on the assumption that some erupted magma type represents the primary partial melt produced in equilibrium with a known or assumed residual upper mantle compositions, are liable to yield results whose significance can easily be misinterpreted if the evolved magma hypothesis is accepted. O'HARA (1968a, p. 93) has criticised GREEN and RINGWOOD's (1967a, p. 104 para 2 of abstract; p. 148–150; p. 156, para 5; p. 158, fig. 10; p. 166, l. 23–26; p. 168 foot) concept of orthopyroxene fractionation as the prime control of magma evolution in the upper mantle on these grounds. Although orthopyroxene is

---

* The diopside-ilmenite eutectics of the Monastery mine (WILLIAMS, 1932) should, however, be kept in mind.

indeed observed as liquidus crystalline phase in the experimentally studied natural magma compositions within certain pressure ranges, no plausible mechanism has been suggested whereby these actual compositions, which are not simultaneously saturated with olivine, can be generated in an olivine-rich environment where partial melt liquids must, by definition, be saturated with olivine, and the required orthopyroxenite accumulates are not in evidence, a point appreciated elsewhere by GREEN and RINGWOOD (1967a).

There is now ample experimental evidence O'HARA, 1968a, pp. 83–85 and figs. 4–6) that the compositions of the majority of erupted magmas cannot be derived directly by the partial melting of any peridotite source material at pressures greater than a few kilobars under dry conditions. The low volatile content of the tholeiitic magmas does not permit evasion of this conclusion by appeals to wet melting; the geothermal gradients required to produce partial melting at the shallow depths implied for primary origin are too steep to be acceptable if the pressure is lithostatic, and pressure release due to open fracturing to greater depths is plausible only until the fracture is filled with liquid whereafter further melting will proceed at the pressure imposed by the liquid column. The primary partial melts of a peridotite mantle at high pressures have been demonstrated by experiment to be picritic in composition (O'HARA and YODER, 1963, 1967; O'HARA, 1963, 1965, 1968a; DAVIS and SCHAIRER, 1965; GREEN and RINGWOOD, 1967; ITO and KENNEDY, 1967) and, independent of any eclogite fractionation which may or may not occur under isobaric high pressure conditions, substantial volumes of olivine must be lost during the ascent of these magmas to the surface under conditions of polybaric crystallisation (O'HARA, 1968a, pp. 78, 97 and fig. 7A and B). This conclusion applies even to the very primitive looking ocean-floor tholeiites, whose erupted compositions are such that they cannot seemingly be derived by direct partial melting of an orthopyroxene-bearing source rock *at any pressure* (i.e. from any of the four mantle models discussed above) or from any olivine-rich source rock (such as the four models discussed above) at pressures greater than 10, or in most cases 5–7 kb (O'HARA, 1968b). The only alternative to these conclusions is to abandon a peridotite model of upper mantle compositions altogether, which, however, raises more problems than it solves.

Geochemical models based on erupted magma compositions (e.g. GAST, 1968; MURTHY and GRIFFIN, 1969) must, therefore, be in error to some undetermined extent. This is particularly true where the calculations of the initial partial melt compositions are based on the assumption that all the major minerals contribute in positive amounts to the liquid composition, when in fact it has been established that either orthopyroxene or olivine is in reaction relationship with the liquid over most of the pressure region of interest (O'HARA, 1968a, pp. 74–78) and the contribution of these phases to the liquid composition must, therefore, be reckoned as a negative quantity.

The choice of answer to the third question is governed by the following considerations. Eclogite fractionation from static magma bodies within the upper mantle involves the extraction of material whose major oxide concentrations (SiO_2, CaO, MgO, FeO, Al_2O_3, Na_2O) differ only slightly from those of the liquid phase. Consequently the volume of a residual liquid can be greatly reduced without too drastic a change in major element chemistry. Eclogite, however, does not extract any appreciable amount of NiO, K_2O, Rb, Ba, TiO_2, P_2O_5 volatiles and a long list of "incompatible" elements whose concentrations are very low in the ocean-floor tholeiites, range up to quite high values in other tholeiites, but reach their highest concentrations in the nepheline-normative magmas. Eclogite fractionation, which can occur at depths of greater than ca. 50 km can, therefore, explain the range of values and the concentrations of the incompatible elements in erupted lavas by processes operating in a closed magma body. If the mechanism really operates, it predicts that substantial volumes of bimineralic eclogite will be present in the upper mantle. If 15% of magma is produced in the initial stages (O'HARA, 1968a, fig. 8, stages 2–3) of partial melting of the mantle, then between 0 and 15% eclogite may be present in any upper mantle column. Its distribution within the column will depend upon where the isobaric fractionation conditions occur. PRESS (1968) has shown that substantial volumes of eclogite may indeed be present in the sub-oceanic mantle between 100 and 300 km depth. These results imply that the depth of magma generation may be greater than is currently believed. The eclogite fractionation mechanism also predicts that the senile stages of large volcanoes, when the probability of prolonged isobaric fractionation of magma at depth is higher, will be marked by the eruption of small amounts of nepheline-normative magma, as is in fact observed in the Hawaiian chain.

Orthopyroxene fractionation involves the extraction of material rich in SiO_2, MgO and FeO, poor in Al_2O_3, but almost devoid of CaO, Na_2O, K_2O etc. from liquids relatively rich in all these constituents. Large reductions of residual liquid volume can only be accomplished at the expense of dramatic changes in residual liquid major element composition. Only small amounts of orthopyroxene extraction (ca. 30%) are permissible if basic igneous chemistry is to be preserved at all. This mechanism, therefore, predicts the presence of relatively small amounts of orthopyroxenite cumulus within the upper mantle, whose density will be relatively low and certainly cannot be held to explain the existence of the high density zone revealed by PRESS' (1968) results. The small volume of extract is insufficient to explain the concentrations and concentration variations of the incompatible elements, hence it is essential to combine this hypothesis with that of zone refining or wall-rock reaction to extract the required trace elements from the surrounding mantle. While this combination of hypotheses, like the eclogite mechanism, predicts the production of late nepheline-normative lavas from large volcanoes, the previous eruption of huge volumes of lava also, moderately enriched in incompatible elements from the same region of mantle, leads one to predict that these late lavas should be poor, rather than rich, in incompatible elements.

The origin of andesite is a problem with too many possible solutions in the present state of knowledge. At least six hypotheses are reasonably well supported by available experimental data (O'HARA, 1968a, pp. 97–102) and further data bearing on two of these hypotheses have been published by GREEN and RINGWOOD (1968) and KUSHIRO et al. (1969). There does, however, seem to be a growing body of opinion that accession of water to the system is the dominant factor, either by modifying the initial partial melting product of peridotite, or by modifying the course of evolution of originally dry basic magma. If this trend of opinion is well founded, the present-day localised origin of andesite may be controlled either by the down-dragging of serpentinised oceanic crust in the orogens, or by the accession of water into basic magmas rising through the overlying wet sediments. The presence of substantial volumes of water in the mantle at an early stage of the Earth's development would, however, have led to the presence of a supercritical fluid phase at the higher pressures (>25 kb) or a low melting water-rich liquid phase at lower pressures. Removal of this to the surface may have been the dominant factor in the origin of the continental crust and the oceans, and this leads to the concept of a uniform continental cover over a uniform mantle. Subsequent events may or may not lead to fundamental differences between sub-oceanic and sub-continental mantle, depending upon one's attitude, positive or negative, towards the hypotheses of deep or shallow mantle convection, and Earth expansion. The nature of any difference to be expected is also controlled by these same attitudes and by one unknown factor which is the effect of pressure up to 1 megabar upon phase equilibria, particularly the fluid phase composition in the peridotite-water system. All conclusions based upon assumptions about one or more of these factors must, for the present be treated with caution, and it is not yet possible to use the andesites to limit the parameters of composition in the underlying mantle, in the way which appears to be becoming possible for basaltic rocks.

3. Summary of conclusions

Garnet-lherzolite nodules in kimberlite provide the best available model of the primitive upper mantle composition and the suite of garnet-lherzolite, garnet harzburgite, harzburgite and eclogite nodules in kimberlite provide the best available means of estimating the relative abundance of primitive and residual mantle, and igneous precipitates within the top 150 km of mantle.

Erupted lavas are not the original liquids produced by partial melting of the upper mantle, but are residual liquids from processes which have left behind complementary eclogite accumulates in the upper mantle. Geochemical inferences about upper mantle composition based upon the assumption that erupted liquids are primary magmas, and that no reaction relationships exist, are invalid.

The site of partial melting in the upper mantle may be deeper than envisaged in recent hypotheses and residual character may increase from the bottom upwards.

The origin of the continental crust is bound up with

the availability and distribution of water in the mantle, during geological history.

References

Davis, B. T. C. and J. F. Schairer (1965) Melting relations in the join diopside–forsterite–pyrope at 40 kilobars and at one atmosphere, Carnegie Inst. Wash. Yearbook **64**, 123–126.

Gast, P. W. (1968) Trace element fractionation and the origin of tholeiitic and alkaline magma types, Geochim. Cosmochim. Acta **32**, 1057–1086.

Green, D. H. and A. E. Ringwood (1963) Mineral assemblages in a model mantle composition, J. Geophys. Res. **68**, 937–945.

Green, D. H. and A. E. Ringwood (1967a) The genesis of basaltic magmas, Contrib. Mineral. Petrol. **15**, 103–190.

Green, D. H. and A. E. Ringwood (1967b) The stability fields of aluminous pyroxene peridotite and garnet-peridotite and their relevance to upper mantle structure, Earth Planet. Sci. Letters **3**, 151–160.

Green, D. H. and A. E. Ringwood (1968) Genesis of the calc-alkaline igneous rock suite, Contrib. Mineral. Petrol. **18**, 105–162.

Griffin, W. L. and V. R. Murthy (1968) Distribution of K, Rb, Sr and Ba in some minerals relevant to basalt genesis (preprint).

Ito, K. and G. C. Kennedy (1967) Melting and phase relations in a natural peridotite to 40 kilobars, Am. J. Sci. **265**, 519–539.

Kushiro, I., Y. Syono and S. Akimoto (1968) Melting of a peridotite nodule at high pressures and high water pressures, J. Geophys. Res. **73**, 6023–6029.

O'Hara, M. J. (1963a) Melting of garnet-peridotite at 30 kilobars, Carnegie Inst. Wash. Yearbook **62**, 71–76.

O'Hara, M. J. (1965) Primary magmas and the origin of basalts, Scottish J. Geol. **1**, 19–40.

O'Hara, M. J. (1967) Mineral parageneses in ultrabasic rocks, in: P. J. Wyllie, ed., *Ultramafic and Related Rocks* (Wiley, New York) 393–403.

O'Hara, M. J. (1968a) The bearing of phase equilibria studies on the origin and evolution of basic and ultrabasic rocks, Earth Sci. Rev. **4**, 69–133.

O'Hara, M. J. (1968b) Are ocean floor basalts primary magma? Nature **220**, 683–686.

O'Hara, M. J. and H. S. Yoder, Jr. (1963) Partial melting of the mantle, Carnegie Inst. Wash. Yearbook **62**, 67–71.

O'Hara, M. J. and H. S. Yoder, Jr. (1967) Formation and fractionation of basic magmas at high pressures, Scottish J. Geol. **3**, 67–117.

Press, F. (1968) Earth models obtained by Monte Carlo inversion, J. Geophys. Res. **73**, 5223–5234.

Ross, C. S., M. D. Foster and A. T. Myers (1954) Origin of dunites and olivine-rich inclusions in basaltic rocks, Am. Mineralogist **39**, 693–737.

Taylor, S. R. and A. J. R. White (1965) Geochemistry of andesites and the growth of continents, Nature **208**, 271–273.

Williams, A. F. (1932) *The Genesis of the Diamond* (Benn, London) 1: 352 pp., 2: 284 pp.

Wilshire, H. G. and R. A. Binns (1961) Basic and ultrabasic xenoliths from volcanic rocks of New South Wales, J. Petrol. **2**, 185–208.

DISPERSED ELEMENTS IN OCEANIC VOLCANIC ROCKS

PAUL W. GAST

Lamont-Doherty Geological Observatory Palisades, New York, U.S.A.

Dispersed elements (elements existing as solid solutions in most natural occurrences) abundances in oceanic volcanic rocks clearly distinguish mid-ocean ridge basalts from basalts found on seamounts and intra-oceanic volcanic islands. Igneous rocks that originate within oceanic-ridge systems (abbreviated herein to OR basalts and gabbros and sometimes designated oceanic tholeiites or abyssal tholeiites) are characterized by

1. chondritic or light depleted rare earth patterns, 2. relatively high abundances of the heavy REE, 3. occasional negative europium anomalies, 4. very low abundances of K, Rb, U, Th and Ba, 5. Sr contents between 80 and 180 ppm, 6. K/Rb ratio ranging from 300 to 1500, 7. covariation of Al and Ti contents (Al_2O_3 ranges from 19–14 weight percent), and 8. Ni abundances from 50-200 ppm that correlate with MgO contents.

Basaltic rocks from oceanic islands and seamounts have

1. REE patterns enriched in light REE and depleted in heavy REE relative to OR basalts, 2. no europium anomalies, 3. high abundances of K, Rb, U, Th and Ba, and 4. a wide range of Sr contents, 200–1200 ppm.

Hawaiian tholeiitic rocks can be clearly distinguished from OR basalts in terms of dispersed element abundances. It is suggested, that Hawaiian basalts have closer affinities to other basaltic rocks from central vent eruptions than to OR basalts.

The origin of the different dispersed element abundance patterns may be a function of

1. the phase composition of the mantle materials from which the observed volcanic materials separated, 2. the extent of partial melting involved in forming the observed liquids and 3. the extent of crystallization that liquids have undergone during the ascent and 4. the dispersed element composition of the mantle.

The abundances of the dispersed elements in oceanic basaltic rocks suggest that the present upper mantle is three to four times richer in the REE, Sr, Ba and U than chondritic meteorites, it is, furhermore, depleted in Ba and Rb relative to Sr and K respectively.

On the basis of the dispersed element characteristics summarized above, the bulk element chemistry, and the tectonic setting of the OR and central vent volcanics, it is suggested that the two magma types described here originate by quite different mechanisms. The OR liquids are produced by extensive melting (20–40%) of mantle material rising in the lithosphere under the mid-ocean ridges. The separation of this liquid from the surrounding mantle takes place at shallow depths, i.e. 25–15 km, after which it ascends rapidly with negligible fractional crystallization. The basaltic liquids from oceanic islands, e.g. St. Helena, Tahiti, the Azores, etc. probably are formed at great depths within a partially molten ($\sim 3\%$) low velocity zone. The extent of partial melting in these regions is probably buffered by the abundance of H_2O and CO_2. Even though these liquids may undergo significant fractional crystallization at shallow depths, many of their dispersed element characteristics are essentially those of the original deep seated liquids, particularly their REE abundances.

Acknowledgment. Work reported here was supported by the National Science Foundation contract NSF-GA-1188; and by the National Aeronautics and Space Administration contract NGL-33-008-012. Lamont-Doherty Geological Observatory Contribution No.1434

EXPERIMENTAL DUPLICATION OF CONDITIONS OF PRECIPITATION OF HIGH-PRESSURE PHENOCRYSTS IN A BASALTIC MAGMA

D. H. GREEN and W. HIBBERSON

Department of Geophysics and Geochemistry, Australian National University, Canberra

An olivine basalt from the Auckland Islands contains partially resorbed "xenocrysts" of orthopyroxene, clinopyroxene and minor olivine. Electron probe microanalyses of these crystals confirm their similarity to near-liquidus crystals obtained experimentally in tholeiitic and alkali olivine basalts at high pressures. A high pressure experimental study of the host olivine basalt demonstrates that orthopyroxene and clinopyroxene are near-liquidus phases at 11–18 kb but the degree of solid solution between the pyroxenes crystallized from the dry magma is much greater than that observed in the natural pyroxenes. Addition of water to the experimental runs results in lowering of the liquidus of the basalt and in the appearance of orthopyroxene, clinopyroxene and olivine as near-liquidus phases at 13–18 kb, 1130–1230 °C. A close correspondence between chemical compositions of the natural "xenocrysts" and the experimental near-liquidus pyroxenes and olivine is obtained and the conditions of the precipitation of the "xenocrysts" from their host magma are inferred to be near 14–16 kb and 1200 °C. The host magma contained $\sim 2\%$ H_2O at these conditions to produce the requisite depression of the liquidus. The "xenocrysts" are regarded as high pressure phenocrysts giving natural evidence of high pressure magmatic fractionation controlled largely by separation of orthopyroxene.

1. Introduction

Recent experimental studies on the melting behaviour of natural basalts at high pressures have led to the formulation of hypotheses of magma generation and fractionation at depths in the mantle. To test such hypotheses we require evidence from the natural basalts themselves on their pre-eruption history. Many basalts have suffered low pressure fractionation to various degrees, so that their chemical compositions bear little relation to the compositions of liquids originally existing at depth. This is particularly true of tholeiitic basalts. However, among the undersaturated basalts, some alkali olivine basalts, olivine basanites, olivine nephelinites etc. contain high density lherzolite and pyroxenite xenoliths of high pressure mineralogy. These liquids have clearly not fractionated by crystal settling at shallow depths (less than those from which the xenoliths were acquired) and their rapid movement to the surface provides minimal opportunity for contamination by digestion of crustal material. Many of these lherzolite-bearing magmas also contain large single crystals ("xenocrysts") including clinopyroxene, olivine, orthopyroxene, spinel and amphibole. These "xenocrysts" (with the exception of olivine) were out of equilibrium with the magma during the final stages of crystallization and show evidence of resorption and reaction. It is a reasonable hypothesis that these crystals are the liquidus or near-liquidus phases of their host magma at depths greater than or similar to those at which the magmas picked up the lherzolite xenoliths. This hypothesis can be experimentally tested for any given magma and "xenocryst" assemblage and if it can be demonstrated that there are unique P, T conditions at which the host magma has liquidus or near-liquidus phases which match the observed xenocrysts in mineralogy and chemical composition, then these P, T conditions may be those at which the magma batch was held, however briefly, allowing some crystallization before final rapid movement to the surface. Although this characterization of P, T environment for the "magma-xenocryst" assemblage does not identify the initial source-region of the magma batch, it does demonstrate a compositional point and a fractionation trend and process which was operating at depth in the particular magma batch. In this way evidence from the natural rocks can be used to evaluate and test rival hypotheses of magma generation and fractionation within the mantle.

An olivine basalt from Mt. Eden plug, Auckland Id., Southern Ocean, was collected and examined by Dr. J. B. Wright* and the presence of both orthopyroxene and clinopyroxene as large, vitreous, corroded crystals was noted. The basalt also contains lherzolite inclusions with the typical olivine, enstatite, clinopyroxene and pale brown spinel assemblage but the "xenocrysts" contrast in size and colour with crystal fragments detached from these xenoliths.

2. Chemical compositions of xenocrysts and host magmas

Several features of the chemical and normative composition (table 1) of the host magma are worthy of comment. The norm contains a small amount of hypersthene but falls within the alkali basalt field using the

TABLE 1

Chemical composition of olivine basalt, Mt. Eden plug, Auckland Id. Otago University No. 19594, A.N.U. No. 2900. Analyst E. Kiss, A.N.U.

SiO_2	46.55		
TiO_2	3.18	CIPW Norm	
Al_2O_3	12.70	Or	5.6
Fe_2O_3	2.98	Ab	24.8
FeO	9.72	An	18.9
MnO	0.17	Di	17.7
MgO	10.63	Hy	2.7
CaO	8.66	Ol	18.8
Na_2O	2.95	Ap	1.3
K_2O	0.95	Ilm	6.1
P_2O_5	0.60	Mt	4.4
H_2O^+	0.67		
CO_2	0.24	Normative feldspar	
Cr_2O_3	0.06	$Or_{11}Ab_{51.5}An_{37.5}$	
NiO	0.04	D.I. = 30.4	
CoO	0.01		
	100.11		

$$\frac{100\ Mg}{Mg + Fe^{++}}\ (\text{atomic ratio}) = 66.2$$

criteria of POLDERVAART (1964). Petrographically it is an alkali olivine basalt in that olivine occurs as both phenocrysts and crystallites, co-precipitating with the clinopyroxene and with no evidence of a reaction relationship with the liquid. The composition also lies within the Hawaiian alkali olivine basalt field on an $Na_2O + K_2O$ vs SiO_2 diagram. The basalt is an example illustrating the continuity of composition between the nepheline-normative alkali olivine basalts and hy-

* Ahmadu Bello University, Zaria, Nigeria.

persthene normative olivine tholeiites (YODER and TILLEY, 1962 p. 353). It lies close to the "plane of critical undersaturation" of the latter authors.

The normative plagioclase composition is andesine $(Ab_{58}An_{42})$ and is more typical of hawaiite composition than alkali olivine basalt if a comparison is made with Hawaiian lavas (MACDONALD and KATSURA, 1964). However the low SiO_2 content and high $Mg/(Mg + Fe^{++})$ ratio of the basalt is typical of alkali olivine basalts and alkali picrites. Except for normative feldspar composition, the host basalt falls close to the alkali olivine basalt point in all the criteria used by TILLEY and MUIR (1964) in their characterization of members of the alkali olivine basalt to trachyte magma series. The term "olivine basalt" is used in this paper for the magma but affinities to alkali olivine basalts are recognized. The points of similarity to some hawaiites are noted and may be linked to data (unpublished) showing that some hawaiites are magmas formed within the deep crust or mantle ($P \geq 8$ kb) and that not all hawaiites can be regarded as products of alkali olivine basalt fractionation in shallow magma chambers.

The compositions of the large phenocrysts present in the magma have been determined using the electron probe microanalyzer and empirical calibration curves based on analyzed minerals and synthetic glasses (cf. GREEN and RINGWOOD, 1967a). Analyses of several crystals and of different areas within one crystal demonstrated small but real variations in composition – the most magnesian and most iron rich compositions obtained for both pyroxenes are listed in table 2. The orthopyroxene and clinopyroxene (table 2) have a high degree of mutual solid solution (i.e. high CaO in orthopyroxene, low CaO in clinopyroxene). They differ in this respect from the co-existing pyroxenes of lherzolite inclusions in basalts (Ross, FOSTER and MYERS, 1954) and are also more iron-rich than the lherzolite assemblages (olivines, enstatites and clinopyroxenes with $100\ Mg/(Mg + Fe^{++}) = 92$–$89$).

The contrast between the clinopyroxene "xenocryst" compositions and the composition of the recrystallised outermost rim, presumed to be in equilibrium with the basalt magma during crystallization at or near the surface, is clearly shown in table 2. The recrystallized rim is markedly different in TiO_2 content, CaO content and hypersthene solid solution, and shows smaller differences in Na_2O and Al_2O_3 contents.

TABLE 2

Chemical compositions of orthopyroxene, clinopyroxene and olivine xenocrysts in olivine basalt. Analyses by electron probe microanalyzer

	Orthopyroxene		Clinopyroxene		Recrystallized rim
SiO_2	56.0	56.0	54.6*	52.9*	50.5
TiO_2	0.2	0.3	0.45	0.65	2.2
Al_2O_3	3.0	3.6	3.4	4.3	2.8
Fe_2O_3	—	—	—	—	1.3*
FeO	8.1	8.6	5.9	6.5	6.7
MgO	31.6	31.0	20.7	19.1	14.2
CaO	2.1	2.1	16.0	15.6	21.9
Na_2O	<0.1	<0.1	0.7	0.7	0.5
	101.1	101.7	101.7	99.7	100.1
Molecular ratios Ca	4	4	32	33	46
Mg	84	83	59	56	41
Fe	12	13	9	11	13
$\frac{100\ Mg}{Mg+Fe^{++}}$	87.4	86.5	86.2	84.0	79.1

Partial analyses of olivines
a) Olivine partially enclosed by orthopyroxene xenocryst ($Ca_4Mg_{84}Fe_{12}$)

	FeO	CaO	Al_2O_3	$\frac{100\ Mg}{Mg+Fe^{++}}$
1. Part against orthopyroxene	12.8	0.1	0.1	86.9
2. Zoned edge against basalt	28.9	0.3	0.6	67.2

b) Olivine phenocryst – continuously zoned, strongly zoned on edge

| 1. Centre | 18.8 | — | — | 80.0 |
| 2. Outer edge | 29.6 | — | — | 66.5 |

* Values calculated from other elements assuming normal pyroxene molecules in solid solution.

A single olivine grain was observed to be partially enclosed by an orthopyroxene "xenocryst" and has closely similar $Mg/Mg+Fe^{++}$ value to the orthopyroxene (table 2). The crystal is not significantly zoned except at the margin contacting the basalt where a sharp compositional gradient leads to more iron-rich olivine with a minimum $Mg/(Mg+Fe^{++})$ value of 67.2. An euhedral olivine phenocryst has a core composition of Fo_{80} with very sharp marginal zoning to at least $Fo_{66.5}$ at the outer edge.

3. Experimental study of the crystallization behaviour of the olivine basalt

Anhydrous conditions: The experimental methods used have been described previously (GREEN and RINGWOOD, 1967a, 113–117). The starting material for the runs was a glass prepared from the analysed basalt and rechecked after fusion for FeO and Fe_2O_3 content. The effect of Fe-loss to the platinum capsule, previously shown to be of minor importance (op. cit. p. 115–117) and not significantly affecting the sample mineralogy, has been further minimised by shorter run times. As a further confirmatory measure, some runs were carried out in graphite capsules demonstrating the same sequence of appearance of phases at 13.5 kb. Runs in both platinum and graphite at 13.5 kb and at 1330 °C and 1320 °C all yielded orthopyroxene+clinopyroxene+liquid. The similarity of degree of crystallization and of the 100 $Mg/(Mg+Fe^{++})$ values (83.0) of the clinopyroxenes in both 1330 °C and 1320 °C runs in graphite suggests that there is no actual difference in degree of crystallization between these runs. The more magnesian compositions of the pyroxenes from the run in platinum at 1330 °C may be due to this run being nearer the liquidus or to some iron loss. The latter effect may also have caused the appearance of orthopyroxene (100 $Mg/(Mg+Fe^{++})$ = 86.4) alone in the one hour run at 1325 °C. Microdetermination of FeO (10.3%) and Fe_2O_3 (0.97%) in the 30 min. run at 13.5 kb 1320 °C in platinum confirms the relatively small change in chemical and normative composition produced by iron loss within the run times used. Microprobe methods and the accuracy of analyses are as previously described (GREEN and RINGWOOD, 1967a).

The crystallization behaviour of the Auckland Island olivine basalt is almost identical to that of the olivine basalt studied by GREEN and RINGWOOD (1967a). Comparison of the compositions shows higher normative olivine and Al_2O_3 in the previous composition but otherwise very similar chemical and normative compositions. Details of experimental runs are given in table 3. Olivine is the liquidus phase at 9 kb and is joined by clinopyroxene as the second phase. At 13.5 kb orthopyroxene and clinopyroxene occur together near the liquidus. Orthopyroxene is the major phase, or possibly the only phase on the liquidus, but at lower temperatures, clinopyroxene is more abundant. At 18 kb clinopyroxene appears to be the liquidus phase and orthopyroxene probably appears over a very restricted lower temperature interval. In comparison with the previous data on olivine basalt (GREEN and RINGWOOD, 1967) the present composition has a slightly more re-

TABLE 3

Experimental crystallization of Auckland Island olivine basalt at various pressures and temperatures

Pressure (kb)	Temperature (°C)	Time (mins)	Sample capsule	Results
A. Dry conditions				
9.0	1280	30	Pt	Olivine+glass. Very near liquidus
9.0	1260	30	Pt	Olivine+clinopyroxene+glass
11.3	1320	30	Pt	Above liquidus
11.3	1310	30	Pt	Clinopyroxene+rare orthopyroxene+minor olivine. Clinopyroxene with rare parallel growth of orthopyroxene
13.5	1350	30	Pt	Above liquidus
13.5	1330	30	Pt	Uncommon orthopyroxene and clinopyroxene+glass. Very near liquidus
13.5	1330	60	Graphite	Clinopyroxene+orthopyroxene+glass Clinopyroxene > orthopyroxene
13.5	1325	60	Pt	Orthopyroxene+glass
13.5	1320	30	Pt	Clinopyroxene+rare orthopyroxene+glass Cpx ≫ Opx
13.5	1320	60	Graphite	Clinopyroxene+orthopyroxene+glass Cpx > Opx
13.5	1300	30	Pt	Clinopyroxene+minor orthopyroxene+glass (~30% crystallization)
18.0	1380	60	Graphite	Above liquidus
18.0	1370	60	Graphite	Clinopyroxene+glass. Cpx may be quench
18.0	1360	60	Graphite	Clinopyroxene+possible rare orthopyroxene+glass
B. "Wet" conditions				
13.5	1200	30	Pt	Olivine+glass. Very near liquidus
13.5	1190	30	Pt	Olivine+orthopyroxene+glass. Opx > ol
13.5	1180	30	Pt	Olivine+orthopyroxene+glass. Opx ≫ ol
13.5	1160	30	Pt	Olivine+orthopyroxene+clinopyroxene+glass. Opx > Cpx. Minor olivine. Possible amphibole
13.5	1150	30	Pt	Olivine+orthopyroxene+clinopyroxene+glass
13.5	1130	30	Pt	Olivine+orthopyroxene+amphibole+glass. Clinopyroxene not certain
18.0	1260	30	Pt	Orthopyroxene+glass. Very near liquidus
18.0	1240	30	Pt	Orthopyroxene+clinopyroxene+glass. Opx > Cpx
18.0	1200	30	Pt	Orthopyroxene+clinopyroxene+glass. Orthopyroxene and clinopyroxene intergrowths well developed Opx ≃ Cpx
C. Controlled Water Contents				
15.3	1200	20	Pt	With 3% H_2O. Above liquidus
14.4	1200	30	Pt	With 2% H_2O. Above liquidus
14.4	1170	30	Pt	With 2% H_2O. Orthopyroxene+clinopyroxene+rare olivine. Opx common
15.3	1200	20	Pt	With 2% H_2O. Rare large orthopyroxene, no definite olivine.
15.3	1170	30	Pt	With 2% H_2O. Common orthopyroxene and possible rare olivine

stricted field of orthopyroxene crystallization. The analytical data on the pyroxene compositions (table 4) demonstrate a very high degree of hypersthene solid solution in the clinopyroxene and somewhat lower Al_2O_3 contents in both pyroxenes than those observed in the previous olivine basalt. The clinopyroxenes have compositions suggestive of very magnesian pigeonites but comparison with the sub-calcic (9–10% CaO) clinopyroxenes previously obtained experimentally does not provide any evidence as yet for a compositional break between augites, sub-calcic augites and compositions near to pigeonite. The coexistence of orthopyroxenes with slightly varying CaO content with this range of clinopyroxene compositions suggests that we are dealing with the "roof" of the two-pyroxene miscibility gap, the orthopyroxene side being "steep" (i.e. CaO content varies only slightly with temperature and with $Mg/(Mg+Fe^{++})$ ratio) while the clinopyroxene side in contrast is "shallow" and the hypersthene solid solution rapidly increases for small increases in temperature

TABLE 4

Compositions of phases crystallized from the olivine basalt under dry conditions in sealed platinum capsules or in graphite capsules with "dry" (talc+boron nitride) furnace assemblies

Pressure (kb)	11.3		13.5		13.5	
Temperature (°C)	1310		1330		1320	
Sample capsule	Platinum		Platinum		Graphite	
Time	30 mins		30 mins		60 mins	
Phases present	Ol+Cpx+Glass		Opx+Cpx+Glass		Opx+Cpx+Glass	
Analyzed phase	Olivine	Clino-pyroxene	Ortho-pyroxene	Clino-pyroxene	Ortho-pyroxene	Clino-pyroxene
SiO_2	40.0*	53.5*	55.0*	54.4*	54.0*	53.0*
TiO_2	—	(0.6)+	0.4	0.5	0.4	0.7
Al_2O_3	<0.2	3.9	4.1	3.8	4.0	4.0
Fe_2O_3	—	—	—	—	—	—
FeO	14.2	9.4	8.4	8.1	10.3	9.9
MgO	45.5*	27.2*	29.9*	27.9*	28.9*	27.0*
CaO	0.3	5.3	2.2	5.1	2.4	5.2
Na_2O	—	(0.2)+	<0.05	0.2	<0.05	0.25
Molecular proportions Ca	0.4	10.5	4.5	10.0	4.5	10.5
Mg	84.8	75.0	82.5	77.5	79.7	74.5
Fe	14.8	14.5	13.0	12.5	15.8	15.0
$\frac{100\ Mg}{Mg+Fe^{++}}$	85.1	83.6	86.3	86.0	83.3	83.0

* Calculated value.
+ Value assumed from analyses of similar clinopyroxenes.

or decreases in $Mg/(Mg+Fe^{++})$ ratio. A similar effect is evident in the simple system diopside-enstatite (DAVIS and BOYD, 1966) and is reported for the system hedenbergite-ferrosilite at 20 kb by LINDSLEY (this volume).

"*Wet*" *Conditions*: Apparatus for sealing of water in the small high pressure capsules was not available when the majority of the experiments were carried out so that the techniques of running with an unknown but fairly reproducible water contents were followed (BULTITUDE and GREEN, 1968). Approximately 1 mgm (~5%) of water was added to the sample and the platinum capsule crimped but not welded. The high pressure furnace assembly used a simple talc cylinder (see GREEN and RINGWOOD, 1967b fig. 1) in which dehydration of the talc and use of open (non-welded) platinum capsules had previously (GREEN and RINGWOOD, 1964) produced a depression of the liquidus of an olivine tholeiite composition by approximately 100 °C. Use of these techniques produced a depression of the liquidus of the olivine basalt of about 130 °C at 13.5 kb and 100 °C at 18 kb. It might be anticipated that runs with this technique would yield rather random points between liquidus and solidus but in fact the careful repetition of the same run procedure produced a series of runs at 13.5 kb and 18 kb in which the order of decreasing temperature is also the order of increasing degree of crystallization as illustrated by the regular sequence of appearance of phases, and the regular decrease of the ratio $100\ Mg/(Mg+Fe^{++})$ in the phases. These data suggest that the technique produces similar but unknown activity of water in the experimental runs. Several later experiments carried out with sealed Pt capsules and known water contents yielded crystallization products consistent with the previous work and showing that the observed depression of the liquidus required water contents of 2–3% (table 3). The earlier technique of running with appreciable but unknown water activity has been criticized by O'HARA (1968) and KUSHIRO (1968) but the data obtained on this particular composition give no grounds for suspecting change of sample composition by transport of components in an aqueous phase either moving into or out of the sample capsule.

Table 5

Compositions of analyzed crystals obtained in crystallization of olivine basalt at 13.5 kb and 18 kb under "wet" conditions i.e. lowering of liquidus by ~120 °C by the addition of water.

Phase	Temperature (°C)	100 Mg / (Mg+Fe^{++}) (atomic) ratio	Molecular proportions			Weight per cent	
			Ca	Mg	Fe	Al$_2$O$_3$	CaO
A. At 13.5 kb							
Olivine	1200	86.9	0.2	86.7	13.1	<0.2	0.2
Olivine	1190	86.9	0.2	86.7	13.1	<0.2	0.2
Orthopyroxene		90.0	3.4	87.1	9.6	1.7	1.8
Olivine	1180	84.8	0.2	84.6	15.2	<0.2	0.2
Orthopyroxene		88.1	3.6	85.2	11.3	1.7	1.9
Olivine		81.4	0.2	81.2	18.6	<0.2	0.2
Orthopyroxene	1160	84.3	4.0	81.2	14.8	4.4	2.0
Clinopyroxene		no consistent analyses					
Olivine		80.0	0.3	79.8	19.9	<0.2	0.3
Orthopyroxene	1150	83.3	4.0	80.0	16.0	4.3	2.0
Clinopyroxene		83.3	32.0	56.6	11.3	6.1	≥15.2
Olivine		81.0	0.2	80.8	19.0	<0.2	0.2
Orthopyroxene	1130	82.8	4.0	79.5	16.5	4.5	2.0
Amphibole		—	not analyzable				
B. At 18 kb							
Orthopyroxene	1260	89.3	3.8	85.9	10.3	2.4	2.0
Orthopyroxene	1240	86.1	3.9	82.8	13.3	2.5	2.0
Clinopyroxene		83.5	30.4	58.0	11.6	5.5	≥14.0
Orthopyroxene	1200	85.6	3.9	82.3	13.8	3.1	2.0
Clinopyroxene		83.2	35.5	53.4	11.1	5.9	16.2

Coexisting orthopyroxene and clinopyroxene

	13.5 kb 1150 °C		18 kb 1240 °C		18 kb 1200 °C	
	Opx	Cpx	Opx	Cpx	Opx	Cpx
SiO$_2$	51.1*	51.7*	55.3*	52.8*	54.9*	52.3*
TiO$_2$	—	0.6	0.4	0.7	0.4	0.7
Al$_2$O$_3$	4.5	6.1	2.5	5.5	3.1	5.9
FeO	10.4	6.9	8.8	6.8	9.1	6.4
MgO	29.4*	19.3*	30.9*	19.2*	30.5*	17.5*
CaO	2.0	≥15.2	2.0	≥14.0	2.0	16.2
Na$_2$O	—	0.3	—	1.0	—	1.0

* Calculated values.

The depression of the liquidus of the olivine basalt to 1200 °C at 13.5 kb resulted in appearance of olivine as the liquidus phase, closely followed by orthopyroxene and joined by clinopyroxene as the third phase. At 1130 °C, amphibole may replace clinopyroxene as the latter was not definitely identified. It may also be noted that the olivine coexisting with amphibole at 1130 °C is more magnesian than that at higher temperature. Olivine is present throughout the crystallization interval studied but remains a minor phase and does not perceptibly increase in abundance, in contrast with the large increase in abundance of the orthopyroxene and clinopyroxene. The orthopyroxene analyses (table 5) show CaO contents consistently lower than those obtained from orthopyroxenes at 1320–1330 °C. Al$_2$O$_3$ contents in the higher temperature orthopyroxenes are low but increase in the 1160–1130 °C runs. The clinopyroxene analyzed at 1150 °C has higher Al$_2$O$_3$ content than coexisting orthopyroxene and, most significantly, has a very much higher CaO content than the clinopyroxenes analyzed from 1320 °C and 1330 °C runs.

The experimental runs at 18 kb yielded orthopyroxene as the liquidus phase, joined by clinopyroxene at lower temperatures. Orthopyroxene is slightly more magnesian than coexisting clinopyroxene and has lower Al_2O_3 content (cf. GREEN and RINGWOOD, 1967a). The clinopyroxene compositions have CaO contents of 14–16%. The most reliable clinopyroxene analyses obtained from the runs were those at 18 kb, 1200 °C; in most runs the crystals were smaller than orthopyroxene and with quench outgrowths and were analyzed with difficulty. The Na_2O contents of the clinopyroxenes in equilibrium with the olivine basalt liquid increase with increasing pressure – thus in both the dry and "wet" runs at 13.5 kb, the clinopyroxenes contained 0.2–0.3% Na_2O with apparently slightly lower contents in the higher temperature clinopyroxenes. At 18 kb however the clinopyroxenes contain 1.0% Na_2O, probably in jadeite solid solution.

A puzzling feature of the compositions listed in table 5 is that olivine has a consistently lower 100 $Mg/(Mg+Fe^{++})$ ratio than co-existing orthopyroxene in the "wet" runs. This differs from the pattern previously obtained at 9 kb, 1290–1250 °C (GREEN and RINGWOOD 1967a) where olivine has essentially the same 100 $Mg/(Mg+Fe^{++})$ ratio as co-existing orthopyroxene. Except for the 13.5 kb, 1150 °C run, co-existing pyroxene pairs have clinopyroxene with lower 100 $Mg/(Mg+Fe^{++})$ ratio than coexisting orthopyroxene. The differences are very small at 13.5 kb, 1320–1330 °C i.e. where the compositional differences between the two pyroxenes are relatively small, but are larger in the 18 kb, 1240–1200 °C runs. The partition coefficients for Fe and Mg between pyroxene pairs in natural rocks are such that the clinopyroxene is relatively enriched in magnesium although for magnesian igneous pyroxene pairs the coefficients approach unity. The apparent reversal of this trend in the experimental runs may in part be due to appreciable Fe^{+++} contents (not determinable by microprobe analyses) in the clinopyroxenes or may be a real feature of the high temperature of equilibration of these assemblages.

Comparison of natural xenocryst compositions with experimentally crystallized phases: The experiments under dry conditions show that olivine, orthopyroxene and clinopyroxene (the three natural xenocryst phases) occur together on the liquidus at 11.3 kb, 1310 °C; the two pyroxenes occur without olivine on the liquidus at 13.5 kb and olivine occurs on the liquidus at 9 kb. The analyzed synthetic phases are similar or slightly lower in 100 $Mg/(Mg+Fe^{++})$ value than the natural xenocryst phases and it is probable that accurate matching of this parameter could be obtained at 11–12 kb and appropriate temperature (1310–1320 °C). However, the analyzed clinopyroxene at 11.3 kb and the pyroxene pairs at 13.5 kb are very different in composition from the natural pyroxenes, particularly in the extremely sub-calcic nature of the clinopyroxene and the slightly higher CaO content of the synthetic orthopyroxene.

Comparison of tables 2 and 5 shows that the natural pyroxene compositions are very closely matched by analyzed pyroxenes from "wet" runs at 13.5 kb and 18 kb. CaO contents in both pyroxenes, TiO_2 contents, 100 $Mg/(Mg+Fe^{++})$ ratios and Na_2O contents are closely matched. In detail the Na_2O content of the natural clinopyroxenes (0.7%) is between the values for the 13.5 kb (0.3) and 18 kb (1.0) clinopyroxene; the CaO content of the natural orthopyroxene is slightly greater and the CaO content of the more Fe-rich natural clinopyroxene slightly lower than the 18 kb, 1200 °C pyroxene pair. These differences suggest that a pressure of 14–16 kb and a temperature slightly above 1200 °C would yield near-liquidus pyroxenes identical to the natural xenocrysts. It may be noted also that while olivine precedes orthopyroxene in the crystallization sequence at 13.5 kb, and is absent at 18 kb, it should occur together with orthopyroxene on the liquidus at about 15 kb. The natural pyroxenes encompass a small range in 100 $Mg/(Mg+Fe^{++})$ ratio and thus represent crystal/liquid equilibria over a small range of P, T conditions such as might be anticipated in a cooling magma chamber or in a static or slow moving feeder dyke. The range of experimental conditions in the wet melting experiments appears to exceed the range required to produce the observed compositional variations.

4. Conclusions

The experimental study of the crystallization of the Auckland Id. olivine basalt has shown that the olivine basalt can precipitate, as near-liquidus phases, crystals which closely match the observed partly resorbed "xenocrysts" within the basalt. This matching of liquidus and xenocryst phases in chemical composition and in paragenesis can only be achieved experimentally over a very small P, T range. In particular, the possible temperature of precipitation is rather closely fixed by the observed degree of solid solution between co-existing pyroxenes and the load pressure at precipitation may be deduced from the relative roles of pyroxene or olivine as liquidus phases and by the Na_2O content of the clinopyroxene. It is considered that the "xenocrysts" are not of accidental origin but are cognate, high-pressure phenocrysts precipitated from the host olivine basalt at depths of around 50–55 kms, (load pressure 14–16 kb) and a temperature close to 1200 °C. These conditions of precipitation require that the basalt liquidus was depressed about 130 °C below the dry liquidus – this is consistent with the presence of approximately 2% water within the magma at the depth of precipitation of the crystals.

The inferred high pressure phenocrysts have lower $100\ Mg/(Mg+Fe^{++})$ values than that inferred for parental mantle material or those values present in lherzolite or garnet peridotite xenoliths ($100\ Mg/(Mg+Fe^{++})$ ~ 88–92). Thus the composition of the Auckland Id. olivine basalt is not considered to be that of a direct partial melt from mantle peridotite but rather to be a liquid produced from some more primitive parent magma by fractional crystallization at depths ≥ 50 km. While it is not possible to unequivocally deduce the nature of this parent magma, the Auckland Id. olivine basalt magma provides natural evidence of a process of fractional crystallization operating at about 50 km depth and "quenched" by the rapid eruption of both liquid and precipitating phases. This fractionation trend is dominated by pyroxene separation and in particular, by orthopyroxene. Although olivine accompanies the near-liquidus pyroxenes in the experiments, it is volumetrically a minor percentage of the precipitated material i.e. the precipitated material would be olivine-poor pyroxenite and not peridotite mineralogy. In chemical composition, the estimated crystal extract at 14–16 kb is very similar to that deduced for the olivine basalt and alkali olivine basalt by GREEN and RINGWOOD (1967a). The calculated fractionation trend at 14–16 kb, ~ 1200 °C, deduced for the Auckland Id. olivine basalt predicts more hypersthene-normative, tholeiitic compositions as parental to the observed host basalt and predicts nepheline-normative alkali olivine basalts as lower temperature derivative liquids from the host basalt. The data support and extend the conclusions of GREEN and RINGWOOD (1964, 1967a) on the dominant role of orthopyroxene and orthopyroxene + clinopyroxene crystallization at 13–18 kb in producing a spectrum of basaltic liquids from olivine tholeiite to basanite. It has been demonstrated that it is possible to characterize in a rather unequivocal manner, the P, T conditions and water content of some magmas within the upper mantle, prior to rapid extrusion.

Acknowledgements

We thank Dr J. B. Wright for originally bringing the Auckland Id. basalt and its xenocrysts to our attention. The technical assistance of Mr E. H. Pedersen in the preparation of microprobe mounts is gratefully acknowledged.

References

BULTITUDE, R. J. and D. H. GREEN (1968) Earth Planet. Sci. Letters **3**, 325.
DAVIS, B. T. C. and F. R. BOYD (1966) J. Geophys. Res. **71**, 3567.
GREEN, D. H. and A. E. RINGWOOD (1964) Nature **201**, 1276.
GREEN, D. H. and A. E. RINGWOOD (1967a) Contr. Mineral. Petrol. **15**, 103.
GREEN, D. H. and A. E. RINGWOOD (1967b) Geochim. Cosmochim. Acta **31**, 767.
KUSHIRO, I. (1968) Upper Mantle Committee Symposium, Abstr. Proc. XXIII Internat. Geol. Congr. Prague. 11.
LINDSLEY, D. H. (in press).
MACDONALD, G. A. and T. KATSURA (1964) J. Petrol. **5**, 82.
O'HARA, M. J. (1968) Earth Sci. Rev. **4**, 69–133.
POLDERVAART, A. (1964) Geol. Soc. Am. Bull. **75**, 229.
ROSS, C. S., M. D. FOSTER and A. T. MYERS (1954) Am. Mineralogist **39**, 693.
TILLEY, C. E. and I. D. MUIR (1964) Geol. Foren. Stockholm Forh. **85**, 434.
YODER, H. S. and C. E. TILLEY (1962) J. Petrol. **3**, 342.

GABBROIC AND ULTRAMAFIC NODULES: HIGH LEVEL INTRACRUSTAL NODULAR OCCURRENCES IN ALKALIC BASALTS AND ASSOCIATED VOLCANICS FROM KENYA, DESCRIBED AND COMPARED WITH THOSE OF HAWAII

G. J. H. McCALL

Reader in Geology, University of Western Australia, Nedlands 6009, Western Australia

Nodules in alkalic olivine basalt and associated volcanic rocks from Kenya include gabbroic, dioritic, syenitic and eclogitic material. Nodules of ijolitic and uncompahgritic material are found in the nephelinite volcanics. All but the eclogitic nodules of the Chyulu Hills are considered to be of high level, intracrustal provenance: the eclogitic nodules described by Saggerson may be of upper mantle provenance. These occurrences are compared with gabbroic and peridotite/dunite nodule occurrences of Hualalai and Mauna Kea, Hawaii, which are probably largely of intracrustal cumulate derivation. The mechanism of gas-coring that is believed to be responsible for the phenomenon of nodular inclusions is tentatively related to an immiscible carbonate fraction, separated in alkalic and alkaline magmas, but not in tholeiitic magmas. A brief discussion of the whole spectrum of nodular occurrences, with special reference to their level of provenance, concludes this account.

1. Introduction

The discovery of gabbroic nodules* in profusion on and within the uppermost lava flows of the Katenmening Basalt sequence, one of the last erupted members of the pre-caldera eruptive sequence of Silali Volcano (McCALL, 1958a, b) is described. This occurrence has been coupled with previous observations by the writer of syenite nodules and blocks within the last erupted, pre-caldera phonolitic trachyte tufflavas ("froth flows") of Menengai and Kilombe volcanoes situated to the south (McCALL, 1957a, b, 1963, 1964a, b, 1967, 1968b), in the same Cainozoic volcanic province.

The author has been prompted by these discoveries and by a recent visit to Hawaii to relate these obviously high-level (crustal) enclave occurrences to the wider problem of the occurrence of ultrabasic and basic nodules in alkalic basalts, a problem that has for some years assumed critical importance in upper mantle studies.

The Silali occurrence is described in some detail, syenite enclaves occurring within trachytic volcanics on Menengai, Kilombe and Silali (and in older, similar volcanics in the foundation beneath that volcano) are briefly mentioned. Nodules of high level provenance in nephelinites and eclogite nodules of possible deep provenance in alkalic basalts are also briefly mentioned, followed by a general discussion of the level of provenance and mechanism of production of such nodules in alkalic volcanic suites.

2. Silali volcano

Silali volcano, situated (fig. 1) at the boundary between Baringo and Turkana districts, Kenya, on the floor of the eastern trough of the Gregory Rift Valley, is a composite caldera volcano. First visited by the writer in 1965 for a month, it was again visited in 1967 (for two months), in association with the East African Geological Research Unit of Professor B. C. King, Bedford College, London University.

This is one of the most spectacular caldera volcanoes in the world, and it presents many problems to the volcanologist and petrologist (McCALL, 1968a, b). An account of the general features of the geology of this volcano and its environs is completed in manuscript form, and a petrological description of the component volcanic rocks is in preparation – some 130 chemical analyses are being carried out by XRF methods at Leeds University. This account deals solely with the unusual enclave occurrences within the basalt flows.

---

* The term nodules strictly denotes rounded bodies: both the intracrustal and sub-crustal (upper mantle) enclaves considered in this paper tend to be angular, facetted masses broken off a rock mass, though rounded individual masses also occur. The term nodule, by custom applied to these enclaves, is thus used *sensu lato*.

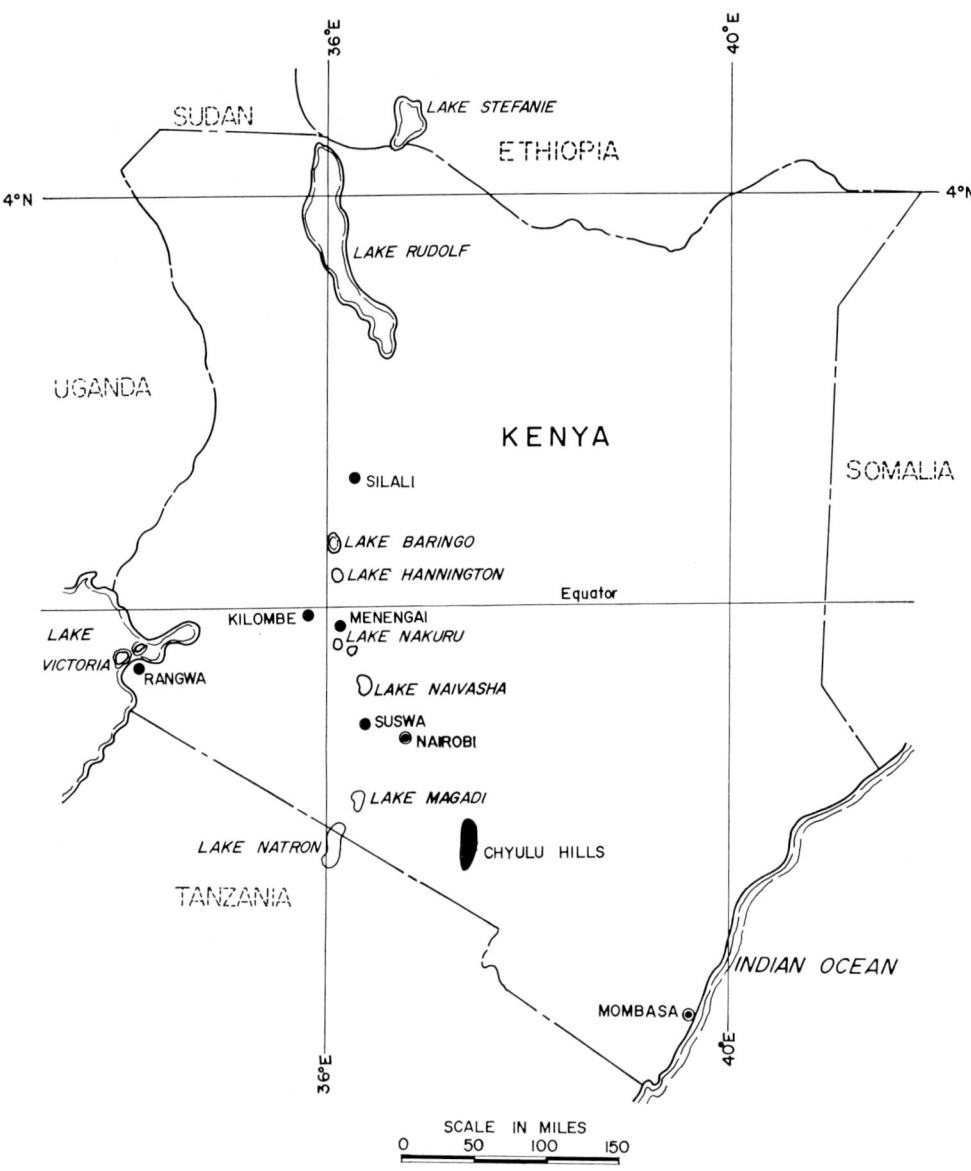

Fig. 1. Sketch map showing the location of Silali, Menengai and Kilombe volcanoes.

3. Gabbroic nodules of Silali

These occurrences are restricted to the top flow (or possibly flows?) of a thick sequence of individually thin (<10 feet thick) basalt flows known as the Katenmening Basalts. These flows were erupted subsequent to the main phase of intermediate lava and tuff emission that built up the 100 square mile central volcano of Silali during Pleistocene times. The probable date of commencement of the initial eruptions of phonolitic trachyte lavas was during the Middle Pleistocene, that is about 500 000 years ago, at the time of the Kanjeran deposition in the Nakuru Basin to the south (MCCALL et al., 1967). However, it must be noted that in the light of the now widely-accepted concept of an extended Pleistocene, not necessarily glacially defined, the time referred to is actually well up towards the top of the Pleistocene time range. These basalts were erupted in the manner of lava plateaux, from innumerable closely spaced, fissure aligned vents situated within a

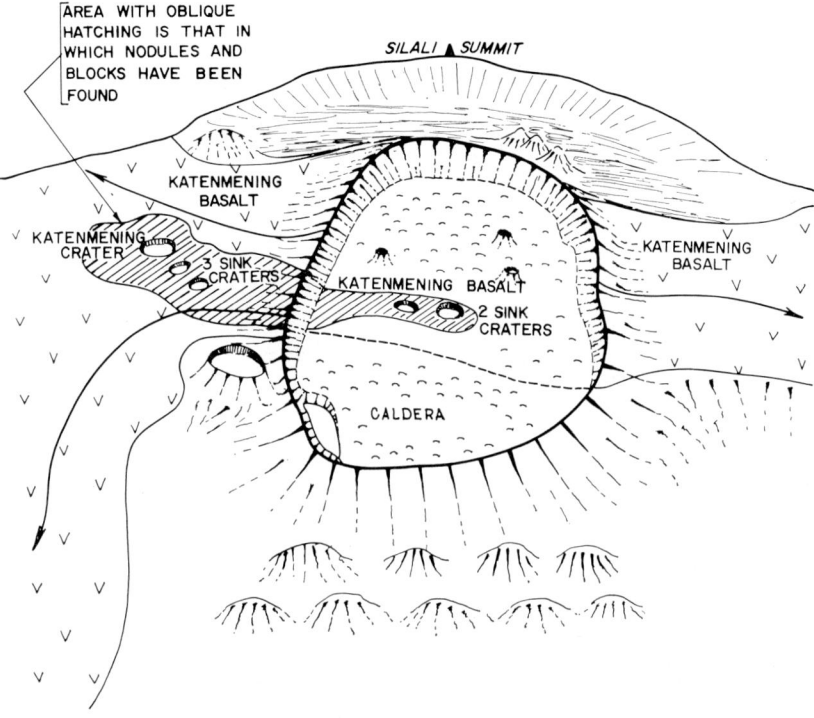

Fig. 2. Sketch of Silali from the south-east.

median sag zone that traverses the volcano meridionally. In this sag zone there developed, shortly after the cessation of effusion, a multitude of closely spaced meridional faults and joints, forming a very closely spaced grid system. Small cones of cinders and lapilli scoria, together with welded spatter, loose bombs and rare agglomerate, are aligned along the faults, and largely restricted to the area of outcrop of the basaltic flows. Trains of small blowholes along the faults testify to the fact that the Katenmening Basalt eruptive sequence terminated in a phase of gas emission. The small cones probably represent the vents from which the lava flows were emitted.

The subordinate tuffs and agglomerates intercalated in the lava sequence and forming the cones may display a cement of carbonate. There are a number of cylindrical sink craters situated within the pile of basalt flows: of these the Katenmening Crater is the largest (figs. 2, 3) situated in a chain of three such craters on the southern slope of the volcano, about two miles south of the caldera rim. It is about half a mile in diameter. These sink craters expose the tiers of thin flows in their cliffed walls (fig. 4), and an even greater number of thin basalt flows is exposed in a fault scarp section to the east. However, the flows that emanated from vents near the pre-caldera summit of Silali were few in number and only a handful of flows are superimposed on one another in the caldera rim section. The increase in number downslope is due to augmentation by flows emanating from the numerous vents situated down the slope from the summit.

The nodules and blocks are found on the surface of, and within the uppermost flow. They have been recovered as far north as the vicinity of the two sink craters within the down faulted surface of the Katenmening Basalt which forms the caldera floor, three miles north of Katenmening crater. Here, however, they are very small in size and few in number, while even at the caldera rim, to the south, they are small and relatively scarce: it is downslope, towards the Katenmening crater and its two smaller neighbours, that they increase in both size and abundance. Around the Katenmening Crater for a distance of about a mile on all sides the surface of the basalt is littered with a profusion of blocks and bombs, and these are far more conspicuous than outcrops of the basalt. All the really large blocks

Fig. 3. Katenmening sink-crater from the north rim.

Fig. 4. Tiered basalt flows of the Katenmening sequence exposed in a small sink-crater, north east of the crater shown above.

(fig. 5) are situated within a mile of Katenmening crater.

The blocks and bombs take various forms. They may be blocks of angular shape, uncoated by basalt: they may be partly veneered with a thin coating of tachylitic basalt; or they may be entirely enclosed in such a basaltic veneer. Though the almost completely and completely basalt veneered forms have the character of volcanic bombs, they lack the spindle, banana and triangular shapes typical of the so-called "aerodynamic" bombs, and more closely resemble the type known as the "toffee-apple bomb", carried along with a moving flow and not ejected from a vent in a parabolic trajector (fig. 6, 7, 8). The bomb form is restricted to the smaller individuals, the larger ones being angular and uncoated (fig. 5). Though some of the nodules are rounded due to the glass coating, the actual crystalline fragments forming the cores are mostly angular, as if broken off along brittle fractures from a rock mass.

Some of the medium sized blocks show a partial veneering (fig. 9), and it is apparent that both blocks and bombs have the same origin.

The most common mode of occurrence is in the form of loose blocks and bombs, weathered out of the lava, but in rare cases blocks and bombs can be found welded into the basalt of the flow (figs. 10, 11) revealing the immediate provenance of the nodular material. Small angular chips of gabbro, flow aligned in swirling trains, occur within the uppermost basalt flow on the northern rim of Katenmening Crater, confirming that the immediate provenance is from within this flow (fig. 12).

The largest of the blocks is more than 20 feet in diameter (fig. 5), its diameter being apparently in excess of the thickness of the flow upon which it was carried to its present site!

There is no certainty that similar nodular material is not contained in lower flows of the Katenmening sequence, but no other occurrences have been recorded

Fig. 5. The largest single block of gabbro discovered, south of Katenmening sink-crater.

Fig. 7. The same bomb, showing the gabbroic core, punctuated by vesicles and bounded by the dark, narrow tachylite selvage.

Fig. 8. Another bomb showing a vesicular core.

Fig. 6. Bomb, entirely coated with tachylite displaying small phenocrysts and shrinkage cracks.

in traverses across the vast expanse of outcrop of these flows, nor in cliff sections through the sequence. This seems to suggest that nodules were restricted to the terminal phase of the eruptive sequence, but it is probable that closer examination of some of the as yet unvisited localities in which similar terminal eruptions occurred may reveal other nodular occurrences.

Mechanisms that could conceivably have transported this material to its present site are:

1) Explosive ejection from a crater to fall back onto the surface of a still moving and partly molten lava flow. The Katenmening crater and other sink craters seem to be the only possible source craters, and they do seem to have a geographic grouping close to the occurrences of nodules. Yet they seem to be quietly-

Fig. 9. Partial tachylitic coating covering a rather larger, crystalline enclave.

Fig. 10. A bare-surfaced block of gabbro welded into the basalt flow.

formed collapse structures closely resembling Halemaumau, Hawaii, rather than explosion craters of the maar type.

2) Eruption of a flow of basalt heavily charged with volatiles: this has carried up blocks and bombs from the depths. The character of the bombs (i.e. "toffee-apple" form) supports this explanation. The provenance of the included material could be:

a) earlier cooled crystalline rocks from the underlying magma chamber, product of crystallisation at an intermediate level during rise of magma through the crust. The rocks crystallised were later disrupted and

Fig. 11. A tachylite-coated bomb welded into the basalt flow.

detached by gas-coring during a terminal phase involving ascent of a volatile-charged magma.

b) patches more in the nature of autoliths than xenoliths, areas of coarse crystallisation in an inhomogeneous, volatile charged flow. While such a process is believed to account for some of the quite sharply defined crystalline enclaves in tufflavas ("froth flows") from this volcanic province (McCall, 1964b), the possibility of autolithic origin seems remote in this case, in view of the size and angular block character of the enclaves.

3) Destruction of a cone consisting largely of exotic-cored bombs and blocks by a later flow during its overland passage, the material gathered up being transported with the flow. The lack of "aerodynamic" shapes to the bombs, the lack of non-cored bombs and the lack of cinder and spatter detritus as enclaves in the host flow virtually eliminate this explanation.

In spite of the great size of some blocks, explanation 2a) is preferred.

4. Petrography of the host basalt flow

The host basalt is not in any way unusual, being the common *alkalic* olivine basalt rock type of this volcanic province. The phenocrysts include iddingsitised olivine (var. chrysolite), titanaugite and labradorite (commonly displaying a marginal zone of andesine composition). They are set in a fine base in which titanaugite, magnetite and plagioclase are conspicuous. Tachylite glass, dusted with magnetite, may be a minor or abundant component of the base. These basalts display abundant vesicles, mostly unfilled giving the lava the appearance of black bread. Vesicles infilled with carbonate and zeolite are also a feature of these basalts.

5. Petrography of the nodules

Unlike the host lava, these are of most unusual character. They are composed of a rock type that has the essential character of gabbro, though more felsic varieties have been recovered from close to the caldera rim and are of dioritic rather gabbroic character. The gabbro is anomalous in that some thin sections display interstitial fields of tachylite or sideromelane, enclosing small rounded vesicles (fig. 13). Empty vesicles are clearly visible on fresh cut surfaces across cores to bombs (figs. 7, 8), and large, lenticular vesicles completely infilled by a greenish mineral (zeolite?) were noted on the bare surface of some large blocks. Interstitial glass and vesicles are not features in any way typical of gabbro, and these rocks are clearly transitional between normal, holocrystalline gabbros characteristically crystallised in a magma chamber environment and surface basalt lavas. Their character is compatible with crystallisation within a very shallow magma chamber beneath Silali volcano.

It is significant that the mineralogy of the gabbroic nodules is, broadly speaking, that of the host basalt: olivine (var. chrysolite), titanaugite, magnetite + interstitial glass. It seems to be an inescapable conclusion that these nodular rocks are the deep seated (magma chamber) equivalent of the surface basalts. Some specimens do exhibit a dioritic rather than gabbroic character, biotite taking the place of olivine and the feldspar being andesine (fig. 14); these are interpreted as products of slight magma chamber differentiation of the basaltic magma.

The glassy veneer coating the nodules does not display exactly the same mineralogical species in the isomorphous series as the crystalline core, and so the veneer does not appear to simply represent a product of marginal melting and resolidification (like a meteorite fusion crust). However, broadly speaking, the olivine, titanaugite and labradorite phenocrysts of this selvage are very similar to the minerals of the nodules, just as are those of the host basalt flow. They are set in a magnetite dusted tachylite glass base, and the laths of plagioclase display a concentric flow alignment

Fig. 12. Swirling trains of gabbro fragments within the top flow of the Katenmening sequence, on the northern rim of Katenmening crater.

Fig. 13. Interstitial sideromelane glass, punctured by small circular vesicles, within an olivine-gabbro nodule (photomicrograph, ×60, plane polarised light).

Fig. 14. Dioritic enclave within alkalic olivine basalt: biotite, augite and andesite are the constituents (photomicrograph, ×6.6, plane polarised light).

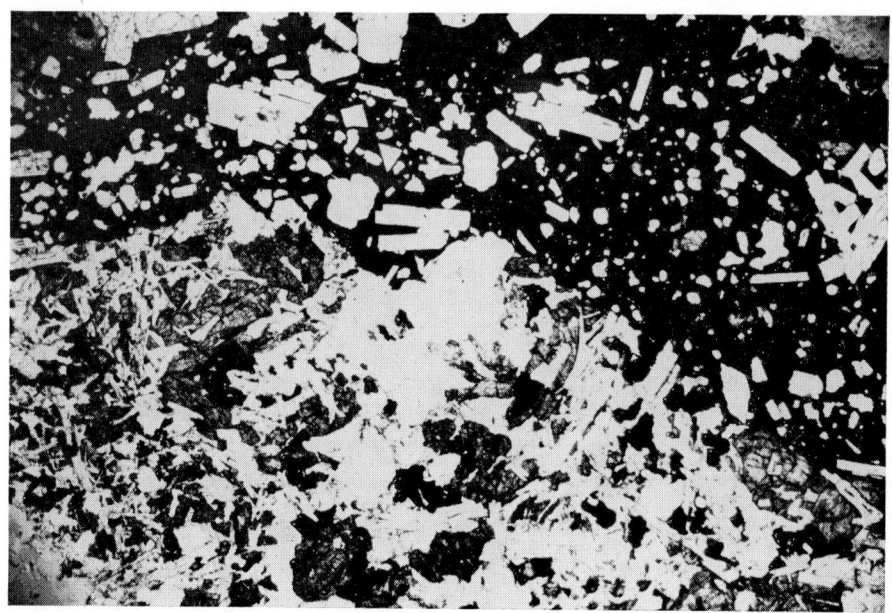

Fig. 15. Tachylite selvage to a gabbro enclave (below) (photomicrograph, ×6.6, plane polarised light).

around the margins of the crystalline cores (fig. 15). Chemical analyses of two nodules and their cores are compared in table 1. The textures of the cores to the bombs and the blocks range from gabbroic to doleritic, but for convenience the nodules have been referred to as gabbroic.

6. Other occurrences of crystalline enclaves in the Cainozoic volcanics of Kenya

It seems relevant to briefly mention other occurrences of enclaves in Kenya. Nodules are very common in the nephelinitic suites of this Cainozoic volcanic province,

TABLE 1

Comparison of the chemical analyses of two nodules and their cores. Analysis by XRF method, Dr. J. Graham, C.S.I.R.O., Floreat Park, Western Australia

	56214		56215	
	a Nodule (%)	b Glassy coating (%)	a Nodule (%)	b Glassy coating (%)
SiO_2	45.16	48.06	45.93	48.17
Al_2O_3	15.02	16.70	15.70	16.09
Fe_2O_3	6.96	5.23	7.01	2.85
FeO	9.47	6.81	8.29	10.28
MgO	4.84	7.07	4.87	4.95
CaO	9.78	10.95	9.56	11.06
Na_2O	1.40	2.50	1.92	3.20
K_2O	0.86	0.51	0.94	0.78
H_2O^+ †	2.10	0.78	2.04	0.65
H_2O^-	1.10	0.27	0.76	0.20
TiO_2	3.69	1.84	3.33	2.98
MnO	0.26	0.19	0.23	0.20
Total	100.64	100.91	100.58	101.41

CIPW norm*

Q	6.3*	—	4.9*	—
Qr	5.1	3.0	5.6	4.6
Ab	11.8	21.1	16.2	25.9
An	32.1	32.8	31.4	21.2
Ne	—	—	—	0.6
Di	13.3	17.2	12.9	22.9
Hy	11.7 (7.8 En) (3.8 Of)	13.8 (10.5 En) (3.2 Of)	10.3 (7.6 En) (2.7 Of)	—
Ol	—	0.8 (0.6 Fo) (0.2 Fa)	—	9.6 (4.6 Fo) (4.9 Fa)
Mt	10.1	7.6	10.2	4.1
Il	7.0	3.5	6.3	5.7

Niggli values

Al	21.16	22.04	22.23	21.93
Fm	49.22	45.52	47.24	42.31
C	25.05	26.27	24.61	27.41
Alk	4.55	6.15	5.91	8.32
Si	107.96	107.63	110.36	111.44
K	0.28	0.11	0.24	0.13
Mg	0.46	0.64	0.50	0.45
Q_e	−10.25	−16.99	−13.29	−21.86
1) MgO/(MgO+FeO)	0.726	0.809	0.787	0.577
2) Total iron as FeO	15.73	11.51	14.59	12.84
3) FeO/(FeO+Fe_2O_3)	0.57	0.56	0.54	0.78
4) Total alkalis	2.26	3.01	2.86	3.98
5) Na_2O/K_2O	1.62	4.90	2.84	4.10
6) Larsen function	−14.43	−13.00	−12.77	−12.01
7) Solidification index	20.56	31.96	21.14	22.43

† May include some CO_2.
* These rocks include two quartz normative, one olivine normative and one nepheline normative specimens. They have characteristic alkalic mineralogy. They straddle the line of saturation on the total alkali/silica plot. The high iron and titania, and low silica values are atypical of continental tholeiites. There is possibly a transition between alkalic basalts and olivine tholeiites of oceanic affinities.

but are mostly of ijolitic or related petrographic types, and are clearly derived from the immediately underlying magma chamber, for, in some cases, such as Rangwa (McCall, 1958), the actual source rocks are laid bare by a freak pattern of dissection which reveals the underworks of the volcano. No other occurrence of nodules in alkalic basalts had been recognised by the author during several years field experience of these rocks, but Saggerson (1968) has recorded the discovery of sparse nodules of eclogite in Quaternary basalts of the Chyulu Hills in southest Kenya. These occur within scoria which forms a very young cone: he has argued the case for upper mantle provenance for this material. The extreme rarity of nodular occurrences in Kenya despite the copious development of such basalts is worth emphasising. Conditions that permit gas-coring at depth, and transport up through the crust, of nodules, whatever their level of provenance, are exceptional and not the norm, and this is true on a world wide basis. Furthermore, such conditions seem to be achieved in the course of late stage or terminal phases in particular eruptive sequences in the case of basaltic eruptions, though they seem to be a quite common feature of nephelinitic and trachytic eruptions, occurring wherever pyroclastic rocks are encountered. The rarity in basaltic suites may, of course, be related to the comparative scarcity of basaltic pyroclasts. The syenite blocks and nodular inclusions associated with phonolitic trachyte lavas, and tufflavas of the Cainozoic volcanic province of Kenya also tend to be restricted to terminal phases of eruptive sequences. Thus in the case of Silali, they are found in the uppermost phonolitic trachyte tufflava flow, beneath the Katenmening Basalt in the south-east caldera wall, though they are only sparsely represented. This tufflava is, incidentally, unusual in that it contains patches of basalt. Similar enclaves are quite common in the older phonolitic trachyte lavas, tufflavas and tuffs that form the subvolcanic foundation beneath Silali, but the most spectacular occurrences are seen on the outer slopes of the caldera volcanoes of Menengai and Kilombe, situated a hundred miles to the south (McCall, 1957a, b, 1964a, 1967, 1968b). Syenite blocks (fig. 16) litter the surface of the last erupted pre-caldera phonolitic trachyte tufflava flow in both cases. The blocks are actually found in situ, in the form of syenitic nodules within the flow in the case of Kilombe (figs. 17, 18), and their

Fig. 16. Syenite block from the uppermost tufflava flow of phonolitic-trachyte composition, Kilombe volcano. Note the patches of pegmatoid texture.

Fig. 17. Syenite nodule within the tuff-lava host.

immediate provenance is certain, though both here and on the slopes of Menengai they occur mostly in the form of a rubble of weathered out blocks, individually up to six feet diameter in the case of Kilombe. The occurrence of the syenite blocks welded into the tufflava (a flow of heterogeneous eutaxite autobreccia) provides evidence for the tufflava origin of this flow rather than an origin in the air fall / lithostatic load compaction / welding process (McCall, 1964b), for the blocks have come up with the flow and been transported with it to a site several miles from the source vent.

Once again, transport with the moving flow of material brought up from magma chamber level is indicated, no other mechanism of eruption being tenable: the large blocks are situated many miles from any

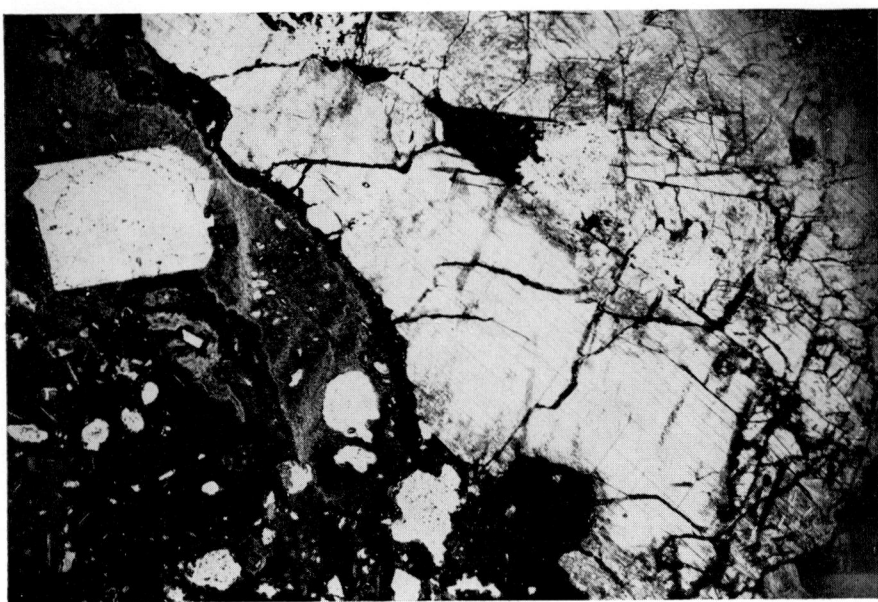

Fig. 18. The same as fig. 17: Boundary with host glassy trachytic material seen in a photomicrograph (×6.6, plane polarised light).

possible explosion vent and projection into the air followed by air-fall onto the moving flow, though remotely tenable, seems less likely than actual emission within the flow, considering the repetition of this phenomenon on Menengai. The blocks were erupted just prior to the pumice mantle of Menengai, while on Kilombe no such mantle was formed and there was no intervening eruption between eruption of the last nodular flow and the collapse process that formed the caldera. The flow containing blocks of syenite exposed within the south-east caldera wall section of Silali is a terminal flow of a sequence: the enclaves within the Katenmening Basalts were erupted just prior to the pumice mantle of Silali before the collapse of the caldera floor formed the caldera. Added to this there is another occurrence on Suswa volcano, where blocks of phonolite litter the surface of the phonolite flow erupted immediately prior to the formation of the inner caldera by subsidence (McCall and Bristow, 1965). The field evidence for a common association of nodular inclusions in lavas and tufflavas with terminal phases of eruptive sequences seems quite strong. It is a common pattern for sequences of eruptivity to commence and terminate explosively, with a main phase of quiet effusion separating these explosive phases, and this pattern, which is by no means the only pattern known, may reflect an initial apical concentration of volatiles in the rising magma "column", and a residual concentration of volatiles at the end of the eruptive sequence. It seems from the evidence available in the Kenya Cainozoic volcanic province that the phenomenon of nodular flows is related to a terminal concentration of volatiles, after emission of a number of perfectly normal lava flows. The pattern may be world wide. In some cases the terminal eruptions may take the form of fragmental emission, of scoria and bombs, with which the nodules are found associated, or of pyroclasts containing blocks of crystalline material.

The syenite nodules of Kilombe have exactly the same form and mode of occurrence as the gabbroic blocks of Silali, but there are no cored bombs: the syenite nodules within the flow show only a dark staining on their surface. Intermediate lavas rarely form volcanic bombs, of course. Both these nodular occurrences must represent accidental incorporation of material from a high level in the crust: the intermediate host tufflava appears unlikely to stem from a process of generation at extreme depth in the crust, though deep provenance cannot be entirely discounted in view of modern evidence of possible deep sources for andesitic magmas. The contained syenite nodules certainly represent high level crustal crystallisation. And the gabbro nodules of Silali, with their vesicles and glassy interstices, must likewise stem from a high level. Indeed, unusually shallow magma chamber (*intermediate stage, intracrustal*) crystallisation is indicated, and the patches of pegmatoid textures in the syenite nodules (fig. 16) suggest a cupola environment, where volatiles are concentrated. We are not dealing with upper mantle here – the only possible occurrence of such material known in Kenya is the Chyulu Hills occurrence – but these intracrustal nodules are of some significance in discussion of the provenance of the ultrabasic nodules, from which they cannot and must not be entirely divorced.

7. Discussion

These high-level, intracrustal nodules do not resemble those ascribed to the upper mantle and at first sight there may appear to be little connection. Yet the gabbroic nodules of Silali occur within alkalic basalts just like the classic nodules of Hawaii and their mode of occurrence is astonishingly similar. In particular they closely match the nodules of Mauna Kea and the 1801 flow of Hualalai, Hawaii, not only in their mode of occurrence but also on the character of the gabbros. Like them they are associated with a late phase of eruption.

The process by which such nodules could be detached from their position in the deep-seated rock mass, whether it be at upper mantle or intracrustal level, must first be considered. The process of gas-coring, in which a combination of a *fluid phase* and a *liquid magma* is invoked, has been well described by CLOOS (1941), in studies of the Swabian tuff-pipes (vulkanembryonnen) of South Germany. A slight yielding of the roof is envisaged, allowing agitation of the fluid phase and boiling off of uprushing gas streams which penetrate narrow cracks. These cracks are persistently widened by the erosive power of the rising gas/particle streams until massive blocks of rock become isolated, stoped off and engulfed, to subside into the melt beneath.

Now such gas-coring effects are most commonly associated with the uprise of the rare, extremely alkaline magmas, such as kimberlite, melilitite and nephelinite. In these magmas extreme dispersion of isolated patches of silicate melt, detached blocks or crystals in a carbonate matrix may occur, and it is considered likely that an immiscible carbonate fraction, separated from the rising magma, is the actual fluid phase. It is in fact, the melilitic magmas that Cloos was dealing with in the case of the Swabian tuff pipes. The problem of the mode of origin of nodular inclusions within basaltic magmas is in all probability not unrelated to the gas-coring process described above. It seems certain, considering the angular nature of many of the fragmental enclaves and their polymictic character, that gas-coring is the process involved. It is indeed very difficult to envisage any other process by which a rising magma could become choked with *an assortment of fragments of crystalline rock*.

The fluid, immiscible phase in the case of the extreme alkaline magmas, of kimberlite, melilitite and nephelinite composition, may be so abundant as to outweigh the silicate fraction, and in this case we may obtain carbonatite melts, either virtually pure or with subordinate silicate minerals contained in them. A less abundant carbonate fraction may give rise to carbonate-based tuffs and agglomerates such as are a common feature of these extremely alkaline suites (fig. 19), and it appears significant that similar carbonate-based agglomerates and tuffs are a feature of the alkalic basalt suites, though not abundantly developed. A carbonate-based tuff outcrops two miles south of Katenmening and a carbonate-based agglomerate some miles to the south-east (fig. 20): both are late members to the Katenmening Basalt eruptive sequence.

It seems reasonable to suggest that the alkalic basalt magmas do, in some cases, have a concentration of an immiscible carbonate fluid fraction associated with them. This suggestion is entirely compatible with the suggestion of TOMKEIEFF (1942) that extreme amygdaloidal developments of carbonate in basalt may be due to a primary, immiscible carbonate fraction. Such a fluid fraction might well tend to become residually concentrated in the terminal phases of eruptive sequences, and so, only in these terminal phases could the basaltic magma, rising through the crust, develop the power to perform gas-coring detachment of exotic blocks. M. J. O'HARA (verbal communication) suggests that the concentration of such a residual phase is the expectation of eclogite fractionation after partial melting, and envisages carbonatites as derived in this way. It is noteworthy too that the nephelinitic suites of the Cainozoic volcanic province of Kenya, as it were *take the place of* the alkalic basalts occurring in exactly the same manner – in plateau-like, multiple sequences of

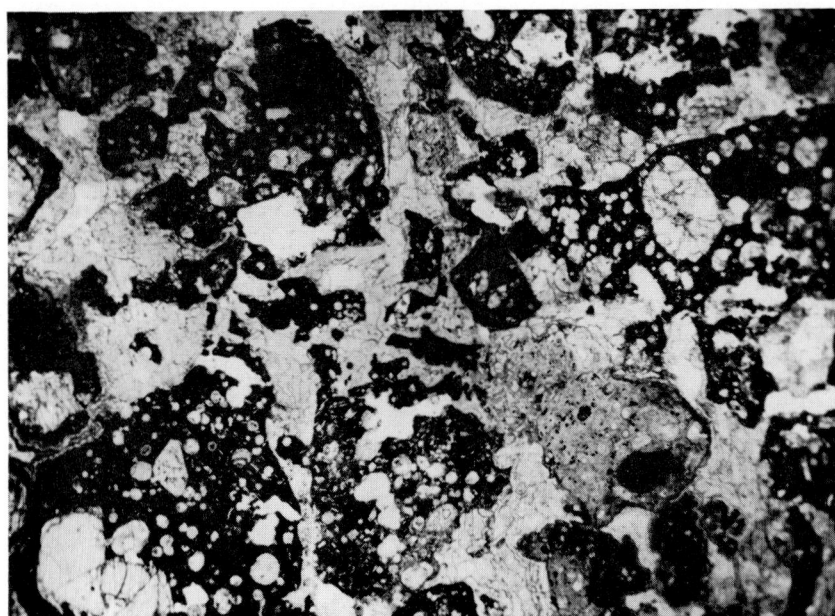

Fig. 19. Carbonate-based nephelinite tuff from Ikoro, Kisingiri Volcano, South West Kenya (photomicrograph, ×25, plane polarised light).

TABLE 2

Volcanic rock types of the Cainozoic province of Kenya
Nodular occurrences

	Periphery of the Rift Valley upswell culminated on Nakuru and Naivasha			Culmination of the Rift Valley upswell centered on Nakuru and Naivasha	Chyulu Hills* (S.W. Kenya)
Nodule type	—	Ijolite Uncompahgrite	Diorite Olivine gabbro	Syenite	Eclogite
Mode of occurrence of residual carbonate fraction	Carbonatite	Carbonate-based tuffs and agglomerates common		Carbonate-based tuffs and agglomerates rare, and mostly restricted to terminal phases of eruptive sequences (?)	?
Main volcanic rock types	—	Nephelinite Melilitite	——(basanite)—— Geographical and statistical hiatus	Alkalic olivine basalt Picrite-basalt Picrite Limburgite	Alkalic olivine basalt
Intermediate differentiate volcanic rock types	—	Phonolitic nephelinite		Phonolite Phonolitic trachyte Quartz-trachyte (Comendite)	—

* reported by E. P. Saggerson, 1968.

thin flows congregated around immense central volcanoes, which are situated at the periphery of the structural culmination of the Rift Valley upwarping, in contrast to the alkalic basalts which tend to occur closer to the culmination (table 2). The two types of suite seem to be antipathetic, and, though there are transitional basanites, a distributional and statistical hiatus separates the two types of suite. It seems reasonable to regard the alkalic (weakly alkaline, basanitoid) suites as representative of a moderate development of the trend towards alkalinity, of which the melilitites and nephelinites are the extreme expression, together with the carbonatites. It is quite logical, therefore, for the same tendency to separate out an immiscible carbonate fraction at great depth to be present in the alkalic basalts, whereas tholeiitic melts do not tend to separate such an immiscible phase, perhaps because of the essentially drier nature of the melt or because a quite different fractionation pattern gives rise to silicic rather than carbonate residues. This seems a likely explanation for the paucity of nodular occurrences in tholeiitic basalt suites: only three occurrences, all in olivine tholeiites are recorded by FORBES and KUNO (1967), while more than 200 nodular occurrences of peridotite alone are known in alkalic basalts and associated volcanic rocks, basanitoids, basanites, olivine–melilite basalts, olivine–nepheline–melilite basalts, limburgites, mugearites and hawaiites. It seems likely to the writer that the nature of the residual fraction alone limits the occurrence of nodules virtually entirely to this related group of alkalic rock types, the restriction being purely mechanical, being simply due to the ability of the mobile, volatile charged, immiscible carbonate fraction to perform gas-coring detachment from the rock mass through which the magma passes on its ascent to the surface. The limiting factor may be the dryness or viscosity of the tholeiitic melts or a combination of both.

Accepting that gas-coring by the agency of an immiscible, residual carbonate fluid fraction separates the nodular material from the rock mass in depth, it is apparent that this process can produce a variety of types of nodular occurrence at the surface: the possible surface manifestations are listed in table 3.

The provenance of nodules contained in alkalic basalts and associated lavas has been the subject of much

Fig. 20. Carbonate-based agglomerate from Jebunbun, at the south-eastern margin of the outer slope of Silali volcano.

TABLE 3

Mode of occurrence of nodules at the surface

1) Enclaves (nodules, blocks) in vent agglomerates and breccias.
2) Enclaves in tuffs and agglomerates dispersed at some distance from the vent, especially flood tuffs etc.
3) Enclaves in the form of blocks and cores to "aerodynamic" bombs in cinder cone deposits. Loose fragments in scoria cones.
4) Enclaves in the form of "non-aerodynamic" toffee-apple bombs, blocks and chips, transported within and contained in lava flows, from which, however, they become readily detached by weathering to form a loose rubble.

controversy. It has been suggested by various authorities that they are:

1) Cognate intracrustal cumulate accretions of normal basalt magmas formed while temporarily arrested in reservoirs or at conduit margins.

2) Accidentally incorporated products of earlier intermediate stage, intracrustal (magma-chamber) crystallisation (of cumulate or non-cumulate type).

3) Accidentally incorporated material from the upper mantle.

4) Cognate partial melting residua from the upper mantle representing what is left behind after extraction of a partial melting fraction.

5) Primary mantle material.

It seems probably that alkalic basalts and associated lavas contain enclaves of all at least the first four types, for it is apparent that both intracrustal and upper mantle derived material are represented: and cognate cumulate accretions and cognate patches of early intracrustal (intratelluric) crystallisation are likely to be contained in basaltic lavas, though their mode of occurrence is rather different to the nodular associations, which appear to be mostly of provenance 2 or provenance 3 and 4 or a combination of both.

The occurrence of nodules of supposed upper mantle provenance from Salt Lake Crater, Oahu (where the nodules are of eclogite/peridotite and the host rock a tuffaceous nepheline-basalt) and of similar nodules in pipes in New South Wales, is well summarised by WYLLIE (1967) and FORBES and KUNO (1967).

It is emphasised by FORBES and KUNO (1967) that

ultramafic material is *not* the most abundant of known rock types in nodular assemblages. The order of decreasing abundance is gabbro, peridotite, basic granulite, garnet peridotite and eclogite. Thus it is obvious that intracrustal material predominates in nodular associations, and mantle derived material is subordinate: calcic plagioclase, an essential component of gabbro, is not stable at upper mantle pressure/temperature conditions. The conclusion of Kuno (1967) that nodular populations are likely to represent increments from various levels is an entirely reasonable one, though it seems that many nodular occurrences represent increments from a restricted depth range (table 4).

All nodular associations are, in a sense accidental increments, except cognate cumulate accretions or cognate partial melting residues actually representing the complement to the host magma: it is likely that by far the greater part of the known nodular associations consist of nodules that are not cognate partial melting residues actually complementary to the host magma. Accidental incorporation of unrelated ultramafic (eclogite, peridotite) material from the sub-volcanic basement will, as O'Hara (1967) observes only rarely confuse the picture, though such a relationship has been demonstrated by drilling in the case of eclogite nodules from Czechoslovakia. Most of the very abundant eclogite and peridotite nodules in alkalic basalts must be related to the genesis of the alkalic magma of a particular volcanic province, but the relationship is unlikely, in most cases, to be a direct, cognate residual relationship.

It is still far from certain which of the numerous reported ultramafic nodule occurrences are truly upper-mantle derived associations, for the data on high pressure or low pressure mineralogy (involving the aluminous nature of the olivine and pyroxenes) is still scanty. The focussing of world-wide attention on the eclogite/peridotite occurrence in tuffaceous nepheline basalt at Salt Lake Crater in Oahu, and related occurrences, has tended to obscure the real facts concerning nodular occurrences. Very spectacular nodular associations in the island of Hawaii, show considerable similarity to the Silali high-level, intracrustal occurrence, and appear to be, at least in part, representative of intracrustal, magma chamber material. One such occurrence is of prolific nodules forming cores, of commonly angular form, to "aerodynamic" volcanic bombs on the flanks of small parasitic cinder cones near to the summit of Mauna Kea. Another occurrence is in the form of flow-aligned trains of angular and rounded, polymictic fragments within the 1801 flow of basalt from Hualalai, on the east side of the island near Captain Cook. These nodular associations appeared to the author, on a brief visit in 1964, to be at least in part derived from intracrustally crystallised material, even though, unlike the representation so far made from Silali, dunite and peridotite nodules accompany the gabbroic material. The peridotite nodules were noted to display an alternation of pyroxenic and olivinic bands, evocative of magma chamber cumulates. Though the author suspected that these nodules represented a complete gradation between dunitic, peridotitic, felspathic peridotite and pyroxenite, and gabbroic nodules reflecting a single cumulate differentiation series, Jackson (verbal communication) has carried out detailed studies of these nodules and regards them as a mixture of magma chamber cumulate and deeper, upper mantle derived material. The dunites show high pressure mineralogy. This occurrence is significant because some of the peridotite, as well as the gabbro, appears to be of intracrustal, magma chamber provenance.

This brief mention of Hawaiian occurrences is introduced purely to effect a comparison with the Kenya occurrences, a comparison made in table 4.

The evidence suggests then that both high level, intracrustal and deep level, upper mantle derived nodules are brought up within alkalic basalt magmas, gabbroic nodules being restricted to the former provenance whereas ultramafic nodules can stem from either provenance.

If we accept the reality of deep level nodular provenance, we must also accept that gas-coring can operate at extreme depths. How can a gas phase be liberated at upper mantle depth? Such a liberation seems to the author to be a pre-requisite for the detachment of angular blocks from the rock mass: a search through the literature has revealed that angular blocks are characteristic of nodular associations attributed to upper mantle provenance: the angularity, and polymictic character of these nodular associations seems to be of critical significance. The occurrence of diamonds and eclogite, inclusions apparently crystallised at extreme pressure, within Kimberlite diatremes has been taken as an indication that these pipes do emanate directly

TABLE 4

Relationship of nodular occurrences from different levels of provenance+ †

		Host	Enclaves
1) Intracrustal, high level, magma chamber (cupola?)	Kenya type 1 (Silali, Menengai, Kilombe)	Phonolitic trachyte	Syenite (Pegmatoid textures)
2) Intracrustal, high level, magma chamber, non-cumulate	Kenya type 2 (Silali)	Alkalic basalt	Olivine gabbro (Diorite) (Glass, vesicle)
3) Intracrustal, high level, magma chamber (cupola?)	Kenya type 3 (Kisingiri, Tinderet)	Nephelinite Melilitite (mostly agglomerate)	Ijolite Uncompahgrite (Pegmatoid textures. Source rock bodies actually exposed)
4) Intracrustal, slightly deeper level, magma chamber, cumulate	Hawaii type 1 (Mauna Kea, 1801 Flow of Hualalai)	Alkalic basalt	Gradational* series dunite (?), through peridotite to gabbro and anorthosite
5) Subcrustal	Hawaii type 2 (Salt Lake Crater)	Tuffaceous nepheline basalt	Eclogite/ peridotite
6) Subcrustal	Kenya type 4 (Chyulu Hills)	Alkalic basalt	Eclogite

† Of obvious high level provenance are eclogite nodules in Czechoslovakia where drilling has revealed a crystalline basement provenance, sedimentary nodules in alkalic basalts in Oregon, and nodules of gneissic material in alkalic basalts of northern Kenya.
* E. D. Jackson, in a discussion after this paper was delivered, remarked that the dunites have appeared to be distinct from the peridotite–gabbro–anorthosite cumulate association, and derived from a deeper level.
+ This table is in no way comprehensive, several hundred nodular occurrences are known. Certain occurrences familiar to the author are compared to provide a picture of the various types of associations of contrasting provenance. Except possibly for 4) above these appear to be associations derived from one level, and not mixtures of material derived through a long vertical section of crust and subcrust.

from the upper mantle. The whole nature of kimberlites is suggestive of liberation of a gas phase at such extreme depths, under certain conditions.

There is a further problem involved for gas-coring does require the existence of a system of fractures within the rock mass, and these extreme depths are surely those at which we would not expect long term stress to produce brittle fracturing. However, the answer to this problem might well be found in short term stresses related to the uprise of the magma column itself, producing brittle fracturing even at that extreme depth.

The weight of evidence does seem to favour gas-coring detachment of nodular material from a range of depth levels extending through the crust down to the Upper Mantle, some nodular associations showing an even representation of this depth range whereas others represent a very limited depth-range section of Crust or Mantle. Variations from one occurrence to another of this sort are compatible with the fact that gas-coring will commence to operate at widely different levels during the ascent of the host magma, and may continue to operate over a wide depth-range or operate only at a single, very restricted depth level.

References

CLOOS, H. (1941) Bau und Tätigkeit Tuffschloten, Untersuchungen an den Schwabischen Vulkan, Geol. Rundschau **32**, 708.

FORBES, R. B. and H. KUNO (1967) Peridotite inclusions and basaltic host rocks, Geol. Rundschau **57**, 328.

KUNO, H. (1967) Mafic and ultramafic nodules from Itinome-Gata, Japan, in: P. J. Wyllie, ed., *Ultramafic and related rocks* (Wiley, New York) 337.

MCCALL, G. J. H. (1957a) The Menengai Caldera, Kenya Colony, Proc. 20th Intern. Geol. Congress (Mexico) **1**, 55.

MCCALL, G. J. H. (1957b) The geology and groundwater conditions of the Nakuru area, Kenya Colony, Hydraulic Branch Ministry of Works, Kenya, Tech. Rept. **3**, 55 p.

MCCALL, G. J. H. (1958) Geology of the Gwasi area, Geol. Surv. Kenya Rept. **45**, 88 p.

McCall, G. J. H. (1963) Classification of Calderas; Krakatoan and Glencoe types, Nature **197**, 136.

McCall, G. J. H. (1964a) Kilombe Caldera, Kenya, Proc. Geologists' Assoc. Engl. **75**, 563.

McCall, G. J. H. (1964b) Froth flows in Kenya, Geol. Rundschau **54**, 563.

McCall, G. J. H. (1965) Possible meteorite craters; Wolf Creek, Australia and analogs, Ann. N.Y. Acad. Sci. **123**, 970.

McCall, G. J. H. (1967) Geology of the Nakuru, Thomsons Falls, Lake Hannington area, Geol. Surv. Kenya Rept. **78**, 122 p.

McCall, G. J. H. (1968a) Silali, another major caldera volcano in the Rift Valley of Kenya, Proc. Geol. Soc. London **1644**, 267.

McCall, G. J. H. (1968b) The five caldera volcanoes of the Central Rift Valley of Kenya, Proc. Geol. Soc. London **1647**, 54.

McCall, G. J. H. and C. M. Bristow (1965) An introductory account of Suswa volcano, Kenya, Bull. Volcanol. **28**, 1.

McCall, G. J. H., B. H. Baker and J. Walsh (1967) Late Tertiary and Quaternary sediments of the Kenya Rift Valley, in: W. W. Bishop and J. D. Clark, eds., *Background to evolution in Africa* (University of Chicago Press) 191.

O'Hara, M. J. (1967) Crystal–liquid equilibria and the origins of ultramafic nodules in basic igneous rocks, in: P. J. Wyllie, ed., *Ultramafic and related rocks* (Wiley, New York) 346.

Saggerson, E. P. (1968) Eclogite nodules associated with alkaline olivine basalts, Kenya. Geol. Rundschau **57**, 890.

Tomkieff, S. I. (1942) The Tertiary lavas of Rum, Geol. Mag. **79**, 1.

Wyllie, P. J. (1967) Mafic and ultramafic nodules: introduction, in: P. J. Wyllie, ed., *Ultramafic and related rocks* (Wiley, New York) 327.

CHEMISTRY OF ULTRAMAFIC NODULES AND THEIR BEARING ON THE ORIGIN OF BASALTIC MAGMAS

HISASHI KUNO and KEN-ICHIRO AOKI

Geological Institute, University of Tokyo, Tokyo, Japan
Institute of Mineralogy, Petrology, and Economic Geology, Tohoku University, Sendai, Japan

Bulk compositions of lherzolite nodules in basaltic rocks range widely in MgO/ΣFeO. With the increase of this ratio, the lower melting oxide components increase and the higher melting components (MgO and Cr_2O_3) decrease. Websterite nodules associated with the lherzolite show a similar variation in oxide components, but differ in the content of some oxides if rocks of the same MgO/ΣFeO are compared. These two groups of rocks are of different origin. The major feature of the lherzolite variation can be explained as due to different degrees of partial melting of a postulated primordial material which is represented by some members of lherzolite having lower MgO/ΣFeO. The magmas thus generated also vary in composition from oceanite to olivine tholeiite. The lherzolite variation could also be explained as due to successive crystal settling in a series of magmas from oceanite to olivine tholeiite. For various reasons, the partial melting is a more likely process. The websterite variation is probably caused by crystal settling in basaltic magmas. The postulated primordial material has a composition similar to the silicate phase of the average bronzite chondrite.

Contents

1. Introduction 273
2. Acknowledgments 274
3. Description of nodules and their origin . . . 274
   3.1. Nodules from Itinome-gata 274
   3.2. Nodules from Hawaii 280
   3.3. Nodules from Dreiser Weiher 280
   3.4. Nodules from other localities 285
4. Variation in lherzolite composition 285
5. Vatiation in websterite-pyroxenite composition 285
6. Origin of lherzolite variation 288
   6.1. Partial melting 289
   6.2. Crystal settling 295
7. Discussion 297
8. Summary 300

1. Introduction

Since the contribution by Ross *et al.* (1954), ultramafic nodules have attracted attention of petrologists and geochemists, as the nodules may represent possible upper mantle materials. Ross *et al.* emphasized a striking uniformity of the mineral assemblage and chemistry of individual minerals of nodules throughout the world.

Nearly 200 localities of such nodules are now known (FORBES and KUNO, 1965). It has been shown that a fairly large variety of rock types are represented. They are associated with gabbro, granulite, and eclogite nodules. Thus in the Hawaiian Islands, groups of lherzolite, dunite-wehrlite, eclogite, and gabbro are distinguished (WHITE, 1966; JACKSON, 1966; KUNO, 1969a). Although JACKSON (1966) and JACKSON and WRIGHT (1969) found a close correlation between the types of nodules and of their host basalt in Hawaii, KUNO (1969a) stated that no such correlation appears to exist for other localities of the world.

Among the ultramafic nodules of the world, lherzolite is the most common type, wehrlite and pyroxenite are the next, and eclogite occurs only at several localities, being invariably associated with lherzolite and garnet lherzolite. Dunite and harzburgite may occasionally be present in association with lherzolite and wehrlite.

Various hypotheses have been proposed for the origin of nodules. They may represent 1) primordial mantle material, 2) residue of partial melting of the primordial mantle to yield basalt magmas, 3) product of crystal settling in magmas in some earlier periods, 4) product of crystal settling in magmas which are represented by the lavas or pyroclastic rocks surrounding the nodules, and 5) fragments of intrusive masses in the crust.

In view of the great variety of rock types and texture found among nodules, any single hypothesis cannot apply to all the nodule types. Thus, the dunite-wehrlite and eclogite of Hawaii and some other localities are interpreted as products of crystal settling (hypothesis 3 or 4) (WHITE, 1966; KUNO, 1969a, b), whereas lherzolite in most localities appear to be mantle fragments (hypothesis 1 or 2) as originally suggested by ROSS et al. (1954). KUNO (1969a, b) showed that a wide and systematic variation exists in bulk composition of lherzolite nodules and suggested that those members rich in lower melting components are probably close to the primordial mantle material and therefore a potential source of basalt magmas.

As it has been realized that a greater number of analyses of lherzolite nodules are needed for understanding the exact nature of variation of their compositions, such nodules were collected from various localities of the world, including those in oceanic regions, island arcs, and stable continental regions. They were examined under the microscope and only those free from alteration were analysed.

Pyroxenite nodules are found together with the lherzolite nodules in Itinome-gata crater, Oga Peninsula, Akita Prefecture, Japan (KUNO, 1967) and in Dreiser Weiher crater, Eifel district, West Germany (FRECHEN, 1948; 1963). These were also studied in order to know whether they have the common origin as that of the lherzolite.

2. Acknowledgments

Our thanks are due to Mr. H. Haramura of the University of Tokyo for many chemical analyses used in this paper, to Professor J. Frechen of the University of Bonn for helping one of us (H. K.) in collecting the nodules of Dreiser Weiher, to Drs. R. M. Honca and L. A. Warner of the University of Colorado, to Dr. I. Kushiro of the University of Tokyo, and to Mr. H. Hayashi of the Government Chemical and Industrial Research in Nagoya for the gifts of nodules from New Mexico, Antarctica, Dreiser Weiher, and Itinome-gata respectively, and to Dr. D. H. Green of the Australian National University for some information on mineralogy and chemistry of the Australian nodules. Some part of the costs of this study, including those for the analysis and the high-temperature and high-pressure experiments, was defrayed by the Government Fund allotted to the research for the Japanese Upper Mantle Project, which is greatly appreciated.

3. Description of nodules and their origin

3.1. Nodules from Itinome-gata

The mode of occurrence and mineralogy of the nodules have been described and a discussion on their origin and their significance in the upper mantle and crustal structure has been given by KUNO (1967). Only a brief mineralogical description is given here together with a discussion on the origin of the mineral assemblages.

The nodules occur in air-fall lapilli tuff and explosion breccia of alkali basalt from Itinome-gata crater or maar. The nodule types are lherzolite, websterite mostly containing olivine, hornblendite, hornblende gabbro, and gabbro. There are also andesite and partly fused granodiorite fragments in the pyroclastics, the former constituting the direct basement of the pyroclastics. Thus the nodules and fragments may represent a nearly complete succession of rocks of the upper mantle and the crust. The lherzolite and websterite are the upper mantle rocks, the hornblendite and gabbros the lower crustal rocks, and the granodiorite (probably upper Cretaceous) and andesite (lower Miocene) the upper crustal rocks.

The bulk compositions of the lherzolite and websterite nodules are given in tables 1 and 9 respectively. The rock analyses are arranged in the decreasing order of $MgO/\Sigma FeO$ in these and other tables. The mineral assemblages and analyses of the constituent minerals are given in tables 12, 13, 14, and 15.

As seen in table 12, most lherzolite and garnet lherzolite contain a small amount of calcic plagioclase, and some of them contain straw-yellow, pargasitic hornblende (pargasite $Na_2Ca_2Mg_5Al_2Si_6O_{22}(OH)_2 + Ca_2Mg_3Al_4Si_6O_{22}(OH)_2$ 87%, tschermakite $Ca_2Mg_3Al_4Si_6O_{22}(OH)_2$ 7%, and cummingtonite 6% in mol. as calculated from no. 87-HO, table 13). A similar hornblende has been described from the lherzolite of St. Paul's Rocks (TILLEY, 1947; MELSON et al., 1967). Garnet has been completely replaced by symplectite of diopside, enstatite, and spinel. This aggregate has been further transformed along its margin to aggregate of calcic plagioclase, olivine, and spinel.

The transformation from garnet, passing through

TABLE 1

Chemical compositions of lherzolite nodules from Itinome-gata, Oga Peninsula, Akita Prefecture, Japan

Rock name	Lh[a]	Lh	Lh	Lh	Lh	Gl[a]	Gl	Lh	Gl	Gl	Gl	Lh	Lh	Lh	Lh	Lh
No.	1[b]	2	3	4	5	6	7	8	9	10	11	12	13	14	15	16
SiO_2	43.54	44.02	43.44	44.08	43.34	44.31	44.59	44.28	44.33	44.71	44.42	45.28	44.70	44.45	46.76	42.80
Al_2O_3	0.54	1.03	0.91	1.39	0.28	2.36	2.98	2.92	2.48	2.75	2.92	2.88	2.53	2.10	0.89	2.47
Fe_2O_3	1.70	0.83	1.42	1.55	1.73	1.46	1.68	1.71	2.02	1.64	1.52	1.96	1.68	2.00	2.20	1.67
FeO	6.15	7.06	7.08	6.92	7.61	6.86	6.83	6.88	6.63	6.87	7.28	6.72	7.36	7.69	7.71	8.64
MgO	46.94	46.99	45.93	44.68	46.33	41.18	41.10	40.73	40.48	40.00	40.37	39.43	39.41	39.88	38.62	39.16
CaO	0.22	tr.	0.29	0.44	0.16	2.46	2.22	2.47	2.38	2.56	2.65	2.37	2.88	2.80	2.14	3.61
Na_2O	0.06	0.05	0.15	0.10	0.07	0.15	0.22	0.20	0.21	0.20	0.14	0.25	0.17	0.10	0.06	0.14
K_2O	0.04	tr.	0.07	0.04	0.03	0.04	0.05	0.01	0.03	0.04	0.00	0.06	0.03	0.06	0.03	0.04
$H_2O(+)$	0.23	0.09	0.25	0.30	0.18	0.27	0.09	0.28	0.32	0.31	0.21	0.14	0.29	0.44	0.32	0.40
$H_2O(-)$	0.00	0.00	0.00	0.02	0.00	0.00	0.00	0.00	0.07	0.04	0.03	0.05	0.07	0.02	0.07	0.07
TiO_2	0.08	tr.	0.08	0.14	0.09	0.18	0.06	0.14	0.17	0.19	0.14	0.17	0.17	0.12	0.15	0.23
P_2O_5	<0.01	tr.	<0.01	<0.01	<0.01	<0.01	0.01	0.01	<0.01	<0.01	0.00	<0.01	<0.01	0.01	0.01	<0.01
MnO	0.13	0.11	0.12	0.13	0.13	0.13	0.17	0.13	0.14	0.14	0.13	0.14	0.14	0.14	0.17	0.17
Cr_2O_3	0.32	0.42	0.61	0.33	0.27	0.33	0.26	0.41	0.36	0.38	0.36	0.39	0.37	0.37	0.67	0.41
Total	99.96	100.60	100.36	100.13	100.23	99.74	100.26	100.17	99.63	99.84	100.17	99.85	99.81	100.18	99.80	99.82
Analyst	Aoki	Haramura	Aoki	Aoki	Aoki	Aoki	Harmura	Aoki	Aoki	Aoki	Aoki	Aoki	Aoki	Aoki	Aoki	Aoki
$MgO/FeO + Fe_2O_3 \times 0.9$	6.11	6.02	5.49	5.37	5.05	5.04	4.88	4.84	4.79	4.79	4.67	4.65	4.44	4.20	3.99	3.86

a) Lh: lherzolite, Gl: garnet lherzolite; b) Sample no. in the Univ. Tokyo collection: 1 – HK64081205d; 2 – HK63111602a; 3 – HK63111602b; 4 – HK50061803b; 5 – HK64081205c; 6 – HK64081206g; 7 – HK50061804g; 8 – HK67051202; 9 – HK67051201; 10 – HK64081206b; 11 – HK64081206c; 12 – HK64081206j; 13 – HK64081206e; 14 – HK64081205a; 15 – HK64081206d; 16 – HK64081205b.

TABLE 2

Chemical compositions of lherzolite nodules from 1801 flow of Hualalai, Hawaii Island (No. 17), and from Salt Lake, Oahu Island (Nos. 18–27) (KUNO, 1969a)

Rock name	Lh[a]	Lh	Lh	Lh	Lh	Lh	Lh	Lh	Lh	Lh	Lh
No.	17[b]	18	19	20	21	22	23	24	25	26	27
SiO_2	44.08	43.40	43.29	44.48	44.39	44.23	44.94	48.02	43.40	49.76	42.93
Al_2O_3	0.47	2.29	2.36	2.73	3.26	2.63	2.84	4.88	2.00	5.35	4.08
Fe_2O_3	1.29	1.00	1.23	1.12	2.02	1.12	1.26	1.94	2.02	2.30	2.31
FeO	7.25	7.74	7.20	7.58	7.06	7.89	9.86	8.15	10.17	7.77	10.96
MgO	45.32	42.98	40.91	40.46	40.79	39.62	38.44	32.35	39.08	30.51	29.77
CaO	0.76	1.60	2.84	2.12	0.94	2.78	1.16	2.97	1.93	1.94	7.88
Na_2O	0.14	0.17	0.33	0.42	0.35	0.29	0.40	0.66	0.69	0.41	0.74
K_2O	<0.05	<0.03	0.05	0.03	0.04	<0.03	0.16	0.07	0.07	0.04	0.05
$H_2O(+)$	0.65	0.41	0.93	0.53	0.13	0.41	0.45	0.79	0.46	0.56	0.23
$H_2O(-)$	0.10	0.00	0.07	0.05	0.08	0.00	0.00	0.32	0.25	0.63	0.06
TiO_2	0.06	0.15	0.20	0.19	0.13	0.15	0.46	0.22	0.10	0.26	0.45
P_2O_5	<0.05	tr.	0.03	0.02	0.15	tr.	0.04	0.07	0.11	0.05	0.18
MnO	0.13	0.13	0.14	0.13	0.13	0.11	0.15	0.14	0.14	0.15	0.15
Cr_2O_3	0.32	0.42	0.27	0.42	0.46	0.48	0.18	0.25	0.35	0.30	0.14
Total	100.57	100.29	99.85	100.28	99.93	99.71	100.34	100.83	100.77	100.03	99.93
Analyst	Haramura	Haramura	Haramura	Haramura	Saito	Haramura	Haramura	Asari	Asari	Asari	Saito
MgO/FeO+ $Fe_2O_3 \times 0.9$	5.39	4.97	4.92	4.71	4.59	4.45	3.50	3.27	3.26	3.10	2.28

a) Lh: lherzolite; b) Sample no. in the Univ. Tokyo collection: 17 – HK61030402; 18 – HK61030313a; 19 – HK57101204d 20 – HK61082601b; 21 – HK61082601a; 22 – HK61082601c-A; 23 – HK57101204a; 24 – HK66101703; 25 – HK66101702; 26 – HK66101701; 27 – HK61030312e.

TABLE 3

Chemical compositions of lherzolite and wehrlite nodules from Dreiser Weiher, West Germany (Analyst, Aoki)

Rock name	Lh[a]	Lh	Lh	Lh	Lh	Lh	Lh	Lh	Lh	Lh	Lh	We[a]	Lh	Lh	Lh
No.	28[b]	29	30	31	32	33	34	35	36	37	38	39	40	41	42
SiO_2	42.84	43.77	43.81	44.21	43.78	45.02	42.51	43.85	44.26	44.84	44.42	39.74	43.90	41.81	39.96
Al_2O_3	1.42	0.29	1.12	2.36	1.27	2.71	0.59	1.62	2.85	2.75	2.59	1.51	1.88	1.15	0.20
Fe_2O_3	1.22	1.06	1.63	1.62	1.37	1.91	1.71	1.72	1.26	1.86	2.07	0.98	1.51	1.59	1.98
FeO	6.13	6.53	6.53	6.77	7.40	6.34	7.49	7.47	7.48	7.03	7.16	9.68	7.99	9.76	12.00
MgO	47.51	47.70	44.94	41.82	43.60	40.44	44.89	42.38	40.28	40.16	40.50	46.84	40.83	41.27	45.11
CaO	0.15	0.07	1.32	2.42	1.56	2.68	1.80	1.74	2.82	2.55	2.67	0.57	2.50	3.76	0.41
Na_2O	0.10	0.07	0.13	0.20	0.12	0.22	0.13	0.15	0.23	0.22	0.18	0.22	0.24	0.57	0.08
K_2O	0.03	0.02	0.04	0.01	0.01	0.02	0.04	0.02	0.02	0.04	0.00	0.11	0.02	0.05	0.24
$H_2O(+)$	0.19	0.34	0.21	0.21	0.28	0.22	0.28	0.19	0.26	0.19	0.15	0.21	0.22	0.22	0.29
$H_2O(-)$	0.08	0.06	0.07	0.00	0.00	0.00	0.00	0.03	0.05	0.00	0.00	0.05	0.01	0.00	0.02
TiO_2	0.10	0.09	0.14	0.17	0.11	0.19	0.19	0.21	0.22	0.23	0.22	0.24	0.20	0.26	0.17
P_2O_5	0.00	0.00	0.00	0.00	0.00	0.00	0.00	0.00	0.00	0.00	0.01	0.03	0.00	0.00	0.00
MnO	0.11	0.11	0.13	0.13	0.14	0.13	0.15	0.14	0.13	0.13	0.14	0.15	0.13	0.14	0.18
Cr_2O_3	0.42	0.36	0.38	0.35	0.29	0.37	0.38	0.33	0.37	0.32	0.37	0.27	0.46	0.08	0.02
Total	100.30	100.47	100.45	100.27	99.93	100.25	100.16	99.85	100.23	100.32	100.48	100.60	99.89	100.66	100.66
MgO/FeO +$Fe_2O_3 \times 0.9$	6.57	6.38	5.62	5.08	5.05	5.02	4.97	4.70	4.68	4.62	4.49	4.44	4.37	3.69	3.27

a) Lh: lherzolite, We: wehrlite; b) Sample no. in the Univ. Tokyo collection: 28 – HK61092801a; 29 – HK61092801c; 30 – HK61092801b; 31 – HK67112004; 32 – HK61092801g; 33 – HK67112003; 34 – HK67112002; 35 – HK61092801d; 36 – HK61092801e; 37 – HK67112007; 38 – HK67112008; 39 – HK61092802c; 40 – HK61092802a; 41 – HK61092802d; 42 – HK61092801h.

pyroxene-spinel symplectite, to plagioclase-olivine (-spinel) aggregate indicates decrease of pressure (KUSHIRO and YODER, 1966).

Some of the hornblende-free lherzolite (table 12) shows distinct lineation due to concentration of spinel and orthopyroxene in tiny strings. YOSHINO (oral communication, 1969) found preferred crystal lattice orientation of olivine in this type of lherzolite. These features were probably produced by flowage in solid state and recrystallization. Exsolution lamellae in ortho- and clinopyroxenes are common in the lherzolite with and without garnet.

From these observations, it is evident that the lherzolite is not merely a product of crystal accumulation in magmas but had been deformed, recrystallized, and then slowly cooled before they were captured by the magma to errupt to the surface.

From the analyses given in table 1, atomic ratios $(Al+Fe^{+3}):Mg:Fe^{+2}$, $(Al+Fe^{+3}):Ca:(Mg+Fe^{+2})$, and $Ca:Mg:Fe^{2+}$ were calculated. It was found that

TABLE 4

Chemical compositions of lherzolite nodules from Vesuvius, Antarctica, and New Mexico (Analyst, Aoki)

Rock name	Lh[a]	Lh	Lh	Lh
No.	43[b]	44	45	46
SiO_2	43.14	44.42	44.67	44.79
Al_2O_3	0.34	2.86	3.38	3.18
Fe_2O_3	1.66	1.69	1.91	1.67
FeO	7.21	6.76	7.30	7.68
MgO	45.95	39.73	39.03	39.24
CaO	1.01	3.16	2.88	2.52
Na_2O	0.34	0.37	0.23	0.25
K_2O	0.13	0.02	0.01	0.02
$H_2O(+)$	0.17	0.26	0.18	0.15
$H_2O(-)$	0.00	0.05	0.00	0.00
TiO_2	tr.	0.07	0.20	0.05
P_2O_5	0.01	0.01	0.00	0.00
MnO	0.12	0.14	0.13	0.14
Cr_2O_3	0.25	0.50	0.38	0.39
Total	100.33	100.04	100.30	100.08
MgO/FeO+ $Fe_2O_3 \times 0.9$	5.28	4.80	4.33	4.27

[a] Lh: lherzolite; [b] No. 43. Lherzolite (HK55091501) from Vesuvius, Italy. No. 44. Lherzolite (HK61040701) from Volcano 116, Edsel Ford Range, West Antarctica (FENNER, 1938). The sample collected and gifted by L. A. Warner of the University of Colorado. No. 45. Lherzolite (HK61081701) from Afton, New Mexico, U.S.A. No. 46. Lherzolite (HK67112024) from Kilbourne Hole near Gallop, New Mexico, U.S.A. The sample gifted by R. M. Honca of the University of Colorado.

the garnet lherzolite is isochemical with some of the garnet-free lherzolite (nos. 8, 12, 13, 14, and 15). Lherzolite nos. 1, 2, 3, 4, and 5 are higher in $Mg:Fe^{+2}$ ratio and poorer in Al. It is concluded that the garnet lherzolite was originally located at a level deeper than that of the garnet-free lherzolite, and that the former was successively brought up to higher levels, probably by convection, resulting in the transformation of garnet to pyroxene-spinel symplectite and then to plagioclase-olivine-spinel aggregate.

Thus, before the convective overturn, the Moho was underlain successively by layers of the following assemblages disregarding the presence of hornblende: 1) olivine + orthopyroxene + clinopyroxene + spinel + plagioclase, 2) olivine + orthopyroxene + clinopyroxene + spinel, and 3) olivine + orthopyroxene + clinopyroxene + spinel + garnet. The layer 3) may have been underlain by a layer with the assemblage 4) olivine + orthopyroxene + clinopyroxene + garnet, although no direct evidence is available. The depth of the boundary between each pair of the two successive layers can be roughly estimated by the subsolidus phase relations determined experimentally or inferred for rocks nos. 24 (table 2) and 67b (table 8) and for the hypothetical primordial mantle material "pyrolite".

The temperatures underneath Itinome-gata are assumed to be a little higher than those of the oceanic geotherm (CLARK and RINGWOOD, 1964), having the value of 700 °C at the Moho which lies at about 30 km. HORAI (1964) estimated the temperature at the Moho of this region as between 816 and 1146 °C, but this value would require partial melting of the hornblende gabbro (YODER and TILLEY, 1962).

According to the hypothetical diagram of ITO and KENNEDY (1968) for rock no. 67b, whose composition is close to those of the higher MgO/ΣFeO members of the Itinome-gata lherzolites, the boundaries between the layers of assemblages 1) and 2), 2) and 3), and 3) and 4) would lie at 30 km (9 kb), 50 km (14 kb), and 74 km (21 kb) respectively.

According to GREEN and RINGWOOD's diagram (1967b) for the phase relation of pyrolite (MgO/ΣFeO = 4.43), which is close to the lower MgO/ΣFeO members of the Itinome-gata lherzolites in composition, those boundaries would lie at 25 km (7 kb), 75 km (22 kb), and 75 km, the range for the layer 3) being very narrow.

Kushiro et al. (1968a) determined the phase relations on the solidus for rock no. 24 which is a little lower in MgO/ΣFeO than any of the Itinome-gata lherzolites. Assuming the inclinations dP/dT of the boundary lines between the stability fields of assemblages 2) and 3) and of 3) and 4) as 22.2 b/°C (Ito and Kennedy, 1968), the boundaries between the layers of these assemblages would lie at 35 km (10 kb) and 150 km (46 kb respectively. This is rather surprising because the garnet-spinel assemblage is stable even at the Moho.

As the stability field of garnet would be enlarged toward the lower pressure region with the decrease of MgO/ΣFeO ratio, the most garnet lherzolite of Itinome-gata was originally located at depths between 50 and 150 km, and the garnet-free lherzolite was above this horizon. The plagioclase-bearing assemblage was attained when the above two rock types were brought up to the level just below the Moho.

In the websterite, the amount of olivine is usually subordinate to that of pyroxenes. In some nodules, thin layers comparatively rich in olivine alternate with those poor in olivine. The amount of augite invariably exceeds that of orthopyroxene. The exsolution lamellae is common in pyroxenes, indicating that the rocks had been cooled slowly before they were captured by the magma as nodules. Bytownite (An$_{70}$; no. 74-PG, table 14), dark spinel, and pale brown or straw-yellow pargasitic hornblende (pargasite 84%, tschermakite 7%, and cummingtonite 9% in mol. as calculated from no. 74-HO, table 14) are accessories.

Compositional relations between co-existing ortho- and clinopyroxenes of lherzolites from various localities, of the Itinome-gata websterites, and of the eclogites from Salt Lake are shown in figs. 1 and 2.

As seen in fig. 1, the clinopyroxenes from the websterite are more Ca-rich than most lherzolite and eclogite clinopyroxenes, and the orthopyroxenes are less Ca-rich than most lherzolite and eclogite orthopyroxenes. According to Davis and Boyd (1966), the shape of the solvus in the system diopside-enstatite is little

TABLE 5

Chemical compositions of lherzolite and harzburgite nodules from Australia quoted from literature

Rock name	Lh[a)]	Lh(?)	Lh	Lh	Ha[a)]	Lh	Lh	Lh	Lh	Lh
No.	47[b)]	48	49	50	51	52	53	54	55	56
SiO_2	45.22	45.33	44.97	44.34	44.11	43.43	44.20	39.13	41.20	43.25
Al_2O_3	2.99	2.75	0.97	1.82	1.93	2.74	2.95	3.48	2.85	2.55
Fe_2O_3	0.20	0.57	2.69	0.62	2.07	1.01	1.16	1.83	1.67	2.09
FeO	7.17	7.33	5.53	7.38	6.21	7.21	7.70	7.58	7.23	7.33
MgO	40.89	41.22	41.60	41.35	41.14	40.98	40.01	42.15	38.04	37.77
CaO	1.93	1.34	1.77	2.22	0.91	2.12	2.65	0.07	1.29	2.07
Na_2O	0.55	tr.	0.06	tr.	0.06	0.19	0.16	—	tr.	0.20
K_2O	0.21	tr.	0.02	0.11	0.02	0.006	0.07	—	abs.	0.006
$H_2O(+)$	0.20	1.18	0.62	0.18	2.96	0.74	0.17	2.80	2.65	2.58
$H_2O(-)$	0.05	0.60	abs.	0.82	0.00	0.68	0.29	0.80	0.92	0.07
TiO_2	0.14	tr.	0.21	0.10	0.05	0.12	0.12	0.16	tr.	0.28
P_2O_5	0.20	abs.	0.04	0.68	0.03	nil	nil	—	—	0.01
MnO	0.28	—	0.13	0.11	0.13	0.12	0.14	0.21	0.26	0.12
Cr_2O_3	0.20	—	0.53	0.14	0.34	0.25	0.31	0.20	0.21	0.35
CO_2	abs.	—	0.44	—	0.28	nil	0.06	3.05	3.40	nil
NiO	—	—	—	—	—	0.23	0.27	—	—	0.24
Total	100.41	100.32	99.58	99.87	100.24	99.826	100.26	101.50[c)]	100.17[d)]	98.916[e)]
MgO/FeO+$Fe_2O_3\times0.9$	5.56	5.27	5.23	5.21	5.10	5.05	4.58	4.57	4.36	3.70

a) Lh: lherzolite, Ha: harzburgite; b) No. 47: Lherzolite from Mt. Gambier, S. A. Analyst, Stanley (Joplin, 1963, p. 302); No. 48: Lherzolite(?) from Aberfeldy, Victoria, Analyst, Field (Joplin, 1963, p. 302); No. 49: Lherzolite from Eucumbene-Tumut Tunnel, Snowy Mts., N.S.W., Analyst, Easton (Joplin, 1963, p. 302); No. 50; Lherzolite from Eucumbene-Tumut Tunnel, Snowy Mts., N.S.W., Analyst, Avery and Anderson (Joplin, 1963, p. 302); No. 51: Harzburgite from Delegate pipe, N.S.W., Analyst, Eaton (Lovering and White, in press). No. 52: Lherzolite from Drogheda Trigonometrical Survey Station, N.S.W., Analyst, Herdsman (Wilshire and Binns, 1961); No. 53: Lherzolite from Armidale, N.S.W., Analyst, Pyle (Wilshire and Binns, 1961); No. 54: Lherzolite from Dundas near Sydney, N.S.W., Analyst, Benson (Joplin, 1963, p. 303). No. 55: Lherzolite from The Basin, Nepean River, N.S.W., Analyst, Osborne (Joplin, 1963, p. 302). No. 56: Lherzolite from Kiama, N.S.W., Analyst, Herdsman (Wilshire and Binns, 1961); c) Including 0.04% NiO+CoO; d) Including 0.45% NiO+CoO; e) The total is given as 99.916 in the original paper.

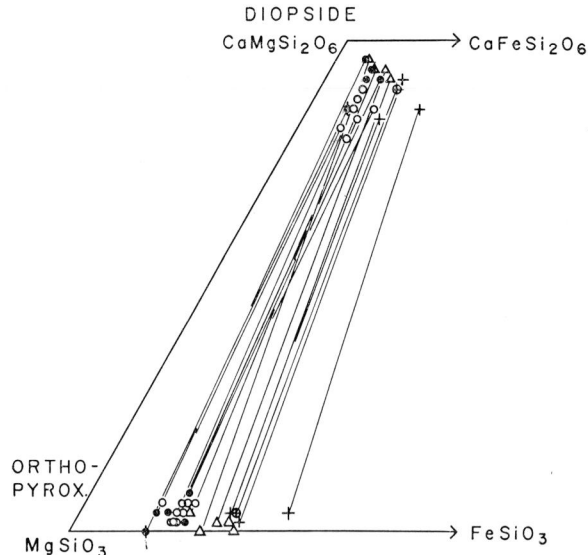

Fig. 1. A part of $CaSiO_3$-$MgSiO_3$-$FeSiO_3$ diagram for ortho- and clinopyroxenes of lherzolite nodule (solid circle with vertical bar) and garnet lherzolite nodules (solid circles) from Itinome-gata, of lherzolite nodules from other localities (open circles), of websterite nodules from Itinome-gata (triangles), of pyroxenite nodule (open circle with cross) and eclogite nodules (crosses) from Salt Lake. The points for co-existing pyroxenes are connected by a straight line.

affected by pressure up to 30 kb, if the temperature is below 1400 °C. If this also holds for a system containing a little iron and Al, it may be concluded that the websterite pyroxenes were equilibrated at temperatures somewhat lower than those for the lherzolite and eclogite pyroxenes. The websterite orthopyroxenes are lower in $Mg:Fe^{+2}$ than the lherzolite orthopyroxenes, but the clinopyroxenes are similar in this ratio to the lherzolite clinopyroxenes. The websterite ortho- and clinopyroxenes are higher in this ratio than the eclogite pyroxenes. Thus the lines connecting the points for the co-existing pyroxenes of websterite intersect those for the lherzolite pyroxenes but are parallel to those for the eclogite pyroxenes. This fact implies that the websterite pyroxenes were equilibrated under a condition somewhat different from that for the lherzolite pyroxenes. However, it is not certain whether this difference is simply related to the temperature or to the set of temperature and pressure.

In fig. 2, it is seen that Al–2Ti contents of the lherzolite pyroxenes vary through a wide range, and that those of the pyroxenes from the websterites and from pyroxenite associated with the eclogites of Salt Lake

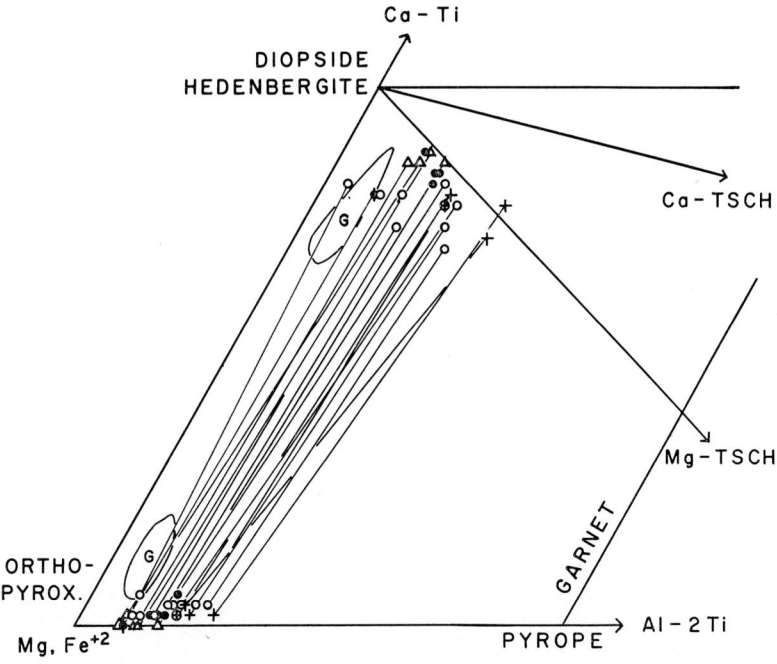

Fig. 2. A part of (Ca–Ti) - (Mg+Fe^{+2}) - (Al–2Ti) diagram for ortho- and clinopyroxenes plotted in fig. 1. The marks are same as those in fig. 1. The points for co-existing pyroxenes are connected by a straight line. The closed curves G enclose points for the two pyroxenes of gabbros. Ca–Ti and Al–2Ti are taken at the corners of the triangle instead of Ca and Al in order to eliminate the effect of $CaTiAl_2O_6$ component in the pyroxenes (KUNO, 1964).

have Al–2Ti within the range of the lherzolite pyroxenes. However, the websterite pyroxenes have lower Al–2Ti than the pyroxenite pyroxenes, suggesting their lower pressure of formation than the pyroxenes of the pyroxenite and eclogites.

In fig. 2, the composition ranges of clinopyroxenes and associated orthopyroxenes from gabbros are indicated by closed curves. The gabbro clinopyroxenes are poorer in Ca–Ti and Al–2Ti than the websterite clinopyroxenes, and gabbro orthopyroxenes are richer in Ca–Ti and poorer in Al–2Ti than the websterite orthopyroxenes. The gabbros were formed in the upper part of the crust and have the two-pyroxene-plagioclase assemblage like that of the websterite. The above-mentioned difference is probably ascribable to higher temperature and lower pressure of formation of the gabbros than that of the websterites.

TABLE 6

Chemical compositions of lherzolite from St. Paul's Rocks quoted from MELSON et al. (1967) (Nos. 57a and b, 59a and b, and 60a and b) and from TILLEY (1947) (No. 58) where b means recalculated from a as water-free

Rock name	Lh[a]	Lh	Lh	Lh	Lh	Lh	Lh
No.	57a	57b	58	59a	59b	60a	60b
SiO_2	43.80	44.50	43.97	42.22	44.69	44.35	44.80
Al_2O_3	2.40	2.44	2.89	4.42	4.68	3.41	3.44
Fe_2O_3	1.41	1.43	1.04	2.86	3.03	1.19	1.20
FeO	6.22	6.32	6.89	4.45	4.71	7.07	7.14
MgO	42.13	42.81	41.11	34.61	36.64	38.88	39.28
CaO	1.13	1.15	2.35	3.92	4.15	2.77	2.80
Na_2O	0.14	0.14	0.07	0.43	0.46	0.17	0.17
K_2O	0.07	0.07	nil	0.11	0.12	0.05	0.05
$H_2O(+)$	1.54	—	0.35	5.73	—	1.16	—
$H_2O(-)$	0.10	—	0.20	0.19	—	0.13	—
TiO_2	0.07	0.07	0.17	0.30	0.32	0.08	0.08
P_2O_5	<0.05	<0.05	n.d.	0.05	0.05	<0.05	<0.05
MnO	0.14	0.14	0.13	0.13	0.14	0.15	0.15
Cr_2O_3	0.54	0.55	0.50	0.50	0.53	0.53	0.54
NiO	0.32	0.33	0.21	0.27	0.29	0.25	0.25
CoO	—	—	tr.	—	—	—	—
CuO	—	—	nil	—	—	—	—
Cl	0.05	0.05	—	0.20	0.21	0.09	0.09
NaCl	—	—	0.09	—	—	—	—
Total	100.06	100.00	99.97	100.39	100.02	100.28	99.99
Analyst	Not mentioned		Vincent	Not mentioned			
MgO/FeO $+Fe_2O_3$ $\times 0.9$	—	5.63	5.25	—	4.92	—	4.26

a) Lh: lherzolite.

TABLE 7

Chemical compositions of lherzolite nodules from California, England and Madeira quoted from literature

Rock name	Lh(?)[a]	Lh	Lh
No.	61[b]	62	63
SiO_2	44.35	43.82	42.42
Al_2O_3	2.97	2.26	1.32
Fe_2O_3	0.67	3.15	4.27
FeO	7.59	5.85	6.96
MgO	40.80	39.83	40.80
CaO	2.55	2.44	1.19
Na_2O	0.20	0.22	0.72
K_2O	0.01	tr.	0.45
$H_2O(+)$	0.06	1.17	0.70
$H_2O(-)$	0.03	0.56	
TiO_2	0.14	tr.	0.30
P_2O_5	0.02	0.03	0.10
MnO	0.13	0.11	n.d.
Cr_2O_3	0.41	0.30	0.40
CO_2	nil	—	—
NiO	0.31	0.32	—
S	—	—	0.04
Total	100.24	100.06	99.67
$MgO/FeO+Fe_2O_3 \times 0.9$	4.98	4.58	3.78

a) Lh: lherzolite; b) No. 61: lherzolite(?) from Ludlow, California, U.S.A., Analyst, Ellestad (HESS, 1955); No. 62: Lherzolite from Calton Hill, Derbyshire, England Analyst, Hamad (HAMAD, 1963); No. 63. Lherzolite-dunite from Parto Monitz, Madeira, Analyst, Klüss (GAGEL, 1912).

3.2. Nodules from Hawaii

The descriptions of various types of nodules and the genetical discussion thereof have been given by WINCHELL (1947), ROSS et al. (1954), WHITE (1966), JACKSON (1966), GREEN (1966), JACKSON and WRIGHT (1969), and KUNO (1969a). In the present paper, we are concerned only with the lherzolite nodules. Of the eleven analyses quoted in table 2, one is from alkali basalt flow of Hualalai volcano and ten from olivine nephelinite air-fall tuff of Salt Lake. Their mineral compositions are shown in table 12.

It was already suggested that the eclogite nodules of Salt Lake were formed at depths between 40 and 60 km (KUNO, 1969a). One of the eclogite nodules encloses a lherzolite fragment (no. 22, table 2), indicating that these two types of nodules were derived from the same general depth.

3.3. Nodules from Dreiser Weiher

The nodules occur in air-fall pumice tuff of basanite

TABLE 8

Chemical compositions of saxonite and lherzolite nodules in kimberlite pipes of South Africa

Rock name	Sa[a)]	Sa	Lh[a)]	Lh
No.	64a[b)]	64b	65a	65b
SiO_2	45.57	47.36	44.77	46.86
Al_2O_3	1.43	1.49	0.84	0.88
Fe_2O_3	1.30	1.35	1.70	1.78
FeO	4.42	4.59	3.99	4.18
MgO	42.75	44.42	40.03	41.90
CaO	0.02	0.02	1.76	1.84
Na_2O	none	none	0.12	0.13
K_2O	0.04	0.04	0.94	0.98
$H_2O(+)$	3.53	—	4.07	—
$H_2O(-)$	0.40	—	0.56	—
TiO_2	0.03	0.03	0.55	0.58
P_2O_5	none	none	none	none
MnO	0.09	0.09	0.10	0.10
CO_2	0.09	0.09	0.15	0.16
ZrO_2	none	none	0.02	0.02
Cl	0.03	0.03	0.03	0.03
S	0.03	0.03	0.10	0.10
F	—	—	0.02	0.02
Cr_2O_3	0.14	0.15	0.14	0.15
V_2O_3	0.01	0.01	0.01	0.01
NiO	0.28	0.29	0.27	0.28
Total	100.16	99.99	100.17	100.00
$MgO/FeO+Fe_2O_3 \times 0.9$	—	7.65	—	7.25

a) Sa: saxonite, Lh: lherzolite, Gl: garnet lherzolite.

TABLE 8 (continued 1)

Rock name	Gl[a)]	Gl	Gl
No.	66	67a	67b
SiO_2	45.15	43.86	45.83
Al_2O_3	2.27	1.96	2.05
Fe_2O_3	0.27	0.69	0.72
FeO	6.35	6.21	6.49
MgO	42.21	39.82	41.60
CaO	2.08	1.68	1.76
Na_2O	0.24	0.14	0.15
K_2O	0.00	0.08	0.08
$H_2O(+)$	0.65	4.06	—
$H_2O(-)$	0.12	0.06	—
TiO_2	0.15	0.39	0.41
P_2O_5	0.03	0.03	0.03
MnO	0.12	0.14	0.15
Cr_2O_3	0.21	0.46	0.48
NiO	—	0.25	0.26
Total	99.85	99.83	100.01
$MgO/FeO+Fe_2O_3 \times 0.9$	6.41	—	5.83

of younger Pleistocene age which was erupted from a maar named Dreiser Weiher in Eifel district, West Germany. They have been studied by FRECHEN (1948, 1963), ROSS et al. (1954), and AOKI and KUSHIRO (1968).

Aoki and Kushiro found nodules of clinopyroxenite with or without hornblende or phlogopite, hornblendite, wehrlite with or without hornblende or phlogopite, harzburgite, and lherzolite. They concluded that the clinopyroxenite and wehrlite are crystal accumulates in alkali basalt magmas near the base of the crust whereas lherzolite represents fragments of the upper mantle.

Bulk analyses of the lherzolite and wehrlite nodules and those of olivine-bearing websterite and clinopy-

TABLE 8 (continued 2)

Rock name	Gl	Gl	Gl	Gl
No.	68a	68b	69a	69b
SiO_2	46.00	46.98	42.30	44.42
Al_2O_3	2.78	2.84	2.87	3.01
Fe_2O_3	1.91	1.95	2.46	2.58
FeO	5.07	5.18	5.25	5.51
MgO	38.86	39.69	40.01	42.02
CaO	1.88	1.92	1.75	1.84
Na_2O	0.19	0.19	0.18	0.19
K_2O	0.34	0.35	0.06	0.06
$H_2O(+)$	2.28	—	4.49	—
$H_2O(-)$	0.31	—	0.44	—
TiO_2	0.06	0.06	0.18	0.19
P_2O_5	none	none	0.04	0.04
MnO	0.12	0.12	0.12	0.13
CO_2	0.09	0.09	—	—
Cl	tr.	tr.	—	—
F	0.01	0.01	—	—
S	0.03	0.03	—	—
Cr_2O_3	0.31	0.32	—	—
V_2O_3	0.03	0.03	—	—
NiO	0.24	0.25	—	—
Total	100.51	100.01	100.15	99.99
$MgO/FeO+Fe_2O_3 \times 0.9$	—	5.72	—	5.36

b) Nos. 64a and b: Saxonite (enstatite peridotite) in kimberlite from Bultfontein, South Africa, Analyst, Theobald (HOLMES, 1936). No. 64b, recalculated as H_2O-free. Nos. 65a and b: Lherzolite with phlogopite in kimberlite from Wesselton, South Africa, Analyst, Theobald (HOLMES, 1936). No. 65b, recalculated as H_2O-free. No. 66: Garnet lherzolite in kimberlite from Thaba Putsoa, Basutoland, South Africa, Analyst, Kerr (NIXON et al., 1963). Nos. 67a and b: Garnet lherzolite in kimberlite from Dutoitspan mine, South Africa, Analyst, Onuki (ITO and KENNEDY, 1967). No. 67b, recalculated as H_2O-free. Nos. 68a and b: Garnet lherzolite with phlogopite in kimberlite from Wesselton, South Africa, Analyst, Theobald (HOLMES, 1936). No. 68b, recalculated as H_2O-free. Nos. 69a and b: Garnet lherzolite in kimberlite from Farm Lauwrencia, South West Africa. Analyst, Kerr (NIXON et al., 1963). No. 69b, recalculated as H_2O-free.

TABLE 9

Chemical compositions of websterite nodules from Itinome-gata (Analyst, Aoki)

Rock name	Ow[a]	Ow	Ow	Ow	Ow	Wb[a]	Ow	Ow	Ow	Ow
No.	70[b]	71	72	73	74	75	76	77	78	79
SiO_2	49.86	50.50	52.15	53.91	51.78	52.28	49.24	50.18	52.25	48.60
Al_2O_3	5.11	2.01	3.72	2.45	5.06	3.19	2.71	5.33	2.93	2.93
Fe_2O_3	1.56	1.85	2.16	2.33	1.06	1.74	1.65	1.71	2.66	1.89
FeO	4.27	4.22	4.03	5.06	4.28	4.85	6.06	4.30	5.13	5.97
MgO	21.82	22.45	21.80	25.51	18.10	22.11	24.86	19.00	23.24	23.22
CaO	14.55	16.96	14.12	9.61	16.68	13.87	13.81	16.53	12.07	15.34
Na_2O	0.61	0.27	0.40	0.24	0.82	0.39	0.35	0.65	0.31	0.27
K_2O	0.02	0.04	0.06	0.02	0.05	0.02	0.04	0.05	0.04	0.04
$H_2O(+)$	0.52	0.47	0.31	0.39	0.51	0.38	0.59	0.63	0.48	0.52
$H_2O(-)$	0.06	0.08	0.02	0.00	0.09	0.03	0.06	0.11	0.03	0.09
TiO_2	0.48	0.20	0.40	0.25	0.52	0.23	0.31	0.53	0.28	0.30
P_2O_5	0.01	0.01	<0.01	<0.01	0.01	0.00	<0.01	<0.01	0.01	0.01
MnO	0.14	0.14	0.15	0.17	0.13	0.16	0.17	0.13	0.20	0.18
Cr_2O_3	0.90	0.62	0.85	0.76	0.70	0.80	0.61	0.85	0.74	0.66
Total	99.91	99.82	100.18	100.71	99.79	100.05	100.47	100.01	100.37	100.02
$MgO/FeO+Fe_2O_3 \times 0.9$	3.85	3.81	3.65	3.56	3.46	3.44	3.29	3.25	3.09	3.03

a) Ow: olivine-bearing websterite, Wb: olivine-free websterite.
b) Sample no. in the Univ. Tokyo collection: 70 – HK50061803d; 71 – HK64081211; 72 – HK59020401; 73 – HK67051501; 74 – Aoki collection E; 75 – HK63111602b; 76 – HK64081208; 77 – HK64081209; 78 – HK63111602c; 79 – HK50061804f.

TABLE 10

Chemical compositions of pyroxenite nodules from Dreiser Weiher (Analyst, Aoki)

Rock name	Ow[a]	Ow	Ow	Ow	Cs[a]
No.	80[b]	81	82	83	84
SiO_2	49.06	50.80	51.89	45.81	41.78
Al_2O_3	3.39	3.54	2.50	9.11	10.90
Fe_2O_3	1.60	1.35	1.59	4.48	5.23
FeO	5.06	4.44	4.36	5.84	6.64
MgO	22.60	18.68	18.10	11.34	6.12
CaO	15.68	17.69	18.21	17.86	18.23
Na_2O	0.67	0.79	1.06	1.66	1.96
K_2O	0.03	0.05	0.08	0.58	0.57
$H_2O(+)$	0.33	0.41	0.37	0.79	2.71
$H_2O(-)$	0.00	0.10	0.05	0.13	0.22
TiO_2	0.59	0.59	0.60	2.13	3.42
P_2O_5	0.00	0.01	0.00	0.01	1.93
MnO	0.13	0.13	0.12	0.17	0.25
Cr_2O_3	0.77	1.01	0.70	0.04	0.00
Total	99.91	99.59	99.63	99.95	99.96
$MgO/FeO+Fe_2O_3 \times 0.9$	3.48	3.30	3.13	1.15	0.54

a) OW: olivine-bearing websterite, Cs: clinopyroxenite with scapolite; b) Sample no. in the Univ. Tokyo collection: 80 – HK67112009, 81 – HK61092802b, 82 – HK61092801i, 83 – HK61092801j, 84 – HK61092801k.

TABLE 11

Chemical compositions of pyroxenite nodules from Australia quoted from literature

Rock name	Py[a]	Di[a]
No.	85[b]	86
SiO_2	50.09	50.36
Al_2O_3	11.31	2.46
Fe_2O_3	1.91	4.26
FeO	3.09	4.41
MgO	13.83	20.76
CaO	15.63	6.15
Na_2O	1.92	0.51
K_2O	tr.	0.09
$H_2O(+)$	1.06	4.07
$H_2O(-)$	0.61	2.21
TiO_2	0.95	tr.
P_2O_5	—	tr.
MnO	0.06	0.29
Cr_2O_3	0.21	0.04
CO_2	0.16	4.44
NiO	—	0.31
BaO	—	0.02
Total	100.83	100.38
$MgO/FeO+Fe_2O_3 \times 0.9$	2.88	2.52
Analyst	Benson	Mingaye

a) Py: pyroxenite, Di: diallage rock; b) No. 85: pyroxenite from Dundas, near Sydney, N.S.W. (JOPLIN, 1963); No. 86: Diallage rock from the Basin, Nepean River, N.S.W. (JOPLIN, 1963).

TABLE 12

Mineral compositions of nodules

Loc.	Rock name, no.	Ol.	Opx.	Cpx.	Sp.	Ho.	Ga.	Pl.	Remarks
Itinome-gata	Lherzolite Nos. 1, 2, 3, 4, 5, 12, 14, 15, 16	+	+	+	+	−	−	±	Lineation
	Lherzolite Nos. 8, 13	+	+	+	+	+	−	+	
	Garnet lherz. Nos. 6, 7, 9, 10, 11	+	+	+	+	+	+	+	
	Websterite Nos. 70, 71, 72, 73, 74, 75, 76, 77, 78, 79	+	+	+	±	+	−	±	Ol. absent in No. 75 and Ho. absent in No. 72
Hawaii	Lherzolite Nos. 17–27	+	+	+	+	−	−	−	Cpx. and Sp. absent and Ho. present in No. 23
Dreiser Weiher	Lherzolite Nos. 28, 29, 30, 31, 32, 33, 34, 35, 36, 37, 38, 40, 41, 42	+	+	+	+	−	−	−	Sp. absent in Nos. 29 and 41 and biotite present in No. 34
	Wehrlite No. 39	+	−	+	+	−	−	−	
	Websterite Nos. 80, 81, 82, 83, 84	+	+	+	−	−	−	−	Biotite present in No. 83 and magnetite present in No. 84
Vesuvius etc.	Lherzolite Nos. 43, 44, 45, 46	+	+	+	+	−	−	−	Plagioclase present in Nos. 45 and 46

TABLE 13

Chemical compositions of olivine (OL), orthopyroxene (OP), clinopyroxene (CP), hornblende (HO), and spinel (SP) from lherzolite nodules from Itinome-gata

No.	2-OL[a]	2-OP	2-CP	87-OL	87-OP	87-CP	87-HO	88-OL	88-OP	88-CP	88-SP	89-OP	89-CP	90-OP	90-C
SiO_2	40.55	56.10	52.07	40.50	54.73	51.45	44.21	40.30	54.78	51.41	—	54.47	51.43	53.39	50.66
Al_2O_3	tr.	2.06	2.57	0.43	3.47	4.11	16.70	0.25	3.50	4.69	53.43	3.51	4.46	4.13	5.13
Fe_2O_3	0.19	1.09	1.14	0.02	1.56	1.24	1.44	0.00	0.00	0.69	0.00	2.24	0.83	1.96	0.85
FeO	8.40	4.73	2.39	9.21	4.91	1.74	3.22	10.26	6.86	2.64	13.15	4.99	2.53	6.03	2.94
MgO	50.67	35.71	19.08	50.03	34.20	16.90	18.78	48.60	33.28	16.32	18.29	32.94	16.41	31.56	16.28
CaO	tr.	tr.	21.17	tr.	0.72	22.89	11.88	0.07	0.62	21.63	0.60	0.98	21.61	1.70	21.38
Na_2O	none	tr.	0.33	tr.	0.06	0.44	3.18	0.04	0.07	0.72	—	0.08	0.56	0.02	0.49
K_2O	none	none	0.03	tr.	0.02	0.02	<0.1	0.03	0.03	0.04	—	0.01	0.08	0.01	0.08
$H_2O(+)$	0.58	0.0	0.35	0.15	0.31	0.25	0.54	0.34	0.13	0.11	—	n.d.	0.25	n.d.	0.35
$H_2O(-)$	0.03	0.00	0.07	0.00	0.12	0.10	0.07				—	n.d.		n.d.	
TiO_2	tr.	0.06	0.15	tr.	0.12	0.50	0.76	0.15	0.15	0.33	1.39	0.39	0.32	0.39	0.68
P_2O_5	tr.	tr.	0.01	tr.	tr.	tr.	0.05	n.d.	n.d.	n.d.	—	n.d.	0.09	0	n.d.
MnO	0.11	0.11	0.07	0.13	0.12	0.06	0.10	0.09	0.13	0.11	0.11	0.11	0.09	0.11	0.10
Cr_2O_3	0.017	0.40	0.91	n.d.	0.45	0.87	n.d.	0.03	0.55	1.03	12.23	0.50	0.98	0.44	0.83
Total	100.547	100.26	100.34	100.47	100.79	100.57	100.93	100.56[b]	100.20[c]	99.81[d]	99.20	100.12	99.55	99.74	99.77
Analyst	Haramura			Haramura				Foster[e]				Onuki[f]		Onuki[f]	

a) 2-OL, OP, CP: rock no. 2 of table 1; 87-OL, OP, CP, HO: garnet lherzolite HK58012403, 88-OL, OP, CP, SP: garnet lherzolite HK36072601c, 89-OP, CP: garnet(?) lherzolite Ip-1 (Onuki), 90-OP, CP: garnet(?) lherzolite Ip-2 (Onuki); b) Including 0.41% NiO; c) Including 0.01% V_2O_5 and 0.09% NiO; d) Including 0.05% V_2O_5 and 0.04% NiO; e) Ross et al. (1954); f) ONUKI (1965).

TABLE 14

Chemical compositions of olivine (OL), orthopyroxene (OP), and clinopyroxene (CP), hornblende (HO), and plagioclase (PG) from olivine-bearing websterite nodules from Itinome-gata

No.	70-OL	70-OP	70-CP	72-OP	72-CP	74-OP	74-CP	74-HO	74-PG	91-OP	91-CP	92-OP
SiO_2	40.02	55.01	51.11	54.26	50.72	55.14	51.63	43.32	49.36	54.91	52.27	54.74
Al_2O_3	0.18	2.31	3.77	3.40	4.91	2.51	3.94	14.74	31.96	2.43	3.00	2.33
Fe_2O_3	1.01	2.03	1.50	0.88	1.33	0.55	1.25	2.79	0.34	0.21	0.68	0.30
FeO	12.17	6.78	2.17	8.01	2.24	8.80	2.62	3.47	n.d.	9.27	3.48	9.60
MgO	46.25	32.28	16.68	32.08	16.02	31.40	16.07	16.98	0.16	31.51	16.64	31.46
CaO	0.08	0.76	22.18	0.11	22.12	0.40	21.98	11.35	14.02	0.35	22.02	0.08
Na_2O	0.08	0.08	0.58	<0.1	0.68	0.05	0.55	2.79	3.34	0.03	0.23	0.10
K_2O	0.03	tr.	tr.	<0.1	<0.1	0.02	0.02	0.20	0.03	0.01	0.02	0.02
$H_2O(+)$	0.26	0.33	0.35	0.46	0.27	n.d.	n.d.	2.34	0.51	0.35	0.31	0.35
$H_2O(-)$	0.00	0.05	0.00	0.05	0.12	n.d.	n.d.	0.00	n.d.			
TiO_2	tr.	0.14	0.58	0.10	0.42	0.19	0.56	0.87	tr.	0.16	0.25	0.15
P_2O_5	0.07	0.02	0.03	<0.05	<0.05	n.d.	n.d.	n.d.	n.d.	n.d.	n.d.	n.d.
MnO	0.16	0.17	0.09	0.22	0.12	0.22	0.13	0.09	n.d.	0.23	0.14	0.24
Cr_2O_3	0.05	0.44	0.96	0.60	0.88	0.40	0.83	0.79	n.d.	0.42	0.72	0.37
Total	100.36	100.40	100.00	100.42	99.83	99.68	99.58	99.73	99.72	99.88	99.76	99.74
Analyst		Haramura		Haramura			Aoki			Aoki		Aoki

a) 70-OL, OP, CP: rock no. 70 of table 9, 72-OP, CP: rock no. 72 of table 9, 74-OP, CP, HO, PG: rock no. 74 of table 9, 91-OP, CP: Aoki collection B, olivine-bearing websterite, 92-OP: Aoki collection F, websterite.

TABLE 15

Atomic proportions of pyroxenes on the basis of 6 oxygens

No.	2-OP		2-CP		87-OP		87-CP	
Si	1.925	2.000	1.896	2.000	1.887	2.000	1.870	2.000
Al	0.075		0.104		0.113		0.130	
Al	0.008		0.006		0.029		0.046	
Fe^{+3}	0.028		0.031		0.041		0.034	
Cr	0.011		0.026		0.012		0.003	
Fe^{+2}	0.136		0.073		0.142		0.053	
Mn	0.003	2.013	0.002	2.027	0.003	2.018	0.002	1.990
Mg	1.825		1.035		1.756		0.915	
Ca	—		0.826		0.027		0.891	
Na	—		0.023		0.004		0.031	
K	—		0.001		0.001		0.001	
Ti	0.002		0.004		0.003		0.014	

TABLE 15 (continued 1)

No.	88-OP		88-CP		89-OP		89-CP	
Si	1.898	2.000	1.879	2.000	1.885	2.000	1.885	2.000
Al	0.102		0.121		0.115		0.115	
Al	0.039		0.081		0.028		0.078	
Fe^{+3}	—		0.018		0.058		0.023	
Cr	0.016		0.031		0.014		0.029	
Fe^{+2}	0.200		0.081		0.144		0.077	
Mn	0.004	2.020	0.002	2.011	0.003	1.997	0.003	2.007
Mg	1.730		0.895		1.699		0.896	
Ca	0.023		0.846		0.036		0.848	
Na	0.004		0.048		0.005		0.040	
K	—		—		0.000		0.004	
Ti	0.004		0.009		0.010		0.009	

TABLE 15 (continued 2)

No.	90-OP		90-CP		70-OP		70-CP	
Si	1.869	2.000	1.858	2.000	1.919	2.000	1.874	2.000
Al	0.131		0.142		0.081		0.126	
Al	0.039		0.080		0.014		0.037	
Fe^{+3}	0.052		0.023		0.053		0.041	
Cr	0.012		0.024		0.012		0.028	
Fe^{+2}	0.176		0.090		0.198		0.067	
Mn	0.003	2.003	0.003	2.008	0.005	1.998	0.003	2.015
Mg	1.646		0.890		1.679		0.911	
Ca	0.064		0.840		0.028		0.871	
Na	0.001		0.035		0.005		0.041	
K	0.000		0.004		—		—	
Ti	0.010		0.019		0.004		0.016	

TABLE 15 (continued 3)

No.	72-OP		72-CP		74-OP		74-CP	
Si	1.900	2.000	1.862	2.000	1.936	2.000	1.891	2.000
Al	0.100		0.138		0.064		0.109	
Al	0.041		0.075		0.040		0.061	
Fe^{+3}	0.023		0.037		0.014		0.034	
Cr	0.016		0.022		0.011		0.024	
Fe^{+2}	0.245		0.069		0.258		0.080	
Mn	0.007	2.024	0.004	2.019	0.007	1.996	0.004	1.997
Mg	1.673		0.876		1.642		0.877	
Ca	0.004		0.870		0.015		0.862	
Na	0.007		0.049		0.003		0.039	
K	0.005		0.005		0.001		0.001	
Ti	0.003		0.012		0.005		0.015	

TABLE 15 (continued 4)

No.	91-OP		91-CP		92-OP	
Si	1.934	⎫ 2.000	1.919	⎫ 2.000	1.933	⎫ 2.000
Al	0.066	⎭	0.081	⎭	0.067	⎭
Al	0.035	⎫	0.049	⎫	0.030	⎫
Fe^{+3}	0.005		0.019		0.008	
Cr	0.012		0.021		0.010	
Fe^{+2}	0.273		0.107		0.283	
Mn	0.007	2.004	0.004	2.000	0.007	2.009
Mg	1.653		0.910		1.655	
Ca	0.013		0.866		0.003	
Na	0.002		0.016		0.008	
K	0.000		0.001		0.001	
Ti	0.004	⎭	0.007	⎭	0.004	⎭

roxenite are given in tables 3 and 10 respectively and their mineral compositions in table 12.

3.4. Nodules from other localities

Lherzolite nodules in alkali basalt lavas from Vesuvius, Antarctica, and New Mexico were also analysed with the result given in table 4. Their mineral compositions are shown in table 12.

Bulk analyses of lherzolite, harzburgite, and pyroxenite nodules, sometimes with garnet, from alkali basalts and kimberlite, and also those of lherzolite from St. Paul's Rocks on the Mid-Atlantic Ridge, all quoted from literature, are given in tables 5, 6, 7, 8, and 11.

4. Variation in lherzolite composition

Contrarily to the currently accepted concept of the uniformity of lherzolite nodule composition, tables 1 to 7 show that the lherzolite in basaltic rocks and from St. Paul's Rocks vary through a wide range in the contents of individual oxides. These variations are discussed in reference to $MgO/\Sigma FeO$ ratio given at the base of each column of analyses. This ratio would be a reliable measure of the degree of partial melting or of fractional crystallization, because selective concentration of olivine or pyroxenes which might occur during crystal settling, or selective removal of some particular minerals in the analysed powder during crushing, would not significantly affect this ratio.

In most ultramafic nodules, K_2O contents are too low to obtain reliable values by flame photometric analysis methods. K_2O per cents determined by the neutron activation (WAKITA et al., 1967), which would give more reliable figures, are 0.010 and 0.062 for some Itinome-gata garnet lherzolite and lherzolite respectively, 0.003 for the Hualalai lherzolite (no. 17), 0.004 to 0.011 for the Salt Lake lherzolite (nos. 21, 24, 25, and 26), 0.009 for the Dreiser Weiher lherzolite (no. 38), and 0.022 for the New Mexico lherzolite (no. 45).

A systematic variation of the oxide components of the nodules is best demonstrated by plotting their weight per cents against $MgO/\Sigma FeO$ (fig. 3). With the decrease of this ratio, Al_2O_3, FeO, CaO, Na_2O+K_2O, TiO_2, and MnO generally increase, whereas MgO and Cr_2O_3 decrease. Thus the lower melting components increase with the decrease of this ratio.

Harzburgite (H) and wehrlite (W) fall generally within the limit of variation of lherzolite, although wehrlite shows a slight departure from the lherzolite variation in some oxides.

Even among the nodules from a single crater, $MgO/\Sigma FeO$ varies through a wide range, accompanied with variations of individual oxide components in the same way as mentioned above (see fig. 7, and also fig. 3 of KUNO's paper, 1969a). Thus, $MgO/\Sigma FeO$ varies from 6.11 to 3.86 in the Itinome-gata lherzolite (table 1) from 4.97 to 2.28 in Salt Lake (table 2), and from 6.57 to 3.27 in Dreiser Weiher (table 3). The wide variation seen among the nodules from a single crater strongly indicates that the upper mantle is considerably heterogeneous in major components even within a small region through which the magma penetrates in ascending to the surface.

The oceanic lherzolites such as those of Salt Lake and Madeira show a slight tendency to have lower $MgO/\Sigma FeO$, and therefore to be more enriched in the lower melting components, than the continental lherzolites. This might imply that the continental upper mantle has been more depleted of the material that formed the crust.

In fig. 3, anhydrous compositions of the nodules from kimberlite (table 8) are also plotted. They are higher in $MgO/\Sigma FeO$ (7.65–5.36) than most lherzolite nodules from basalts, and are generally higher in SiO_2 and CaO and lower in ΣFeO and MgO if rocks of the same $MgO/\Sigma FeO$ are compared. It appears that the nodules in kimberlite and those in basalts belong to slighly different genetical lineage.

5. Variation in websterite-pyroxenite composition

The compositions of the websterite and pyroxenite nodules (tables 9, 10, and 11) are plotted also in fig. 3.

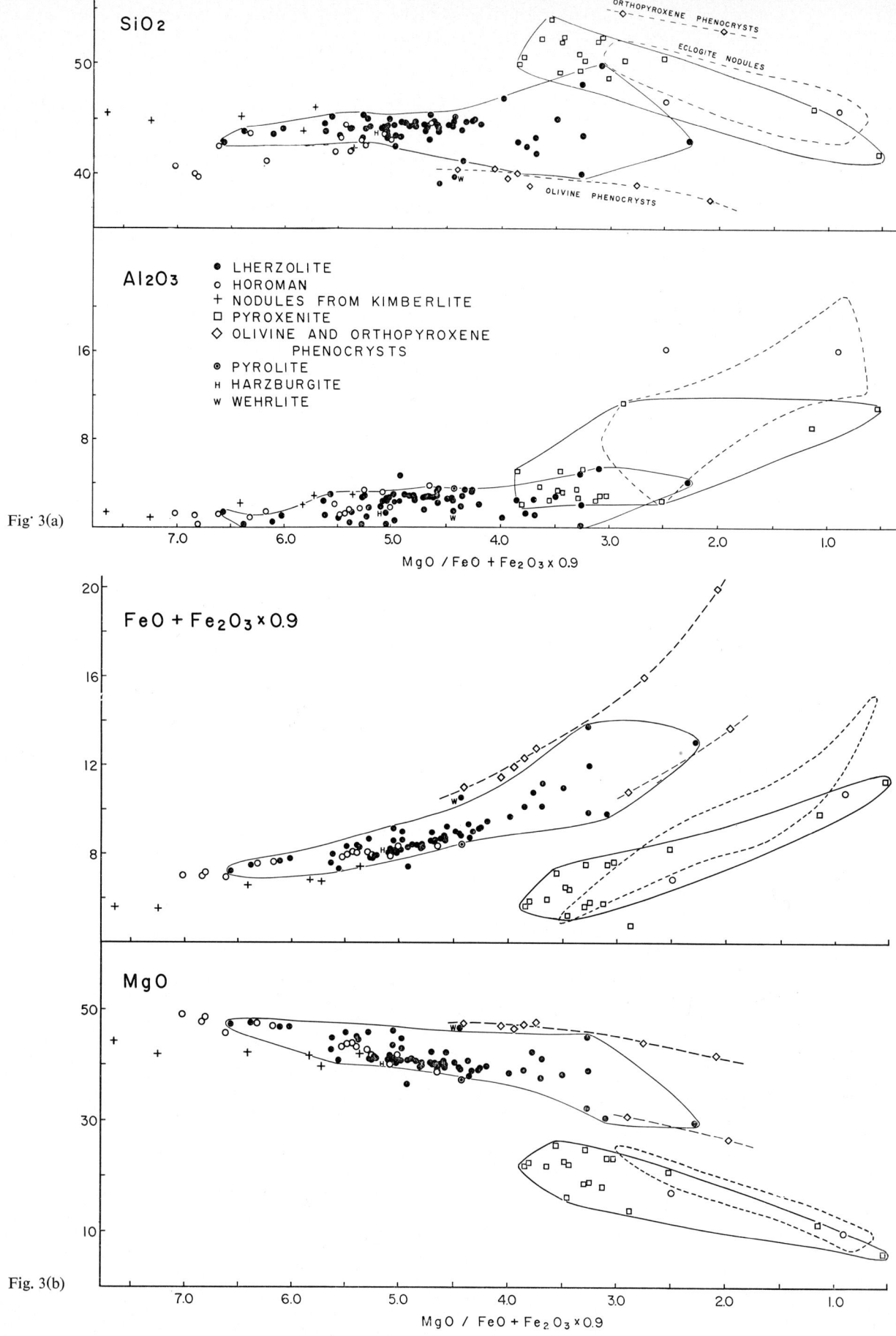

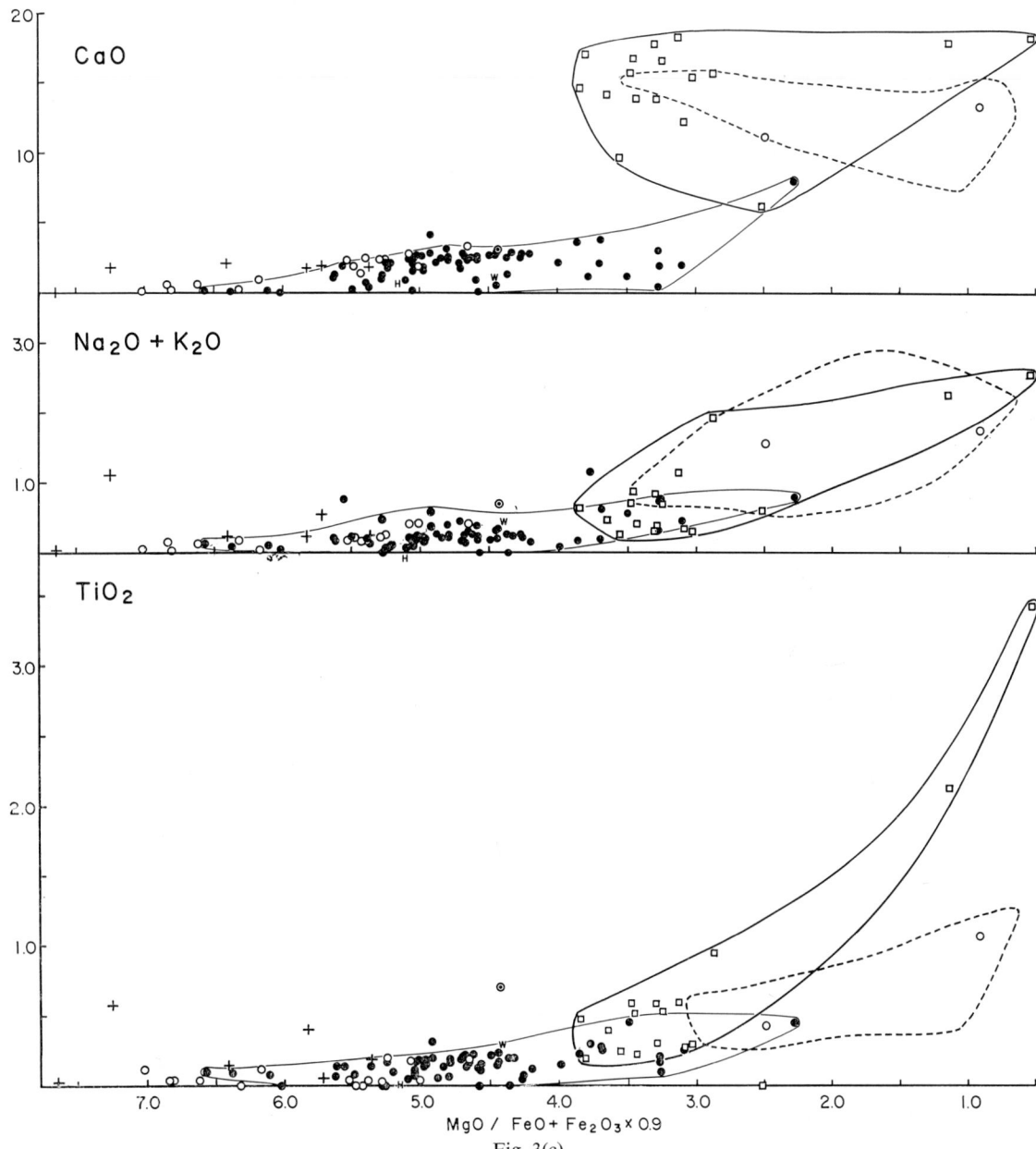

Fig. 3. Weight per cents of oxide components of ultramafic nodules and an intrusive plotted against MgO/ΣFeO (weight). Solid circles: lherzolite nodules. Squares: websterite and pyroxenite nodules. H: harzburgite nodule. W: wehrlite nodule. Crosses: lherzolite and garnet lherzolite nodules from kimberlite, South Africa. Open circles: layered ultramafic complex of Horoman, Hokkaido, Japan (data from MOTOJIMA, 1965; ONUKI, 1965; NAGASAKI, 1966). Double circle: the hypothetical primordial mantle material "pyrolite" (RINGWOOD, 1966). Diamonds: olivine and orthopyroxene phenocrysts in Hawaiian olivine tholeiite and oceanite. The two olivines between MgO/ΣFeO 3.0 and 2.0 are those from crystal accumulates (allivalite and olivine eucrite) in tholeiite of Izu-Hakone region, Japan (KUNO, 1952). The closed solid curves mark the general limits of variations of lherzolite and websterite-pyroxenite nodules. The closed broken curves mark the limit of variation of eclogite nodules from Salt Lake, Hawaii (KUNO, 1969a).

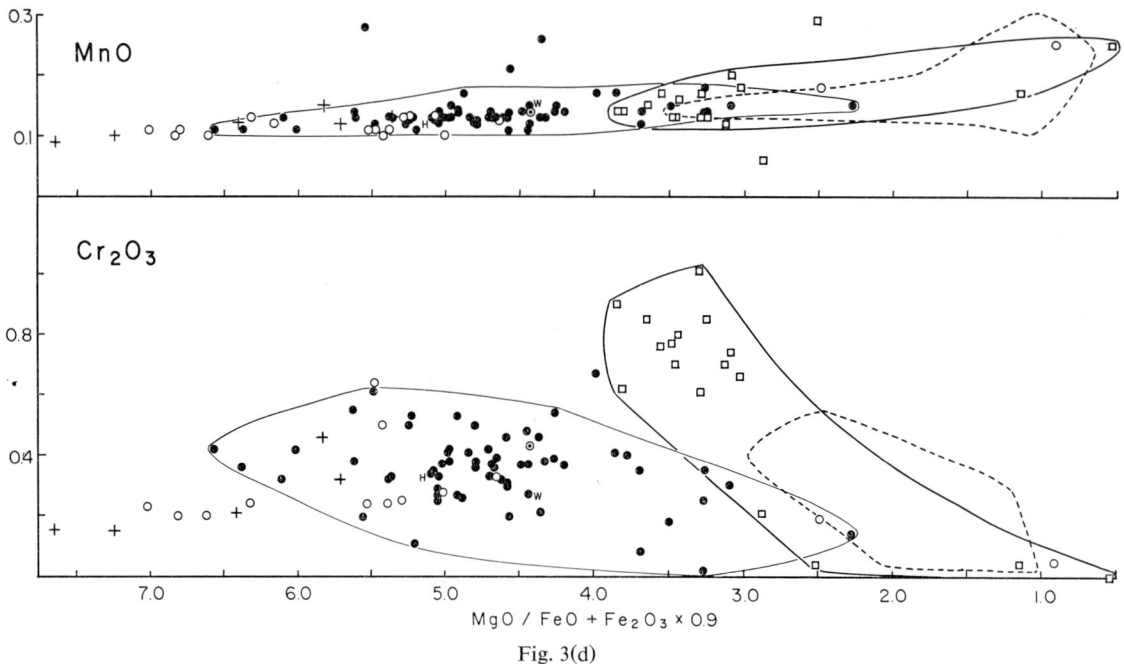

Fig. 3(d)

Fairly wide variation in MgO/ΣFeO is seen in the Dreiser Weiher nodules.

With the decrease of MgO/ΣFeO, Al_2O_3, ΣFeO, Na_2O+K_2O, TiO_2 and MnO increase whereas SiO_2, MgO, and Cr_2O_3 decrease. The variation trends of SiO_2, ΣFeO, MgO, CaO, and Cr_2O_3 are distinctly different from those in the lherzolite nodules. Thus these two types of nodules are of different origin, as is also suggested by the difference in pyroxene compositions.

In the same figure, the limits of variation of the Hawaiian eclogite nodules are shown by broken lines. According to KUNO (1969a), the eclogites were formed by successive crystal settling in magmas of picrite-basalt (oceanite) to alkali basalt compositions at depths of 40 to 60 km. The websterite-pyroxenite variation is very similar to the eclogite variation, suggesting their similar origin. The presence of compositional layering in some Itinome-gata nodules indicates that they were formed also by crystal settling. AOKI and KUSHIRO (1968) suggested that the wehrlite (including some olivine-bearing websterite?) and clinopyroxenite nodules of Dreiser Weiher originated by crystal accumulation in alkali basalt magma.

The Itinome-gata websterite was probably formed at a shallower depth (possibly 30–40 km) than the depth of formation of the Hawaiian eclogite, as suggested by the lower Al–2Ti content in the Itinome-gata pyroxenes (fig. 2).

However, more chemical analyses should be available before any detailed discussion on the origin of websterite-pyroxenite nodules can be made.

6. Origin of lherzolite variation

In this section, the origin of the major chemical variation seen in the lherzolite nodules in basaltic rocks is discussed. The discussion given in the earlier papers of KUNO (1969a, b) were based on a smaller number of analyses and are revised here.

The whole variation cannot be ascribed to a primary inhomogeneity of the mantle caused by accretion of different types of chondrite. The different chondrite types such as hypersthene-, bronzite, and enstatite chondrites are believed to have been originated by different degrees of reduction of iron. The reduction would result simply in the increase of MgO/ΣFeO in the silicate phases accompanied with increase of all the oxide components other than ΣFeO. This is not the case in the lherzolite variation. Therefore the variation is here considered as resulting from either partial melting or fractional crystallization in the mantle.

6.1. Partial melting

As seen in fig. 3, the lherzolites with MgO/ΣFeO lower than 4.0 are much fewer in number and more diverse in various oxide contents than those with higher MgO/ΣFeO. It must be noted that those with MgO/ΣFeO lower than 4.0 are all contained in pyroclastic rocks except for no. 56. This would imply that the nodules rich in lower melting components are more easily disintegrated when they are included in lavas than in pyroclastics because of the slower transportion to the surface by the lavas. It is assumed therefore that those lherzolites near the lowest MgO/ΣFeO end of the variation represent primordial mantle material, or are at least close to it, and that the diversity of their composition is due to the inhomogeneity of the primordial material. Those with higher MgO/ΣFeO may be residue formed by successive partial melting of this primordial material. This postulated primordial material is lower

TABLE 16

Average compositions of lherzolite nodules grouped on the basis of MgO/FeO+Fe$_2$O$_3\times 0.9$

No.	A	B	C	D
MgO/ΣFeO	6.99–6.00	5.99–5.00	4.99–4.00	3.99–3.00
No. of analyses	4	15	25	10
SiO$_2$	43.54	44.08	44.26	44.31
Al$_2$O$_3$	0.82	1.63	2.59	2.37
Fe$_2$O$_3$	1.20	1.44	1.64	2.13
FeO	6.47	6.95	7.25	8.84
MgO	47.29	43.10	40.55	38.31
CaO	0.11	1.53	2.44	2.12
Na$_2$O	0.07	0.13	0.23	0.39
K$_2$O	0.02	<0.04	<0.03	0.12
H$_2$O(+)	0.21	0.39	0.31	0.67
H$_2$O(−)	0.04	0.17	0.06	0.16
TiO$_2$	0.07	0.12	0.15	0.24
P$_2$O$_5$	0.00	<0.06	<0.01	<0.04
MnO	0.12	0.13	0.14	0.15
Cr$_2$O$_3$	0.38	0.35	0.38	0.30
Total	100.34	100.02	100.00	100.11
MgO/FeO+Fe$_2$O$_3\times 0.9$	6.26	5.22	4.64	3.56

in MgO/ΣFeO than the "pyrolite" of RINGWOOD (1966) as seen in fig. 3.

In order to show the mean variation trend of the lherzolite, average compositions for groups of rocks lying within different MgO/ΣFeO intervals are calculated with the result given in table 16 and are plotted in fig. 4. In this calculation, wehrlite (no. 39), harzburgite (no. 51), and lherzolites (nos. 27, 47, 54, 55, 57a, b, 59a, b, and 60a, b) are not included. These lherzolites are either partly serpentinized as seen from the high H$_2$O or show marked departure from the general variation in some oxide contents (no. 47) or has MgO/ΣFeO lower than 3.00 (no. 27).

In this figure, the oxide per cents are plotted against MgO, because in the MgO/ΣFeO diagram, the addition and subtraction relations cannot be shown by straight lines.

Point PL in fig. 4 represents an assumed composition of the average primordial lherzolite. It is located so as to lie on the extension of the general trend line for each oxide and its MgO/ΣFeO is assumed to be 3.00. Its composition obtained graphically is given in table 17.

In fig. 4, rock no. 27, which has the lowest MgO/ΣFeO of the lherzolite variation, some representative basalts from Hawaii and Japan having highest MgO/ΣFeO among aphyric lavas, average oceanite of Hawaii, olivine phenocrysts in Hawaiian oceanite and olivine tholeiite and in Japanese tholeiite, and orthopyroxene phenocrysts in Hawaiian olivine tholeiite are also plotted.

If we suppose that partial melting of rock PL produces an initial liquid of some basaltic composition, for example the Pele's hair of Hawaii (PH of fig. 4 and table 17), further melting would shift the composition

TABLE 17

Estimated compositions of residues at successive stages of partial melting and of co-existing magmas

No.	A[a]	S-1[a]	PL[a]	L-2[a]	L-1[a]	AO[a]	PH[a]
SiO$_2$	43.54	44.4	45.0	45.9	46.4	46.41	48.82
Al$_2$O$_3$	0.82	2.6	4.1	5.9	7.2	8.53	13.42
ΣFeO	7.55	10.2	11.0	12.7	12.6	12.04	11.43
MgO	47.29	38.8	33.0	26.0	21.5	20.81	9.00
CaO	0.11	2.3	4.0	5.9	7.3	7.38	11.32
Na$_2$O+K$_2$O	0.09	0.38	0.68	0.96	1.30	1.90	2.83
TiO$_2$	0.07	0.19	0.36	0.50	0.69	1.98	2.77
MnO	0.12	0.15	0.16	0.18	0.18	0.15	0.18
Total	100.34[b]	99.02	98.30	98.04	97.17	99.65[b]	100.22[b]
MgO/FeO+Fe$_2$O$_3\times 0.9$	6.26	3.80	3.00	2.05	1.70	1.73	0.79

a) PL: Postulated primordial mantle lherzolite; L-1 and S-1: Magma (L-1) produced by 33% melting of PL, co-existing with the residue S-1; L-2: Magma produced by 66% melting of PL, co-existing with the residue A (Table 16); AO: Average of 14 oceanites of the Hawaiian Islands (MACDONALD and KATSURA, 1964); PH: Pele's hair of 1959 eruption of Kilauea Iki (MACDONALD and KATSURA, 1961); b) Including minor components.

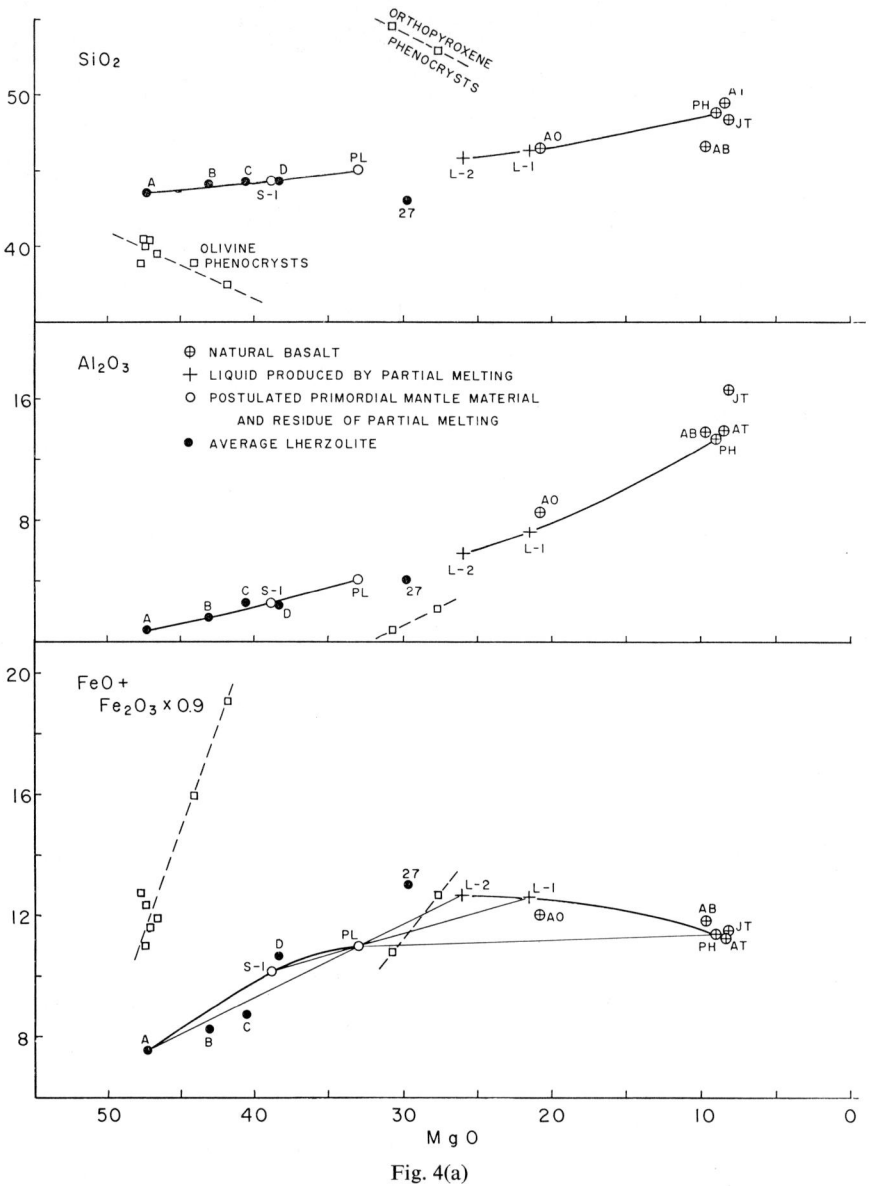

Fig. 4(a)

Fig. 4. Average compositions of lherzolite nodules, compositions of some representative basaltic rocks, and those of liquids produced by successive partial melting of the lherzolites plotted against weight per cent of MgO. A, B, C, and D: average compositions of lherzolite nodules in basaltic rocks given in columns A, B, C, and D respectively of table 16. PL: the postulated primordial mantle lherzolite given in table 17. L-2 and L-1: liquids (table 17) produced by successive partial melting of PL. S-1: solid residue (table 17) produced by partial melting of PL. AO: average oceanite of Hawaii (table 17). PH: Pele's hair (table 17, olivine tholeiite) of 1959 eruption of Kilauea Iki. AT: average tholeiite of Hawaii (MACDONALD and KATSURA, 1964). AB: average alkali basalt of Hawaii (KUNO et al., 1957). JT: average tholeiite of Izu-Hakone region, Japan (KUNO, 1966). Diamonds are for olivine and orthopyroxene phenocrysts of fig. 3. Rock no. 27 of table 2 is plotted in order to show the least magnesian end of the lherzolite variation. Points for co-existing liquids and solids during partial melting are connected by solid lines.

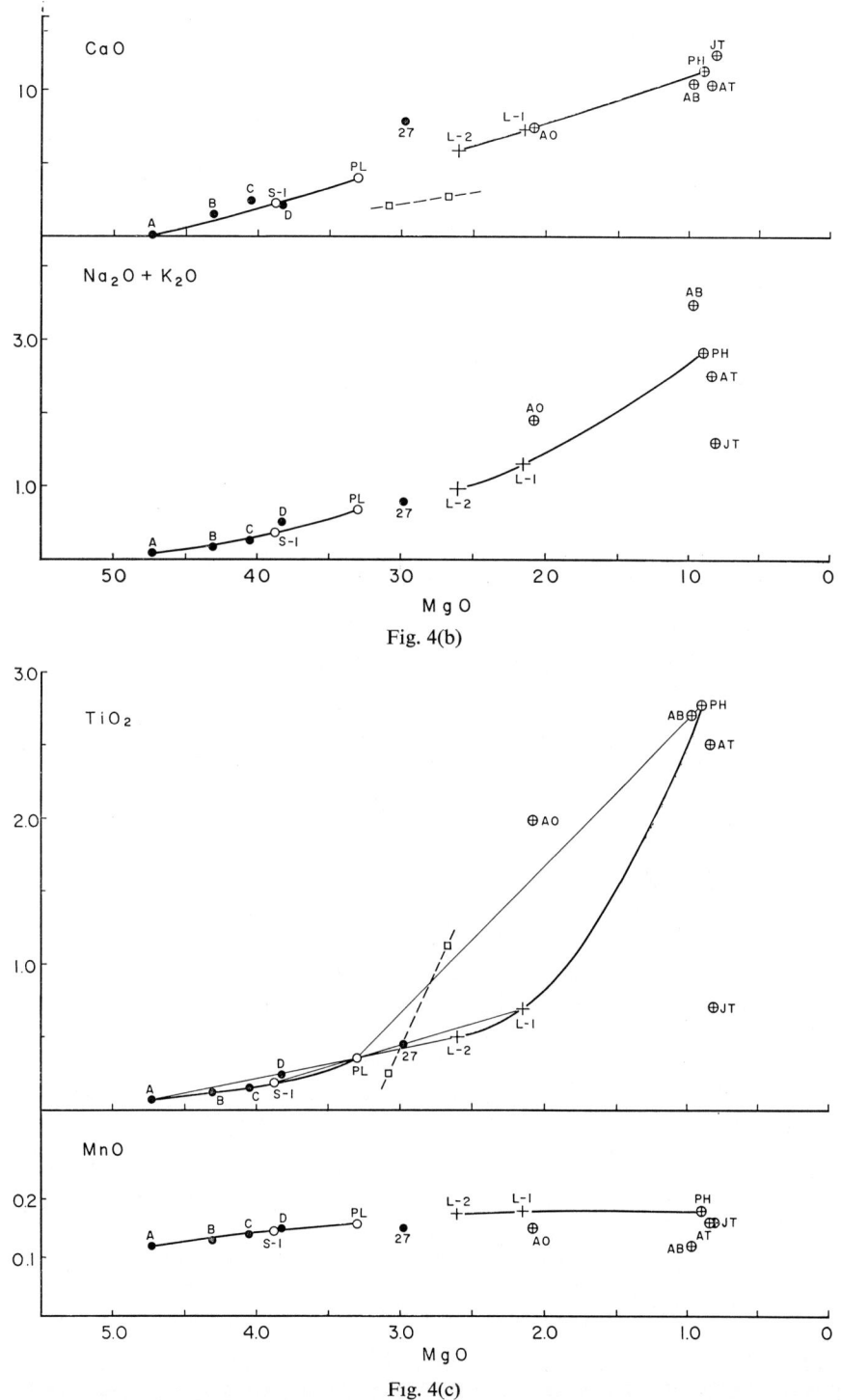

Fig. 4(b)

Fig. 4(c)

of the solid residue from PL to A along a line passing close to points D, C, and B, while the liquid composition also changes from PH to a point lying on the MgO-rich side of PH. Any Hawaiian lavas having compositions more magnesian than the Pele's hair (PH) are invariably porphyritic with olivine and rarely orthopyroxene phenocrysts; they are oceanite and olivine-rich tholeiite. This implies that, if any liquids more magnesian than the Pele's hair are ever formed at some depth, they separate much olivine and perhaps some orthopyroxene upon rising to shallower depth, changing successively their composition to that of the Pele's hair.

MACDONALD and KATSURA (1964) showed that the chemical variation of the Hawaiian rock series from oceanite to olivine tholeiite (Pele's hair) is solely controlled by olivine fractionation. TILLEY et al. (1964) demonstrated experimentally that olivine is the dominant phase crystallizing from liquids ranging in composition from the Hawaiian oceanite to olivine tholeiite. Thus any liquids produced by successive partial melting of the primoridal lherzolite PL would lie between the points for the Hawaiian olivine phenocrysts of fig. 4 and point PH, namely on a line, or lines depending on the pressures at which the partial melting takes place, passing close to the point for the average oceanite (AO of fig. 4 and table 17).

Then, it is necessary to know the approximate compositions of liquids co-existing with solids PL and A.

GREEN and RINGWOOD (1967a) determined the ratio

$$K = \frac{Mg \times 100}{Fe^{+2} + Mg} \text{ for crystal and } \frac{Mg \times 100}{Fe^{+2} + Mg} \text{ for liquid}$$

for co-existing pairs olivine-liquid and orthopyroxene-liquid, both on the liquidus, at various pressures. They found that K varies little for different pairs and at different pressures, its average value being 1.27. Taking this value, the liquid co-existing with solid PL should have MgO 10.0% and FeO 9.1%, MgO 9.0% and FeO 8.1%, or MgO 8.0% and FeO 7.3% (all in weight). These values are very close to those of Pele's hair and other basalts plotted in fig. 4. In the same way, the liquid co-existing with solid A should have MgO 25.0% and FeO 16.5% or MgO 20.0% and FeO 13.2%. The latter pair is close to that of L-1 of table 17 which represents an estimated liquid formed by partial melting as will be mentioned later.

Another way of approach to the problem is to compare the solidi of various lherzolite with the liquidi of basaltic rocks determined at various pressures.

The solidi for lherzolites nos. 2, 24, and 67b and the pyrolite were determined by AKIMOTO and ARAMAKI (1966), KUSHIRO et al. (1968a), ITO and KENNEDY (1967) and GREEN and RINGWOOD (1967b) respectively as reproduced in fig. 5 by solid lines. The liquidi for synthetic picrite (oceanite) and olivine-rich tholeiite (ITO and KENNEDY, 1968), for olivine tholeiite (COHEN et al., 1967), and for quartz eclogite of tholeiite composition (YODER and TILLEY, 1962) are also reproduced in fig. 5 by broken lines.

It is seen in fig. 5 that lherzolite no. 2, having MgO/ΣFeO close to that of average lherzolite A, would produce by initial melting a liquid of the olivine-rich tholeiite composition (MgO/ΣFeO = 1.56) at 20 to

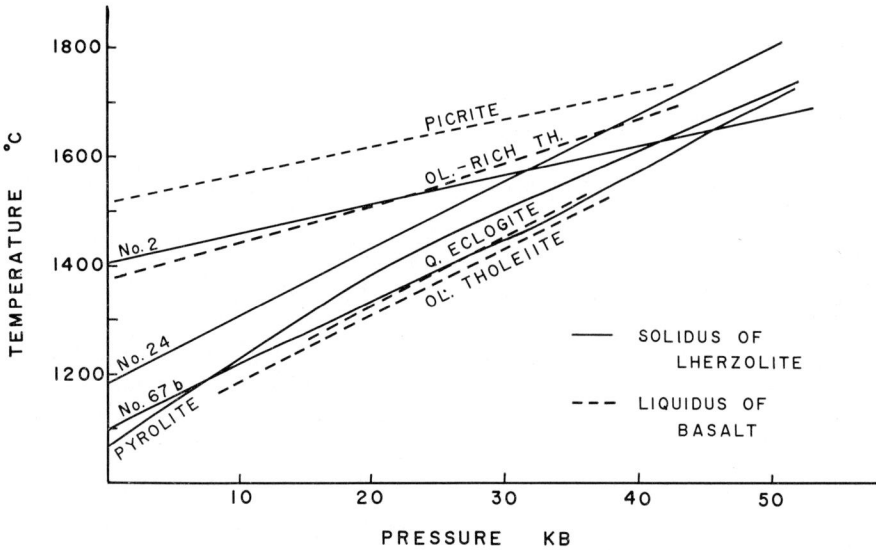

Fig. 5. Solidi of lherzolite (solid lines) and liquidi of basalt and picrite (broken lines). Nos. 2, 24, and 67b refer to those in tables 1, 2, and 8 respectively. See also text.

30 kb, and by further melting would produce a liquid of the picrite composition (MgO/ΣFeO = 2.65) at any pressures up to 40 kb. Lherzolite no. 24, having MgO/ΣFeO close to that of the postulated primordial lherzolite PL, and the pyrolite (MgO/ΣFeO = 4.43) cannot produce by melting liquids of the olivine tholeiite (MgO/ΣFeO = 0.55-0.67 depending on the analyses by different analysts, YODER and TILLEY, 1962) and quartz eclogite (MgO/ΣFeO = 0.47) compositions at 20 to 30 kb. Probably, the initial melting of lherzolite no. 24 would produce an olivine tholeiite liquid with MgO/ΣFeO a little higher than 0.67, namely a liquid close to the Pele's hair (table 17) and some other basalts plotted in fig. 4.

The temperature relation between the solidi of the four lherzolites studied is not consistent with their MgO/ΣFeO. It is rather surprizing to see that garnet lherzolite no. 67b has a low temperature solidus and appears to yield a liquid of the quartz eclogite composition by initial melting.

From these considerations, it is likely that the primordial lherzolite PL and the average lherzolite A are in equilibrium with liquids of some olivine tholeiite composition such as the Pele's hair and of oceanite composition respectively, possibly at pressures between 10 and 30 kb.

The above discussion is based on the assumption that the melting takes place under anhydrous condition. KUSHIRO et al. (1968a) showed experimentally that a drastic decrease of solidus temperatures takes place by the presence of water in the system. However, the liquidus temperatures of basalts may equally be lowered by the presence of water. Therefore the above discussion may hold also for hydrous systems.

According to KUSHIRO et al. (1968a) the four solid phases of lherzolite no. 24 are present on the solidus at 20 kb, and therefore lherzolite PL would maintain the same four phases when partial melting starts at 20 kb. During further melting, the same phases may co-exist with the successive liquids until the solid has the composition A, if the temperature difference between the initial melting and the stage of solid A is small. However, it is also possible that, during the successive melting, clinopyroxene or clinopyroxene and spinel may disappear (ITO and KENNEDY, 1968) and the solid phases in equilibrium with the liquid are olivine and orthopyroxene. The last two minerals form a crystal aggregate having the bulk composition of A after separation from the liquid. Upon cooling to a temperature on the relevant geotherm, they may recrystallize to form the four phases of lherzolite. This implies that the orthopyroxene co-existing with the liquid contains clinopyroxene and spinel components in solid solution. Such an orthopyroxene composition is possible.

If the partial melting takes place at pressures from 30 to 50 kb, garnet would appear on the solidus and would persist upon cooling to the relevant geotherm. This does not appear to be the case in the garnet-free lherzolite.

Thus the most likely process to produce the observed lherzolite variation is the partial melting at about 20 kb with liquid co-existing with the four solid phases at least in the earlier stage of the melting. The liquid would remain mostly on the isobaric univariant line and then may leave this line to move on a divariant surface.

The compositional relation between the solids and co-existing liquids during partial melting is illustrated by the hypothetical phase diagram of the system SiO_2-MgO-FeO (fig. 6). It is assumed that at 20 kb olivine and orthopyroxene have a cotectic relation. Even at 20 kb, magnesian olivine may bear a reaction relation to orthopyroxene under hydrous condition, as has been shown by KUSHIRO et al. (1968b). However, the following discussion still holds in principle even if such a reaction relation exists.

In the upper diagram of fig. 6, line L_1-L_3 is a boundary between the orthopyroxene and olivine fields. If solid S_1, having a composition lying on the extension of the line from L_1 to L_3, undergoes partial melting, the first liquid to form is L_1. Upon further melting, the liquid moves along the boundary line from L_1 to L_2 and then to L_3. As liquids L_2 and L_3 are in equilibrium with solids S_2 and S_3, the solid residue changes successively in composition from S_1 to S_2 and then to S_3.

The variation of SiO_2 and FeO for liquids from L_1 to L_3 and solids from S_1 to S_3 are plotted against MgO in the lower diagrams of the figure. It is seen that the composition of the solid changes along a line lying approximately on the extension of the liquid variation line. If separation of some liquid from the solid takes place during the partial melting, namely partial melting with fractionation, the composition of the solid moves toward the extension of the solid variation line as shown by the arrow.

If the initial solid has the composition of S_4 which is a little removed from the extension of the boundary line L_1-L_3 of the upper diagram, the first liquid to form is again L_1. Upon further melting, the liquid changes from L_1 to L_3, while the solid residue changes from S_4 to S_6 passing through S_5 along a curved course, provided a complete reaction obtains between the liquid and solid throughout the melting. If fractionation takes place when solid S_5 is formed, the solid composition moves toward the arrow head. The SiO_2 and FeO variations for liquids from L_1 to L_3 and solids from S_4 to S_6 are shown in the lower diagrams. The pair of points for the co-existing liquid and solid are connected by straight lines. In contrast to line S_1-S_2-S_3, line S_4-S_5-S_6 shows a distinct departure from the extension of liquid variation line L_1-L_2-L_3. The fractionation at the stage of S_5 shifts the solid variation toward the arrow head. If such fractionation occurs continuously throughout the partial melting, the solid variation would be represented by a line which is close to the extension of the straight line connecting L_1 and S_4, whereas the liquid still remains on the boundary line.

If the initial solid is S_7, the solid variation is from S_7 to S_8, showing more departure from the liquid variation line L_1-L_2 than does line S_4-S_6. Thus the degree of departure depends on the compositional relation of the initial solid with the univariant line along which the liquid moves during partial melting. In the natural system, whether the solid variation line is close to the extension of that of the liquid or shows some departure therefrom may differ in different oxide components.

Applying the above principle, the liquid variation which would result in the mean lherzolite variation for each oxide is shown by a line from PH to L-2 passing through L-1 of fig. 4. The pairs of points for the co-existing solid and liquid are connected by straight lines where possible without causing any confusion of lines in the diagram. The liquid line PH to L-2 is drawn so as to pass between the points for the Hawaiian olivine phenocrysts and the Hawaiian tholeiite and also close to the average oceanite point AO.

It must be remembered that this is not a unique solution; we could draw some other liquid variation lines for other magma suites, for example, a line passing through the Japanese tholeiite point JT. In addition, the location of point PL is somewhat flexible, and one could assume different amounts of melting to produce the residue A. In spite of these allowances, it was found that the possible location of the liquid variation line is still limited within a narrow range.

The compositions of the successive liquids, L-1 and L-2, and of solid S-1, which is in equilibrium with liquid L-1, as obtained by this graphical method, are listed in table 17. The total of the oxide per cents for each inferred composition is close to 100, indicating that the location of the liquid variation line for each oxide is quite reasonable.

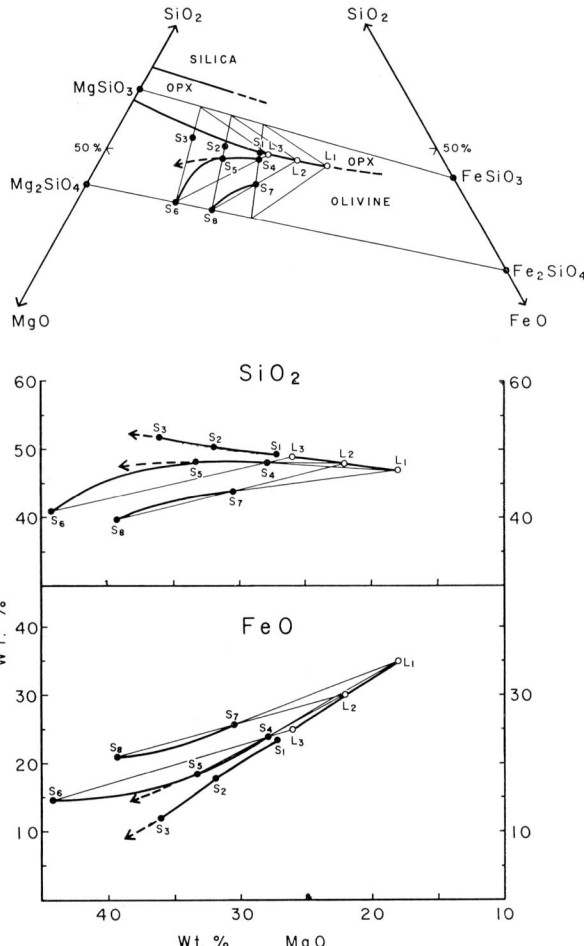

Fig. 6. A hypothetical diagram of a part of the system SiO_2 – MgO – FeO at 20 kb (upper diagram) and an oxide variation diagram for the liquids (L_1, L_2, and L_3) and solids (S_1, S_2, S_3, S_4, S_5, S_6, S_7, and S_8) plotted against MgO (lower diagram), showing the compositional relations between liquids and solid residues produced by partial melting. Curve L_1 to L_3 of the upper diagram is the boundary between the fields of orthopyroxene and olivine solid solutions. In the lower diagram, points for the co-existing liquids and solids are connected by straight lines where they do not cause any confusion of lines.

The above graphical solution is based on the assumption of a complete reaction between the solid and liquid throughout the partial melting. If fractionation takes place at successive stages of partial melting, the solid variation would be represented by a line passing close to the extension of the straight line from PH to PL *for every oxide*, or the liquid variation line would lie on the extension of the solid variation line from A to PL. Either of these two cases does not apply to the variations for ΣFeO and TiO_2. Therefore, fractionation does not appear to have taken place.

Because of the extremely low content of K_2O in the lherzolite, as mentioned before, the liquids produced should also be low in this component. Some process other than partial melting might operate in producing magmas with comparatively high K_2O contents, such as some trachybasalt and leucite-bearing basaltic rocks.

OXBURGH (1964) also pointed out that the K_2O contents in ultramafic nodules are too low to produce basalt magmas and suggested the existense of K-bearing amphibole in the peridotite of the upper mantle. As mentioned before, pargasitic hornblende is often found in lherzolite and websterite nodules, but its K_2O content is also low (87-HO of table 13 and 74-HO of table 14).

Although the major feature of the compositional variation of lherzolite can be explained as originating from a single partial melting process, this does not exclude the possibility that the individual lherzolite compositions have also been affected by repeated partial melting in minor degrees.

For example, if the compositions of the Itinome-gata lherzolites are plotted as in fig. 7, together with those of the representative basalts, it is seen that the lherzolites fall into two groups (nos. 1 to 5 and nos. 6 to 16) differing in the content of the lower melting components. Those poorer in these components can be produced by subtracting about 17% basaltic fraction from the others. This basaltic fraction should be poor in Na_2O+K_2O and TiO_2. The Japanese tholeiite JT fits this requirement.

6.2. *Crystal settling*

In fig. 3, compositions of early olivine and orthopyroxene crystals from Hawaiian oceanite and olivine tholeiite and from Japanese tholeiite are plotted. The lherzolites with lower $MgO/\Sigma FeO$ could be formed by settling of these crystals, but the other lherzolites are too high in $MgO/\Sigma FeO$ to be formed from these crystals.

In the same figure, compositions of lherzolite and plagioclase-bearing peridotite of the layered Horoman ultramafic intrusion, Hokkaido, Japan are plotted. According to NAGASAKI (1966), the successive layers were formed by crystal settling in some basaltic magma. The compositional variation of the Horoman rocks is similar to that of the lherzolite nodule variation, although a slight difference is noted in ΣFeO, MgO, and CaO variation. This similarity may suggest that the lherzolite nodule variation is also produced by crystal settling.

Let us first assume that the crystallization takes place at 20 kb under anhydrous condition. The liquid may move along an isobaric univariant line precipitating the four solid phases of lherzolite. If the univariant line is in effect nearly straight for some oxide components, the solid variation should follow a line lying on the extension of the liquid variation line, like the relation between line L_3-L_1 and line S_3-S_1 of fig. 6. This relation does not change according to whether a complete reaction is maintained or an extreme fractionation takes place. Therefore, so far as the variation of SiO_2, Al_2O_3, CaO, Na_2O+K_2O, and MnO is concerned, solids A, S-1, and PL of fig. 4 may have been precipitated from liquids L-2, L-1, and PH respectively.

If the univariant line is curved, the relation between the liquid and solid variation lines is like that between line L_1-L_2-L_3-L_4 and line S_1-S_2-S_3-L_1 of fig. 8, if a complete reaction is maintained. This relation applies to the lherzolite variation for ΣFeO and TiO_2 as shown in fig. 9 which is a mere reproduction of fig. 4.

In this case, PL is again assumed as representing the primordial material which was once completely melted. Upon crystallization, it changes to L-2 which precipitates solid A. Liquid L-2 then moves to L-1, and finally to PH where the crystallization is complete, PL being the final crystallization product.

However, crystal settling would inevitably cause fractionation. In the case of extreme fractionation, liquids L_2, L_3, and L_4 of fig. 8 precipitate solids S_4, S_5, and S_6 respectively, lines S_4-L_2, S_5-L_3, and S_6-L_4 being tangent to the liquid variation line at L_2, L_3, and L_4 respectively. Applying this principle, the solid variation line from A to S-2 passing through S-1 of fig. 9 is drawn

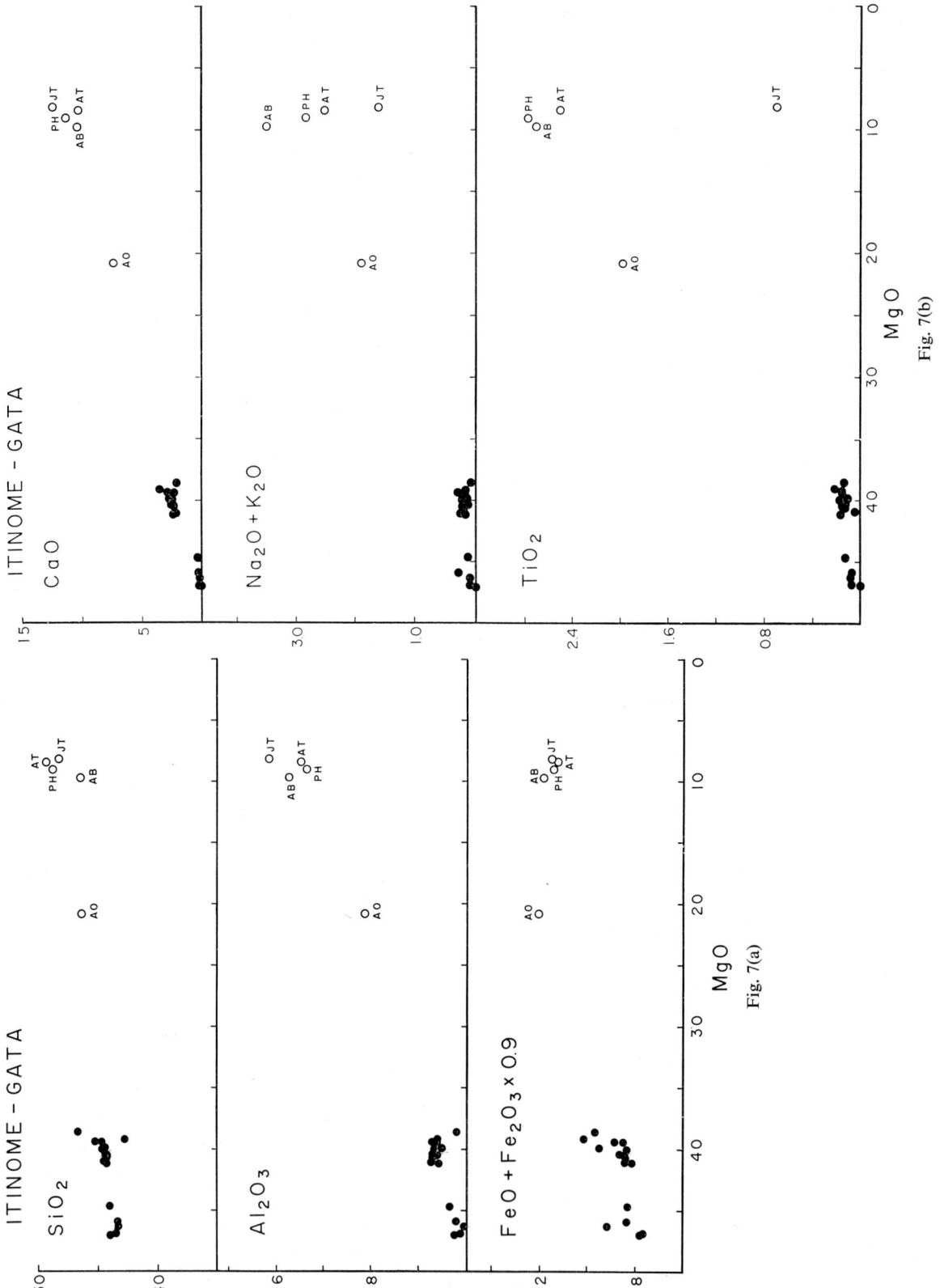

Fig. 7(a)

Fig. 7(b)

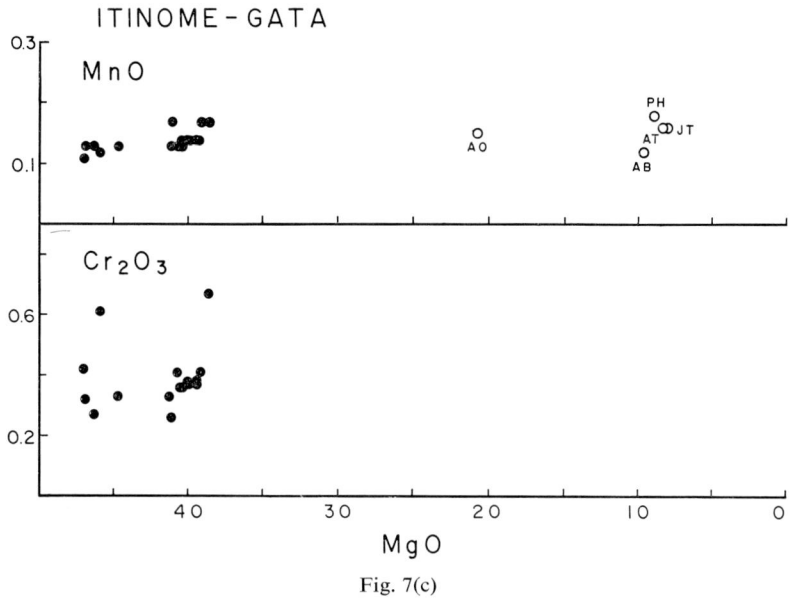

Fig. 7. Compositions of lherzolite nodules (solid circles) from Itinome-gata, Japan, plotted against MgO. Open circles are for the representative basaltic rocks plotted in fig. 4.

as representing the lherzolite variation caused by precipitation with extreme fractionation from liquid changing from L-3 to PH passing through L-4. Liquid L-3 may be derived by partial melting of some solid material more magnesian than A. It is also possible that liquid L-3 is formed by partial melting of solid PL or by crystallization of liquid PL at a physical condition different from that now under consideration.

As we assume that A is a fixed point and also as there is little allowance in locating points S-1 and S-2 for given MgO per cents, the line from L-3 to PH cannot be drawn closer to the line from L-2 to PH for both ΣFeO and TiO$_2$. Therefore, as compared with the ΣFeO and TiO$_2$ values for points L-2 and L-1 given in table 17, ΣFeO values for L-3 and L-4 are lower by 1.6 and 1.2 per cents respectively and TiO$_2$ values are higher by 0.8 and 0.9 per cents. Thus the total of all the oxides becomes a little too low.

However, as the extreme fractionation does not appear to be a realistic process, the most likely liquid variation line would pass somewhere between the line from L-2 to PH and that from L-3 to PH.

If some water is present, olivine may bear a reaction relation to orthopyroxene even at 20 kb total pressure (Kushiro et al., 1968b). In the case of equilibrium crystallization, the liquid remains on the univariant line until the crystallization is complete. The solid consists of the four phases throughout the crystallization. The relation between the solid and liquid variations is essentially the same as that shown in fig. 4.

If extreme fractionation takes place, the liquid would leave the univariant line; olivine no longer precipitates and no more lherzolite is formed. Therefore, the only possibility is that the discontinuous reaction is complete but the continuous reaction is somewhat incomplete so that the liquid still remains on the univariant line. In this case, the solid and liquid variations are essentially the same as those shown in fig. 4.

7. Discussion

As mentioned above, a more reasonable liquid variation line can be drawn on the assumption of complete reaction between the solid and liquid rather than on the assumption of strong fractionation both in the cases of partial melting and of crystal settling.

The partial melting starts with a small amount of liquid which gradually increases in proportion as the melting proceeds. In this case, any strong fractionation may not occur. On the contrary, the crystallization starts with a large proportion of liquid in which crystal settling is likely to occur, resulting in strong fractionation. Thus, the partial melting appears to be a more likely process which produced the observed lherzolite variation.

This problem is also related to the thermal history of the earth and the validity of the assumption on the primordial mantle material.

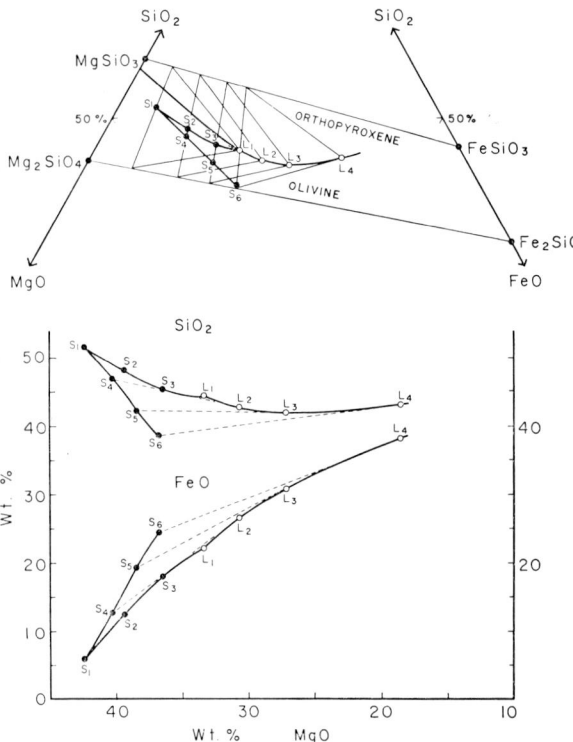

Fig. 8. A hypothetical diagram of a part of the system $SiO_2 - MgO - FeO$ at 20 kb (upper diagram) and an oxide variation diagram for the liquids (L_1, L_2, L_3, and L_4) and solids (S_1, S_2, S_3, S_4, S_5, and S_6) plotted against MgO (lower diagram), showing the compositional relation between liquids and solids produced by crystal settling. Curve L_4 to L_1 of the upper diagram is the boundary line between the fields of orthopyroxene and olivine solid solutions. In the lower diagram, points for the co-existing liquids and solids are connected by broken lines where they do not cause any confusion of lines.

If the earth has experienced a stage of steep thermal gradient by which a bodily melting or a great amount of partial melting was caused, the crystal settling would have been the major process which produced the observed lherzolite variation. It is also possible that such a great amount of melting took place in different parts of the earth in different geologic times.

If the thermal gradient was not sufficiently high to cause this kind of melting throughout the earth's history, the partial melting would have been the direct process to produce the lherzolite variation.

The magmas of oceanite to olivine tholeiite compositions could produce upon crystallization quartz- or olivine-normative tholeiite (and high-alumina basalt) or alkali basalt, depending on the total pressure and water pressure. However, the same types of magmas would also be produced directly by partial melting of the primordial mantle material under different pressure conditions. In this case, no marked variation in the solid material would result.

Even after the variation of the mantle material such as shown in fig. 3 has been formed, the initial melting of the portion of the mantle still retaining lower melting components would also produce those different types of basalt magmas. If sufficient amount of heat is supplied, melting of the more refractory lherzolite would take place, resulting in olivine-rich tholeiite and oceanite magmas.

In fig. 3, the primordial mantle material "pyrolite" as postulated by RINGWOOD (1966) is plotted. It is seen that the pyrolite lies in the middle of $MgO/\Sigma FeO$ variation of the lherzolite nodules, and that its oxide contents fall within the range of the lherzolite variation except for $Na_2O + K_2O$ and TiO_2. The high $Na_2O +$

TABLE 18

Average compositions of hypersthene chondrite and bronzite chondrite (MASON, 1965)

	Hyp. ch.	Hyp. ch. silicate phase	Bronz. ch.	Bronz. ch. silicate phase
SiO_2	39.87	46.17	36.33	47.83
Al_2O_3	2.51	2.91	2.66	3.50
FeO	14.66	16.98	9.61	12.65
MgO	25.16	29.14	23.53	30.98
CaO	1.89	2.19	1.76	2.32
Na_2O	0.95	1.10	0.86	1.13
K_2O	0.15	0.17	0.12	0.16
TiO_2	0.15	0.17	0.15	0.20
P_2O_5	0.27	0.31	0.26	0.34
MnO	0.29	0.34	0.25	0.33
Cr_2O_3	0.45	0.52	0.42	0.55
Total	86.35	100.00	75.95	99.99
MgO/FeO	—	1.72	—	2.45
Fe	6.28	—	16.79	—
Ni	1.10	—	1.63	—
Co	0.06	—	0.09	—
FeS	6.07	—	5.27	—

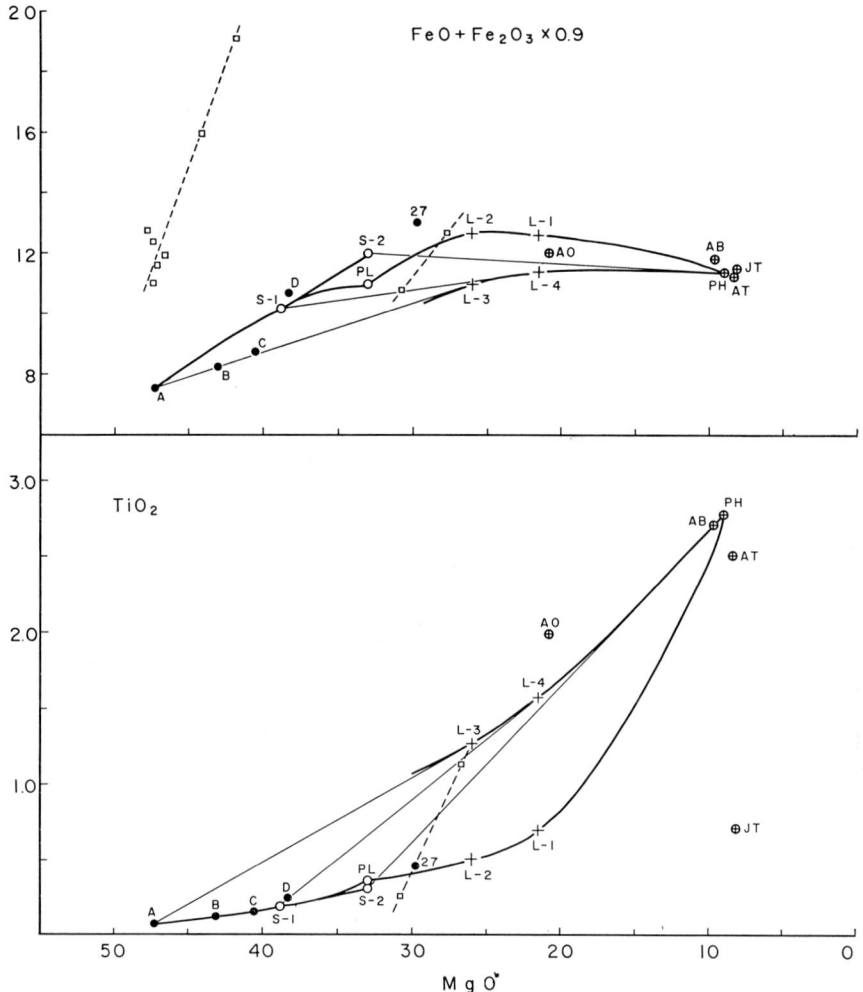

Fig. 9. ΣFeO and TiO$_2$ plotted against MgO for the average lherzolite nodules (A, B, C, and D of table 16), compositions of some representative basaltic rocks (AO, PH, AT, AB, and JT, see fig. 4), and those of liquids (L-1, L-2, L-3, and L-4) and of solids (S-1 and S-2) produced by crystal settling. Lherzolite no. 27 of table 2 is also plotted. Points for the co-existing solids and liquids are connected by straight lines where they do not cause any confusion of lines. See also text.

K$_2$O and TiO$_2$ of the pyrolite is due to the fact that, in calculating this composition, the analysis of the Pele's hair (PH of table 17) is used. The Pele's hair has a little higher contents of Na$_2$O+K$_2$O and TiO$_2$ than other tholeiites of the world with similar MgO/ΣFeO (compare Na$_2$O+K$_2$O and TiO$_2$ of PH with those of JT and AT of fig. 4). Thus the pyrolite composition could be modified so as to fit completely the lherzolite variation by using an average world tholeiite in the calculation.

However, there is some doubt as to whether the pyrolite can be accepted as the primordial mantle material, because many lherzolite nodules have higher MgO/ΣFeO than the pyrolite. The pyrolite may represent the average upper mantle composition of the present time.

We prefer PL of table 17 as the average primordial mantle material. Its composition closely resembles that of the silicate phase of bronzite chondrite (table 18) recalculated from the average composition of this type of meteorite including metal phase as given by MASON (1965). The silicate phase of hypersthene chondrite (table 18) has a composition slightly different from those of PL and of any lherzolite nodules.

It is likely that the earth was formed by accretion of particles having the composition of bronzite lherzolite.

Its silicate phase was separated from the metal phase either before or after the accretion.

8. Summary

1) Mineralogy of lherzolite and websterite nodules from Itinome-gata, Japan, from Hawaiian Islands, from Dreiser Weiher, West Germany, and several other localities is described. From the mineral assemblage and chemistry of pyroxenes, it is suggested that garnet lherzolite of Itinome-gata was originally located at some depth between 50 and 150 km and was overlain by garnet-free lherzolite. The garnet lherzolite was later brought up to the level of the garnet-free lherzolite probably by the mantle convection. The websterite of Itinome-gata was probably formed at the top of the mantle.

2) By plotting the bulk compositions of these nodules against their $MgO/\Sigma FeO$, it was found that the lherzolite shows a wide and systematic variation in their oxide components which is distinctly different from the websterite variation. The latter is similar to the variation of eclogite nodules from Salt Lake, Hawaii. It is suggested that the websterite was formed by crystal settling.

3) In the lherzolite, the lower melting components generally increase with the decrease of $MgO/\Sigma FeO$. Assuming a primordial mantle material having a composition of average nodule lherzolite with low $MgO/\Sigma FeO$, the lherzolite variation can be explained as resulting from different degrees of partial melting of this primordial material, the magma thus produced changing in composition from olivine tholeiite to oceanite. A complete reaction between the solid and liquid should be maintained during the parial melting. The variation can also be explained as resulting from crystal settling in magma which changes in composition from oceanite to olivine tholeiite, provided a complete reaction is maintained. From various considerations, the partial melting appears to be a more likely process.

4) The oceanite to olivine tholeiite magmas could produce upon crystallization quartz- or olivine-normative tholeiite (and high-alumina basalt) or alkali basalt, depending on the total pressure and water pressure. However, the same types of magmas would also be produced directly by partial melting of the primordial mantle material under different pressure conditions.

5) The lherzolite with low $MgO/\Sigma FeO$, being assumed as representing the primordial mantle material, has a composition close to the silicate phase of the average bronzite chondrite, suggesting that the earth was originally formed by accretion of this type of chondrite.

References

Akimoto, S. and S. Aramaki (1966) Quoted in second progress report on the upper mantle project of Japan, Nat. Comm. UMP, Sci. Council of Japan, 46.

Aoki, K. and I. Kushiro (1968) Some clinopyroxenes from ultramafic inclusions in Dreiser Weiher, Eifel. Contr. Mineral. Petrol. **18**, 326–337.

Clark, S. P. and A. E. Ringwood (1964) Density distribution and constitution of the mantle, Rev. Geophys. **2**, 35–88.

Cohen, L. H., K. Ito and G. C. Kennedy (1967) Melting and phase relations in an anhydrous basalt to 40 kilobars, Am. J. Sci. **265**, 475–518.

Davis, B. T. C. and F. R. Boyd (1966) The join $Mg_2Si_2O_6$-$CaMgSi_2O_6$ at 30 kilobars pressure and its application to pyroxenes from kimberlites, J. Geophys. Res. **71**, 3567–3576.

Fenner, C. N. (1938) Olivine fourchites from Raymond Fosdick Mountains, Antarctica, Bull. Geol. Soc. Am. **49**, 367–400.

Forbes, R. B. and H. Kuno (1965) Peridotite inclusions and basaltic host rocks, in: *Upper mantle symposium New Delhi 1964* (Intern. Union of Geol. Sci., Copenhagen) p. 161–179.

Frechen, J. (1948) Die Genese der Olivinausscheidungen vom Dreiser Weiher (Eifel) und Finkenberg (Siebengebirge), Neues Jahrb. Mineral. Abhand. **79A**, 317–406.

Frechen, J. (1963) Kristallisation, Mineralbestand, Mineralchemismus und Förderfolge der Mafitite vom Dreiser Weiher in der Eifel, Neues Jahrb. Mineral. Monatsh. 205–225.

Gagel, C. (1912) Studien über den Aufbau und die Gesteine Madeiras, Z. Deut. Geol. Ges. **64**, 344–491.

Green, D. H. (1966) The origin of the "eclogites" from Salt Lake Crater, Hawaii, Earth Planet. Sci. Letters **1**, 414–420.

Green, D. H. and A. E. Ringwood (1967a) The genesis of basalt magmas, Contr. Mineral. Petrol. **15**, 103–110.

Green, D. H. and A. E. Ringwood (1967b) The stability fields of aluminous pyroxene peridotite and garnet peridotite and their relevance in upper mantle structure, Earth Planet. Sci. Letters **3**, 151–160.

Hamad, S. el D. (1963) The chemistry and mineralogy of the olivine nodules of Calton Hill, Derbyshire, Mineral. Mag. **33**, 483–497.

Hess, H. H. (1955) The oceanic crust, J. Marine Res. **14**, 423–439.

Holmes, A. (1936) A contribution to the petrology of kimberlite and its inclusions, Trans. Geol. Soc. S. Africa **39**, 379–427.

Horai, K. (1964) Studies of the thermal state of the earth, The 13th paper: terrestrial heat flow in Japan, Bull. Earthquake Res. Inst. Tokyo Univ. **42**, 93–132.

Ito, K. and G. C. Kennedy (1967) Melting and phase relations in a natural peridotite to 40 kilobars, Am. J. Sci. **265**, 519–538.

Ito, K. and G. C. Kennedy (1968) Melting and phase relations in the plane tholeiite-lherzolite-nepheline basanite to 40 kilobars with geological implications, Contr. Mineral. Petrol., **19**, 177–211.

JACKSON, E. D. (1966) Xenoliths in Hawaiian basalts, Program 1966 Ann. Meeting, Geol. Soc. Am., 101–102.

JACKSON, E. D. and T. L. WRIGHT (1969) The magmas and xenoliths of the Honolulu volcanic series, Abstracts of papers presented to UMC symposium on phase transformations, Canberra.

JOPLIN, G. A. (1963) Chemical analyses of Australian rocks, Part 1: igneous and metamorphic, Bureau of mineral resources, Commonwealth of Australia, Bull. no. 65, 302–303.

KUNO, H. (1952) Explanatory text of the Geological Map of Japan "Atami". Geol. Surv. Japan, 1–141 (in Japanese).

KUNO, H. (1964) Aluminian augite and bronzite in alkali olivine basalt from Taka-sima, North Kyusyu, Japan, in: *Advancing Frontiers in Geology and Geophysics*, dedicated to Dr. Krishnan (India) 205–220.

KUNO, H. (1966) Lateral variation of basalt magma type across continental margins and island arcs, Bull. Volcanol. **29**, 195–222.

KUNO, H. (1967) Mafic and ultramafic nodules from Itinomegata, Japan, in: P. J. Willey, ed., *Ultramafic and Related Rocks*, (Wiley, New York) 337–342.

KUNO, H. (1969a) Mafic and ultramafic nodules in basaltic rocks of Hawaii, Geol. Soc. Am. Memoir. **115**, 189–234.

KUNO, H. (1969b) Volcanic inclusions and the nature of the upper mantle, Am. Geophys. Union Monograph **13** (in press).

KUNO, H., K. YAMASAKI, C. IIDA and K. NAGASHIMA (1957) Differentiation of Hawaiian magmas, Jap. J. Geol. Geography **28**, 179–218.

KUSHIRO, I. and H. S. YODER, JR. (1966) Anorthite-forsterite and anorthite-enstatite reactions and their bearing on the basalt-eclogite transformation, J. Petrol. **7**, 337–362.

KUSHIRO, I., Y. SYONO and S. AKIMOTO (1968a) Melting of a peridotite nodule at high pressures and high water pressures, J. Geophys. Res. **73**, 6023–6029.

KUSHIRO, I., H. S. YODER, JR. and M. NISHIKAWA (1968b) Effect of water on the melting of enstatite, Bull. Geol. Soc. Am. **79**, 1685–1692.

LOVERING, J. F. and A. J. R. WHITE (1969) Granulitic and eclogitic inclusions from basic pipes at delegate, Australia (in press).

MACDONALD, G. A. and T. KATSURA (1961) Variations in the lava of the 1959 eruption in Kilauea Iki, Pacific Sci. **15**, 358–369.

MACDONALD, G. A. and T. KATSURA (1964) Chemical composition of Hawaiian Lavas, J. Petrol. **5**, 82–133.

MASON, B. (1965) The chemical composition of olivine-bronzite and olivine-hypersthene chondrites, Am. Museum Novitates, No. 2223, 1–38.

MELSON, W. G., E. JAROSEWICH, V. T. BOWEN and G. THOMPSON (1967) St. Peter and St. Paul Rocks: a high temperature, mantle-derived intrusion, Science **155**, 1532–1535.

MOTOJIMA, K. (1965) Average chemical compositions of rocks, The progress report on the UMP of Japan (1964–65), compiled by R. Takahashi, 24.

NAGASAKI, H. (1966) A layered ultrabasic complex at Horoman, Hokkaido, Japan, J. Fac. Sci. Univ. Tokyo, Sect. II, **16**, 313–346.

NIXON, P. H., O. VON KNORRING and J. M. ROOKE (1963) Kimberlites and associated inclusions of Basutoland: a mineralogical and geochemical study, Am. Mineralogist **48**, 1090–1132.

ONUKI, H. (1965) Petrochemical research on the Horoman and Miyamori ultramafic intrusives, northern Japan, Sci. Rep. Tohoku Univ., Ser. III, **9**, 217–276.

OXBURGH, E. R. (1964) Petrological evidence for the presence of amphibole in the upper mantle and its petrogenetic and geophysical implications, Geol. Mag. **101**, 1–19.

RINGWOOD, A. E. (1966) The chemical composition and origin of the earth, P. M. Hurley, ed. *Advances in Earth Science* (M.I.T.) 287–365.

ROSS, C. S., M. D. FOSTER and A. T. MYERS (1954) Origin of dunites and olivine-rich inclusions in basaltic rocks, Am. Mineralogist **39**, 693–737.

TILLEY, C. E. (1947) The dunite-mylonites of St. Paul's Rocks (Atlantic), Am. J. Sci. **246**, 483–491.

TILLEY, C. E., H. S. YODER, JR. and J. F. SCHAIRER (1964) New relations on melting of basalts, Carn. Inst. Washington Year Book **63**, 92–101.

WAKITA, H., H. NAGASAWA, S. UYEDA and H. KUNO (1967) Uranium, thorium and potassium contents of possible mantle materials, Geochem. J. **1**, 183–198.

WHITE, R. W. (1966) Ultramafic inclusions in basaltic rocks from Hawaii, Contr. Mineral. Petrol. **12**, 245–314.

WILSHIRE, H. G. and R. A. BINNS (1961) Basic and ultrabasic xenoliths from volcanic rocks of New South Wales, J. Petrol. **2**, 185–208.

WINCHELL, H. (1947), The Honolulu series, Oahu, Hawaii, Bull. Geol. Soc. Am. **58**, 1–48.

YODER, H. S., JR. and C. E. TILLEY (1962) Origin of basalt magmas: an experimental study of natural and synthetic rock systems, J. Petrol. **3**, 342–532.

GEOCHEMICAL EVIDENCE FOR THE ORIGIN OF SOME ULTRAMAFIC INCLUSIONS FROM VICTORIAN BASANITES

J. D. KLEEMAN and J. A. COOPER

Department of Geophysics and Geochemistry, Australian National University, Canberra

The ultramafic inclusions occurring in some undersaturated Victorian basanites are shown to be accidental xenoliths. U, Th and K abundances and lead isotope measurements in inclusions and hosts, and uranium distribution studies in the phases of the inclusions all indicate that there is no genetic relationship between the ultramafics and the host basanites. The xenoliths are thought to be fragments of an inhomogeneous upper mantle of peridotitic composition. The lead isotope data on the lherzolites are consistent with a two-stage differentiation process from a primeval source, and the uranium concentration (mean 0.3 p.p.m.) in the primary clinopyroxene suggests that this mineral may have this uranium abundance in a peridotitic upper mantle.

1. Introduction

Ultramafic inclusions have widespread distribution in undersaturated basalts, and the abundance and distribution of trace elements and their isotopes is an important consideration when discussing their origin and history. This paper summarizes works already in press as separate papers: KLEEMAN et al. (1969), COOPER and GREEN (1969). For complete references, the reader is referred to those papers.

2. Previous investigations

GREEN et al. (1968) investigated the U, Th and K abundances of six inclusions and six basanite host rocks. The lherzolites have consistently higher Th/U ratios when compared to their hosts, and they found much larger distinctions between them in the K/Th and K/U ratios. Both these indexes showed much lower values in the lherzolites compared with the basanites. Although they considered the possibility of a very distinctive partition relationship between inclusion and host, the authors favoured the interpretation that the inclusions were accidental xenoliths preserving their own geochemical characteristics. It was noted that 1% contamination by introduction of basanite liquid would more than double the K content of the inclusions, and produce element ratios halfway between those of inclusions and hosts. This is particularly important because two of the inclusions contained patches of glass: these results indicate that this glass is not chemically related to the basanite.

3. Uranium distribution studies

3.1. *Uranium distribution in minerals and glasses*

The same inclusions used in the U, Th and K study, and three more in addition, were studied for uranium distribution using fission-track analysis. The fission-tracks were registered in Lexan plastic prints, and their densities compared to those produced simultaneously in a standard by the same thermal neutron dose. The results are summarized in table 1.

The lherzolites have the typical four-phase primary assemblage of olivine, orthopyroxene, clinopyroxene and spinel. In addition one has apparently primary phlogopite, another has hornblende, and two have primary apatite. They range from dunitic, with little other than olivine, to lherzolitic, and rich in clinopyroxene. In five of the inclusions the primary clinopyroxene (I) has a high uranium content (mean 0.30 p.p.m. U), and in most cases this is partially melted to a modified phase (cpx Ia) with lower U content, which has the same optic orientation, but slightly different chemistry. This readjustment takes place in bands across grains, or on rims, and it is easily recognized by its more turbid nature, caused by the presence of glass blebs. Where the clinopyroxene is in contact with spinel, there is a partial melting reaction involving them, and a liquid

TABLE 1

Uranium abundance in phases of lherzolite inclusions

Sample no.	Total rock	Clinopyroxene			Apatite		Olivine	Ortho-pyroxene	Spinel	Other	Glass at reaction sites	Interior glass veinlets	Glass veinlets near contact†
		I	Ia	II	I	II	I	I	I				
2640	0.0030	—	—	—	—	—	0.0002	0.0007	0.0002	Phlogopite 0.0005	—	—	0.22–0.52
2728	0.0097	0.0057	—	—	—	—	0.0002	0.0006	0.0005	—	—	1.08–1.25	—
										Hornblende			
2642	0.0180	0.042	—	—	—	—	0.0005	0.0031	0.0002	0.0005	—	0.36	—
2604	0.0212	0.35	—	—	—	—	0.0007	0.0049	0.0006	—	0.68	0.12–0.70	—
2700	0.1136	0.28	0.047	—*	38.4	—	0.0006	0.0035	0.0007	—	1.44	0.76–1.32	—
2669	0.0400	0.37	0.037	—	—	—	0.0005	0.0039	0.0003	—	2.45–3.90	1.73–3.84	—
2659	n.a.	0.33	0.035	0.014	—	—	0.0007	0.0022	0.0005	—	2.19–4.26	0.29–2.22	—
2683	n.a.	0.16	0.016	0.012	—	—	0.0005	0.0005	0.0005	—	0.72–2.14	0.41–0.79	—
2638	n.a.	—	0.013	0.016	32.1	5.8	0.0002	—	—	—	2.26–6.06	2.26–6.06	1.85–3.70
Error	—	3%	10%	20%	3%	6%	20–40%	15–40%	25–40%	100%	8–13%	11–20%	11–20%

* Insufficient resolution due to small size.
† Only 2640 and 2638 included contact zone in fission-track specimen.
All figures p.p.m. uranium.

is produced, now preserved as a glass. In this glass there are euhedra of olivine II, spinel II, plagioclase, apatite II and clinopyroxene II, and this secondary clinopyroxene has a lower uranium content again (mean 0.014 p.p.m. U).

In all but one of the inclusions there is no continuity between glass formed at these partial reaction sites, and the glassy groundmass of the host basanite. Such continuity would have been immediately obvious from the plastic print, due to the relatively high uranium content (0.4–6.0 p.p.m. U) of the glass. Inclusion 2638 did, however, have physical continuity between glass formed internally, and the groundmass of the host, although there is a gradual change in chemical content. This inclusion also contained abundant apatite, and where the glass is in contact with it, the turbid primary (I) phase has a clear, recrystallised rim, with euhedral outlines against the glass. This rim of apatite II has much lower uranium content (5.8 p.p.m.) than the primary phase (32.1 p.p.m. U).

3.2. *Crystal–liquid partitions*

Where clinopyroxene crystals are growing out of the glass, and where the apatite rims are clear, and have euhedral outlines against the glass, it can be assumed that there is at least local equilibrium between the glass (liquid) and these two phases. On this basis, partition coefficients have been calculated between the uranium contents of the glass and crystals at that point. The results of these calculations are shown in table 2. The U content of the glass is between 100 and 250 times as concentrated as the clinopyroxene crystallites growing

TABLE 2

Uranium partition between coexisting phases and liquids

Sample no.	Glass veinlets/cpx Ia	Glass patches/cpx II	Glass patches/Ap II	Cpx I/Opx I	Cpx I/Ol I
2604	—	—	—	71	500
2700	16–28	—	—	80	470
2669	46–106	—	—	95	740
2659	8–63	240–253–270	—	150	470
2683	37–72	97–101–105	—	320	320
2638i	—	194–224–250	0.50–0.84–1.45	—	—
2638ii	—	—	0.86–0.99–1.24	—	—

All figures are ratios between the uranium contents of the minerals indicated.

out of it, and between 0.5 and 1.2 that of the apatite rims. The partition between glass veinlets and recrystallizing clinopyroxene (Ia) is considered to be dynamic, and so does not enter the present argument.

These partitions apply to lower pressure conditions when compared to the primary assemblage, which was formed at pressures greater than 8 kb, but provided there are only slight compositional changes, the effect of moderate pressures, of the order of 10 to 15 kb, should be slight. It is considered that there have been no major changes in the oxidation state between high and low pressure conditions, especially since the Fe^{++}/Fe^{+++} ratios are high in both inclusions and hosts.

These partition coefficients are now used to calculate the uranium content of a hypothetical liquid that could have been in equilibrium with the primary clinopyroxene I and apatite I. These calculations derive that this liquid would have had 28–35 p.p.m. U to have been in equilibrium with apatite with 35 p.p.m. U, and between 30–75 p.p.m. U to have been in equilibrium with clinopyroxene containing 0.30 p.p.m. U. These levels are a factor of ten to a hundred times higher than those observed in basalts, even the very undersaturated nephelinitic magmas. It is therefore concluded that these inclusions have not formed as an accumulate from the magma in which they now occur, or in any other natural basaltic magma. This data also implies that the lherzolite material could not have acted as source rock for the present host, or for a basaltic magma, provided that equilibrium was maintained between liquid and source material. This conclusion follows if it is assumed that any liquid formed by melting is completely removed from the source material. If only part of the liquid generated in such lherzolite material escaped, leaving some trapped, then on later cooling the melting process would be essentially reversed, and the partially extracted material would revert to an assemblage with lower abundance of "basalt" components, including U, and clinopyroxene. This clinopyroxene would have a uranium content similar to the pre-melting assemblage, but would be present in lower modal amounts, compared to the primary assemblage.

4. Lead isotope measurements

4.1. *Measurements*

The isotopic composition of lead has been measured in a suite of inclusions and hosts from the western Victorian basanites. The vacuum volatilization technique makes it possible to extract lead from large quantities of sample, and up to 300 grams were used. Despite this, only seven out of twenty lherzolite samples gave sufficient lead for isotopic measurements. The lherzolite lead content ranged from 0.3 p.p.m. to less than 0.01 p.p.m., based on volatilization yield. Very low lead content correlates with very low uranium abundance. The basanite lead content was much higher: 2–5 p.p.m.

The results are presented in tables 3 and 4, and plotted on figs. 1 and 2. The diagram insets show the basanite results after isotopic fractionation has been removed by double spiking. It is not possible to do

TABLE 3

Lead isotope data and lead content of Lherzolites which yielded sufficient lead for isotopic analysis (samples not double spiked)

Sample no.	Pb extracted (p.p.m.)	(µg)	206/204	206/207	206/208	207/204	208/204	Load composition
2639 Mt. Leura	0.3	60	18.42	1.170	0.4708	15.73	39.11	oxalate
2642 Mt. Leura	0.2	40	18.12	1.155	0.4694	15.68	38.60	oxalate
2669 Mt. Shadwell	0.2	50	17.87	1.147	0.4685	15.58	38.14	oxalate
2903 Mt. Noorat	0.09	25	16.70	1.084	0.4594	15.40	36.35	sulphide
2904 Mt. Noorat	0.13	35	16.23	1.049	0.4499	15.47	36.08	sulphide
2697 Mt. Noorat	0.25	100	18.50	1.173	0.4720	15.77	39.20	oxalate
2700 Mt. Noorat	0.2	15	16.6	1.086	0.461	15.3	36.1	sulphide (multiplier)

TABLE 4

Lead isotope data and approximate lead content of host basanite. Lower figures displaced to the right are figures corrected through double spiking using NBS 981 standard lead as reference calibrator.

Sample no.	Pb (p.p.m.)	206/204		206/207		206/208		207/204		208/204	
2642	5	18.65_1		1.183_7		0.4764		15.75_7		39.15	
Mt. Leura			18.53_8		1.187_4		0.4793		15.61_2		38.68
2679	2	18.49_9		1.183_7		0.4776		15.62_8		38.73	
Mt. Shadwell			18.43_3		1.185_8		0.4793		15.54_5		38.46
2693	2	18.72_9		1.183_5		0.4754		15.82_4		39.40	
Mt. Noorat			18.51_7		1.190_3		0.4808		15.55_8		38.51
2740	2	18.52_1		1.177_4		0.4736		15.73_1		39.17	
Mt. Gambier			18.34_7		1.183_0		0.4773		15.51_0		38.44
2909 (JC45)	3.5	18.49_2		1.174_3		0.4733		15.74_7		39.07	
Mt. Schank			18.38_4		1.177_7		0.4761		15.60_9		38.61

this in the case of the inclusions, due to the smaller amounts of lead extracted from the samples. Most of the discussion is based on the un-normalized data.

4.2. Discussion

The basanite points fall in a small group with similar ratios to modern lead, but the lherzolites are different, all being B type anomalous, forming a sub-linear pattern with various degrees of deficiency of ^{206}Pb. If the lherzolite material was genetically related to the basanites, then each group must have the same lead isotope ratios. From this it can be said that the lherzolites are not residual material after the extraction of the basanite host magma, nor are they crystal differentiates from that magma. They are therefore accidental xenoliths, and their mineralogical characteristics place their source within the upper mantle.

5. Lead isotope evolution in the lherzolites

If is assumed that these inclusions are accidental xenoliths of an inhomogeneous upper mantle, some useful discussion follows regarding the processes which gave rise to the present spread of ratios. There are two reference lines on fig. 1. One is the lead growth curve for a μ value of 9.0 (^{238}U/^{204}Pb). The straight dotted line is the line of best fit drawn through a large number of world-wide volcanics. It is not a regression line through the small number of points plotted from the present study, although it is not in disharmony with them.

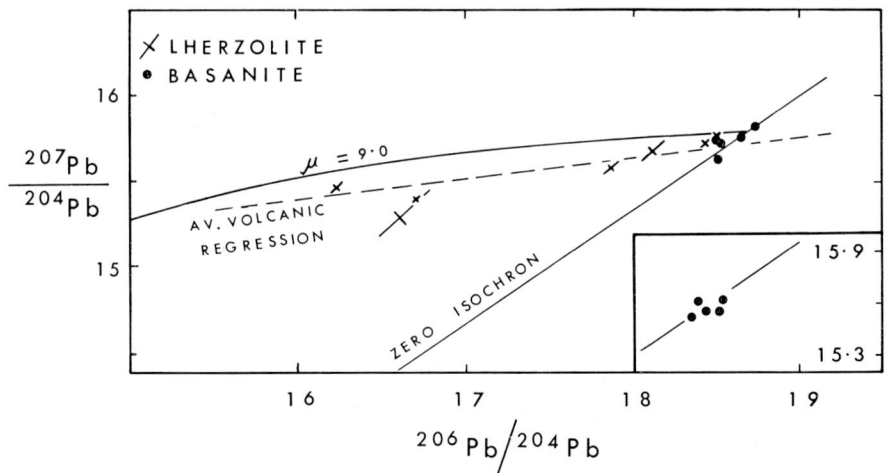

Fig. 1. ^{206}Pb/^{204}Pb–^{207}Pb/^{204}Pb plot of unspiked lead isotope ratios of whole rock lherzolite and host basanite samples. Also in plot is $\mu = 8.99$ growth curve and meteoritic zero isochron passing through the Primordial Point 9.56, 10.42. The inset shows the basanite values corrected for fractionation by double spiking calibrated to N.B.S. 981 value.

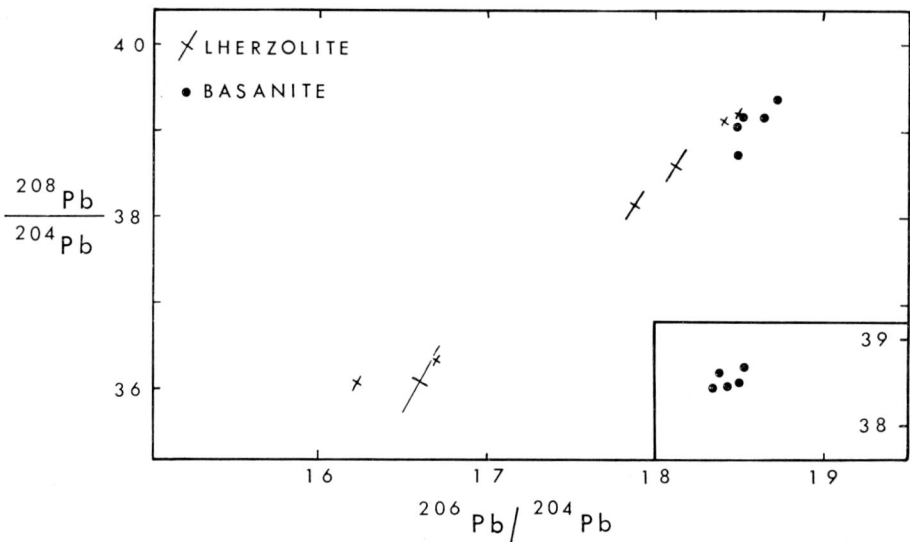

Fig. 2. $^{208}Pb/^{204}Pb-^{206}Pb/^{204}Pb$ plot of unspiked lead isotope ratios of whole rock lherzolite and host basanite samples. The inset shows the basanite values corrected for fractionation by double spiking calibrated to N.B.S. 981 value.

A linear distribution of the lherzolite results could be explained in three ways:

1. A modern mixing of lead from two end members. For example one of basanite content, and one poorer in radiogenic lead than the lherzolites. This seems unlikely from the U, Th and K results.

2. All the lherzolite samples had similar Pb isotopic compositions about 2 b.y. ago or earlier, and the present spread is due to different proportions of radiogenic lead added to each by different U/Pb environments, all apparently less than or equal to that of the basanites.

3. The lherzolites formed later from some other material which had this history.

Physically this means that all the lherzolite material formed at this time in the past as a mass with the same isotope ratios, or an older lherzolite mass or its parent had its lead isotopes homogenized about 2 b.y. ago. The data do not exclude the possibility that some of the samples were formed at subsequent times, or that the lherzolites were residual or accumulates from earlier basaltic magmas. The U, Th contents and uranium distribution data do not support the latter alternatives.

Fig. 2 demonstrates that the lherzolite lead has varying degrees of ^{208}Pb deficiency proportionate to ^{206}Pb deficiency. This indicates that the lherzolite material has a limited range of Th/U ratios despite large fluctuations in U/Pb ratio.

The lead isotope composition of the lherzolites could be satisfied by a two stage differentiation process from a primeval mantle source. A knowledge of t_1, the time of differentiation, would enable the $^{238}U/^{204}Pb$ (i.e. μ values) of each stage of each sample to be calculated. If it is assumed that the $^{232}Th/^{238}U$ ratio of the first stage (k_1 value) is equal to the primordial value of 3.9, then k_2, the $^{232}Th/^{238}U$ for the next stage, can also be calculated for each sample (GAST, 1969). Table 5 shows the calculated μ_1, μ_2 and k_2 values if t_1 is assumed to be 1.5, 2.0, 2.5 and 3.0 b.y.. All samples require that μ_2 is lower than μ_1 for all values of t_1. This suggests loss of a component of higher μ value at this time. It is considered that the small range of calculated μ_1 (8.6–9.1) justifies the use of an invariant primordial k_1 value to calculate approximate k_2 values. Since it has already been concluded that the lherzolite has not passed through a modern chemical fractionation, the calculated k_2 values can be compared with the direct measurements of Th/U (= 3.7–5.0) reported by GREEN et al. (1968). If t_1 equals 2.0 or 2.5 b.y. there is a much more satisfactory agreement than the older or younger selections of t_1 for the group as a whole, although some individual samples are not restricted by this range.

6. Uranium distribution in the upper mantle

The uranium distribution data are consistent with a hypothesis that the inclusions are accidental xenoliths of inhomogeneous upper mantle material, and the lead

TABLE 5

Calculated μ_1, μ_2 and k_2 values for lherzolite lead isotope composition assuming a two stage history, 4 different values for t_1, $k_1 = 3.9$, and the promordial lead isotope ratios of MURTHY and PATTERSON (1962), and following GAST (1969)

Sample no.	t_1 (b.y.)	μ_1	μ_2	k_2
2639	1.5	8.9	8.3	5.4
2642		8.9	7.2	5.3
2669		8.7	7.1	4.7
2903		8.6	2.6	4.3
2904		8.8	0.2	26.0
2697		9.0	8.3	5.4
2700		8.4	2.8	3.4
2639	2.0	8.9	8.5	5.0
2642		8.9	7.7	4.9
2669		8.7	7.2	4.7
2903		8.6	4.1	4.0
2904		8.9	2.4	5.2
2697		9.0	8.5	5.0
2700		8.5	4.2	3.6
2639	2.5	8.9	8.5	4.7
2642		8.9	7.9	4.6
2669		8.8	7.5	4.5
2903		8.8	5.0	3.9
2904		9.1	3.6	4.5
2697		9.0	8.5	4.8
2700		8.6	5.0	3.7
2639	3.0	9.0	8.6	4.6
2642		9.0	8.0	4.5
2669		8.9	7.7	4.3
2903		9.0	5.6	3.9
2904		9.5	4.4	4.2
2697		9.0	8.6	4.6
2700		8.9	5.6	3.7

isotope data suggest that all of these samples have been reacted geochemically in the past. This implies that they are not primordial upper mantle material, and it is doubtful if such primordial mantle material still exists.

6.1. *Uranium abundance in the upper mantle*

A major point of this aspect of study is that there is an important amount of uranium in an essential phase of the lherzolite inclusions; clinopyroxene would account for 15–20% of the modal content of a pyroxene pyrolite zone in the upper mantle. If this clinopyroxene had a mean abundance of 0.3 p.p.m. U, then this alone would account for 0.045–0.060 p.p.m. in such a rock. The possibility that there may also be up to around 0.2% modal apatite with about 35 p.p.m. U means there may be locally an additional 0.07 p.p.m. available, if apatite is present.

These estimates compare very favourably with those from heat flow considerations.

6.2. *Basalt genesis*

The primary assemblages of at least six of the inclusions have the potential for producing basalt liquid of normal uranium content, and therefore a model upper mantle based on them would be very suitable from the point of view of the geochemistry of uranium. The first infinitesimal drop of liquid formed by partial melting would necessary be of very high uranium content, since it would be in equilibrium with the primary clinopyroxene. However the observed clinopyroxene/liquid partition coefficients imply that for slightly higher degrees of melting, the uranium is rapidly partitioned into the liquid, producing liquids (nepheline normative on other grounds) with high U content, but with the U content decreasing with higher degrees of melting, as low-U clinopyroxene (Ia), enstatite and olivine continue to enter the liquid.

Using the pyrolite model as a basis for a model for uranium distribution in the upper mantle, as described above, the uranium abundances of basalts can be calculated from the degrees of melting inferred from major element and phase relationship studies. The results of these calculations are consistent with observed abundances in basalts.

7. Conclusion

The primary conclusion from the three studies summarized in this paper, is that the lherzolite inclusions from the newer volcanics of western Victoria are not accumulates from their host basanites, and they have not acted as source rock for the generation of that magma in which they now occur.

The inclusions are interpreted as accidental xenoliths of an inhomogeneous peridotitic upper mantle. As such, speculations on the evolution of the lead isotopes indicate that they were geochemically fractionated at about 2.0–2.5 b.y.. A model for the uranium distribution in a peridotitic upper mantle, based on the high-uranium primary assemblages in the lherzolites, is found to satisfy predictions from heat-flow considerations, and provide a suitable source rock from which to extract basalt magmas of typical uranium content.

Acknowledgements

These studies were initiated at the suggestion of Dr. D. H. Green, who also criticised this manuscript. The cost of neutron irradiations for the uranium distribution studies was met by a grant from the Australian Institute of Nuclear Science and Engineering.

References

COOPER J. A. and D. H. GREEN (1969) Earth Planet. Sci. Letters **6**, 69.

GAST, P. W. (1969) Earth Planet. Sci. Letters **5**, 353.

GREEN, D. H., J. W. MORGAN and K. S. HEIER (1968) Earth Planet. Sci. Letters **4**, 155.

KLEEMAN J. D., D. H. GREEN and J. F. LOVERING (1969) Earth Planet. Sci. Letters **5**, 449.

MURTHY, V. R. and C. C. PATTERSON (1962) J. Geophys. Res. **67**, 1161.

CRYSTALLISATION OF AN ALKALI-OLIVINE BASALT UNDER CONTROLLED P_{O_2}, P_{H_2O} CONDITIONS

R. W. NESBITT and D. L. HAMILTON

Department of Geology, University of Adelaide, S. Aus., and Department of Geology, University of Manchester, U.K.

An alkali-basalt from Hualalai, Hawaii, previously examined by YODER and TILLEY (1962) has been crystallised under controlled P_{O_2} at 2 kb total pressure. Significant differences, in the order of appearance of the phases result from lowering the P_{O_2} from $10^{-0.7}$ to about 10^{-10} atm. In particular magnetite, present at 1100 °C in the unbuffered runs does not crystallise until below 900 °C in the oxygen controlled runs. There is a drop of up to 50 °C on the liquidus of the system if the oxygen is controlled at the quartz-fayalite-magnetite buffer. Calcium amphiboles crystallised under such conditions are nepheline normative and rich in titanium and aluminium. Decreasing the water content but maintaining total pressure at 2 kb results in an increase in the melting point of the amphibole. There is no data in the runs to support the concept of a primary amphibole gabbro crystallising from a basaltic melt of this composition. Rather, the amphibole would form by reaction of the pyroxene with the liquid, or direct growth at lower temperatures from a basaltic mineralogy.

1. Introduction

The genesis of basaltic magmas is a fundamental problem of igneous petrology. Recently, the experimental approach to the problem has been completely revised by the use of natural basalts as starting materials (YODER and TILLEY, 1962; GREEN and RINGWOOD, 1967). In such experiments, the parameters, temperature and pressure are controlled but there is poor control of chemical composition. This is in part resolved by duplicating the experiments with several basalts of varying composition, but in such cases little is known of the effects of minor constituents. A further chemical complication is the oxidation of the iron bearing minerals in the basalts during the heating experiments. It is the lack of control of this oxidation parameter which casts doubt on the usefulness of such experiments in interpreting natural phenomena. This investigation is specifically concerned in evaluating the effects of controlling this parameter. To do this, an alkali olivine basalt, previously used by and reported on by YODER and TILLEY (1962, Cambridge No. 65992) was used as starting material. Data on temperature and order of appearance of major phases in the water saturated basalt could thus be compared with results of the previously reported unbuffered runs.

A further aim of the investigation was to obtain data on liquidus temperatures in buffered runs in which known amounts of water were added. Thus for each P-T condition, varying amounts of water were added to the capsules such that saturated through to dry conditions were obtained.

2. Control of oxidation in experimental runs

KENNEDY (1948) and OSBORN (1959, 1962), FUDALI (1965) and ROEDER and OSBORN (1966) have pointed out the influence of PO_2 in the crystallisation trends of basalts. However, high pressure – high temperature experiments on basalts have either been conducted under oxidising conditions (YODER and TILLEY, 1962) or under reducing conditions in which P_{O_2} was unknown (GREEN and RINGWOOD, 1967). The use of buffers (EUGSTER and WONES, 1962) in this type of experimentation is handicapped by the rapid reaction rate of the buffer, particularly at high temperatures. This problem has been discussed by HAMILTON et al. (1964), who concluded that the introduction of fixed amounts of hydrogen into the argon mixture effectively controlled P_{O_2} pressures. In all of the runs reported in this paper a calculated P_{H_2} was introduced to the pressure vessel so that at the desired P-T conditions of the run the P_{O_2} imposed on the water vapour saturated samples would be approximately equivalent to a quartz-fayalite-magnetite-water buffer.

3. Experimental method

An alkaline olivine basalt, from Hualalai, Hawaii, described by YODER and TILLEY (1962) as Specimen 20 (Cambridge No. 65992) was used. The analysis of this material taken from YODER and TILLEY (1962) is given in table 1.

TABLE 1

Alkali Basalt, Hualalai, Hawaii (Cambridge No. 65992) (analyst J. H. Scoon [YODER and TILLEY, 1962])

SiO_2	46.53		
Al_2O_3	14.31	Qz	—
Fe_2O_3	3.16	Or	5.28
FeO	9.81	Ab	20.04
MnO	0.18	Ne	2.20
MgO	9.54	An	23.63
CaO	10.32	Di	20.89
Na_2O	2.85	Hy	—
K_2O	0.84	Ol	18.48
H_2O^+	0.08	Mt	4.53
H_2O^-	nil	Il	4.41
P_2O_5	0.28	Ap	0.67
TiO_2	2.28	Ct	—
Cr_2O_3	0.06	Rest	0.14
	100.24		100.27

Some of the finely ground basalt was initially fused in a platinum envelope suspended in an open furnace. Material around the margin of the resultant slug was discarded and the remainder checked optically to ensure freedom from crystalline products. All runs were carried out with both crystalline and glassy starting materials. About 30–40 mg material was added to a pre-weighed platinum tube, a known weight of water added and the tube welded. All the hydrothermal experiments were carried out in internally heated pressure vessels. The bombs were first flushed with hydrogen which was then preset at a calculated P_{H_2}, dependent on the temperature of the run. Argon was then pumped into the bomb and the vessel heated to the required temperature.

Temperature gradients along the length of the specimen holder were monitored by thermocouples at either end. Final control was achieved by differential power ratings on a doubly wound furnace. The thermocouples were calibrated in situ against the melting point of sodium chloride at atmospheric pressure and also at 2 kb. The melting temperature at pressure was taken from CLARK (1959).

Charges were examined optically for the presence of glass and the mineralogy determined from thin sections. This enabled a positive distinction between quench and primary products to be made. In many cases, X-ray diffraction was used to identify small quantities of clinopyroxene and plagioclase.

Microprobe analyses were carried out on a four channel ACTON microprobe at the University of Yale. Chemically analysed amphiboles provided calibration lines for eight elements. An analysing beam of 5 micron diameter, was used. In the case of titanium and aluminium, no standards were available in the range of the unknown. Consequently values for these elements were obtained by extrapolation. Pulse height analysis techniques were used, but no correction for dead time applied. The use of material of approximately similar composition results in smaller discrepancies due to matrix interferences.

4. Results

Results of the 2 kb runs on the fused rock powders are given in table 2. The charges were water saturated and the P_{O_2} controlled at approximately that of the quartz–fayalite–magnetite buffer. The data of YODER and TILLEY (1962) on the water saturated, unbuffered runs is also given in table 2.

Several significant differences exist between the two sets of data.
1) Magnetite, present as a primary liquidus phase in the unbuffered runs, (YODER and TILLEY, 1962) is observed to crystallise only in the final stages under controlled P_{O_2} (below 900 °C). Hence the temperature difference between unbuffered and buffered is 200 °C.
2) The liquidus is lowered by at least 15 °C, possibly as much as 50 °C.
3) The assemblage amphibole-clinopyroxene-olivine is stable over a wide temperature interval whilst in the unbuffered runs, this assemblage is not found. At 900 °C, the assemblage amphibole+clinopyroxene +plagioclase+olivine+glass is observed. Since the starting material was a glass, the olivine cannot be a residual from the original rock. However, it is possible that the length of the run (4 days) was insufficient to produce complete reaction of an early olivine phase.

TABLE 2

A Crystallisation of alkali basalt Hualalai, Hawaii, (Cambridge No. 65992) at 2 kb. Excess water present on completion of the run, and P_{O_2} at quartz-fayalite-magnetite buffer. Rock initially fused in air at 1400 °C.

B Data from table 30, YODER and TILLEY (1962), 2 kb runs.

	Run no.	Temp.	Time	Products (optical identification)	% H_2O
A	Fused basalt				
	139	1095 °C	41 hrs	Glass+Q Amph+Q Mi	6%
	174	1086	23 hrs	Ol+Glass+Q Amph	5½%
	107	1066	6 hrs	Ol+Q Cpx+Glass	6%
	208	1035	40 hrs	Ol+Cpx+Glass+Q Amph	5%
	241	1010	23 hrs	Ol+Cpx+Glass+Q Amph	10%
	131	975	2½ days	Amph+Cpx+Ol+Glass	5%
	153	930	2 days	Amph+Cpx+Ol+Glass	3.5%
	193	906	4 days	Amph+Cpx+Ol+Plag+Glass	5.5%
	220*	860	5 days	Amph+Plag+Mt+Glass	2.0%
B	Finely ground basalt				
		1150 °C	½ hr	Glass	
		1100	1 hr	Ol+Mt+Glass	
		1050	2 hrs	Cpx+Ol+Mt+Glass	
		1000	24 hrs	Amph+Cpx+Mt+Glass	
		900	24 hrs	Amph+Plag+Mi+Mt+Glass	
		850	168 hrs	Amph+Plag+Mi+Mt	

Q = quench; Amph = amphibole; Mi = mica; Ol = olivine; Cpx = clinopyroxene; Plag = plagioclase; Mt = magnetite.
* Data on unfused rock powder.

The differences between the results obtained by YODER and TILLEY (op. cit.) and the oxygen buffered water saturated and "dry" runs are shown in fig. 1.

As well as comparisons with the water saturated unbuffered runs we can comment on the amphiboles produced in the buffered experiments. The amphibole appears as a primary phase at 975 °C, and occurs as euhedral, strongly pleochroic hornblende. This is some 25 °C below that observed by YODER and TILLEY (op. cit.). Later amphiboles form at the expense of olivine and clinopyroxene which react with the liquid. At temperatures less than 900 °C, amphibole is the only mafic phase present.

Microprobe analysis of amphiboles crystallised at 930 °C are given in table 3. Apart from the Mg/Fe ratio, there is little difference between amphiboles crystallised in water saturated conditions and in which water was not added. The ratio 100 Mg/(Mg+Fe) is slightly lower in the water saturated runs. Mineral formulae (based on 24 anions) have been calculated assuming a 2% H_2O value (table 3). Fig. 2 is a plot of tetrahedrally co-ordinated aluminium against total alkali (after DEER et al., 1963) and indicates a large tschermakitic component of the analyses.

C.I.P.W. norm calculations (table 4) on these amphiboles, assumes that the mineral analysis can be expressed in terms of a basaltic composition (BOWEN, 1928, p. 271 and YODER and TILLEY, 1962, p. 455). All the norms contain nepheline and hence as basalts would be classified as alkali basalts.

The results of the "anhydrous" runs are complicated by the appearance of amphibole. This indicates either the presence of water in the powder or its production during the run. In the dry runs, the liquidus is raised about 45 °C and the melting interval decreased by 235 °C to about 150 °C (fig. 3). Amphibole appears at 1035 °C along with plagioclase and possibly magnetite and the solidus is above 975 °C but below 1010 °C.

Runs with varying quantities of water, below saturation level, confirm the differences observed between "dry" and saturated. Such a technique does allow rough values to be assigned to water solubility in the liquids at varying temperatures. For example, at 1100 °C (above the liquidus) 6% water is required to produce excess water on opening of the capsule. At lower temperatures, the presence of crystalline material, particularly amphibole makes such measurements of little quantitative value. Saturation levels at 900 °C and 1000 °C

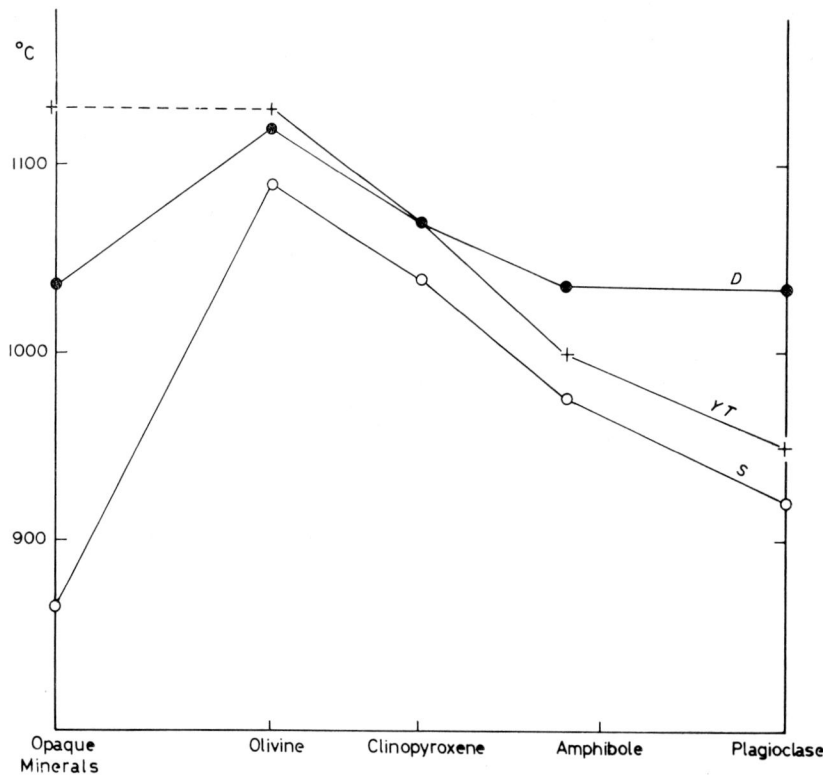

Fig. 1. Temperature of crystallisation of the major phases of the alkali basalt at 2 kb. D: under "dry" conditions (see text); S: under conditions of water saturation. Both D and S represent runs under controlled oxygen fugacity; YT: data from YODER and TILLEY (1962) water saturated runs with no control of oxygen fugacity.

appear to be at 2.5% and 3.5% water respectively. One further complication in these estimates is the presence of hydrogen during the run.

5. Discussion

5.1. *Magnetite stability*

Our data, limited to one isobaric section, are nevertheless useful in outlining the changes which occur between experiments with and without oxygen buffering.

Undoubtedly the absence of magnetite from the primary assemblage, observed in the oxygen buffered runs, is the most important observable difference. The experimental data of ROEDDER and OSBORN (1966) demonstrates the diminution of the magnetite phase field in an anhydrous synthetic system. Our data on a natural system supports this observation and demonstrates that this diminution also occurs in hydrous systems. This effect, discussed by OSBORN (1959), will alter the course and composition of liquids derived by fractional crystallisation. In the buffered runs, iron will be enriched in the later liquids since the magnetite stability field is decreased.

This effect of the suppression of magnetite is evident in many of the large gabbroic intrusions of the world. In the Skaergaard intrusion, magnetite appears as a primary cumulate in the upper part of the hypersthene-olivine gabbro zone (WAGER and DEER, 1939, p. 93). This according to the estimates of Wager and Deer on the thickness of the hidden layered series, would indicate that cumulus magnetite does not appear until over half the magma has crystallised. Intrusions of the Giles Complex, particularly Mt. Davies (NESBITT and KLEEMAN, 1964), Ewarara (GOODE and KRIEG, 1967) and Gosse Pile (MOORE, 1969) are characterised by the absence of iron oxide phases except in the upper most parts of the intrusions.

Intrusions, such as the Bushveld and Stillwater Complexes, are characterised by persistent chromite horizons in their basal portions, but not magnetite. It is

TABLE 3

Partial analyses of amphiboles, crystallised from alkali basalt (Hualalai) 930 °C, 2 kb. P_{O_2} approx. $= 10^{-13}$ 2 days*. (Powder initially fused in air at 1400 °C)

	I	II	III	IV
SiO_2	43.0	42.5	40.0	43.5
Al_2O_3	14.5	14.5	15.5	14.5
TiO_2	3.5	4.0	4.4	3.2
FeO	11.0	11.0	12.0	8.0
MgO	14.0	13.0	13.0	13.0
CaO	10.7	11.0	11.0	11.5
K_2O	0.3	0.5	0.5	0.5
Na_2O	2.0	2.0	2.0	1.5
	99.0	98.5	98.4	95.7
Si	6.16 ⎱ 8.00	6.14 ⎱ 8.00	5.84 ⎱ 8.00	6.36 ⎱ 8.00
Al	1.84 ⎰	1.86 ⎰	2.16 ⎰	1.85 ⎰
Al	0.62	0.61	0.50	0.64
Ti	0.37 ⎱	0.44 ⎱	0.48 ⎱	0.35 ⎱
Fe	1.32 ⎰ 5.31	1.34 ⎰ 5.19	1.47 ⎰ 5.27	1.00 ⎰ 4.8
Mg	3.00	2.80	2.82	2.83
Ca	1.64 ⎱	1.71 ⎱	1.73 ⎱	1.81 ⎱
K	0.03 ⎰ 1.97	0.04 ⎰ 2.03	0.05 ⎰ 2.06	0.05 ⎰ 2.07
Na	0.30	0.28	0.28	0.21
(OH)	1.92	1.94	1.96	1.90
$\frac{Mg100}{Mg+Fe}$	69	68	66	74

I Run 152: Random grains (Av. 6);
II Run 153: Random grains (Av. 6);
III Run 153: Single grain (7 measurements);
IV Run 154: Random grains (Av. 4).
* The following amounts of water were added to the capsules Run 152: 8%; Run 153: 3.5%; Run 154: 0.0%.

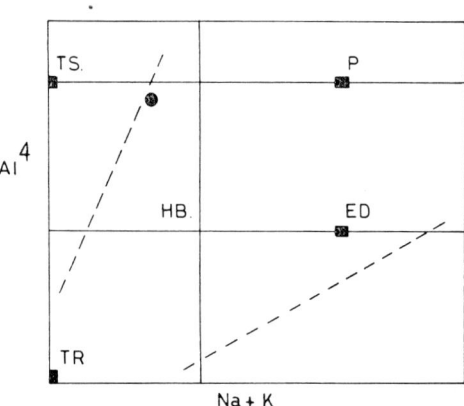

Fig. 2. Plot of analyses of amphibole (large solid circle) crystallised at 930 °C (table 2) on a total alkali–tetrahedrally co-ordinated aluminium, diagram. P: Pargasite; HB: Hornblende; ED: Edenite; TR: Tremolite; TS: Tschermakite.

TABLE 4

C.I.P.W. norms on amphiboles crystallised from the Hualalai alkali basalt (numbers as for table 3)

		I	II	III	IV
Or		1.8	3.0	2.7	3.0
Ab		7.9	6.3	—	11.1
An		29.4	28.8	31.6	32.2
Lc		—	—	0.2	—
Ne		4.9	5.8	9.2	0.8
Di	Wo	9.7	10.7	9.6	11.0
	En	6.3	7.0	6.1	7.8
	Fs	2.6	3.0	2.8	2.3
Ol	Fo	19.8	17.6	18.2	17.8
	Fa	9.1	8.2	9.2	5.8
Il		6.5	7.6	8.4	6.2
		98.0	98.0	98.0	98.0

therefore pertinent to enquire the role of oxygen fugacities in controlling the crystallisation of chromite. Since it is such ubiquitous mineral in the early products of basic magmas of the Bushveld type, it is probable that much lower oxygen fugacities would be required together with an alternative site for the chromium.

Its absence in the gabbros of the Giles Complex cannot be ascribed to initial low chromium content, since the clinopyroxenes are rich in the element. We suggest its absence may be due to the very low oxygen fugacities developed and also to the preponderance of pyroxene over olivine. This latter feature is in part controlled by depth of crystallisation (GREEN and RINGWOOD, 1964, 1967).

5.2. *Amphibole stability*

Microprobe analyses of the amphiboles crystallised at 930 °C (table 3) demonstrate the alkali nature of this mineral. C.I.P.W. norms of the analyses (table 4) all contain nepheline. It has been suggested that selective resorption of amphiboles is capable of producing undersaturated basalts (BOWEN, 1928). Bowen's concept was one of partial resorption and his calculations involved the selective removal of normative nepheline, orthoclase and leucite into the liquid and the precipitation of the olivine and anorthite components of the norm. The amphiboles crystallised in the oxygen buffered runs appear first as primary minerals, crystallised directly from the liquid. Our data suggests that later amphiboles result from a reaction between liquid and olivine or clinopyroxene. This indicates that partial melting of the amphibole would occur below 970 °C. Temperatures in excess of this would result in total

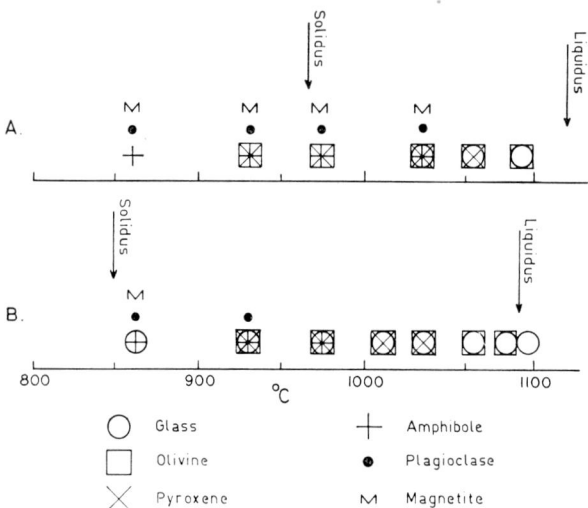

Fig. 3. Comparison of mineral phases crystallised from the alkali basalt at 2 kb. A: data from runs in which no water was added; B: data from water saturated runs. Oxygen fugacity controlled at quartz-fayalite-magnetite buffer.

melting, a factor which would not increase the alkaline nature of the basalt.

If P_{H_2O} is lowered, then the data of fig. 3 suggest the breakdown temperature of the amphibole would be increased. This observation is not in accord with the suggestion of YODER and TILLEY (1962) who believed that decreasing water content would decrease the stability of the mineral if P_{total} remained constant. However, there is no doubt that our data indicate amphibole at higher temperatures in the water undersaturated systems than in the saturated runs.

The presence of amphibole in the dry runs is probably due to the reaction of the introduced hydrogen with the oxygen in the silicate melt. This would result in the production of (OH) molecules which would stabilise the amphibole structure. The reduction of ferric iron (present because of the earlier fusion of the rock) by the hydrogen would result in H$^+$ ions which would then be available for bonding to free oxygen in the melt.

$$Fe^{3+} + \tfrac{1}{2}H_2 \rightarrow Fe^{2+} + H^+$$
$$H^+ + \tfrac{1}{2}O_2 \rightarrow (OH)^-$$

Hence none of the experimental runs could be termed "dry". However, the failure to produce an hydrous conditions under controlled P_{O_2} was not entirely unrewarding. The data indicates, that amphibole appears at temperature of 1035 °C, in systems where hydrogen is available. This is some 70 °C above the temperature at which the mineral appears in hydrous conditions. We take this data as further confirmation of the increasing stability of amphibole in situations where $P_{H_2O} < P_{total}$.

The primary nature of amphibole in hornblende gabbros is not supported by our data. In all cases, olivine and clinopyroxene precede the crystallisation of the amphibole, even under fully water saturated conditions. The only examples of direct amphibole crystallisation we observed involved low temperature crystallisation of the basalt without the production of much liquid. This "amphibolitisation" which is a solid state process, must be a common phenomenon in the metamorphism of gabbroic rocks.

5.3. *Plagioclase stability*

The appearance of plagioclase at low temperatures in the runs, results in slow reaction rates and hence precise data are lacking. However, it appears that there is a direct relationship between water-saturated systems and the late appearance of plagioclase. For example, in the dry system, at one atmosphere, plagioclase crystallises before pyroxene (YODER and TILLEY, 1962). At 2 kb P_{H_2O}, there is a difference of 110 °C in the crystallisation of the two minerals, with plagioclase being the least stable. This feature has been commented on by YODER (1954) OSBORN (1959, 1963) the latter with particular emphasis on the production of andesites. However, if the correlation is real, then it may be useful in evaluating the original P_{H_2O} of the system. Thus an assessment of the order of crystallisation of cumulus phases (WAGER et al., 1961; JACKSON, 1967) in particular the clinopyroxene-plagioclase relationship may give an estimate of the P_{H_2O} during crystallisation.

6. Summary

Our data confirm the conclusions reached on the experimental work on synthetic systems with regard to the importance of oxygen fugacity. Although there are no differences in final mineralogy of the Hualalai alkali basalt between the water saturated unbuffered and buffered runs, there are important differences in the order of appearance. The major mineral in this respect is magnetite whose stability field is considerably decreased by low oxygen pressures. Hence this high iron bearing phase crystallises late in the sequence, an indication that the late liquids are rich in iron. The absence of magnetite in the early cumulate phases of large gabbro bodies is probably due to this controlling factor. We are unable to comment definitively on the extensive chromite deposits associated with such intrusions. However, it has been suggested (IRVINE, 1965) that oxygen fugacity is also a controlling factor. If such is the case, a distinction may be drawn between those gabbros and ultrabasic intrusions in which the chronium appears in the spinel phase ("high" P_{O_2}) and in the pyroxene phase ("low" P_{O_2}). Two examples of these categories would be Bushveld Complex (JACKSON, 1967) and the Giles Complex (NESBITT et al., 1969).

Amphibole crystallised from the alkali basalt during runs with controlled oxygen fugacity are all nepheline normative. Increase in oxygen fugacity results in an increase in their stability (raises their reaction point). However, primary amphibole always appears at lower temperatures than olivine or pyroxene. This suggests that amphibole gabbros are formed by reaction rather than direct precipitation of the hydrous phase.

As a result of the mixing of the introduced hydrogen with oxygen present in the run, water was produced. This stabilised the amphibole structure so that the mineral crystallised at higher temperatures. This was also true for those runs in which the water added was insufficient to saturate the liquid. Thus the breakdown of an hydroxylated phase in the mantle (e.g. a pyroxene, SCLAR, 1970) may result in the fixing of the hydrogen by amphibole at higher and cooler regions of the mantle.

Finally we note the depression of the melting point of plagioclase in water saturated systems. This may be a possible means of assessing the original P_{H_2O} during the crystallisation of the rocks.

Acknowledgements

We wish to thank Professor C. E. Tilley for his assistance in the examination of the run products and Professor W. S. MacKenzie for his advice and encouragement. One of us (R.W.N.) acknowledges financial support from the Royal Society and Nuffield Foundation. Dr. R. B. Scott, kindly assisted in the microprobe analyses.

References

BOWEN, N. L. (1928) *The evolution of the igneous rocks* (Princeton Univ. Press).
CLARK, S. P. (1959) Effect of pressure on the melting points of eight alkali halides, J. Chem. Phys. **31**, 1526–1531.
DEER, W. A., R. A. HOWIE and J. ZUSSMAN (1963) *Rock-forming minerals*, Vol. 2 (Longmans, London).
FUDALI, R. F. (1965) Geochim. Cosmochim. Acta **29**, 1063–1075.
GOODE, A. D. T. and G. W. KRIEG (1967) J. Geol. Soc. Australia **14**, 185–194.
GREEN, D. H. and A. E. RINGWOOD (1967) Contr. Mineral. and Petrol. **15**, 103–190.
GREEN, D. H. and A. E. RINGWOOD (1964) Nature **201**, 4926, 1276–1279.
HAMILTON, D. L., C. W. BURNHAM and E. F. OSBORN (1964) J. Petrol. **5**, 21–39.
IRVINE, T. N. Chromian spinel as a petrogenetic indicator (1965) Pt. I. theory, Can. J. Earth Sci. **2**, 648–672. (1967) Pt. II. petrologic applications, Can. J. Earth Sci. **4**, 71–103.
JACKSON, E. D. (1967) Ultramafic cumulates in the Stillwater, Great Dyke, and Bushveld intrusions, in: P. J. Wyllie, ed., *Ultramafic and related rocks* (Wiley, New York) 20–38.
KENNEDY, G. C. (1948) Am. J. Sci. **246**, 529–549.
MOORE, A. C. (1969) The petrology, geochemistry and structure of the Gosse Pile Intrusion, Tomkinson Ranges, South Australia, Ph.D. thesis, University of Adelaide (in preparation).
NESBITT, R. W. and A. W. KLEEMAN (1964) Nature **203**, 391–393.
NESBITT, R. W., A. D. T. GOODE, A. C. MOORE and T. P. HOPWOOD (1969) The Giles Complex, Central Australia; a stratified sequence of basic and ultrabasic intrusions, in: Geol. Soc. S. Africa, *Symposium on the Bushveld igneous complex and other layered mafic intrusions* (Pretoria).
OSBORN, E. F. (1959) Am. J. Sci. **257**, 609–647.
OSBORN, E. F. (1962) Am. Mineralogist **47**, 211–226.
OSBORN, E. F. (1963) Estudios Geol. **19**, 1–7.
ROEDER, P. L. and E. F. OSBORN (1966) Am. J. Sci. **264**, 428–480.
SCLAR, C. G. (1970) High pressure studies in the system MgO–SiO$_2$–H$_2$O (Abst.) Phys. Earth Planet. Interiors **3**, 333.
WONES, D. R. and H. P. EUGSTER (1962) J. Petrol. **3**, 82–125.
WONES, D. R. and H. P. EUGSTER (1965) Am. Mineralogist **50**, 1228–1272.
WAGER, L. R., M. G. BROWN and W. J. WADSWORTH (1960) J. Petrol. **1**, 73–85.
YODER, H. S., JR. and C. E. TILLEY (1962) J. Petrol. **3** (3) 342–532.

MELTING IN THE DEEP CRUST AND UPPER MANTLE AND THE NATURE OF THE LOW VELOCITY LAYER

I. B. LAMBERT* and P. J. WYLLIE

Department of Geophysical Sciences, University of Chicago, Chicago, Illinois, 60637, U.S.A.

It is possible that there are minor amounts of water in the upper mantle. This water could significantly lower melting temperatures and may give rise to melts of different composition to those formed by melting under anhydrous conditions.

This paper summarizes our experimental study of the influence of water on phase relations in natural gabbro and tonalite samples at pressures from 10 to 25 kb. Our results are used in conjunction with recently published data for other rock-water systems to discuss melting in the upper mantle and lower continental crust. It is suggested that the low velocity zone in the upper mantle could be a layer of peridotite containing interstitial melt, which may be overlain by a layer of peridotite containing interstitial "water". The minimum possible temperatures for the generation of andesite and basalt liquids are discussed, and it is noted that the initial melt formed from wet peridotite may be andesitic. Finally, granulite facies metamorphism in the lower continental crust and the generation of "granites" are considered.

1. Introduction

For a long while it has been generally accepted that the mantle consists of anhydrous peridotite. However, there has been some interest recently in the possibility that small amounts of water may be present, being mainly the result of degassing of the earth's interior. A number of people have suggested that the hydrous minerals amphibole and/or phlogopite may be present as minor phases in the upper mantle (RINGWOOD, 1962; OXBURGH, 1964; GREEN and RINGWOOD, 1967; KUSHIRO et al., 1967; ERNST, 1968), and a glance at the papers in this volume shows that several laboratories are at present engaged in experimental studies of phase relations in the presence of water to upper mantle pressures.

We have carried out experimental investigations of natural gabbro and tonalite samples at pressures from 10 to 25 kb, in the presence of excess water. In this paper we discuss our data, and use them in conjunction with other high pressure data, for granite-water (BOETTCHER and WYLLIE, 1968a), and peridotite-water (KUSHIRO et al., 1968) to consider a model for the low seismic-velocity zone, some aspects of andesite and basalt generation, and metamorphism and melting in the deep continental crust.

2. Experimental methods and apparatus

The gabbro we used is a high alumina olivine tholeiite from the island of Soay, near Skye. The tonalite is a typical, intermediate, calc-alkaline plutonic rock from the Sierra Nevada batholith. The chemical analyses, CIPW norms and approximate modes of these rocks are given in table 1. Our tonalite is chemically similar to the andesite composition studied by GREEN and RINGWOOD (1968).

Both rocks were ground to pass through 200 mesh and sealed into platinum capsules with 25–35 weight % water. All runs were carried out in a $\frac{1}{2}''$ piston-cylinder apparatus of the design of BOYD and ENGLAND (1960), which was calibrated as described by BOETTCHER and WYLLIE (1968b). All runs were done with the piston advancing. Pressures are precise to $\pm 3\%$, and are corrected for -8% friction. Temperatures were measured with a chromel–alumel thermocouple, with no correction for pressure effect on emf, and are precise to $\pm 5\,°C$. Phases were identified by optical and X-ray diffraction studies.

We did not control oxygen fugacity because the small dimensions of the sample furnace make this difficult. Hydrogen can pass through the platinum sample capsules, causing some oxidation of the samples, and in

TABLE 1

Chemical analyses, CIPW norms and approximate modes of the rocks studied

	Gabbro	Tonalite
(1) *Chemical analyses*		
SiO_2	45.91	59.14
TiO_2	0.94	0.79
Al_2O_3	17.19	18.23
Fe_2O_3	2.33	2.32
FeO	7.67	3.62
MnO	0.22	0.11
MgO	7.48	2.50
CaO	13.54	5.92
Na_2O	1.63	3.81
K_2O	0.14	2.19
H_2O^+	1.78	0.82
H_2O^-	1.26	0.04
P_2O_5	0.04	0.30
CO_2	nil	0.01
(2) *CIPW norms*		
Quartz	—	11.7
Orthoclase	0.83	12.9
Albite	13.78	32.2
Anorthite	39.14	26.2
Diopside	22.59	0.9
Hypersthene	6.95	9.4
Olivine	8.55	—
Magnetite	3.39	3.4
Ilmenite	0.78	1.5
Apatite	0.10	0.7
(3) *Approximate modes*		
Quartz	—	13
Alkali feldspar	—	4.5
Plagioclase	47	59
Augite	47	—
Biotite	—	12.5
Hornblende	—	9
Olivine	3	—
Opaques	3	2

addition, iron can move from the sample into solid-solution in the Pt capsule, but this effect should be minor at the temperatures of our runs. As a result the stability fields we have defined for the iron bearing minerals will differ by uncertain amounts from the stability fields in the natural environment. Most previous experiments on rock-water systems suffer from the same short-comings. However, it is unlikely that these effects will be of such a large extent as to invalidate the deductions we have drawn from the data.

3. Results

The gabbro-water phase relations are shown in fig. 1; below 10 kb we have drawn on the results of YODER and TILLEY (1962). The tonalite-water phase relations are shown in fig. 2; below 10 kb we have extrapolated to the 1, 2 and 3 kb runs on the same rock by PIWINSKII (1968).

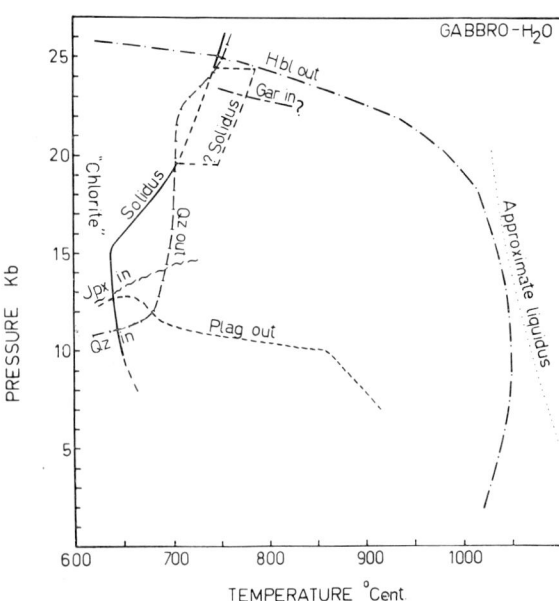

Fig. 1. Pressure-temperature projection of phase boundaries for the high alumina tholeiitic gabbro in the presence of 25 to 35 weight % water. Data below 10 kb are taken from YODER and TILLEY (1962). Abbreviations: Gar = garnet, Hbl = hornblende, Jpx = jadeitic pyroxene, Plag = plagioclase, Qz = quartz. Note that the location of the solidus is uncertain in the vicinity of 20 to 24 kb (see text).

In fig. 3 we have compared the solidus curves for granite, tonalite, gabbro and peridotite. It is apparent that above 5 kb or so, granite, tonalite and gabbro begin to melt at similar temperatures in the presence of water. All three solidi exhibit reversals of slope which appear to be related to the incoming of dense jadeitic pyroxene at the expense of plagioclase. With increasing anorthite molecule in the sequence granite-tonalite-gabbro, the reversal becomes more gradual and occurs at lower pressures. The peridotite-water solidus is at considerably higher temperatures than the other solidi. It exhibits a gentle reversal of slope which coincides with the incoming of garnet.

We have also plotted in fig. 3 some reactions which have been defined in simple systems and which we consider may be relevant to the tonalite-water and gabbro-water systems.

3.1. Gabbro water

Comparison of our results with the anhydrous experiments of GREEN and RINGWOOD (1968) shows that in the presence of water the gabbro solidus is lowered by some 500° to 600 °C at deep crust and upper mantle

remains, and there is common jadeitic pyroxene. This is the result of the reaction:

$$\text{Albite} \rightleftharpoons \text{jadeite} + \text{quartz}.$$

Quartz is present at 12.5 kb and melts just above the solidus; at pressures near 20 kb, quartz disappears

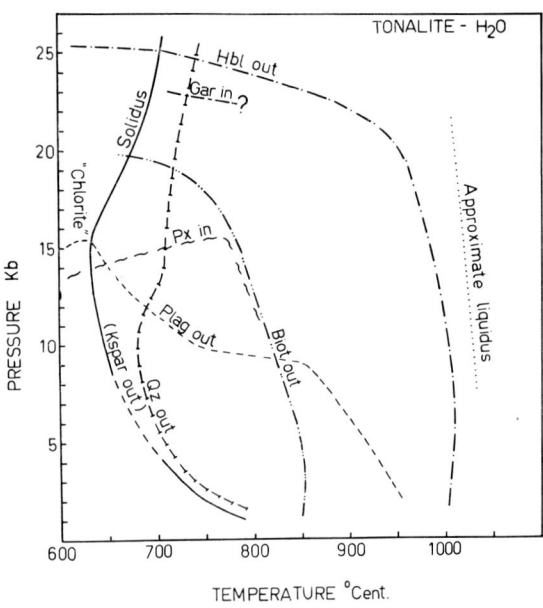

Fig. 2. Pressure-temperature projection of phase boundaries for the tonalite in the presence of 25 to 35 weight % water. Data below 10 kb are extrapolated to the 1, 2 and 3 kb runs on the same rock by PIWINSKII (1968). Abbreviations as for fig. 1, plus Biot = biotite, Kspar = alkali feldspar, Px = pyroxene. We could not recognise alkali feldspar in our run products, because it is only a minor constituent of the tonalite; we infer that it melts close to the solidus, as it does in the granite-water system (BOETTCHER and WYLLIE, 1968a).

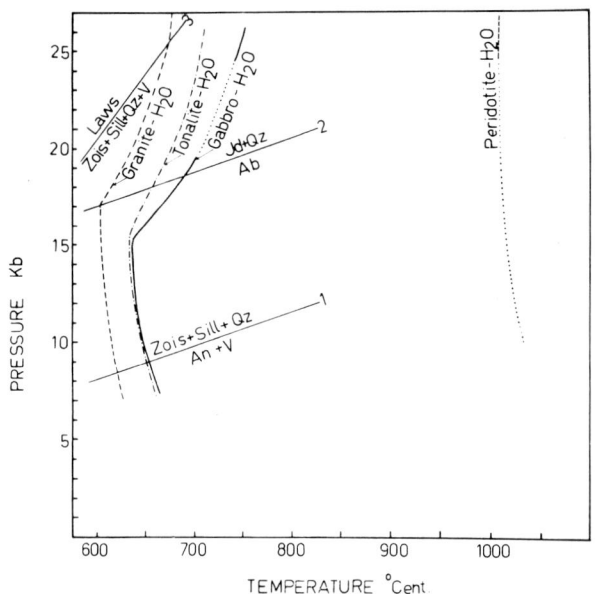

Fig. 3. Comparison of beginning of melting curves (solidi) for granite-water (BOETTCHER and WYLLIE, 1968) tonalite-water, gabbro-water, and peridotite-water (KUSHIRO et al., 1968). Small dots along gabbro and peridotite curves indicate that the curves are extrapolated at these places. These solidi are several hundred degrees below those for the anhydrous rocks. Also shown are reactions that appear relevant to the gabbro-water and tonalite-water systems. Abbreviations are as for fig. 1, plus Ab = albite, Laws = lawsonite, Sill = sillimanite, Zois = zoisite, V = vapour. Reactions 1 and 3 are from NEWTON (1966) and reaction 2 is from BIRCH and LE COMTE (1960).

pressures. At these pressures the wet gabbro melts over a temperature interval of about 400 °C, which is much larger than the 100 °C interval from dry gabbro.

In the presence of excess water, most of the pyroxene of our starting rock was converted to hornblende at run conditions, so at 10 kb the subsolidus assemblage was dominantly plagioclase plus hornblende, with some pyroxene. At 10 kb, plagioclase is completely melted near 850 °C, and hornblende is completely melted near 1000 °C (based on the data in YODER and TILLEY, 1962). In our runs at 12.5 kb, much of the anorthite molecule of the plagioclase has broken down, leaving minor amounts of relatively albite-rich plagioclase, which melts soon above the solidus. By 15kb, no plagioclase

just below the solidus (see below). Because quartz and feldspar melt near the solidus, the initial liquid produced must be silica rich.

There is no evidence for the presence of kyanite and zoisite, which might be expected from the reaction:

$$\text{Anorthite} + \text{vapour} \rightleftharpoons \text{quartz} + \text{kyanite} + \text{zoisite}.$$

It is possible that their place is taken by a sheet silicate phase with a major peak near 14 Å which is abundant in the 600 °C runs between 10 and 20 kb, and decreases in amount with increasing temperature, finally disappearing in the vicinity of the solidus. This seems to be some sort of chlorite or montmorillonite. As the sheet silicate goes out, quartz decreases and hornblende increases, suggesting a reaction of the sort:

"Chlorite" + quartz ⇌ hornblende.

YODER and TILLEY (1962) also noted a sheet silicate phase in their lower temperature runs at 5 and 10 kb on basalt-water compositions. They regarded it as a metastable phase, possibly quenched from the vapour. This could be so, but one would then expect it to occur in all runs, or at least a random assortment of runs. We consider it might be a stable phase in some glaucophane schist facies type assemblage, possibly coexisting with lawsonite, which is hard to identify with certainty because its main X-ray peaks overlap those of hornblende.

There is very little hornblende in our subsolidus runs at 25 kb, and none at all in those above the solidus. It appears that hornblende starts to break down by subsolidus reaction between 17.5 and 20 kb, and finally disappears along a curve with a shallow negative slope, not far above 25 kb. At the solidus, the hornblende stability limit must take on a slightly steeper negative gradient. We have not pinpointed the hornblende stability limit below 25 kb, but it must change gradient at the incoming of garnet and at the incoming of jadeite, by which time it has a steep positive slope and is in the vicinity of 1000 °C at 10 kb (YODER and TILLEY, 1962).

Garnet was not recognized until the 25 kb runs, which is higher than recorded in anhydrous basic rocks (e.g. GREEN and RINGWOOD, 1967). The appearance of garnet in our excess water runs is probably delayed by the existence of hornblende, but it would be expected to form as soon as hornblende begins to break down. It is probable that garnet nucleates very sluggishly at the relatively low temperatures of our runs, and seeding our gabbro starting material should cause garnet to grow at lower pressures.

In the few runs we did at 20 and 22.5 kb we lost track of the solidus. Some more runs are needed at these pressures, but our present interpretation is as follows: From just above 10 kb to just below 20 kb the quartz field extends to temperatures just above the solidus - thus we are picking up the solidus for a hydrous quartz eclogite. At about 19 kb the quartz stability limit crosses the solidus, and we are now looking at the solidus for a quartz-free, hydrous eclogite, which should occur at higher temperatures (just how much higher has not been determined). By 25 kb, most of the hornblende has broken down, yielding quartz, and the quartz field extends to higher temperatures, and we again pick up the hydrous quartz eclogite solidus.

3.2. Tonalite water

At deep crustal and upper mantle pressures our tonalite-water solidus is some 500 to 600 °C lower than the anhydrous solidus for the andesite (of similar composition) determined by GREEN and RINGWOOD (1968). At these pressures the wet tonalite melts over a temperature interval of some 400 °C, compared with about 100 °C under anhydrous conditions.

A feature of the melting behaviour of wet tonalite is that the granite components - quartz, alkali feldspar and albitic plagioclase, melt just above the solidus, whilst calcic plagioclase biotite and hornblende do not melt until much higher temperatures.

The subsolidus assemblage at 10 kb is that of the original tonalite -quartz, plagioclase, hornblende, biotite, with the exception that the few percent alkali feldspar is probably dissolved in the aqueous fluid phase. At 12.5 kb plagioclase has started to break down, yielding additional quartz. Minor albitic plagioclase is present at 15 kb, but melts just above the solidus. Jadeitic pyroxene is common in the 15 kb runs.

The sheet silicate phase with the major peak near 14 Å was again present in the 600 °C runs between 10 and 20 kb, and decreased in amount with increasing temperatures, disappearing near the solidus. The stability field of hornblende in the tonalite seems similar to that of hornblende in the gabbro. There is very minor hornblende in the subsolidus runs at 25 kb and no hornblende in any of the super-solidus runs. In a run at 900 °C/15 kb most of the tonalite had melted but there was very minor hornblende coexisting with slightly larger amounts of pyroxenes. Biotite was not noted in the runs above 20 kb. It melts at considerably lower temperatures than hornblende. Garnet again did not appear until the 25 kb runs, presumably because of sluggish nucleation - we consider it should appear about 17.5 kb, when hornblende breakdown commences.

4. Discussion and geological applications

Our experimental conditions are obviously not

directly applicable to the lower crust and upper mantle. In these environments there can only be minor amounts of water, the aqueous phase is unlikely to be pure water, and P_{H_2O}(water pressure) is probably less than P_{total} (total pressure).

The temperature for beginning of melting depends on P_{H_2O}, and is independent of the amount of aqueous phase present. However, the volume of melt produced is proportional to the amount of water available, and only traces of liquid should form where mantle peridotite begins to melt.

When P_{H_2O} is not much less than P_{total}, the hydrous minerals will break down at slightly lower pressures, but their high temperature stability limits are determined by melting reactions and will therefore extend to higher temperatures than they do for $P_{H_2O} = P_{total}$. At sufficiently low P_{H_2O} the tonalite solidus will have moved to such high temperatures that the hydrous minerals break down completely by subsolidus reaction, and at still lower P_{H_2O} their stability limits will move to lower temperatures.

Bearing these factors in mind we will consider the relevance of the experimental data to melting in the deep crust and upper mantle. We will discuss a model for the low velocity zone, some aspects of the generation of andesitic and basaltic magmas, and granite formation during high grade metamorphism in the deep crust.

4.1. *The nature of the low velocity zone*

We assume that the upper mantle is dominantly peridotite and that small amounts of water are present. In the uppermost mantle much of this water will be bound in hydrous minerals, presumably hornblende and/or phlogopite. The hydrous minerals may be stable to the peridotite-water solidus, but they should break down fairly rapidly at higher temperatures, and the water will enter the melt phase. Alternatively, the hydrous minerals may become unstable at temperatures lower than the peridotite-water solidus, resulting in a zone of mantle where all the water is in an interstitial aqueous phase, immediately above the zone of partial melting.

We do not have sufficient data to distinguish between these possibilities because the stabilities of hornblende and phlogopite in peridotite have not been determined. The only information we have on hornblende stability is that KUSHIRO et al. (1968) did not record it in their peridotite-water run at 980 °C and 26 kb (the lowest P–T conditions they studied). On the basis of this limited information it appears that the stability field of hornblende in peridotite cannot be much more extensive than in the systems gabbro-water and tonalite-water. Knowledge of phlogopite stability at upper mantle pressures is limited to the preliminary results of KUSHIRO et al. (1967), which indicate that phlogopite in the absence of other minerals is stable to significantly higher pressures than the hornblende in our runs. But the phlogopite field would be much more restricted in the presence of mantle pyroxenes (LUTH, 1967).

We feel that the limited stabilities of these hydrous minerals in the mantle provide a satisfactory explanation for the zone of low seismic velocity. Our model for the low velocity zone is summarized in fig. 4. Small amounts of interstitial melt phase must lower the seismic velocity of the peridotite, and a zone of interstitial water phase should also lower it slightly. The "floor" of this low velocity zone may be the result of water once more entering hydrous minerals, and thus not being available for melting. The synthesis of hydroxylated pyroxenes by SCLAR et al. (1967) has suggested this is possible. Another explanation might be that the "floor" is where the concentration of water in the mantle falls off to effectively zero.

The low velocity zone has previously been correlated

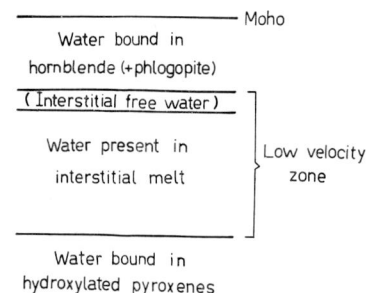

Fig. 4. Model for the zone of low seismic velocity in the upper mantle. If hornblende does not breakdown before mantle peridotite starts to melt, the zone of interstitial free water will not occur. The "floor" of the low velocity zone may alternatively be where the water content drops off to effectively zero.

with partial melting in the upper mantle, but under dry conditions, and this required a mantle with unexpectedly high temperatures at moderate depths. We published a note on our wet melting model in September, 1968.

KUSHIRO et al. (1968), in their paper published in the same month, also mentioned that partial melting of wet peridotite could explain the low velocity zone, but they did not discuss the probable importance of hydrous minerals in the upper mantle.

By making various simplifying assumptions, we can discuss the depth of the low velocity zone. Consider, for instance, that $P_{H_2O} = P_{total}$, and that the stability fields of hornblende and phlogopite in mantle perido-

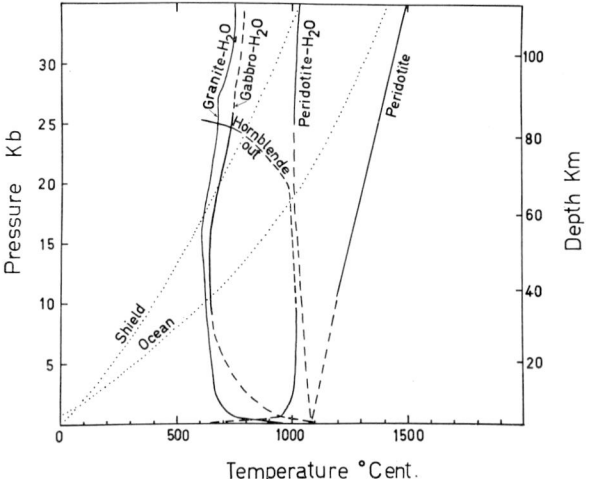

Fig. 5. Diagram showing the positions of the average geotherms under oceanic and continental shield regions (CLARK and RINGWOOD, 1964), in relation to the curves for beginning of melting in granite-water, gabbro-water, peridotite-water and anhydrous peridotite, and to the stability limit of hornblende in gabbro-water.

tite are both similar to the hornblende field we have outlined for gabbro-water. Fig. 5 shows that for these assumptions, beginning of melting should occur at about 60 km for normal oceanic geothermal gradients, and at about 110 km for normal shield geotherms. Under the oceans there should be a thin layer of peridotite incorporating interstitial water, and this should have slightly lower seismic velocities than the hornblende and/or phlogophite bearing peridotite above it. Under the shields this layer of interstitial water would be much thicker, beginning at about 80 km.

Of course, these are greatly simplified conditions and there are probably significant departures from them. For instance, the presence of CO_2 would lower the P_{H_2O}, and thus would raise the temperatures (depths) for beginning of melting of the peridotite, and lower the breakdown temperatures (depths) for the hydrous minerals, if these minerals break down by subsolidus reaction.

4.2. Aspects of the production of andesitic and basaltic magmas

GREEN and RINGWOOD (1968) showed experimentally that partial melting of basaltic rocks was a possible mechanism for generating andesitic magmas. Their experiments were conducted under conditions of $P_{H_2O} = 0$, and $P_{H_2O} \ll P_{total}$. At higher P_{H_2O} andesite liquid would form at markedly lower temperatures. Our gabbro-water solidus approximates the lowest possible temperatures for beginning of melting of basic rocks, and our tonalite-water liquidus approximates the minimum temperatures at which andesite melts could form.

It is usually considered that basaltic magma is the first product of partial melting of peridotite. In the presence of minor amounts of water, basaltic magma should develop at significantly lower temperatures than from anhydrous peridotite. The minimum possible temperatures for basalt formation are given by the gabbro-water liquidus.

It has recently been established that enstatite melts incongruently in the presence of water at upper mantle pressures, yielding a silica enriched liquid (SCLAR et al., 1967; KUSHIRO, et al., 1968). It is thus conceivable that the *initial* liquid produced when peridotite melts in the presence of water, is silica rich. This could mean that the minor amounts of interstitial melt in the low velocity zone are andesitic, and it raises the question of whether mantle peridotite could be the source of some of the andesitic rocks which are so commonly erupted in active regions near continent-ocean boundaries. If it is ever demonstrated that considerable amounts of water are available in the upper mantle under these regions, this mechanism would have to be considered. (The water may be dragged down when laterally spreading oceanic crust turns downwards, it may move down major fault zones, or it may move up from deeper in the earth if higher than normal rates of degassing characterize these active regions).

Free water beneath the layer of hydrous minerals in the uppermost mantle, and water liberated where these hydrous minerals break down, would probably contain strongly lithophile elements such as K, U, Th, and Rb, because these are not easily accommodated in the anhydrous mineral lattices in the mantle. Therefore the contents of these elements in magmas derived in the mantle should be controlled in part by incorporation of the aqueous phase.

4.3. *Metamorphism in the deep crust and the generation of batholithic "granites"*

Pyroxene granulite facies terrains are fairly uncommon at the surface. They frequently contain mineral assemblages which indicate they must have been metamorphosed at considerable depths in the continental crust. LAMBERT and HEIER (1967, 1968) noted the general lack of acid rocks in these terrains, which are frequently dominated by intermediate rocks. Lambert and Heier also confirmed the depletions of U, Th and Rb in these granulites that had previously been suggested by HEIER (1964), and HEIER and ADAMS (1965). They considered that there was partial melting during deep-seated metamorphism, followed by upward migration of the anatectic melts, which preferentially incorporated the granitophile elements and water.

The experimental data summarized in the present paper support this conclusion. These indicate that if water is available in the lower crust, rocks of acid, intermediate and basic compositions would be partially melted at the elevated temperatures which are generally thought to characterize orogenic episodes. Melting would cease when all the available free water was used up, and if the melt phase could move away, the water pressure in the residual rocks would become effectively zero. Under these conditions, the hydrous minerals would be unstable, and anhydrous granulite facies mineral assemblages would develop. In deep crustal environments where the free aqueous phase has diffused away completely before the climax of metamorphism, there would be no anatexis preceding the breakdown of the hydrous minerals, but there should be some subsequent melting.

The most common rocks in "granite" batholiths are adamellite (quartz monzonite), granodiorite and tonalite. Many geologists have argued that these rocks formed from magmas generated within the crust. The experimental work herein shows that the minimum temperatures required to generate a melt of tonalite composition within the crust is about 1000 °C, $(P_{H_2O} = P_{total})$. There is likely to be appreciable CO_2 in any fluid phase in the lower crust, because a number of carbonate minerals are unstable at the elevated P–T conditions. Therefore temperatures considerably higher than 1000 °C could be needed, and it is unlikely that such temperatures are attained during crustal evolution. However, mixture of melts of acid composition with mafic materials could give rise to batholithic "granites" at lower temperatures. We feel that acid anatectic melts formed during high grade metamorphism, digest varying amounts of refractory crystals and unmelted rock from the region of melting. The chemical and petrographic features of "granites" seem consistent with the operation of this process during slow cooling and concomitant fractionation.

References

BIRCH, F. and P. LE COMTE (1960) Am. J. Sci. **258**, 209.
BOETTCHER, A. L. and P. J. WYLLIE (1968a) J. Geol. **76**, 235.
BOETTCHER, A. L. and P. J. WYLLIE (1968b) J. Geol. **76**, 314.
BOYD, E. R. and J. L. ENGLAND (1960a) J. Geophys. Res. **65**, 741.
ERNST, W. G. (1968) *Amphiboles* (Springer, New York).
GREEN, D. H. and A. E. RINGWOOD (1967) Geochim. Cosmochim. Acta **31**, 767.
GREEN, T. H. and A. E. RINGWOOD (1968) Contrib. Mineral. Petrol. **18**, 105.
HEIER, K. S. (1964) Nature **202**, 477.
HEIER, K. S. and J. A. S. ADAMS (1965) Geochim. Cosmochim. Acta **29**, 53.
KUSHIRO, I., Y. SYONO and S. AKIMOTO (1967) Earth Planet. Sci. Letters **3**, 197.
KUSHIRO, I., Y. SYONO and S. AKIMOTO (1968) J. Geophys. Res. **73**, 6023.
KUSHIRO, I., H. S. YODER and M. NISHIKAWA (1968) Bull. Geol. Soc. Am. **79**.
LAMBERT, I. B. and K. S. HEIER (1967) Geochim. Cosmochim. Acta **31**, 377.
LAMBERT, I. B. and K. S. HEIER (1968) Lithos **1**, 30.
LAMBERT, I. B. and P. J. WYLLIE (1968) Nature **219**, 1240.
LUTH, W. C. (1967) J. Petrol. **8**, 372.
NEWTON, R. C. (1966) Am. J. Sci. **264**, 204.
OXBURGH, E. R. (1964) Geol. Mag. **101**, 1.
PIWINSKII, A. J. (1968) J. Geol. **76**, 548.
RINGWOOD, A. E. (1962) J. Geophys. Res. **67**, 4473.
SCLAR, C. B., L. C. CARRISON and O. M. STEWART (1967) Program Ann. Meeting, Geol. Soc. Am. 198.
YODER, H. S., JR. and C. E. TILLEY (1962) J. Petrol. **3**, 342.

RARE EARTH ABUNDANCES IN ALPINE ULTRAMAFIC ROCKS

FRED A. FREY

Department of Earth and Planetary Sciences, Massachusetts Institute of Technology, Cambridge 02139, Mass., U.S.A.

The rare-earth (RE) abundances of some alpine ultramafic rocks have been determined. A variety of RE distributions were observed. The high-temperature peridotites are markedly depleted in the light RE. The high-temperature peridotite at Lizard, Cornwall was studied in detail. The results are interpreted as providing evidence that the Lizard peridotite is a residue rock left after basaltic magma formation. The RE content of the marginal portions of the peridotite appear to have been contaminated by the country rock. Mineral RE abundances imply that the RE are located in structural sites of the host minerals. The observed mineral RE distributions can be used to develop a partial melting model for basalt generation. The residue mineralogy of this model is not consistent with residue mineralogies proposed to explain basaltic major element composition.

1. Introduction

The abundances of rare-earth elements (RE) in chondritic meteorites is rather constant. Analyses of twenty individual meteorites have indicated that the average deviation from the mean for each RE is approximately 20% (SCHMITT et al., 1963, 1964; HASKIN et al., 1966). The relative abundances of one RE to another in chondrites varies by approximately 11% (HASKIN et al., 1966). The RE abundances appear to be independent of the mineralogical and chemical variations which distinguish the various chondrite types.

Surveys of the RE in terrestrial rocks have demonstrated that in terrestrial rocks large variations in both RE absolute and relative abundances commonly occur (HASKIN and FREY, 1966). Specifically RE abundances in basaltic rocks vary from 4 to 100 times the chondrite abundances. The relative RE distribution in basalts is normally enriched in the light RE (low atomic number) compared to a chondritic average (see fig. 1).

There is evidence to support the hypothesis that terrestrial matter overall has the same relative distribution of RE as chondrites. This evidence arises from the similarity in relative abundance of the heavy RE in terrestrial rocks and chondrites (fig. 1). Also the RE distribution observed in sub-alkaline (oceanic tholeiites) basalts dredged from oceanic ridges is very similar to the chondritic distribution (FREY et al., 1968). The consistent light RE enrichment observed in continental and oceanic island basalts implies that this enrichment may be characteristic of the processes responsible for formation of basaltic material in the mantle. Residual

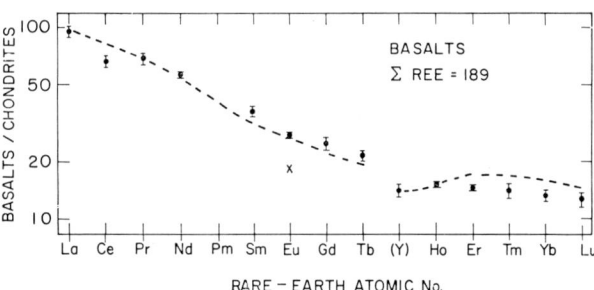

Fig. 1. Comparison diagram for the composite basalt with the composite of 9 chondrites. Dashed line represents the RE abundances for a composite of North American Shales. × represents Eu abundance of the shale composite.

ultramafic matter left after basalt generation would be expected to be depleted in the light RE. If the chondritic model is valid for the RE, then undifferentiated mantle material would be expected to have relative RE abundances as in chondrites.

Therefore, the RE abundances of terrestrial ultramafic rocks may provide information pertaining to the geologic history of ultramafic bodies presently exposed in the crust. In the research discussed in this paper several ultramafic bodies have been analyzed for RE.

TABLE 1

RE contents of Lizard peridotites* (ppm)

Element	Primary peridotites		Recrystallized anhydrous peridotites		Recrystallized hydrous peridotite	
	90683‡	90681	90689	90686	90691	Amphibole from 90691
Y	2.0 ± 0.1	1.6 ± 0.1	3.7	3.3	5.0	17
La	0.01	0.0092 ± 0.0008	0.161	0.157	0.44	2.2
Ce	0.02	—	0.92	0.68	1.6	6.1
Pr	0.014	0.0045 ± 0.0003	0.127	0.107	0.245	0.90
Nd	0.13	0.047	0.87	0.66	1.4	4.9
Sm	0.11 ± 0.01	0.052 ± 0.001	0.329	0.285	0.51	1.65
Eu	0.0450 ± 0.0003	0.0237 ± 0.0004	0.137	0.112	0.203	0.751
Gd	0.25 ± 0.01	0.132 ± 0.006	0.54	0.44	0.78	2.59
Tb	—	0.046 ± 0.006	0.113	0.099	0.164	0.46
Ho	0.09 ± 0.01	0.0690 ± 0.0007	0.146	0.139	0.225	0.708
Er	0.302 ± 0.002	0.213 ± 0.007	0.41	0.39	0.61	2.0
Tm	0.05 ± 0.02	0.038 ± 0.003	0.071	0.065	0.11	0.33
Yb	0.33 ± 0.04	0.28 ± 0.04	0.42	0.40	0.64	1.9
Lu	0.045	0.037 ± 0.005	0.054	0.052	0.095	0.34
ΣREE†	3.5	2.6	8.0	6.9	12.0	42
La/Yb	0.03	0.03	0.38	0.39	0.69	1.2

* States errors are mean deviations for samples analyzed in duplicate. For other samples see results and discussion in text.
† ΣREE includes the Y abundance and estimates for elements not determined.
‡ Sample numbers in all tables refer to Green's samples (GREEN, 1964a, 1964b).

Particular emphasis has been given to the RE abundances in the Lizard Intrusion, Cornwall, England.

2. Results and discussion

The neutron activation procedures used are described in HASKIN et al. (1968). Because of the low RE concentrations in most ultramafic rocks, the precision is somewhat poorer than the ±5% normally obtainable. The precision for the data presented can be estimated from the results of duplicate and triplicate analyses for two peridotites. Table 1 indicates the deviations from the mean for these analyses. The accuracy of the data are believed to be equal to the precision.

The comparison diagrams (CORYELL et al., 1963; HASKIN et al., 1966) are based on normalization to a composite of nine chondritic meteorites (HASKIN et al., 1968). Values for Y are plotted at atomic number 66 (Dy). Although the behavior of Y usually corresponds to that of a heavy lanthanide, it is somewhat variable and does not correspond exactly to any one of the heavy lanthanides.

3. Alpine peridotites

Crustal occurences of ultramafic rocks can be divided into several characteristic types. One type, the alpine peridotites, typically occur in orogenic zones among regionally metamorphosed rocks. These peridotites have some features in common (THAYER, 1967), but it is probable that rocks included within this type reflect several different geneses of ultramafic rocks.

Locations of alpine peridotites which have been analyzed for RE are:

1. Addie-Webster, North Carolina,
2. Canyon Mountain, Oregon,
3. Lac de Lherz, Pyrenees,
4. Lizard, Cornwall,
5. Mt. Albert, Quebec,
6. Shikoku, Japan,
7. Tinaquillo, Venezuela,
8. Twin Sisters, Washington.

Rare-earth abundances in the rocks analyzed range from 0.1 to 2 times the chondritic average. Definite correlations with mineralogy exist. The low RE containing rocks are dominantly olivine or orthopyroxene. When clinopyroxene becomes a dominant mineral the RE abundances are similar to chondrites. The relative RE distribution also varies considerably. Figure 2 indicates the kinds of variations typically found. Clearly some alpine ultramafics are enriched in light RE, some are depleted in light RE, and others have a nearly chondritic RE distribution. Detailed discussions of RE

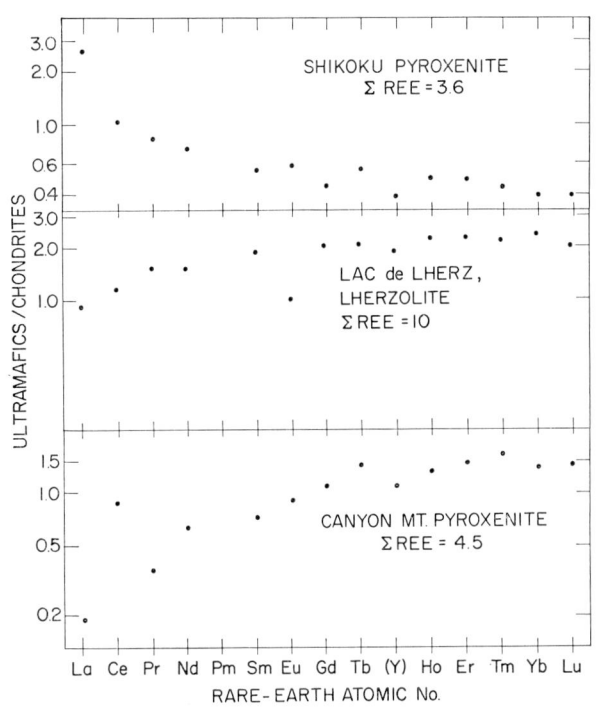
Fig. 2. Comparison diagram for three alpine ultramafic rocks.

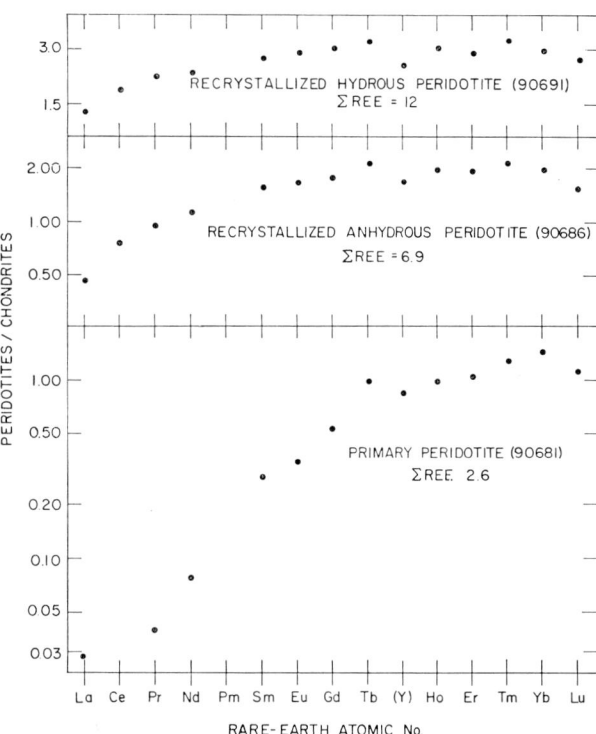
Fig. 3. Comparison diagram for Lizard peridotites.

in the alpine peridotites listed above are given by FREY and HASKIN (1969). The wide variety of RE abundances reinforces the previous conclusion that special tectonic conditions necessary for their intrusion may be the only feature which alpine peridotites have in common (GREEN, 1964; O'HARA, 1967).

GREEN (1964) observed that three alpine peridotite areas (Lizard, Mt. Albert, Tinaquillo) are unique in having caused extensive contact metamorphism. THAYER (1967) does not interpret the aureole rocks as a result of contact metamorphism. Nevertheless these three peridotites were found to be consistently severely depleted in the light RE relative to chondrites (fig. 3, also HASKIN and FREY, 1966). The light RE depletion in these ultramafics exceeds that of the other alpine ultramafics analyzed. A mass balance can be made by calculating what the mantle RE content must be if the crustal RE abundances developed from a source with chondritic RE abundances. The result of this calculation is that 25% of the mantle would have to have RE abundances as in these high-temperature peridotites. This implies that if the high-temperature peridotite RE abundances are characteristic of large mantle regions, 25% of the mantle has been involved in the formation of crustal material. This is a reasonable percentage. Because of the evidence for a high-temperature intrusion and since this is the RE distribution expected of residual mantle material, one of these peridotite bodies was studied in detail.

4. Lizard peridotite

The field and petrologic relations of the Lizard peridotite were studied in detail by GREEN (1964). The intrusion consists of a coarse-grained core enclosed within a recrystallized marginal shell. The major element composition of the primary core and recrystallized margin is constant. The mineralogy of the primary core consists of olivine, spinel, high-alumina enstatite, and high-alumina diopside. The recrystallized margins contain olivine, spinel, low-alumina enstatite, low-alumina diopside, plagioclase and accessory hornblende (termed recrystallized anhydrous assemblage). In some localized areas within the marginal region pargasitic amphibole becomes a dominant phase (termed recrystallized hydrous assemblage). Serpentinization commonly affects up to 50% of the rocks. The effects of serpentinization on the RE in the Lizard intrusion have been discussed by FREY (1969). The constant chemical composition with differing mineralogy has been interpreted as representing adjustment to equilibrium conditions in different pressure-temperature environments. O'HARA (1967) and GREEN (1967) have provisionally suggested that the primary peridotite pyroxene compositions indicate equilibration pressures of 15–20 kb and temperatures in excess of 1000 °C.

Five whole rocks have been analyzed for RE (table 1). The RE abundances and distribution are rather constant within each mineral assemblage but differ quite markedly between assemblages (fig. 3). The total RE content increases and the relative abundance of light RE increases in the sequence primary peridotite-recrystallized anhydrous peridotite-recrystallized hydrous peridotite. This result contrasts with the constant major element composition of these assemblages.

Pargasite separated from the recrystallized hydrous peridotite was analysed for RE (table 1). The pargasite forms about one-third of the rock and it is the principal host phase for RE in this rock. The presence of amphibole in the recrystallized peridotite appears to be responsible for the different RE abundances. The RE trend observed in the different assemblages is that expected if the primary peridotite has been contaminated with typical light RE enriched crustal RE abundances. The peridotite was intruded into amphibolite country rock which has a RE content typical of crustal material (FREY, 1969). The contact area contains a significant amount of pyroxene granulite which GREEN (1964) has interpreted as a result of contact metamorphism of the amphibolite. RE analyses of the metamorphic contact indicate that there is a relative loss of light RE in going from the amphibolite to the pyroxene granulite (FREY, 1969). Frey proposed the following sequence of events to explain the RE trends of the Lizard peridotite. Intrusion of the peridotite caused breakdown of country rock amphiboles into pyroxenes plus an aqueous phase. The observed deficiency of light RE in the pyroxene granulite implies that this aqueous phase would be enriched in light RE. This aqueous phase then permeated the peridotite margins where it was later incorporated into amphiboles upon crystallization. Because of the likely crustal contamination of the recrystallized peridotite, further analyses were done only on the primary peridotite.

5. Rare earths in ultramafic minerals

Separated minerals (olivine, spinel, clinopyroxene, orthopyroxene) from two typical primary peridotites were analyzed for RE. The pyroxenes were also separated and analyzed from a pyroxene rich band in the primary peridotite. The results are indicated in table 2 and fig. 4. If the Lizard primary peridotite is to be interpreted as a residue material, this figure should use the coexisting liquid for the comparison plots. Since the true nature of this coexisting liquid is unknown the chondritic average has been used as a reference. The shape of the RE distributions would have the same trends if various basaltic RE abundances were used as references.

An important problem in utilizing trace element data is to determine whether the trace elements are actually present in a structural position of a host mineral. The characteristic RE distributions observed for each mineral implies that the RE do substitute into structural positions. The RE distribution in the pyroxenes is consistent with simple crystal chemical predictions. The three enstatites are similar in RE distribution and exhibit increasing depletion of RE as the ionic size increases from Lu to La. The M_1 and M_2 cation sites in orthopyroxene are smaller than any of the RE ions, and it is expected that it should be increasingly difficult for larger ions to be incorporated into these sites. This is exactly the observed RE distribution for the enstatites (fig. 4). A quantitative theory has been formulated which expresses the energetics involved when a large trace element ion substitutes for a smaller host ion (NAGASAWA, 1966). This theory appears to be consistent with orthopyroxene RE data (ONUMA et al., 1968).

In diopside the M_2 site (cn = 8) is larger than the M_1 site (cn = 6). The M_2 site ideally corresponds to an ionic size which occurs in the middle of the lanthanide series. Crystal chemical considerations imply that ions smaller than M_2 site can be accomodated about equally well, but larger ions should be increasingly discriminated against. The distribution observed in the three diopsides is in accord with this idea. The distribution is flat from La to Tb or Gd but increasing depletion occurs for larger RE ions (fig. 4). Similar results were found by ONUMA et al. (1968) who analyzed for RE in the orthopyroxene and clinopyroxene phenocrysts of an alkaline olivine basalt.

TABLE 2

RE contents of minerals in Lizard primary peridotites* (ppm)

Element	Primary peridotite (90683)				Primary peridotite (90681)				Primary Peridotite (90684)	
	Diopside	Enstatite	Olivine	Spinel	Diopside	Enstatite	Olivine†	Spinel	Diopside	Enstatite
Y	20.8	2.7	0.24	0.167	12.2	1.91	0.065	0.90	17.2	3.1
La	0.0073	0.0030	0.045	0.75	0.0183	0.0039	0.078	0.48	0.088	0.012
Ce	0.21	0.022	0.11	1.3	0.063	0.015	0.26	0.96	—	—
Pr	0.126	0.0052	0.0109	0.152	0.0159	0.0019	0.0194	0.121	0.095	0.0059
Nd	1.70	0.062	0.048	0.54	0.29	0.031	0.112	0.54	0.90	0.048
Sm	1.32	0.063	0.0132	0.097	0.43	0.0259	0.0136	0.090	0.97	0.062
Eu	0.645	0.038	0.0041	0.0143	0.191	0.015	0.0011	0.0190	0.388	0.029
Gd	2.8	0.18	0.0216	0.076	1.23	0.099	—	0.119	2.1	0.15
Tb	0.60	0.047	0.0043	0.0085	0.255	0.028	0.0025	0.020	0.46	0.051
Ho	0.855	0.109	0.0094	0.0066	0.471	0.075	0.0027	0.038	0.813	0.147
Er	2.3	0.33	0.034	0.024	1.49	0.28	0.011	0.15	2.7	0.50
Tm	0.35	0.077	0.0055	0.0032	0.22	0.052	0.0020	0.021	0.40	0.088
Yb	2.1	0.50	0.045	0.018	1.28	0.44	0.012	0.13	1.79	0.62
Lu	0.34	0.098	0.0105	0.0054	0.20	0.069	0.0044	0.021	0.35	—
ΣREE	34.2	4.3	0.60	3.3	18.4	3.0	0.60	3.6	28.6	5.0
La/Yb	0.003	0.006	1.0	42	0.014	0.009	6.5	3.7	0.049	0.019
Impurity‡	serp.	serp.	sp.0.2% pyx.2.5%	none	serp.	serp.	sp.2.5% pyx.2.5%	none	serp.	serp.

* For estimated errors see Results and Discussion in text.
† Spinel impurity contributions have been subtracted.
‡ All pyroxene separates contain some grains which show incipient serpentinization designated in table as "serp".

The RE distributions in olivine and spinel are not easily explained by crystal chemical reasoning. The cation sites in olivine are smaller than all of the RE ions, and a RE distribution as in orthopyroxene is predicted. This is not observed. A low RE content is observed, but the light RE are not increasingly discriminated against. Instead a "V" shaped pattern is observed with the middle of the lanthanide series having the lowest relative abundance (fig. 4). The olivine separates were examined microscopically, and this pattern can not be explained by contamination with other minerals. There is considerable evidence accumulating that olivine characteristically has this type of RE distribution. For example, MASUDA (1968) and SCHMITT et al. (1964) have observed similar distributions for olivines in pallasite meteorites. Also SCHNETZLER and PHILPOTTS (1968) have obtained similar results for olivine phenocryst-matrix distribution coefficients. Several alpine dunites also have this distribution (FREY and HASKIN, 1969). No simple explanation for this distribution is apparent. It does not appear to be an absorption phenomena since acid washed olivine (MASUDA, 1968) has the same distribution.

The two spinels analyzed are enriched in the light RE. The RE distributions are quite different because one spinel (90683) is relatively enriched in the light RE more than the other spinel (90681). This is a result of different heavy RE abundances.

The variation of RE distribution among coexisting minerals in a single intrusion can be seen clearly when distribution coefficients are calculated for coexisting minerals (table 3). The three pairs of pyroxenes have distribution coefficients which vary by no more than a factor of three. For the elements Sn through Lu the variation from the mean averages at 12%. The similar relative RE distributions observed in the two primary peridotites plus the similarity of the clinopyroxene/orthopyroxene distribution coefficients imply a reasonably close approach to an equilibrium situation. Distribution coefficients calculated from pyroxene basaltic phenocryst data differ in magnitude but exhibit similar trends (table 3). This may be a result of different pyroxene major element composition. Also the pressure and temperature regime of phenocryst formation was probably quite different from that where the primary peridotite was equilibrated.

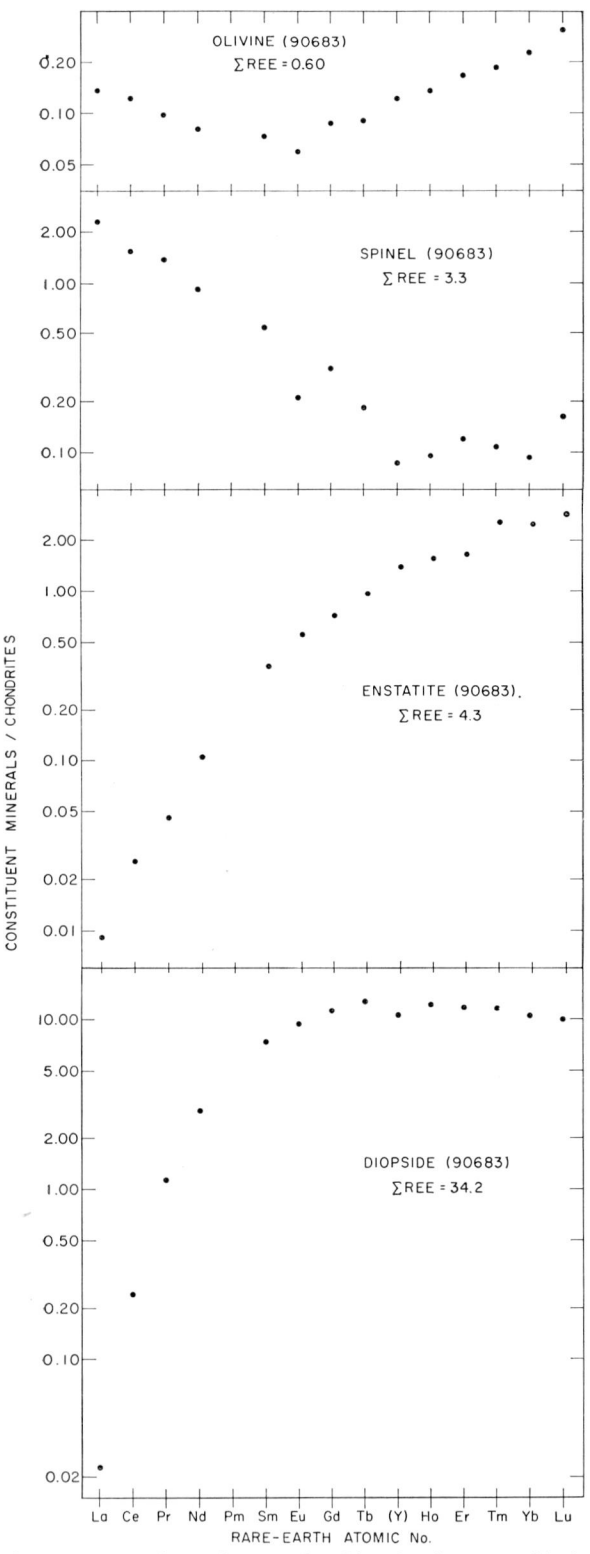

Fig. 4. Comparison diagram for Lizard primary peridotite minerals.

Similar data for olivine/spinel distribution coefficients indicate that the values are very different in the two primary peridotites. Spinel always contains more RE than the olivine, but in one case (90681) the olivine appears to preferentially incorporate the light RE

TABLE 3
RE distribution coefficients

Element	Clinopyroxene/Orthopyroxene				Olivine/Spinel	
	Lizard 90681	Lizard 90683	Lizard 90684	Basalt phenocrysts*	Lizard 90681	Lizard 90683
La	4.7	2.4	7.3	—	0.16	0.06
Ce	4.2	9.5	—	64.7	0.27	0.08
Pr	8.4	24.2	16.1	—	0.16	0.07
Nd	9.4	27.4	18.8	—	0.21	0.09
Sm	16.6	21.0	15.6	55.6	0.15	0.14
Eu	12.7	17.0	13.4	33.1	0.06	0.29
Gd	12.4	15.6	14.0	—	—	0.28
Tb	9.1	12.8	9.0	21.3	0.13	0.51
Y	6.4	7.7	5.5	—	0.07	1.4
Ho	6.3	7.8	5.5	—	0.07	1.4
Er	5.3	7.0	5.4	—	0.07	1.4
Tm	4.2	4.5	4.5	—	0.10	1.7
Yb	2.9	4.2	2.9	9.4	0.09	2.5
Lu	2.9	3.5	—	8.5	0.21	1.9

* Data from ONUMA et al. (1968).

whereas in the other sample (90683) the heavy RE are preferentially incorporated by olivine. Obviously an equilibrium situation is not established. A possible explanation may be that at the initial equilibration pressure and temperature of the peridotite spinel was not a stable phase. Spinel may have formed upon subsequent cooling of the peridotite solid. This process is suggested by the data of GREEN and RINGWOOD (1967a). At this lower temperature the small amount of spinel formed (1–2%) may not have equilibrated in trace element content with all the coexisting minerals.

6. Applications to basalt genesis

This topic has been extensively discussed by FREY (1969). The following comments are a summary of that discussion.

The RE distribution of the primary Lizard peridotite suggests that the peridotite material is a residue left after basalt generation by partial melting. The minerals forming the Lizard peridotite are similar to the residue minerals expected at depths of 40–60 km. FREY et al.

(1968) and SCHILLING (1966) have demonstrated that basalts have a variety of RE distributions. The RE distributions (relative to chondrites) in basalts can be generalized into three broad types:

1. A steadily increasing light RE enrichment from Lu to La (e.g. alkali oceanic island basalts).
2. A distribution which is flat from Lu to Eu and then is increasingly enriched in the light RE (e.g. continental tholeiitic basalts).
3. A distribution which is fairly flat from Lu to Nd with slight depletions of Pr, Ce, and La. A broad maximum may occur in the middle of the lanthanide series (eg. oceanic ridge sub-alkaline basalts).

The RE distributions observed in the Lizard orthopyroxenes, clinopyroxenes, and olivines are generally complementary to these basaltic RE distributions. This result suggests that the ultramafic residual minerals left after partial melting formation of basaltic magma could be very important in determining the RE distribution of the generated basalt. Intervening crystallization processes in the magma before crustal intrusion must also be considered, but GAST (1968) has demonstrated that for trace elements partial melting effects are likely to dominate effects of fractional crystallization.

If the Lizard minerals are residue minerals, then solid–liquid RE distribution coefficients may be calculated by utilizing various kinds of basalt as the coexisting liquid. FREY (1969) found that RE distribution coefficients calculated in this manner were similar to those determined from basaltic phenocryst-matrix data (SCHNETZLER and PHILPOTTS, 1968; ONUMA et al., 1968). The same trends were observed although the Lizard data indicated significantly smaller clinopyroxene-liquid distribution coefficients for the light RE. The overall similarity of distribution coefficients calculated in different ways suggests further that the Lizard peridotite is a residual body.

GREEN and RINGWOOD (1967b) have suggested the partial melting residue mineralogies which are necessary to explain basaltic major element compositions. Utilization of the RE solid–liquid distribution coefficients of either the Lizard data or the phenocryst data in the partial melting processes suggested by Green and Ringwood does not result in liquids which have the light RE enriched distributions observed in basalts. Furthermore, if chondritic abundances are assumed for the source material, the absolute RE abundances in the liquid are too low for basalts. This discrepancy can be alleviated if the basaltic source material has an RE abundance 4 times chondrites. This suggests a possible zoning of trace elements within the mantle.

The residual mineralogy proposed by Green and Ringwood does not cause light RE enrichment in the coexisting liquid because clinopyroxene is not an important residual mineral in explaining basaltic major element abundances. Clinopyroxene is the most important residue mineral for causing the RE fractionation observed in basalts. Garnet can fractionate the RE more than clinopyroxene (HASKIN and FREY, 1966; GAST, 1968), but garnet is thought to be an important residual phase only in generation of very silica undersaturated magmas (GREEN, 1970).

In summary there is evidence that the Lizard primary peridotite is a residue material left after basalt generation. The RE distribution observed in the Lizard minerals are complementary to the observed basaltic RE distributions. This result implies that a solid–liquid fractionation process is important in determining basaltic RE distributions. A knowledge of basaltic RE abundances may provide information concerning the original formation of the basaltic magma.

Partial melting processes appear to be important in determining the trace element content of basalts (GAST, 1968). There are discrepancies in matching the residue mineralogy which explains basaltic major element composition with that which explains RE abundances. Additional data for trace element abundances among coexisting minerals pertinent to basalt generation are needed. Also some changes in the Green and Ringwood model may occur as additional information is obtained about the stability of ultramafic minerals at upper mantle conditions. It seems feasible that the present discrepancy may be resolved by additional data. If the discrepancy continues, then processes in addition to partial melting are important in causing the RE abundances observed in basalts.

Acknowledgements

I thank Dr. D. H. Green for providing the Lizard peridotite samples studied. The neutron irradiations were made at the Massachusetts Institute of Technology nuclear reactor.

This work was completed with the support of the National Science Foundation under grant GA-4463. I thank the Australian Academy of Science whose travel grant enabled me to participate in this symposium.

References

CORYELL, C. D., J. W. CHASE and J. W. WINCHESTER (1963) J. Geophys. Res. **68** (2), 559.

FREY, F. A. (1969) Geochim. Cosmochim. Acta, in press.

FREY, F. A. and L. A. HASKIN (1969), in preparation.

FREY, F. A., M. HASKIN, J. POETZ and L. HASKIN (1968) J. Geophys. Res. **73** (18), 6085.

GAST, P. W. (1968) Geochim. Cosmochim. Acta **32** (10), 1057.

GREEN, D. H. (1964) J. Petrol. **5** (1), 134.

GREEN, D. H. (1967) High-temperature peridotite intrusions, in: P. J. Wyllie, ed., *Ultramafic and related rocks* (Wiley, New York) 212.

GREEN, D. H. (1970) Phys. Earth Planet. Interiors **3**, 221.

GREEN, D. H. and A. E. RINGWOOD (1967a) Earth Planet. Sci. Letters, 151.

GREEN, D. H. and A. E. RINGWOOD (1967b) Contrib. Mineral. Petrol. **15**, 103.

HASKIN, L. A. and F. A. FREY (1966) Science **152**, 299.

HASKIN, L. A., F. A. FREY, R. A. SCHMITT and R. H. SMITH (1966) Phys. Chem. Earth **7**, 167.

HASKIN, L. A., M. A. HASKIN, F. A. FREY and T. R. WILDEMAN (1968) Relative and absolute terrestrial abundances of the rare earths, in: L. H. Ahrens, ed., *Origin and distribution of the elements* (Pergamon Press, Oxford) p. 889.

HASKIN, L. A., T. R. WILDEMAN and M. A. HASKIN (1968) J. Radioanal. Chem. **1**, 337.

MASUDA, A. (1968) Earth Planet. Sci. Letters **5** (1), 59.

NAGASAWA, H. (1966) Science **152**, 767.

O'HARA, M. J. (1967), Mineral paragenesis in ultrabasic rocks, in: P. J. Wyllie, ed., *Ultramafic and related rocks* (Wiley, New York) 393.

ONUMA, N., H. HIGHUCHI, H. WAKITA and H. NAGASAWA (1968) Earth Planet. Sci. Letters **5** (1), 47.

SCHILLING, J. G. and J. W. WINCHESTER (1966) Science **153**, 867.

SCHMITT, R. A., R. H. SMITH, J. E. LASCH, A. W. MOSEN, D. A. OLEHY and J. VASILEVSKIS (1963) Geochim. Cosmochim. Acta **27**, 577.

SCHMITT, R. A., R. H. SMITH and D. A. OLEHY (1964) Geochim. Cosmochim. Acta **28**, 67.

SCHNETZLER, C. C. and J. A. PHILPOTTS (1968) Partition coefficients or rare-earth elements and barium between igneous matrix material and rock forming phenocrysts, in: L. H. Ahrens, ed., *Origin and distribution of the elements* (Pergamon Press Oxford) 929.

THAYER, T. P. (1967) Chemical and structural relations of ultramafic and feldspathic rocks in alpine intrusive complexes, in: P. J. Wyllie, ed., *Ultramafic and related rocks* (Wiley, New York) 222.

HYDROTHERMAL MELTING RELATIONSHIPS IN SILICATE-H_2O SYSTEMS AT VAPOR PRESSURES GREATER THAN 10 KBARS

A. L. BOETTCHER

Department of Geochemistry and Mineralogy Pennsylvania State University, University Park, Pennsylvania, U.S.A.

The effect of H_2O in lowering the temperatures of beginning of melting of rocks and rock-forming silicates is well known for pressures below 10 kbars. The solidus in each such system extends to lower temperatures with increasing vapor pressure, in contrast to the positive dP/dT slopes of anhydrous silicate systems. Insufficient data are available to predict fusion curves under high-pressure hydrothermal conditions, but experiments at vapor pressures above 10 kbars pressure reveal that incongruent melting, solid-phase transformations, and critical phenomena greatly influence phase relationships for many petrologically important systems.

In the system $CaO-SiO_2-CO_2-H_2O$, the solidus terminates in a second critical end-point at about 32 kbars. Similar phenomena are inferred for other univariant reactions in this and in more complex systems pertinent to kimberlites and carbonatites.

In the system $NaAlSiO_4-SiO_2-H_2O$, there are phase transformations involving jadeite, coesite, and other high-pressure modifications, which result in pronounced, abrupt changes in the dP/dT slopes of at least six univariant melting reactions. As a consequence of incongruent melting of albite-H_2O at about 17 kbars, the solidus for compositions on this join extends with a flat, negative dP/dT slope to a low-temperature invariant point involving quartz and jadeite, proceeding with a positive slope at higher pressures. Critical phenomena occur only for very SiO_2-rich compositions.

The appearance of dense phases, including kyanite (and sillimanite) zoisite, grossularite, and corundum, create similar complications in the system $CaO-Al_2O_3-SiO_2-H_2O$, for example in changing the dP/dT slope of the solidus for anorthite-SiO_2-H_2O compositions from negative to positive. Incongruent melting, together with the formation of zoisite, drives the solidus for anorthite-H_2O compositions to a low-temperature, quartz-bearing invariant assemblage.

These three systems provide a basic pattern for melting relationships in simple silicate-H_2O systems, and results on granite, together with work in progress on basalts, disclose similar complications in the hydrothermal melting of rocks.

THERMODYNAMIC PROPERTIES OF WATER-BEARING MAGMAS

C. WAYNE BURNHAM and N. F. DAVIS

Department of Geochemistry and Mineralogy, Pennsylvania State University, University Park, Pennsylvania, U.S.A.

The $P-V-T$ relations in the system $NaAlSi_3O_8-H_2O$ have been determined experimentally up to 8.5 kb and 900 °C in order to gain insight into the thermodynamic behaviour of water-bearing magmas. The data, in the form of mean specific volumes as functions of P, T and X_w, were obtained in the phase fields $S+V$, $S+L$, $L+V$ and L.

Partial molar volumes ($\bar{V}_w$) of water in solution, obtained graphically from the experimental data in the all-liquid (L) field, range from 25.0 cm³ mole⁻¹ at 1 kb, 900 °C to 15.5 cm³ mole⁻¹ at 10 kb, 700 °C. $\bar{V}_w$ is independent of X_w and various linearly with T (0.007 cm³ mole⁻¹ °C⁻¹).

Fugacities (f_w/f_w^0) of water in solution (L), calculated from integrated expressions for $\bar{V}_w$, were found to vary linearly with X_w^2. Activity coefficients ($\gamma_w = f_w/f_w^0 X_w^2$) obtained from the slopes of $(f_w/f_w^0)_{P,T}$ vs. X_w^2 at 900 °C range from 5.7×10^3 at 1 kb to 3.3×10^4 at 10 kb. The corresponding coefficients at 800 °C are slightly lower. The fact that $(f_w/f_w^0)_{P,T}$ is directly proportional to X_w^2, and not X_w, implies that water in the melt is dissociated, probably according to the following reaction: $H_2O_v + O_1^{2-} = 2 OH_1^-$.

The $P-V-T$ data also were used to compute the ΔV_r of the reaction: albite (S) + H_2O (V) = melt (L), along the S–L–V boundary. The ΔV values, together with the experimentally determined $P-T$ coordinates of this boundary, were used to calculate a ΔH_r of reaction which was found to pass through a minimum of 19 cal gm⁻¹ at 3 kb.

HIGH PRESSURE STUDIES IN THE SYSTEM MgO-SiO_2-H_2O

C. B. SCLAR

Department of Geological Sciences, Lehigh University, Bethlehem, Pennsylvania, U.S.A.

(i) Synthesis and stability of hydroxylated pyroxenes: Three quenchable hydroxylated pyroxenes with the ideal formula $Mg_2SiH_4O_6$ have been synthesized. Charge compensation is achieved by substitution of $(H_4O_4)^{-4}$ for $(SiO_4)^{-4}$. Two of these polymorphs are isostructural with orthorhombic enstatite and monoclinic enstatite and are stable above 40 kb, 1200 °C and 100 kb, 700 °C respectively. The third is not analogous to any known pyroxene and is stable above 50 kb and 500 °C. Each of these hydroxylated pyroxenes is related to forsterite by the reaction

$$Mg_2SiO_4 + 2H_2O \rightleftharpoons Mg_2SiH_4O_6,$$

which provides an experimental and crystallochemical basis for storing or releasing large amounts of water in the upper mantle in accord with vertical movements of the geoisotherms.

(ii) The effect of water vapor on the melting of forsterite involves a dramatic decrease in the melting point from 2000 °C at 20 kb (dry) to 1330 °C at 20 kb (wet). Although enstatite melts congruently at 20 kb (dry) it melts incongruently to forsterite plus a silica-rich liquid at 20 kb in the presence of water vapor. At 20 kb (wet), forsterite crystallizes from a charge of enstatite composition between 1220 °C and 1160 °C, below which forsterite reacts with the liquid to yield enstatite in the interval 1160 °C – 975 °C. These results preserve the possibility of magmatic differentiation by fractional crystallization at or near the continental crust-mantle interface, and prompt reconsideration of the possible existence of water-rich peridotitic magmas at temperatures compatible with those believed to prevail in the upper mantle and near the continental crust-mantle interface.

ELECTRICAL CONDUCTIVITY OF BASALTS AT HIGH T-P AND PHASE TRANSITIONS UNDER UPPER MANTLE CONDITIONS

N. I. KHITAROV, A. B. SLUTSKY and V. A. PUGIN

V.I. Vernadsky Institute of Geochemistry and Analytical Chemistry, USSR Academy of Sciences, Moscow, USSR

The conductometric method was used to determine the temperatures of phase transitions of tholeiitic basalts. Values of conductivity and capacity changes were determined under various conditions.

Analysis of the capacity and conductivity curves, which were obtained under isobaric conditions, and results of the study of quenching products show that in the range of pressures from 12 000–14 000 kg/cm^2 to 30 000 kg/cm^2 the transition from melt to the solid state in the cooling process is characterised by a successive change of mineral associations. These minerals are in equilibrium with the liquid phase in the liquidus–solidus temperature interval of about 150 °C at low pressures and up to 200 °C at high pressures.

Firstly at 28 000 kg/cm^2 a pyroxene mineral association is formed from the melt, followed by the gabbroic mineral association, which changes to the granulitic. The process results in full crystallization with the formation of eclogite.

At pressures below 10 000–12 000 kg/cm^2 the basalts studied behave in the process of melting and crystallization as mixtures close to eutectic. The liquidus–solidus temperature interval lies between 50–80 °C. The interval from liquidus to solidus is considerably greater with increase of pressure; beginning at 10 000–12 000 kg/cm^2 favourable conditions for the development of mineral associations exist, alternating with each other successively in the form of separate zones.

1. Introduction

Vast geological materials having been accumulated to the present time together with geophysical and geochemical data have created a favourable base for the development of many general considerations on the characteristics of the upper mantle structure, the nature and evolution of its matter on the elucidation of its interrelation with the crust of the Earth.

In this connection the data on the behaviour of basalts under different conditions play an important role and therefore during recent years a large number of experimental investigations have been carried out with the aim of obtaining additional data in this field. Already as a result some regularities have been revealed, but the domain of the unknown is still very large.

Many problems connected with basalt are still poorly understood and there is extreme scantiness of data on regularities in the development of zones of transition from solidus to liquidus under conditions where temperature and pressure increase with depth. At the same time it seems important to know of these regularities for the development of many genetic ideas, and the present investigation was carried out for this purpose.

2. Methods

For the determination of the temperatures of phase transitions the conductometric method was used. The most widely used method of studying the state of matter under conditions of simultaneous action of pressure and temperature – the method of quenching – does not allow the determination of the state of matter directly during the experiment. Therefore the method of quenching was used as a subsidiary one (KHITAROV and SLUTSKY, 1965).

The investigation was carried out in the region of temperatures up to 1500 °C and pressures up to 30 000 kg/cm^2. In this range of P–T values the electroconductivity and the change of the capacity of quartz tholeiite (QT), olivine tholeiite (OT), and high-aluminous tholeiite (AT) were determined. The analyses are given in table 1.

The study was carried out on a specially designed high pressure cylinder-piston apparatus. The heating of the sample is carried out by a graphite heater placed inside the chamber. A measuring cell is mounted within the heater (fig. 1).

To ensure reliable results of the temperature meas-

TABLE 1

Chemical composition of the basalts used in the experimental work

	Quartz tholeiite (%)	Olivine tholeiite (%)	High-aluminous tholeiite (%)
SiO_2	52.05	49.58	48.41
TiO_2	0.82	1.71	0.23
Al_2O_3	15.16	14.62	19.56
Fe_2O_3	4.48	3.03	1.35
FeO	5.16	7.80	4.36
MnO	0.10	0.16	0.41
MgO	9.49	7.34	8.19
CaO	9.29	11.32	15.42
Na_2O	2.46	3.19	1.36
K_2O	0.89	0.18	0.00
P_2O_5	0.13	0.04	0.00
H_2O^-	0.03	0.34	0.18
H_2O^+	0.06	0.26	0.53
	100.12	99.57	100.00
$\frac{100\,Mg}{Mg+Fe^{2+}}$	77	62	72
Q	1.92	—	—
Or	5.56	1.11	—
Ab	20.97	27.66	11.33
Ne	—	—	—
An	27.54	25.15	47.03
Di	14.97	25.04	22.92
Hy	21.04	7.10	12.04
Ol	—	5.89	0.60
Ilm	1.52	3.19	4.05
Mt	6.48	4.40	2.03
Normative olivine	—	Fe_{80}	Fo_{80}
Normative plagioclase	An_{57}	An_{47}	An_{81}

urement, three Pt–Pt 10% Rh thermocouples were introduced into the chamber, one of which was present in the sample itself and the two others allowed to follow the temperature gradient on both sides of the measuring cell. The influence of pressure on the E.M.F. of the thermocouples was not taken into account. The error of temperature measurements at 1400 °C is estimated about 25 °C. The pressure in the chamber was calibrated according to the phase transition in Bi and Tl. Pressures lower than 25 000 kg/cm² were determined by the resistance of manganin wire. A special method was used to determine the influence of temperature on the pressure determination. It was established that at 800 °C the friction in the pressure assembly is half that at room temperature. The total error in the pressure determination in the range between 14 000–30 000 kg/cm² is 4–5%. In regions of lower pressures these errors are about 15–20%.

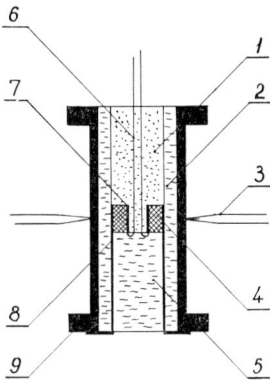

Fig. 1. Schematic diagram of the coaxial cell for electrical conductivity measurements. 1 = Insulator of alumina, 2 = Pyrophyllite layer, 3 = Side thermocouples, 4 = Sample, 5 = pyrophyllite, 6 = Internal thermocouple and electrical leads, 7 = internal electrode, 8 = Grounded external electrode, 9 = Heater.

For the measurement of conductivity the sample, in the form of a finely crushed powder, was pressed into the cell mounted in the chamber of high pressure. After reaching the necessary pressure the automatic equipment was switched on which ensured the heating at a rate of 120–200 °C/hr. The cooling was, as a rule, carried out at the same rate.

If it was necessary to fix the state of the sample at some definite temperature it was quenched with a speed of 400–500 °C/sec at a constant pressure and the sample was subjected to a microscopic and X-ray study.

The measurement of the resistance was carried out during heating or cooling of the sample under conditions of constant pressure. During the measurement the temperature was not changed until a stable resistance value was obtained. All the measurements were carried out in the frequency range of 1–10 kHz. Contemporary with the resistance measurement the values of the capacity were recorded. The bridge allowed measurement of resistances up to $2 \times 10^6\,\Omega$ with a sensitivity of the order of 0.1%. The cell was calibrated according to the resistances of NaCl and KCl.

3. Measurement of conductivity; results; discussion

The conductivity of the samples under investigation was studied at constant pressure and varying temperature. For all three basalts curves of $\log_{10} \sigma = f(1/T)$ (T in °K) have been drawn for different isobars from 2500 to 28 000 kg/cm². Conductivity curves have been obtained both when the sample was heated and cooled from melt. These curves consist of separate straight lines, which may be expressed by the equation $\sigma = \sigma_0 \cdot e^{-E/kT}$, where σ_0 is a constant, E the activation energy, T the absolute temperature, k Boltzmann's constant.

Breaks in the conductivity curves are, as a rule, connected either with a change of the conductivity mechanism without change of phase or with the appearance of a new phase.

Sharp breaks in the capacity curves also support the appearance of the liquid phase. This feature of the capacity curves was shown in the preliminar experiments on the determination of the influence of pressure on the temperature of albite melting (KHITAROV and SLUTSKY, 1965).

The joint examination of conductivity and capacity curves allow us to fix (at the moment of a sharp break) the temperature of the beginning and the end of melting of the sample under investigation. In fig. 2 curves of the conductivity (top) and of the capacity of quartz tholeiite are given. In fig. 3 conductivity curves of three basalts at a pressure of 28 000 kg/cm² are shown. At this pressure in the whole temperature range olivine tholeiite has the highest conductivity and high-aluminous tholeiite the lowest. The conductivity of quartz tholeiite occupies an intermediate position.

The cause of the differences may be the considerable alumina and calcium oxide content and the small content of iron in the high aluminous tholeiite and the relatively substantial content of iron and alkalis in olivine tholeiite. In this relation quartz tholeiite occupies an intermediate position.

It has to be noted that the difference in conductivity of the basalts is larger in the region below 800 °C. At higher temperatures the difference in conductivity of the basalts is not so substantial. The comparison shows that FILLOUX's (1967) conclusions based on our data for quartz tholeiite are also correct for other basalts in the considered T and P region. Based on the results of magneto-telluric soundings in the NE part of the Pacific at a distance of 630 km from the shore of Central California, Filloux has determined a conductivity value of $0.4 \, \Omega^{-1} m^{-1}$ for a 30 km depth from the ocean floor

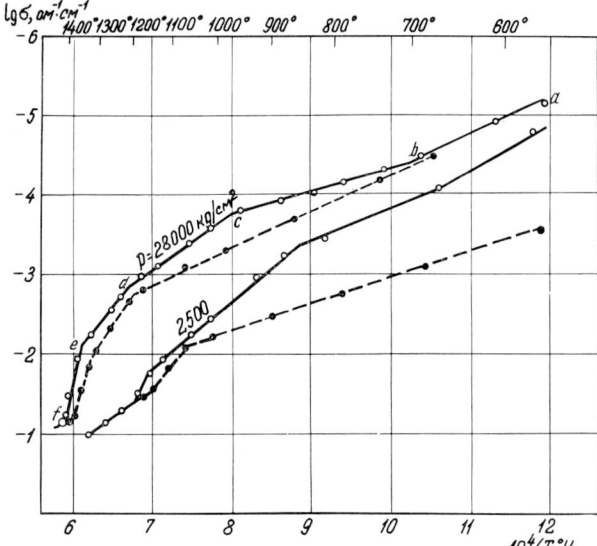

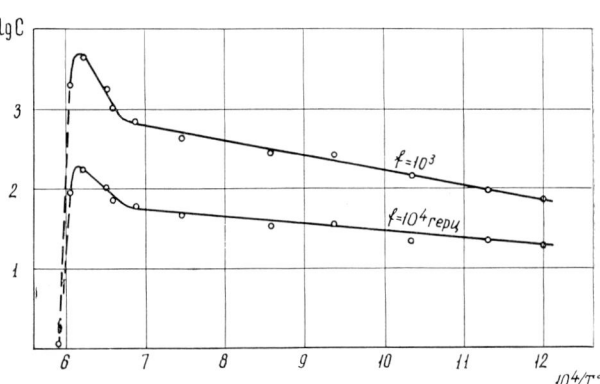

Fig. 2. Electrical conductivity and capacity versus temperature for quartz tholeiite. Dashed lines were obtained during cooling.

and compared the result with some experimental data.

As is seen in fig. 4, the conductivity level established by him, intersects at the points E and D of the conductivity curves of quartz tholeiite obtained in our work at $P = 28\,000$ and 2500 kg/cm². At a pressure corresponding to 30 km depth, the conductivity of $0.4 \, \Omega^{-1} m^{-1}$ is observed approximately at 1070 °C. This result will be considered below.

4. Phase transformations

In addition to conductometric data for the study of phase transformations which basaltic matter undergoes under conditions of high P and T the quenching products of the samples were investigated. A series of quenching experiments were carried out for basalts

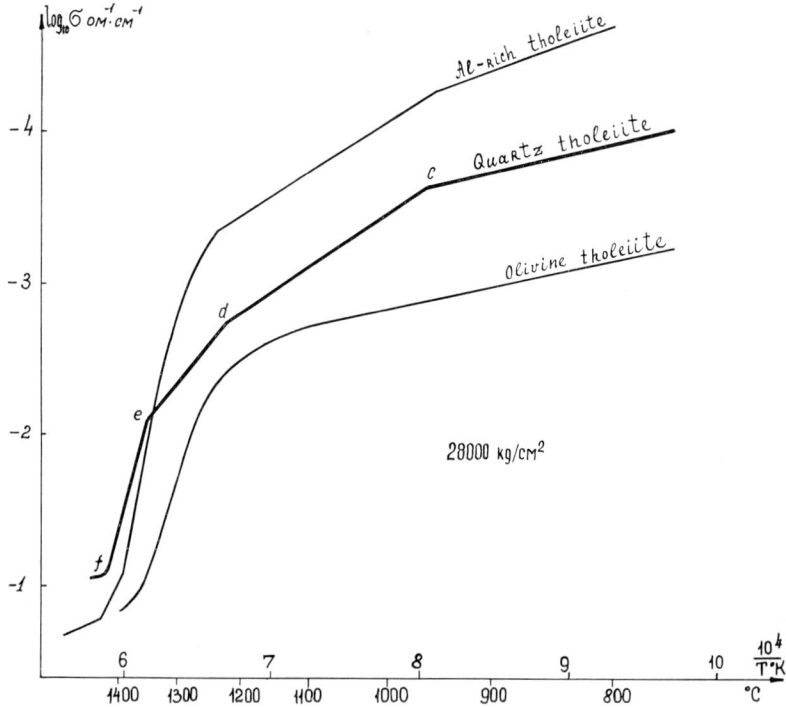

Fig. 3. Electrical conductivity versus temperature for the studied basalts.

which had been exposed for a certain time to different temperatures under isobaric conditions. Quenching experiments were bound to a separate part of the conductivity curves, which were limited by breaks on this curves.

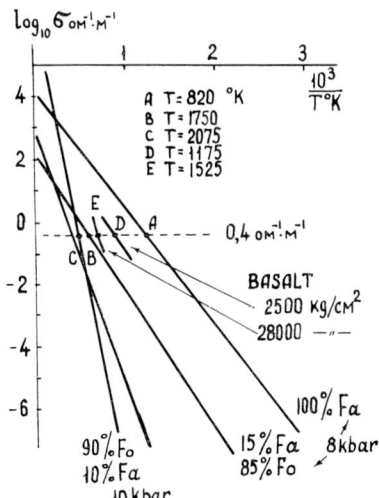

Fig. 4. Dependence of the electrical conductivity on temperature for different samples. Fo = forsterite, Fa = faylite (summarized by FILLOUX, 1967).

From this analysis of the conductivity and capacity curves as well as from the results of optical and X-ray study of quenching products, the temperature conditions of different phase transitions at pressures from 2500 to 28 000 kg/cm² were determined (see tables 2 and 3). Quartz tholeiite is characterized in more detail by quenching experiments. Samples of olivine and high-aluminous tholeiite were used for the elucidation of general regularities in the process of melting and crystallization of basalt compositions.

4.1. *Quartz tholeiite*

Experiments have shown that the melt of quartz tholeiite composition is transformed into a completely crystallized rock after passing through a series of reactions. At 28 000kg/cm² and a temperature below the liquidus temperature (1420 °C) the first C_{px}, O_{px} and Mt crystallize from the melt. In some experiments it has been noted that after C_{px} olivine is isolated. At a further temperature drop below 1350 °C plagioclase and some olivine appear a partial dissolution of orthopyroxene being observed. It is seen under the microscope that the gabbroid mineral association forming is submerged into a glassy mass. With the increase of the number of mineral phases and the growth of the crystals the amount of liquid phase decreases, though it remains present. The two mentioned domains of crystalline phase stability being at equilibrium with the melt pass one into the other at about 1350 °C.

A further temperature drop leads to a change of the mineral composition of the gabbroid association to the formation of a phase of high pressure garnet of pyrop composition with approximately Mg > Ca > Fe. In the 1300–1240 °C temperature range there is no olivine and at equilibrium with the melt are C_{px}, O_{px}, Pl, Gr and Mt and in some cases quartz. The O_{px} amount abruptly decreases and plagioclase gets a more acid composition (An_{30}). According to GREEN and RINGWOOD's (1966) example this mineral association may be attributed to the garnet–pyroxene granulite association. It is at equilibrium with the liquid phase, the amount of which is small. It may be assumed that the composition of the liquid phase in this case is mainly determined by constituents of minerals being dissolved – of orthopyroxene and the anorthite part of plagioclase from which garnet is then formed.

A further temperature drop below 1240 °C leads to a transformation of the complicated granulite association of minerals into a simpler eclogitic one, being represented by garnet and pyroxene of augite–jadeite composition (C_{px_2}). During the formation of this association an increase of the liquid phase amount is preserved. An explanation of this fact may be found in the appearance of a residue enriched in silica and alkalis. It has to be noted that, if during the garnet formation in the preceding zone, the supply of components due to the dissolution of the initial minerals O_{px} and Pl took place gradually in a section somewhat expanded by temperature, and in connection with this relicts of Ol, O_{px} and basic plagioclase were observed. The dissolution of acid plagioclase and quartz proceeds intensively in a narrower temperature range and with a noticeable increase of the liquid phase. The appearance of the latter at a certain ratio of the quartz-feldspar mixture may occur at a sufficiently low temperature. In the melting process apparently magnetite also participates, which disappears by transition into an eclogitic association of minerals.

Thus by the transition into the eclogite association of minerals at a temperature about 1220 °C a considerable fusion of the quartz-feldspar magnetite part occurs which joins the melt which existed at equilibrium with the granulite association of minerals. The formed melt participates in the formation of the eclogitic association of minerals together with the unfused components of the granulite association of minerals.

At 14000 kg/cm² only phases of the gabbroid association crystallize from the quartz tholeiite melt. Beginning at 1315 °C in a small temperature range clinopyroxene prevailingly develops, and then from 1260 °C develops an other components gabbroid association which has been traced up to 1000 °C.

Results of the experiments at 2500, 7000 and 12 000 kg/cm² also show crystallization of only the gabbroid association of minerals.

4.2. *High-aluminous tholeiite*

The melt of this basalt at 28 000 kg/cm² crystallizes to the first stage with the formation of C_{px}, O_{px}, Ol, Pl and Mt. As the temperature drops, garnet appears, which coincides with the disappearance of Ol and O_{px}, and then a granulite association of minerals is formed. At low pressures of 3000 and 7000 kg/cm² only the gabbroid association crystallizes.

4.3. *Olivine tholeiite*

At the initial crystallization stage of the olivine tholeiite melt at 28 000 kg/cm², C_{px} and Mt are formed. At lower temperatures Pl and Ol appear. A formation of the granulite association of minerals has not been noted and the complete crystallization is finished by C_{px_2} and Gr. Thus, during crystallization of the olivine tholeiite melt, a direct transition of the gabbroid association into the eclogitic is traced, though there is some doubt about this owing to the small number of experiments.

As a result, on the basis of analysis of the capacity and conductivity curves of different basalts, obtained under isobaric conditions, as well as of the study of quenching products, it is found that in the pressure range from 12 000–14 000 kg/cm² to 30 000 kg/cm² the transition from the melt to the solid state in the course of cooling is characterized by an alternation of mineral associations. The latter are present in equilibrium with the liquid phase in the temperature range up to 150 °C at low pressures and up to 200 °C at high pressures. Depending on the chemical composition of the basalt

TABLE 2

Quartz tholeiite

The mineral associations which coexist with the melt in the temperature interval between solidus and liquidus at various pressures

P (kg/cm^2)	Temperature interval (°C)	Mineral phases	Mineral associations
28000	1420°–1350°	Cpx, Opx, Mt, L$_1$	Pyroxenitic
	1350°–1300°	Cpx, Opx, (Ol), Pl, Mt, L$_2$	Gabbroidic
	1300°–1240°	Cpx, Opx, Pl, Gr, (Q), Mt, L$_3$	Granulitic
	1240°–1220°	Cpx$_2$, Gr, L$_4$	Eclogitic
	<1220°	Cpx$_2$, Gr, Q	Eclogite
14000	1315°–1265°	Cpx, Mt, L$_1$	Pyroxenitic
	1265°–1165°	Cpx, Opx, (Ol), Pl, Mt, L$_2$	Gabbroidic
12000	1265°–1190°	—	
	1190°–1165°	Cpx, Opx, Pl, Mt, L$_2$	Gabbroidic
7000	1200°–1155°	Cpx, Opx, Pl, Mt, L$_2$	Gabbroidic
2500	1200°–1150°	Cpx, Pl, Mt, L$_2$	

melt and on pressure, one or several zones of mainly pyroxenite, gabbroid or granulite composition may develop, which at the completion of crystallization pass into eclogite.

The experimental results lead to the conclusion that the investigated basalt melts under a pressure of 28 000 kg/cm^2 show non-eutectic compositions, and can transform in complex ways in rather wide intervals of temperature.

5. Position of liquidus and solidus boundaries

The estimation of the temperature range determining the position of liquidus and solidus for basalt compositions at atmospheric pressure was made by YODER and TILLEY (1962). They came to the conclusion that this range is determined for different compositions by temperatures from 135 °C to 195 °C. This estimation is close enough to one given by HESS (1941), who indicated that the maximum temperature of this interval is about 185 °C.

According to experimental data the liquidus–solidus temperature interval decreases with increasing pressure. A diminution of the temperature interval which determines the joint presence of the liquid and solid phases in equilibrium is observed when going from the melt to the eclogitic association of minerals. Thus, according to YODER and TILLEY's data (1962), the liquidus–solidus temperature range at 28 000 kg/cm^2 is in this region 80–90 °C and according to data given by COHEN et al. (1966) it is 50–60 °C. A diminution of the temperature range at high pressures is also observed in some data of GREEN and RINGWOOD (1966).

The task of determining the positions of the solidus–liquidus boundaries in a wide pressure range is of special interest in connection with the existence of a wide region of phase transformations of basalt matter, which was distinctly shown at 28 000 kg/cm^2 by investigations of quartz tholeiite. For this purpose quartz tholeiite was studied in more detail by the conductometric method in combination with the method of quenching. With the aim of revealing the most general picture of the solidus–liquidus boundary location high-aluminous and olivine tholeiites were investigated, mainly by conductometry. Results of the study of liquidus–solidus temperatures are listed in table 3 and in fig. 5 for quartz tholeiite. In fig. 6 a general scheme for three tholeiites is given.

For quartz tholeiite the study was carried out in five isobaric profiles (sections). The solidus–liquidus interval changes considerably with increasing pressure. From 2500 kg/cm^2 to 10 000–12 000 kg/cm^2 it is 50–100 °C. Then at 12 000–28 000 kg/cm^2 it is noticeably broadened and at 28 000 kg/cm^2 reaches 200 °C. For the two other tholeiites – the olivine and the high-aluminous ones – a broadening of the solidus–liquidus range in the region of high pressures and a narrowing at 3000–5000 kg/cm^2 have also been established. Owing to limited data, a general trend of boundaries for these basalts has been drawn conventionally by contours, but upper and lower values of the liquidus–solidus interval are clearly determined.

As a result, a very important feature of possible basalt melt evolution is revealed. At pressures below 10 000–12 000 kg/cm^2 in a relatively narrow temper-

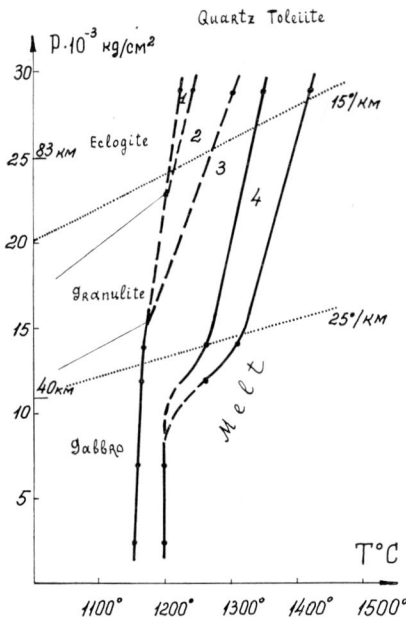

Fig. 5. *T–P* phase diagram for quartz tholeiite. Mineral associations: 1 = Eclogitic, 2 = Granulitic, 3 = Gabbroic, 4 = Pyroxenic.

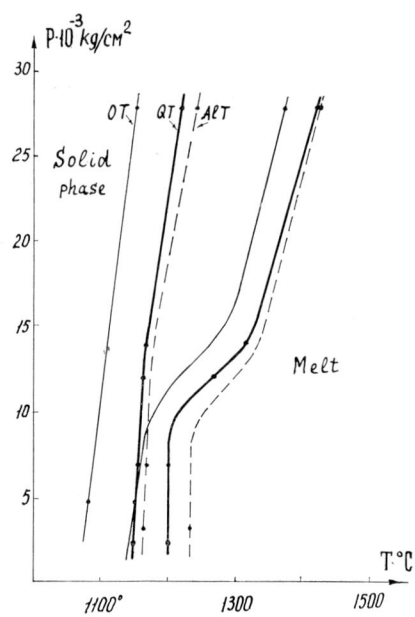

Fig. 6. The liquidus–solidus boundaries for three basalts.

TABLE 3

Liquidus and solidus temperatures of basalts at various pressures

	P (kg/cm^2)	Temperature (°C)		
		liquidus	solidus	ΔT
Quartz tholeiite	2 500	1200	1150	50
	7 000	1200	1155	45
	12 000	1265	1160	105
	14 000	1315	1165	150
	28 000	1420	1220	200
High-aluminous tholeiite	3 000	1235	1160	75
	7 000	—	1170	—
	28 000	1430	1240	190
Olivine tholeiite	5 000	1155	1075	80
	28 000	1375	1150	225

ature range the studied compositions behave in the process of melting and crystallization as mixtures close to eutectics. The region of such a behaviour appears to change and this change will be connected with the basalt composition. At higher pressures the transition from solidus to liquidus is considerably broadened and as a result favorable conditions for the development of mineral associations which successively alternate in the form of separate zones are created.

6. Discussion

As shown by the investigation, the interval of solidus–liquidus is considerably broadened with pressure and a number of regions are traced in it in which the solid phase is at equilibrium with the liquid one. The composition of the liquid phase will at first depend on temperature.

If it is assumed that the source of basaltic magma is eclogite, then only its complete melting can produce a magma of eclogite composition. It has to be noted that at levels deeper than 40 km the magma must be considerably overheated in comparison with the melting conditions in less deep districts, as the liquidus curve at high pressures is steeper than at low pressures. This fact must favour the appearance on the Earth's surface of magma in liquid state, even in a case of its relatively slow uprise.

The most probable process of magma formation in deeper parts is the process of melting out. The uncrystallized part in the thin sections of quenching products of different zones of quartz tholeiite, arising with a temperature change at 28 000 kg/cm^2, consists of glasses. The refractive indices of these glasses are characterized by a noticeable lowering towards eclogite, as can be seen from the data given in table 4.

TABLE 4

Zone	Refractive index
Melt	1.604
Pyroxenite	1.572–1.590
Gabbroid	1.572–1.576
Granulite	1.537–1.539
Eclogite	1.537–1.548

According to these data the most interesting melting out region for quartz tholeiite is of a pyroxene–garnet–granulite association of minerals, being at equilibrium with the melt at the transition to the eclogitic association on the one hand and to the gabbroid one on the other. In this region the melting out of the quartz-feldspar constituent takes place, and this appreciably increases the amount of melt in the system. The products of melting out in the form of a liquid phase may be considered as acid in relation to the initial material.

If an acid melt, approaching the eutectic composition, is assumed, the chemical composition of the crystalline mass in equilibrium with the melt may be approximately estimated. It will be a rough estimate, but a general regularity of the chemical separation into a liquid and a solid part may be revealed.

If from the quartz tholeiite melt, fully crystallized under conditions of a granulite zone, is taken the amount of acid melt arising from orthoclase, a part of the albite and quartz (according to the model composition) will approximately make up 18%. The chemical composition of the crystalline residue will in case of complete melting correspond to a melt – magma of an olivine tholeiite type, almost devoid of potassium and somewhat enriched with alumina. That means that during separation into an acid and a still more basic composition, potassium completely or almost completely goes away with the acid melt.

It has to be noted that the formation of acid and intermediate magma is possible only from eclogites containing excess of silica, i.e. corresponding to the composition of quartz tholeiites.

Concerning the question of the possibility of granulite transition from gabbro to eclogite in the lower parts of the crust and in the upper mantle, on the base of our data it can only be said that the granulite mineral association in equilibrium with the melt is only a transitional one from solid state to melt. The region of its existence is for the present drawn conventionally on the basis of Green and Ringwood's (1966) boundaries of granulite abundance in the solidus region.

The formation of a magmatic series may be explained either by complete melting of a gabbro type basic rock in a region of low pressure where these rocks practically melt as eutectic mixtures, or by a partial melting out at high pressures below liquidus temperature from eclogite type mineral associations with the formation of acid and alkaline melts and ultrabasic residues. It has been found that the pressure necessary for the formation of acid magmas and, consequently, the depth of their formation must be considerably greater than for the formation of basic magmas.

On the part of the liquidus–solidus zone up to a depth of the order of 35–45 km, the pressure drop due to tectonic disturbances and the slackening of general tensions may lead to the appearance of basaltic melts at a minimum temperature of about 1200 °C. At greater depths melting out must occur. There an emergence of melts of derivative composition and their moving to upper horizons is possible. In extreme cases, at local high pressures and relatively high rates of upward movement, granulite compositions may be intruded into the lower parts of the crust at a temperature of about 1200 °C.

Zones of different mineral associations, coexisting with melts similar to those shown for quartz tholeiite, may reach considerable thickness. At a geothermal gradient of 15 °C/km at a level corresponding to 28 000 kg/cm^2 the thickness of separate zones may reach several kilometres and abruptly increase at lower gradient values. These zones must apparently be of a stratified character and perhaps their existence is detectable by precise geophysical measurements.

The abrupt change of conductivity in passing from the solid phase to the liquid one, must assist in the solution of some geophysical problems. In this connection the results of Filloux (1967), who, on the basis of geophysical measurements carried out by him, notes a sharp jump of conductivity at depths of about 30 km (fig. 4) are of interest. Using his conductivity data and ours on quartz tholeiite, the temperature at this depth below the bottom of the Pacific at 630 km from the shores of Central California is 1070 °C. As the composition of the basalt assumed at this depth is not known, it is possible that the jump established by him corresponds to the transition to the liquid phase. This is possible, as may be seen from fig. 6, if there is olivine tholeiite at this depth.

The determination of the solidus–liquidus boundaries at high pressures for rocks, the existence of which is assumed at depth, is an important but complicated problem. The available data are limited, and it is hoped that investigators will further study this region using special and modern methods.

References

COHEN, L. H., K. ITO and G. C. KENNEDY (1966) Geol. Soc. Am. Spec. Papers **87**.

FILLOUX, J. H. (1967) Doctoral dissertation, University of California.

GREEN, D. H. and A. E. RINGWOOD (1966) Petrology of the upper mantle, Australian Nat. Obs. Publ. **4444**.

HESS, H. H. (1941) Am. Mineralogist **26**, 573.

KHITAROV, N. I. and A. B. SLUTSKY (1965) Geochemistry **12**, Moscow.

YODER, H. S. JR. and C. E. TILLEY (1962) J. Petrol. **3** (3).

INFLUENCE OF WATER ON MELTING OF SILICATES AT HIGH PRESSURE

A. A. KADIK and N. I. KHITAROV

V. I. Vernadsky Institute of Geochemistry and Analytical Chemistry, USSR Academy of Sciences, Moscow, USSR

The conditions of solidus in water-silicate systems are analysed from thermodynamic considerations. In the region of low H_2O pressures a molar volume of water is considerably greater than that of silicates. It is shown that this fact must determine the inevitable temperature decrease of silicate melting in the presence of water. A substantial decrease of the water molar volume with pressure increase leads to the fact that thermodynamic conditions are created under which the melting temperature of water-silicate systems will augment, i.e. with respect to pressure the water-silicate systems will behave similarly to dry systems. This fact is of considerable petrological interest for the solution of problems of Earth material transitions to the molten state.

When passing from low pressures to high ones the change of the dT/dP sign from negative to positive may occur either in the point of extremum for temperature or in the invariant point which lies on the intersection of the T–P curve of the monovariant reaction of melting with the line of solid phase transformations.

1. Introduction

Invariant equilibria

$$[\text{melt}]'' = N''[\text{vapor}]'' + \sum_i N_i'''[\text{crystal}]_i''' \quad (1)$$

determine the beginning of melting of silicates in the presence of water. They are of great importance, because they provide information on the conditions of transition of the Earth substance to the melt state. T–P conditions of such melting reactions are determined by changes of the thermodynamic properties and the solubility of H_2O with pressure and temperature.

At low pressure the molar volume of the water vapour phase is considerably greater than that of silicates, but it decreases with pressure (fig. 1a). The molar entropy of the water vapour phase is in general lower than that of silicates and this relationship does not change with pressure (fig. 1b). That determines a number of regularities of the water solubility and the melting of silicates at low pressures and indicates some principle changes at high pressures (KADIK and KHITAROV, 1963).

2. Solubility of water

The isobaric and isothermal solubility of water in silicate melts may be described by the following equations:

$$\left(\frac{dN''_{H_2O}}{dP}\right)_T = \frac{V' - V'' - (N'_{H_2O} - N''_{H_2O})(\partial V''/\partial N''_{H_2O})}{(N'_{H_2O} - N''_{H_2O})(\partial^2 Z''/\partial N''^2_{H_2O})_{P,T}} \quad (2)$$

and

$$\left(\frac{dN''_{H_2O}}{dT}\right)_P = \frac{S' - S'' - (N'_{H_2O} - N''_{H_2O})(\partial S''/\partial N_{H_2O})_{P,T}}{(N'_{H_2O} - N''_{H_2O})(\partial^2 Z''/\partial N''^2_{H_2O})_{P,T}}. \quad (3)$$

In eqs. (2), (3) V' and S' are the molar volume and the molar entropy, V'' and S'' are the same for H_2O-saturated silicate melt, Z is the isobaric–isothermal potential of the melt, N' is the molar fraction of H_2O in the vapour phase, N'' is the same value in the melt. At low pressure V' and S' are determined practically by the molar volume and the molar entropy of H_2O, i.e. $V' \approx V_{H_2O}$, $S' \approx S_{H_2O}$. This assumption is supported by the relatively low solubility of the silicate components in the water vapour phase.

In the region of low pressure

$$V_{H_2O} \gg V' - (N'_{H_2O} - N''_{H_2O})\left(\frac{\partial V''}{\partial N''_{H_2O}}\right)_{P,T}.$$

As a result of this the solubility of water is described by the equation

$$\left(\frac{dN''_{H_2O}}{dP}\right)_T \approx \frac{V_{H_2O}}{(N'_{H_2O} - N''_{H_2O})\left(\frac{\partial^2 Z''}{\partial N''^2_{H_2O}}\right)_{P,T}}. \quad (4)$$

343

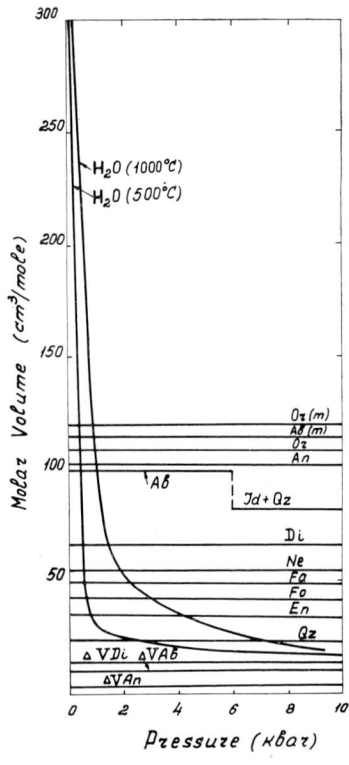

Fig. 1a. Molar volumes of H_2O and of some silicates at different pressures.

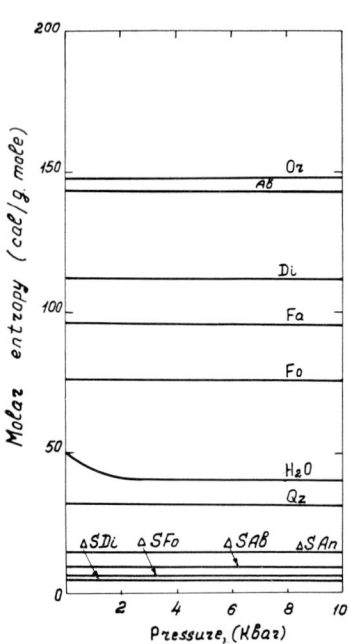

Fig. 1b. Molar entropies of H_2O and of some silicates at different pressures.

Ab = albite, An = anorthite, Di = diopside, En = enstatite, Fo = forsterite, Fa = fayalite, Ne = nepheline, Or = orthoclase, Jd = jadeite, qz = quartz;

(m) = melt ΔV = change of volume at melting ΔS = change of entropy at melting.

Since in a broad range of pressure and temperature $(N'_{H_2O} - N''_{H_2O}) > 0$, the condition (5) determines the increase of the water solubility in the silicate melts with pressure. In addition the shape of the isothermal curve of the water solubility must repeat the shape of the curve describing the change of the molar volume of water with pressure (fig. 2). At high pressure the occurrence of pressure maxima on the isotherms of the solubility of H_2O is probable. For these reasons the solubility of water should decrease with increasing pressure. The necessary condition for the maxima is

$$V' - V'' - (N'_{H_2O} - N''_{H_2O})\left(\frac{\partial V''}{\partial N_{H_2O}}\right) = 0. \quad (5)$$

Unlike eq. (2), the prevalence of any terms in the numerator of eq. (3) determining an effect of temperature on water solubility, do not occur. This does not allow to say something about the obligatory character of the sign of $(dN''_{H_2O}/dT)_P$.

It is to be expected that the sign of $(dN''_{H_2O}/dT)_P$ will

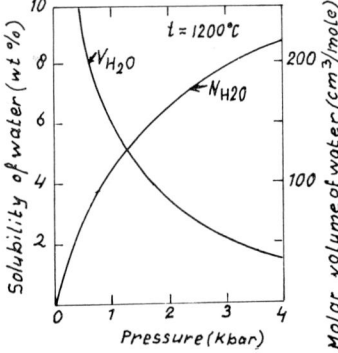

Fig. 2. The correlation between the magnitude of the molar volume of H_2O at different pressures and the shape of the curve of the solubility isotherm of water in albite melt (KADIK and LEBEDEV, 1968).

depend on a specific composition of silicates and at low pressures silicate systems may even exist in which the solubility of water increases with temperature. Experimental data which confirm this conclusion are given by KHITAROV et al. (1963), KHITAROV et al. (1968), KADIK and LEBEDEV (1968).

For example, the solubility of water in the system albite–water decreases with temperature at pressures up to 5000 b at 800–1400 °C and at pressures above 5000 b inversion takes place and the solubility of water begins to increase with temperature (KADIK and LEBEDEV, 1968).

3. Melting

T–P conditions of the reaction of melting are well known for a variety of water–silicate systems. Experiments show that a considerable fall of the melting temperature of silicates with water pressure is a general law in the region of low pressures, and thermodynamic analysis explains such general dependence.

In the region of low pressure, where the solubility of silicates in the water vapour phase is very small, the invariant reaction of the melting (1) can be described by an equation similar to that of Clausius–Clapeyron:

$$\left(\frac{dT}{dP}\right) = \frac{V'' - (N''_{H_2O} V_{H_2O} + \sum_i N''_i V'''_i)}{S'' - (N''_{H_2O} S_{H_2O} + \sum_i N''_i S'''_i)} = \frac{\Delta V}{\Delta S}. \quad (6)$$

In this equation V'' is the molar volume of the water–silicate melt, V_{H_2O} is that of the water vapor phase, V'''_i that of the crystal i, S'' is the molar entropy of the water–silicate melt, S_{H_2O} is the molar entropy of the water vapor phase, S''_i is the molar entropy of the crystal i, N''_{H_2O} and N_i are the molar fractions of the phases [vapor]' and [crystal]$'''_i$ in the liquid phase [melt]'', respectively.

In the region of low pressures the molar volume of water is considerably greater than that of the silicates (fig. 1) because ΔV of the invariant reaction of the melting is practically determined by the product $N''_{H_2O} \cdot V_{H_2O}$.

Hence in the region of low pressures the influence of the water pressure on the melting temperature is practically determined by the equation:

$$\left(\frac{dT}{dP}\right) = -\frac{N''_{H_2O} V_{H_2O}}{\Delta S}. \quad (7)$$

Since in a broad range of pressure $\Delta S > 0$ (KADIK and KHITAROV, 1963), the condition (7) determines the fall of the melting temperature of silicate as the water pressure increases ($dT/dP < 0$).

As the numerator in eq. (3) depends considerably more on pressure than the denominator, the shape of the T–P curve of melting must be determined by the change of the molar volume of water with increasing pressure and with decreasing melting temperature. This supposition is confirmed by experimental data. The shape of the P–T curve of melting repeats the shape of the curve describing the change of the molar volume of water with temperature and pressure of the reaction (1) (fig. 3).

With increase in pressure the molar volume of H_2O decreases considerably. As a result, at pressures of 500–5000 atm it becomes smaller than the molar volume

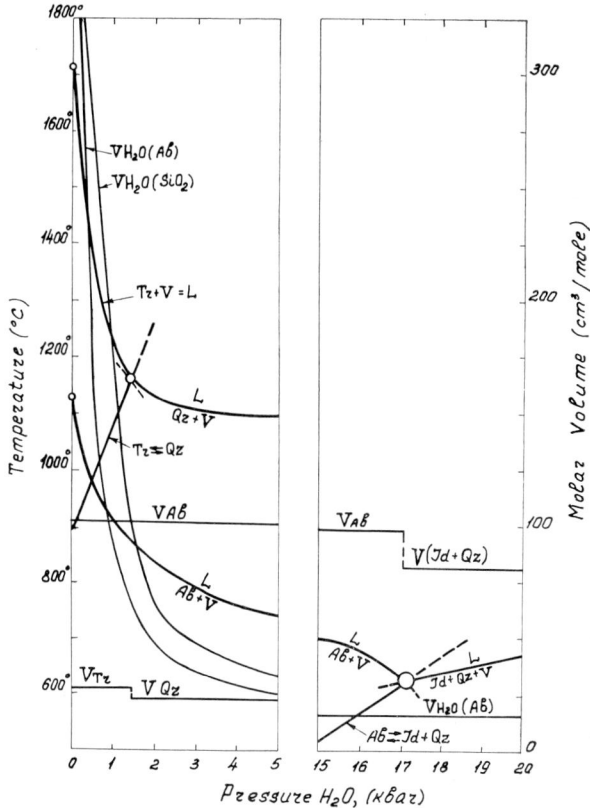

Fig. 3. The molar volume of water and of the crystals at T–P conditions of melting in the systems SiO_2–H_2O (KENNEDY et al., 1962) and $NaAlSi_3O_8$–H_2O (BOETTCHER and WYLLIE, 1967). $V_{H_2O(SiO_2)}$ = molar volume of water at T–P conditions of melting in the system SiO_2–H_2O, $V_{H_2O(Ab)}$ = that at T–P conditions of melting in the system of albite–H_2O, Tr = tridymite, L = melt, V = vapor.

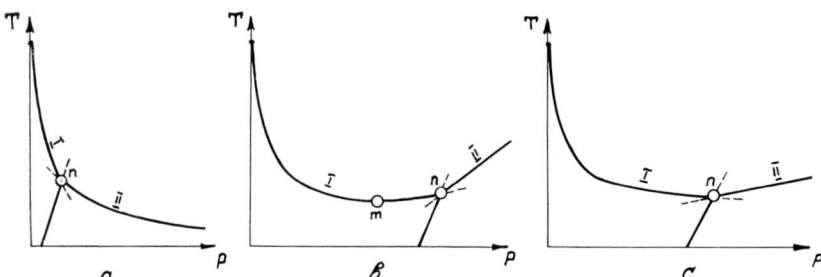

Fig. 4. The basic variants of T–P dependence of the reactions of melting in the water-silicate systems at high pressure.

of the silicates and is commensurable with the magnitude of the change of the volume during the melting or the solid phase transition of silicates. In other words, if at low pressures the thermodynamic conditions of melting are

$$V_{H_2O} \gg V'', \tag{8}$$

$$V_{H_2O} \gg V_i''', \tag{9}$$

and

$$V_{H_2O} \cdot N_{H_2O}'' > (V'' - \sum_i N_i'' V_i'''), \tag{10}$$

then at high pressures they are

$$V_{H_2O} \gtrless V'', \tag{11}$$

$$V_{H_2O} \gtrless V_i, \tag{12}$$

and

$$V_{H_2O} \cdot N_{H_2O}'' \gtrless (V'' - \sum_i N_i'' V_i'''). \tag{13}$$

It follows from conditions (11), (12) and (13) that at high pressures we are dealing with an essentially new region, in which the melting temperature of silicates may increase with increasing pressure in the presence of water, i.e. in the high pressure region a water–silicate system in which the dependence of the melting temperature on the pressure becomes analogous to that in the dry silicate system.

It is obvious that the change in the sign of dT/dP may be a consequence of two causes: the first of them is conditioned by a change of the relation between the members of the numerator in eq. (6), which causes an appearance of a temperature minimum on the P–T melting curve.

The thermodynamic conditions of such a critical point is given by

$$V'' - (N_{H_2O}'' \cdot V_{H_2O} + \sum_i N_i'' V_i''') = 0. \tag{14}$$

A change of the law of the water solubility with pressure in a system (the pressure maximum eq. (5)) must promote the achievement of this minimum (eq. (14)) of the melting.

At a high enough pressure, at which the molar volume of water and the product $N_{H_2O}'' \cdot V_{H_2O}$ are commensurable with the change of volume during the solid phase transition of the silicate, the change in the sign of dT/dP may be caused by the achievement of an invariant point such as n in fig. 4b. This invariant point is the intersection of the melting curve with the line of solid phase transition. In this case a large volume decrease of silicate crystals can lead to an increase of $(V'' - \sum_i N_i V_i)$. This must make for the appearance of the invariant melting reaction with a positive sign of dT/dP.

By virtue of the thermodynamic analysis, the following types of T–P melting curves, which are complicated by the solid phase transition, can be expected:

The first type corresponds to the low pressure region. The temperature of the invariant melting reaction decreases with increasing pressure. The occurrence of the temperature minimum on the P–T melting curve and the change of the dT/dP sign due to the solid phase transition is improbable (fig. 4a). This is due primarily to the fact that at low pressures the molar volume of water is incommensurably large compared with the molar volume of the silicates and with the quantity of the change of the volume during the solid phase transition.

For example, in the system SiO_2–H_2O (KENNEDY et al., 1962) the polymorphous transition tridymite–quartz does not lead to a change of the sign of dT/dP.

The decrease of the silica molar volume at the T–P conditions of the invariant point ($P_{H_2O} = 1500$, $T = 1160$ °C) is very much smaller than the water molar volume ($-\Delta V \approx 3.5$ cm^3/mole, $V_{H_2O} = 90$ cm^3/mole).

In the second type of system the temperature minimum on the T–P melting curve occurs before the solid phase reaction which is accompanied by a decrease of the molar volumes of the crystals (fig. 4b).

The third type of system corresponds to such a case, when at high pressures the solid phase reaction, changing the sign of dT/dP is accomplished before the possible temperature minimum on the T–P melting curve (fig. 4c), as it occurs in the composition join NaAlSi$_3$O$_8$–H$_2$O (BOETTCHER and WYLLIE, 1967). This join is effectively binary with a negative slope dT/dP of the invariant curve for the reaction

$$[\text{Albite} + \text{vapor} = \text{melt}]^{\text{I}}$$

up to 17 kb. At pressures above 17 kb albite breaks down to jadeite and quartz and the composition join are involved in the melting reaction

$$[\text{jadeite} + \text{quartz} + \text{vapour} = \text{melt}]^{\text{II}},$$

which has a positive slope dT/dP (fig. 3).

The phase transition albite = jadeite + quartz involves a volume decrease ($-\Delta V \approx 17.5$ cm^3/mole), which is larger than the molar volume of water at the conditions of the invariant point. One can suppose that the sharp volume decrease of the solid phase causes just the change of the sign of dT/dP in the invariant point.

An analogous explanation is given by BOETTCHER and WYLLIE (1967, 1968) but they simultaneously suppose that a change of sign of dT/dP established by them is not a result of a change in the properties of water in the vapour and in the equilibrium melt. However, more accurate definition is necessary for this assertion since without that the assertion may be incorrect.

As explained above, the great size of the molar volume of water at low pressures determines the sign of dT/dP and the form of the melting curve. It is particularly the progressive change of the molar volume of water with increasing pressure which at high pressures creates the situation in which the appearance of water–silicates systems with a positive value of dT/dP is possible.

The change of water properties with pressure is a necessary condition for a similar phenomenon. The transition of dP/dT from negative to positive at a point of temperature minimum or at a point of solid phase transition are only variants by which the change of the melting law can be realized when the molar volume of the water vapour phase will be sufficiently small.

4. Summary

Qualitative thermodynamic analysis in combination with experimental data shows that in the high pressure region (corresponding to the upper mantle) a water–silicate system in which a melting temperature of silicates increases with pressure can be expected. The occurrence of such an effect is improbable at low pressure (corresponding approximately to the earth's crust). This fact is of considerable petrological interest for the solution of problems of the transition of crustal and mantle material to the molten state.

References

BOETTCHER, A. L. and P. I. WYLLIE (1967) Nature **216**, 572.
BOETTCHER, A. L. and P. I. WYLLIE (1968) J. Geol. **76**.
KADIK, A. A. and N. I. KHITAROV (1963) Geokhimiya, 917–936.
KADIK, A. A. and E. B. LEBEDEV (1968) Geokhimiya, 1444–1955.
KENNEDY, G. C., G. I. WASSERBURG, H. C. HEARD and R. C. NEWTON (1962) Am. J. Sci. **260**, 501–521.
KHITAROV, N. I., E. B. LEBEDEV and A. A. KADIK (1963) Geokhimiya, 992.
KHITAROV, N. I., A. A. KADIK and E. B. LEBEDEV (1967) Geokhimiya, 1274–1283.
KHITAROV, N. I., A. A. KADIK and E. B. LEBEDEV (1968) Geokhimiya, 763–771.

SHOCK COMPRESSION OF POWDERED SiO_2, Mg_2SiO_4, $ZrSiO_4$ AND OTHER MATERIALS

N. L. DOBRETSOV, A. A. DERIBAS and V. I. MALY

*Institute of Geology and Geophysics and Institute of Hydrodynamics
Siberian Branch of the USSR, Academy of Sciences, Novosibirsk, USSR*

Phase-transformation in shock-compressed solids has been studied by many scientists in the last few years. Geologists hoped to apply these investigations to the theory of solid-state transformation in the Earth's mantle (DOBRETSOV et al., 1968; DERIBAS et al., 1966; RINGWOOD et al., 1967; McQUEEN et al., 1967) but the specificity of shock compression prevents the realization of this idea at present. However, experiments in shock compression of geological materials are interesting in themselves. These experiments may be particulary useful for the general theory of solid-state transformation and for meteorite problems.

1. Methods of shock compression

The methods of shock compression with shocked material preservation have been described in BATSANOV et al. (1965) and DERIBAS et al. (1967). Theoretical investigations and direct observations show the possibility of the existence of a three-shock configuration in the axial part of the compressed material in the experiments with cylindrical containers (ADADUROV et al., 1967). A photograph of shock configuration obtained by the optical method in (ADADUROV et al., 1967) is shown in fig. 1. The existence of a steady three-shock configuration allows the determination of the pressure in the axial zone. For this case the Hugoniot's curve determines the pressure behind the plane shock wave propagating with the detonation velocity. The corresponding values of pressure for the two types of explosive used are given in table 1.

Evidently, the possibility of the existence of a three-shock configuration depends on the correlation of the sizes of explosive charge and that of the container, as well as of the composition and density of powder and other factors. In the case when there is no steady plane shock wave in shocked powder, the determination of pressure, temperature and density become very complicated.

Usually in shock-compression processes the pressure increases in a time of about 10^{-8} sec and decreases in

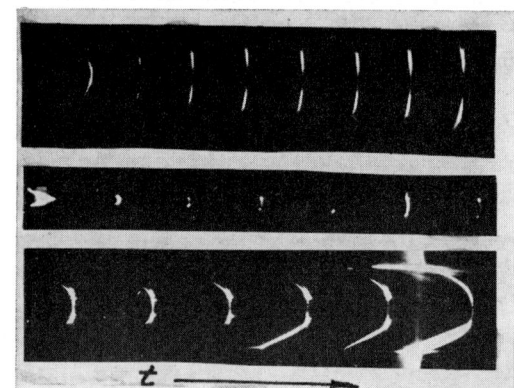

Fig. 1. Formation of three-wave configuration in cylindrical case.

TABLE 1

Pressure (in kbars) in the axial zone of containers in front of the head wave for different substances

Explosive	Explosive density g/cm³	Rate of detonation km/sec	Substances and their density (g/cm³)					
			SiO_2 (quartz) 2.60	SiO_2 (glass) 2.20	SiO_2 (powder) 1.6	Mg_2SiO_4 (powder) 3.05	$MgSiO_3$ (powder) 2.71	TiO_2 (crystalline) 4.25
Hexogene	1.1	6.6	480	490	420	—	510	—
Trotile/hexogene 50/50	1.6	7.6	680	680	—	660	—	800

a time of about 10^{-6} sec. The temperature decreases slower than the pressure in accordance with the thermoconductivity process.

2. Orthosilicates

2.1. *Forsterite*

The typical appearance of shocked powder is shown in fig. 2. The distinct clear zone is noticeable here in the axial part of the specimen. The dark and intermediate zones are observable on the periphery of the specimen. The insignificant changes of the initial powder are fixed in this zone. The observation of the specimens after compression is insufficient, evidently, for the single-significant determination of shock-wave configuration. Nevertheless, it is possible to compare the axial zone with the place of presumably three-shock configuration. No new phases are found in the experiment shown in fig. 2, but the structure of material in the axial zone suggests the formation of neogenic forsterite from the phase other than the initial one. The broadening of lines, the appearance of two new weak lines and the dissapearence of several weak lines, were observed in the X-ray pattern in the intermediate zone. The refractive index in the axial zone for natural olivine ($N_g = 1.686$, from kimberlite) was found to be equal to the pure forsterite ($N_g = 1.670$). Olivine in the dark border of the intermediate zone had a higher refractive index ($N_g = 1.693$). A redistribution of Fe from the axial zone into the intermediate one is probable in this process.

2.2. *Zircon*

The transformations in the powder of natural zircon ($ZrSiO_4$) are shown in fig. 3 (DOBRETSOV *et al.*, 1968). Distinctive axial zone 1, intermediate zone 2 and zone of the insignificant changes of initial powder 3 were observed in this experiment. The existence and width of these zones depend somewhat on the weight of explosive, but the composition of the material in zone 1 changes with weight of explosive. The relics of $(ZrSiO_4)^1$ with the destroyed lattice similar to the natural metamict zircon were found in zone 1. The amorphous glass-like phase of SiO_2, monoclinic ZrO_2 and some rhombic modification of ZrO_2, that is stable under high static pressure (BENDELIANI *et al.*, 1967), were also found in zone 1. It is necessary to emphasize that the last phase was definitely established in the experiments with the large charges of explosive. The quantity of "metamict" zircon $(ZrSiO_4)^1$ (lines in fig. 3 are marked by x), was decreased with the increase of the explosive weight.

Investigations of zone 1 by microsonde (fig. 4) confirmed the existence of the isolations of pure SiO_2. The particles of SiO_2 and ZrO_2 were very small in the ground part of zone 1 (less than 1–2 μ). These sizes are

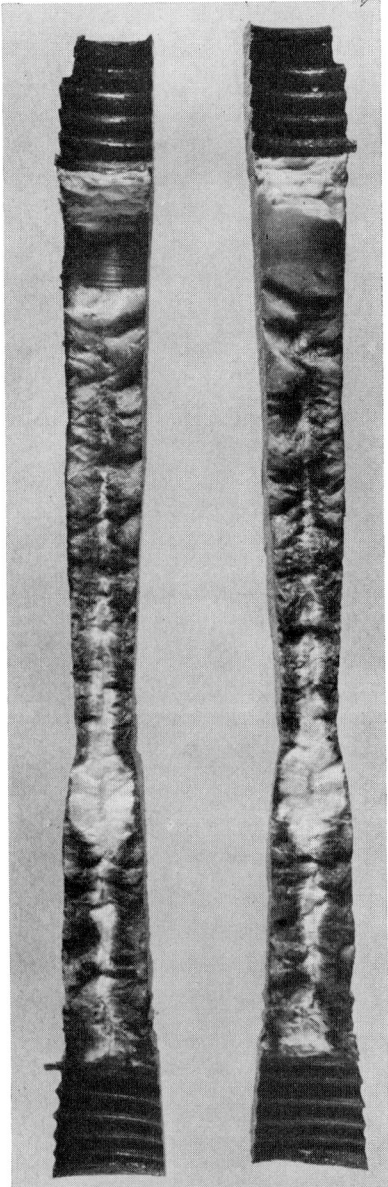

Fig. 2. Shocked powder of Mg_2SiO_4.

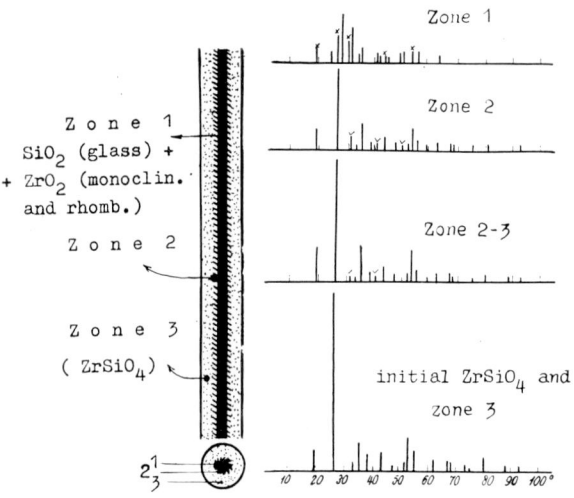

Fig. 3. Shocked powder of ZrSiO$_4$.

comparable with the accuracy of the microsonde, and the parts with a mixture of particles of SiO$_2$ and ZrO$_2$ are not distinguised in X-rays of SiKα and ZrLα from the relics of "metamict" zircon. These particles can be distinguished in an optical microscope.

The partial destruction of the lattice (broadening of the lines in the X-ray pattern and disappearance of the weak lines) was observed in zone 2. Sometimes the weak lines of ZrO$_2$ (marked by V in fig. 3) were found. The EPR-spectrums of zone 2 (fig. 5) reveal the characteristic effects similar to the partially metamict zircon. The broadening of line of EPR-spectra from zone 3 to zone 1, and the 10-fold decrease of its intensity were established.

In general, the behaviour of zircon powder is similar to the natural zircon with metamict destruction. The basic difference is that the high-pressure phase (rhombic ZrO$_2$) appears in shocked zircon and no intermediate states between the partially metamict zircon in zone 2 and the completely dissociated into oxides in zone 1 were found.

3. Framework silicates and SiO$_2$

As distinct from orthosilicates, we obtained the glass-like amorphous phases in framework silicates, often with heightened density and without the indication of melting. The appearance of shocked framework silicates differs from that of the othosilicates. The distinctive axial zone has not been observed in this case. Possibly it depends on the shock waves propagating in the frame-

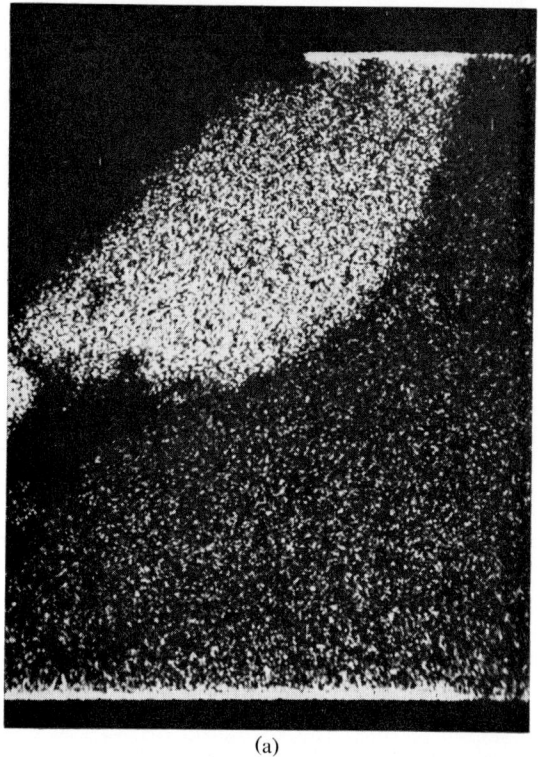

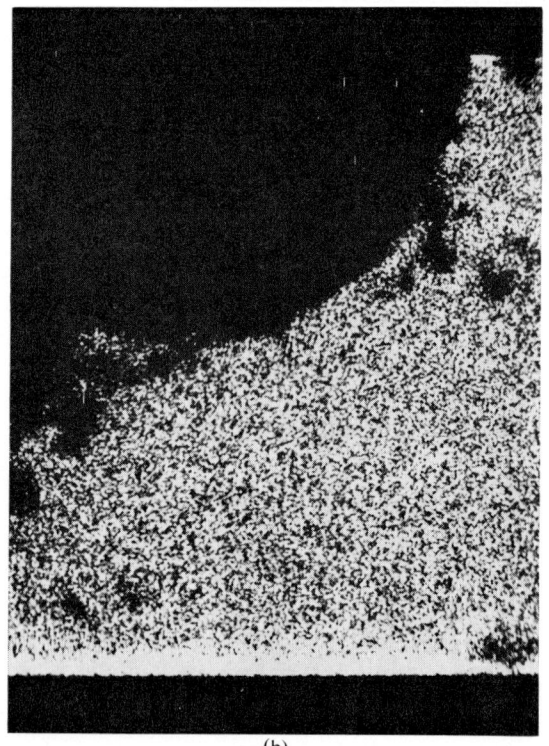

Fig. 4. Photograph (0.1 × 0.1 mm) of central zone in X-ray radiation (a) SiK$_\alpha$, (b) ZrL$_\alpha$.

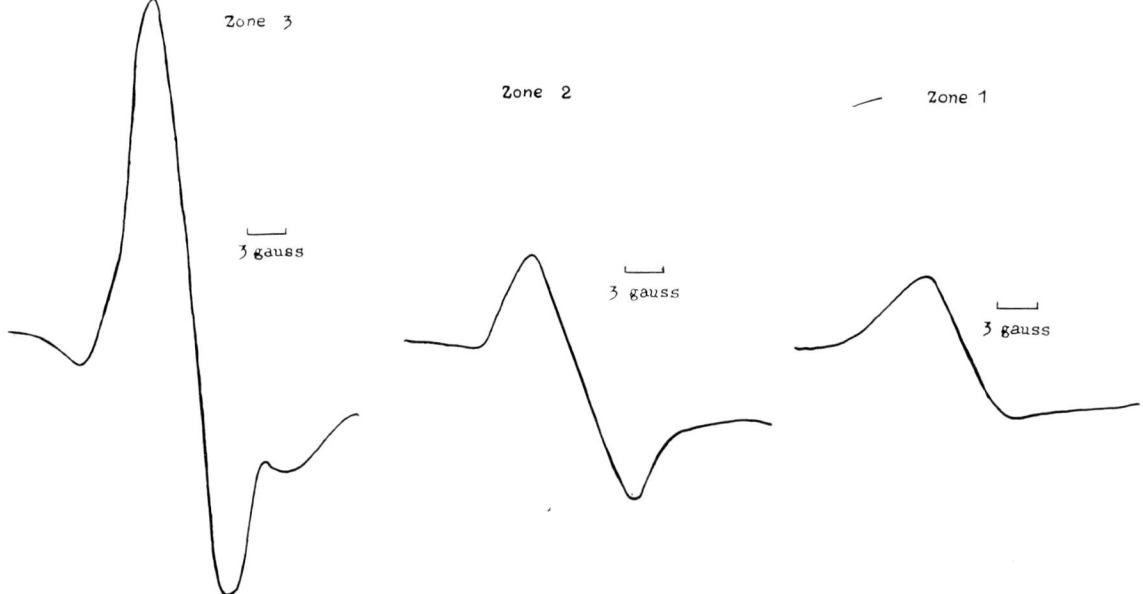

Fig. 5. EPR (electron paramagnetic resonance) spectra of the paramagnetic defect in shocked zircone of zone 3, 2 and 1.

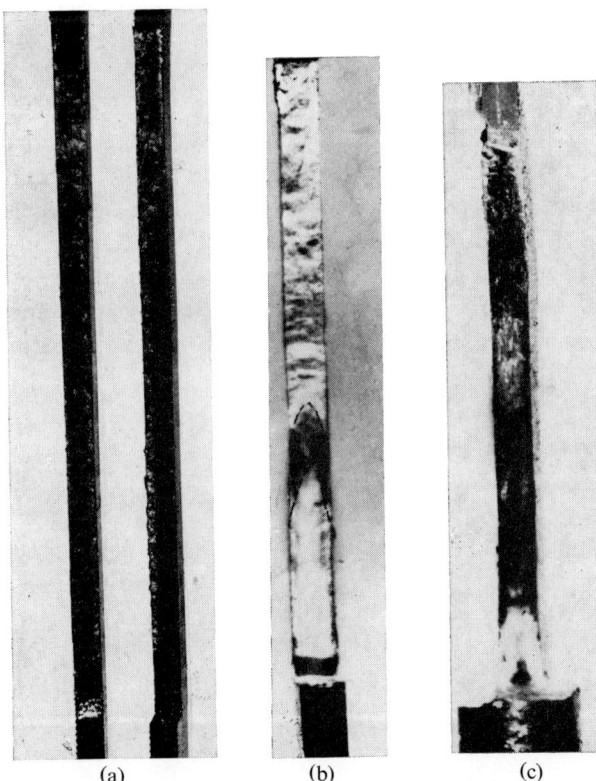

Fig. 6. Shocked powder of SiO$_2$ under different conditions: a – long container and large charge, size 0.5 of real size, b – short container and small charge, c – double shocked powder, band c being twice their original size.

work silicates without formation of a three-shock configuration (fig. 6). Several cone-shaped zones are distinguishable in this case. The sizes of these zones essentially depend on the sizes of containers, explosive charge and initial density of powder.

It is possible to assume, that the absence of the three-shock configuration is connected with the "friable" structure of these silicates in comparison with the orthosilicates.

The situation and composition of phases are shown in fig. 7 for SiO$_2$ and KAlSi$_3$O$_8$ shocked in the equivalent conditions. The major part of the shocked products is the glass of normal density containing the great number of smallest bubbles in the lower part of a container. (Detonation moves from above.) The intermediate zones 2 and 3 are most interesting.

3.1. SiO$_2$

The fragments of grains of powder SiO$_2$ transit gradually with the preservation of form into the glass-like phase with the heightened density (N = 1.510 instead of N = 1.460 for normal SiO$_2$ glass). Simultaneously the relics of quartz grains acquired the lower refractive index (N up to 1.520) and the low double-refraction (0.001–0.003), the broadening of lines and disappearing of weak lines in X-ray patterns were observed. The appearance of the X-ray patterns indicates

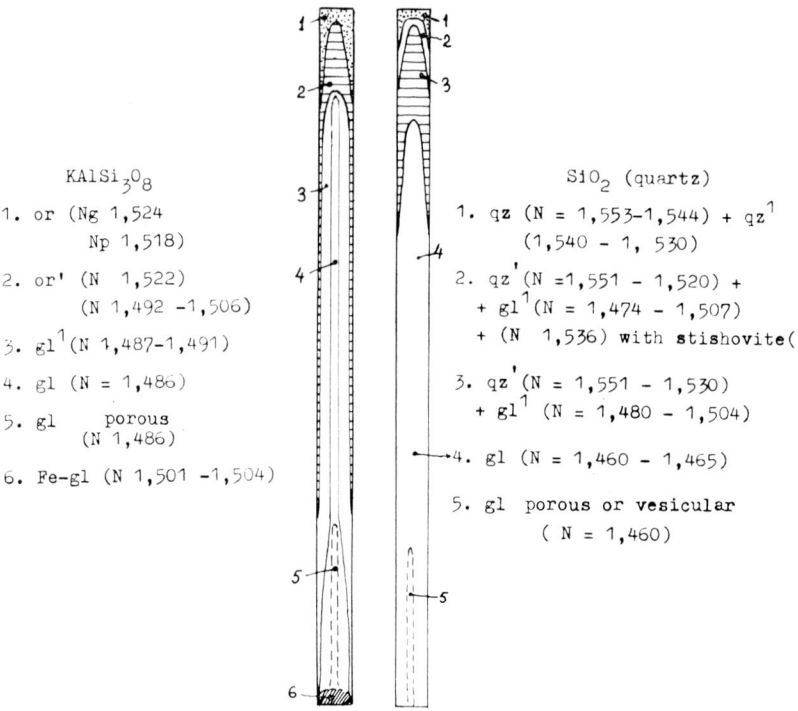

Fig. 7. Shocked powder of SiO₂ and KAlSi₃O₈ with zones of variable density.

the partial destruction of the lattice. The small quantity of the very dense glass (N = 1.536, constant density) was found in zone 2. This glass contained the needle-shaped grains which were hardly observed in the microscope, with these grain being possibly a coesite or stipoverite (stishovite). The stipoverite was found in SiO_2 after shock compression after treatment by a solution of HF and HNO_3 (DERIBAS et al., 1967; DE CARLI and MILTON, 1965) and the coesite was found in some of our previous experiments (DERIBAS et al., 1966). In the present experiments with large explosive charges (fig. 7) some coesite, tridimite (hexagonal) (coesite ≥ tridimite) and sometimes (in experiments with initial glass of SiO_2) neogenic quartz were found in zones 1, 2, 3. Some tridimite, traces of coesite and sometimes (in experiments with initial glass of SiO_2) α-cristobalite were found in zones 4–5. We suppose that the formation of tridimite and cristobalite is connected with the high residual temperature after shock compression. Zones 2 and 3 moved along the detonation and their sizes increased with explosive charge. Zones 1 and 2 disappeared with the largest explosive charges. Simultaneously, the appearance of stipoverite instead of coesite is possible.

According to this phenomenon, we considered in the previous paper (DERIBAS et al., 1966) three different regimes of stability of quartz, coesite and stipoverite according to the weight of explosive charge. The present investigation shows that the detection of different zones in one experiment is more correct. It is necessary to note that the appearance of coesite instead of stipoverite depends on many factors little understood at present.

3.2. $KAlSi_3O_8$

The behaviour of $KAlSi_3O_8$ in shock experiments (fig. 7) is similar to that of SiO_2 in general. The glass-like phase (gl') appeared in zones 2 and 3 with higher variable density (N up to 1.505). The relics of potassic feldspar with strongly destroyed lattice were observed. The presence of leucite and the variations of composition of glass were not found. The comparison of these data with the phase-diagram of a system K_2O–Al_2O_3–SiO_2 (SCHAIRER and BOWEN, 1955) shows that these effects are not simple melting at high temperature. The new high-pressure phase of $KAlSi_3O_8$ recently discovered (RINGWOOD et al., 1967) was not found definitely. Probably a very small quantity of this phase was formed in our experiments.

3.3. Other framework silicates

A similar situation in shock experiments with other framework silicates was observed. Some effects were determined for albite ($NaAlSi_3O_8$) similar to those for $KAlSi_3O_8$. A small quantity of jadeite was also found in shocked albite. The paper (MILTON and DE CARLI, 1963) contains a description of the transformation of anorthite $CaAl_2Si_2O_8$ into an X-ray-amorphous phase, with preservation of the shape of grains and even cleavage, similar to mascelenite in stone meteorites.

4. Discussion

It is possible to distinguish two types of phase-transformation in silicates (table 2). Type I "without axial zone" corresponds to framework silicates and SiO_2, type II "with axial zone" is characteristic of orthosilicates and some oxides. Type II may be divided into two subtypes: IIa "with decomposition", IIb "without decomposition". Naturally, this division is connected essentially with the conditions of shock compression, but in framework silicates and SiO_2 the axial zone has never been observed.

An intermediate picture was found in some other silicates. For instance, $MgSiO_3$ decomposed into Mg_2SiO_4 and X-ray-amorphous SiO_2, and the formation of glass $MgSiO_3$ with variable density was observed in the axial zone. In ferro-magnesian micas after shock compression, magnesian micas with the destroyed lattice, magnetite, native Fe and glass near to potassic felspatic composition were found.

Some transformation in shocked oxides are shown in tables 2 and 3. These transformations correspond to the high-temperature modifications (α-Al_2O_3), and to the high-pressure phases (Zr_2O_2, PbO). No transformations were found in some oxides (TiO_2) (BATSANOV et al., 1967). However, a defect structure with change of colour was observed in this case. This type of transformation may be singled out as type III, and the relaxation and disappearance of X-ray lines were usually found in these experiments.

Broadening of Laue reflections in shocked single crystals of several materials was found in some experiments. It shows that the single crystals transform into a fine grained powder at a definite shock pressure.

The similarity between the shocked and "metamict"

TABLE 2

Transformations of silicate and oxide powders under shock wave treatment in the cylindrical case

Type	Characteristic features	Examples	
		Initial minerals	New phases*
I	1. Indistinct or absent axial zone, i.e. unstable three-shock configuration	SiO_2 (quartz, glass)	Destroyed "quartz", s.r.o. phase of high density, traces of stishovite, rarely of coesite
	2. Formation of glass-like phases of variable density (without fusion)	Framework minerals: $KAlSi_3O_8$ (orthoclase) $NaAlSi_3O_8$ (albite) $CaAl_2Si_2O_8$ (anortite)	Destroyed "orthoclase", s.r.o. phase of high density, traces of high pressure phase?) s.r.o. phase, jadeite+SiO_2 s.r.o. phase (maskelinite)
IIa	1. Distinct axial zone, corresponding to Mach's three shock configuration	Silicates: $ZrSiO_4$ (zircone)	Destroyed (metamictic) zircone, SiO_2+ZrO_2 (monoclin.), glass.
	2. No glass-like phases, with partial or complete lattice deformation	$MgSiO_3$ (enstatite) $K(M,Fe)_3AlSi_3O_{10}(OH)_2$	SiO_2 (s.r.o.)+Mg_2SiO_4, glass Destroyed Mg-mica+$FeFe_2O_4$ or Fe+SiO_2+glass
IIb	a. with decomposition to constituents	Non-complex silicates and oxides: Mg_2SiO_4 (forsterite)	Fine-grade fracturing and partial deformation of the lattice	Traces of new phase high pressure (?)
	b. with polymorphic transformations	Al_2O_3 (α and γ)		α-Al_2O_3
III	No phase transformations; partial lattice deformation	TiO_2		No new phases

* Destroyed – phase with partially or completely destroyed lattice; s.r.o. – short-range order (glass-like) phase.

TABLE 3

Comparison of the shock wave transformations and transformations caused by irradiative treatment (LASTMAN, 1963)

Minerals	Type of irradiative treatment	Type of transformation caused by shock wave treatment	Analogy with shock treatment
SiO_2 (quartz, glass)	Fast neutrons up to 2.10^{20} neutr./cm²	I	Line broadening, decrease in SiO_2 density up to glass formation. Increase in glass density
$NaAlSi_3O_8$ $CaAl_2Si_2O_8$	Fast neutrons	I	Decrease in density up to formation of glass
ZrO_2 (monoclin.)	Fast neutrons up to 6.10^{19} neutr./cm²	IIb	Decrease in the density and transformation to a new modification (to high temperature, cubic one with irradiative treatment; high pressure, rhombic one with the shock wave treatment)
$ZrSiO_4$	α-particles up to 3.10^{-4} α-particles/atom	IIa	Lattice deformation up to the X-ray amorphous state with decrease in density, decomposition to SiO_2 (X-ray amorphous) and ZrO_2 (various modifications)
	Fast neutrons up to 3.10^{20} neutr./cm²	IIa	Decrease in density, disappearance of the far-order lines
Mg_2SiO_4	Fast neutrons	IIb	No observable change (except for disappearence of the weak lines)

$ZrSiO_4$, formed by natural radiation, was mentioned above. The same similarity can be stated for the other materials at conditions of shock compression (table 3). The radiation produced the basic effects of destruction of the lattice (LASTMAN, 1963). The main distinction of shock compression from radiation is the formation of high-pressure phases in shock experiments. It is possible to assume that the destruction of lattice in shock front is similar to that one produced by radiation.

Many investigations show that the basic processes characteristic of shock compression (destruction of lattice, formation of high-pressure phases, polymerisation) proceed in the short time of the existence of high pressure (about 10^{-6} sec) (BETSANOV et al., 1965; DERIBAS et al., 1967; ALTSHULER et al., 1967; ADADUROV et al., 1965). This is evidence of the abnormal speed of transformations in shock waves, exceeding by several orders the speed of the same processes under normal conditions. There is no common explanation of this anomaly in spite of some attempts in this direction (ALTSHULER et al., 1967; ADADUROV, 1965).

Our conception is that the lattice is destroyed completely by a shock wave with energy exceeding a definite critical value depending on the properties of the powder.

In this case the material transforms into some "state of activation" similar to a strongly compressed gas (GLASSTONE et al., 1941). This "state of activation" transforms into the glass-like phase for the framework silicates and SiO_2 and into the mixture of fine-grained crystalline phases for other materials under condition of high residual temperatures. In this case, destruction of the lattice and mixing of its elements creates the conditions for the formation of high-pressure and other phases. These phases form, possibly, in small quantities and transform partly into initial or metastable phases under the action of high residual temperature. As a rule, only the relics of these phases are observed, and the search of them is very complicated. The absence of high-pressure phases in the axial zone may be explained by the influence of residual temperature. Contrary to the axial zone, the intermediate zones are of the most interest. Possibly the using of oblique shock wave and the organisation of effective cooling will be useful for the increasing quantity of high-pressure phases after shock compression.

The absence of thermodynamic equilibrium in shock waves TROFIMOV, 1967 is the reason for the limited application of these experiments to geological problems, except possibly to the problem of meteorites. However, the principal possibility of obtaining diamonds, stipoverite, spinel-like modification of Mg_2SiO_4 and other high-pressure minerals by shock compression is very interesting. The destruction of the lattice, abnormal velocity of transformation and other phenomena of shock compression are also of interest in the general theory of solid phase transformations.

References

ADADUROV, G. A., A. N. DREMIN, G. I. KANEL and S. V. PERSHIN (1967) Phys. Combust. Explosion **2**, 281–285 (in Russian).

ADADUROV, G. A., I. M. BARKALOV, V. I. GOLDANSKY et al. (1965) Dokl. Akad. Nauk SSSR **165**, N 4 (in Russian).

ALTSHULER, L. V., N. N. PAVLOVSKY and V. P. DRAKIN (1967) Zh. Experim. i Teor. Fiz. **52**, Vyp. 2, 400–408 (in Russian).

BATSANOV, S. S., A. A. DERIBAS and E. V. DULEPOV et al. (1965) Phys. Combust. Explosion **4**, 78–82 (in Russian).

BATSANOV, S. S., G. K. BORESKOV, G. V. GRIDASOVA et al. (1967) Kinetika i Kataliz **8**, Vyp. 6 (in Russian).

BENDELIANI, N. A. et al. (1967) Geochem. N **6**, 677 (in Russian).

CHAO, E. C. T., J. I. FAHEY and I. LITTLER (1961) Science **133**, N 3456.

DERIBAS, A. A., N. L. DOBRETSOV, V. M. KUDINOV and N. I. ZYUZIN (1966) Dokl. Akad. Nauk SSSR **165**, N 4, 665–668 (in Russian).

DERIBAS, A. A., N. L. DOBRETSOV et al. (1967) Symposium H.D.P., Paris.

DE CARLI, P. S. and D. I. MILTON (1965) Science **147**, N 3654, 144.

DOBRETSOV, N. L., I. L. DOBRETSOVA, V. S. SOBOLEV and V. I. MALY (1968) Dokl. Akad. Nauk SSSR, N **4**, 910–913 (in Russian).

GLASSTONE, S., K. J. LAIDLER and H. EYRING (1941) *The theory of rate processes* (McGraw Hill, New York) 611 p.

LASTMAN B. (1963) Irradiation effects in uranium dioxide, in: I. Belle, ed., *Uranium dioxide, properties and nuclear application*, (U.S. Atomic Energy Commission).

McQUEEN, R. G., S. P. MARCH and J. N. FRITZ (1967) J. Geophys. Res. **72**, N 20, 4999–5036.

MILTON, D. I. and P. S. DE CARLI (1963) Science **140**, N 3567, 670–71.

RINGWOOD, A. E., A. F. REID and A. D. WADSLEY (1967) Acta Cryst. **23**, 1093–1095.

SCHAIRER, J. F. and N. L. BOWEN (1955) Am. J. Sci. **253**, 681–746.

TROFIMOV, V. S. (1967) Phys. Combust. Explosion **3**, N 4, 573–584.

PETROLOGY OF THE UPPER MANTLE AND LOWER CRUST

MINERALOGY OF PERIDOTITIC COMPOSITIONS
UNDER UPPER MANTLE CONDITIONS

D. H. GREEN and A. E. RINGWOOD

Department of Geophysics and Geochemistry, Australian National University, Canberra

An experimental study of the stability fields at high pressure of plagioclase peridotite, aluminous pyroxene (±spinel) peridotite and garnet peridotite, has been carried out in compositions matching estimates of the average undifferentiated upper mantle (pyrolite). The appearance of garnet at high pressures results from either of two complex reactions:

a. spinel + pyroxene ⇌ olivine + garnet,
   (solid solutions)

b. aluminous pyroxene ⇌ garnet + pyroxene (lower Al_2O_3).
   (solid solution)

In the model pyrolite composition, garnet appears at 21 kb (1100 °C) to 24 kb (1300 °C) by reaction (a) but at temperatures above 1300 °C spinel is absent from the low pressure assemblage and garnet develops by reaction (b) at pressures of 24 kb (1300 °C) to 31 kb (1500 °C). Reactions (a) and (b) have very different P, T gradients. In the garnet peridotite field, the amount of garnet is inversely proportional to the alumina content of the pyroxenes. This is confirmed by microprobe analyses of orthopyroxene (ranging from 6.0 to 2.2% Al_2O_3, 0.7–0.9% Cr_2O_3) coexisting with garnet, olivine and clinopyroxene. Coexisting clinopyroxenes have higher R_2O_3 content and exhibit a variable, temperature dependent enstatite solid solution.

The experimental data are applied to estimate mineralogical, density and seismic velocity variations along oceanic and continental geothermal gradients in a pyrolite upper mantle. Although the mineralogical variation may produce a low velocity zone, the alternate explanation of incipient or minor partial melting is preferred. Experimental data on the maximum stability limit of amphibole in pyrolite (with 0.1–0.2% water) are presented showing that amphibole breaks down to olivine + pyroxenes + garnet + fluid phase at approximately 29 kb, 1000 °C. The instability of amphibole at depths >80–90 km is considered to result in a zone of incipient or minor melting in a pyrolite upper mantle with 0.1–0.2% water. The nature of the liquid in this zone is considered to be highly undersaturated olivine basanite or olivine nephelinite.

1. Introduction

The uppermost region of the Earth's mantle acts as the immediate source region for basaltic magmas and is the one part of the mantle "sampled" by magmatic or tectonic processes and, in certain circumstances, exposed for direct petrological and geochemical investigation. The properties of the upper mantle can also be rather closely defined by geophysical methods, particularly by seismological and gravity studies. The weight of geological and geophysical evidence on upper mantle composition favours an overall peridotitic composition and arguments against an alternative of eclogitic composition have previously been presented (RINGWOOD and GREEN, 1966; RINGWOOD, 1966a). In seeking to define an upper mantle composition more closely the petrologist must interpret the variety of basaltic magmas, of peridotitic xenoliths in magmas and vent breccias, high-temperature peridotites, and other peridotite occurrences in terms of parental mantle compositions, partial melting products, crystal residues and crystal accumulates. No unique synthesis of these data is possible at this stage but in previous papers we have used the term "pyrolite" for parental mantle capable of yielding basaltic magmas by partial melting and advanced arguments for specific pyrolite compositions (GREEN and RINGWOOD, 1963; RINGWOOD, 1966a). Other authors (ITO and KENNEDY, 1967; KUSHIRO and KUNO, 1963; KUSHIRO et al., 1968; HESS, 1964; HARRIS et al., 1967) have derived possible upper mantle compositions on other grounds or have selected specific samples of peridotite xenoliths. These compositions also have essential Al_2O_3, CaO and Na_2O and, like the pyrolite compositions, may crystallize in four different mineral assemblages within the P, T conditions of the upper mantle (minor accessory minerals such as ilmenite, apatite, phlogopite are omitted):

a. Olivine + amphibole ± enstatite ("ampholite"),

b. Olivine + pyroxenes + plagioclase + chromite, ("plagioclase pyrolite")
c. Olivine + aluminous pyroxenes ± spinel, ("pyroxene pyrolite")
d. Olivine + pyroxenes + garnet ("garnet pyrolite").

2. Previous investigations

Experimental investigations of the stability relationships of these assemblages have been undertaken by several authors using either complex systems or simple, end member systems. Reactions between olivine and plagioclase to yield aluminous pyroxenes + spinel have been studied for the system Fo–An (KUSHIRO and YODER, 1966) and for the systems Fo–An, olivine (Fo_{92})-labradorite ($Ab_{41}An_{59}$), and for pyrolite composition (GREEN and HIBBERSON, 1969). These investigations show that plagioclase pyrolite will not occur as the stable mantle assemblage along typical geothermal gradients in either continental or oceanic regions but may be present in areas of particularly high heat flow and thin crust, e.g. if near-solidus temperatures are reached at depths of 25–35 km beneath mid-oceanic ridges. More interest centres on the conditions for the incoming of garnet and the role of garnet pyrolite and pyroxene pyrolite assemblages in upper mantle mineralogy.

MACGREGOR (1964, 1965) studied the reaction

$$\text{pyroxene} + \text{spinel} \rightleftharpoons \text{garnet} + \text{olivine} \quad (1)$$

in the system $MgO-Al_2O_3-SiO_2$ and in the system $MgO-CaO-Al_2O_3-SiO_2$. In the latter system MACGREGOR (1965) used pyroxene proportions (enstatite/diopside) of 75/25, 25/75 and 0/100. KUSHIRO and YODER (1965) studied the analogous reaction for enstatite:diopside = 50:50 and established a boundary for the incoming of garnet close to that found by Macgregor for the enstatite + diopside bearing mixtures. MACGREGOR (1964) did not report olivine on the low pressure side of the reaction (1) in the system $MgAl_2O_4 + 4MgSiO_3$ but did observe olivine in the low pressure assemblages (Ol + Al-pyroxene(s) + spinel) in the systems $2MgAl_2O_4 + CaMgSi_2O_6 + 6MgSiO_3$ and $2MgAl_2O_4 + 3CaMgSi_2O_6 + 2MgSiO_3$. The appearance of olivine results from a reaction

$$\text{pyroxene} + \text{spinel} \rightleftharpoons \text{aluminous pyroxene} + \text{olivine} \quad (2)$$

$$mMgSiO_3 + MgAl_2O_4 \rightleftharpoons (m-2)MgSiO_3 \cdot MgAl_2SiO_6 + Mg_2SiO_4 \quad (2a)$$

$$mCaMgSi_2O_6 + nMgSiO_3 + MgAl_2O_4 \rightleftharpoons mCaMgSi_2O_6 \cdot MgAl_2SiO_6 + (n-2)MgSiO_3 + Mg_2SiO_4 \quad (2b)$$

The apparent absence of reaction (2a) (MACGREGOR, 1964) contrasts with other data showing high solubility of Al_2O_3 in enstatite and with the presence of reaction (2b) (or an analogous reaction involving ($CaAl_2SiO_6$) solid solution) in the Ca-bearing system. The study of reaction (1) in the system $MgO-CaO-Al_2O_3-SiO_2$ by MACGREGOR (1965) utilized mixes containing both high and low pressure assemblages and the position of the reaction boundary was determined by changes in intensity of garnet and pyroxene reflections. However metastable growth of garnet from aluminous pyroxene (e.g. reaction (3))

$$mMgSiO_3 \cdot nMgAl_2SiO_6 \rightleftharpoons Mg_3Al_2Si_3O_{12} + (m-2)MgSiO_3 \cdot (n-1)MgAl_2SiO_6 \quad (3)$$

may have proceeded more rapidly than reaction of the aluminous pyroxene with olivine to yield less aluminous pyroxene and spinel. A difficulty involving metastable growth of garnet may account for some results of KUSHIRO and YODER (1966, fig. 2) in which, using a mix of An + Fo seeded with Ga + Fo + Cpx, they obtained some runs in which garnet and plagioclase disappeared in favour of Cpx + Opx + Fo + Sp but others at lower pressure in which the seed phases, Ga + Fo + Cpx, increased at the expense of An + Fo. Although there are no true reversals, it would seem that the boundary for reaction (1) should lie at higher pressures at 1100°–1300 °C, than shown by KUSHIRO and YODER (1966).

Other apparent discrepancies exist in the experimental data attempting to define the conditions for reaction (3), i.e. the breakdown of aluminous enstatite to garnet and less aluminous pyroxene. BOYD and ENGLAND (1964) and MACGREGOR and RINGWOOD (1964) carried out experiments homogenizing different garnet + pyroxene mixtures to a single pyroxene phase using pure pyrope and enstatite and natural pyrope-rich garnet and enstatite respectively. The natural enstatite in equilibrium with garnet contained 1.5–2 mol % more $(Al, Cr, Fe)_2O_3$ in solid solution than the enstatite in the MgO–Al$_2$O$_3$–SiO$_2$ system. RINGWOOD et al. (1964) and RINGWOOD (1966) integrated the data from the enstatite + pyrope studies with the pyrolite (RINGWOOD 1966a) composition* to infer stability fields for the assemblages olivine + pyroxenes + spinel, olivine + aluminous pyroxenes, and olivine + pyroxenes + garnet. Direct experimental study (GREEN and RINGWOOD, 1967a) of the pyrolite III revealed a much lower solubility of Al$_2$O$_3$ in orthopyroxene than that predicted from the study of the simple systems—this resulted in persistence of spinel to higher temperature ($\approx$1200 °C, 11 kb to $\approx$1300 °C, 23 kb) than the 900 °C predicted by RINGWOOD et al. (1964) and in appearance of garnet from breakdown of aluminous pyroxene at lower pressures ($\approx$26.5 kb at 1400 °C) than previously predicted ($\approx$39 kb at 1400 °C). These surprisingly large differences imply that the presence of coexisting olivine or clinopyroxene considerably alters the solubility of Al$_2$O$_3$ in orthopyroxene—this can readily occur if the coexisting phases can enter solid solution in either garnet or orthopyroxene. The addition of clinopyroxene to the assemblage obviously establishes new partition relations involving grossular solid solution in garnet and diopside solid solution in enstatite. However MACGREGOR and RINGWOOD (1964) found higher solubilities of garnet in orthopyroxene for the natural enstatite (0.82% CaO) + garnet (5.21% CaO) mixes than for the pure enstatite + pyrope system. MACGREGOR and RINGWOOD (1964) homogenized mixtures of 50% garnet, 50% enstatite to a pyroxene phase at 30 kb, 1500 °C implying that the pyroxene contained 3.0% CaO (by weight)—this either exceeds or is very close to saturation in diopside solid solution for these P–T conditions (DAVIS and BOYD, 1966; GREEN and RINGWOOD, 1967b). Thus it does not appear that diopside solid solution greatly inhibits Al$_2$O$_3$ solubility in enstatite. If the addition of olivine to garnet–enstatite assemblages is the factor inhibiting Al$_2$O$_3$ solubility then it is implied that olivine has a significant solid solubility in enstatite. This possibility requires further investigation but it may be noted that KUSHIRO (1964) reported olivine solid solution in diopside and BOYD and ENGLAND (1960) noted that natural enstatites from peridotites had values of the molecular ratio $M^{2+}/Si > 1$. The observation of spinel exsolution lamellae within ensta-

* This composition is denoted as pyrolite III in the rest of this paper (see table 1).

tite of lherzolite nodules and some intrusive peridotites also implies non-stoichiometric composition of the orthopyroxene (cf. reaction (2)).

MACGREGOR (1967) briefly reported data on mixtures of natural orthopyroxene + clinopyroxene + spinel using separated minerals from lherzolite inclusions. Two assemblages differing mainly in Cr$_2$O$_3$/Al$_2$O$_3$ ratio were used and the preliminary data indicate that high Cr$_2$O$_3$/Al$_2$O$_3$ ratios of spinel will have a marked effect on stabilizing the olivine + orthopyroxene + clinopyroxene + spinel assemblages to higher pressures. ITO and KENNEDY (1967) presented data on the solidus and melting relations in a garnet lherzolite nodule from kimberlite. They obtained solidus temperatures approximately 100 °C lower than those of GREEN and RINGWOOD (1967a) but the pressure obtained for the incoming of garnet ($\approx$23 kb, 1320 °C) agrees with that obtained by GREEN and RINGWOOD (1967a) on pyrolites I and II*. ITO and KENNEDY (1967) did not obtain a field of olivine + aluminous pyroxenes (without spinel) although this would be anticipated from a comparison of their composition with that of GREEN and RINGWOOD (1967a) for pyrolites I, II and III. Ito and Kennedy used a natural garnet lherzolite as starting material—this presents difficulties in representative sampling of the initial coarse-grained rock and in the avoidance of sample composition change during the extremely fine-grinding needed to ensure some chance of reactions proceeding to equilibrium.

From the preceding summary of published data, it is apparent that there is as yet no general agreement on the P, T coordinates of reactions (1), (2) and (3), either in simple systems or in complex natural peridotitic compositions. More detailed work is required, attempting to eliminate the possibility of metastable reactions or other experimental uncertainties. From the data already obtained, it is clear that the reaction (1) is one in which the pyroxene will vary in composition along the reaction boundary. This boundary is actually the locus of intersections of reactions (2) and (3) which define the curves of constant Al$_2$O$_3$ content of pyroxene in peridotitic compositions. The position of reaction (1) will also depend sensitively on the spinel composition

* This agreement may be fortuitous as GREEN and RINGWOOD (1967a) applied a (-10%) pressure correction while ITO and KENNEDY (1967) applied no pressure correction. It should also be noted that MACGREGOR (1964, 1965, 1967) did not apply a pressure correction.

(Cr_2O_3/Al_2O_3 ratio) in complex compositions and this will vary due to coupled reaction (2) with pyroxene.

With the complexity of the reactions in mind, there is much to be said for the direct experimental testing of model pyrolite compositions. In the following sections the data obtained on the model pyrolite compositions of GREEN and RINGWOOD (1963) and RINGWOOD (1966a) are summarized and emphasis is placed on the attempts to establish reaction boundaries by unequivocal reversals.

3. Experimental data on the incoming of garnet in pyrolite

3.1. Experimental methods

The experimental compositions (table 1) were prepared from AR grade chemicals, carefully ground and reacted together under high temperature reducing conditions and then analyzed for FeO and Fe_2O_3 contents. The initial mixes were extremely finegrained (≤ 1 μ) and consisted of olivine, clinopyroxene, orthopyroxene, plagioclase and minor chromite and ilmenite. To facilitate identification of minor phases and of the presence or absence of small degrees of partial melting, compositions were prepared which are equivalent to the pyrolite compositions of table 1 after extraction of 50% olivine ($Mg_{91.5}Fe_{8.0}Ni_{0.4}Mn_{0.1}$) from pyrolites I and II and after extraction of 40% olivine ($Mg_{91.6}Fe_{8.1}Ni_{0.2}Mn_{0.1}$) from pyrolite III. In all experiments conducted with these modified pyrolite compositions, excess olivine was present. Hence the above procedure did not affect in any way the equilibrium relationships which are discussed below.

Crystallization of the experimental compositions was carried out using a single-stage, piston-cylinder apparatus and a pressure correction of -10% was applied to the nominal pressure for all runs. Samples were run in both platinum capsules and in graphite capsules. The run conditions using platinum capsules were such that iron loss to the Pt capsule was less than 25% of the amount present. Analyses of samples after experimental runs are listed in table 1a, the average iron content is 5.7% FeO, 0.7% Fe_2O_3 yielding a normative composition with approx. 33% Ol and 30% enstatite. The worst examples (in table 1a) give normative olivine of approximately 31% and 33% enstatite.

Microprobe analyses across a polished sample (36 kb,

TABLE 1

Chemical compositions and CIPW norms of model pyrolite compositions used in experimental runs. Pyrolite I and pyrolite III refer to the model compositions calculated by GREEN and RINGWOOD (1963) and RINGWOOD (1966a) respectively. They differ principally in their MgO/SiO_2 ratios and thus in pyroxene/$(Al, Cr)_2O_3$ and pyroxene/olivine ratios. Pyrolite II is a composition intermediate between the two in which the enstatite/olivine ratio of pyrolite I was increased without appreciable change in the R_2O_3 content

	Pyrolite I	Pyrolite II	Pyrolite III	Pyrolite III less 40% olivine
SiO_2	43.20	43.95	45.20	47.84
TiO_2	0.58	0.57	0.71	1.18
Al_2O_3	4.01	3.88	3.54	5.90
Cr_2O_3	0.42	0.41	0.43	0.72
Fe_2O_3	0.35	0.75	0.48	0.80
FeO	7.88	7.50	8.04	8.21
MnO	0.13	0.13	0.14	0.13
NiO	0.39	0.39	0.20	0.18
MgO	39.54	39.00	37.48	28.73
CaO	2.67	2.60	3.08	5.14
Na_2O	0.61	0.60	0.57	0.95
K_2O	0.22	0.22	0.13	0.22
100 Mg/$(Mg+Fe^{2+})$ atomic ratio	89.9	90.3	89.2	86.5
CIPW Norm				
Or	1.1	1.1	0.8	1.3
Ab	5.2	5.2	5.0	8.3
An	7.5	7.3	6.6	11.0
Di	4.6	4.3	6.8	11.3
Hy	3.8	9.4	15.8	26.4
Ol	75.6	69.8	62.5	37.5
Ilm	1.1	1.1	1.3	2.2
Mt	0.5	1.2	0.7	1.1
Chr.	0.6	0.6	0.6	1.0

TABLE 1a

Effect on FeO, Fe_2O_3 contents of pyrolite III less 40% olivine of loss of iron to platinum capsules during experimental runs (E. Kiss, analyst)

P (kb)	T (°C)	Time (hrs)	% FeO	% Fe_2O_3	% total Fe as FeO
27	1500	0.33	5.62	0.38	5.96
29.3	1500	0.33	5.18	0.78	5.88
30.4	1500	0.33	5.11	1.03	6.04
29.3	1400	1.0	5.10	0.54	5.59
30.4	1400	1.0	5.56	0.75	6.24
24.8	1300	2.0	5.21	0.56	5.71
20.3	1200	2.0	6.89	1.01	7.80
22.5	1100	4.0	7.17	0.75	7.70

1500 °C, 20 mins, pyrolite III less 40% olivine) showed average Fe as FeO content of 6.1% (i.e. sample 100 Mg/(Mg+Fe) = 89.5) at the edge against platinum and Fe as FeO content of 7.0% (i.e. sample 100 Mg/(Mg+Fe) = 88) in the centre of the sample. Microprobe analyses of olivines and orthopyroxenes from runs in Pt capsules yield values of 100 Mg/(Mg+Fe) = 90–91.5, whereas from subsolidus runs in graphite capsules the olivine and orthopyroxene had 100 Mg/(Mg+Fe) = 87–88. Runs across the pyroxene pyrolite/garnet pyrolite boundary at 1350 °C and a reversal on the boundary at 1400 °C were carried out in graphite capsules. The use of graphite capsules in longer runs allowed access of water producing depression of the solidus and the presence of amphibole in runs at 1100 °C, 22.5 kb and 20.5 kb. These effects were eliminated by repeating these runs in sealed Pt capsules. To summarize, the use of sealed Pt capsules appears necessary for reliable determination of the solidus and prevention of growth of amphibole at low temperatures. The use of platinum causes slight iron loss changing the overall sample 100 Mg/(Mg+Fe) ratio by 2–3% but this does not perceptibly affect the position of the phase boundary marking the appearance of garnet as subsolidus runs in graphite and platinum are mutually consistent. The iron loss also causes a small increase in normative Hy/Ol ratio and thus in pyroxene/R_2O_3 ratio. It may be noted that the effect of Fe increase due to experiments in Fe capsules by ITO and KENNEDY (1967) appears to be similar in magnitude (modal olivine $\approx Fo_{95}$ but experimental olivines in subsolidus assemblages range from Fo_{88}–Fe_{91}) but has the opposite effect in decreasing normative Hy/Ol and pyroxene/R_2O_3 ratios.

Samples were examined optically and by X-ray powder diffraction. Orthopyroxene forms quite large tabular porphyroblasts; garnet is commonly subhedral but contains many inclusions at lower temperatures and olivine and clinopyroxene form small anhedral grains. Spinel occurs as small, equant, green, isotropic grains and differs from ilmenite (ilmenite+geikielite solid solution) in that the latter is commonly elongate, brown, translucent and with high birefringence. The orthopyroxene crystals are readily analyzed by electron microprobe techniques and some data have been obtained on olivine, clinopyroxene and garnet compositions. The presence of small amounts of amphibole required detection by X-ray powder diffraction photographs.

In the experiments on the stability of amphibole, the methods used should outline the *maximum* stability of amphibole at 1000 °C and were designed to specifically test the model that pyrolite contains 0.1–0.2% H_2O at shallow mantle depths. The use of a very small water content also avoids the problems of large degrees of melting and depression of the solidus, and of the possibility of solubility of components in a vapour phase. The starting material for these runs was prepared by adding 0.3% H_2O to the "pyrolite III less 40% olivine" mix. This mix was run at 10.8 kb, 900 °C for $4\frac{1}{2}$ hrs yielding a fine-grained assemblage of olivine, pyroxenes and amphibole ($\approx 15\%$). This was then rerun at 1000 °C and at pressures of 18 kb to 29 kb and examined carefully by optical and X-ray methods for the disappearance of amphibole and the appearance of garnet.

3.2. *Experimental results*

The results of the determination of the synthesis fields for garnet pyrolite and pyroxene pyrolite are presented in fig. 1. The data points denote the phase assemblages present in pyrolite III composition. Plagioclase pyrolite is stable under dry conditions on the low pressure side of AB. Between AB and ELF, both garnet and plagioclase are absent and the mineral assemblage is dominated by aluminous pyroxenes. Within this field, spinel is present as a minor phase at temperatures below the line marked K but only olivine, aluminous pyroxenes and accessory ilmenite are present at higher temperatures. Garnet first appears in trace amounts along the line ELF and, at a given temperature, steadily increases in abundance as pressure increases.

The data points for pyrolites I and II are not shown in fig. 1 but in these compositions the first appearance of garnet is along EJ and spinel remains up to solidus temperatures on the low pressure side of EJ. Garnet and spinel co-exist together over a very small pressure interval on the high pressure side of EJ. The triangular P, T field FLJ is one in which the olivine+aluminous pyroxenes assemblage occurs in pyrolite III but olivine+aluminous pyroxenes+garnet occurs in pyrolites I and II.

The boundary for the appearance of garnet was apparently defined by different reactions at low and high temperatures and it was desirable to obtain reversals

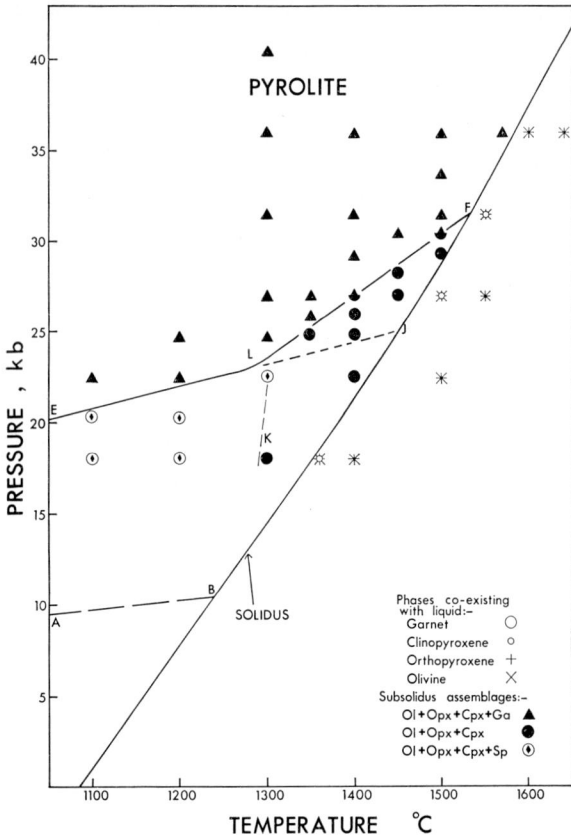

Fig. 1. Experimental runs on pyrolite III composition. Garnet is absent on the low pressure and present on the high pressure side of ELF. Spinel is absent on the high temperature side of the line K. In pyrolites I and II garnet is present on the high pressure side of ELJ. (From Green and Ringwood, 1967a).

across these boundaries as previous data (Boyd, 1960) had demonstrated difficulty in nucleation of magnesian garnet near its low pressure stability limit. Reversals involving complete disappearance of garnet on the low pressure side of the boundary, were achieved at 1400 °C and 1200 °C (table 2). The starting mix for the 1400 °C reversals was a large capacity run at 36 kb, 1000 °C, 3 hrs yielding fine grained olivine and pyroxene and very poikilitic garnet. For the 1200 °C reversal, the starting material was a 50:50 mix of runs carried out at 1200 °C, 18 kb, 2 hrs and 1200 °C, 27 kb, 2 hrs. The reversals at 1400 °C and 1200 °C confirm the positions of the boundaries established from the olivine + pyroxenes + plagioclase + chromite starting mixtures. The analysis of the roles of spinel and aluminous pyroxenes gives a theoretical explanation for the change in slope of the boundary marking the appearance of garnet in pyrolite III and the absence of such a change in slope of the boundary in pyrolites I and II. Microprobe analyses of orthopyroxene (table 4) show regular variations in Al_2O_3 and $CaSiO_3$ solid solution which are in themselves very good evidence that the experimental assemblages closely approach equilibrium.

In the runs (table 3) on the olivine + amphibole + pyroxenes mix, amphibole breaks down finally between 27 kb and 28.8 kb at 1000 °C. There is a transitional assemblage of olivine + pyroxenes + garnet + amphibole from 23.5 kb to 28 kb. In this assemblage the amount of garnet is less than in lower pressure runs. Experiments on the stability of amphibole in basaltic compositions (Essene et al., 1970) have demonstrated that amphibole may form readily from (glass + H_2O) mixtures and persist metastably at high pressures and at 700–800 °C beyond its stability field as defined by growth in garnet + pyroxene + amphibole + water mixtures. This possibility has not been fully tested in the

TABLE 2

Results of experimental runs aimed at reversal of the boundary for appearance of garnet from olivine + aluminous pyroxenes and from olivine + spinel + pyroxenes assemblages

Run no.	Capsule	Pressure (kb)	Temperature (°C)	Time (hrs)	Starting material	Products
1096	Graphite	27.0	1450	1	Ol + Opx + Cpx + Ga	Ol + Opx + ?Cpx + Melt
1098	Graphite	28.1	1450	1	Ol + Opx + Cpx + Ga	Ol + Opx + Cpx + Melt
1099	Graphite	25.9	1400	1	Ol + Opx + Cpx + Ga	Ol + Opx + Cpx
1055	Pt	25.9	1400	1	Ol + Opx + Cpx + Plag + Chromite	Ol + Opx + Cpx
1100	Graphite	27.0	1400	2	Ol + Opx + Cpx + Ga	Ol + Opx + Cpx + rare Ga
1016	Pt	27.0	1400	1	Ol + Opx + Cpx + Plag + Chromite	Ol + Opx + Cpx + rare Ga
2289	Pt	20.7	1200	4	50% (Ol + Opx + Cpx + Ga) 50% (Ol + Opx + Cpx + Sp)	Ol + Opx + Cpx + rare Spinel
2290	Pt	22.5	1200	4	50% Ol + Opx + Cpx + Ga 50% Ol + Opx + Cpx + Sp	Ol + Opx + Cpx + Ga (minor garnet but with euhedral form)

TABLE 3

Experiments on the stability of amphibole in pyrolite with low water content. All runs used "pyrolite III less 40% olivine" mix containing 0.3% H_2O and all were carried out in *sealed* Pt capsules

Run no.	Pressure (kb)	Temperature (°C)	Time (hrs)	Starting material	Products
2170	18	1100	7	Ol+Amph+Px	Ol+Amph+Px
2173	19.8	1000	15	Ol+Amph+Px	Ol+Amph+Px
2168	22.5	1000	17	Ol+Amph+Px	Ol+Amph+Px
2189	24.8	1000	12	Ol+Amph+Px	Ol+Amph+Px+rare Ga
2338	25.2	1000	26	Ol+Amph+Px	Ol+Amph+Px+rare Ga
2339	27.0	1000	24	Ol+Amph+Px	Ol+Amph+Px+minor Ga
2344	28.8	1000	24	Ol+Amph+Px	Ol+Px+common Ga+ trace amphibole
2347	25.2	1000	48	90% Ol+Amph+Px 10% Ol+Px+Ga	Ol+Px+minor Ga+minor amphibole

present experiments although the higher temperature of the runs and the persistence of amphibole in a 48 hr run seeded with garnet+pyroxene+olivine at 25.2 kb, 1000 °C suggest that the effect may be small. Another difference from the experiments on basaltic amphibolites is that the experiments on pyrolite are devised with a specific model i.e. mantle composition with 0.1–0.2% water (RINGWOOD, 1969), in mind and the problem of partial melting at low temperatures for $P_{H_2O} = P_{total}$ has been avoided. KUSHIRO et al. (1968) reported data on melting of a lherzolite nodule under anhydrous and hydrous conditions and found the solidus to be ≈ 1000 °C for $P_{H_2O} = P_{total}$ at 20–30 kb. They did not report amphibole at 26 kb, 980 °C but apparently used optical methods only in identification of phases and minor amphibole may have been missed.

Partial analyses by electron probe micro-analyser are listed in table 4. The orthopyroxenes, because of their porphyroblastic habit, could be analysed with similar accuracy to those in partial melting runs (GREEN and RINGWOOD 1967b) but the finer grain size of olivine and clinopyroxene and the poikilitic nature of the garnet made analysis more difficult. The figures given in table 4 include only those grains where two or more consecutive analyses (2–3 μ steps) gave consistent analytical data. The data show a very clear pattern of decreasing Al_2O_3 solubility in orthopyroxene with increasing pressure within the garnet pyrolite field. The Al_2O_3 content of orthopyroxene from pyrolite III at the incoming of garnet (boundary LF, fig. 1) is 6.0 ± 0.2%. Coexisting clinopyroxene contains approximately 7.7% Al_2O_3 and the Cr_2O_3 contents are 0.9% Cr_2O_3 (orthopyroxene) and ≥1.0% Cr_2O_3 (clinopyroxene). The higher $Al_2O_3+Cr_2O_3$ content of the clinopyroxene is due to $NaR^{3+}Si_2O_6$ solid solution. The minimum Al_2O_3 content of analyzed orthopyroxene is 2.2% at 40.5 kb, 1300 °C. Clinopyroxene contains 4.6% at 36 kb, 1300 °C in equilibrium with orthopyroxene with 2.6% Al_2O_3. The degree of mutual solid solution between pyroxenes shows a consistent pattern although the difficulty of analysing the small clinopyroxenes prevents any confident statement on limits of pyroxene solid solution in the complex composition. Orthopyroxene ($\approx$ Ens_{90}) coexisting with clinopyroxene contains approximately 2.4% CaO at 1500 °C, 2.2% CaO at 1400 °C, 2.0% CaO at 1350 °C and 1–5% CaO at 1300 °C. Coexisting clinopyroxene contains ≥11.4% CaO at 1400 °C and ≥14.5% CaO at 1350 °C – this is a higher CaO content than in clinopyroxene obtained at similar temperatures at 9–18 kb from more iron-rich basaltic compositions (GREEN and RINGWOOD, 1967b, p. 136–8). The garnets are more Fe-rich than coexisting clinopyroxene and olivine and contain 4.5–5.0% CaO and $\approx 1.7\%$ Cr_2O_3. Higher values of CaO (6.9% and 5.6%) are probably due to small clinopyroxene inclusions.

The experimental data on pyrolite III composition are summarized in fig. 2 in which we have plotted curves showing maximum solubility of Al_2O_3 (with $\approx 0.8\%$ Cr_2O_3 in addition) in orthopyroxene coexisting with garnet, olivine and clinopyroxene. Data are inadequate to draw similar curves in the olivine+py-

TABLE 4

Partial analyses, using electron probe microanalyser, of crystalline phases from pyrolite

Temperature (°C)	Pressure (kb)	Time (hrs)	Sample capsule	Phase analysed	Analysed constituents					Garnet content of assemblage	Remarks
					Al_2O_3	Cr_2O_3	FeO	CaO			

Pyrolite III less 40% olivine

1500	22.5	0.33	Pt	Opx	3.5	—	4.0	2.2	Nil	≈50% melting, Ol+Opx+Quench, Opx > Ol
1500	27.0	0.33	Pt	Opx	5.4	—	5.2	2.4	Nil	Minor melting, Ol+Opx+Cpx+Quench
1500	29.3	0.33	Pt	Opx	5.7	0.9	5.2	2.4	Nil	No identifiable melting, Ol+Opx+Cpx
1500	30.4	0.33	Pt	Opx	5.9	0.8	5.3	2.4	Trace	No identifiable melting, Ol+Opx+Cpx+Ga
				Ol	≤0.5		8.5	≤0.5	Trace	
1500	31.5	0.33	Pt	Opx	5.6	0.7	5.7	2.4	Minor	Ol+Opx+Cpx+Ga
				Ga	23.0	1.7	—	4.9		
1500	33.8	0.33	Pt	Opx	4.8	—	5.2	2.3	Minor	Ol+Opx+Cpx+Ga
1500	36.0	0.33	Pt	Opx	3.9	—	5.5	2.3	Common	Ol+Opx+Cpx+Ga
				Ga	>16	—	≈8.0	≈4.3		
				Whole sample-centre	6.1	—	7.0	5.2		
1450	27.0	1.0	Pt	Opx	5.4	—	4.2	2.3	Nil	No identifiable melting, Ol+Opx+Cpx
				Ol	0.1	—	8.6	0.3	Nil	
1400	22.5	0.66	Pt	Opx	3.9–5.1	—	4.8–5.7	2.4	Nil	Local patchy melting, Ol+Opx+Cpx
1400	24.8	1.0	Pt	Opx	4.3	—	5.1	2.4	Nil	Minor melting ?water access, Ol+Opx+Cpx
1400	25.9	1.0	Pt	Opx	5.0	0.8	5.6	2.2	Nil	No identifiable melting, Ol+Opx+Cpx
				Cpx	≥5.1	>1.0	—	≥10.0		
1400	27.0	1.0	Pt	Opx	5.5	—	5.8	2.1	Trace	No identifiable melting, Ol+Opx+Cpx+Ga
				Cpx	6.6	—	4.8	11.4		
				Ga	22.0	1.7	6.8	5.3		
1350	24.8	2.0	Graphite	Opx	6.1	—	7.0	2.1	Nil	No garnet, no melting, Ol+Opx+Cpx
				Cpx	7.7	—	5.8	14.5		
				Ol	≤0.2	—	11.6	≤0.2		
1350	25.9	2.0	Graphite	Opx	5.5	—	7.1	2.1	Minor	No melting, Ol+Opx+Cpx+Ga
				Cpx	7.3	—	5.6	14.5		
				Ol	≤0.6	—	12.0	≤0.4		
				Ga	≥20.2	—	7.9	6.9		
1350	27.0	2.0	Graphite	Opx	4.9	—	7.1	1.9	Minor	No melting, Ol+Opx+Cpx+Ga
				Cpx	6.7	—	5.4	≥14.1		
1300	36.0	4	Graphite	Opx	2.6	—	7.2	1.5	Common	Ol+Opx+Cpx+Ga
				Cpx	4.6	—	4.7	≥13.9		
				Ol	≤0.3	—	12.5	≤0.2		
				Ga	23.5	—	8.2	4.3		
1300	40.5	4	Graphite	Opx	2.2	—	7.2	1.5	Common	Ol+Opx+Cpx+Ga
				Ga	≥20.9	—	8.7	5.6		

Pyrolite I less 50% olivine

| 1400 | 22.5 | 1 | Pt | Opx | 3.0 | — | 4.2 | 2.2 | Nil, uncommon spinel | Minor melting, Ol+Opx+Cpx+spinel |
| 1400 | 24.8 | 1 | Pt | Opx | 5.9 | — | 5.0 | 2.1 | Trace garnet, uncommon spinel | No melting, Ol+Opx+Cpx+Ga+Spinel |

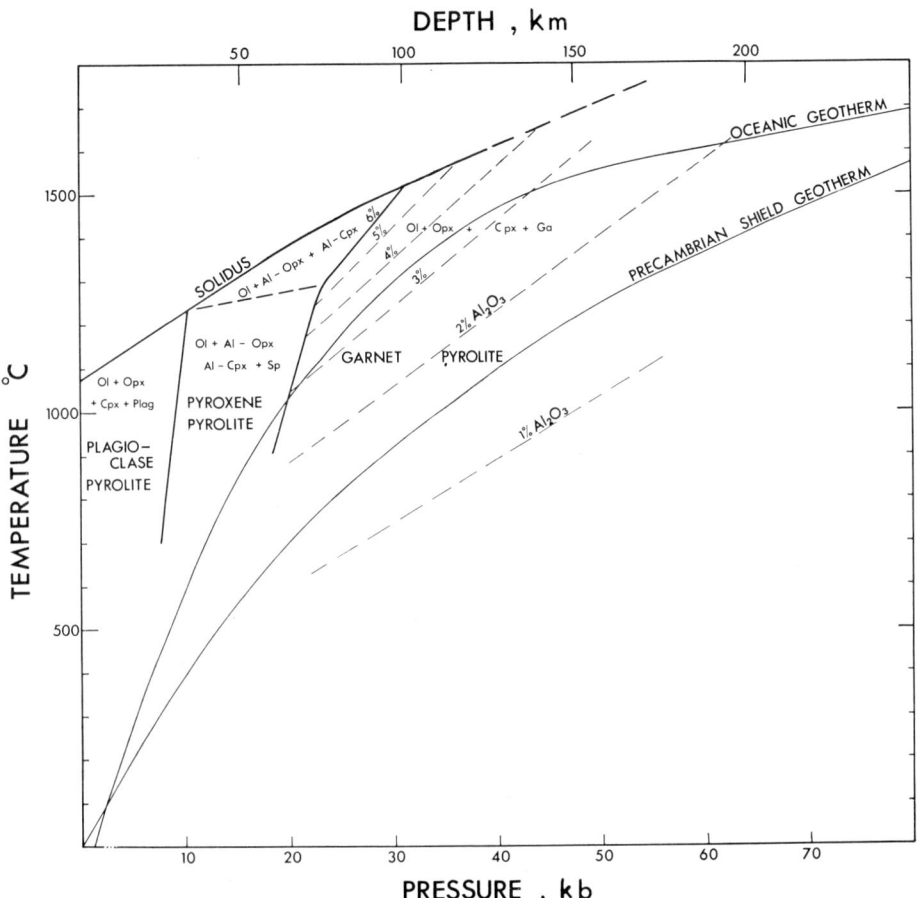

Fig. 2. Diagram illustrating the P, T fields of different mineral assemblages in pyrolite III composition. The figures 1%Al₂O₃, 2%Al₂O₃ etc. refer to the Al₂O₃ content of orthopyroxene in equilibrium with garnet in the garnet pyrolite field. The oceanic and Precambrian shield geotherms are those illustrated by RINGWOOD et al. (1964). (From GREEN and RINGWOOD, 1967a.)

roxenes + spinel field but such curves, combined with data on the pyroxene miscibility gap, will ultimately provide a very useful means of estimating the P, T conditions of equilibration of natural spinel and garnet lherzolite (O'HARA, 1967). With variation in bulk composition, the conditions for appearance of garnet will vary; compositions with higher Al₂O₃/pyroxene ratio will yield spinel up to solidus temperatures and garnet will appear at 23–24 kb at the solidus. For a bulk composition such that, with all Al₂O₃ in pyroxenes, the orthopyroxene would contain only 3.0% Al₂O₃, then spinel would be expected to disappear above 1000 °C, 10–18 kb and there would be a greatly expanded field of olivine and aluminous pyroxenes. Garnet would not appear on the solidus of such a composition until pressures of approximately 60 kb.

4. Mineralogy in a pyrolite upper mantle

In the following discussion, it is assumed that the average composition of the upper mantle beneath oceanic and geologically "young" regions is that of pyrolite III. In continental regions and particularly in Precambrian shield regions, there is probably a much greater proportion of refractory residual peridotite at shallower levels and the upper mantle may be chemically zoned. Mineralogical variation in the upper oceanic mantle is thus determined by the intersection of geothermal gradients with the stability fields of fig. 1. The variation of temperature in the continental and oceanic crusts and upper mantle has been discussed by CLARK (1962) and CLARK and RINGWOOD (1964). It is clear from these papers that there is a large difference between oceanic and stable shield geotherms and that this difference takes the form illustrated in fig. 2. Nevertheless, the specific form of the curves, particularly their "convexity" and closeness of approach to the pyrolite solidus, depend sensitively on knowledge of the magnitude of radiative heat transfer in the upper mantle. No unique solution for geothermal gradients is currently possible but a general evaluation of the role of mineralogical zoning in the upper mantle can best be obtained by consideration of the two examples in fig. 2.

Along the Precambrian shield geotherm, the probability of chemical zoning would limit the possible phase assemblages. If rocks of composition approaching pyrolite occur locally, then there may be an extremely limited zone near the base of the crust where these would crystallize to olivine + orthopyroxene (1% Al_2O_3) + clinopyroxene + spinel assemblages. Similar compositions below about 35–40 km would yield olivine + orthopyroxene (1–2% Al_2O_3) + clinopyroxene + garnet assemblages and it may be noted that the garnet content would be relatively high (e.g. 12% garnet in pyrolite III composition). No regular change in mineralogy would occur for rocks of pyrolite composition along the Precambrian shield geotherm.

The picture is very different along the oceanic geotherm. The olivine + orthopyroxene + clinopyroxene + spinel assemblage is stable in pyrolite composition to depths of 60–70 km. Within this interval the amount of spinel would decrease and the Al_2O_3 content of pyroxenes would increase with increasing depth (reaction (2)). Prior to the incoming of garnet at 60–70 km and about 1000 °C, aluminous spinel would coexist with orthopyroxene containing about 3% Al_2O_3. An intersection of the geotherm with the phase boundary at a higher temperature than that illustrated in fig. 2 would yield assemblages with less spinel and with orthopyroxene of higher Al_2O_3 content (4–5% Al_2O_3).

An extremely steep geothermal gradient, possibly realized only in regions actively producing basaltic magmas, would be required to enter the olivine + aluminous enstatite (6% Al_2O_3) + aluminous clinopyroxene field. In regions of partial melting and magma generation, such gradients must be attained and it may be noted that garnet does not appear on the pyrolite III (anhydrous) solidus until depths > 100 kms are reached.

At depths of 60–70 km on the oceanic geotherm, garnet appears from reaction (1) and is in equilibrium with orthopyroxene containing about 3% Al_2O_3. It is estimated that about 6% garnet would appear in pyrolite III composition at 60–70 km on the geothermal gradient illustrated. If the geothermal gradient intersected the boundary at a higher temperature, the amount of garnet appearing would be correspondingly less, e.g. about 3–4% garnet coexisting with orthopyroxene containing about 4.5% Al_2O_3.

The incoming of garnet due to reaction (1) at about 60–70 km in the oceanic mantle probably occurs over a relatively small depth interval (5–15 km). With further penetration along the geotherm into the garnet pyrolite field the amount of garnet may actually *decrease* – this will occur for temperature gradients steeper than the lines of constant Al_2O_3 content of orthopyroxene shown in fig. 2. For the gradient shown, pyrolite III will contain about 5% garnet at depths between 90 and 120 km and the mineralogy will remain constant over this interval. At depths greater than 120 km the geothermal gradient becomes increasingly transgressive to the lines of constant Al_2O_3 content for orthopyroxene. Thus, along this part of the geothermal gradient, the aluminous pyroxenes will gradually break down to yield an increasing garnet content and low-alumina pyroxenes. At depths of 200–250 km the assemblage of pyrolite III will probably contain 11–12% of garnet.

The transition from aluminous pyroxenes + spinel pyrolite to garnet pyrolite at depths of 60–70 km in the oceanic mantle is in agreement with MACGREGOR's (1964) data and with the conclusions of ITO and KENNEDY (1967) on the stability of spinel and garnet-bearing peridotite. Although the present work does not support earlier conclusions (RINGWOOD et al., 1964; MACGREGOR and BOYD, 1964) that garnet pyrolite would not in general be stable until depths of 120–150 km, it provides excellent confirmation of the importance at this depth interval of the breakdown of aluminous pyroxenes to garnet + low alumina pyroxenes, the reactions on which these earlier conclusions were based. It should be pointed out that for pyrolite-like compositions with higher pyroxene/(Al, Cr)$_2$O$_3$ ratios than pyrolites I, II and III, the field of garnet pyrolite may not be entered until depths in excess of 120 km are reached – this is particularly relevant for mantle regions from which basaltic fractions have been removed (cf. MACGREGOR, 1967). Geothermal gradients steeper than that

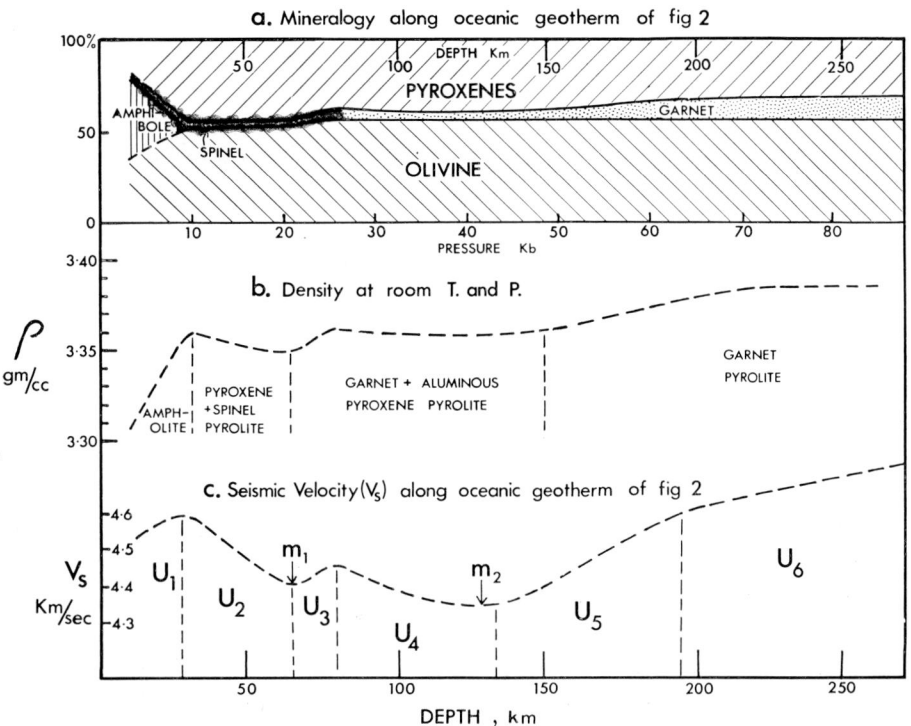

Fig. 3. Diagram illustrating the changes in mineralogy, density (room T and P) and the relative changes in seismic velocity (V_s) along the oceanic geotherm. The mantle is assumed to be of pyrolite III composition and the critical gradient $(\partial T/\partial P)_{V_s} = 4.5$ °C/km. (From GREEN and RINGWOOD, 1967a.)

illustrated will have a similar effect in diminishing the amount of garnet produced at 60–70 km depth by reaction (1) and increasing the amount of garnet produced at deeper levels from reaction (3).

This brief discussion serves to illustrate the application of our experimental data to a more flexible range of mantle compositions and geothermal gradients than implied in fig. 2. The effects of the mineralogical variation along the oceanic geothermal gradient in causing variations of seismic velocity have been previously discussed by GREEN and RINGWOOD (1967a) and fig. 3 is reproduced from this paper.

The previous discussion and the data presented in fig. 3 illustrate the principle that both mineralogical changes and P, T effects on V_s are of primary importance in determining seismic velocity distributions in the upper mantle and may in themselves produce velocity minima in the upper 200 km of the mantle. Variations from the seismic velocity distributions illustrated in fig. 3 may be set up from fig. 2 by assuming various geothermal gradients and including the complexity of chemical variation from pyrolite to refractory peridotite and dunite.

5. Amphibole stability, partial melting and the low velocity zone

The mineralogical variations illustrated in fig. 3 may produce V_s variations of approximately 2% but these are inadequate to account for a low velocity zone with a V_s minimum of 4.1 km/sec below the uppermost mantle "lid" with $V_s \approx 4.6$ km/sec (ANDERSON, 1967; BERRY and KNOPOFF, 1967). Other analyses of seismic data have inferred high values of seismic attenuation in the low velocity zone (ANDERSON, 1962, 1967), or have inferred very low values of shear modulus for the low velocity zone (HALES and DOYLE, 1967). These considerations have led a number of authors to suggest that the low velocity zone of the mantle owes its existence to partial or incipient melting (ANDERSON, 1962, 1967; OXBURGH and TURCOTTE, 1968; AKI, 1967). LAMBERT and WYLLIE (1968) carried out experiments aimed at establishing the high pressure stability limit of amphibole for $P_{H_2O} = P_{total}$ and suggested that the existence of interstitial amounts of water-rich fluid phase or water-rich silicate melts at depths below the stability limit of amphibole would lead to low seismic velocities. At greater depths, incorporation of water into high pressure hydrous magnesian silicates and hydroxylated pyroxenes (RINGWOOD and MAJOR, 1967; SCLAR et al., 1967) would produce a lower depth limit to this feature. The temperature of formation of magmas in the zone of amphibole instability and free water would be considerably below the anhydrous pyrolite solidus.

The data on maximum amphibole stability in pyrolite presented herein support this hypothesis and suggest that amphibole would be stable along the shield geotherm of fig. 2 to depths of approximately 100–120 km and along the oceanic geotherm to depths of 80–90 km. The latter estimates derive from the data of table 3 and appearance of amphibole in pyrolite runs at 1200 °C, 18 kb and 22.5 kb, unless stringent precautions against drying of furnace components are observed.

RINGWOOD (1969) inferred a water content for the upper mantle of around 0.1% and CO_2 contents may also be significant. The presence of a fluid phase, even if $P_{H_2O} < P_{total}$, will depress the beginning of melting of pyrolite, probably by as much as 200 °C. RINGWOOD (1969) has presented arguments that a limited water content of around 0.1% will have the effect of buffering a condition of incipient melting ($\approx 1\%$ melting) so that comparatively large changes in temperature will have little effect on the amount or nature of the melt formed – this condition would not apply for melting of anhydrous pyrolite (GREEN and RINGWOOD, 1967a). The nature of hydrous magmas in equilibrium with olivine, orthopyroxene, clinopyroxene and garnet has been shown to be highly undersaturated olivine-rich basanite (GREEN, 1969, 1970) or possibly olivine-rich nephelinite and olivine melilite nephelinite (BULTITUDE and GREEN, 1968). Thus it is inferred that regions with a relatively shallow (80–120 kms) low velocity zone with proportionately large velocity decrease owe these characteristics to the presence of very small amounts of partial melting. The liquids formed by degrees of melting up to about 5% are highly undersaturated basanites and nephelinites and extraction of such liquids will leave regions of the low velocity zone selectively depleted in elements strongly partitioned into the initial liquid.

Movement of magma batches through overlying mantle may lead to breakdown of hydrous phases and selective element depletion by "wall rock reaction" processes (GREEN and RINGWOOD, 1967b; GREEN, 1970). Models of mantle evolution such as those of OXBURGH and TURCOTTE (1968) and RINGWOOD (1969) would also lead to regions of the mantle overlying the low velocity zone which were partially or completely depleted in basaltic components.

6. Conclusions

Reactions producing garnet or garnet+olivine from assemblages with aluminous pyroxenes or pyroxenes+spinel are complex, involving sensitively P, T dependent solid solution of R_2O_3 components in pyroxene. There is a need for further experiments to confirm data presented for several simple systems and natural assemblages and to evaluate the role of Cr_2O_3 and other components on the phase boundaries. The experimental study of the relations between pyroxene pyrolite and garnet pyrolite for the model mantle composition of RINGWOOD (1966a) has established reversals for this boundary at key temperatures and enables application of these studies to evaluation of upper mantle mineralogy.

Pyroxene+spinel pyrolite is stable to depths of 60–70 km along an estimated sub-oceanic geothermal gradient. At this depth spinel and pyroxene react to form garnet and olivine. The amount of garnet formed from this reaction depends sensitively upon the temperature at which the geotherm intersects the spinel+pyroxene $\rightleftharpoons$ garnet+olivine boundary. At temperatures in excess of 1000 °C, less than half the potential garnet in the pyrolite compositions forms by this reaction, the rest remaining in solid solution in aluminous orthopyroxene ($>3\%$ Al_2O_3) and aluminous clinopyroxene. It has been demonstrated that mineralogical variation along geothermal gradients, particularly in oceanic regions, may be expected to strongly influence the seismic velocity distribution in the upper mantle. No unique model of mineralogical or seismic velocity distribution in the upper mantle is presented, rather it is argued that regional variations in chemical composition (from pyrolite to refractory peridotite), and in geothermal gradients will produce significant, regional differences in seismic velocity distributions.

Although mineralogical variation in anhydrous pyrolite may cause the existence of a low velocity zone, the alternate explanation of a zone of incipient or minor partial melting is preferred. The presence of this zone is inferred to result from the presence of small ($\approx 0.1\%$) quantities of water in the mantle, and from the instability of the hydrous phase amphibole at depths greater than about 80 km. The presence of a fluid, water-rich phase in limited quantity produces incipient melting well below the anhydrous pyrolite solidus. Liquids produced in this way are highly undersaturated basanites and nephelinites.

References

AKI, K. (1967) J. Geophys. Res. **73**, 585.
ANDERSON, D. L. (1962) Sci. Am. July, 2–9.
ANDERSON, D. L. (1967) in: T. Gaskell, ed., *The Earth's Mantle* (Academic Press, London) 355.
BERRY, M. J. and L. KNOPOFF (1967) J. Geophys. Res. **72**, 3613.
BOYD, F. R. and J. L. ENGLAND (1960) Carnegie Inst. Wash. Yearbook **59**, 47.
BOYD, F. R. and J. L. ENGLAND (1964) Carnegie Inst. Wash. Yearbook **63**, 157.
BOYD, F. R. and I. D. MACGREGOR (1964) Carnegie Inst. Wash. Yearbook **63**, 152.
BULTITUDE, R. J. and D. H. GREEN (1968) Earth Planet. Sci. Letters **3**, 325.
CLARK S. P. (1962) in: C. M. Herzfeld, ed., *Temperature, its measurement and control in science and industry* 3 (R. Reinhold, New York) 779.
CLARK, S. P. and A. E. RINGWOOD (1964) Rev. Geophys. **2**, 35.
DAVIS, B. T. C. and F. R. BOYD (1966) J. Geophys. Res. **71**, 3567.
ESSENE, E. J., B. J. HENSEN and D. H. GREEN (1970) Phys. Earth Planet. Interiors **3**, 378.
GREEN, D. H. (1969) Proc. Prague Upper Mantle Symp., Tectonophysics **5/6**, in press.
GREEN, D. H. (1970) Phys. Earth Planet. Interiors **3**, 221.
GREEN, D. H. and W. HIBBERSON (1969) Lithos, in press.
GREEN, D. H. and A. E. RINGWOOD (1963) J. Geophys. Res. **68**, 937.
GREEN, D. H. and A. E. RINGWOOD (1967a) Earth Planet. Sci. Letters **3**, 151.
GREEN, D. H. and A. E. RINGWOOD (1967b) Contrib. Mineral. Petrol. **15**, 103.
HALES, A. L. and H. A. DOYLE (1967) Geophys. J. **13**, 403.
HARRIS, P. G., A. REAY and I. G. WHITE (1967) J. Geophys. Res. **72**, 6359.
HESS, H. H. (1964) in: *A study of serpentinite*, U.S. Natl. Res. Council Publ. **1188**, 169.
ITO, K. and G. C. KENNEDY (1967) Am. J. Sci. **265**, 519.
KUSHIRO, I. (1964) Carnegie Inst. Wash. Yearbook **63**, 101.
KUSHIRO, I. and H. KUNO (1963) J. Petrol. **4**, 75.
KUSHIRO, I., Y. SYONO and S. AKIMOTO (1968) J. Geophys. Res. **73**, 6023.
KUSHIRO, I. and H. S. YODER, JR (1966) J. Petrol. **7**, 337.
LAMBERT, I. B. and P. J. WYLLIE (1968) Nature **219**, 1240.
MACGREGOR, I. D. (1964) Carnegie Inst. Wash. Yearbook **63**, 157.
MACGREGOR, I. D. (1965) Carnegie Inst. Wash. Yearbook **64**, 126.
MACGREGOR, I. D. (1967) Ann. Rept. South West Centre Adv. Studies **66–67**, 6.
MACGREGOR, I. D. and A. E. RINGWOOD (1964) Carnegie Inst. Wash. Yearbook **63**, 161.
O'HARA, M. J. (1967) in: P. J. Wyllie, ed., *Ultramafic and related rocks* (Wiley, New York) 393.
OXBURGH, E. R. and D. L. TURCOTTE (1968) J. Geophys. Res. **73**, 2643.
RINGWOOD, A. E. (1966a) in: P. M. Hurley, ed., *Advances in Earth science* (M.I.T. Press, Cambridge, Mass.) 287.
RINGWOOD, A. E. (1966b) in: P. M. Hurley, ed., *Advances in Earth science* (M.I.T. Press, Cambridge, Mass.) 357.
RINGWOOD, A. E. (1969) Upper Mantle Committee Monograph, in press.
RINGWOOD, A. E., F. R. BOYD and I. D. MACGREGOR (1964) Carnegie Inst. Wash. Yearbook **63**, 147.
RINGWOOD, A. E. and D. H. GREEN (1966) Tectonophysics **3**, 383.
RINGWOOD, A. E. and A. MAJOR (1967) Earth Planet. Sci. Letters **3**, 130.
SCLAR, C. B., L. C. CARRISON and O. M. STEWART (1967) Trans. Am. Geophys. Union Abstr. **48**, 226.

THE EFFECT OF CaO, Cr_2O_3, Fe_2O_3 and Al_2O_3 ON THE STABILITY OF SPINEL AND GARNET PERIDOTITES

IAN D. MACGREGOR*

Southwest Center for Advanced Studies, Dallas, Texas, U.S.A.

The reaction pyroxene+spinel ⇌ garnet+olivine defines the boundary between low pressure spinel peridotite and high pressure garnet peridotite. Experimental determination in the four component system MgO-CaO-Al_2O_3-SiO_2 indicates that the CaO content is critical in defining the position of the reaction boundary. Where orthopyroxene is the only pyroxene present, increasing the CaO content decreases the pressure at which the reaction occurs; where both pyroxenes are present changing the CaO content has no effect; and where clinopyroxene is the only pyroxene present, increasing the CaO content increases the pressure of the reaction.

Experiments on analyzed clinopyroxenes, orthopyroxenes and spinels from ultramafic nodules, indicate that the reaction boundary moves to higher pressures with increasing Cr_2O_3 and Fe_2O_3, and decreasing Al_2O_3 contents. Variations in the trivalent oxide ratio may result in changes of the equilibrium reaction by as much as 10 kilobars.

The results indicate that the composition of the mantle is critical in defining the position of the spinel-to-garnet-peridotite boundary, and leaves open the possibility that spinel and garnet peridotite may coexist throughout a wide depth within the earth's upper mantle.

1. Introduction

Water free ultramafic rocks found on the earth's surface fall into three main mineral assemblages (fig. 1). The mineralogy for all three assemblages is similar except for the aluminous phases (plagioclase, spinel and garnet) which define the temperature and pressure, or metamorphic facies, at, or within which the rocks have finally equilibrated. Olivine, orthopyroxene and clinopyroxene occur either with plagioclase, spinel or garnet to form phase assemblages that are characteristic of the plagioclase peridotite, spinel peridotite or garnet peridotite facies (MACGREGOR, 1968), respectively. Their geological environments indicate that plagioclase, spinel and garnet peridotites have formed at successively increasing pressures. Experimental determination of the equilibrium reactions between the different ultramafic facies in pressure-temperature space (MACGREGOR, 1965; KUSHIRO and YODER, 1966) confirm the geological conclusions, and allow a more quantitative estimate of the temperature and pressures at which the reactions occur. The experimental data also show that the phase assemblages stable in a mantle of ultramafic composition lie in the spinel and garnet peridotite facies. The position of the boundary between these two phase assemblages is thus of interest to geologists and geophysicists interested in the density structure of the upper mantle.

The variables that define the equilibrium position of this reaction are pressure, temperature and composition. Since all the minerals in ultramafic rocks form complex solid solutions, it will be necessary to evaluate the compositional variables. This paper deals particularly with the effect of variations in the CaO, Cr_2O_3, Al_2O_3 and Fe_2O_3 content of the assemblage, on the

ULTRAMAFIC MINERAL ASSEMBLAGES

MINERAL ASSEMBLAGE	ROCK NAME
(1) OLIVINE + ORTHOPYROXENE ± CLINOPYROXENE ± PLAGIOCLASE	PLAGIOCLASE PERIDOTITE OR LHERZOLITE
(2) OLIVINE + ORTHOPYROXENE ± CLINOPYROXENE ± SPINEL	SPINEL PERIDOTITE OR LHERZOLITE
(3) OLIVINE + ORTHOPYROXENE ± CLINOPYROXENE ± GARNET	GARNET PERIDOTITE OR LHERZOLITE

Fig. 1. Phase assemblages of water free ultramafic rocks.

---

* New address: Dept. of Geology, University of California, Davis, California, 95616, USA.
Contribution Southwest Center for Advanced Studies, P.O. Box 30365, Dallas, Texas.

position of the reaction. The experimental data supporting the conclusions in this paper have not been published, but will be submitted for publication in the near future.

2. The spinel to garnet peridotite reaction

Typical modes for a spinel and garnet peridotite with a possible primary mantle composition (CARTER, 1966) are given in table 1. The reaction defining the

TABLE 1

Modal proportions of a spinel and garnet peridotite for a possible upper mantle composition (CARTER, 1966)

Mineral Assemblage	Volume Percent	
	Low Pressure	High Pressure
Olivine $(Mg,Fe)SiO_4$	55 ± 10	55 ± 10
Orthopyroxene $(Mg,Fe)_2Si_2O_6$	27 ± 5	19 ± 4
Clinopyroxene $Ca(Mg,Fe)Si_2O_6$	15 ± 3	12 ± 2
Spinel $(Mg,Fe)(Al,Cr,Fe)_2O_4$	3 ± 1	0
Garnet $(Mg,Fe)_3(Al,Cr,Fe)_2Si_3O_{12}$	0	13 ± 3

boundary between these two phase assemblages results in the combination of pyroxene and spinel, on the low pressure side, to form garnet and forsterite, on the high pressure side. The specific reactions that define the boundary between spinel and garnet peridotite are as follows:

$$4MgSiO_3 + MgAl_2O_4 = Mg_2SiO_4 + Mg_3Al_2Si_3O_{12}, \quad (1)$$

$$2CaMgSi_2O_6 + MgAl_2O_4 = Mg_2SiO_4 + Ca_2MgAl_2Si_3O_{12}, \quad (2)$$

$$2xMg_2Si_2O_6 + 2yCaMgSi_2O_6 + (x+y)MgAl_2O_4 = (x+y)Mg_2SiO_4 + [(x+y/3)Mg_3Al_2Si_3O_{12} + (2y/3)Ca_3Al_2Si_3O_{12}]. \quad (3)$$

Reaction (1) defines the boundary for the three component system $MgO-Al_2O_3-SiO_2$ and may be considered as a simplified peridotite. Reaction (2) defines the boundary for the reaction between diopside and spinel to form olivine and garnet, and represents one of the critical reactions in the four component system $CaO-MgO-Al_2O_3-SiO_2$. A more general statement is given in reaction (3), which takes into account the effect of varying CaO content. Similar reactions may be written to include the effect of varying amounts of Cr, Al and Fe^{++}. It should be added, that because of Al, Cr and Fe^{+++} solid solution in the pyroxenes, which is also a function of temperature and pressure, the actual reaction involves pyroxenes of varying composition.

3. The system $MgO-Al_2O_3-SiO_2$

Looking first at the three component system $MgO-Al_2O_3-SiO_2$, reaction (1) is defined by a univariant equilibrium boundary (MACGREGOR, 1964) that would define a unique pressure for the phase boundary at any temperature (fig. 2).

4. The system $CaO-MgO-Al_2O_3-SiO_2$

With the addition of CaO to the three component system the reactions become more complex. The variation of the CaO content results in a variation of the diopside/enstatite ratio of the pyroxene entering into reaction with the spinel. These reactions have been experimentally determined (MACGREGOR 1965; KUSHIRO and YODER, 1966) and are shown in fig. 2. The reaction boundary (1) refers to that for the simple three component system $MgO-Al_2O_3-SiO_2$; the reaction boun-

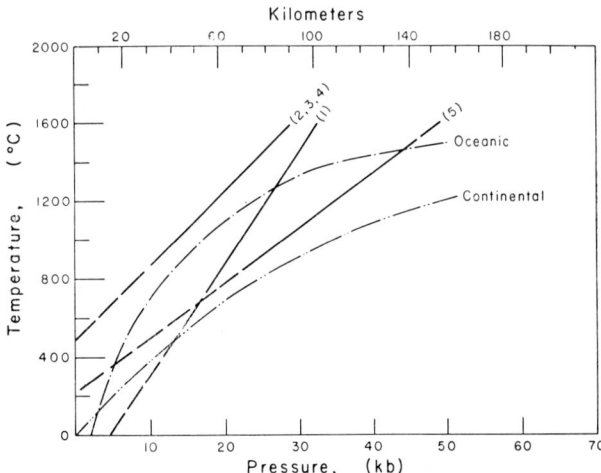

Fig. 2. Equilibrium boundaries for the reaction, pyroxene + spinel = garnet + olivine ((1) for reaction where enstatite constitutes pyroxene phase; (2), (3) and (4) for reaction where both enstatite and diopside comprise pyroxene phase, reactions (2), (3) and (4) refer to experiments having diopside/enstatite weight ratios of 25:75, 50:50 and 75:25, respectively; (5) for reaction where diopside is only pyroxene phase). Geothermal gradients from CLARK and RINGWOOD (1963).

daries (2–4) refer to experiments using diopside/enstatite weight ratios of 25:75, 50:50 and 75:25, respectively; reaction (5) is for pure diopside. It can be seen that the addition of CaO results in a significant change in the spinel to garnet peridotite boundary. With increasing CaO the boundary initially moves to lower pressures and subsequently back to higher pressures. The geothermal gradients superimposed (CLARK and RINGWOOD, 1964) on this diagram illustrate how the composition of a peridotite will drastically affect the depth at which it transforms from a spinel to a garnet peridotite.

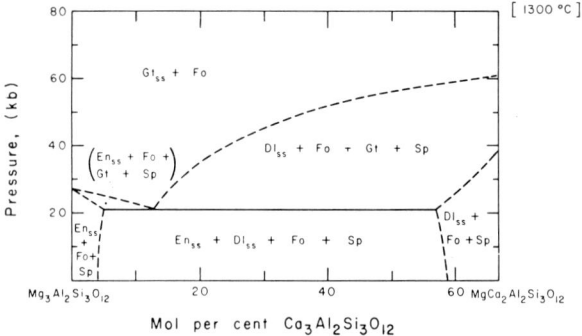

Fig. 3. Pressure-composition projection at 1300 °C along the join pyrope-grossulanite.

The effect of CaO may be better seen in a pressure-composition projection (fig. 3) which represents the phase relations at 1300 °C. Fig. 3 gives a projection of part of the join, pyrope-grossular, and illustrates the pressure at which the reaction occurs for a specific diopside/enstatite ratio. At pressures less than 20 kb all peridotite compositions will exist as one of the three following spinel-bearing assemblages:

a) forsterite + enstatite$_{ss}$ + spinel;
b) forsterite + enstatite$_{ss}$ + diopside$_{ss}$ + spinel, and
c) forsterite + diopside$_{ss}$ + spinel.

With increasing pressure, compositions within the field enstatite + diopside + forsterite + spinel are the first to react to the high pressure assemblage. Moreover, for all four phase compositions there are 5 phases on the equilibrium curve which defines a univariant boundary in the four component system, and changes in the CaO content will not change the pressure of reaction. For those compositions sufficiently CaO-poor or -rich to lie in the two bounding 3 phase fields

a) forsterite + enstatite$_{ss}$ + spinel or
c) forsterite + diopside$_{ss}$ + spinel,

the reaction surface is divariant and changes of the CaO content results in a change of the pressure of reaction. In the CaO-poor field increasing CaO content decreases the pressure of reaction, while in the CaO-rich field increasing the CaO content increases the pressure of reaction.

5. The effect of Cr_2O_3, Al_2O_3 and Fe_2O_3

The next step is to examine the effect of the compositional variables Cr_2O_3, Fe_2O_3 and Al_2O_3. Since these elements substitute for each other in similar structural sites in the crystal lattice of both garnet and spinel, it is convenient to think of these variables as ratios of the trivalent oxides. To check the effect of these variables an experiment was designed using natural minerals which covered a wide range of trivalent oxide composition. In order to limit the effect of CaO as a variable, all reactants were made up to fall in the 4 phase field olivine + orthopyroxene + clinopyroxene + spinel. Thus, the natural assemblage is being treated as a 4 component system in which olivine, orthopyroxene, clinopyroxene, and spinel solid solutions are treated as components. Variations of the orthopyroxene/clinopyroxene ratio thus should not seriously affect the reaction boundary.

With this restriction in mind, a set of compositions were made up using analyzed, coexisting sets of orthopyroxenes, clinopyroxenes and spinel from peridotite xenoliths from the Kilbourne Hole basalt, New Mexico (CARTER, 1965). The bulk chemistry of the reaction assemblages, using a molecular ratio of 5 orthopyroxene to 3 clinopyroxene, which lies close to an average model ratio for the Kilbourne Hole samples, while still having four phases, is shown in table 2.

The chemistry of the assemblages listed in table 2 show no systematic variation in SiO_2, TiO_2, FeO, MgO, CaO, NiO, MnO and Na_2O. There are systematic variations for the three trivalent oxides Al_2O_3, Cr_2O_3 and Fe_2O_3, which is reflected in the trivalent oxide ratios. Although the fayalite content of the coexisting olivine varies systematically with the trivalent oxides this is not reflected in the FeO/FeO + MgO ratio of the reactants. Thus the major chemical variations in the suite of reactants is in the trivalent oxides, and one may

TABLE 2

Chemistry of the reaction assemblage used in experiments

(5 $Mg_2Si_2O_6$ + 3 $CaMgSi_2O_6$ + 4 $MgAl_2O_4$ ⇌ 4 Mg_2SiO_4 + $Ca_3Mg_9Al_8Si_{12}O_{48}$)

SAMPLE	104	2	109	89	129	114
OXIDE						
SiO_2	39.82	39.55	40.19	40.01	37.70	40.17
Al_2O_3	20.77	18.96	17.38	14.13	12.80	7.96
TiO_2	0.85	0.27	0.21	0.11	0.19	0.21
Cr_2O_3	0.81	2.87	3.69	7.21	11.02	12.35
FeO	5.51	6.10	5.64	6.15	6.52	4.92
Fe_2O_3	0.40	1.11	1.32	1.32	2.21	3.48
MgO	25.29	25.32	24.95	24.50	23.94	25.16
CaO	5.86	5.27	5.85	5.89	5.10	6.04
NiO	0.13	0.14	0.14	0.11	0.08	0.10
MnO	0.10	0.09	0.09	0.15	0.06	0.09
Na_2O	0.42	0.46	0.48	0.46	0.27	0.35
TOTAL	99.96	100.14	99.94	100.04	99.99	100.83
FeO/FeO + MgO	17.89	19.14	18.17	20.06	21.40	16.3
Al_2O_3/ΣR_2O_3	94.50	82.60	77.62	62.36	49.10	33.4
Cr_2O_3/ΣR_2O_3	3.69	12.50	16.48	31.81	42.30	51.9
Fe_2O_3/ΣR_2O_3	1.82	4.80	5.90	5.83	8.40	14.6
Mol. % Fa in olivine (coexisting)	12.0	10.1	9.4	9.8	8.2	7.2

suspect that variations in the temperatures and pressures at which the reaction occurs will be related to these compositional variables.

Fig. 4 shows a pressure-composition section at 1200 °C; the pressure at which the reaction occurs is shown as a function of the trivalent oxide ratio. Increasing the $Al_2O_3/\Sigma R_2O_3$ ratio of the reactants decreases the pressure of reaction, while increasing the $Cr_2O_3/\Sigma R_2O_3$ and $Fe_2O_3/\Sigma R_2O_3$ ratios results in an increase of the pressure of reaction, the $Cr_2O_3/\Sigma R_2O_3$ ratio having a more profound effect than the $Fe_2O_3/\Sigma R_2O_3$ ratio. There is no systematic variation with the FeO/FeO + MgO ratio (fig. 4) or for that matter with

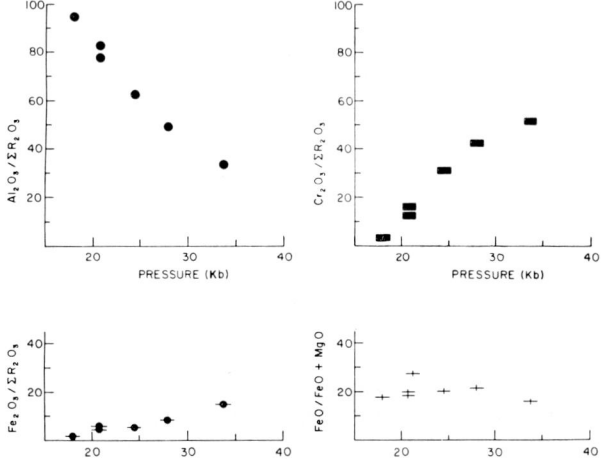

Fig. 4. Pressure-composition plots at 1200 °C showing pressure at which spinel converts into garnet peridotite for particular compositional ratios.

any other component. Thus, within the range of compositional variation of four phase ultramafic rocks, it appears that the trivalent ions, which occupy a combination of tetrahedral and octahedral lattice sites, are critical compositional variables in defining the relative stability of spinel and garnet peridotite.

A pressure-temperature section of the same results (fig. 5) indicates the wide range of pressures over which the reaction may occur. Reaction assemblages from typical garnet lherzolites from kimberlite occurences have a range of $Al_2O_3/\Sigma R_2O_3$ from 85 to 90 (HARRIS et al., 1967). The primary anhydrous Lizard assemblage has an $Al_2O_3/\Sigma R_2O_3$ of approximately 90 (GREEN, 1964). Both these sets of rocks would have a reaction boundary similar to that for the lower pressure curves (fig. 5). Similarly the boundary for the 4 component system $CaO-MgO-Al_2O_3-SiO_2$, which has an $Al_2O_3/\Sigma R_2O_3$ ratio of 100, coincides approximately with the lowest reaction boundary. In contrast, the spinel peridotites from Alpine occurrences generally have low $Al_2O_3/\Sigma R_2O_3$ ratios, varying from 10 to 40 (HARRIS et al., 1967), which suggests that reaction boundaries should extend to even higher pressures than those shown in the figure. Indeed, reaction boundaries for the Ca-poor, diopside-free harzburgitic assemblages probably extends to even higher pressures. The data illustrate that the composition of the ultramafic rock, in particular the ratio of the trivalent oxides, must be specified prior to predictions of the temperature and pressure at which that rock converts from a spinel- to a garnet-bearing assemblage.

It is worth noting that for compositions with $Cr_2O_3/\Sigma R_2O_3$ ratios in excess of a value between 16 and 32 (fig. 5) both garnet and spinel appear initially on the high pressure side of the boundary. The width of the 5 phase transition zone has not been experimentally defined but presumably expands with increasing Cr_2O_3 content.

In fig. 6 the data are summarized in a more general perspective. The range of pressures over which different groups of ultramafic rocks may be expected to convert from the low pressure spinel to the high pressure garnet peridotite is shown. In general, the more primitive compositions such as a pyrolite (GREEN and RINGWOOD, 1963), the primary Lizard composition, and garnet lherzolites from diamondiferous kimberlite pipes have a reaction boundary which lies at the lowest pressure

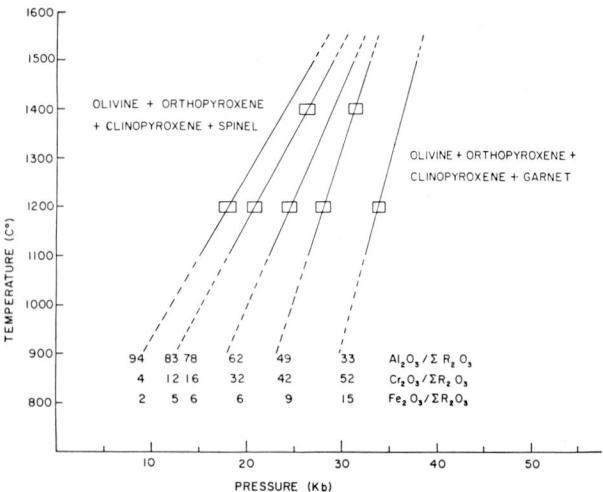

Fig. 5. Pressure-temperature section illustrating the boundary between a spinel and garnet peridotite for the compositions given in table 1.

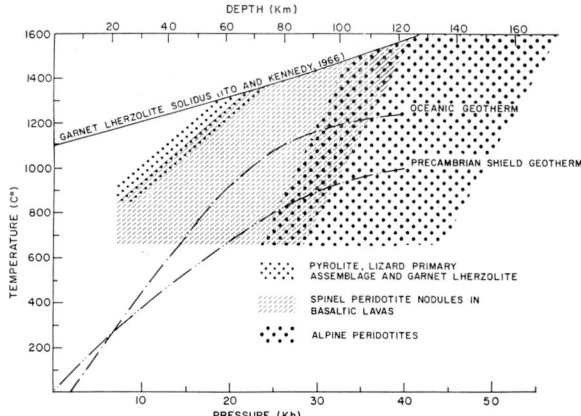

Fig. 6. Pressure-temperature section showing general range of pressures over which spinel to garnet peridotite reaction may occur, for different groups (compositions) of ultramafic rocks.

range. The spinel peridotite nodules from basaltic lavas (CARTER, 1965) have a wide compositional range, with the reaction occurring at increasingly higher pressures for the more refractory Cr_2O_3-rich assemblages. Alpine peridotites, which generally represent a more highly differentiated sequence than the xenolithic suites, have compositions which indicate reaction boundaries that overlap the xenolithic samples and extend to considerably higher pressures. This conclusion would be reinforced for the Ca-poor, diopside free, spinel harzburgites which convent to the higher density assemblage at higher pressures than the two pyroxene assemblages. Superimposed on this P-T section are an average oceanic and Precambrian shield geotherm (CLARK and RINGWOOD, 1964) and a lherzolite solidus determined by ITO and KENNEDY (1967).

6. Conclusions

These results can be related to the distribution of natural ultramafic rocks in geological occurrences. Large Alpine spinel peridotites commonly found in the cores of old mountain belts generally have high $Cr_2O_3/\Sigma R_2O_3$ ratios. They would be expected to lie within the spinel peridotite stability field down to depths of approximately 80 and 95 km beneath average shield and oceanic regions, respectively. Garnet peridotites in high grade metamorphic terraines such as the Norwegian Caledonides have low $Cr_2O_3/\Sigma R_2O_3$ ratios (CARSWELL, 1968) which places them within the garnet peridotite stability field. It should be pointed out that in the same metamorphic terrain ultramafic rocks with different trivalent oxide or diopside/enstatite ratios may stably coexist as spinel and garnet peridotites. Garnet peridotite xenoliths from kimberlite pipes have $Cr_2O_3/\Sigma R_2O_3$ ratios which place them well within the garnet stability field for both oceanic and shield geotherms. The high temperature peridotites such as the Lizard generally have low $Cr_2O_3/\Sigma R_2O_3$ ratios and a high pressure garnet peridotite assemblage may be expected. However, the evidence of local high temperatures well above the normal geothermal gradients, and, in the Lizard intrusion, the presence of plagioclase peridotites indicate conditions of final equilibration to the lower pressure side of the Cr_2O_3-poorest assemblages.

The data from the Kilbourne Hole basalt allow the specific conclusion that this suite of spinel peridotites represents a final equilibration at depths less than 55 kn. Since they are interpreted as crystals in equilibrium with a basaltic liquid (CARTER, 1965), this represents the maximum depth of the magma chamber. The absence of plagioclase peridotites in the suite allows the use of the reaction, anorthite + forsterite = pyroxene + spinel, to place an upper limit on the depth of final equilibration of these nodules. For the four component system $CaO-MgO-Al_2O_3-SiO_2$ KUSHIRO and YODER (1966) indicate that the anorthite + forsterite reaction places an upper limit of 25 km on the spinel peridotite assemblage. This information agrees well with the experimental result with natural minerals and suggests that, if the Kilbourne Hole xenoliths were in equilibrium with the magma, the magma chamber in which these rocks have formed lies in the depth zone 25 to 55 km.

A second area to which these data apply is the question as to whether seismic discontinuities in the lower crust and upper mantle are related to phase changes from the lower density spinel to the higher density garnet peridotite. Depending on ones choice of composition a wide variety of possible results may ensue. To make any prediction of the present phase distribution we need a far more sophisticated knowledge of the compositional distribution of the upper mantle than is presently available.

Two interesting points which emerge are as follows: Firstly, it is possible that seismic discontinuities caused by phase changes in the upper mantle may be related to small compositional variations within the overall ultramafic composition and: Secondly, if we assume that the crust has evolved from the mantle beneath it, we may expect an increasingly more refractory mantle both in depth and time. This would result on the average in an increase of the stability field of the spinel peridotites to greater depths with time.

Acknowledgements

The experimental data on the system CaO-MgO-Al_2O_3-SiO_2 were collected while the author was a fellow at the Geophysical Laboratory, Washington. The author is indebted to Dr. F. R. Boyd for the use of the high pressure equipment, and for his advice and assistance during the course of the experiments. The final experiments were conducted at the Southwest Center for Advanced Studies with support from the United States Army Research Office (Contract: DA-31-124-ARO-D-463). The criticisms and comments of my colleagues, Dr. H. Brueckner, Dr. J. L. Carter and Dr. D. C. Presnall, did much to improve the presentation of this paper.

References

CARSWELL, D. A. (1968) Lithos **1**, 322.
CARTER, J. L. (1965) Ph. D. Dissertation, Rice University, Houston.
CARTER, J. L. (1966) Ann. Meeting Geol. Soc. Am. 35 (Abs.).
CLARK, S. P. and A. E. RINGWOOD (1964) Rev. Geophys. **2**, 35.
GREEN, D. H. and A. E. RINGWOOD (1963) J. Geophys. Res. **68**, 937.
GREEN, D. H. (1964) J. Petrol. **5**, 134.
HARRIS, P. G., A. REAY and I. G. WHITE (1967) J. Geophys. Res. **72**, 6359.
ITO, K. and G. C. KENNEDY (1967) Am. J. Sci. **265**, 519.
KUSHIRO, I. and H. S. YODER (1966) J. Petrol. **7**, 337.
MACGREGOR, I. D. (1964) Ann. Rept. Dir. Geophys. Lab. Carnegie Inst. Wash. **63**, 157.
MACGREGOR, I. D. (1965) Ann. Rept. Dir. Geophys. Lab. Carnegie Inst. Wash. **64**, 126.

EXPERIMENTAL STUDY OF AMPHIBOLITE AND ECLOGITE STABILITY

E. J. ESSENE, B. J. HENSEN and D. H. GREEN

Department of Geophysics and Geochemistry, Australian National University, Canberra

The amphibolite-eclogite transition is often observed in metamorphic rocks and may be expected to occur in the lower crust or upper mantle if $P_{H_2O} \approx P_S$. Many simple amphiboles become unstable at high water pressures due to backbending of dehydration reactions or to solid–solid transitions involving sheet silicates. Contrary to YODER and TILLEY (1962), eclogite is expected to be stable relative to amphibolite at sufficiently high water pressures because of the formation of dense garnet and the high compressibility of water vapor. Experiments in progress on several basaltic compositions have produced eclogite at the expense of amphibolite at high water pressures. Seeding has been found to be important as hornblende crystallizing from glass will persist at least 4 kb above its stability field defined by reversals with garnet-pyroxene-amphibole mixtures. In reconnaissance experiments on a quartz tholeiite and an alkali olivine basalt, hornblende disappears at high water pressures between 15–25 kb and 700–900 °C. Biotite appears to be stable in the alkali olivine basalt to at least 40 kb at these temperatures.

Basaltic amphibolites do not appear to be stable very far into the low-velocity zone even at $P_{H_2O} = P_S$ confirming LAMBERT and WYLLIE's (1968) earlier experiments. Biotite however seems to be stable deeper in the mantle (at least for potassic basalts) than previously considered possible if $P_{H_2O} \approx P_S$ and may be a source of water for magmas when partially melted.

1. Introduction

Eclogite and amphibolite relationships have long been of interest to the petrologist as both rocks are closely associated in metamorphic terranes. The classical petrological studies of HEZNER (1903), BRIÈRE (1920), ESKOLA (1921) and TILLEY (1936, 1937) showed that early-formed eclogites have often become partially hydrated to amphibolites. ESKOLA (1939) originally placed the eclogite facies at the highest pressures and temperatures, but it is now recognized that eclogites may form at relatively low pressures as well. If eclogites form at relatively low temperature and low P_{H_2O}/P_T^* then increase of water pressure could hydrate the eclogites to amphibolites (or glaucophane schists) as suggested by SAHLSTEIN (1936), BEARTH (1959), GREEN and RINGWOOD (1967) and others. For $P_{H_2O} \ll P_T$, either other gases must be very abundant to keep $P_F \approx P_T$, or there can be no vapour phase as in the water-deficient region (YODER, 1955). A third possibility is that the thermodynamic properties of intergranular films of impure water are markedly different from those of pure water. Most eclogites (except for rare scapolite-bearing varieties) show little evidence of abundant gases other than water, though of course they need not be evident in the preserved solid phases. One may also question the possibility of crystallizing a coarse-grained eclogite from an initial olivine-pyroxene-feldspar rock in the absence of a vapour phase without leaving traces of the original basaltic minerals, though BEARTH (1959, 1965) has observed all steps of the basalt–eclogite transition for some unusual eclogites. The possibility that the partial pressure of impure water is considerably lower "in solution" along grain boundaries than pure water vapour at the same P–T cannot be properly evaluated at present. To avoid the apparent difficulties in these explanations ESSENE and FYFE (1967) postulated that eclogites are stabilized relative to amphibolites at high water pressures and showed that density data alone confirm this prediction. Similarly, GREEN and RINGWOOD (1967, p. 805–6) showed that amphiboles, including the simple end member tremolite, will break down to anhydrous products and water vapour at sufficiently high P_{H_2O} and argued that this would considerably restrict the stability of amphibole in the upper mantle. Experiments are clearly needed to resolve the role of water pressures in the formation of eclogites.

* The notation used is $P_T = P_{Total}$, $P_F = P_{Fluid}$, $P_S = P_{Solid}$.

2. Previous experiments

YODER and TILLEY (1962) showed that the stable subsolidus assemblage for many basaltic rocks is amphibolite for $P_T = P_{H_2O} > 2$ kb, < 10 kb, T > 600 °C, and concluded that eclogite is unstable in the presence of water for *all* water pressures. However the beginning of melting could develop a positive slope and the disappearance of amphibole a negative slope at high water pressures due to the formation of garnet and the high compressibility of water. The combination of these two effects leads to the stabilization of eclogite relative to amphibolite at sufficiently high water pressures (fig. 1).

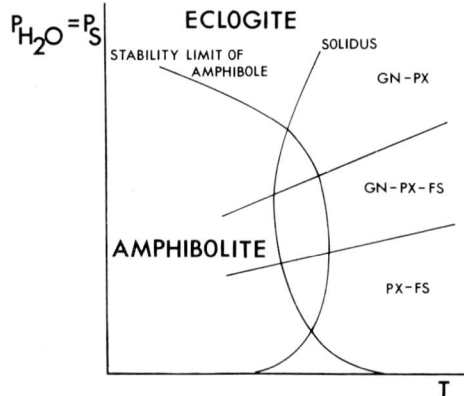

Fig. 1. A possible geometry for the beginning of melting and the stability of amphibole in basaltic rocks. The solid–solid transformation of feldspar to garnet and the high compressibility of water combine to stabilize eclogite at high water pressures. Fs = Feldspars, Px = Pyroxenes, gn = garnet.

This geometry has been confirmed in preliminary synthesis experiments on hornblende stability under its own composition and in a gabbro by LAMBERT and WYLLIE (1968). HENSEN and GREEN (unpublished) established the synthesis limit of hornblende in an alkali olivine basalt* with similar results also for $P_{H_2O} \approx P_T$ (fig. 2). These experiments show that eclogite is stable at high water pressures but represent syntheses mostly in the supersolidus region.

3. Thermodynamic calculations

Many simple amphiboles can be shown to be unstable at high water pressures by thermodynamic cal-

---
* The composition of this alkali-olivine basalt as well as the albite-rich quartz tholeiite used in following experiments is listed in GREEN and RINGWOOD (1967).

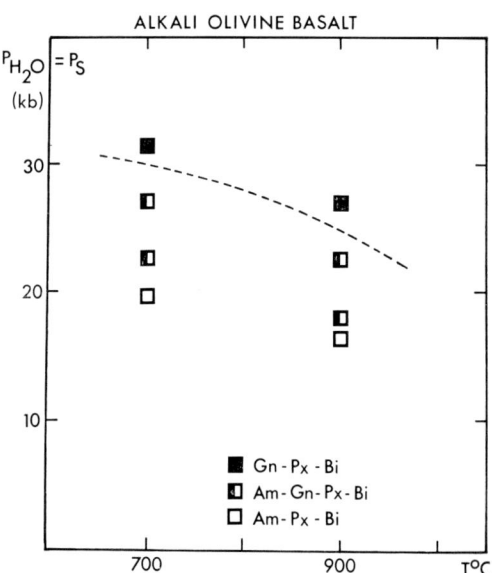

Fig. 2. Stability of amphibole for an alkali olivine basalt as determined by syntheses in cold-sealed gold capsules. The restriction of garnet to relatively high pressures is probably a reflection of the lack of initial garnet nuclei. This stability of amphibole compares well with the hot-sealed synthesis runs of fig. 3, showing that water pressures were not significantly lowered by chance leaks during the run. Am = Amphibole, Bi = Biotite, Px = Pyroxenes.

culations which permit extension of experimental data to higher water pressures (see Appendix 1 for thermodynamic data and procedures)*. The simple dehydration of tremolite (BOYD, 1959) will develop a negative slope at high water pressures due only to the rapid compressibility of water vapor, as predicted by GREEN and RINGWOOD (1967). At high enough water pressures the tremolite = diopside + enstatite + quartz + water curve will cross the talc = enstatite + quartz + water curve and a solid–solid decomposition of tremolite = talc + diopside is found instead (fig. 3). A similar solid–solid transformation predicted for anthophyllite by GREENWOOD (1963) is also shown in fig. 3, though calculations using ROBIE et al.'s (1967) volume data would move this curve to at least 50 kb. Glaucophane might similarly be expected to break down to talc + jadeite at high pressures (shown schematically in fig. 3), and the curve could be located more exactly if dependable experiments were available for the stability of glaucophane. Amphiboles may also break down at

---
* These calculations ignore the possibility of partial melting at high water pressures.

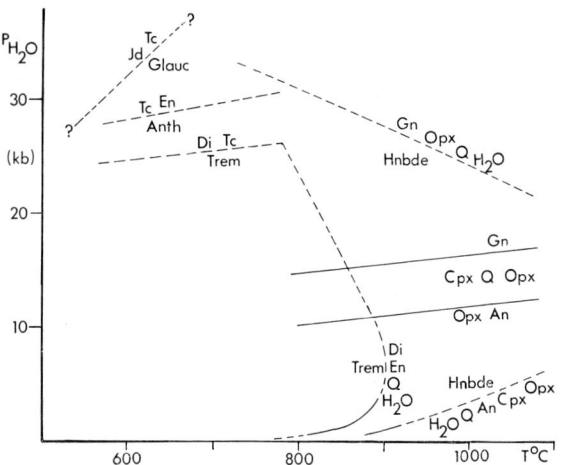

Fig. 3. Stability of various amphiboles estimated at high water pressures. Trem = Tremolite, Di = Diopside, En = Enstatite, Q = Quartz, Hnbde = "Hornblende", Gn = Garnet, Opx = Orthopyroxene, Cpx = Clinopyroxene, An = Anorthite, Anth = Anthophyllite, Tc = Talc, Jd = Jadeite, Gl = Glaucophane.

high water pressures because the products themselves react to form denser phases, as shown in fig. 3 for a "hornblende" – $Ca_2Mg_4Al_2Si_7O_{22}(OH)_2$. The initial point for the curve "hornblende" = anorthite+diopside+enstatite+water is taken at 900 °C and 1 kb* and when the feldspar reacts with pyroxenes to yield garnet the stability of this amphibole is sharply back-bent (fig. 3). While the true stability of this "hornblende" may be represented instead by some reaction as

hornblende$_A$ = hornblende$_B$+pyroxenes+feldspars
+quartz+water vapor,

the effect of high water pressures when garnet is produced may be qualitatively represented as in fig. 3**.

The instability of these amphiboles at high water pressures disagrees markedly with similar estimates by O'HARA (1967), who extended the stability of amphiboles vertically regardless of reactions among the producuts giving garnet at the expense of feldspars. O'Hara's schematic amphibole stability curves are unrealistic and

* This is an estimate taken from BOYD's (1959) experiments. When Boyd attempted to determine the stability of pargasite+ quartz the pargasite broke down to form an amphibole close to $Ca_2Mg_4Al_2Si_7O_{22}(OH)_2$ stable to ~900 °C, 1 kb.
** Following preparation of this manuscript, GILBERT's (1969) note on the experimental stability of end-member amphiboles appeared in the Ann. Rept. Geophys. Lab. Carnegie Inst. Wash. Gilbert's preliminary experiments on tremolite and "hornblende" are generally consistent with the calculations shown in fig. 3.

should be disregarded. Extrapolation of simple amphibole stabilities to high water pressures demonstrates that they may decompose by increasing water pressure but bear little on the stability of complex amphiboles which may first react with other phases in natural basaltic compositions.

4. Present experiments

The writers attempted to establish the upper stability limit of amphibole in an alkali olivine basalt and a quartz tholeiite for $P_{H_2O} = P_S$ by sealing glassy and crystalline samples with water and holding them in a piston cylinder apparatus at P and T for >24 hours (see appendix II for details of experimental procedure). Amphibole disappeared rather sharply within a 2 kb interval (figs. 4, 5) and in the case of the alkali olivine basalt, compared rather well with earlier experiments with cold-sealed capsules (fig. 3). When the crystallized mixtures of the earlier runs (amphibolite and eclogite) were rerun, amphibole was found to be unstable in the region where it had previously been synthesized, and it did not perceptibly grow at the expense of the garnet and pyroxene until the pressure had been lowered 5–10 kb below the original synthesis limit. While the reversal limits are rather wide in these preliminary experiments, they illustrate the danger of accepting short-time synthesis experiments as representing equilibrium at these relatively low temperatures, even if the syn-

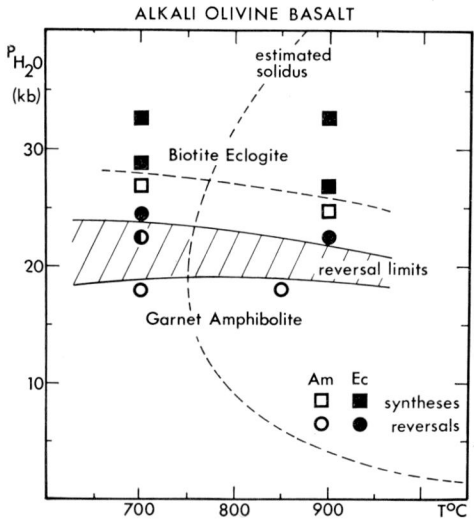

Fig. 4. Stability limit of amphibole in an alkali olivine basalt, comparing synthesis runs with reversals. Am = Amphibolite, Ec = Eclogite.

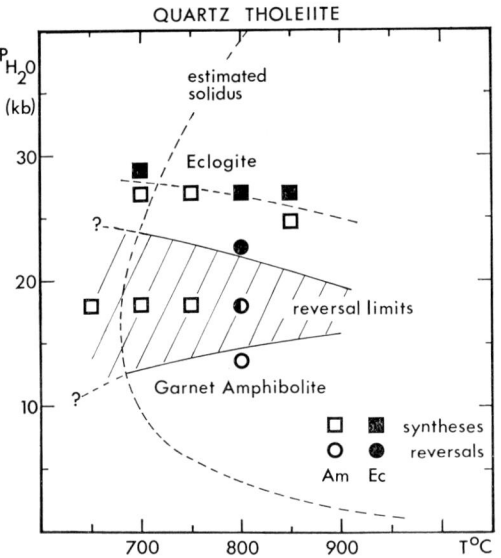

Fig. 5. Stability limit of amphibole in an albite-rich quartz tholeiite, comparing synthesis runs with reversals.

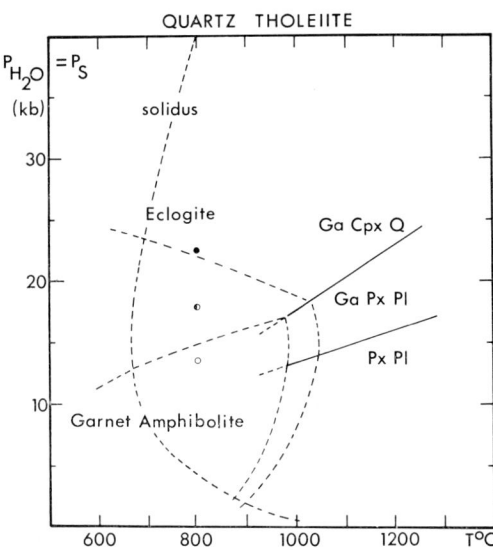

Fig. 6. Combination of present data for the alkali olivine basalt with that of GREEN and RINGWOOD (1967). Sa = Sanidine, Ga = Garnet, Pxs = Pyroxenes, Pl = Plagioclase, Ol = Olivine.

thesis boundary appears to be rather sharp and reproducible.

The solidus was difficult to locate for these basaltic compositions due to the formation of quench amphibole and the formation of only a small amount of glass near the solidus. It was estimated from the runs plotted in figs. 4 and 5 and by comparison with YODER and TILLEY's (1962) and LAMBERT and WYLLIE's (1968) data. The solidus for the alkali olivine basalt with 26% normative olivine is thought to be 50–75° higher than that for the quartz-normative tholeiite.

The upper stability of amphibole in supersolidus runs will involve reaction with melt as well as vapor, but these data have been included to place gross limits on the possible slopes for the supersolidus curves. GREEN and RINGWOOD (1967) found amphibole between 10.1 and 16.8 kb at 1100 °C with $P_{H_2O} < P_T$ in the alkali olivine basalt and this is used in fig. 6 to tentatively fix the amphibole stability. Yoder and Tilley's data on the stability of amphibole in various basalts have been consulted to estimate the amphibole stability in the quartz tholeiite (fig. 7).

The subsolidus reaction of amphibolite to eclogite is not likely to be a sharp transition but a gradual reaction among amphibole, garnet and pyroxenes (and feldspars?). It is best thought of as a sliding reaction with a number of substitutions in each phase and may be

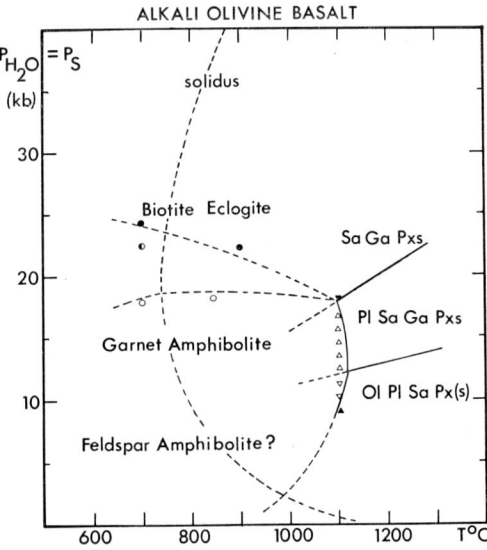

Fig. 7. Combination of the present data for the quartz tholeiite with that of GREEN and RINGWOOD (1967) and data for the amphibole stability from the Silberbach eclogite as determined by YODER and TILLEY (1962).

generalized as:

$$hornblende_A + garnet_B + pyroxene_C = hornblende_D + garnet_E + pyroxene_F \mp quartz \pm olivine + water.$$

The hornblende etc. on each side of the reaction will form a solid solution shifting with $\Delta P, T$ to any of a large number of end members. The area over which this sliding reaction will occur has not been determined experimentally and much longer run times (≈ 1 month) may be needed to establish this zone.

5. Interpretation

The present data suggest that eclogite will form at the expense of amphibolite when pressures are somewhere in the region of 10–30 kb at medium temperatures for basaltic rocks at high water pressures. This generally confirms LAMBERT and WYLLIE's (1968) experiments, but the difference between syntheses and reversals found here suggests that their synthesis experiments may be in error and should not be used for any specific argument as to magma generation in the upper mantle. Stability of one amphibole in a basaltic rock or under its own composition is no guide to the behavior of another amphibole in an ultramafic rock. Certainly their incoming of garnet at $P > 18$ kb (LAMBERT and WYLLIE, 1970) is open to doubt as no garnet nuclei were available in the starting mix. Tighter reversals are needed before evaluating the possibility that crustal eclogites form at high water pressures.

Garnet amphibolites lacking feldspar are expected to have higher densities (3.2–3.4 g/cm^3) and higher estimated seismic velocities ($V_p = 7.6$–8.0 km/s) than previously thought possible for hydrated basic rocks, and may be important in the lower crust or upper mantle (< 70 km), if water pressures are high. Garnet amphibolites lacking feldspar have been found in high-pressure crustal rocks (ESSENE, 1967) and may be widespread in the lower crust.

The apparent stability of biotite in the potassic alkali olivine basalt suggests that trioctahedral mica may be an important hydrous phase in the upper mantle. The decomposition of several amphiboles at high water pressures (fig. 3) apparently involves formation of another sheet silicate (talc) at high water pressures. SCLAR (1970) has synthesized a 10 Å micaceous phase in the system MgO–SiO$_2$–H$_2$O at $P > 30$ kb and $T < 500$ °C but has not yet shown whether this phase is stable by reversals nor reported its relation to other phases of ≈ 10 Å – as attapulgite. KUSHIRO et al. (1967) have examined phlogopite at high pressures and claim it is stable to at least 100 kb, but they have located a solidus in the system outside their stability field for phlogopite, whereas LUTH (1967) has shown that phlogopite breakdown curve intersects the solidus at only 2 kb water pressure. This suggests that Kushiro et al. may have had $P_{H_2O} \ll P_T$ in their runs; they may have also had trouble with quench phlogopite. In a later series of experiments, YODER and KUSHIRO (1969) confirmed Luth's data and showed that phlogopite remains stable up to 1200 °C and 40 kb.

Appendix I

The thermodynamic calculations in fig. 3 were all based on experimental data, and extended to higher pressures and temperatures with the well-known relations

$$\left(\frac{\delta \Delta G}{\delta T}\right)_P = -\Delta S, \quad \left(\frac{\delta \Delta G}{\delta P}\right)_T = \Delta V.$$

The free energy data for water were taken from BURNHAM et al. (1968) below 10 kb, and above 10 kb from SHARP (1962). Measured high-temperature entropies ($S_T - S_{298}$) are not available for talc, tremolite or "hornblende" and were estimated by summing the high-temperature entropies of other silicates:

talc = 2 serpentine − 3 brucite*,
tremolite = talc + 2 diopside**,

and the entropy of "hornblende" was assumed equal to tremolite. From these estimates, the high-temperature entropies of talc and tremolite were fitted to the quadratic equation of the form

$$S_T - S_{298} = a \ln T + b 10^{-3} T + c 10^5 T^{-2} + d,$$

where the values of a, b, c, d are given in table 1.

TABLE 1

	a	b	c	d
Talc	+126.26	−5.06	+24.89	−745.82
Tremolite	+232.00	+10.02	+38.63	−1365.52

The molar volumes (V^0) of the silicates were taken from ROBIE et al. (1967) and the standard entropies (S^0_{298}) from ROBIE (1966); these data are not available for "hornblende" and were estimated: $V^0 = 270.5$ cm^3, $S^0_{298} = 131.6$ e.u. In making the calculations the com-

---

* Entropy data for serpentine and brucite are from KING et al. (1967).
** High-temperature entropy data for diopside are from KELLEY (1960).

TABLE 2

	amphibole	garnet	pyroxene	biotite	rutile	opaque	quartz
Alkali olivine basalt							
amphibolite	× ×	×	×	×	tr	tr	—
eclogite	tr	× ×	× ×	×	tr	—	—
Quartz tholeiite							
amphibolite	× ×	×	×	—	—	tr	—
eclogite	tr	× ×	× ×	—	tr	—	tr

× × = major component; × = minor component; tr = trace.

pressibility of the solids was neglected, i.e. ΔV_{solids} = constant.

Because of the approximations involved in the estimation of volumes and entropies at high pressures and temperatures, the slopes and locations of the solid-solid reactions can only be regarded as approximate. Negative slopes for these curves cannot be ruled out at present.

Appendix II: Details of experimental procedure

The starting material for the synthesis runs were glasses of the appropriate composition prepared as described by GREEN and RINGWOOD (1967). Weighed amounts of glass (10–40 mg) were sealed in Au, Pt or Ag–Pd capsules with 5–10 wt% water added with a microburette. The capsules were sealed with a carbon arc-welder and reweighed to ensure against loss of the water. For reversals, products of earlier runs containing amphibolite and eclogite were mixed, weighed with additional water and rerun. The capsules were enclosed with a talc assembly which fitted into the standard graphite furnace and external talc sheath of the Boyd piston-cylinder apparatus. Pt-(Pt+10% Rh) or chromel–alumel thermocouples were used to measure and control temperatures to ±10 °C. Pressure (10–40 kb) was first applied then the cell was brought to temperature (600–900 °C) with the piston remaining on the compression stroke; a −10% pressure correction was applied to all runs. Synthesis run times were generally 12–24 hours, but successful reversals required at least 2–3 days. At temperatures >800 °C large pressure and/or temperature drops often occurred due to talc dehydration, and boron nitride was substituted for the talc internal to the graphite furnace. After the run was quenched in ≈30 s by cutting the power input, the capsule was removed, cleaned and reweighed. It was then punctured and examined for excess water, usually seen only when >2 mg water had initially been added. Runs were then heated to ≈120 °C for at least 1 hour and reweighed to measure the water loss insuring that excess water had been present during the run. The sample was then optically examined and powder photographs taken. Iron loss to the Pt capsule at these relatively low temperatures was probably restricted to a narrow 2–5 μ bleached zone observed at the edge of the sample after a run. Reducing conditions of an uncontrolled nature were guaranteed by the presence of the graphite furnace external to the sample.

Appendix III: Phases observed in the experimental runs

The phases observed in the alkali olivine basalt and the quartz tholeiite are listed in table 2. Feldspar was not positively identified in low pressure runs though small amounts of glass are difficult to distinguish from these phases. In synthesis runs on the quartz tholeiite at 27 kb and 700 °C, large weakly birefringent crystals with a low refractive index were tentatively identified as quartz. When amphibole is a major constituent, other minor phases (except for biotite) are difficult to distinguish with powder photographs because of the overlap with amphibole lines, and an optical identification is necessary. Small amounts of pyroxene can be distinguished from amphibole by their equant habit, and garnet by its characteristic shape, refringence and lack of birefringence. The trace amounts of amphibole in the eclogite may be quench material. For the alkali olivine basalt, some of the pyroxene may be orthopyroxene though much has probably reacted with the K-feldspar component to give biotite. The eclogitic pyroxene of the quartz tholeiite has ≈20% Jd and that of the alkali olivine basalt ≈ 10% Jd, as determined by the $\bar{2}21$ d-spacing (see ESSENE and FYFE, 1967).

References

Alderman, A. R. (1936) J. Geol. Soc. London **92**, 488–530.
Bearth, P. (1959) Schweiz. Mineral. Petrog. Mitt. **39**, 267–286.
Bearth, P. (1965) Schweiz. Mineral. Petrog. Mitt. **45**, 179–188.
Boyd, F. R. (1959) in: P. H. Abelson, ed., *Researches in geochemistry* (Wiley, New York).
Briere, P. Y. (1920) Bull. Soc. Franç. Mineral. **43**, 72–222.
Burnham, C. Wayne, J. R. Holloway and N. F. Davis (1968) College Earth Mineral. Sci. Penn. State Univ. preprint 18–19.
Essene, E. J. (1967) Ph.D. thesis, Dept. of Geology, Univ. of Calif., Berkeley.
Essene, E. J. and W. S. Fyfe (1967) Contrib. Mineral. Petrol. **15**, 1–23.
Green, D. H. and A. E. Ringwood (1967) Geochim. Cosmochim. Acta **31**, 767–833.
Greenwood, A. J. (1963) J. Petrol. **4** (3), 317–351.
Hezner, L. (1903) Tschermarks Mineral. Petrog. Mitt. **22**, 437–471, 505–573.
Kelley, K. K. (1960) U.S. Bur. Mines Bull. **584**, 232 pp.
King, E. G., R. Barany, W. W. Weller and L. B. Pankratz (1967) U.S. Bur. Mines Rept. Invest. **6962**, 1–18.
Kitahara, S., S. Takenouchi and G. C. Kennedy (1966) Amer. J. Sci. **264**, 223–233.
Kushiro, I., Y. Syono and S. Akimoto (1967) Earth Planet. Sci. Letters **3**, 197–203.
Lambert, I. B. and P. J. Wyllie (1968) Nature **219**, 5160, 1240–1241.
Lambert, I. B. and P. J. Wyllie (1970) Phys. Earth Planet. Interiors **3**, 316.
Luth, W. C. (1967) J. Petr. **8**, 372–416.
O'Hara, M. J. (1967) in: P. J. Wyllie, ed., *Ultramafic and related rocks* (Wiley, New York, London, Sydney).
Robie, R. A. (1966) in: S. P. Clark, Jr, ed., *Handbook of physical constants*, Geol. Soc. Am. Mem. **97**.
Robie, R. A., P. M. Bethke and K. M. Bardsley (1967) U.S. Geol. Surv. Bull. **1248**, 1–87.
Sahlstein, T. G. (1935) Medd. Gronland **95** (5), 1–43.
Sclar, C. B. (1970) Phys. Earth Planet. Interiors **3**, 333.
Sharp, W. E. (1962) Univ. of Calif. Rad. Lab. UCRL Rept. **7118**, 1–50.
Tilley, C. E. (1936) Mineral. Mag. **24**, 422–432.
Tilley, C. E. (1937) Mineral. Mag. **24**, 158, 555–568.
Yoder, H. S. (1955) Geol. Soc. Amer. Spec. Papers **62**, 505–524.
Yoder, H. S. and I. Kushiro (1969) Carnegie Inst. Wash. Geophys. Lab. Ann. Rept. 1967–1968, 161–167.
Yoder, H. S. and C. E. Tilley (1962) J. Petrol. **3** (2), 342–532.

EXPERIMENTAL DUPLICATION OF MINERAL ASSEMBLAGES IN BASIC INCLUSIONS OF THE DELEGATE BRECCIA PIPES

A. J. IRVING and D. H. GREEN

Department of Geophysics and Geochemistry, Australian National University, Canberra, Australia[*]

Subsolidus phase relationships have been determined with a piston-cylinder apparatus for compositions representing a garnet clinopyroxenite, a garnet-plagioclase clinopyroxenite and a spinel-garnet websterite from the Delegate basic breccia pipes. Modes of experimentally produced and natural assemblages have been compared to deduce conditions of final equilibration for these inclusions of 14–16 kb, 1050–1100 °C. Some important features of the genesis of the inclusions have been clarified, particularly the role of solid state exsolution phenomena. A relatively high-temperature origin (possibly as basaltic accumulates) is likely for all garnet pyroxenite inclusions in nephelinitic basalts and breccias.

1. Introduction

The garnet-bearing basic inclusions of the Delegate pipes in southeast Australia (LOVERING and WHITE, 1969) are intensively studied examples of a small class of inclusions (rich in pyroxenes, garnet and, in some cases, plagioclase) which have been also described from nephelinitic breccias and basalts of Hawaii (YODER and TILLEY, 1962; WHITE, 1966; JACKSON, 1966), New Zealand (DICKEY, 1968), Algeria (GIROD, 1967), Kenya (SAGGERSON, 1968) and elsewhere in eastern Australia (LOVERING and WHITE, 1964).

This paper is a report of an experimental study undertaken with a view to (a) delimiting conditions of equilibration of the Delegate inclusions and (b) elucidating some features of their genesis.

2. Procedure

Examples of three different inclusion types were selected for study from the material analysed by LOVERING and WHITE (1969):

Garnet clinopyroxenite R392: homogeneous aggregate of clinopyroxene (56%), garnet (35%) with narrow rims of fine clinopyroxene-spinel-"clay" aggregate (8.5%), and a trace of pyrite. Assignment of secondary phases to garnet implies primary weight percent proportions of approximately clinopyroxene 53 garnet 47.

Garnet-plagioclase clinopyroxenite R130: nonhomogeneous aggregate of clinopyroxene, garnet (rimmed by turbid, fibrous material and fine clinopyroxene, orthopyroxene, plagioclase and opaque) and plagioclase ($Ab_{61}An_{36}Or_3$) with accessory rutile, apatite and scapolite. The primary weight percent proportions of phases computed from total rock and mineral data are approximately clinopyroxene 45, garnet 31, plagioclase 22, accessories 2.

Spinel-garnet websterite R394: aggregate of very coarse clinopyroxene with finer interstitial orthopyroxene, scattered large spinel grains (rimmed by garnet) and a trace of olivine and sulphides. The clinopyroxene contains abundant oriented orthopyroxene lamellae, garnet blebs and minor spinel. The computed weight percent proportions are approximately clinopyroxene 68, orthopyroxene 10, garnet 12, spinel 10.

It is important to note that a mineralogical gradation exists between the Delegate garnet and garnet-plagioclase clinopyroxenites; the clinopyroxenes and garnets have very similar compositions in all these inclusions (LOVERING and WHITE, 1969), and certain finely-banded examples display both mineral assemblages in adjacent bands.

The experimental runs were carried out on homogeneous glasses prepared by fusing the natural rock

---
[*] Correspondence address: Department of Geophysics and Geochemistry, Australian National University, Box 4, P.O., Canberra, A.C.T. 2601, Australia.

powders in an argon atmosphere. The glasses were seeded with 10% of clinopyroxene + garnet of the appropriate composition. Chemical analyses of the starting materials (recalculated volatile-free from rock analyses incorporating redetermined Fe data) are presented in table 1. R392 and R394 are quite similar except for the Mg/Fe^{2+} ratio and both show affinities with an alkali-poor olivine tholeiite composition studied by GREEN and RINGWOOD (1967a); R130 resembles an alkali olivine basalt composition but has high-alumina character.

TABLE 1

Chemical composition of starting materials for experimental study

	Garnet-plagioclase clino-pyroxenite (R130)	Garnet clino-pyroxenite (R392)	Spinel-garnet websterite (R394)
SiO_2	46.26	46.55	45.12
TiO_2	0.56	0.44	0.59**
Al_2O_3	17.09	15.19	14.69
Cr_2O_3	0.01	0.16	0.05
Fe_2O_3	2.26*	0.95*	0.91*
FeO	9.67*	5.54*	4.47*
MgO	8.03	15.23	17.98
MnO	0.17	0.11	0.14
CaO	12.04	14.74	15.21
Na_2O	3.23	0.83	0.74
K_2O	0.21	0.22	0.06
P_2O_5	0.47	0.04	0.04
$100\ Mg/(Mg+Fe^{2+})$	60	83	88

* Redetermined by wet chemical analysis of glass (E. Kiss, analyst). Other values recalculated from analyses of LOVERING and WHITE (1969).
** Calculated from mineral analyses and mode (since the value given by LOVERING and WHITE (1969) is inconsistent with mineral data).

Subsolidus phase relationships were determined for each composition with piston-cylinder apparatus (piston-in technique) using a dry furnace assembly with boron nitride sleeve (cf. GREEN and RINGWOOD, 1967a). In general, graphite sample capsules were employed for runs below 1100 °C and sealed platinum capsules used to establish the solidi (although the latter were used exclusively for runs on R394). Run times of 1 hour (1400 °C, 1300 °C), 2.5 hours (1200 °C), 6 hours (1100 °C) and 18 hours (1000 °C) were used.

Relative proportions of garnet and clinopyroxene were estimated (especially for the garnet clinopyroxenite runs) by visual comparison of X-ray powder photographs with those of accurate garnet–clinopyroxene mixtures. Proportions of garnet, clinopyroxene and plagioclase in runs on the garnet-plagioclase clinopyroxenite composition proved more difficult to determine and were estimated from powder mounts and, in some cases, from thin-sections of charges. Electron microprobe analysis of selected run products is in progress.

3. Results

The three phase diagrams determined experimentally are described briefly below and essential features are shown on a composite diagram (fig. 1). Reported pressures include a -10% correction to the nominal load pressure calculated for the $\frac{1}{2}$-inch piston (GREEN et al., 1966).

3.1. Garnet clinopyroxenite R392

In order of increasing pressure, the subsolidus mineral assemblages observed for this composition are

clinopyroxene + plagioclase + olivine (below a),
clinopyroxene + plagioclase + spinel (below b),
clinopyroxene + plagioclase + garnet + spinel (below c),

and

clinopyroxene + garnet (above c, 13kb at 1100 °C). Minor amounts of amphibole occur in runs below 1100 °C and 18 kb. Within the field of clinopyroxene + garnet, subparallel lines of equal garnet percentages are oblique to the solidus; the minimum proportion of garnet is about 20% on the solidus at the low pressure boundary of the clinopyroxene + garnet field.

3.2. Garnet-plagioclase clinopyroxenite R130

At the investigated pressures (above 12 kb) this composition shows a broad field of clinopyroxene + garnet + plagioclase in varying proportions bounded above e (20.5 kb at 1100 °C) by a field of clinopyroxene + garnet (+ quartz?). Minor amphibole is again present in lower temperature and pressure runs.

3.3. Spinel-garnet websterite R394

Although the websterite composition closely resembles the garnet clinopyroxenite (except for a higher Mg/Fe^{2+} ratio), the phase diagram for R394 is quite different in having a subsolidus field of clinopyroxene + spinel + *orthopyroxene* bounded on the low temper-

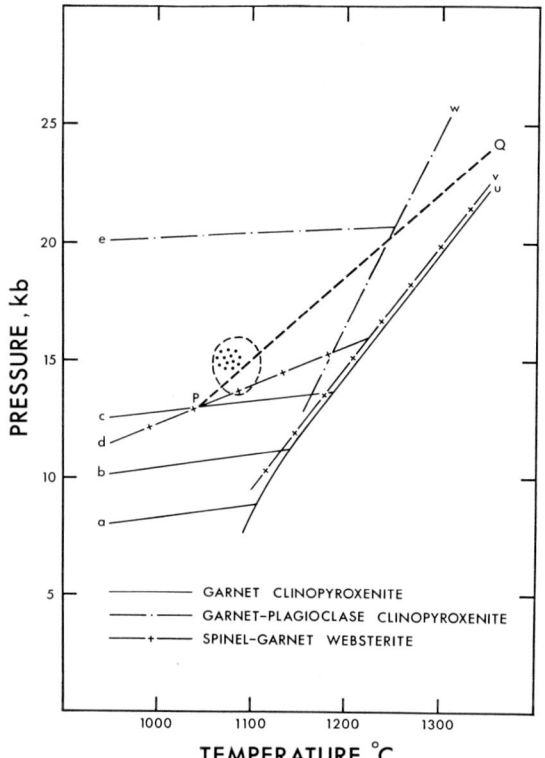

Fig. 1. Composite diagram illustrating phase relationships for the three Delegate compositions studied experimentally. Lines a, b, c, d, e define subsolidus phase fields (see text for description); lines u, v, w are the solidi. The line PQ (garnet percentage = 47), the dotted area, and the line d represent the ranges of conditions under which respectively the garnet clinopyroxenite (R392), the garnet-plagioclase clinopyroxenite (R130), and the spinel-garnet websterite (R394) are duplicated experimentally. The deduced conditions of final equilibration of these inclusions are within the field enclosed by dashes.

ature, high pressure side (along d) by a field of clinopyroxene + garnet + spinel.

4. Discussion

4.1. *Genesis of the Delegate inclusions*

Despite several isotope studies (e.g. LOVERING and TATSUMOTO, 1968), the relationships of the Delegate inclusions to associated basaltic materials has not been unambiguously ascertained. From the present study few deductions can be made concerning the original status of the inclusions, but some insight can be gained into the processes of development of the present mineral assemblages.

4.1.1. *Garnet clinopyroxenites and garnet-plagioclase clinopyroxenites.* As an alternative to formation by direct crystallisation of clinopyroxene and garnet, LOVERING and WHITE (1969) suggested that the Delegate garnet clinopyroxenites may be the end products of exsolution of garnet at high pressure from an original complex pyroxene aggregate, a hypothesis also proposed (GREEN, 1966) for garnet websterite inclusions in olivine nephelinite tuff at Salt Lake Crater, Hawaii. The fact that there is a minimum of 20% garnet 80% clinopyroxene on the solidus of R392 precludes an origin *wholly* by exsolution processes for the garnet of these rocks. It is possible that 20% of the observed garnet of R392 is primary and a further 27% has exsolved from pyroxene, however there is no textural evidence of exsolution in this or any other of the Delegate garnet clinopyroxenites examined.

For reasons outlined above, it seems likely that the garnet-plagioclase clinopyroxenites have had a similar type of origin to the garnet clinopyroxenites, and the occurrence of banded examples of both types (particularly composite examples) is at least consistent with their origin as accumulates from a magma. It should be noted however that the banded rocks do not show cumulus textures.

4.1.2. *Spinel-garnet websterite.* In the spinel-garnet websterite R394 there is excellent textural evidence for the exsolution of garnet, most orthopyroxene and minor spinel from original clinopyroxene as well as for the reaction of spinel with clinopyroxene to yield garnet. From the experimentally determined phase diagram, it is apparent that these changes could take place in an initial high-temperature pyroxene(s)–spinel(–olivine) assemblage as it cooled approximately isobarically to enter the clinopyroxene + garnet + spinel field (cf. GREEN, 1966). Entirely analogous textures have been figured for garnet websterites from Hawaii (100 Mg/(Mg+Fe^{2+}) = 85; YODER and TILLEY, 1962), Algeria (100 Mg/(Mg+Fe^{2+}) = 87; GIROD, 1967) and Kenya (100 Mg/(Mg+Fe^{2+}) = 84; SAGGERSON 1968), and have been observed in some inclusions from Kakanui, New Zealand (J. F. LOVERING, personal communication).

4.2. *Conditions of equilibration of the Delegate inclusions*

Possible *P–T* fields of formation for each of the three rocks studied could be delimited using knowledge of the modes of both experimentally produced and natural assemblages (see fig. 1). Although each of the assemblages could have crystallised over quite a wide range of physical conditions, there is a restricted P–T field (14–16 kb, 1050–1100 °C) within which *all* three could have equilibrated. Textural and mineral chemical evidence strongly suggests that the garnet and garnet-plagioclase clinopyroxenites formed under similar conditions (near 15 kb, 1100 °C), and it is possible for the spinel-garnet websterite to have also finally equilibrated under very similar conditions (near the boundary d of the garnet- and orthopyroxene-bearing fields of fig. 1), although this assemblage requires a history involving higher temperatures (perhaps up to 1200 °C).

The suggested range of conditions of final equilibration 14–16 kb, 1050–1100 °C implies an origin for these inclusions in the uppermost part of the Earth's mantle at a depth of about 50 km. It is notable that the deduced temperature is considerably above that anticipated along an average continental geothermal gradient (CLARK and RINGWOOD, 1964). Available data for garnet pyroxenites from Delegate, Hawaii, Kakanui and Algeria show that for this group partition coefficients of Fe^{2+}/Mg between coexisting garnet and clinopyroxene (cf. BANNO and MATSUI, 1965) are consistently low with a small range (1.9–2.8). By comparison with data for eclogites from glaucophanitic terranes, amphibolite–granulite terranes and inclusions in kimberlite (LOVERING and WHITE, 1969; BANNO and MATSUI, 1965), these data can be used to deduce a similar high-temperature origin for all garnet pyroxenites in nephelinitic basalts and breccias.

From available experimental data (GREEN and RINGWOOD, 1967b) it appears possible to obtain garnet clinopyroxenites and/or garnet-plagioclase clinopyroxenites as near-solidus accumulates from an alkali olivine basalt magma at about 15 kb. However, the inferred conditions of final equilibration of these inclusions are ca. 100 °C below the dry alkali olivine basalt solidus. Thus, *if* these inclusions do indeed represent local precipitates and the mineralogical banding is of accumulative origin, then a subsequent recrystallisation on cooling to ca. 1100 °C (15 kb) is required (explaining the absence of cumulus textures). The layering may alternatively represent flow banding made up of schlieren of early crystallised phases in a residual liquid finally crystallising as pockets of relatively Fe-rich garnet-plagioclase clinopyroxenite.

The rare spinel-garnet websterite with its high Mg/Fe^{2+} ratio could represent an original near-liquidus pyroxene(s)–spinel (–olivine) precipitate from an olivine basalt magma. However, precipitation from a dry magma would require conditions at least 70 °C *above* the websterite solidus, and it seems necessary in this scheme to invoke depression of the dry basalt liquidus (most likely by the presence of water). Such liquidus depression could also have occurred in magmas postulated as precipitating the garnet and garnet-plagioclase clinopyroxenites, particularly since hornblende occurs in many Delegate examples.

5. Conclusions

5.1. The inferred conditions of final equilibration of the Delegate garnet-bearing basic inclusions are in the P–T range 14–16 kb, 1050–1100 °C. Very similar high temperature conditions of formation are indicated for other garnet pyroxenites from nephelinitic basalts and breccias.

5.2. Certain features of these Delegate inclusions are consistent with an origin as local high pressure precipitates from a basaltic magma. In order to reconcile experimental basalt data it is necesssary to propose a hydrous parent magma and/or post-accumulation solid-state changes.

5.3. The Delegate garnet and garnet-plagioclase clinopyroxenites may be products of direct crystallisation of clinopyroxene, garnet and plagioclase. The garnet of these rocks cannot have originated entirely by exsolution from clinopyroxene. If solid state exsolution processes played a role in the development of these assemblages, evidence of their operation has been obliterated by recrystallisation.

5.4. The Delegate spinel-garnet websterite is inferred to be the end product of approximately isobaric cooling of a high temperature pyroxene(s)–spinel (–olivine) aggregate involving exsolution of garnet, orthopyroxene (and spinel) and the reaction of spinel with clinopyroxene. To judge from documented occurrences of similar rocks, such an origin (suggested by textural evidence) is indicated for those garnet pyroxenites characterised chiefly by 100 Mg/(Mg+Fe^{2+}) $\geq$ ca. 85.

Acknowledgements

The authors wish to thank Drs. J. F. Lovering and A. J. R. White for the samples of analysed Delegate rocks and for helpful discussions during the course of the study. The technical assistance of Mr. W. Hibberson is gratefully acknowledged.

References

BANNO, S. and Y. MATSUI (1965) Proc. Japan Acad. **41**, 716.
CLARK, S. P. and A. E. RINGWOOD (1964) Rev. Geophys. **2**, 35.
DICKEY, J. S. (1968) Amer. Mineralogist **53**, 1304.
GIROD, M. (1967) Bull. Soc. Franç. Minéral. Crist. **90**, 202.
GREEN, D. H. (1966) Earth Planet. Sci. Letters **1**, 414.
GREEN, D. H. and A. E. RINGWOOD (1967a) Geochim. Cosmochim. Acta **31**, 767
GREEN, D. H. and A. E. RINGWOOD (1967b) Contrib. Mineral. Petrol. **15**, 103.
GREEN, T. H., A. E. RINGWOOD and A. MAJOR (1966) J. Geophys. Res. **71**, 3589.
JACKSON, E. D. (1966) U.S. Geol. Survey Profess. Paper **550-D**, 151.
LOVERING, J. F. and M. TATSUMOTO (1968) Earth Planet. Sci. Letters **4**, 350.
LOVERING, J. F. and A. J. R. WHITE (1964) J. Petrol. **5**, 195.
LOVERING, J. F. and A. J. R. WHITE (1969) Contrib. Mineral. Petrol. **21**, 9.
SAGGERSON, E. P. (1968) Geol. Rundschau **57**, 890.
WHITE, R. W. (1966) Contrib. Mineral. Petrol. **12**, 245.
YODER, H. S. and C. E. TILLEY (1962) J. Petrol. **3**, 342.

GRANULITIC AND ECLOGITIC INCLUSIONS FROM BASIC PIPES IN EASTERN AUSTRALIA

J. F. LOVERING

Department of Geophysics and Geochemistry, Australian National University, Canberra, Australia

Several composite basic breccia and massive alkali basalt pipes in eastern Australia contain virtually identical suites of basic two-pyroxene granulite, garnet granulite, fassaite eclogite, spinel pyroxenite and rare ultramafic inclusions. Mineralogical and chemical studies are consistent with the hypothesis that the two-pyroxene granulites represent lower crustal material while the garnet granulites, fassaite eclogites and spinel pyroxenites represent uppermost mantle materials from beneath eastern Australia.

THE CHEMISTRY OF CLINOPYROXENES AND GARNETS OF ECLOGITE AND PERIDOTITE XENOLITHS FROM THE ROBERTS VICTOR MINE, SOUTH AFRICA

IAN D. MACGREGOR* and J. L. CARTER

Southwest Center for Advanced Studies, Dallas, Texas, U.S.A.

The eclogite xenoliths from the Roberts Victor pipe may be divided into two groups. The first group consists of coarse grained, subhedral to rounded garnets in a matrix of anhedral to interstitial clinopyroxene, and they may be subdivided on the basis of their mineralogy into kyanite and rutile-bearing eclogites. The second group consists of irregular, anhedral garnet and clinopyroxene with a tightly interlocking fabric. The garnets of the first group have greater contents of pyrope, and, for equivalent diopside: jadeite ratios, the clinopyroxenes have a greater enstatite-ferrosilite content than those of the second group. In addition, both the garnets and clinopyroxenes of the first group have greater K_2O and Cr_2O_3 and lower CaO contents than those of the second group. FeO and MnO are higher in the clinopyroxenes, and MgO, NiO, Li_2O and Na_2O higher in the garnets, of the first than the second group. Both groups appear to have formed at similar pressures, but distribution coefficients indicate equilibrium over a range of temperatures. Experimental data indicate that at pressures in the eclogite stability field, mafic liquids that precipitate crystals of eclogite mineralogy can also recrystallize to the same bimineralic assemblage. It is interpreted that the second group of eclogites are the sets of liquids, and that the first group of eclogites are the sets of crystals that have formed as the result of fractional crystallization in a high pressure magma chamber.

1. Introduction

Ultramafic and mafic xenoliths, with high pressure phase assemblages, are commonly found as xenoliths in the kimberlite pipes of South Africa (WILLIAMS, 1932; NIXON et al., 1963). Of fundamental importance is the question as to whether the nodules are accidentally related to, or cognate with the kimberlitic host magma. If accidental, there remains the problem of whether the samples may at least be related to each other or are entirely different groups of samples coming from different levels or regions in the mantle; if cognate, the nature of the association is important. The following represents a preliminary report of how the mineral chemistry of eclogites from the Roberts Victor Mine may relate to the above questions.

2. Roberts Victor xenoliths

2.1. General

Ultramafic and mafic xenoliths are common in the kimberlite pipe of the Kimberley area, South Africa. In nearly all the pipes ultramafic xenoliths are much more abundant than the mafic xenoliths. However, the Roberts Victor pipe is exceptional in this regard in that the exposed portion of the pipe has xenoliths which are mainly mafic in composition. Because of the erratic distribution of xenoliths in these pipes it is difficult to place a quantitative value on the ratio of mafic to ultramafic xenoliths in the different pipes, but a rough estimate would indicate a mafic to ultramafic ratio of approximately 5:95 for all pipes, except for the Roberts Victor occurrence, which has a mafic to ultramafic ratio of approximately 80:20.

2.2. Ultramafic xenoliths

The ultramafic xenoliths in the Roberts Victor mine are made up of harzburgite, garnet harzburgite and garnet lherzolite. The olivines have compositions ranging from 88 to 92 mole % forsterite, and the orthopyroxenes from 90 to 95 mole % enstatite. The chemistry of coexisting garnets and clinopyroxenes from garnet lherzolites is shown in figs. 4, 5 and 7.

2.3. Eclogite xenoliths

a) General

The eclogites form a suite of coarse grained granular rocks made up, by definition, of garnet and clinopy-

---

* New address: Department of Geology, University of California, Davis, California, 95616.
Contribution No. 93, Southwest Center for Advanced studies.

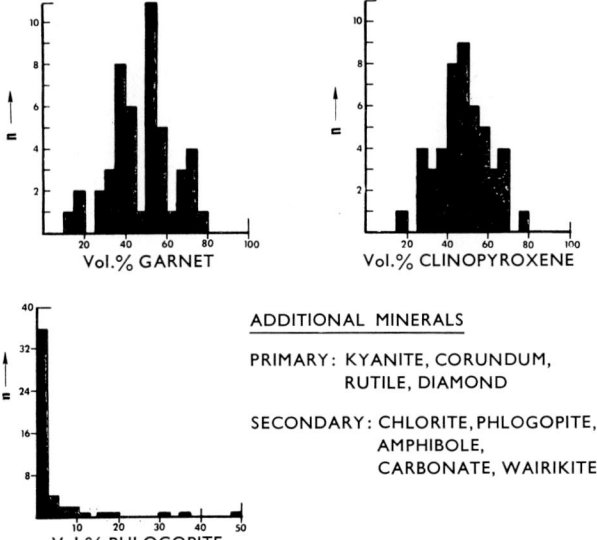

Fig. 1. Distribution of minerals in Roberts Victor eclogites.

modal variation of garnet and clinopyroxene from layer to layer. In addition some 5% of the eclogite xenoliths have a crude planar fabric marked by the roughly planar arrangement of elongated garnet and clinopyroxene grains.

It has been found that well defined mineralogical layering and the patchy distribution of phases is only associated with the group I eclogites, while the planar fabric is only found in xenoliths characterized as group II eclogites. Both group I and II eclogites also include samples that show no fabric or structure.

c) Textures

Eclogites of the first group (I) are composed of coarse grained, subhedral to rounded garnets set in anhedral to interstitial matrix of clinopyroxene (fig. 2). Occasionally, large poikilitic clinopyroxene crystals enclose subhedral garnets. Fine exsolution lamellae of another pyroxene are commonly seen in the clinopyroxene. Kyanite, rutile corundum and rarely diamond are primary accessory minerals (fig. 1). Chlorite, wairakite, carbonates and amphibole are common alteration products. Phlogopite is present in many samples (fig. 1); but whether it is a primary and/or a secondary mineral is still uncertain. The modal abundance of unaltered garnet and clinopyroxene is normally distributed (fig. 1), with most of the samples having approximately 50 volume % of each phase. Extreme variations from 50% of each phase often come from layered eclogites. Secondary reaction and alteration is a ubiquitous feature of all the eclogites. This is reflected by the presence of kelyphytic rim around the garnets, and the cloudy appearance of most clinopyroxenes (figs. 2 and 3).

It has been possible to divide the eclogites into two major groups (I and II). The group I eclogites may also be further subdivided on the basis of their accessory minerals (kyanite, rutile, corundum). The distinguishing characteristics of each group is given in later sections.

b) Structures

A number of different types of structures were observed. Most of the nodules (45%) have an even grained granular texture with a uniform distribution of phases. About 20% exhibit a patchy distribution of phases, and about 30% have a well defined layering, marked by the

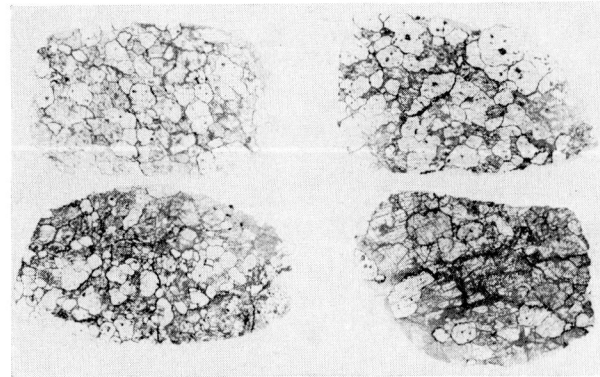

Fig. 2. Characteristic textures of group I eclogites (clear garnets, dusty clinopyroxenes).

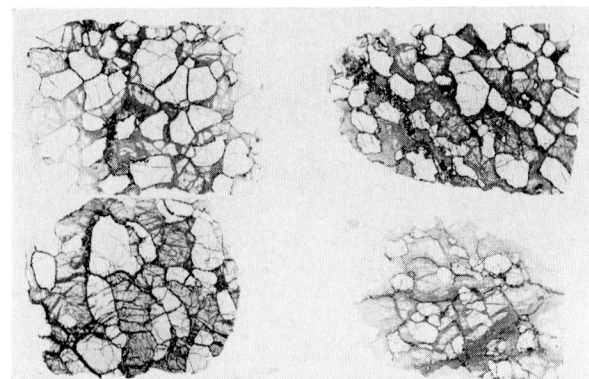

Fig. 3. Characteristic textures of group II eclogites (clear garnets, dusty clinopyroxenes).

roxenes. There is often a patchy distribution of the phases, the clinopyroxenes are more altered, and the garnets significantly more cloudy than those of the group II eclogites.

The group II eclogites consist of irregular, anhedral garnet and clinopyroxene grains which have a tightly interlocking texture (fig. 3). There is significantly less alteration of the clinopyroxene of the group II than the group I eclogites, and the group II garnets are generally clear and unaltered. Oriented inclusions of

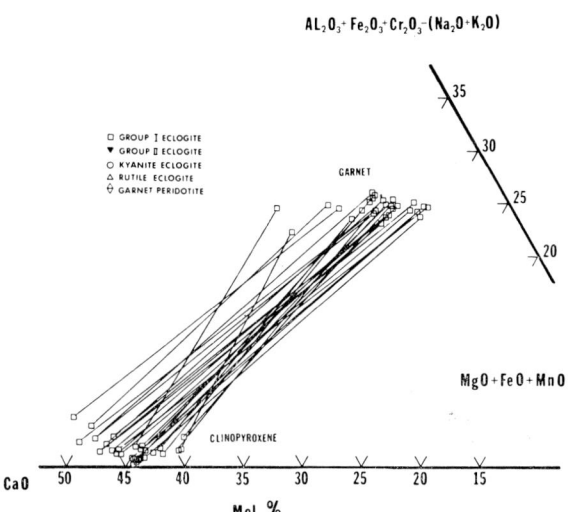

Fig. 6. ACF diagram showing the projected chemistry of coexisting phases from group I eclogites.

rutile in both the clinopyroxenes and garnet are a characteristic feature of the group II eclogites.

d) Chemistry*

The chemistry of the garnets may be summarized in the ternary plot, almandine – grossular – pyrope (fig. 4). The garnets show a fairly systematic trend from the pyrope-rich garnets, in the ultramafic rocks, to the almandine-rich eclogitic garnets. In this progression the grossular content increases. In some group I eclogites, garnets and clinopyroxenes of variable composition were found. Extreme compositions were separated and analyzed.

The chemistry of the clinopyroxenes is summarized in fig. 5. The clinopyroxenes vary in composition from being jadeite-poor in the ultramafic xenoliths, to jadeite-rich in the eclogite xenoliths. In addition, for equivalent diopside: jadeite ratios, the clinopyroxenes of the group II eclogites are systematically poorer in the enstatite-ferrosilite molecule, than the group I eclogites.

The chemistry of the coexisting phases is shown in the ACF diagrams (figs. 6 and 7). The group I eclogites (fig. 6) show a general correlation of increasing CaO content in the garnet with increasing CaO content and the A component in the clinopyroxenes. The kyanite and rutile subdivisions of the group I eclogites show a

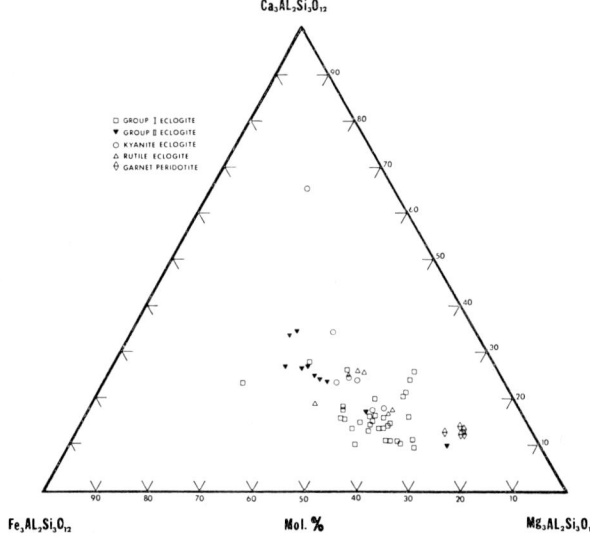

Fig. 4. Almandine: pyrope: grossular ratios of garnets from Roberts Victor eclogites and peridotites.

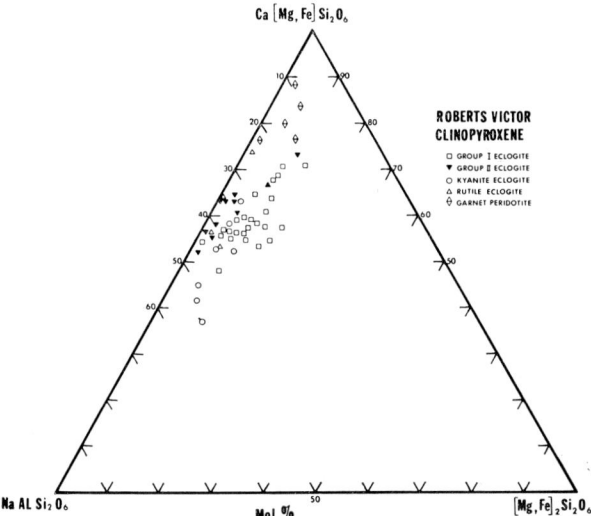

Fig. 5. Diopside: jadeite: enstatite-ferrosilite ratios of clinopyroxenes from Roberts Victor eclogites and peridotites.

* Minerals analyzed by atomic absorption spectrophotometer, Fe^{2+} and Fe^{3+} ratios calculated from atomic formulae.

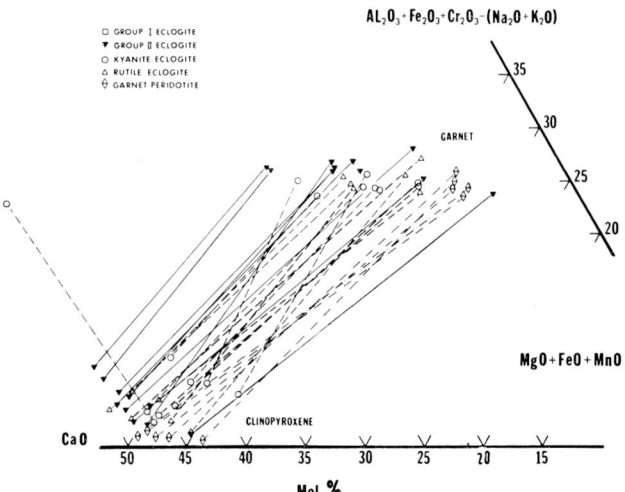

Fig. 7. ACF diagram showing the projected chemistry of coexisting phases from group II, kyanite and rutile bearing group I eclogites and garnet peridotites.

garnets and clinopyroxenes of the group II eclogites generally have a lower Cr_2O_3 and K_2O, and higher CaO content than the group I eclogites. FeO and MnO are higher in the clinopyroxenes, and MgO, NiO, LiO and Na_2O higher in the garnets of the group I, than the group II eclogites. It has been noticed that there is generally a good correlation between the chemistry of coexisting mineral pairs, indicating that the eclogites are equilibrium assemblages. Variation diagrams com-

continued increase of the CaO content in the garnets, with a concommitant increase in the CaO and A-components of the clinopyroxenes. The group II eclogites have garnets and clinopyroxenes which cover the whole range of chemistry exhibited by the group I eclogites.

A summary ACF diagram (fig. 8) shows the Roberts Victor eclogites in comparison with grosspydites from the Zagadochnya pipe (SOBOLEV et al., 1968). Fig. 8 shows a continuous variation of the garnet and clinopyroxene composition throughout the compositional range of eclogites from kimberlite pipes. Compared with the Roberts Victor eclogites, the continued trend of CaO enrichment in the garnets, and CaO and A-component enrichment in the clinopyroxenes of the eclogites from the Zagadochnya pipe is evident.

The ACF diagram may also be used to project the whole rock chemistry. For an essentially bimineralic rock, such as an eclogite, the mode may be used to plot the primary whole rock chemistry of each coexisting mineral pair (fig. 9). The group I eclogites form a cluster at lower C values, while the group II eclogites form a trend at A values of about 15 with a variation of the $C/(A+C+F)$ ratio from 25 to 70. Within the group I eclogites the kyanite and rutile-bearing eclogites are generally clustered at higher A values.

The mineral chemistry also reflects the different eclogite groups seen in the structures, textures and whole rocks recalculations on the ACF diagram (fig. 9). The

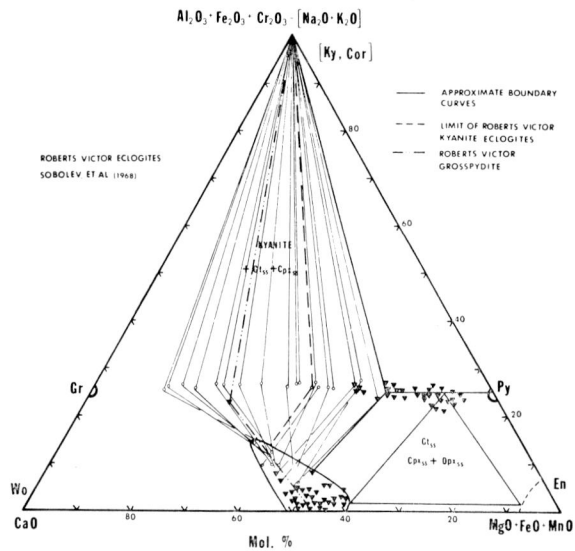

Fig. 8. Summary ACF projection showing data from Roberts Victor and Zagadochnya pipe (SOBOLEV et al., 1968).

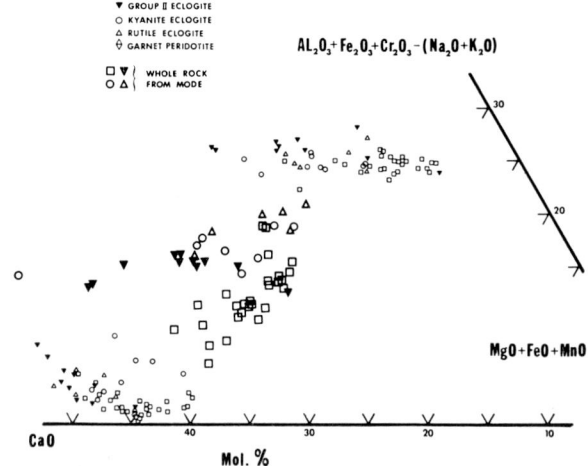

Fig. 9. ACF diagram showing the projection of the chemistry of garnets and clinopyroxenes from eclogites. Whole rock chemistry plotted from modal data on coexisting pairs.

paring mineral chemistry with a differentiation index show very little correlation of mineral chemistry to a differentiation index for group I eclogites, but a good correlation for the group II eclogites.

e) *Temperatures and pressures of formation*

The eclogite mineralogy of these mafic compositions indicate that the xenoliths have formed within the eclogite stability field (GREEN and RINGWOOD, 1966); the presently known limit of garnet and clinopyroxene compositions for grosspydites from the Roberts Victor pipe is comparable with the chemistry of coexisting phases, experimentally defined for similar compositions at 1100 °C and 36 kilobars (GREEN, 1967). There is little evidence which indicates whether the eclogite xenoliths have formed over a wide or narrow pressure range within the eclogite stability field. However, some indication of their relative temperatures of formation may be obtained by looking at distribution coefficients. For example, assuming that pressure and composition have a negligible effect, the distribution of Mg and Fe^{2+} (KUSHIRO and AOKI, 1968), and Mn^{2+} and Fe^{2+} (BANNO and MATSUI, 1965) between coexisting garnet and clinopyroxenes (figs. 10 and 11) indicate that the Roberts Victor eclogites have formed over a range of temperatures. This applies to both the group I and group II eclogites.

2.4. *Interpretation*

Structural, textural and chemical data indicated that there are two main groups of eclogites in the Roberts Victor pipe. The problem now arises as to whether they are accidentally or genetically related to each other, or to the kimberlite host magma. No data have been found which offer any comment on the nature of the relationship to the host magma. However, the data do show groups of eclogites that have similar characteristics and are probably internally related. The interrelation of the two eclogite groups is a further question. We propose that the group I and group II eclogites are related by the processes of crystal-liquid equilibria, and are portions of a high pressure magma, which has been intruded and cooled within the eclogite stability field. Specifically, the group II eclogites are interpreted as the set of successive mafic liquids that have formed by continuous fractional crystallization through gravitative accumulation of garnet and clinopyroxene crystals,

that now comprise the group I eclogites. The factors leading to this interpretation are listed as follows.

1) The association of two groups of eclogites in one kimberlite pipe is compatible with the hypothesis that they are genetically related.

2) Layered structures characteristic of crystal cumulates are only seen in the group I eclogites. The planar foliation seen in the group II eclogites is not reminiscent of settling structures, but rather of a metamorphic or flow structure.

3) The textures of the group I eclogites are compatible with an igneous cumulate origin. Subhedral to anhedral garnets and poikilitic clinopyroxenes are sug-

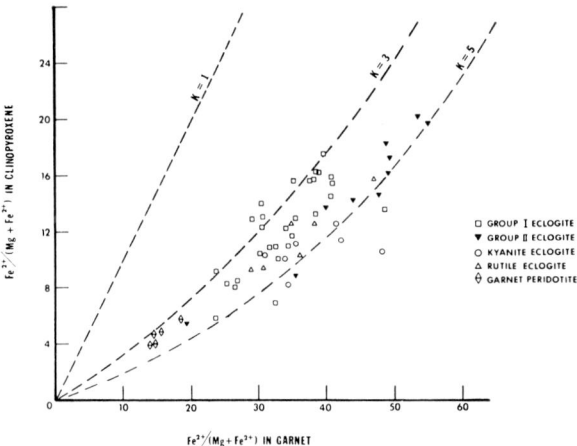

Fig. 10. Distribution of Mg and Fe^{2+} between coexisting garnets and clinopyroxenes.

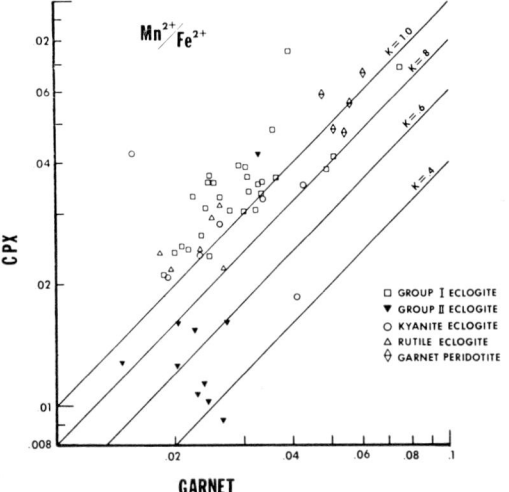

Fig. 11. Distributions of Mn^{2+} and Fe^{2+} between coexisting garnets and clinopyroxenes.

gestive of igneous cumulate textures. The crystals of the group II eclogite have no well defined shape and the textures are more compatible with a metamorphic origin.

4) Variation diagrams of the whole rock (CHEN, personal communication) and mineral chemistry generally show that the group II eclogites fall on simple curvilinear trends, while there is a considerably greater scatter of points for the group I eclogites. The different distribution is compatible with the interpretation that the group II eclogites are liquids, and the group I eclogites cumulates (BOWEN, 1928, p. 122).

5) Distribution coefficients (figs. 10 and 11) indicate that both groups of eclogites have formed over a range of temperatures; an unlikely result if each group is unrelated and derived from two different depths within the mantle. Further there is an overlap in their apparent equilibrium temperatures, and when plotted against a differentiation index such as the $C/(A+C+F)$ ratio (fig. 12) the group II eclogites show a general decrease of their apparent equilibrium temperatures; a result compatible with the hypothesis that the group II eclogites are successive sets of liquids.

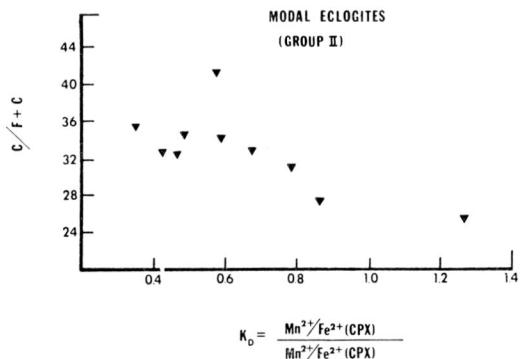

Fig. 12. Variation of the distribution of Mn^{2+} and Fe^{2+} between coexisting garnets and clinopyroxenes, and a whole rock differentiation index ($C/(A+C+F$, fig. 9).

6) Experimental data in the system CaO–MgO–Al_2O_3–SiO_2 (DAVIS, 1964; DAVIS and SCHAIRER, 1965) show that at high pressures (40 kb) fractionation of liquids, derived as partial melts from probable upper mantle compositions (garnet peridotite), result in movement of the liquids to a reaction point, P (fig. 13), where liquid + garnet + orthopyroxene + diopside coexist. Further fractional crystallization at temperatures lower than the reaction point yields crystal precipitates that are mixtures of garnet and clinopyroxene, and liquids that also crystallize to a mixture of clinopyroxene and garnet. Both the liquid and crystal phases finally yield rocks with an eclogite mineralogy. The interpretation of group I and group II eclogites is summarized in the ACF plot in fig. 13. Superposition of the experimental

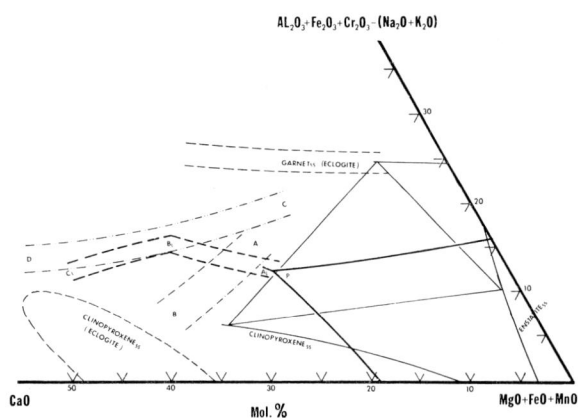

Fig. 13. Interpretive ACF projection showing relationship of group II eclogites (A_L-B_L-C_L; "liquids"), and group I eclogites (A-B, C-D, "crystals") to experimental data at 40 kilobars (DAVIS, 1964). P is experimental reaction point liquid + diopside + enstatite + garnet. (Solid lines – experimental data; dashed lines – interpretation of eclogites; heavy lines – liquid trends; light lines – crystal trends.)

data (DAVIS, 1964) on this plot (fig. 13) illustrates that, at least in the projection shown, the group II eclogites may be satisfactorily classifie das liquids. There are no experimental data on the proportions of crystal phases crystallizing from the liquids, and if the interpretation offered is correct it is possible that the Roberts Victor eclogites give a rough indication of the crystal trend. The systematic projection of kyanite and rutile-bearing eclogites to points above the liquid trend (figs. 9 and 13) has led to the distinction of the two separate crystal trends. It is proposed that the crystal trend A-B (fig. 13) represents the sets of garnets and clinopyroxenes that are precipitating from the liquid trend A_L–B_L; at B_L, it is interpreted that kyanite appears as a primary phase and the liquid changes direction towards C_L precipitating garnet + clinopyroxene + kyanite as crystal phases along the crystal trend C-D (fig. 13). It is apparent that the above interpretation still needs much further support to establish it as a valid model. At this state it is presented as a working hypothesis.

3. Summary and conclusions

It is proposed that the Roberts Victor eclogites may be divided into two groups; group I and II. The group I eclogites may be further subdivided on the basis of their mineralogy into kyanite and/or rutile-bearing eclogites and eclogites without any additional phases. The division is based on textures, structures mineral chemistry and recalculated whole rock chemistry.

The two groups are subsequently interpreted as being genetically related as a set of cumulates (group I eclogites) which have precipitated from a set of successively cooled liquids (group II eclogites) in a high pressure magma chamber. There is no evidence which points to either a cognate or accidental association of the eclogite xenoliths to the kimberlite host magma.

Acknowledgements

We are grateful to the Anglo American Corporation of South Africa and the DeBeers Consolidated Mines for the opportunity of making the collection which forms the basis of this study. Their generous cooperation and assistance has been critical to this study and is very much appreciated. In particular we would like to thank A. E. Waters, D. Hallam, C. Oosterveldt and B. Hawthorne for their time, local knowledge and hospitality.

This study has been supported by a National Science Foundation Grant (GP-5142). Their financial assistance is gratefully acknowledged.

References

BANNO, S. and Y. MATSUI (1965) Proc. Japan Acad. **41**, 716.
BOWEN, N. L. (1928) *The evolution of igneous rocks* (Dover Pub. Inc.).
DAVIS, B. T. C. (1964) Ann. Rept. Dir. Geophys. Lab. Carnegie Inst. Wash. **63**, 165.
DAVIS, B. T. C. and J. F. SCHAIRER (1965) Ann. Rept. Dir. Geophys. Lab. Carnegie Inst. Wash. **64**, 123.
GREEN, T. H. (1967) Contrib. Mineral. Petrol. **16**, 84.
GREEN, D. H. and A. E. RINGWOOD (1966) Geochim. Cosmochim. Acta **71**, 3589.
KUSHIRO, I., and K. AOKI (1968) Am. Mineralogist **53**, 1347.
NIXON, P. H., O. VON KNORRING and J. M. ROOKE (1963) Am. Mineralogist **48**, 1090.
SOBOLEV, N. V., JR., I. K. KUZNETSOVA and N. I. ZYUZIN (1968) J. Petrol. **9**, 253.
WILLIAMS, A. F. (1932) *The genesis of the diamond* (E. Benn Ltd., London).

ECLOGITES AND PYROPE PERIDOTITES FROM THE KIMBERLITES OF YAKUTIA

N. V. SOBOLEV

Institute of Geology and Geophysics, Siberian Branch of the USSR Academy of Sciences, Novosibirsk, U.S.S.R.

During the 15 years since the kimberlite pipes of Yakutia were discovered, Russian geologists achieved considerable progress in studying the composition of the kimberlites and xenoliths. A number of monographs are devoted to the petrography and mineralogy of these rocks as well as to geology of diamond-bearing occurrences (BOBRIEVICH et al., 1959; MILASHEV et al., 1963; BOBRIEVICH et al., 1964); there are a series of papers devoted to this subject as well. A number of general problems of metamorphism, such as eclogitization of hypersthene crystalline schists, were exemplified by the Yakutia kimberlite zenoliths (BOBRIEVICH and SOBOLEV, 1957) and granatization of spinel peridotites (SOBOLEV and SOBOLEV, 1964; LUTSS, 1965). A new type of deep-seated rock, grospydite, has been discovered (BOBRIEVICH et al., 1960) and a continuous series of pyrope-grossullarite garnets was found for the first time in natural conditions (SOBOLEV et al., 1966). For the first time diamond-bearing eclogites were investigated in greater detail (BOBRIEVICH et al., 1959; SOBOLEV and KUZNETSOVA, 1966) and also corundum eclogite (SOBOLEV, 1964; SOBOLEV and KUZNETSOVA, 1965).

The evidence available at present permits one to make some generalizations. On 1) the chemical composition of xenoliths of some deep-seated rocks; 2) the correlation between the kimberlite clinopyroxene composition and conditions of their formation; 3) on the nature of the effect of Cr_2O_3 admixture on the equilibrium parameters of pyrope peridotites; 4) on the nature of the distribution of the Fe/Fe+Mg ratio in coexisting garnets and pyroxenes, particularly in relation to the Na effect.

On the basis of the detailed study, xenoliths of pyrope-bearing rocks can be divided into several types. These are presumably pyrope peridotites and pyroxenites, being widespread in the pipes.

With respect to Cr distribution in the rocks and associated garnets and pyroxenes, one can distinctly distinguish the magmatic peridotites among them, very similar in general to peridotites of the Czech massif (FIALA, 1965), and the rocks with complex crystallization history formed by granatization of spinel peridotites (SOBOLEV and SOBOLEV, 1964, 1967).

Garnet-pyroxene assemblages with variable content of Ca in garnets and Na in pyroxenes can be related to typical eclogites.

Kyanite eclogites together with corundum eclogites and grospydites must be distinguished as an individual type characterized by distinct common features under the conditions of their formation, which are very close to P,T conditions, and by the Na effect produced on the stability of the continuous series of pyrope-grossularite garnets (SOBOLEV et al., 1966).

Moreover, diamond-bearing eclogites, distinguished as an individual type by the presence of diamonds, are characterized by distinct typomorphic peculiarities of the minerals, in particular, pyroxenes, with high Na_2O (4.40–6.57%), and by the presence of clinoenstatite molecules (SOBOLEV, 1968).

By their chemical composition, pyrope-bearing xenoliths form a continuous series from ultrabasic to basic varieties (SOBOLEV, 1968b). On the diagram of fig. 1, one can see a gradual increase in the Ca content, with localization of the separate field of kyanite eclogites and then grospydites. In these rocks one can also observe an increase in the Al_2O_3 content, in which they are very close to the undersaturated SiO_2 anorthosites. Broad variations of the quantitative-mineralogical compositions of the ultrabasic xenoliths and especially the

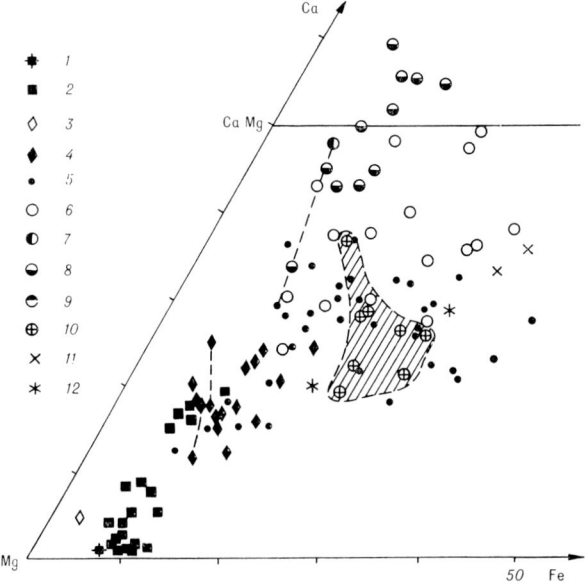

Fig. 1. Ca:Mg:(Fe^{2+}+Fe^{3+}) plot of the compositions of pyrope bearing xenoliths: olivinites (1); lehrzolites (2); enstatitite (3); websterites (4); eclogites (5); kyanite eclogites (6), corundum eclogite (7); grospydites (8); kyanite pyroxenite (9); diamond-bearing eclogites (10); average compositions of olivine tholeites after KUTOLIN (1968) – (11): average compositions of olivine basalt after NOCKOLDS (1954) – (12). The area plot of diamond-bearing eclogites is shaded. The plot of complex composition xenoliths are joined by dotted lines.

presence of unique xenoliths of complex composition (in fig. 1 the dots standing for various zones of such xenoliths are joined by a straight line) suggest a composite composition of the substrate from which these xenoliths were extracted, i.e. the Upper Mantle.

As shown by various authors (COLEMAN et al., 1965; SMULIKOWSKI, 1965), all the eclogites so far known can be divided by the garnet f-value ($f = \text{FeO}/(\text{FeO}+\text{MgO})$) into three large groups: 1. those associated with ultrabasic rocks ($f = 15–40\%$); 2. eclogites in gneisses ($f = 35–75\%$); 3. eclogites associated with glaucophane schists ($f = 75–90\%$).

The garnets from eclogitic xenoliths vary with respect to the f-value within a narrower range than would be expected for the eclogites as a whole (SOBOLEV, 1964), with f-value variations of 15%–57%. However, their compositions are not within the range of the 1st group, as suggested by several authors; rather they overlap the magnesian part of the 2nd group. In particular, garnets from the diamond-bearing eclogites can be entirely related to the 2nd group with variations of 33%–57% (BOBRIEVICH et al., 1959; SOBOLEV and KUZNETSOVA, 1966).

Of special interest is the composition of diamond-bearing eclogites. In table 1 are listed the average data based on 8 analyses of these rocks. The average composition turns out to possess a higher f-value than for previous data of eclogites from kimberlite pipes (KUTOLIN, 1968). However, if in counting the mean composition of eclogites without diamonds we exclude the rocks approaching the limiting paragenesis with orthopyroxene and olivine (i.e. garnet pyroxenites), then the same figures as for the diamond-bearing eclogites are obtained, being different only in having a just insignificant decrease in the Na content (table 1).

Hence, eclogites selected in this way from kimberlite pipes are most similar to several of the most magnesian eclogites of the 2nd group, in particular to kyanite eclogites and related rocks of lower and intermediate f-values (GODOVIKOV, 1968).

Corundum eclogite differs sharply (towards enrichment) in Al_2O_3 and CaO together with grospydites and related kyanite eclogites (SOBOLEV et al., 1968). They all are characterized by considerably higher contents of these oxides as compared with kyanite eclogites from the metamorphic complexes (table 1).

The experimental work provides evidence that the Ca/(Ca+Mg) ratios in clinopyroxenes coexisting with orthopyroxenes is almost independent of pressure, and

TABLE 1

Average compositions of eclogite xenoliths from kimberlite pipes

Rocks	N		SiO$_2$	TiO$_2$	Al$_2$O$_3$	Fe$_2$O$_3$	Cr$_2$O$_3$	FeO	MnO	MgO	CaO	Na$_2$O	K$_2$O
Eclogites	21	$\bar{x}$	46.13	0.60	14.34	4.49	0.06	7.15	0.20	13.54	9.60	1.28	0.49
		s	2.12	0.29	2.80	2.11	0.07	1.40	0.15	2.03	1.97	0.86	
Diamond-bearing	8	$\bar{x}$	46.00	0.56	14.08	3.42	0.09	7.68	0.21	13.57	9.50	2.26	0.68
eclogites		s	2.20	0.26	1.90	0.67	0.06	1.67	0.08	2.27	1.50	1.01	0.44
Grospydites (after	10	$\bar{x}$	43.73		27.48	1.67		2.77		7.53	12.14	1.53	0.66
SOBOLEV et al., 1968)			1.74		2.96	1.22		0.70		1.12	2.25	0.41	0.35

N – number of samples, $\bar{x}$ – arithmetical means, s – standard deviation.

is mainly a function of equilibrium temperature (BOYD and SCHAIRER, 1964; DAVIS and BOYD, 1966). This ratio for xenocrysts of diopside from kimberlite and from peridotite xenoliths of the South African pipes were considered by Boyd (BOYD, 1967), who collected data from 40 specimens, chiefly analysed chemically and by the microprobe. It was shown that the temperatures of equilibrium of these diopsides vary in a wide range from 900° to 1300 °C with a maximum between 900°–1000 °C. We have made an attempt to generalize such material with reference to the kimberlites of Yakutia, among them our unpublished data.

On the histogram (fig. 2), plotted with separate symbols, are the data on pyroxenes from the kimberlites of Africa and Yakutia, as well as bedrock occurrences of peridotites. Though the number of analyses is twice as large (90) compared with Boyd's data, the general nature of the distribution is entirely the same. Only a few analyses fall into the high temperature field between 1200° and 1300 °C. Among these are pyroxenes studied by us from the pyrope lehrzolites of the Udachnaya pipe (table 2, analyses 4, 5) with $Ca/(Ca+Mg)$ ratios of 41% and 37%, whose garnets are characterized by increased content of Cr_2O_3 (5.00% and 7.15%).

It should be stressed that pyroxenes from the diamond-bearing eclogites are characterized by a decreased $Ca/(Ca+Mg)$ ratio, attained in the specimens ap-

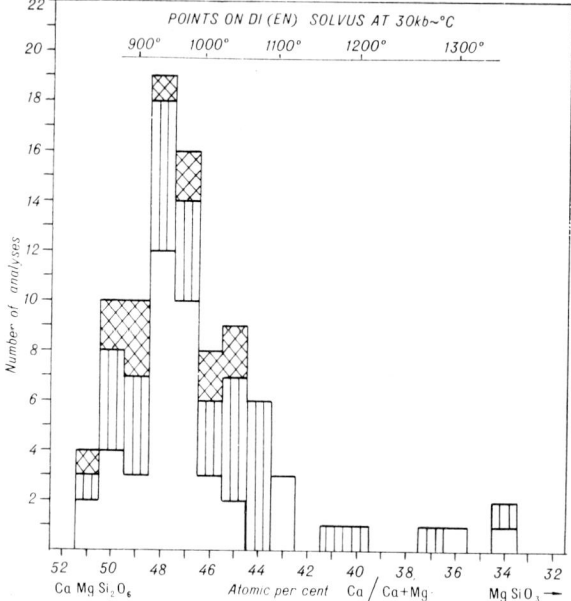

Fig. 2. Histograms of values for the ratio $Ca/(Ca+Mg)$ in diopsidic pyroxenes from African kimberlites – white area (after BOYD, 1967); from Yakutian kimberlites – striped area; from pyrope peridotites known in bedrock occurrence – crossed area.

proaching the limiting association up to 43%, which is their typomorphic peculiarity, distinguishing them from the pyroxenes of many other eclogites.

Apart from this specific feature, the pyroxenes from diamond-bearing eclogites are characterized by high

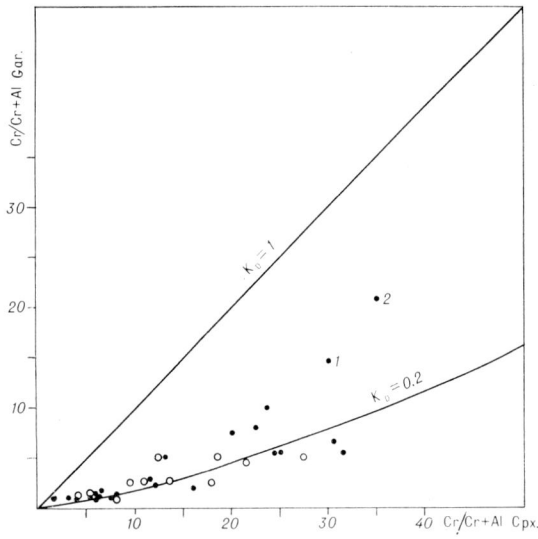
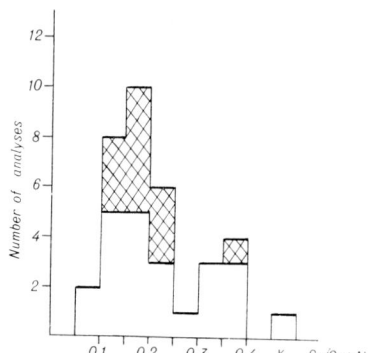

Fig. 3. The distribution diagram for $Cr/(Cr+Al)$ ratio in garnets and clinopyroxenes from peridotites known in a) bedrock occurrences – open circles; b) kimberlite pipes – solid circles. Solid circles numbered 1 and 2 – plot of Ud-3 and Ud-5 analyses pairs (see table 2). Histograms for the K_D value. White area – for xenoliths minerals analyses; crossed area – for outcropped peridotite mineral analyses.

Na (0.31%–0.46%). Taken as a whole, these specific features will enable one to separate them into a special group (SOBOLEV, 1968).

To check the possible effect of Cr_2O_3 on the parameters of equilibrium of the pyrope peridotite minerals, we investigated the nature of the distribution of the ratio $Cr/(Cr+Al)$ in coexisting garnets and pyroxenes. We used data from the literature and our unpublished data on the Cr_2O_3 content in the minerals, down to 0.2% (35 pairs).

Though the Cr_2O_3 content in the clinopyroxenes is lower than that in associated garnets, the $Cr/(Cr+Al)$ ratio is much higher, varying in accordance with its variation in the garnets (SOBOLEV and SOBOLEV, 1967). As shown in fig. 2, the value for the distribution coefficient (K_D) (KRETZ, 1961) for 35 pairs varies in the majority of cases (26 pairs) from 0.05 to 0.25 exceeding 0.3 in only 8 cases.

As shown in the distribution diagram (fig. 3), the ratio $Cr/(Cr+Al)$ in garnets for the majority of the analysed pairs does not exceed 10%. For garnets with high Cr_2O_3 content there is no evidence in the literature. There are only individual analyses of pyropes without those of pyroxene containing about 7% of Cr_2O_3 (NIXON et al., 1963; FIALA, 1965). We found in the pipe "Udachnaya" in Yakutia some xenoliths of pyrope harzburgites and lehrzolites whose garnets are characterized by increased Cr_2O_3 content. In a number of specimens there are all the transitions of the pyrope composition with 5% to 11% of Cr_2O_3, with CaO content typical of all the garnets from peridotites. In the present paper are given the data for two garnet-pyroxene pairs (table 2) which permit one to draw some conclusions. For most high temperature pyroxenes in the ones investigated associated with garnets rich in chromium, the coefficient of the distribution of the $Cr/(Cr+Al)$ ratio is much larger than for most of compared pairs, i.e. 0.36–0.49. Proceeding from the data available, one can assume that entering of chromium, into silicates in particular, in pyrope occurs not only at elevated pressure (SOBOLEV and SOBOLEV, 1967) but also at increased temperatures. The data on the garnet composition in diamonds support this evidence (MEYER, 1967, 1968; FUTERGENDLER, 1958) for which most probably the highest temperatures occur. In the inclusions in the diamonds investigated, including our unpublished data, the majority of pyropes are characterized by high Cr content and high refractive index, attaining 1.790. Apparently, the value of the distribution coefficient of the $Cr/(Cr+Al)$ ratio for associated pairs rich in chromium will approach 0.5 simultaneously with high temperature nature of the pyroxenes.

TABLE 2

Chemical analyses of garnets and clinopyroxenes from some peridotite xenoliths of the kimberlites of Yakutia

	Garnets			Clinopyroxenes		
	Ud-3	Ud-5	M-60	Ud-3	Ud-5	M-60
SiO_2	41.75	41.68	41.35	54.05	53.61	53.82
TiO_2	0.16	0.80	0.19	0.15	0.48	0.64
Al_2O_3	19.66	16.58	22.74	2.66	1.48	7.96
Cr_2O_3	5.07	7.15	0.26	1.61	1.10	0.29
Fe_2O_3	1.99	2.30	2.76	0.90	1.30	2.34
FeO	5.82	5.39	10.26	1.20	1.97	1.13
MnO	0.38	0.27	0.21	tr.	0.06	0.08
MgO	19.24	20.40	18.81	19.11	20.81	12.32
CaO	5.68	5.90	4.34	18.30	16.85	16.66
Na_2O	—	—	—	1.79	1.25	4.26
K_2O	—	—	—	0.25	0.18	0.05
loss ign.	—	—	—	—	0.49	—
Total	99.75	100.47	99.97	100.02	99.67	99.53
Si	2.992	2.956	2.967	1.942	1.942	1.934
Ti	0.008	0.044	0.013	0.004	0.013	0.017
Al^{iv}	1.678	1.436	1.923	0.058	0.058	0.066
Al^{vi}				0.056	0.009	0.276
Cr	0.287	0.414	0.017	0.047	0.035	0.009
Fe^{3+}	0.035	0.123	0.060	0.026	0.035	0.061
Fe^{3+}	0.070	—	0.095	—	—	—
Fe^{2+}	0.352	0.330	0.617	0.037	0.061	0.035
Mn	0.026	0.017	0.013	—	—	0.002
Mg	2.074	2.229	2.009	1.023	1.125	0.659
Ca	0.443	0.462	0.259	0.703	0.653	0.642
Na	—	—	—	0.124	0.086	0.298
K	—	—	—	0.008	0.008	0.002
Excess						
Si	0.030	0.101	—	—	—	—
n	1.755	1.759	—	—	—	—
a_0 (Å)	11.557	11.579	11.495	—	—	—

To roughly estimate the equilibrium temperatures one often uses the diagrams of the f-value distributions (KRETZ, 1961; SOBOLEV, 1964). In the diagram of fig. 4 are listed the extensive data on the associated garnets and pyroxenes (83 pairs) from xenoliths partially representing our unpublished data. The variation of the value of the distribution coefficient for different types of xenoliths is shown in the histograms.

As to peridotites, most of high temperature assemblages E3 and Ud-5 (table 2) are characterized by an

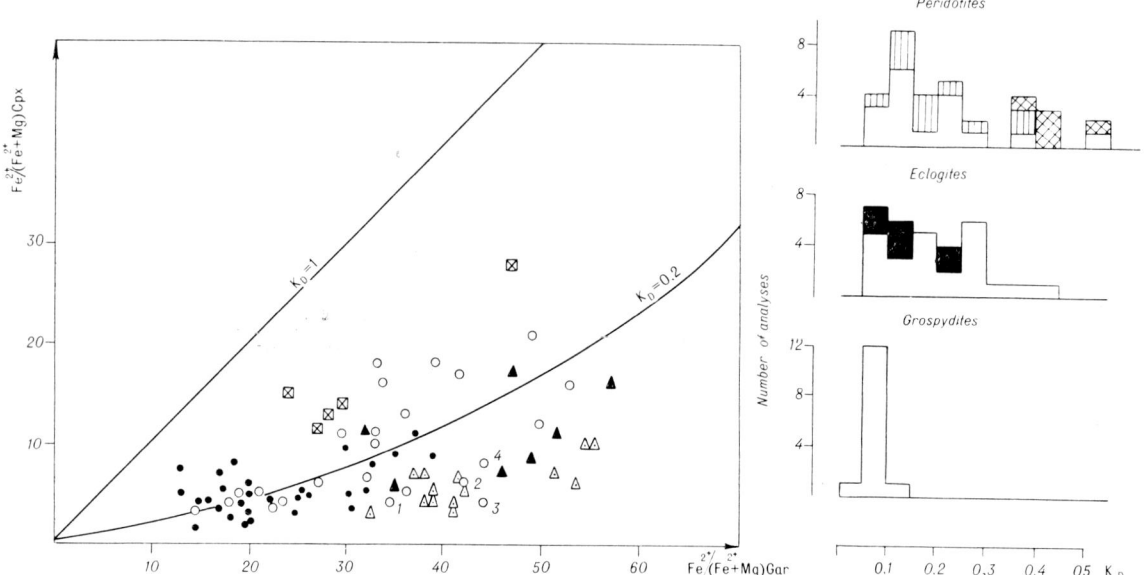

Fig. 4. The distribution diagram and histograms of K_D for the ratio $FeO/(FeO+MgO)$ in xenoliths garnets and clinopyroxenes from: peridotites (solid circles); eclogites (open circles; diamondbearing eclogites (solid triangles); grospydites (open triangles); peridotites form alkali basaltoids (crossed squares); open circles numbered 1–4 – plot of eclogite minerals with high jadeite content in pyroxene (see text). In upper histograms: white area – analyses of peridotite xenoliths minerals; striped area – those of outcropped peridotites; crossed area – analyses of minerals of peridotite xenoliths from alkali basaltoids. In "eclogite" histograms: black area – analyses of diamond-bearing eclogite minerals.

increased distribution coefficient equal to 0.54 and 0.36, respectively. For the main number of analyses of peridotite xenoliths and peridotites known from the bedrock occurences, K_D ranges between 0.1 and 0.25. Extremely high K_D are characteristic of the garnet-pyroxene pairs of peridotite and two-mineral assemblages from xenoliths in the alkali basaltoids of the Hawaiian Islands, Central Sahara and New Zealand (GIROD, 1967; DICKEY, 1968), which along with the low $Ca/(Ca+Mg)$ ratio for pyroxenes (YODER and TILLEY, 1962), confirm their high temperature nature.

For eclogite pairs K_D also strongly varies. Some of the analyses, including almost all the diamond-bearing eclogites, are characterized by a low K_D value (0.05–0.15) which, in part, seem to be connected with higher pressures affecting the side opposite to temperatures. For especially low points of the diagram with minimum K_D, numbered 1–4 in fig. 4, of special importance is the very high jadeite content in the pyroxenes, 41% or 45% (our data). The same holds true for the diamond-bearing eclogites containing 0.31%–0.41% of Na.

High Na content is the most distinct effect produced on the K_D value for 12 grospydites pyroxenes containing 0.31%–0.52% Na (SOBOLEV et al. 1968).

Thus the distribution coefficient for the f-value of garnets and pyroxenes is a function of, at least, three independent variables, P, T and Na concentration, and cannot be applied directly even at a rough determination of the temperature. This is possible only when other conditions one equal.

In addition to the above remarks on the role of Na in pyroxenes, in table 2 are given new analyses of coexisting garnet and pyroxene from xenolith of pyrope websterite from the pipe "Mir", characterized on the one hand by decreased Ca content in the garnet; in place of the 13.3 ± 1.9% of Ca component typical of the average garnets from ultrabasic rocks (SOBOLEV, 1964) we have here only 8.7%, with simultaneous increase of Na content in the pyroxene. The garnet contains 67.3% pyrope, 23.7% almandine, 0.3% spessartine, 4.8% grossularite, 3.1% andradite and 0.8% uwarovite, with $f = 25.2\%$, $N = 1.747$, $a = 11.495$ Å. These analyses confirm the conclusions by Banno (BANNO, 1967) that the Na content should be correlated not only with the change in the Ca content in garnets from kyanite eclogites and grospydites established by us, but also in the assemblage garnet-clinopyroxene-orthopyroxene. We believe that in this case there takes place essentially the same reaction of Ca–Mg garnet decomposition to form clinopyroxene with a jadeite component.

It should be stressed that in the diamond-bearing kyanite eclogites, i.e. higher pressure formations than grospydites and associated kyanite eclogites, the pyroxenes could be distinguished from the latter by higher Na content at constant $FeO/(FeO+MgO)$ ratio.

Conclusions

By assuming the deepseated rocks of the kimberlite pipes to be xenoliths of the Upper Mantle one can make some assumptions on the strong differentiation in the Upper Mantle material both with respect to the main oxides and with respect to Na_2O and FeO. Some rock types found in the pipes have not so far been encountered in the Earth's crust.

It has been confirmed that the temperature of crystallization of clinopyroxenes from the kimberlites ranges between 900° and 1300 °C, to confirm higher temperatures also for the deepest diamond-bearing eclogites.

The correlative nature of the distribution of the $Cr/(Cr+Al)$ ratio in associated garnets and pyroxenes is demonstrated for the higher $Cr/(Cr+Al)$ ratio in the pyroxenes. A tendency to an increase in the distribution coefficient with increasing temperatures has been revealed. This also confirms the tendency to a rise in the crystallization temperature with depth.

The K_D value of the $FeO/(FeO+MgO)$ ratio in the garnet-clinopyroxene pairs varied over a relatively wide range, depending on a number of parameters including P, T and Na concentration, with a relative decrease in K_D value resulting in a possible increase in the Na content.

References

BANNO, S. (1967) Effect of jadeite component on the paragenesis of eclogitic rocks, Earth Planet. Sci. Letters **2**, 249–254.
BOBRIEVICH, A. P. and V. S. SOBOLEV (1957) Eclogitization of pyroxene crystalline schists of Archean complex, Zapiski Vses. Miner. Obshchestva, **86**, N4 (in Russian).
BOBRIEVICH, A. P., G. I. SMIRNOV and V. S. SOBOLEV (1959) The xenolith of eclogite with diamond, Dokl. Akad. Nauk SSSR, **126**, N3 (in Russian).
BOBRIEVICH, A. P., G. I. SMIRNOV and V. S. SOBOLEV (1960) On the mineralogy of xenoliths of the grossular-pyroxene-disthene rocks (grospydite) from kimberlites of Yakutia, Geol. i Geofyz. Akad. Nauk SSSR Sibirsk. Otd., N3 (in Russian).
BOBRIEVICH, A. P. et al. (1959) *The diamond deposits of Yakutia* (Gosgeoltekhizdat) Moscow, 526 p. (in Russian).
BOBRIEVICH, A. P. et al. (1964) *Petrography and mineralogy of the kimberlite rocks of Yakutia* (Nedra, Moscow) 190 p. (in Russian).
BOYD, F. R. (1967) Electron probe study of diopsidic pyroxenes from kimberlites, Carn. Inst. Wash. Yearbook **65**, 252–260.
BOYD, F. R. and J. F. SCHAIRER (1964) The system $MgSiO_3$–$CaMgSi_2O_6$, J. Petrol. **5**, 275–309.
COLEMAN, R. G., D. E. LEE, L. B. BEATTY and W. W. BRANNOCK (1965) Eclogites and eclogites: their differences and similarities, Geol. Soc. Am. Bull. **76**, 483–506.
DAVIS, B. T. C. and F. R. BOYD (1966) The join $Mg_2Si_2O_6$–$CaMgSi_2O_6$ at 30 kb pressure and its application to pyroxenes from kimberlites, J. Geophys. Res. **71**, 3567–3576.
DICKEY, J. S. (1968) Eclogitic and other inclusions in the mineral breccia member of the Deborah volcanic formation at Kakanui, New Zealand, Am. Mineralogist **53**, 1304–1319.
FIALA, I. (1965) Pyrope of some garnet peridotites of the Czech massif, Kristallinikum, N 3.
FUTERGENDLER, S. I. (1958) X-ray study of solid mineral inclusions in diamond, Kristallografiya **3**, 494–496 (in Russian).
GIROD, M. (1967) Données petrographiques sur de pyroxenolites à grenat en enclaves dans les basalts du Hoggar (Sahara central), Bull. Soc. Franc. Mineral. Crist. **90**, 202–213.
GODOVIKOV, A. A. (1968) The CaO : MgO : FeO ratio and mineral composition of eclogites, Doklady Ac. Sci. USSR **179**, 3, 668–671.
KRETZ, R. (1961) Some applications to coexisting minerals of variable composition. Examples: orthopyroxene-clinopyroxene and orthopyroxene-garnet, J. Geol. **69**, N4.
KUTOLIN, V. A. (1968) Petrochemical peculiarities of various formational types basalts and the composition of the Earth's upper mantle, in Repts. of Sov. Geol. XXIII Sess. IGC, Probl. 1. 90–97 (in Russian).
LUTTS, B. G. (1965) Reactions of eclogitization in deepseated rocks, Geol. Ore Deposits, N5 (in Russian).
MEYER, H. O. A. (1968a) Mineral inclusions in diamonds, Carn. Inst. Wash. Yearbook **66**, 446–450.
MEYER, H. O. A (1968b) Chrome pyrope: an inclusion in natural diamond, Science **160** (3835) 1447.
MILASHEV, V. A., M. A. KRUTOYARSKY, M. I. RABKIN and E. N. ERLIKH (1963) Kimberlites and picritic porphyres of Northeastern part of the Siberian platform, Trans. Arctic Inst. (NIIGA), vol. 126 (in Russian).
NIXON, P. H., O. VON KNORRING and J. M. ROOKE (1963) Kimberlites and associated inclusions of Basutoland: a mineralogical and geochemical study, Am. Mineralogist **48**, N9–10.
NOCKOLDS, S. R. (1954) Average chemical composition of some igneous rocks, Bull. Geol. Soc. Am. **65**, N10.
SMULIKOWSKI, K. (1965) Chemical differentiation of garnet and clinopyroxene in eclogites, Bull. Acad. Pol. Sci., Ser. Sci. Geol. Geogr., **13**, N1.
SOBOLEV, N. V. (1964a) *Paragenetic types of garnets* (Nauka) (in Russian).
SOBOLEV, N. V. (1964b) Eclogite xenolith with ruby, Dokl. Akad. Nauk SSSR **157**, N6 (in Russian).
SOBOLEV, N. V. (1968a) Eclogite clinopyroxenes from the kimberlite pipes of Yakutia, Lithos **1**, N1, 54–57.
SOBOLEV, N. V. (1968b) The xenoliths of eclogites from the kimberlite pipes of Yakutia as fragments of the upper mantle substance, XXIII IGC, Vol. 1, 155–163.

Sobolev, N. V. and I. K. Kuznetsova (1965) Recent data on the mineralogy of eclogites from kimberlite pipes of Yakutia, Dokl. Akad. Nauk SSSR **163**, N2 (in Russian).

Sobolev, N. V. and I. K. Kuznetsova (1966) Mineralogy of diamond-bearing eclogites, Dokl. Akad. Nauk SSSR **167**, N6 (in Russian).

Sobolev, N. V., I. K. Kuznetsova and N. I. Zyuzin (1968) The petrology of grospydite xenoliths from the Zagadochnaya kimberlite pipe in Yakutia, J. Petrol. **9**, 253–280.

Sobolev, N. V., N. I. Zyuzin and I. K. Kuznetsova (1966) A continuous range in the series of pyrope-grossularite garnets in grospydites, Dokl. Akad. Nauk SSSR **167**, N4.

Sobolev, V. S. and N. V. Sobolev (1964) The xenoliths in kimberlites of Northern Yakutia and problems of the Earth's Mantle composition, Dokl. Akad. Nauk SSSR **158**, N1 (in Russian).

Sobolev, V. S. and N. V. Sobolev (1967) On chrome and chrome-bearing minerals in deepseated xenoliths of the kimberlite pipes, Geol. Ore Depos., N2 (in Russian).

Yoder, H. S., Jr. and C. E. Tilley (1962) Origin of basalt magmas. An experimental study of natural and synthetic rock systems, J. Petrol. **3**, N3.

CLASSIFICATION OF ECLOGITES IN TERMS OF PHYSICAL CONDITIONS OF THEIR ORIGIN

SHOHEI BANNO

Department of Earth Sciences, Kanazawa University, Kanazawa, Japan

The apparent Fe–Mg distribution coefficient between garnet and clinopyroxene K' is defined by

$$K' = \left(\frac{X_{Fe}}{X_{Mg}}\right)^{ga} \bigg/ \left(\frac{X_{Fe}}{X_{Mg}}\right)^{cpx}.$$

The effects of pressure, temperature, and chemistry of the rocks on K' were examined, using formula volumes of end members of these minerals, and the K' values of natural eclogites. It is shown that K' increases with increasing pressure, and decreases with increasing temperature, and that the effect of chemistry on K' is usually not large.

It is shown that the eclogite types defined by the geological mode of occurrences are also characterized by particular range of K' values, and that the relative temperatures of eclogite crystallization as estimated by K' and by ordinary petrological considerations are usually in harmony with each other. Further, K' can be applied to distinguish the difference in temperatures among the eclogites of some types. The crystallization temperature of eclogites increases in the following order, which is shown by localities: Colombia, Ural and Guatemala, California, Alps and Japan, Bavaria and Spain, Norway, East Sudetes, and granulite facies eclogites. Eclogite inclusions in basalt and in kimberlite represents highest temperature but the former represents higher pressure than the later.

The crystallization temperature of some eclogites which are not included above are also discussed.

1. Introduction

In recent years, the genesis of eclogite, garnet-clinopyroxene rock with basaltic composition, has been discussed by many authors in relation to the status of upper mantle materials. The argument that eclogite is a high pressure modification of basalt may be traced back at least to GRUBENMANN (1904), and classical papers by ESKOLA (1921) and GOLDSCHMIDT (1922) established the basis for later approaches. Recent contributions to this problem may be represented by the papers by GREEN and RINGWOOD (1967) and RINGWOOD and GREEN (1966), who, based on experimental work on the basalt to eclogite transformation, have extensively discussed the petrological, geophysical and tectonophysical aspects of this problem.

Petrologically, it is considered that there are several types of eclogites, each representing a particular mode of occurrence and probably representing a particular field of temperature and pressure of formation. Therefore, there are "eclogites and eclogites" (COLEMAN et al., 1965): some eclogites are crustal basic metamorphics and the others are mantle materials. Not all eclogites came from the mantle.

Geologically, the following mode of occurrences of eclogites or basaltic garnet-clinopyroxene rocks are knwon:

1) Eclogite inclusions in kimberlite;
2) Eclogite inclusions in alkali basalt;
3) Eclogite or pyrope-diopside rock inclusions in peridotite;
4) Granulite facies eclogite or garnet-clinopyroxene granulite;
5) Amphibolite facies eclogite;
6) Low temperature eclogite or eclogite in glaucophanitic metamorphic terranes.

BORG (1956), SMULIKOWSKI (1964), BANNO (1964), WHITE (1964), COLEMAN et al. (1965) and others have shown that each of, or groups of, these eclogite types are characterized by particular compositional range of garnet or clinopyroxene solid solutions. SOBOLEV (1964), COLEMAN et al. (1965), BANNO and MATSUI (1965) and ESSENE and FYFE (1967) have shown that the pattern of the distribution of elements, mainly Fe and Mg, between garnet and clinopyroxene varies from type to type, and that this offers a more sound basis of the petrological classification of eclogites than the compositional range of individual minerals. This view is also the basic standpoint of the present paper, in which the author examines the extent over which the element distribution between garnet and clinopyroxene could be applied as a measure of physical conditions, and examines the possibility of classifying some of the eclogite types in more detail. Throughout this paper, the values of temperature and pressure are not referred to quantitatively. It is the author's opinion that if the equilibrium relations at two temperatures are experimentally determined, the temperature of eclogite crystallization is uniquely determined by the use of the distribution relations to be discussed in this paper.

It is helpful to define the nomenclature to be used. The term eclogite is used in a wide sense, and hence it denotes garnet-clinopyroxene rock with more or less basaltic composition, but the term basaltic is used rather vaguely. FORBES (1965) has mentioned that many, if not all, eclogites are not basaltic in chemical composition. According to this nomenclature, almanine-salite rock in the granulite facies, and pyrope-diopside rock enclosed in peridotite are called eclogite, along with typical eclogite containing pyrope-rich garnet and omphacite. Therefore, we have to accept the existence of eclogites which do not belong to the eclogite facies. The definition of the eclogite facies follows that given by O'HARA (1960), i.e., the eclogite facies is characterized by the assemblage kyanite+garnet+omphacite (or in a wide sense Ca-rich clinopyroxene), or by quartz+garnet+omphacite.

2. Some basic concepts of distribution relations

2.1. *Definition of distribution coefficient*

In later sections, the classification of eclogites will be discussed mainly in terms of the distribution coefficient between garnet and clinopyroxene, and hence this coefficient is defined first.

The apparent distribution coefficient of A and B cations of the same valency between phases α and β, $K'^{\alpha \cdot \beta}_{A \cdot B}$ or simply K' is defined by

$$K'^{\alpha \cdot \beta}_{A \cdot B} = \left(\frac{X_A}{X_B}\right)^\alpha \bigg/ \left(\frac{X_A}{X_B}\right)^\beta, \quad (1)$$

where X^α_A etc. denote the mole fraction of A in phase α etc. This coefficient corresponds to the following exchange reaction:

$$AY^\alpha + BZ^\beta = AZ^\beta + BY^\alpha, \quad (2)$$

or simply expressed by

$$A^\alpha + B^\beta = A^\beta + B^\alpha, \quad (3)$$

where AY, BZ etc. denote the components of solid solutions.

If the solid solutions α and β are ideal, K' defined by eq. (1) is equal to the thermodynamic distribution constant $K^{\alpha \cdot \beta}_{A \cdot B}$ or simply K, as defined by

$$K^{\alpha \cdot \beta}_{A \cdot B} = \left(\frac{a_A}{a_B}\right)^\alpha \bigg/ \left(\frac{a_A}{a_B}\right)^\beta = \exp\left(\frac{\Delta G}{RT}\right),$$

where a^α_A etc. denote the activities of AY in phase α etc., and ΔG denotes the difference in the free energies between the right- and left-hand sides of eq. (2).

In dealing with the distribution relations of trace elements such as Mn and rare earth elements (REE), the distribution coefficients are normalized to Mg for Mn, and to Sm for REE. The distribution relations of trace elements between garnet and clinopyroxene are not based on Nernst's distribution law, but on the exchange reaction such as shown in eq. (2).

Generally speaking, the apparent distribution coefficient K' is a function of temperature, pressure and the chemical composition of the system.

2.2. *Effect of pressure on the distribution coefficient*

The effect of pressure on the distribution constant K as defined by eq. (4) is obtained as

$$\frac{\partial \ln K}{\partial P} = \frac{\partial}{\partial P}\left(\frac{\Delta G}{RT}\right) = \frac{\Delta V}{RT}. \quad (5)$$

Therefore, with a crude approximation that ΔV is independent of pressure, the pressure coefficient of K is obtained as follows:

$$\ln K = \frac{\Delta V}{RT}(P - P_0) + \ln K_0, \quad (6)$$

$$K = K_0 \exp \frac{\Delta V(P - P_0)}{RT}, \quad (7)$$

where K and K_0 are the distribution constants at pressure $P = P$ and $P = P_0$ at given temperature, respectively.

The relationship between the thermodynamic distribution constant K and the apparent distribution coefficient K' as defined by eq. (1) is given as

$$\ln K' = \ln K - \ln \left\{ \left(\frac{\gamma_A}{\gamma_B}\right)^\alpha \middle/ \left(\frac{\gamma_A}{\gamma_B}\right)^\beta \right\}, \quad (8)$$

where γ_A^α etc. denote the activity coefficient of AY in phase α etc. The term containing the activity coefficient in eq. (8) is composition dependent, and hence the pressure coefficient of K' is not easily calculated. We have, however, the following relation:

$$\frac{\partial \ln \gamma_A}{\partial P} = \frac{V_{AY} - V_{AY}^\circ}{RT}, \quad (9)$$

where V_{AY} and V_{AY}^0 denote partial formula volume of AY in phase α and formula of pure AY, respectively. If there is an additivity of formula volume, the activity coefficient is independent of pressure. For neso- and ino-silicate solid solutions, the additivity of formula volume has been examined in detail on several series such as diopside-Ca-Tschermakite (CLARK et al., 1962), olivine and orthopyroxenes for Mg-Fe, Mg-Co and Mg-Ni substitutions (MATSUI and SYONO, 1968; MATSUI et al., 1968), diopside–jadeite series (KUSHIRO, personal communication) and others. Most of these solid solutions have slight volume of mixing, but detailed studies on diopside–hedenbergite, and pyrope–almandine series, with which we are concerned here, have not yet been obtainable to the author. If the volume of mixing is ignored, we have a crude approximation that the pressure affects K' only through the change of K by pressure.

Let us examine the pressure effect on K' in a simplified system where the phase α is ideal and the phase β is non-ideal solid solution. The apparent distribution coefficient K' at pressures P_1 and P_2 at the same temperature are denoted by K_1' and K_2', respectively, and similarly the subscripts 1 and 2 denote the quantities under pressures P_1 and P_2, respectively. We have then

$$\ln K_1' = \ln K_1 + \ln \left(\frac{\gamma_A}{\gamma_B}\right)_1^\beta \quad (10)$$

and

$$\ln K_2' = \ln K_2 + \ln \left(\frac{\gamma_A}{\gamma_B}\right)_2^\beta. \quad (11)$$

From eqs. (10) and (11), we have

$$\ln \frac{K_1'}{K_2'} = \ln \frac{K_1}{K_2} + \ln \left\{ \left(\frac{\gamma_A}{\gamma_B}\right)_1^\beta \middle/ \left(\frac{\gamma_A}{\gamma_B}\right)_2^\beta \right\}. \quad (12)$$

If we fix the composition of the phase β, γ_A^β and γ_B^β are constant under the crude approximation, so that we have

$$\ln \frac{K_1'}{K_2'} = \ln \frac{K_1}{K_2} = \frac{\Delta V}{RT}(P_1 - P_2). \quad (13)$$

Therefore, the difference in K' between rocks formed under different pressures is best examined by comparing the values of K' for the fixed composition of the non-ideal phase, or of the more non-ideally looking phase, because by doing so the pressure coefficient of K' is the same or similar to that of K.

In this paper, we are mainly concerned with the Fe–Mg distribution coefficient between garnet and clinopyroxene K'^{ga-cpx}_{Fe-Mg}, which corresponds to the following exchange reaction:

$$\underset{66.10}{\overset{\text{diopside}}{CaMgSi_2O_6}} + \underset{38.43}{\overset{\text{almandine}}{\tfrac{1}{3}Fe_3Al_2Si_3O_{12}}} =$$
$$= \underset{68.10}{\overset{\text{hedenbergite}}{CaFeSi_2O_6}} + \underset{37.76}{\overset{\text{pyrope}}{\tfrac{1}{3}Mg_3Al_2Si_3O_{12}}} \quad (14)$$
$$\text{(formula volume in cm}^3\text{)}$$
$$\Delta V = 1.33 \text{ cm}^3.$$

The formula volumes given above were taken from the compilation by ROBIE et al. (1966), but the formula volume of hedenbergite given in their table 5–2 appears unreasonably small, so that its formula volume was calculated from the cell constants listed in their table 5–1.

The apparent Fe–Mg distribution coefficient between garnet and clinopyroxene, K'^{ga-cpx}_{Fe-Mg}, is defined as follows, but in many cases, it will be denoted only as K':

$$K'^{ga-cpx}_{Fe-Mg} = \left(\frac{X_{Fe}}{X_{Mg}}\right)^{ga} \middle/ \left(\frac{X_{Fe}}{X_{Mg}}\right)^{cpx}. \quad (15)$$

With increasing pressure, the reaction (14) proceeds from the right- to left-hand sides, so that K increases with increasing pressure. The pressure coefficient of K is shown in fig. 1, in which we see that the pressure coefficient is close to unity in dealing with crustal rocks, but it is not so when we deal with the rocks formed within the mantle. The pressure coefficient of K' may not deviate much from that of K, if we accept MUEL-

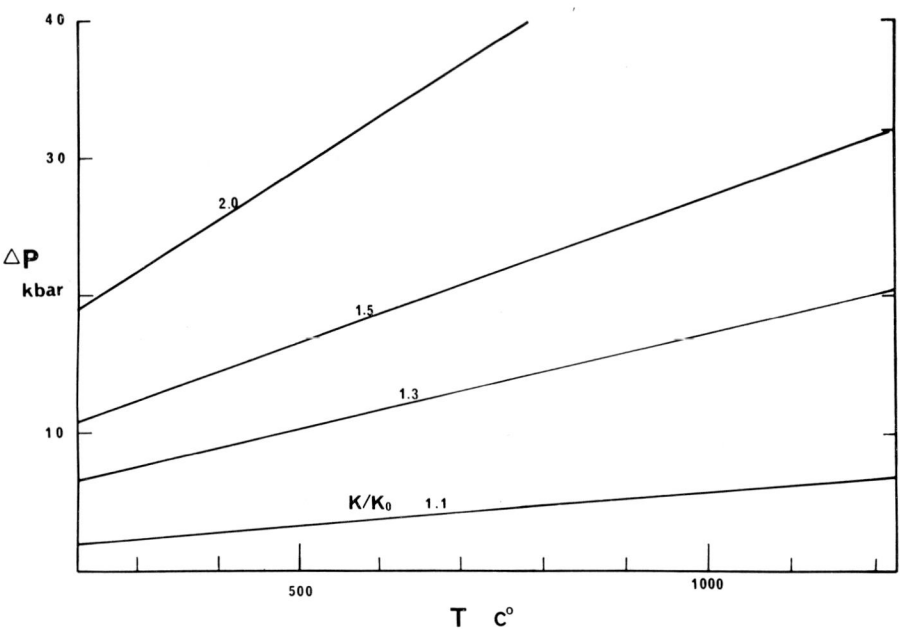

Fig. 1. The pressure coefficient for K'^{ga-cpx}_{Fe-Mg}.

LER's (1962) idea that diopside–hedenbergite series is a nearly ideal solid solution.

2.3. Effect of temperature on distribution coefficient

The effect of temperature on the distribution coefficient K is given by

$$\frac{\partial \ln K}{\partial T} = \frac{-\Delta H}{RT^2}. \qquad (16)$$

Assuming that at sufficiently high temperatures, ΔC_p of the exchange reaction is negligible, we have

$$\ln K = \frac{\Delta H_0}{RT} - \frac{\Delta S_0}{R}, \qquad (17)$$

where the subcript 0 refers to the quantities at $T = T_0$, which is sufficiently high.

It follows that with increasing temperature K approaches to a certain constant, probably close to, but not equal to unity. This obvious conclusion is sometimes mistaken for that K approaches to unity with increasing temperature.

The effect of temperature on K' is seen from eq. (17), too. The temperature effect on the activity coefficient is shown as follows, assuming that the excess specific heat C_p^E is negligible at high temperature,

$$\frac{\partial \ln \gamma}{\partial T} = \frac{H^E}{RT^2}, \quad \ln \gamma = \frac{H^E}{RT} - \frac{S^E}{R}, \qquad (18)$$

where the superscript E denotes the excess thermodynamic quantities of mixing. The activity coefficient for a given composition approaches to a certain constant as the temperature increases, and if S^E/R is small, the non-ideality decreases with increasing temperature.

For the equilibrium between garnet and clinopyroxene, we have no reliable thermodynamic data for calculating ΔG, so that the effect of temperature on K and K' can only be obtained by comparing the values for natural eclogites of distinct mode of occurrences. To determine the sense of $\partial K/\partial T$, the values of K' of three distinct types of eclogites are compared in fig. 2, in which the atomic Fe^{2+}/Mg ratios of garnet are plotted against those of the associated clinopyroxenes. The three representative types are California low temperature eclogites described by COLEMAN et al. (1965), Norwegian amphibolite facies eclogites (metabasites) as compiled by GREEN (in press), and the eclogite inclusions in kimberlite at the Robert Victor mine described by KUSHIRO and AOKI (1968). Geological observations suggest that these rocks were formed under more or less similar physical conditions as the associated non-eclogitic rocks, and that the temperature

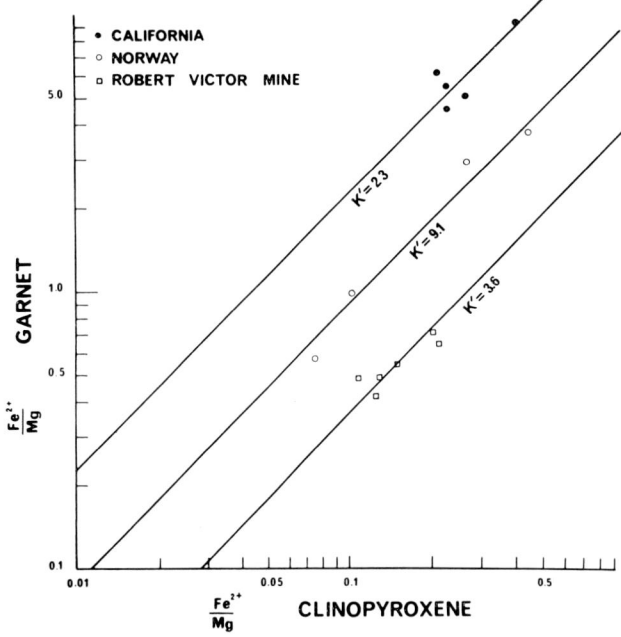

Fig. 2. The Fe–Mg distribution between garnet and clinopyroxene for three distinctive eclogite types: low temperature eclogite, amphibolite facies terrane eclogite, and eclogites inclusions in kimberlite.

of formation increases in the same order as mentioned above. The effect of pressure is negligible in comparing the low temperature and amphibolite facies terrane eclogites, as both are crustal rocks. The distinctly high pressure of the crystallization of the inclusions affects and increases K'. Therefore, it is concluded that the apparent Fe–Mg distribution coefficient between garnet and clinopyroxene decreases with increasing temperature.

2.4. *Effect of chemistry on distribution coefficient*

We have as yet no direct measurement of the activities in garnet and clinopyroxene solid solutions, so that the dependence of K' on the chemistry of the rocks is not clear.

Both garnet and clinopyroxene have only one structural site for Fe, Mg, Mn etc., provided that the clinopyroxene is low in orthopyroxene component, then the non-ideality due to the non-equivalence of lattice sites for the substituting cations (MATSUI and BANNO, 1965, 1968; BANNO and MATSUI, in press) need not be taken into account. The Fe–Mg distribution in pairs involving garnet or clinopyroxene is often approximated by that between the ideal solid solutions, and this can be emphasized as supporting the adequacy of a crude approximation that both these minerals are nearly ideal. We have, however, several reasons to doubt their ideality. The substitution of Mg by Ca in garnet may not be ideal, as their ionic radii are distinctly different from each other and the only common rock-forming mineral which forms continuously a solid solution between the Ca- and Mg-end members is garnet under high pressure (SOBOLEV et al., 1965, 1968).

Clinopyroxene solid solution may not be ideal for the diopside–Ca–Tschermakite series, as there is an excess volume of mixing. The orthopyroxene component necessarily results in the non-ideality of clinopyroxene as it breaks one of the conditions of the ideal solid solution, that the substituting cations should occupy the energetically equivalent site. The contents of Ca-Tschermakite and orthopyroxene components are, however, subordinate in ordinary eclogitic clinopyroxenes, and hence they may be ignored in crude treatments. The diopside–jadeite series has an excess volume of mixing (KUSHIRO, private communication) and an ordered phase is formed between them (CLARK and PAPIKE, 1968). The activity of the jadeite component in the diopside–jadeite series as calculated from the phase equilibrium diagram proposed by KUSHIRO (1965) shows a positive deviation from Raoult's law. The diopside–hedenbergite series was considered by MUELLER (1962) to be nearly ideal, on the basis that the Fe–Mg distribution between clinopyroxene and actinolite is explained as the equilibrium between two ideal solutions.

The existence of non-ideality in various solid solutions series of clinopyroxene, however, does not necessarily lead to a very pessimistic view, because in a ternary solid solution (A, B, C)X, in which the series (A, B)X is ideal, the (A, B, C)X series behaves as if ideal (A, B,C)X solid solution if the concentration of CX is sufficiently low (MATSUI and BANNO, 1968).

It follows that we need not worry too much about the effect of minor components. We have to be careful with the Fe/(Fe+Mg) ratio of minerals, the grossular content of garnet and the jadeite content of clinopyroxene as the possible major sources of the non-ideality.

In the absence of adequate thermodynamic data, the dependence of K' on the chemistry of rocks has to be examined using the data on natural eclogites. For this purpose, we need a set of isofacial eclogites covering a

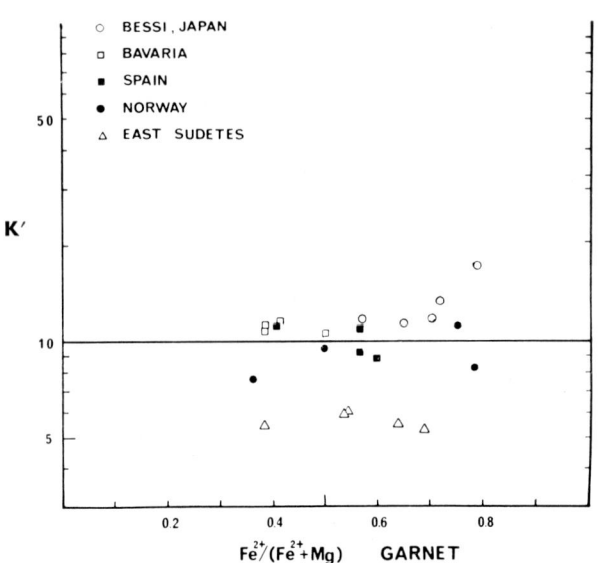

Fig. 3. The relationship between K' and the Fe/(Fe+Mg) ratio of garnet. Data are from the amphibolite facies terrane eclogites, and the low temperature eclogites of the Bessi area, Japan.

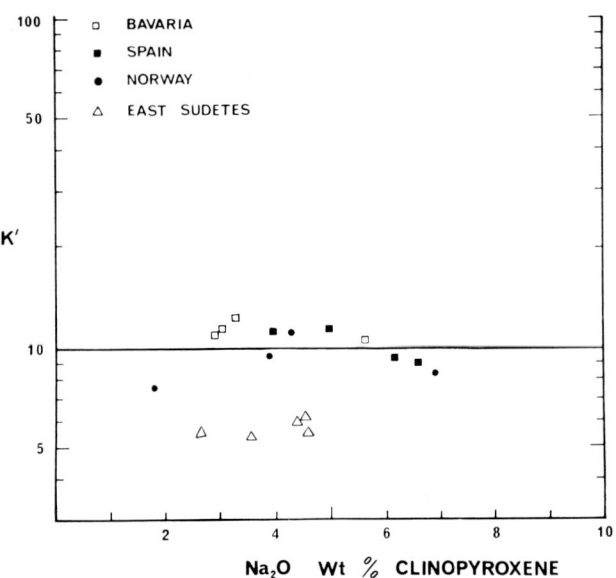

Fig. 4. The relationship between K' and the Na_2O content of clinopyroxene. The data are from the amphibolite facies terrane eclogites.

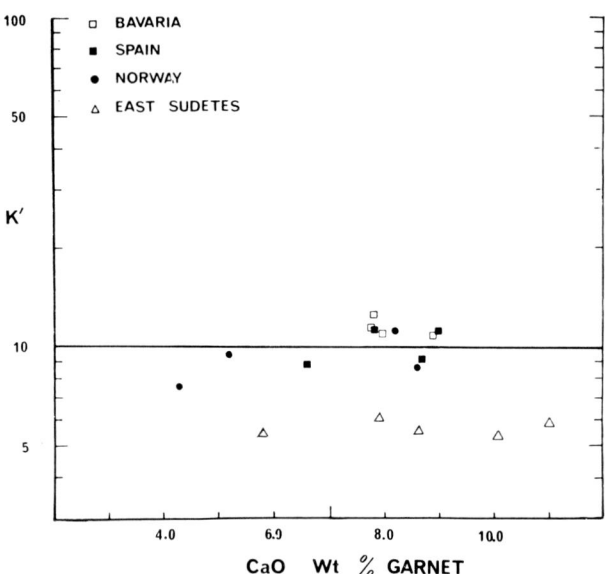

Fig. 5. The relationship between K' and the CaO content of garnet. The data are from the amphibolite facies terrane eclogites.

wide range of chemical composition, but such data are as yet not available to the author. The best data now available are those by GREEN (in press) who described two eclogites from one and the same eclogite lense and obtained two K' values, 7.6 and 11.1, respectively, for the strictly isophysical rocks. Fig. 3 shows the plots of the values of K' of the amphibolite facies terrane eclogites against the Fe/(Fe+Mg) ratio of garnet. Included in the figure are the eclogites from Norway (ESKOLA, 1921; GREEN, in press; K' by Green for both data), East Sudetes (SMULIKOWSKI, 1967), Bavaria (YODER and TILLEY, 1962; BANNO, 1967a), and Spain (VOGEL, 1967), along with the eclogites of the Sanbagawa metamorphic terrane to be referred later. In the text, only ferrous iron is taken into consideration, and thus Fe refers always to ferrous iron. The dependence of K' on the Fe/(Fe+Mg) ratio of garnet cannot be denied for Green's Norwegian data, but it cannot be detected for other eclogites. Similar plots of K' against the Na_2O content of clinopyroxene and the CaO content of garnet as shown in figs. 4 and 5 are not conclusive as to the dependence of K' on these parameters, but rather suggest that K' is nearly independent of these parameters. The absence of clear compositional dependence of K' of the eclogites other than the Green's data might well be due to the fact that the data are from isolated localities and the rocks may not have crystallized strictly under the same physical conditions.

Therefore, it may be concluded that the value of K' is constant within about 20% error or less. The 20% error in K' is not too unsatisfactory because K' is sensitive to temperature and the minerals of eclogites are often heterogeneous. The Fe content of the garnet of the bronzite eclogite described by MATSUI et al. (1966) varies by 20% as revealed by the electron probe study. Heterogeneity of garnet and clinopyroxene of eclogitic rocks have been described by PHILIPSBORN (1930), ESSENE and FYFE (1967), GREEN et al. (1968) and others. Further the determination of FeO by conventional wet chemical analysis is sometimes very difficult on eclogite minerals. A preliminary study of FeO analysis in diopside of the Higasiakaisi eclogite by Mössbauer spectroscopy suggests that the FeO content has been underestimated by the factor of 20% (MATSUI et al., in preparation).

The arguments on the apparent Fe–Mg distribution coefficient between garnet and clinopyroxene, K', are summarized as follows:
1) K' increases with increasing pressure, but the pressure effect is negligible in comparing the crustal eclogites;
2) K' decreases with increasing temperature;
3) There may be the compositional dependence of K', and care must be taken for the effect of the Fe–Mg substitution, the jadeite and orthopyroxene contents of clinopyroxene, and the grossular content of garnet as the possible sources of non-ideality, but in ordinary eclogites, the dependence of K' on chemistry is not large.

In the following, it is considered that the eclogites were formed under more or less similar physical conditions as the associated rocks. The eclogites associated with metamorphic rocks are metabasites belonging to the same or similar metamorphic facies as the enclosing rocks, and the eclogite inclusions are cumulates from basaltic or kimberlitic magmas or the solidified magma, which may have suffered metamorphism at lower temperature than the solidus temperature of the magmas. This assumption does not conflict with the observation that many eclogite lenses have tectonic contact with schists. A more detailed summary of the mode of occurrences of metamorphic eclogites were given elsewhere (BANNO, 1966).

3. Subdivision of and mutual relations among eclogite types

3.1. General statement

In the previous section, it was shown that there is a distinct difference of K' values between three distinctive types of eclogites.

It was pointed out by SMULIKOWSKI (1964), BANNO (1964) and COLEMAN et al. (1965) that the compositional range of garnet is different among some eclogite types, and that the pyrope content increases with increasing temperature. The difficulty of defining the limit of the pyrope content for each eclogite type was noticed, and the reason for this was explained by the consideration of the element distribution as given by COLEMAN et al. (1965) and BANNO and MATSUI (1965). Two methods of analysis of the distribution relations of elements have been proposed, one of which is based on the intersection of tie lines on a Ca–Mg–Fe ternary plot of garnet and clinopyroxene compositions and the other is based on the apparent distribution coefficient. Each of them has its own merit, but it is the author's opinion that the one which possesses the theoretical basis at least for an idealized system, i.e., for the equilibrium of ideal garnet and clinopyroxene solid solutions in our case, is preferable. Fig. 6 illustrates the tie lines for the

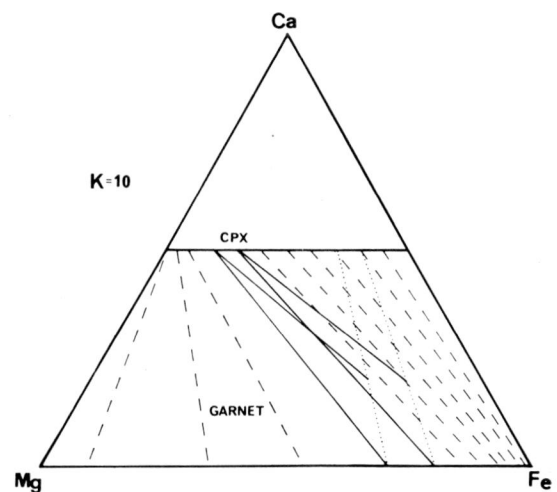

Fig. 6. The Ca–Mg–Fe plot of hypothetical garnet–clinopyroxene equilibrium. Clinopyroxene is ideal for Fe–Mg substitution, and garnet is ideal for Ca–Mg–Fe. The tie lines intersect with each other even if the CaO content of the system differs.

equilibrium of hypothetical ideal grossular-pyrope-almandine, and diopside-hedenbergite solid solutions with $K = 10$. It is seen in the figure that the tie lines intersect with each other even among the pairs being formed under the same physical conditions. This is due to the fact that the Fe–Mg distribution under consideration is not of a ternary but of a quaternary system. Therefore, in this paper, the phase equilibrium relations are discussed mainly in terms of the apparent distribution coefficient between coexisting garnet and clinopyroxene.

3.2. Subdivision of low temperature eclogites

The eclogites included in this type occur mainly in glaucophanitic metamorphic terranes. This type was called ophiolitic eclogite by SMULIKOWSKI (1964) and group "C" by COLEMAN et al. (1965). The term ophiolitic is rather vague, and it is not used here.

The low temperature eclogites have been described from the following localities:

1) Urals: Lawsonite and glaucophane are stably associated (CHESNOKOV, 1960).
2) Colombia: Boulders in a tertiary conglomerate. Lawsonite is considered to have been stable with garnet and clinopyroxene, but it is now changed to zoisite. One of the specimens described contains jadeite-rich pyroxene + quartz assemblage (GREEN et al., 1968).
3) Guatemala: A boulder in a low temperature metamorphic terrane. Lawsonite is associated (MCBIRNEY et al., 1967).
4) California: Glaucophane is stably associated. COLEMAN et al. (1965) considered that the associated lawsonite was formed later than the eclogite minerals, but ESSENE and FYFE (1967) considered it to be stably associated with them. The assemblage jadeite + quartz and aragonite are unstable. The chemical data are from COLEMAN et al. (1965).
5) New Caledonia: The detail of the mode of occurences is not known, but chemical data were described by COLEMAN et al. (1965).
6) Western Alps: According to BEARTH (1966), glaucophane schists are associated, but lawsonite is unstable. The chemical data are from VAN DER PLAS (1959) and BEARTH (1965).
7) Bessi area, Sanbagawa metamorphic belt, Japan: Eclogites occur as layers in an epidote amphibolite mass, and their occurrences appear to be restricted to the neighbourhood of the Higasiakaisi peridotite mass which contains Almklovdalen type eclogites in itself. The chemical data are listed in table 1. The eclogites enclosed in peridotite of glaucophanitic terrane are not included in this type.

The plots of K' values against the Fe/(Fe + Mg) ratios of garnet are shown in fig. 7. In the figure, three groups of the low temperature eclogites are distinguished: the

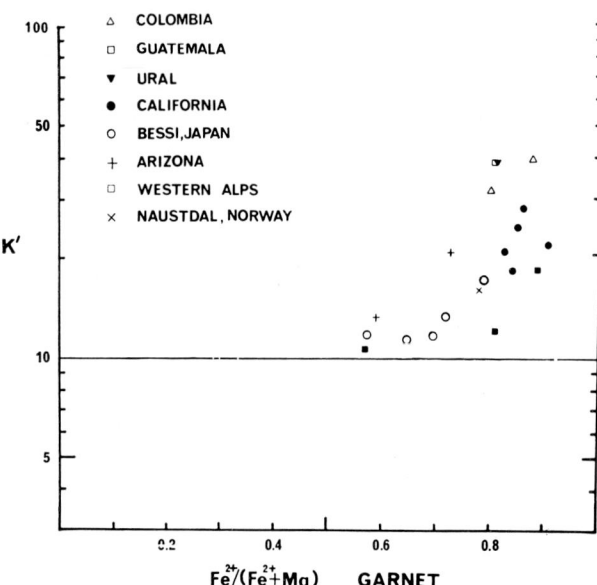

Fig. 7. The relationship between K' and Fe/(Fe + Mg) ratio of garnet for low temperature eclogites.

first group is composed of the Ural, Colombia and Guatemala eclogites, the second group is the Californian ones, and the third group includes the Alpine and Japanese eclogites.

The clinopyroxenes of the low temperature eclogites are rich in Na_2O, as first suggested by BORG (1956). One of the Colombia pyroxenes contains 68 % jadeite and 14 % acmite. The effect of Na_2O content of the pyroxene on K' is examined in fig. 8, in which the values of K' are plotted against the Na_2O content of the pyroxene. There is no clear compositional dependence of K'. The CaO content of garnet generally increases with increasing FeO content, so that it is difficult to separate its effect from that of FeO. It is considered that the observed difference in K' among three groups of the low temperature eclogites reflects the difference in the physical conditions among them.

The Ural, Colombia and Guatemala eclogites are of the lawsonite–glaucophane schist facies, and the Alpine and Japanese ones are of the lawsonite-free glaucophane schist facies and of the albite–epidote amphibolite facies, respectively, and hence the temperature of metamorphism is higher in the latter than in the former. This view is in good agreement with the conclusion obtained from the comparison of K', that the former group has higher values than the latter. The metamorphic facies of the Californian eclogites are not

TABLE 1

The chemical compositions of garnet and clinopyroxene from the Bessi area
(The eclogites enclosed in the Higasiakasi peridotite mass are not included)

	1		2		3		4		5	
	Gar	Cpx	Gar	Cpx	Gar	Cpx	Gar	Cpx	Gar	Cpx
SiO_2	37.15	55.18	39.91	52.9						
TiO_2	0.26	0.38	0.12	0.6						
Al_2O_3	21.08	9.56	20.88	11.7						
Fe_2O_3	1.49	5.14	1.83	6.3		5.04		4.38		4.64
FeO	26.04	2.93	20.26	1.4	20.31*	2.56	24.65*	2.53	22.47*	2.09
MnO	0.87	0.01	0.65	0.02						
MgO	4.23	8.40	8.75	7.2	4.98	7.34	5.34	7.51	6.80	7.12
CaO	8.49	11.26	6.76	12.3						
Na_2O	<0.1	6.52	0.08	5.48						
K_2O	<0.1	<0.01	0.07	0.37						
H_2O^+	0.74	0.43	0.71	1.4						
H_2O^-	0.05	0.05	0.15	0.0						
P_2O_5	0.04	0.03	0.13							
Total	100.44	99.89	100.30	99.67						
K'	17.6		11.9		11.7		13.7		11.3	
$Fe^{2+}/(Fe^{2+}+Mg)$ in garnet	0.78		0.57		0.70		0.72		0.65	

* Total Fe as FeO
1. SBD122 Boulder at the Hodono valley ⎫
2. SB56081103 Gongen shrine ⎬ The corrected analyses given by BANNO (1964).
3. ⎫
4. ⎬ Boulders at Hodono valley.
5. ⎭
Analysts. 1, 2: H. Haramura, 3: Y. Hirano, 4, 5: Y. Oki.

clear, as the presence of lawsonite is interpreted differently by different authors. Petrologically, they are certainly of lower temperature than the Japanese eclogites, and of higher temperature than the Colombia ones, which accompany the jadeite+quartz assemblage. Judging from the element distribution, it is more plausible to consider that the Californian eclogites represent intermediate temperatures between the other two groups, rather than to consider that the difference in the K' values between the Californian and other eclogites is simply due to the poor accuracy of the apparent distribution coefficient to be used as a geological thermometer. Therefore, it is considered that K' can be used to subdivide the eclogite type.

We have three more occurrences of possible low temperature eclogites. BINNS (1967) described an eclogite from Naustdal, Norway, which forms a lens in albite-epidote amphibolite facies terrane. The K' value is 16, which lies within the range for the low temperature eclogite type, and it is distinctly higher than those of other Norwegian eclogites. Geological and phase equilibrium considerations seem to favor the view that the Naustdal eclogite is a metabasite formed under the same or similar physical conditions as the enclosing schists.

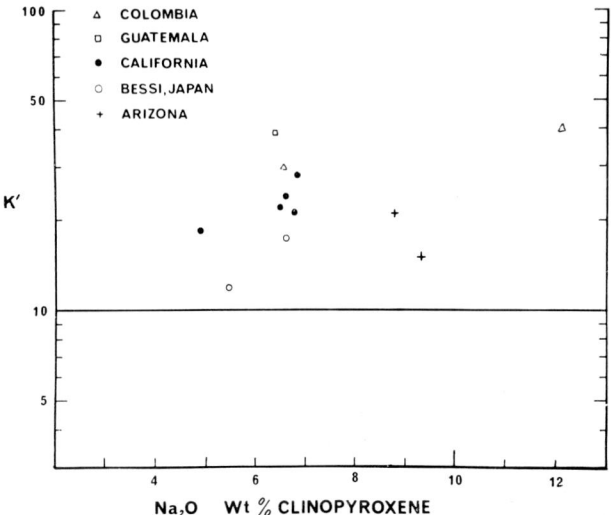

Fig. 8. The relationship between K' and the Na_2O content of clinopyroxene for low temperature eclogites.

SPRY (1958) described an eclogite from Tasmania. The associated metamorphic rocks are of the albite-epidote amphibolite facies. The K' value of this eclogite is 8, which is typical of the amphibolite facies terrane eclogites. More mineralogical data are needed for this eclogite occurrence, but it is not unreasonable to consider that it belongs to an upper albite-epidote amphibolite facies, because the K' values of the eclogites of the albite-epidote amphibolite and amphibolite facies may overlap with each other, for reasons to be discussed later.

O'HARA and MERCY (1966) described the mineralogy of two eclogites found in breccia pipes in Arizona and New Mexico and considered them to be related to basalt or kimberlite activity. BANNO (1967b) mentioned that the extremely low CaO content of the Arizona eclogite garnet is due to the very high Na_2O content of the associated clinopyroxene, and hence it does not indicate physical conditions. He and GREEN et al. (1968) mentioned that the distribution coefficients of the Arizona eclogite are similar to those of glaucophanitic metamorphic terranes. In fig. 8, it is seen that the high Na_2O content of the pyroxene cannot be the sole reason of high K'. GREEN et al. (1968) also mentioned the occurrence of lawsonite-bearing eclogitic rocks in the same pipes (WATSON, 1962), which indicates the presence of low temperature and high pressure metamorphic rocks within the crust of this area. In fig. 9, the distribution of rare earth elements between garnet and clinopyroxene of various eclogites is shown. The factors controlling the distribution of REE are not well known, but the figure shows that the distribution pattern for the Arizona eclogite is distinctly different from those of eclogites that occur as amphibolite facies metamorphics, and the inclusions in basalt. For these reasons, the author favors the view that the Arizona eclogites are crustal rocks belonging to the low temperature eclogite type.

3.3. *Subdivision of amphibolite facies terrane eclogites*

The eclogites of this type form lenticular masses in amphibolite facies terranes. They are typical eclogite,

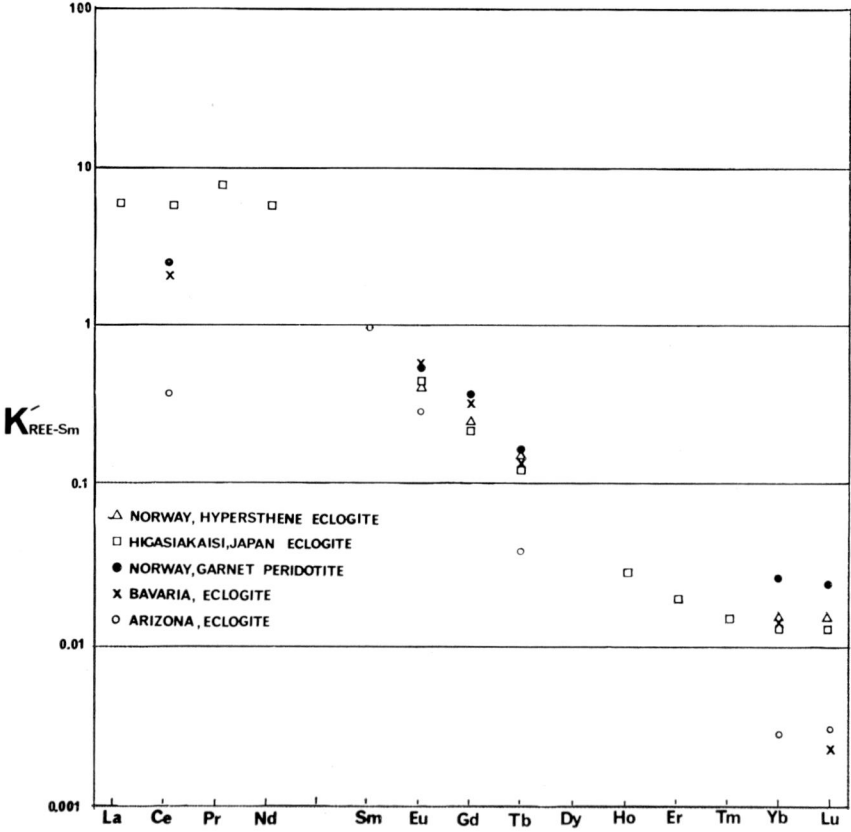

Fig. 9. The rare earth distribution pattern for various eclogites. The apparent distribution coefficient is normalized to Sm.

historically and in phase equilibrium relations. The eclogites enclosed in peridotite of these terranes are excluded again. The known localities of this eclogite type include Fichtelgebirge in Bavaria, Basa' Gneiss region in SW Norway, Glenelg in Scotland, Saualpe, Greenland, the Śnieżnik Mts in East Sudetes, Cabo Ortegal in Spain and others, and sufficient mineralogical data for analysing the distribution relations are obtainable from Bavaria, Norway, East Sudetes and Spain. The plots of K' values of these eclogites against the Fe/(Fe+Mg) ratio of garnet, the Na_2O content of pyroxene, and the CaO content of garnet are shown in figs. 3, 4 and 5, by which it was concluded that the composition dependence of K' is small.

A distinct difference in K' is seen between the eclogites of the East Sudetes, and of the others, thereby suggesting that the Sudetes eclogites were formed at higher temperature than the others. The East Sudetes eclogites have the critical mineral assemblages of the eclogite facies, i.e., kyanite+garnet+omphacite, and quartz+garnet+omphacite, but the described specimens are from scattered localities so that it is not certain to what extent they are isophysical. In discussing the jadeite to Ca-Tschermakite ratio of the eclogitic clinopyroxenes, BANNO and YAMASAKI (in preparation) concluded that the Sudetes eclogites belong to the eclogite facies rather than to the granulite facies. They also suggested that the Sudetes eclogites represent higher temperatures or lower pressures than the typical amphibolite facies terrane ones. Therefore, the conclusions deduced from both K' and compositional range of clinopyroxene are in harmony and suggest that the East Sudetes eclogites are of higher temperature than the ordinary amphibolite facies terrane eclogites. According to SMULIKOWSKI (1967), the associated metamorphic rocks in this area are migmatite and amphibolite.

The subdivision of other eclogites of this type is difficult, but plots of K' values in figs. 3, 4 and 5, as well as the average K' values of each of terranes as shown in table 2 suggest that the Bavarian and Spanish eclogites are of slightly lower temperature than the Norwegian ones, though this is not conclusive.

3.4. *Comparison of low temperature and amphibolite facies terrane eclogites*

Inspection of figs. 3 and 7, in which the plots of K' values against the Fe/(Fe+Mg) ratio of garnet for the low temperature and amphibolite facies terrane eclogites are shown, reveals the fact that the minimum K' of the low temperature eclogites i.e., for Alpine and Japanese ones and the maximum value of K' of the others are rather similar to each other. The overlapping of the K' values between the albite-epidote amphibolite and amphibolite facies eclogites is, however, not against the validity of using K' as a geological thermometer. The pressure-temperature relationships between these two metamorphic facies are shown in fig. 10, which is a

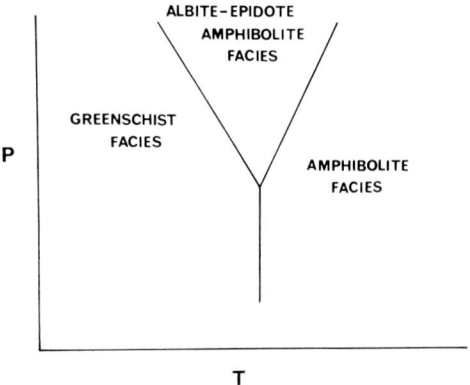

Fig. 10. A schematic diagram showing the pressure-temperature relations among the green schist, albite-epidote amphibolite and amphibolite facies.

qualitative quotation of the diagram given by MIYASHIRO (1961). Most low temperature eclogites occur in glaucophanitic metamorphic terranes, while those of the amphibolite facies terrane probably suffered the kyanite-sillimanite type metamorphism. The pressure of metamorphism is therefore higher in the former than in the latter, so that the temperature of an upper albite-epidote amphibolite facies could well be higher than that of a lower amphibolite facies. As the effect of pressure on K' is negligible in dealing with the crustal eclogites, the overlapping of K' values between these two facies is generally expected. This argument may give some suggestions as to the origin of the problematical Tasmanian eclogite.

3.5. *Granulite facies eclogites and their relations to the amphibolite facies eclogites*

The garnet-clinopyroxene assemblage is stable in the granulite facies, too. In Mg-rich rocks, this assemblage is stable only in SiO_2-undersaturated rocks, while in a Fe-rich environment, it is stable even in SiO_2-saturated

TABLE 2

The values of K' for different garnet–clinopyroxene pairs

Type and locality	Number of samples	K'	References	Remarks
Amphibolite facies eclogites				
Bavaria	4	11.3	1, 2	
Cabo Ortegal, Spain	4	10.3	3	
Norway	4	9.1	4, 5	1
Glenelg, Scotland	2	8.0	1, 6	
Śnieżnik Mts, East Sudetes	5	5.6	7	
Granulite facies eclogites				
Varberg	8	6.3	8	2
Other areas	14	7.3	9, 10	
Eclogite inclusion in basalt				
Hawaii	3	2.5	1, 11	
Others	2	2.7	12, 13	
Eclogite inclusion in kimberlite				
Robert Victor mine	6	3.7	14	
Zagadochnaya	5	10.9	15	3
Basutoland	3	4.9	16	
Eclogites enclosed in peridotite and garnet peridotite intrusives				
Norway	8	5.7	4, 5, 17	
Higasiakaisi	3	8.3	This paper	
Czechoslovakia	3	2.6	18, 19	
Garnet peridotite inclusions in kimberlite				
	4	3.2	5, 16, 17	
Garnet peridotite and eclogite enclosed in peridotite				
Norway	8	5.7	4, 5, 17	4
Higasiakaisi, Japan	3	8.3	This paper	5
Czechoslovakia	3	2.6	18, 19	

Remarks
1) K' as calculated assuming that some Fe_2O_3 of garnet is actually FeO (GREEN, in press).
2) Probe analysis. Reciprocal of K_D given by SAXENA (1968).
3) CaO of garnet is very high.
4) K' neglecting Fe_2O_3.
5) Table 3 of this paper.

Key for references
1. YODER and TILLEY (1962)
2. BANNO (1967)
3. VOGEL (1967)
4. ESKOLA (1921)
5. GREEN (in press)
6. O'HARA (1960)
7. SMULIKOWSKI (1967)
8. SAXENA (1968)
9. SOBOLEV (1964)
10. WARNAARS (1967)
11. KUNO (in press)
12. DICKEY (1968)
13. GIROD (1967)
14. KUSHIRO and AOKI (1968)
15. SOBOLEV et al. (1968)
16. NIXON et al. (1963)
17. O'HARA and MERCY (1963)
18. FIALA (1966)
19. MIKHAIROV and ROVSHA (1966)

ones (O'HARA, 1960; BANNO, 1966; GREEN and RINGWOOD, 1967). SOBOLEV (1964) and ESSENE and FYFE (1967) have shown that the average of K' of the granulite facies eclogites is lower than that of the amphibolite facies terrane eclogites. The plots of K' values of granulite facies eclogites against the Fe/(Fe+Mg) ratio of associated garnet are shown in fig. 11 in which the plots for the Norwegian and East Sudetes eclogites are also shown for comparison. The data on the granulite facies rocks are from SOBOLEV (1964) and WARNAARS (1967). In the figure, it is seen that the K' values of the granulite facies eclogites are lower than those of Nor-

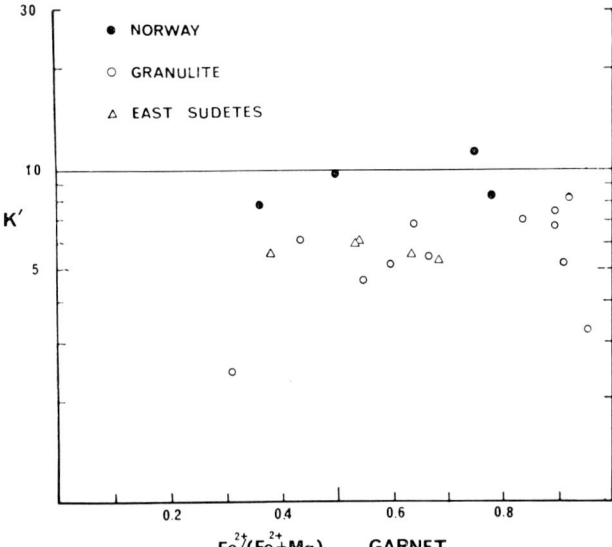

Fig. 11. The relationship between K' and Fe/(Fe+Mg) ratios of the granulite facies eclogites. The data for Norwegian and East Sudetes eclogites are also shown for comparison.

wegian ones, which are the representatives of the amphibolite facies eclogites, but the values of the East Sudetes eclogites are similar to the granulite facies ones. The fact that the granulite facies eclogites generally have lower K' values than the ordinary amphibolite facies eclogites is in harmony with petrological considerations, that the former represents higher temperature than the latter. It has been suggested that the contents of jadeite and Ca-Tschermakite differ between the granulite and amphibolite facies eclogites (WHITE, 1964), but the compositional dependence of K' is negligible for the amphibolite facies eclogites. The agreement of the conclusions by mineral facial considerations and that by K' further supports the adequacy of using K' as a geological thermometer.

The East Sudetes eclogites, however, have similar K' values to the average granulites. They are associated with migmatite, probably of the amphibolite facies, and this is contradictory to the previous view that the granulite facies represent higher temperatures than the amphibolite facies, so that a tentative explanation to this controversy is given below.

The amphibolite-granulite facies boundary is usually defined by the appearance of orthopyroxene in basic metamorphic rocks. The mineral assemblages of pelitic metamorphic rocks are similar to each other between an upper amphibolite and a lower granulite or the hornblende granulite facies. Under high pressures, and within the stability field of the quartz+garnet+clinopyroxene assemblage, the appearance of orthopyroxene in basic metamorphic rocks has to take place at much higher temperatures than under lower pressures, because the following reaction proceeds from the left- to right-hand sides with increasing pressure:

$$\underset{\underset{100.7}{\text{anorthite}}}{CaAl_2Si_2O_8} + \underset{\underset{4\times 31.5}{\text{enstatite}}}{4MgSiO_3} =$$

$$= \underset{\underset{113.3}{\text{pyrope}}}{Mg_3Al_2Si_3O_{12}} + \underset{\underset{66.1}{\text{diopside}}}{CaMgSi_2O_6} + \underset{\underset{23.7}{\text{quartz}}}{SiO_2} \quad (19)$$

(formula volume cm^3)

$$\Delta V = -23.6 \text{ cm}^3$$

where the reaction is expressed in terms of Fe-free end members.

A schematic representation of the pressure-temperature relationships between the amphibolite and granulite facies is shown in fig. 12. In field A of the figure,

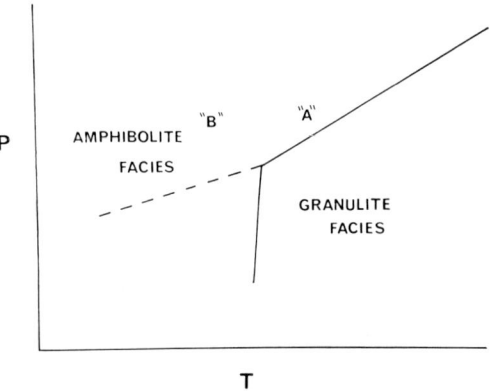

Fig. 12. A schematic diagram showing the pressure-temperature relations between the amphibolite and granulite facies. The boundary between field "A" of the amphibolite facies and the granulite facies is the univariant curve of reaction (19).

eclogite is stable, and orthopyroxene does not coexist with basic plagioclase. The mineral assemblages of pelitic metamorphic rocks are similar between the amphibolite, and the hornblende granulite facies, i.e., the lower granulite facies, and thus they are similar between the fields "A" and "B". The mineral assemblages of basic metamorphic rocks are also similar between the fields A and B, as the appearance of orthopyroxene is depressed by reaction (19) and then the amphibolites between these fields are not easily distinguished. Therefore, the rocks metamorphosed in the field "A" may well be classified as belonging to the amphibolite facies. If we consider that the metamorphic rocks of the East Sudetes belong to field "A", the overlapping of K' values between the Sudetes and granulite facies eclogites is reasonably explained. It is emphasized, however, here that the above discussions are qualitative, and do not propose nor assume that the eclogites are formed under the same water fugacity as the associated metamorphics.

3.6. *Eclogite inclusions in basalt and in kimberlite*

The phase equilibrium relations of the eclogite inclusions in basalt were discussed by LOVERING and WHITE (lecture in this symposium), so they are only briefly mentioned here. The most striking feature of the eclogite inclusions in basalt is the fact that their K' values are lower than those of the eclogite inclusions in kimberlite. The eclogite inclusions in basalt are considered to have been formed at the uppermost mantle or the lowermost crust, probably at the former, while some eclogite inclusions in kimberlite contain diamond (WILLIAMS, 1932; SOBOLEV et al., 1965), and are undoubtedly formed under very high pressures. The shallow origin of the inclusions in basalt is also supported by the presence of plagioclase in some specimens (LOVERING and WHITE, 1964). The melting point of eclogite under dry conditions increases with increasing pressure, so that the inclusions in basalt are considered to have crystallized at lower temperatures than those in kimberlite, but this is contradictory with the fact that the K' values are generally lower in the former than in the latter. This controversy may be avoided by assuming that the magmas are far from being dry, and the water in magmas affected greatly the crystallization temperature of the inclusions, or by assuming that the differences in the chemistry of the host magma as well as that of eclogite itself were so large that the temperature of crystallization has little connection with the depth of the formation and then the inclusions in basalt actually represent higher temperature than the others. Another and more plausible explanation to this controversy is, however, that the pressure affected K'. It was shown in a previous discussion that 20 kb difference in pressure at 1000 °C affects K' by the factor of 1.5, the ratio of K' values between two eclogite types. Judging from the experimental determination of eclogite crystallization as reported by GREEN and RINGWOOD (1967), 20 kb of pressure difference between two types of eclogite inclusions is not an unreasonable estimate.

The dependence of K' on the chemistry of rocks was considered generally to decrease with increasing temperature. However, the ideality of the equilibrium between garnet and clinopyroxene at high temperatures may be affected by the high concentration of orthopyroxene component in clinopyroxene, which necessarily results in the increase of non-ideality. Therefore, a detailed comparison of high temperature eclogites being under consideration in terms of K' is not so reliable as that for the metamorphic eclogites, and only a rough comparison has meaning. It appears that both these inclusions crystallized under more or less similar temperatures to each other.

3.7. *Eclogite enclosed in peridotite and garnet peridotite*

The discussion based on the apparent distribution coefficient of Fe and Mg between garnet and clinopyroxene can be applied in discussing the genesis of eclogite, or pyrope-diopside rock, enclosed in peridotite, and of garnet peridotite. The average K' values of garnet peridotites, which occur as inclusions in kimberlite and as the intrusive mass in metamorphic terranes, as well as those of eclogites enclosed in peridotite, i.e., Almklovdalen type eclogites, are shown in table 2. The data on the garnet peridotite inclusions in kimberlite are rather scanty, but available data show that the K' values are higher for the intrusives than for the inclusions, suggesting that the latter, probably mantle materials, are of higher temperatures than the crustal intrusives. This conclusion is in harmony with that given by O'HARA and MERCY (1963), who discussed this problem on the basis of the compositional range of pyroxenes.

Eclogites consisting of pyrope and diopside are often found to be enclosed in peridotite and garnet peridotite intruded into metamorphic terranes. The known localities include SW Norway, Bohemian massif, Spain and Higasiakaisi, Japan; the analyses for the last are listed in table 3.

The genesis of this eclogite type has not been discussed in detail. Their bulk chemical compositions are not

TABLE 3

Chemical compositions of eclogite, Higasiakaisi, Japan.
(Fe-rich eclogites are not included in this paper, as the consanguinity between Mg-rich and Fe-rich eclogites are in doubt).

	1		2		3	
	Gar	Cpx	Gar	Cpx	Gar	Cpx
SiO_2	39.68	51.86	40.46	53.38	41.20	54.73
TiO_2	0.98	0.20	0.23	0.10	0.13	0.08
Al_2O_3	22.05	1.02	22.72	1.43	22.50	0.84
Fe_2O_3	4.10	0.99	0.49	0.66	0.98	0.95
FeO	11.33	1.73	13.12	2.01	14.03	1.66
MnO	1.57	0.02	0.62	0.07	0.37	0.03
MgO	13.63	17.09	16.28	17.71	14.60	17.19
CaO	6.75	25.24	6.42	24.57	6.33	24.41
Na_2O	0.19	0.08	<0.02	0.37	tr.	0.52
K_2O	0.04	0.06	<0.02	<0.02	tr.	tr.
H_2O^+	0.28	2.13	0.10	0.05	0.00	0.10
H_2O^-			0.00	0.00	0.00	0.00
P_2O_5			0.05	0.12		
Cr_2O_3			0.12	0.08	0.30	0.059
Total	100.60	100.42	100.61	100.55	100.44	100.569

1. MIYASHIRO and SEKI (1958).
2. BANNO and YOSHINO (1965).
3. BANNO and YOSHINO (1965) for garnet. Diopside: new analysis. Analyst: H. Haramura.

strictly basaltic, and very low Na_2O and high CaO contents are noteworthy. The normative mineral assemblage is olivine+plagioclase+diopside (+hypersthene), with colour index more than 50. The hypothesis that they crystallized from basaltic magma within the crust is rejected from the experimental data on the basalt to eclogite transformation, and high pressure solidus phases of basaltic magmas. The possible mechanism for their genesis includes the metamorphism of olivine eucrite in peridotite, or the intrusion of upper mantle materials in essentially solid state.

The K' values of these eclogites are also shown in table 2. They range from 6 to 9, and are within the range of the amphibolite facies terrane eclogites. The plots of the K' values against the Fe/(Fe+Mg) ratio of garnet of this eclogite type are shown in fig. 13, together with the plots of the K' values of the associated garnet peridotite, and of metamorphic eclogites in the same area. The Norwegian and Japanese data are used in the diagram, as they are the only areas which have mineralogical data being capable of being examined by phase equilibrium principles. Inspection of the figure reveals the fact that the values of K' are different between Norwegian and Japanese occurrences, and this difference is accompanied by the difference in the K' values of associated metamorphic eclogites. In Norway, where both eclogite types have comparatively low K' values, the host peridotite is intruded into the amphibolite facies terrane, while in Japan, where prevailing metamorphic rocks are of the albite-epidote amphibolite facies, both eclogite types have comparatively high K' values.

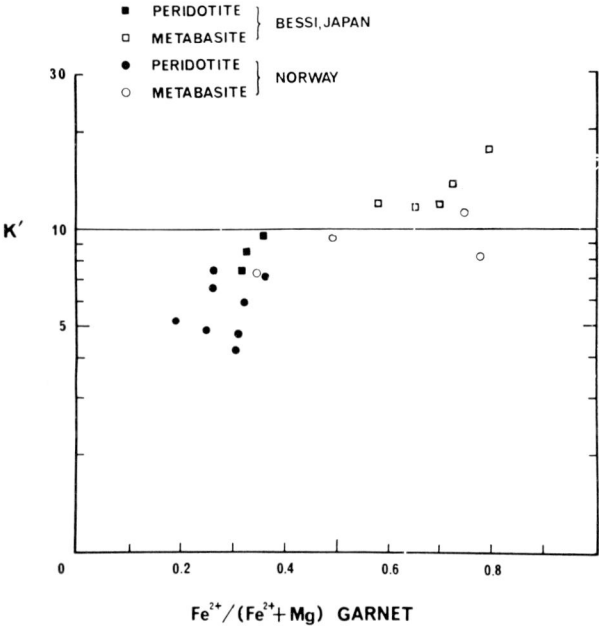

Fig. 13. The relationship between K' and Fe/(Fe+Mg) ratio of garnet for eclogites in peridotite, and garnet peridotite. For comparison, K' for the metamorphic eclogites of the same area is shown.

In this connection, it is worthy of note that in Spain, pyrope-diopside eclogites occur in peridotite intruded into an amphibolite facies terrane, where metamorphic eclogite occurs, while spinel peridotite, which represents lower pressure than the garnet peridotite, occurs in the granulite facies terrane (VOGEL, 1967; WARNAARS, 1967). However, this parallelism between the peridotite types and the metamorphic facies of the enclosing rocks is not necessary, as we have garnet peridotite intrusives in granulite terrane in Bohemia (DUDEK and KOPECKY, 1966).

It is as yet not clear if the similarity of K' between the eclogites enclosed in peridotite and in neighbouring metamorphic rocks represents the compositional dependence of K' among isofacial rocks, or if it is due to the difference in temperatures between them. If the temperatures are considered different, the eclogites in peridotite represent higher temperatures than those in associated metamorphic rocks, and this is geologically reasonable.

It appears that the association of two eclogite types in the same area, and the parallelism of K' between them are not fortuitous, and some genetical connection may exist between them. A possible interpretation of this is that the temperature-pressure conditions of a geosyncline and the underlying upper mantle during regional metamorphism are intimately related with each other, i.e., the upper mantle underneath the glaucophanitic metamorphic terrane was of lower temperature than that underneath the amphibolite facies metamorphic terranes. Another possible interpretation is the metamorphic recrystallization of olivine eucrite and dry basalt under more or less similar depth of a geosyncline, at slightly different temperatures.

4. Distribution of trace elements

BANNO and MATSUI (1965) demonstrated that the apparent Mn–Fe distribution coefficient, K'_{Mn-Fe} between garnet and clinopyroxene varies systematically with K'_{Mg-Fe}. Table 4 shows the examples of the Mn distribution in some representative eclogite types. The values of K' are normalized to Mg, instead of Fe of the previous paper, because the determination of FeO and Fe_2O_3 of eclogite minerals are sometimes questioned because of extreme difficulty in dissolving them in acid. The problem in dealing with the trace element distribution lies in the unnecessarily rounded figures of the Mn concentration of clinopyroxene. For low temperature eclogites, in which a remarkable preferential concentration of Mn into garnet takes place, the published MnO contents of clinopyroxenes are far from being satisfactory.

It was shown in the previous section that K'_{Fe-Mg} is nearly composition independent. If this is accepted it is expected that K'_{Mn-Mg}, K'_{Co-Mg} and K'_{Ni-Mg} are composition independent as well (cf. section 2.4).

A detailed consideration of the trace element distribution in eclogite minerals will be discussed elsewhere.

5. Concluding remarks

In the foregoing discussions it was shown that K'_{Fe-Mg} can be used as a geological thermometer to distinguish various eclogite types, and to subdivide some of the eclogite types. The temperature of crystallization of eclogites is considered to increase in the following order:

1) Ural, Colombia and Guatemala; ⎫
2) California; ⎬ Low temperature eclogites
3) Alps and Japan; ⎭
4) Bavaria and Spain;
5) Norway;
6) East Sudetes and granulites. The former represents higher pressures than the latter at more or less similar temperatures;
7) Inclusions in basalt and in kimberlite.

TABLE 4

Apparent Mn–Mg distribution coefficient for various eclogites

Locality	K'_{Fe-Mg}	K'_{Mn-Mg}	References
California	28	35	21
Spain	10.3	12	3
Norway	9.1 (7.6)*	8.5*	5, 22
East Sudetes	5.6	14	7, 23
		10.5**	
Inclusions in kimberlite	3.6	3.8	14
Inclusions in basalt	2.6	3.7	11, 12
Inclusions in peridotite			
Higasiakaisi	8.3	15	24
			This paper
Norway	5.7	8	5, 17, 25

Numbers in the last column are the source of data as given in table 2. The additional references are:
21. COLEMAN et al. (1965)   22. MATSUI et al. (1966)
23. BAKUN-CZUBAROW (1968)   24. MATSUI (unpublished)
25. GREEN (unpublished)
* K' by (22).   ** K' by (23).

Geological and petrological implication of determining the relative temperature of eclogite crystallization will be discussed elsewhere, as it needs detailed consideration on the relationships between the mineralogy of eclogites and associated rocks. The coexistence of eclogite and basic schists in many metamorphic terranes cannot be explained, if we accept the classical assumption that the chemical potential of water during regional metamorphism was the same or similar within a mineral zone of particular metamorphic terranes. It is also worthy to mention that most of the low temperature eclogites occur in glaucophanitic metamorphic terranes, and no eclogite has been described from the low pressure regional metamorphic terranes. This may require the revision of GREEN and RINGWOOD's (1967) suggestion that the eclogite mineral assemblage may be stable in basic rocks even at very low pressures, if the temperature is low.

Acknowledgements

The author is deeply indebted to Drs. D. H. GREEN and Y. MATSUI for helpful discussions on this problem and the permission to quote their unpublished data. He is also indebted to Prof. H. KUNO, Drs. I. KUSHIRO, Y. OKI, and H. HIGUCHI for permission to use their unpublished data, and to Prof. M. YAMASAKI for the critical reading of the manuscript. The grant from the Australian National University for the visiting appointment from 1965–1967 is also acknowledged.

References

BAKUN-CZUBAROW, N. (1968) Arch. Mineral **28**, 243.
BANNO, S. (1964) J. Fac. Sci. Univ. Tokyo, Sec. II **15**, 203.
BANNO, S. (1966) Japan. J. Geol. Geography, Trans. **37**, 105.
BANNO, S. (1967a) Neues Jahrb. Mineral. Monatsh., 116.
BANNO, S. (1967b) Earth Planet. Sci. Letters **2**, 249.
BANNO, S. (in press) Korzhinskii volume, Moscow.
BANNO, S. and Y. MATSUI (1965) Proc. Japan Acad. **41**, 716.
BANNO, S. and G. YOSHINO (1965) Upper Mantle Symposium, New Dehli, 150.
BEARTH, P. (1965) Schweiz. Mineral. Petrog. Mitt. **45**, 179.
BEARTH, P. (1966) Schweiz. Mineral. Petrog. Mitt. **46**, 13.
BINNS, R. A. (1967) J. Petrol. **8**, 349.
BORG, I. W. (1956) Bull. Geol. Soc. Am. **67**, 1563.
CHESNOKOV, B. V. (1960) Intern. Geol. Rev. **2**, 936.
CLARK, J. R. and J. J. PAPIKE (1968) Am. Mineralogist **53**, 840.
CLARK, S. P., J. F. SCHAIRER and J. DE NEUFVILLE (1962) Yearbook 1961-62 Geophys. Lab. Carnegie Inst. Wash., 59.
COLEMAN, R. G., D. E. LEE, L. B. BEATTY and W. W. BRANNOCK (1965) Bull. Geol. Soc. Am. **76**, 483.
DICKEY, J. S., JR. (1968) Am. Mineralogist **53**, 1304.
DUDEK, A. and P. KOPECKY (1966) Kristallinikum **4**, 7.
ESKOLA, P. (1921) Oslo Vidensk. Skr. Mat.-Naturw. Kl. No. 8.
ESSENE, E. and W. S. FYFE (1967) Contrib. Miner. Petrol. **15**, 1.
FIALA, J. (1966) Kristallinikum **4**, 31.
FORBES, R. B. (1965) J. Geophys. Res. **70**, 1515.
GIROD, M. (1967) Bull. Soc. Franç. Minéral. Crist. **90**, 202.
GOLDSCHMIDT, V. M. (1922) Naturw. **42**, 1.
GREEN, D. H. (1966) Earth Planet. Sci. Letters **1**, 414.
GREEN, D. H. (in press) Korzhinskii volume, Moscow.
GREEN, D. H. and A. E. RINGWOOD (1967) Geochim. Cosmochim. Acta **31**, 767.

GREEN, D. H., A. J. P. LOCKWOOD and E. C. KISS (1968) Am. Mineralogist **53**, 1320.
GRUBENMANN, U. (1904) *Die Kristallinschiefer I* (Borntraeger, Berlin).
KUSHIRO, I. (1965) Yearbook 64–65 Geophys. Lab. Carnegie Inst. Wash., 112.
KUSHIRO, I. and K. AOKI (1968) Am. Mineralogist **53**, 1347.
KUNO, H. Geol. Soc. Am., Mem., in press.
LOVERING, J. F. and A. J. R. WHITE (1964) J. Petrol. **5**, 195.
MATSUI, Y. and S. BANNO (1965) Proc. Japan Acad. **41**, 461.
MATSUI, Y. and S. BANNO (1968) Kagaku-no-Ryoiki **156**, 256 (in Japanese).
MATSUI, Y., S. BANNO and I. HERNES, Norsk Ged. Tidsskr **46**, 364.
MATSUI, Y. and Y. SYONO (1968) Geochem. J. **2**, 51.
MATSUI, Y., Y. SYONO, S. AKIMOTO and K. KITAYAMA (1968) Geochem. J. **2**, 61.
MIYASHIRO, A. (1961) J. Petrol. **2**, 277.
MIYASHIRO, A. and Y. SEKI (1958) Japan. J. Geol. Geography, Trans. **29**, 199.
MUELLER, R. F. (1962) Geochim. Cosmochim. Acta **26**, 581.
NIXON, P. H., O. VON KNORRING and J. M. ROOKE (1963) Am. Mineralogist **48**, 1090.
O'HARA, M. J. (1960) Geol. Mag. **97**, 145.
O'HARA, M. J. and E. L. P. MERCY (1963) Trans. Roy. Soc. Edinburgh **65**, 251.
O'HARA, M. J. and E. L. P. MERCY (1966) Am. Mineralogist **51**, 336.
PHILIPSBORN, H. VON (1930) Chem. Erde **5**, 200.
RINGWOOD, A. E. and D. H. GREEN (1966) Tectonophysics **3**, 383.
ROBIE, R. A., P. M. BETHKE, M. S. TOULMIN and J. L. EDWARDS (1966) in: S. P. Clark, ed., *Handbook of physical constants* (Geol. Soc. Am., Mem. No. 97) p. 30.
SAXENA, S. K. (1968) Am. Mineralogist **53**, 2018.
SMULIKOWSKI, K. (1964) Bull. Acad. Pol. Sci. **12**, 27.
SMULIKOWSKI, K. (1967) Geol. Sudetica **3**, 7.
SOBOLEV, N. V. (1964) *Paragenetic types of garnet* (in Russian, Nauka, Moscow) p. 218.
SOBOLEV, N. V. and I. K. KUZNETZOVA (1966) Dokl. Acad. Nauk SSSR **167**, 1365 (in Russian).
SOBOLEV, N. V., N. I. ZYUZIN and I. K. KUZNETZOVA (1966) Dokl. Acad. Nauk SSSR **167**, 902 (in Russian).
SOBOLEV, N. V., I. K. KUZNETZOVA and N. I. ZYUZIN (1968) J. Petrol. **9**, 253.
SPRY, A. H. (1963) Mineral. Mag. **33**, 589.
VAN DER PLAS, L. (1959) Leidse Geol. Mededel. **24**, 415.
VOGEL, D. E. (1967) Leidse Geol. Mededel. **40**, 121.
WARNAARS, F. W. (1967) Ph.D. Thesis (Univ. Leiden).
WATSON (1960) Bull. Geol. Soc. Am. **71**, 2082.
WHITE, A. J. R. (1964) Am. Mineralogist **49**, 883.
WILLIAMS, A. F. (1932) *The genesis of diamond* (Benn, London) p. 636.
YODER, H. S., JR. and C. E. TILLEY (1962) J. Petrol. **3**, 342.

PRESSURE DEPENDENCE OF CRYSTAL STRUCTURES IN THE SYSTEM MgO-Al$_2$O$_3$-SiO$_2$-H$_2$O AT PRESSURES UP TO 30 KILOBARS

W. SCHREYER and F. SEIFERT

Institut für Mineralogie, Ruhr-Universität, Bochum, Germany

The stability relations, as determined by experiment, of the following quaternary and anhydrous ternary phases of the system MgO–Al$_2$O$_3$–SiO$_2$–H$_2$O are reviewed in the light of their crystal structures: Cordierite, gedrite, aluminous enstatite, boron-free kornerupine, sapphirine, yoderite, Mg-staurolite, pyrope, aluminous talc, and chlorite. With the exception of the last two phases, there is a consistent relationship between density, oxygen packing and coordination of Al on one side, and stability with respect to pressure on the other side: Densely packed phases with only octahedral Al like pyrope and Mg-staurolite are only stable at high pressures, whereas loosely packed cordierite with only tetrahedral Al is only stable at low pressures. The phases gedrite, aluminous enstatite, boron-free kornerupine, sapphirine, and yoderite exhibiting intermediate degrees of packing and (except yoderite) both tetrahedral and octahedral Al are stable at intermediate pressures. The phyllosilicates talc and chlorite are pressure-indifferent within the range considered, although they have low densities and contain Al both in tetrahedral and octahedral coordination.

1. Introduction

The system MgO–Al$_2$O$_3$–SiO$_2$–H$_2$O is regarded as an important model system for pelitic rocks of crustal origin. However, because it contains at least four of the main mineral phases of peridotites, e.g. forsterite, enstatite, spinel and pyrope (cf. fig. 1), it is also of interest in discussions of phase relations within the Earth's mantle.

Experimental studies made in the past within this system have therefore been directed specifically towards these two goals:

Hydrothermal investigations were confined to relatively low pressures and alumina-rich bulk compositions imitating pelites, whereas high-pressure work was done solely in the anhydrous system, and only on bulk-compositions relatively poor in alumina. From the first set of studies the general contributions by YODER (1952), and by ROY and ROY (1955) should be mentioned as well as specific investigations on chlorites by NELSON and ROY (1958), on anthophyllite by GREENWOOD (1963), on cordierite by SCHREYER and YODER (1964), and on the assemblage chlorite+quartz by FAWCETT and YODER (1966). Anhydrous experiments at high pressures by BOYD and ENGLAND (1959, 1960, 1964) were aimed at determining stability relations of pyrope and aluminous enstatites and mixtures thereof. In addition MCGREGOR (1964) studied compatibility relations of the enstatite–spinel and forsterite–pyrope pairs.

It was only most recently that the aluminous portions of the system MgO–Al$_2$O$_3$–SiO$_2$–H$_2$O were also studied at high pressures and in the presence of excess water (SCHREYER and YODER, 1968; SCHREYER, 1968; SCHREYER and SEIFERT, 1969 a, b). The drastic changes in the phase relations observed against those known at lower pressures are mainly due to the instability of cordierite as already expected from earlier work (SCHREYER and YODER, 1960), but also to the appearance of five new crystalline compounds, which had previously not been synthesized or not even been anticipated in this system. These five phases are: *yoderite*, *Mg-staurolite*, *gedrite*, *sapphirine* with the composition Mg$_2$Al$_4$SiO$_{10}$, and a phase isostructural with the rare natural boron-bearing silicate kornerupine, called *boron-free kornerupine*. The approximate compositions of these phases are plotted in fig. 1. In addition more insight was gained into the stability relations of *pyrope* and also into the upper pressure breakdown relations of *cordierite*. More detailed accounts on these topics as well as physical constants of the new phases have already been published in the recent papers mentioned above.

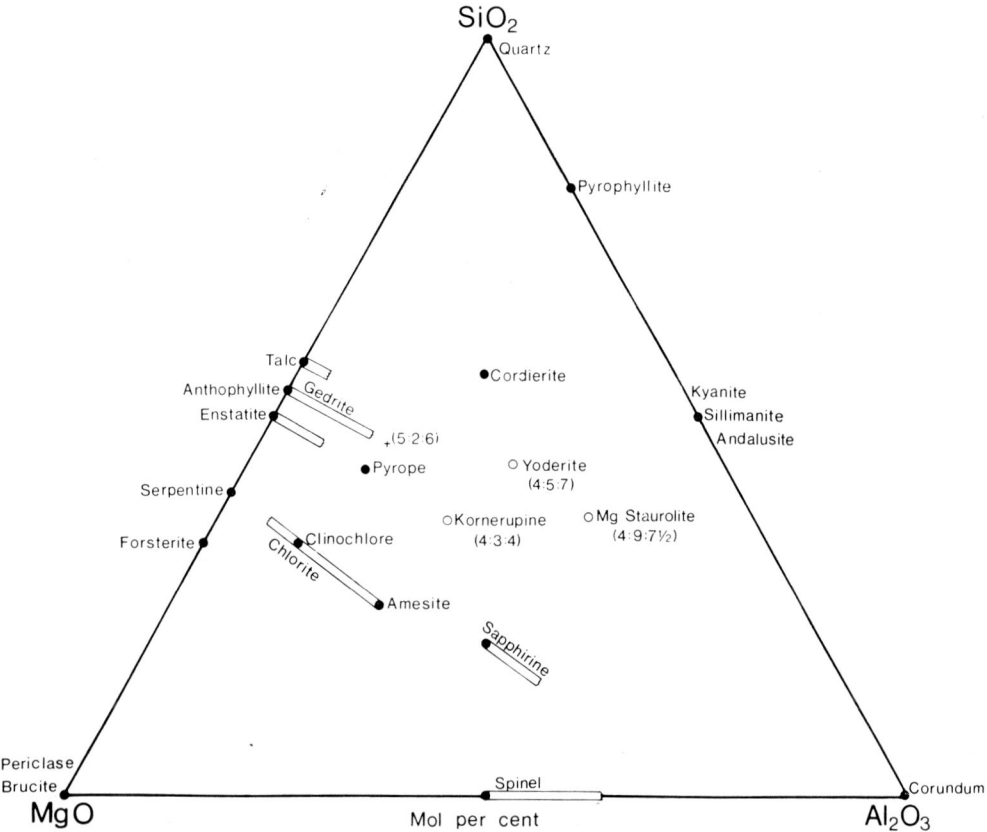

Fig. 1. Crystalline phases in the system MgO–Al_2O_3–SiO_2–H_2O projected from the H_2O apex onto the anhydrous base of the tetrahedron. Phases stable only at low pressures and low temperatures, respectively, have been omitted. Open circles indicate phases of which the compositions are not exactly known as yet. Kornerupine stands for boron-free kornerupine.

It is the purpose of the present paper to review the stability relations, as known at present, of important quaternary and anhydrous ternary phases of the system MgO–Al_2O_3–SiO_2–H_2O in the light of the main features of their crystal structures. Due to the lack of experimental data beyond 25–30 kb the discussion must be limited, however, to still relatively low pressures.

2. Structural properties of crystalline phases

In table 1 the phases discussed in this paper are listed together with their simplified structural formulas, oxide ratios, coordination numbers of the cations Mg and Al, densities as determined from X-ray data, and molar volumes. In addition, a so-called oxygen packing index, i.e. the number of moles of oxygen present in a volume of 1 cm^3 of the various structures was calculated. The higher the values of this packing index the closer is the packing of the anionic framework representing the bulk volume of the structure.

2.1. Cordierite

Cordierite, previously considered to be a ring silicate (BYSTRÖM, 1942), according to more recent structure determinations (e.g. GIBBS, 1966) has all the properties of a tectosilicate with a three-dimensional framework of (Al, Si)O_4-tetrahedra. All the aluminum in cordierite is in tetrahedral coordination. A characteristic structural feature are stacks of six-membered rings forming large open channels, which may accommodate additional ions or, more frequently, foreign molecules such as H_2O (SCHREYER and YODER, 1964). The loosely packed nature of this crystal structure is well demonstrated by its exceptionally low density and its low oxygen packing index (table 1).

2.2. Anthophyllite-gedrite

Amphiboles are represented in the system considered by the orthorhombic solid solution series between an-

TABLE 1

Important quaternary and anhydrous ternary phases of the system $MgO-Al_2O_3-SiO_2-H_2O$ and some of their chemical and crystallographic properties

Mineral	Oxide ratio $MgO:Al_2O_3:SiO_2$	Simplified structural formula	X-ray density (g/cm^3)	Molar volume (cm^3)	Oxygen packing index (mole O/cm^3)
Cordierite	2:2:5	$Mg_2^{VI}Al_4^{IV}Si_5O_{18}$	2.505	233.5	0.0773
Anthophyllite	7:0:8	$Mg_7^{VI}[Si_8O_{22}](OH)_2$	2.954	264.4	0.0908
Gedrite	6:1:7	$Mg_6^{VI}Al^{VI}[Al^{IV}Si_7O_{22}](OH)_2$	3.004	260.4	0.0921
Enstatite	1:0:1	$Mg^{VI}[SiO_3]$	3.198	31.40	0.0964
Al-enstatite	17:3:17	$Mg_{0.85}^{VI}Al_{0.15}^{VI}[Al_{0.15}^{IV}Si_{0.85}O_3]$	3.262	30.83	0.0973
Kornerupine	4:3:4	$Mg_4^{VI}Al_5^{VI}Al^{IV}Si_4O_{20}(OH)_2$	3.214	225.7	0.0975
Sapphirine	2:2:1	$Mg_4^{VI}Al_8^{VI,IV}Si_2O_{20}$	3.493	197.3	0.1014
Sapphirine	7:9:3	$Mg_{3.5}^{VI}Al_9^{VI,IV}Si_{1.5}O_{10}$	3.505	197.0	0.1015
Yoderite	4:5:7	$Mg_{2.28}^{V}Al_{5.72}^{VI,V}Si_4O_{17.72}(OH)_{2.28}$	3.311	194.6	0.1028
Mg-staurolite	4:9:7.5	$Mg_4^{IV?}Al_{18}^{VI}Si_{7.5}O_{44}(OH)_4$	3.535	442.9	0.1083
Pyrope	3:1:3	$Mg_3^{VIII}Al_2^{VI}Si_3O_{12}$	3.559	113.3	0.1059
Talc	3:0:4	$Mg_3^{VI}[Si_4O_{10}](OH)_2$	2.788	136.7	0.0878
Al-talc	—	$(Mg, Al)_3^{VI}[(Al^{IV}, Si)_4O_{10}](OH)_2$	?	?	?
Chlorite	5:1:3	$Mg_5^{VI}Al^{VI}[Al^{IV}Si_3O_{10}](OH)_8$	≈2.65	≈209.7	≈0.0858

thophyllite and gedrite (table 1). In anthophyllite the double chains characterizing the amphibole structure (WARREN and MODELL, 1930) are built up solely of SiO_4-tetrahedra, whereas the planes of cations separating these chains contain only Mg in octahedral coordination. The gedrites being related to anthophyllite through AlAl → MgSi substitution contain Al both in tetrahedral and octahedral coordination. Because there is a marked contraction of the unit cell with increasing Al content (SCHREYER and SEIFERT, 1969a), the density as well as the oxygen packing index is considerably higher for gedrites than for anthophyllite.

2.3. Enstatite

Aluminous enstatites exhibiting a single chain pyroxene structure contain octahedral as well as tetrahedral Al in a very similar fashion as the gedrites. In accordance with the behavior of the anthophyllite–gedrite series, densities and packing indices increase from pure enstatite towards more aluminous members (cf. SKINNER and BOYD, 1964).

2.4. Kornerupine

The crystal structure of kornerupine has only recently been determined by MOORE and BENNETT (1968) using a natural specimen containing boron. However, in the crystal chemical formula derived by these authors boron is not listed as an essential element. It is not so surprising, therefore, that a boron-free kornerupine could be synthesized in the system $MgO-Al_2O_3-SiO_2-H_2O$ (SCHREYER and SEIFERT, 1969a). Although the exact composition of this phase is still unknown, the one given in table 1 is considered most likely. The anionic framework of the kornerupine structure is based on an interrupted close oxygen packing. The cation coordination polyhedra are MgO_6- and AlO_6-octahedra which fuse to form walls, held together laterally by tetrahedral $[T_2O_7]$ doublets and $[T_3O_{10}]$ triplets in which T = Al, Si.

2.5. Sapphirine

According to the recent structure analysis by MOORE (1969) sapphirine may be regarded a chain silicate consisting of complex $(Si, Al)_6O_{18}$ tetrahedral chains interconnecting walls of MgO_6- and AlO_6-octahedra. It is based on a cubic close-packing of oxygens which explains the relatively high packing index (table 1). Its rather high density is mainly due to its low silica content. Sapphirines exhibit solid solution again based on MgSi ⇌ AlAl substitutions. Whereas an Al-rich end member was synthesized at atmospheric pressure by FOSTER (1950), the MgSi-rich end member of the composition $Mg_2Al_4SiO_{10}$ was only recently prepared at 10 kb (SCHREYER and SEIFERT, 1969a). Because, in contrast to the behavior of the gedrites and Al-enstatites, the unit cell volume of the sapphirines synthesized does not change appreciably as a function of composition, density and oxygen packing index remain virtually constant for all sapphirine solid solutions.

2.6. Yoderite

The crystal structure of yoderite (FLEET and MEGAW, 1962) is related to that of the pure Al-silicate kyanite again exhibiting close oxygen packing. As a first approximation yoderite may be derived from Al_2SiO_5 through a partial substitution of Al by MgH. One obvious difference of the yoderite structure against that of kyanite is, however, the presence of MgO_5- and AlO_5-polyhedra in addition to the usual AlO_6-octahedra.

2.7. Mg-staurolite

The crystal structure of Mg-staurolite is not known but will undoubtedly be closely related to that of the iron-rich natural staurolites. It consists in essence of an interlayering of kyanite units with complex units of $AlFe_2^{+2}O_3OH$ composition in which Fe^{+2} occurs in tetrahedral coordination (NÁRAY-SZABÓ and SASVÁRI, 1958; J. V. SMITH, 1968). All the aluminum in staurolite is octahedrally coordinated. According to SCHREYER and SEIFERT (1969a), however, the variation of the b_0 unit cell dimension in the system Fe-staurolite–Mg-staurolite cannot be linear over the entire compositional range. This may possibly indicate a minor structural discontinuity and perhaps even a coordination of Mg in the Mg-rich members other than tetrahedral.

The composition of Mg-staurolite as given in table 1 is based on the data of RICHARDSON (1968) on pure Fe-staurolite and on conclusions by SCHREYER and SEIFERT (1969a) concerning the Mg end member. It seems significant that the oxygen packing index of Mg-staurolite is the highest of all the phases presently discussed in the system $MgO–Al_2O_3–SiO_2–H_2O$.

2.8. Pyrope

Pyrope (GIBBS and SMITH, 1965) displays a normal garnet structure thus having all the aluminum in sixfold, all magnesium in eightfold coordination. Its density is the highest of all the phases considered, although the oxygen packing index is below that of Mg-staurolite.

2.9. Talc

Talc is a typical phyllosilicate with octahedral MgO_6-layers alternating with tetrahedral SiO_4-layers. The Al in aluminous talc enters both of these layers according to the AlAl → MgSi substitution. The extent of this solid solubility is, according to the data obtained by FAWCETT and YODER (1966) at 2 and 10 kb, not very large and practically pressure-insensitive in this pressure range.

There are no data on the variation of the unit cell volume of aluminous talcs. It is obvious from the data of pure talc, however, that both density, and oxygen packing index as determined at atmospheric pressure must be very low for the whole talc series.

2.10. Chlorite

Chlorites are also phyllosilicates consisting of interlayered talc- and brucite-type sheets. In the brucite-type sheet all the cations (Mg, Al) are in octahedral coordination. Thus like aluminous talc the chlorites contain equal amounts of both octahedral and tetrahedral Al and their densities and oxygen packing indices are comparatively low.

3. Stability relations

3.1. Cordierite

Cordierite with its open framework structure and low density is clearly confined to relatively low pressures not higher than 8–11 kb depending on temperature. A series of breakdown assemblages of cordierite was obtained through experiment or deduced theoretically by SCHREYER and YODER (1960, 1964), but more recent experimentation (SCHREYER, 1968) showed that due to metastable crystallization several important assemblages had been omitted. In the pressure temperature plot constructed through Schreinemakers analysis by SCHREYER and SEIFERT (1969b, figs. 1 and 2) the sequence of stable breakdown assemblages is, in the order of increasing temperature:

Chlorite + kyanite + quartz;
chlorite + yoderite + quartz;
talc + yoderite + quartz;
talc + kyanite + quartz;
gedrite + kyanite (or sillimanite) + quartz;
enstatite + sillimanite + quartz;
sapphirine + quartz.

The last two assemblages, which only involve anhydrous minerals, are probably confined to conditions of water pressure being less than total pressure, whereas in the presence of excess water hydrous cordierite would only form hydrous breakdown products or, at higher temperatures, melt incongruently. The incorporation of molecular H_2O into the structural channels of cordierite which must of course increase the density of the mineral (SCHREYER and YODER, 1964) does not seem to have the effect of stabilizing such cordierites towards higher pressures. Most recent yet unpublished data suggest that, quite on the contrary, the anhydrous cordierite breakdown into enstatite+sillimanite+quartz takes place at the highest pressures.

The nature of the hydrous breakdown assemblages of cordierite listed above is also of interest concerning the compatibility relations of this mineral. Thus cordierite can coexist stably over limited pressure temperature ranges with the phases kyanite, yoderite, and gedrite, the stability fields of all these compounds overlapping with the cordierite field. Moreover the assemblage *talc+kyanite* is of considerable importance as it replaces the common low-density assemblage *chlorite+quartz* at pressures in the range of 10–15 kb (SCHREYER, 1968, fig. 25)

3.2. *Anthophyllite–gedrite*

The stability relations of the anthophyllite–gedrite series are extremely complicated and not fully understood as yet. According to the authors' latest, yet unpublished results they vary considerably as a function of composition. It is quite clear however, that the double chain orthoamphibole structure becomes totally unstable at intermediate pressures, that is in the approximate range between 13 and 25 kb.

Pure *anthophyllite* was found to break down to the assemblage enstatite+talc in the pressure range 13–19.5 kb thus closely approaching the breakdown pressure predicted by GREENWOOD (1963) on thermodynamical grounds. Its temperature stability range below this pressure is rather narrow ranging from roughly 700° to 800° C at 13 kb.

Gedrites, on the other hand, are only stable at considerably higher temperatures, roughly between 800° and 900° near 13 kb, that is under conditions at which pure anthophyllite has already broken down to the high temperature assemblage enstatite+quartz. Similarly, the pressure stability range of the gedrites extends to somewhat higher pressures than that of anthophyllite. The phase $Mg_6Al[AlSi_7O_{22}](OH)_2$ synthesized by SCHREYER and SEIFERT (1969a) is stable to at least 20 kb. It is possible that more aluminous gedrites are stable to even higher pressures. Thus it may be concluded that the AlAl → MgSi substitution in orthoamphiboles resulting in a denser packing of the structure stabilizes these amphiboles towards higher pressures as well as temperatures.

The low-temperature breakdown of gedrites yields more hydrous assemblages such as chlorite+talc+corundum. Anhydrous high-temperature breakdown assemblages are, depending on the gedrite composition, aluminous enstatite+kyanite (or sillimanite) ± quartz and aluminous enstatite+kyanite+pyrope. Upper pressure breakdown assemblages of gedrites may be talc+pyrope ± enstatite or ± kyanite, again depending on the gedrite compositions.

The extension of the gedrite stability field towards pressures below 10 kb is not known as yet. It is expected, however, that a minimum pressure in the order of several kilobars is required to form this phase. Thus the only common orthoamphibole stable at very low pressures appears to be anthophyllite (GREENWOOD, 1963).

3.3. *Enstatite*

The basic pyroxene structure possessed by the enstatites is certainly stable over the entire pressure range considered in this paper. However, there is an appreciable pressure-temperature dependence of the amounts of Al to be accommodated in the structure. BOYD and ENGLAND (1964) studying the system enstatite–pyrope have obtained the most aluminous enstatites at intermediate pressures (around 20 kb) and relatively high temperatures. At higher pressures the maximum Al content of the enstatites decreases strongly because this ion is preferentially incorporated in the denser phase pyrope. At lower pressures the aluminous enstatites break down into less dense assemblages such as nearly pure enstatite+cordierite+sapphirine, enstatite+cordierite+spinel, and enstatite+cordierite+forsterite.

3.4. *Kornerupine*

Boron-free kornerupine could be synthesized by SCHREYER and SEIFERT (1969a) at pressures between 8 and 13 kb and at temperatures between 800° and

900 °C. At 10 kb it formed at the expense of the stable phases cordierite and sapphirine and, therefore, must probably be considered a stable phase itself over a limited pressure-temperature range. It becomes clearly unstable, however, at pressures in excess of some 17 kb, when pyrope and corundum form instead. Its lower temperature stability limit at lower pressures may be marked by the breakdown to the hydrous assemblage chlorite + gedrite + corundum whereas its upper temperature stability limit at these same pressures seems to be given by its breakdown to the anhydrous pair enstatite + corundum. It is not known how far the stability field of boron-free kornerupine extends to still lower pressures, i.e. below 8 kb. Because of the relatively high density and packing index of boron-free kornerupine one might expect a minimum pressure of several kilobars to be necessary for the formation of this phase.

3.5. *Sapphirine*

The stability relations of the sapphirines are very complicated indeed, and, similarly as for the gedrites, vary strongly with composition. If one neglects these variations a stability field of sapphirine (regardless of its composition) extends from atmospheric pressure to some 23–28 kb depending on temperature. At higher pressures sapphirine breaks down to the denser anhydrous assemblage pyrope + corundum + spinel (SCHREYER and SEIFERT, 1969a). At intermediate pressures between 8 and 23 kb and temperatures between about 725° and 875 °C it is replaced by the hydrous assemblage chlorite + corundum + spinel. According to current experimental work performed at this Institute by D. ACKERMAND (personal communication) the sapphirines forming at the lowest pressures up to 1 kb are confined to very high temperatures: Below some 1200 °C these sapphirines break down to the assemblage cordierite + spinel + corundum which is not stable at liquidus temperatures (KEITH and SCHAIRER, 1952). This anhydrous low-pressure breakdown curve of sapphirine must have a strongly negative slope as it will join the hydrous low-temperature breakdown curve at higher pressures discussed above. These overall stability relations may explain the rarity of sapphirine in natural rocks, this mineral being confined to environments characterized by relatively high pressures or exceptionally high temperatures or both. The solid solubility of sapphirine for a given temperature and pressure is rather restricted. However, this narrow range is being shifted throughout the entire compositional series $Mg_2Al_4SiO_{10}$ (2:2:1)–$Mg_7Al_{18}Si_3O_{40}$ (7:9:3) as a function mainly of temperature and, to a lesser extent, of pressure: Alumina-poor sapphirines near 2:2:1 composition form at the lowest temperatures ($\approx 750°$–$900°$) and, necessarily, at pressures near 10 kb. With increasing temperatures sapphirines with compositions intermediate between 2:2:1 and 7:9:3 become stable, whereas the most aluminous members are restricted to near-liquidus temperatures at low pressures. Provided these solid solubility relationships are known in detail sapphirines might thus be used as sensitive geothermometers and -barometers.

3.6. *Yoderite*

The preliminary stability field of pure *Mg-yoderite* as worked by SCHREYER and YODER (1968, fig. 22) is characterized by its relatively small width with respect to temperature ($\approx 700°$–850 °C) and its wide pressure range (≈ 8 to at least 25 kb). At low temperatures yoderite was found to break down into the following assemblages arranged in the order of decreasing pressure:

chlorite + pyrope + kyanite;
chlorite + talc + kyanite;
and
chlorite + quartz + kyanite.

The low-pressure end of the yoderite field is probably marked by several breakdown assemblages involving the phase cordierite:

cordierite + kyanite + chlorite;
cordierite + corundum + chlorite;
cordierite + corundum + boron-free kornerupine;
cordierite + corundum + gedrite.

Even at high temperatures yoderite breaks down to form hydrous assemblages, which are probably, in the order of increasing pressure:

gedrite + sillimanite + corundum;
gedrite + sillimanite (or kyanite) + Mg-staurolite;
enstatite + kyanite + Mg-staurolite;
and
pyrope + kyanite + Mg-staurolite.

An upper pressure stability limit of yoderite has thus far not been found by experiment. However, phase theoretical and volume considerations suggest that the univariant low-temperature and high-temperature breakdown curves at high pressures i.e.

chlorite + pyrope + kyanite ⇌ yoderite,
and

yoderite ⇌ Mg-staurolite + kyanite + pyrope,

respectively, may intersect to form an invariant point marking the high-pressure end of the yoderite field. Experimental results obtained at 25 kb seem to indicate that this invariant point may lie at pressures not very much higher than 25 kb. It should be borne in mind, however, that in this pressure range the SiO_2 polymorph *coesite* becomes stable, and, because of the density increase connected with the quartz-coesite transition, the overall compatibility relations in the system $MgO-Al_2O_3-SiO_2-H_2O$ may be changed from those valid in the quartz stability range.

3.7. *Mg-staurolite*

Mg-staurolite is one of the rather unexpected phases in this system, because it has never been found in nature, all natural staurolites being very iron-rich. Its preliminary stability field (SCHREYER, 1968, fig. 23) is roughly wedge-shaped beginning in the vicinity of 12 kb, 800 °C and extending to higher pressures. Thus there is probably no overlap with the cordierite field.

The lower temperature stability limit lies near 750 °C and is marked by the breakdown to the more hydrous assemblage chlorite + kyanite + corundum over the whole pressure range investigated, i.e., up to 25 kb. The upper temperature stability limit of Mg-staurolite shows a pronounced break in slope, where it intersects the lower-pressure stability limit of pyrope (see later). Therefore, at pressures above about 18 kb and at temperatures near 980 °C Mg-staurolite breaks down to pyrope + kyanite + corundum, whereas at slightly lower pressures the breakdown assemblage is enstatite + kyanite (or sillimanite) + corundum. With further decreasing pressures and temperatures the probable breakdown assemblages are gedrite + sillimanite + corundum, and yoderite + kyanite (or sillimanite) + corundum. The latter assemblage implies that the upper temperature stability limits of Mg-staurolite and yoderite intersect at an invariant point involving these two phases plus corundum + gedrite + Al-silicate + vapor.

The low pressure end of the Mg-staurolite field is marked by an invariant point involving the phases Mg-staurolite, yoderite, kyanite, chlorite, corundum and vapor (see SCHREYER and SEIFERT, 1969a, fig. 4). No experimental data are available concerning the extension of the Mg-staurolite stability field to pressures beyond 25 kb. It is clear, however, that in the range 20–25 kb the nucleation of Mg-staurolite is favored by pressure increase. This behavior, which is the opposite of that of yoderite, appears to indicate that the densely packed Mg-staurolite remains stable to considerably higher pressures.

3.8. *Pyrope*

The stability field of pyrope was first determined by BOYD and ENGLAND (1959) under anhydrous conditions. Under hydrous conditions this stability field could be followed to lower temperatures and pressures and approximate limits could be identified (SCHREYER 1969, fig. 24). A low-temperature low-pressure end of the pyrope field under these conditions was found to lie near 800°, 15 kb.

The hydrous low-temperature breakdown of pyrope is given by a series of univariant reaction curves which all appear to have negative slopes thus extending to temperatures near 700 °C with increasing pressures. In the order of decreasing pressure the requisite breakdown assemblages seem to be:

chlorite + talc + kyanite;

chlorite + talc + yoderite;

chlorite + talc + Mg-staurolite;

chlorite + talc + corundum;

and

chlorite + gedrite + corundum.

The latter three phases plus enstatite and pyrope (+vapor) may coexist at an invariant point marking the lowest pressure under which pyrope is stable in the presence of excess water, i.e. about 15 kb at about 800 °C.

The low-pressure breakdown of pyrope at higher temperatures is given by two reaction curves both exhibiting positive slopes. The first curve leading to the breakdown assemblage enstatite + corundum originates from the invariant point just mentioned. The second curve extending at higher temperatures probably up to the melting point of pyrope creates the breakdown assemblage aluminous enstatite + sillimanite + sapphirine, which was already assumed by BOYD and ENGLAND (1959).

It appears possible that the stability field of boron-free kornerupine (see above) overlaps with the lowest-pressure part of the pyrope field just outlined. In this case there may be two additional breakdown assemblages of pyrope stable in the potential range of overlap (15–16 kb, 800°–900 °C): chlorite + gedrite + boron-free kornerupine at relatively low, and enstatite + boron-free kornerupine at relatively high temperatures. Pyrope is stable up to pressures much beyond the range discussed here. As found by RINGWOOD (1967) it even develops solid solubility towards enstatite at pressures above about 100 kb.

3.9. *Talc*

Talc is stable over the entire pressure range from one atmosphere to at least 30 kb (KITAHARA et al. 1966). Its upper temperature stability limit lies near 800 °C, and no lower temperature stability limit has been found as yet. Within the stability field of anthophyllite (see above) the breakdown assemblage is anthophyllite + quartz; at higher pressures it is changed into enstatite + quartz (or coesite) as determined by KITAHARA et al. (1966). Unfortunately, no systematic study is available on the stability of aluminous talcs. However, preliminary results of the authors indicate that these phases can be synthesized over much of the stability field of pure talc.

3.10. *Chlorite*

The general stability field of the chlorites is very similar to that of talc: It was found to extend from one atmosphere to 20 kb by SEGNIT (1963) and, more recently, to at least 25 kb (SCHREYER, 1968). The upper temperature stability limit at 10 kb is 830 °C and the breakdown assemblage was found to be forsterite + enstatite + spinel by FAWCETT and YODER (1966). These authors also came to the conclusion that the chlorite exhibiting the highest temperature stability, that is clinochlore at relatively low pressures (NELSON and ROY, 1958), is becoming more and more aluminous, that is amesitic, with increasing pressure.

4. Summary and conclusions

On the basis of their stability relations the quaternary and anhydrous ternary phases of the system MgO–Al_2O_3–SiO_2–H_2O discussed in the previous section may roughly be subdivided into four main groups:

(1) *Low-pressure phases* that are stable from atmospheric conditions up to pressures in the neighbourhood of 10 kb. Cordierite is the only phase belonging to this group.

(2) *Intermediate-pressure phases* that usually require a minimum pressure of several kilobars for their formation but become unstable at higher pressures ranging from some 15 to 30 kb. The phases gedrite, aluminous enstatite, boron-free kornerupine, sapphirine (as an exception also stable at 1 atm and very high temperature), and, probably, yoderite may be classified in this group.

(3) *High-pressure phases*, the stability fields of which start at pressures above 10 kb and extend much beyond 30 kb. Pyrope and Mg-staurolite represent this group.

(4) *Pressure-indifferent phases* that may form stably over the entire pressure range discussed here, i.e. from atmospheric conditions up to at least some 25 kb. Talc and chlorites are typical phases of this group.

The values of density and oxygen packing index (table 1) yield excellent explanations for the stability relations of all the minerals attributed to groups 1 through 3: Thus cordierite with the lowest numerical values of these properties belongs to group 1, whereas Mg-staurolite with its high values is member of group 3. In a similar way the coordination of cations, especially of Al, has a direct influence on the relative mineral stabilities: Cordierite with only tetrahedral Al is a low-pressure phase; all the intermediate-pressure phases except yoderite with partly five-coordinated Al contain both tetrahedral and octahedral Al; and the high-pressure phases have all their Al in octahedral coordination.

On the other hand, the minerals of group 4 having low densities and packing indices and both tetrahedral and octahedral Al do not follow this pattern. It may not be fortuitous that the two minerals belonging to this group are phyllosilicates which might be particularly pressure-resistant because of their compressible layer structure. KUSHIRO et al. (1967) could even show that the phyllosilicate phlogopite is stable to at least 40 kb and probably some 100 kb.

The list of quaternary phases of the system discussed here may not be complete as yet. There are indications (SCHREYER, 1968) that a phase Mg-chloritoid, $MgO \cdot Al_2O_3 \cdot SiO_2 \cdot H_2O$, may be synthesized at low temperatures (near 500 °C) and intermediate pressures in very long runs. Preliminary attempts were unsuccessful to synthesize Mg-carpholite,

$MgO \cdot Al_2O_3 \cdot 2SiO_2 \cdot 2H_2O$. A mineral close to this end member appears to occur in natural glaucophane schists (DE ROEVER et al., 1967).

The petrological application of the experimental results discussed in this paper can only be outlined briefly at this stage of the investigation. The formation of the *low-pressure phase cordierite* is clearly confined to environments situated within the earth's crust.

The intermediate-pressure phases of group 2 including sapphirine are likely to form in the lower parts of the crust, but also in the uppermost mantle provided the appropriate rock compositions are available there. The common absence of sapphirine from shallow contact aureoles is readily explained because temperatures are not high enough for sapphirine formation in these low pressure environments. The only natural occurrence of *yoderite* discovered thus far (MCKIE, 1959) is in deep-seated pelitic rocks of crustal origin, which also contain very boron-poor *kornerupine* (MCKIE, 1965), *sapphirine* (MCKIE, 1963), and *gedrite* (MCKIE, personal communication 1969). Sapphirine–gedrite–kornerupine assemblages are also known from unusual metamorphic rocks named sakenites (LACROIX, 1939), which because of their aluminous bulk compositions may also be of crustal origin. On the basis of the present results one should expect to find kornerupines even without boron in these rocks.

The *high-pressure-phases* pyrope and Mg-staurolite should only be expected in rocks derived from the earth's upper mantle. *Pyrope* although not of pure end member composition is well known from garnet peridotites of mantle origin, but *Mg-staurolite* has thus far not been detected in any natural rock. This may be due to the lack of sufficiently aluminous bulk composition in the mantle. Yet the recent findings by SOBOLEV et al. (1968) of highly aluminous nodules from kimberlites suggest that Al-enrichment processes may even operate in the upper mantle. The hydrous nature of some of the intermediate- and high-pressure phases discussed in this paper does apparently not preclude their occurrence in the earth's upper mantle: The presence of phlogopite in rocks of mantle origin is evidence for at least local enrichment of this mobile component also in subcrustal environments.

This research is supported by Deutsche Forschungsgemeinschaft, Bad Godesberg, Germany, under the auspices of the International Upper Mantle Committee.

References

BOYD, F. R. and J. L. ENGLAND (1959) Carnegie Inst. Wash. Yearbook **58**, 83–87.
BOYD, F. R. and J. L. ENGLAND (1960) Carnegie Inst. Wash. Yearbook **59**, 49–52.
BOYD, F. R. and J. L. ENGLAND (1964) Carnegie Inst. Wash. Yearbook **63**, 157–161.
BYSTRÖM, A. (1942) Arkiv Kemi Mineral. Geol. **15** B (12).
FAWCETT, J. J. and H. S. YODER, JR. (1966) Am. Mineralogist **51**, 353–380.
FLEET, S. G. and H. D. MEGAW (1962) Acta Cryst. **15**, 721–728.
FOSTER, W. F. (1950) J. Geol. **58**, 135–151.
GIBBS, G. V. (1966) Am. Mineralogist **51**, 1068–1087.
GIBBS, G. V. and J. V. SMITH (1965) Am. Mineralogist **50**, 2023–2039.
GREENWOOD, H. J. (1963) J. Petrol. **4**, 317–351.
KEITH, M. L. and J. F. SCHAIRER (1952) J. Geol. **60**, 181–186.
KITAHARA, S., S. TAKENOUCHI and G. C. KENNEDY (1966) Am. J. Sci. **264**, 223–233.
KUSHIRO, I., Y. SYONO and S. AKIMOTO (1967) Earth Planet. Sci. Letters **3**, 197–203.
LACROIX, A. (1939) Compt. Rend. Acad. Sci. Paris **209**, 609.
MACGREGOR, I. D. (1964) Carnegie Inst. Wash. Yearbook **63**, 156–157.
MCKIE, D. (1959) Mineral. Mag. **32**, 282–307.
MCKIE, D. (1963) Mineral. Mag. **33**, 635–645.
MCKIE, D. (1965) Mineral. Mag. **34**, 346–357.
MOORE, P. B. (1969) Am. Mineralogist **54**, 31–49.
MOORE, P. B. and J. M. BENNETT (1968) Science **159**, 524–526.
NÁRAY-SZABÓ, I. and K. SASVÁRI (1958) Acta Cryst. **11**, 862–865.
NELSON, B. W. and R. ROY (1958) Am. Mineralogist **40**, 147–178.
RICHARDSON, S. W. (1968) Carnegie Inst. Wash. Yearbook **66**, 397–398.
RINGWOOD, A. E. (1967) Earth Planet. Sci. Letters **2**, 255–263.
DE ROEVER, W. P., E. W. F. DE ROEVER, F. F. BEUNK and P. H. J. LAHAYE (1967) Koninkl. Ned. Akad. Wetenschap. Proc. Ser. B **70**, 534–537.
ROY, D. M. and R. ROY (1955) Am. Mineralogist **40**, 147–178.
SCHREYER, W. (1968) Carnegie Inst. Wash. Yearbook **66**, 380–392.
SCHREYER, W. and F. SEIFERT (1969a) Am. J. Sci. **267A**, 407–443.
SCHREYER, W. and F. SEIFERT (1969b) Am. J. Sci. **267**, 371–388.
SCHREYER, W. and H. S. YODER, JR. (1960) Carnegie Inst. Wash. Yearbook **59**, 90–91.
SCHREYER, W. and H. S. YODER, JR. (1964) Neues Jahrb. Mineral. Abhandl. **101**, 271–342.
SCHREYER, W. and H. S. YODER, JR. (1968) Carnegie Inst. Wash. Yearbook **66**, 376–380.
SEGNIT, E. R. (1963) Am. Mineralogist **48**, 1080–1089.
SKINNER, B. J. and F. R. BOYD (1964) Carnegie Inst. Wash. Yearbook **63**, 163–165.
SMITH, J. V. (1968) Am. Mineralogist **53**, 1139–1155.
SOBOLEV, N. V., I. K. KUZNETZOVA and N. I. ZYUZIN (1968) J. Petrol. **9**, 253–280.
WARREN, B. E. and D. I. MODELL (1930) Z. Krist. **75**, 161.
YODER, JR., H. S. (1952) Am. J. Sci. Bowen vol., 569–627.

EXPERIMENTAL DATA ON COEXISTING CORDIERITE AND GARNET UNDER HIGH GRADE METAMORPHIC CONDITIONS

B. J. HENSEN and D. H. GREEN

Department of Geophysics and Geochemistry, Australian National University, Canberra, Australia

An experimental investigation of phase relations in synthetic compositions simulating natural metasediments has been carried out to determine the relative stability of cordierite and garnet as a function of pressure, temperature and chemical composition. Anhydrous solid-solid equilibria have been studied at temperatures of 800–1200 °C and pressures of 0–15 kb. On the basis of the $(MgO+FeO)/Al_2O_3$ ratio (F/A) of the bulk composition silica-saturated cordierite-garnet bearing rocks can be divided into three groups each having different phase assemblages with varying pressure and temperature. At temperatures around 800 °C the following assemblages have been found at low, intermediate and high pressure. Group A $(F/A \gg 1)$: hypersthene-cordierite-quartz/garnet-hypersthene-cordierite-quartz/garnet-hypersthene-quartz. Group B $(F/A > 1)$: cordierite-hypersthene-quartz/garnet-cordierite hypersthene-quartz/garnet-sillimanite-quartz. Group C $(F/A < 1)$: cordierite-sillimanite-quartz/garnet-cordierite-sillimanite-quartz/garnet-sillimanite-quartz. The data indicate that in the intermediate pressure assemblages the $Mg/(Mg+Fe^{2+})$ ratios of the coexisting ferromagnesian phases are controlled by pressure and temperature. Two sliding reactions have been postulated to explain the sequence of assemblages. For groups A and B, $Cd + Hy \rightleftharpoons Ga + Qz$ (1) and for group C, $Cd \rightleftharpoons Ga + Si + Qz$ (8). Above 1000° sapphirine or spinel have been found to be stable with quartz in Mg-rich and Fe-rich compositions respectively. The petrological importance of the sliding reactions as a means to determine P_{total} and T is discussed.

1. Introduction

The occurrence of cordierite in rocks from granulite facies terranes has aroused a great deal of interest among petrologists. Cordierite has been regarded as a low pressure mineral, and therefore as atypical of deepseated regional metamorphism. Similarly the occurrence of almandine in contact aureoles has led to controversy, as almandine-rich garnets were thought to be stable only at relatively high pressure. Coexisting cordierite and garnet are known both from high grade amphibolite and granulite facies terranes and from high level, low pressure, contact aureoles. In the latter, however, garnet occurs in addition to cordierite only in very iron-rich rocks. This has led some authors (e.g. CHINNER, 1962) to stress the importance of the affect of rock-chemistry as well as that of physical parameters on the relative stability of cordierite and garnet. The present experimental study was carried out to determine the stability of cordierite, in a silica-saturated environment, relative to its anhydrous breakdown products, as a function of pressure, temperature and chemical composition.

2. Previous experimental work

BOYD and ENGLAND (1959) determined the breakdown of pyrope to enstatite, sapphirine and sillimanite (?) between 1100 and 1500 °C. The stability of almandine up to 3 kb for a range of oxygen fugacities has been determined by HSU (1968). SCHREYER and YODER (1960) studied the breakdown of anhydrous Mg-cordierite at high pressures. They found that cordierite breaks down to enstatite, sillimanite and quartz at 10 kb from 1000 to 1100 °C and to sapphirine and quartz between 8 and 10 kb at 1200 °C and above. The breakdown of Fe-cordierite has been reversibly determined at 700 and 775 °C by RICHARDSON (1968). HIRSCHBERG and WINKLER (1968) using Ca-poor pelitic compositions, found a field of coexistence of cordierite and garnet together with biotite, sillimanite and quartz. In an iron-rich composition (100 $Mg/(Mg+Fe^{2+})$ = 20–23) cordierite and garnet coexist between 4 and 7 kb at 700 °C. In a more Mg-rich composition (100 $Mg/(Mg+Fe^{2+})$ = 56–61) garnet was found to be unstable up to 7 kb at the same temperature. A short paper summarising some of the results of the current work has appeared recently (HENSEN and GREEN, 1969).

3. Procedure

3.1. *Apparatus and experimental technique*

The experiments have been carried out in a single stage piston cylinder apparatus. The "piston in" procedure has been used throughout and a −10% pressure correction has been applied to all results (GREEN et al., 1967; NEWTON and SMITH, 1967). Maximum pressure uncertainty is thought to be ±0.4 kb.

Graphite sample capsules have been used together with dried furnace assemblies with inner boron nitride sleeves, identical to those described by GREEN and RINGWOOD (1967). By this method essentially dry conditions can be maintained during the runs. As a result no hydration takes place and partial melting is kept to a minimum. The thermocouple bead is separated from the top of the graphite capsule by a ceramic disc.

Temperatures have been controlled with chromel-alumel thermocouples for runs up to 1100 °C. Pt/Pt−10%Rh thermocouples have been used for most runs at 1100 °C and above. Precision of temperature measurements is believed to be ±10 °C.

Because reaction rates are strongly temperature dependent, run duration has been increased with decreasing temperature. Run times vary from around 15 hours at 1100 °C to 96 hours at 800 °C.

3.2. *Attainment of equilibrium*

Close to equilibrium boundaries, reaction rates are extremely low. For this reason true equilibrium cannot be obtained, but it is possible to determine the direction in which reaction is proceeding. Equilibrium is more closely approached in runs further away from reaction boundaries. This is supported by the observation that run time does not influence the results under such conditions.

3.3. *Oxygen fugacity*

Oxygen fugacity has not been directly controlled in the experiments. There is evidence that the fO_2 is low during the experiments:

(a) No magnetite has been observed in any of the runs (transmitted and reflected light observations);

(b) magnetite solid solution in pure iron-hercynite crystallized at 1100 °C in equilibrium with quartz is low (<5 mol%).

Although little water vapour is present, the fact that the sample is surrounded by graphite probably causes the oxygen fugacity to be low.

3.4. *Examination of the sample*

All runs have been examined optically (grain mounts) and by X-ray (powder photographs). Polished sections of selected samples enable a check on the homogeneity of the run and on abundance, size and shape of mineral grains. Changes in the relative abundance of mineral phases have been determined principally by the X-ray powder method. In the case of the incoming of garnet, where only a small change in the abundance of that phase occurs, the interpretation of the X-ray data has been confirmed by examination of polished sections of the runs in question. When a decrease of garnet had been interpreted from the X-ray data, irregular, resorbed grains (seeds) of garnet were observed optically. Rounded grains and an increase in grainsize were found in the runs in which garnet had been growing.

4. Compositions studied

4.1. *Choice of compositions*

Compositions simulating natural metasediments have been used in this work to enable direct comparison with natural rocks. Silica-saturated cordierite-garnet bearing rocks can be divided into three groups on the basis of their $(MgO+FeO)/Al_2O_3$ ratio (F/A). Three series of compositions, labelled A, B and C were chosen for experiment. A and B have $F/A > 1$, and C has $F/A < 1$ (fig. 1). Each series consists of a number of compositions with varying $Mg/(Mg+Fe^{2+})$ ratios. The results reported here are those of experiments on the B70 composition*, the analysis of which is given in table 1.

In the following some reference will be made to results on a few other compositions in the same series and in the C series (fig. 1). Work is in progress on some selected compositions in the $MgO-FeO-Al_2O_3-SiO_2$ system to evaluate the influence of CaO, Na_2O and K_2O on the equilibria studied.

---

* Member of the B series with $100Mg/(Mg+Fe^{2+}) = 70$.

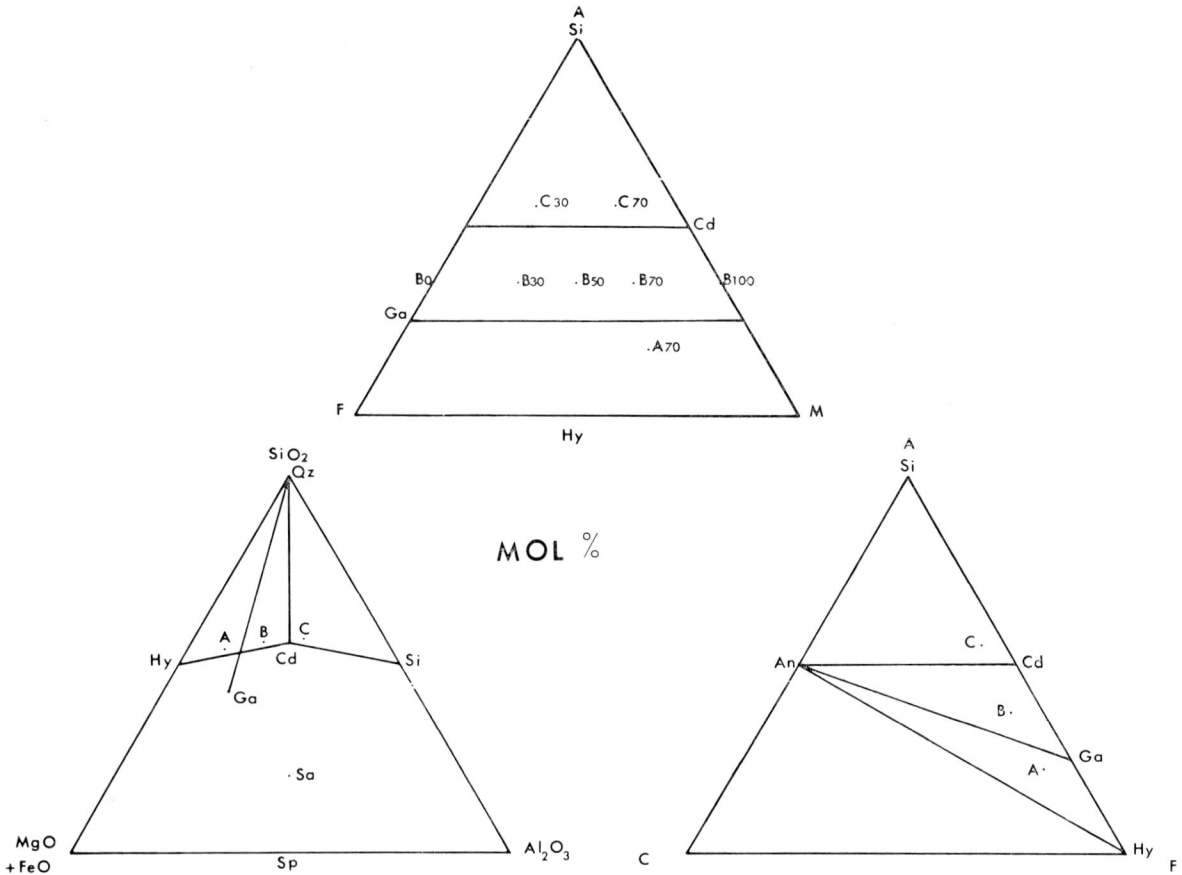

Fig. 1. Position of synthetic compositions A, B and C in various compositional diagrams.
ACF: A = Al_2O_3-(K_2O+Na_2O), C = CaO, F = FeO+MgO.
AFM: A = Al_2O_3-(CaO-K_2O-Na_2O), F = FeO, M = MgO.
SiO_2-(MgO+FeO)-Al_2O_3: SiO_2 = SiO_2-6(K_2O+Na_2O)-2CaO, Al_2O_3 = Al_2O-(K_2O+Na_2O+CaO), MgO+FeO.

TABLE 1

Chemical composition of B70

SiO_2	53.82
Al_2O_3	25.43
FeO	6.83
MgO	8.95
CaO	2.42
Na_2O	1.18
K_2O	1.35
	99.98

4.2. Starting materials

Synthetic glasses have been prepared from A.R. chemicals and fused in an argon atmosphere. Homogeneity has been checked optically with immersion oils. As the use of glass as starting material has proved unreliable owing to the appearance and persistence of metastable low pressure phases (e.g. cordierite) and extremely low nucleation rates of high pressure phases (e.g. garnet) a seeding technique is being used. Finely ground crystalline material, containing both the high and the low pressure assemblages, obtained from large capacity runs on glass, is mixed with glass in a 1:4 ratio. It has been found that the glass shards crystallize in a short time (5–10 minutes) to the low pressure assemblage (at pressures up to 10.8 Kb in the composition studied here). Therefore any increase in the high pressure assemblage takes place at the cost of the low pressure assemblage (and *not* of the glass). A few reversals using crystalline phase only as starting materials have been carried out. For this purpose runs containing the required assemblages were mixed in the desired proportions to serve as starting material.

TABLE 2

Experimental data on B70 composition (seeded glass runs)

Run No.	Temp. °C	Press. kb	Time hrs	Cd	Ga	Hy	Si	Qz	Sa	Sp	Bi	Fsp	Glass
1605	1200	7.2	2	M		tr		M	M			tr	L/M
1606	1200	9	2			tr		M	H			tr	L/M
1675	1150	11.7	4		H	tr	L	M	tr/L			L	L
1592	1100	5.4	19	H		L/M						L	L
1595	1100	6.3	18	H		L/M						L	L
1589	1100	7.2	18	H	L	L/M				tr		L	L
1694	1100	7.2	44	H	L/M	L		tr				L	L
1625	1100	8.1	18	M	M	L		L/M	L/M			L	L
1594	1100	9	18		H	L		M	M	tr		L	L
1669	1100	9.9	13		H	L		M	M			L	L
1673	1100	10.8	20		H	tr/L	L/M	M	L/M			L	L
1627	1100	11.7	15		H	tr	L/M	M	tr			L	L
1622	1100	13.5	12		H		L/M	M				L	L
1664	1050	9	43		L	H	tr	M	tr			L	L
1604[a]	1000	5.4	45	H	tr	M		tr/L				L	tr
1593[b]	1000	6.3	44	H	L	L		L				L	tr
1573	1000	7.2	43	M/H	M/H	L		L				L	tr
1597	1000	9	46	M	H	tr/L		M				L	tr
1616	1000	9.9	45		H	tr/L	L	M				L	tr
1608	1000	10.8	45		H	tr/L	L/M	M				L	tr
1676	950	9.9	72	L/M	H	tr	tr	M				L	tr
1518[c]	900	4.5	18	H		M		tr				L	tr
1536[c]	900	5.4	70	H		M		tr				L	L
1586[c]	900	6.3	72	M/H	M	L/M		L				L	tr/L
1570[c]	900	7.2	72	M	M/H	tr/L		M				L	tr/L
1620[d]	900	10.8	70	tr/L	H	tr	L	M			tr	L	L
1628	900	11.7	72	tr	H	tr	L	M				L	tr
2032[a]	800	4.5	98	H	tr/L	L(?)		L				L	L
H8[e]	750	3.8	12 days	H		(?)		tr			M	tr/L	L
Starting material					M	M	L	L	L	tr		L	M
Reversals													
Starting material						H	tr/L	L/M	M	tr		L	L
1644	1100	9	16			H	tr	tr/L	M	L		L	L
Starting material					tr/L	H	tr	M	M			L	tr
2293	900	9.9	95		L/M	H	tr	tr/L	L/M			tr	L

[a] rare irregular shaped garnet relics. [b] more abundant rounded garnets. [c] runs in Pt-capsules [d] runs in Ag/Pd capsules. [e] hydrothermal run. Ag/Pd capsule. fO_2 between that of the iron-wustite and quartz-fayalite-magnetite buffers.

Relative proportions of phases are indicated by: H = high; M = medium; L = low; tr = trace

5. Results

5.1. Results of seeded glass experiments

The experimental data are given in fig. 2 and table 2*.
At temperatures from 800–1050 °C the following assemblages have been observed with increasing pressure:
 (a) Cd-Hy-Qz,
 (b) Cd-Ga-Hy-Qz,
 (c) Ga-Hy-Si-Qz,
 (d) Ga-Si-Qz.

At temperatures above 1050 °C two more assemblages were encountered:
 (e) Ga-Sa-Hy-Qz,
 (f) Sa-Hy-Qz.

All assemblages also contain feldspars. The observation that feldspars are less abundant in runs at 1100 °C than at lower temperatures suggests that a small amount of partial melting takes place in the highest temperature runs. Several runs contain two of the above assemblages simultaneously and one run even three (see fig. 1). These runs always occur close to the boundaries be-

* Abbreviations used are Cd = cordierite, Ga = garnet, Hy = hypersthene, Si = sillimanite, Sa = sapphirine, Qz = quartz, Sp = spinel, Bi = biotite, Fsp = feldspars.

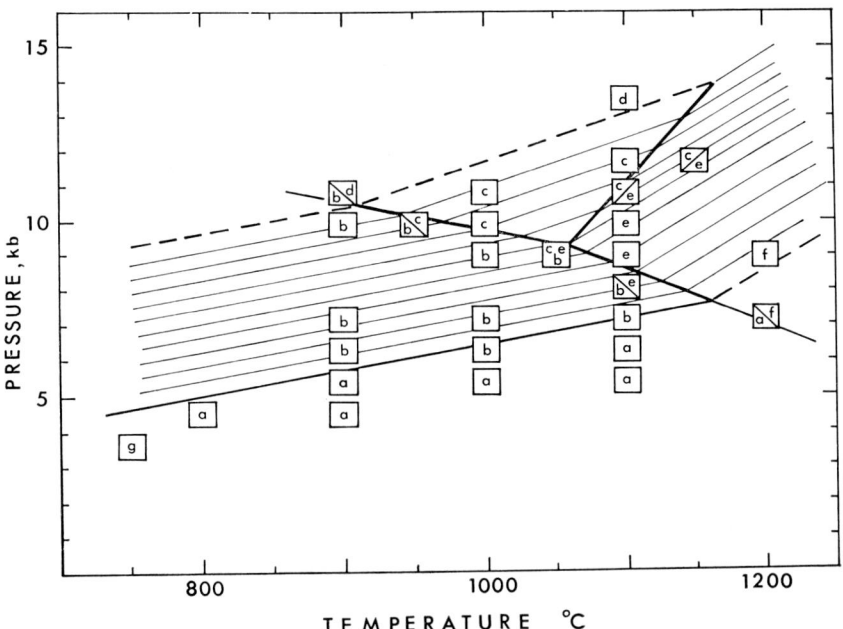

Fig. 2. *P–T* diagram of phase relations in the B70 composition. Phase assemblages are indicated by the letters also used in the text.
    a = cordierite-hypersthene-quartz,
    b = garnet-cordierite-hypersthene-quartz,
    c = garnet-hypersthene-sillimanite-quartz,
    d = garnet-sillimanite-quartz,
    e = garnet-hypersthene-sapphirine-quartz,
    f = hypersthene-sapphirine-quartz,
    g = cordierite-biotite-hypersthene(?)-quartz (hydrothermal run see table 2).
Size of the squares indicates maximum error in *P* and *T*.

tween two (or three) assemblages and are believed to represent disequilibrium due to low reaction rates under such conditions. Mineralogical changes from one assemblage to the next are shown in table 3.

In the interval in which garnet and cordierite coexist the amount of garnet increases while that of cordierite decreases with increasing pressure. The amount of hypersthene shows a small but distinct decrease at the incoming of garnet and decreases very slowly while garnet increases. Under all conditions the $Mg/(Mg+Fe^{2+})$ ratio is highest in cordierite and lowest in garnet. All three iron-magnesian phases are inferred to become more Mg-rich with increasing pressure. (R.I. data show the increase in $Mg/(Mg+Fe^{2+})$ ratio of the garnet.) Quantitative analysis of the crystalline phases using the electron microprobe is in progress.

5.2. *Reversals*

The boundary (a → b) marking the incoming of garnet has been reversed. Garnet seeds break down at the low pressure side of this curve and grow at higher pressure at the cost of the low pressure assemblage (which crystallized from the glass initially).

The possibility existed that the breakdown of cordierite (b → c) in seeded glass experiments occurred according to a temperature dependent rate process rather than an equilibrium reaction. Therefore a run was carried out using crystalline garnet, sillimanite and quartz seeded with a very small amount of cordierite-bearing intermediate pressure assemblage. At 900 °C and 9.9 kb, just below the anticipated equilibrium curve a distinct increase in the amount of cordierite was observed.

TABLE 3

Observed changes in mineral assemblages

From assemblage to assemblage	Appear	Disappear	Increase	Decrease
a → b	Ga	—	Qz	Cd, Hy
b → c	Si	Cd	Ga, Qz, Hy(?)	
c → d	—	Hy	Ga, Qz	Si(?)
d → e	Sa	Si	Qz	Ga, Hy(?)
e → f		Ga	Hy, Sa, Qz	
b → e	Sa	Cd	Hy(?), Ga, Qz	
a → f	Sa	Cd	Qz	

Since cordierite is absent in the seeded runs at 1000 °C, 10.8 kb and 9.9 kb but present in moderately large amount in the 1000 °C, 8.1 kb run the negative slope for the breakdown of cordierite seems sufficiently demonstrated.

Sapphirine could be a metastable breakdown product of cordierite. A cordierite-free mixture of garnet, sillimanite, quartz and very small amounts of hypersthene and sapphirine was run at 1100 °C, 9 kb. An increase of sapphirine could be detected, while the amount of sillimanite decreased. There can be little doubt that sapphirine is stable in the presence of quartz under these conditions.

6. Interpretation of results

6.1. *Reactions and mineral equilibria*

An attempt has been made to interpret the experimentally determined boundaries in terms of reactions. The following reactions express the mineralogical changes listed in table 3.

determined by the $Mg/(Mg+Fe^{2+})$ ratio of the rock (other variables being equal).

Reactions (2), (4) and (6) are of a more complex nature. Five phases are in equilibrium compared to four in the previous reactions. By moving off one of the curves representing these reactions (heavy curves in fig. 2) one of the five phases breaks down. The $Mg/(Mg+Fe^{2+})$ ratios of the coexisting phases change along the curves representing these reactions on the P–T diagram.

In three dimensional compositional space reactions (2) and (6) mark the breakdown of the join cordierite-garnet. In reaction (2) this allows the stable coexistence of hypersthene, sillimanite and quartz and in reaction (6) that of hypersthene, sapphirine and quartz. Reaction (4) most likely represents the disappearance of the join hypersthene-sillimanite allowing the coexistence of garnet, sapphirine and quartz. The volume change of this reaction is probably very small.

Reaction (7) is a degenerate case of a sliding equilibrium. A more detailed treatment of sliding reac-

$$a \to b \qquad Cd+Hy \rightleftharpoons Ga+Qz \qquad \Delta V = (-)ve \qquad (1)$$

$$b \to c \quad Cd+Ga_1+Hy_1 \rightleftharpoons Ga_2+Hy_2+Si+Qz \quad \Delta V = (-)ve \qquad (2)$$

$$c \to d \qquad Hy+Si \rightleftharpoons Ga+Qz \qquad \Delta V = (-)ve \qquad (3)$$

$$d \to e \quad Ga_1+Hy_1+Si \rightleftharpoons Ga_2+Hy_2+Sa+Qz \quad \Delta V = (-)ve? \qquad (4)$$

$$e \to f \qquad Ga \rightleftharpoons Hy+Sa+Qz \qquad \Delta V = (+)ve \qquad (5)$$

$$b \to e \quad Cd+Ga_1+Hy_1 \rightleftharpoons Ga_2+Hy_2+Sa+Qz \quad \Delta V = (-)ve \qquad (6)$$

$$a \to f \qquad Cd \rightleftharpoons Sa+Qz \qquad \Delta V = (-)ve \qquad (7)$$

Reactions (1), (3) and (5) appear to be "sliding reactions", i.e. reactants and products coexist over a limited pressure-temperature interval. For these sliding equilibria the $Mg/(Mg+Fe^{2+})$ ratios of the coexisting ferromagnesian phases are fixed by pressure and temperature. Not until some quantitative data on the composition of the coexisting phases have been obtained will it be possible to determine how composition varies with P and T. Taking one of these equilibria then for each particular bulk composition, four phases, representing a sliding reaction, will coexist over a limited pressure-temperature interval. The width and position of this interval (in terms of P and T) will be

tions, their intersections and resulting topology will be discussed in a separate paper (HENSEN and ESSENE, in preparation).

6.2. *Influence of chemical parameters*

(a) *The $(MgO+FeO)/Al_2O_3$ ratio.* In compositions with an F/A ratio <1 (C series) the incoming of garnet occurs at a higher pressure than in the B series for material with the same $Mg/(Mg+Fe^{2+})$ ratio. Below 900 °C garnet appears and cordierite disappears according to the sliding reaction

$$Cd \rightleftharpoons Ga+Si+Qz \qquad \Delta V = (-)ve. \qquad (8)$$

This reaction has a negative slope.

In compositions with $F/A \gg 1$ (A series) no sillimanite occurs under any conditions. On the basis of the F/A ratio three groups of compositions (A, B and C series in this study) can be distinguished. The low, intermediate and high pressure assemblages of each group at around 800 °C are the following.

Group	A	B	C
low pressure	Hy-Cd-Qz	Cd-Hy-Qz	Cd-Si-Qz
intermediate pressure	Hy-Cd-Ga-Qz	Cd-Ga-Hy-Qz	Cd-Ga-Si-Qz
high pressure	Ga-Hy-Qz	Ga-Si-Qz	Ga-Si-Qz

Under the same physical conditions then $Mg/(Mg+Fe^{2+})$ ratios of coexisting cordierite and garnet are higher in the assemblage Cd-Ga-Hy-Qz than in Cd-Ga-Si-Qz (see section 7).

(b) *The $Mg/(Mg+Fe^{2+})$ ratio*. Both reactions (1) and (8) occur at lower pressure with decreasing $Mg/(Mg+Fe^{2+})$ ratio of the bulk composition i.e. a decrease in this ratio restricts the stability field of cordierite and enlarges that of garnet. In composition B30 garnet appears (by reaction (1)) at 3.6 kb and 1000 °C and at 4.8 kb and 1100 °C. Cordierite disappears at 6.3 kb and 7.2 kb at the same temperatures. Sapphirine does not occur in iron-rich compositions. An iron-rich spinel is found instead. The assemblage Fe-rich olivine, cordierite and quartz occurs at low pressure in the composition B30.

(c) *The Ca-content*. Preliminary results on the system $MgO-FeO-Al_2O_3-SiO_2$ indicate that the presence of calcium influences the position of some of the curves determined with the more complex compositions used in this work. The most important effect expected is the stabilisation of garnet to lower pressure. In many natural rocks and in the experiments on the complex compositions, cordierite, garnet and sillimanite (or hypersthene) coexist with plagioclase. It is inferred that the Ca/Na ratio of the bulk composition has an indirect effect on the relative stability of garnet and cordierite. This also implies that the reactions used above to express the observed changes in mineral assemblages are too simplified (i.e. that plagioclase plays a role in these equilibria).

(d) *Mn- and Ti-content*. As Mn, if present in a rock, tends to concentrate in garnet, its presence will increase the stability field of garnet with respect to the other ferromagnesian phases. The present data are only applicable to rocks with low Mn concentrations. Most garnets from cordierite-garnet bearing rocks have low Mn-contents (MnO <1 weight %). Ti is also very low in these garnets and therefore it can be disregarded as an important factor.

(e) *The influence of oxygen fugacity*. Assuming that Fe^{3+} substitution in garnet, cordierite, hypersthene and sapphirine is negligible, changes in oxygen fugacity will not affect the equilibria described. If this is so, an increase fO_2 could only lead to the formation of magnetite. This would increase the $Mg/(Mg+Fe^{2+})$ ratio of the (remaining) rock and consequently reduce or eliminate the most iron-rich phase in an equilibrium assemblage.

(f) *The presence of additional hydrous ferromagnesian phases*. Most natural rocks having the assemblages described above contain biotite and sometimes amphibole (gedrite-anthophyllite). The presence of these phases does not affect reactions (1) and (8) described here i.e. the assemblages cordierite-garnet-sillimanite-quartz and cordierite-garnet-hypersthene-quartz. At a given pressure and temperature the compositions of the coexisting ferromagnesian phases of these assemblages could only be affected if another component (e.g. Ti or Mn) would substitute in the biotite (or amphibole) and in one or other of the phases of the assemblage and thereby change the distribution of Fe and Mg between them. As has been seen above, Ti is mostly very low in garnet, cordierite and hypersthene, while Mn is very low in biotite (e.g. REINHARDT, 1968). Therefore it is unlikely that the addition of biotite could substantially affect the $Mt/(Mg+Fe^{2+})$ ratios in the above mentioned equilibrium assemblages at constant pressure and temperature. The addition of biotite (by partial hydration) to the assemblages will in most cases change the proportions of coexisting cordierite and garnet etc. (but not their compositions) and it can lead to the breakdown of one of the phases.

(g) *Structural state and water content of cordierite*. The cordierite produced in the experiments is most likely a "high" cordierite as defined by SCHREYER and SCHAIRER (1961). The effect of the structural state on the stability of cordierite is unknown and is not considered in this paper. Although SCHREYER and YODER (1964) concluded that synthetic hydrous cordierites only contain molecular H_2O, the role of water in cordierite is still subject to controversy. The influence of water content on the stability of cordierite cannot be evaluated.

7. Application of results

Reviews of the literature on cordierite-garnet bearing rocks have been given by WYNNE-EDWARDS and HAY (1963) and by DE WAARD (1965). Both the assemblages cordierite-garnet-sillimanite-quartz and cordierite-garnet-hypersthene-quartz (often accompanied by biotite, plagioclase, potash feldspar and opaques) are commonly found in low and intermediate pressure terranes as defined by GREEN and RINGWOOD (1967) on the basis of solid-solid reactions in rocks of basaltic compositions. These authors pointed out the apparent absence of cordierite in high pressure granulites. Cordierite and kyanite are rarely found together and it is usually hard to determine whether or not the two phases are in equilibrium. The experimental data suggest that below 850 °C in relatively dry rocks the stability fields of Mg-rich cordierite and kyanite overlap (using the kyanite-sillimanite boundary as determined by RICHARDSON et al., 1968). SCHREYER (1968) concluded that Mg-cordierite and kyanite are compatible at 800 °C and 10 kb under hydrothermal conditions. In the presence of K-feldspar, however, the assemblage phlogopite-kyanite-quartz is probably stable at high water pressures. The stable coexistence of cordierite and kyanite will therefore be restricted either to rocks which formed under conditions of $P_{H_2O} \ll P_{Load}$ or to rocks without K-feldspar. From the above it can be concluded that the stable occurrence of cordierite in intermediate to high pressure granulite areas is possible and should not be dismissed as retrograde metamorphism without careful investigation. It should also be kept in mind that the absence of cordierite in some areas may well be due to the fact that no rocks of suitable composition i.e. Mg-rich metapelites, occur in those areas.

Two occurrences from W. Ontario, Canada, of garnet-cordierite rocks have been studied in detail by WYNNE-EDWARDS and HAY (1963) and by REINHARDT (1968). The former authors concluded from their chemical data that the relative stability of garnet and cordierite is governed by the $Ca/(Ca+Mg+Fe^{2+})$ ratio as well as the Mg/Fe ratio of the bulk composition, at constant pressure and temperature. REINHARDT (1968) however does not find any systematic relationship between Ca in the rock and the occurrence of garnet or even the Ca-content of the garnet, nor do his data indicate any influence of the Ca/Na ratio (An. % of the plagioclases is 32–60) on the amount of grossular substitution in the garnet. The grossular and pyrope content of the garnet vary independently. Reinhardt attributes the observed shift of compositional (garnet-cordierite-biotite) triangles in an AFM diagram to changes in oxygen fugacity. The following facts which Reinhardt recognized himself do not support this hypothesis:

(a) two of the seven samples fall off the trend of increasing $Mg/(Mg+Fe^{2+})$ ratio in garnet and biotite with increasing oxidation ratio, $2Fe^{3+}/(2Fe^{3+}+Fe^{2+})$;

(b) there is no systematic relation between Fe^{3+} substitution in garnet, cordierite and biotite and the rock oxidation ratio. Increasing fO_2, producing magnetite and in that way increasing the Mg/Fe ratio of the remaining rock cannot change the $Mg/(Mg+Fe^{2+})$ ratios of garnet and cordierite coexisting with sillimanite (or hypersthene) and quartz (see previous paragraph) as suggested by Reinhardt, but can only change their proportions in the rock. It would however have the above-mentioned affect if neither sillimanite nor hypersthene are present. The $Mg/(Mg+Fe^{2+})$ ratios of the coexisting phases of the assemblage garnet-cordierite-biotite are not uniquely fixed by P and T. Their composition is also determined by the bulk composition of the rock. The most Mg-rich garnet coexisting with cordierite occurs in a rock which also contains hypersthene. This would be expected even if the rocks were formed under identical conditions. MARAKUSHEV and KUDRYAVTSEV (1965) describing cordierite-garnet rocks from the Aldan shield, U.S.S.R., made the important observation that garnet coexisting with cordierite, sillimanite and quartz has a lower $Mg/(Mg+Fe^{2+})$ ratio than garnet coexisting with (a more Mg-rich) cordierite and hypersthene in rocks that were most likely formed under the same physical conditions. This is in good agreement with the experimental data (see section 6 of this paper). It is related to the fact that garnet appears at lower pressure in Al-poor $(F/A > 1)$ compositions than in Al-rich compositions having identical $Mg/(Mg+Fe^{2+})$ ratios. The above-mentioned authors also found that the garnet composition at the highest "grade", where it is most pyrope rich, is similar in both assemblages. In terms of the experiments this would mean that we approach reaction (2). On this boundary all phases from the two assemblages coexist.

In this high-grade part of the Aldan-shield the assemblage hypersthene-sillimanite-quartz has been found (KHLESTOV 1964; MARAKUSHEV and KUDRYAVTSEV 1965). This assemblage can only occur when the join cordierite-garnet is broken i.e. when the curve representing reaction (2) is crossed. Unless the stability field of hypersthene-sillimanite-quartz relative to the assemblages garnet-hypersthene-sillimanite-quartz and cordierite-hypersthene-sillimanite-quartz is affected by other factors than pressure and temperature (e.g. fO_2) this assemblage can only be expected to occur in a narrow compositional range. Moreover the position of this range (in terms of the $Mg/(Mg+Fe^{2+})$ ratio) would change with varying P and T. Hypersthene-sillimanite-quartz occurrences have also been described by ESKOLA (1952) from Lapland and by LUTS and KOPANEVA (1968) from the Anabar Massif USSR.

Another very interesting assemblage is that of sapphirine-quartz, which has been recently found in rocks from Antarctica (DALLWITZ, 1968) and from Central Australia and Africa by Prof. A. F. WILSON (private communication). The occurrences of hypersthene-sillimanite-quartz and particularly sapphirine-quartz seem to suggest that in some high grade metamorphic areas extremely high temperatures in the order of 850–1000 °C may have been reached. However until more quantitative chemical data are available both on the rocks and minerals in question and on the experimentally produced phases the above suggestion remains uncertain.

The present data show that intermediate Fe-Mg garnets break down at intermediate pressures to quartz bearing assemblages at high temperatures (i.e. hypersthene-sapphirine-quartz and hypersthene-spinel-quartz). The breakdown of pyrope by BOYD and ENGLAND (1960) at high pressures and that of almandine by HSU (1968) at low pressure both give rise to assemblages without quartz. Both the present data and data of SCHREYER and SEIFERT (this conference) indicate that pyrope will probably break down to enstatite, sapphirine and quartz at very high temperatures and pressures. We also have some evidence to suggest that almandine will go to fayalite-hercynite-quartz at higher P and T.

The formation of almandine in contact metamorphic rocks with a $(MgO+FeO)/Al_2O_3$ ratio > 1 does not require elevated pressures. The interpretation of CHINNER (1962) that in such rocks almandine is only stable in rocks with a low $Mg/(Mg+Fe^{2+})$ ratio is correct. In aluminous rocks with $F/A < 1$ pressures in the order of 4–5 kb are required for the formation of pure almandine ($+$ aluminosilicate $+$ quartz) at temperatures around 600 °C as is also demonstrated by the work of RICHARDSON (1968). Therefore the conclusion of HIRSCHBERG and WINKLER (1968) that: "pressures in excess of 5 kb are necessary for the formation of spessartine-free almandine at 600 °C even in rocks with high FeO/MgO ratios" only applies to rocks with $F/A < 1$. Their statement that "both the assumptions of HARKER (1939) of higher pressure and CHINNER (1968) of high FeO/MgO ratio must be fulfilled for almandine to be a stable phase in contact metamorphic rocks" is incorrect and misleading. (Both quotations translated from German.) The influence of other chemical parameters on the stability of almandine-rich garnets has been discussed by HSU (1968), RICHARDSON (1968) and has been given some attention in this paper (section 6.1).

If it can be demonstrated by microprobe analyses that the phases produced in the experiments have the same compositions as the minerals of natural rocks, direct application of the results to rocks containing homogeneous domains with unzoned minerals in textural equilibrium is justified. The assemblages of sliding reactions (1) and (8) are of particular interest petrologically. The sliding equilibrium $Cd \rightleftharpoons Ga + Si + Qz$ (8) is of special importance because, unlike most other known solid-solid equilibria, it has a negative slope. The analysis of coexisting garnet and cordierite from this assemblage provides a "univariant" line on a P-T diagram. The intersection of this line with other univariant solid-solid equilibria (with positive slopes) should eventually enable a reasonably accurate estimate of P_{total} and T in a metamorphic terrane. Reactions such as reaction (1) in this paper or the solid-solid reactions in basaltic compositions determined by GREEN and RINGWOOD (1967) could be used for this purpose. If P_{total} and T can be determined in this manner it will become possible to evaluate fluid pressures in high grade metamorphic rocks by comparison with experimental data on dehydration, decarbonisation and oxidation-reduction reactions.

Acknowledgements

The authors want to thank Dr. E. J. Essene for his many helpful comments and for critically reading the manuscript.

References

BOYD, F. R. and J. L. ENGLAND (1959) Carnegie Inst. Wash. Yearbook **58**, 108.
CHINNER, G. A. (1962) J. Petrol. **3**, 316.
DALLWITZ, W. B. (1968) Nature **219**, 476.
DE WAARD, D. (1966) Can. Mineralogist **8**, 481.
ESKOLA, P. (1952) Am. J. Sci. Bowen Vol., 133.
GREEN, D. H. and A. E. RINGWOOD (1967) Geochim. Cosmochim. Acta **31**, 767.
GREEN, T. H., A. E. RINGWOOD and A. MAJOR (1966) J. Geophys. Res. **71**, 3589.
HARKER, A. (1939) *Metamorphism* (Methuen and Co., London).
HENSEN, B. J. and D. H. GREEN (1969) J. Geol. Soc. Australia, in press.
HIRSCHBERG, A. and H. G. F. WINKLER (1968) Contrib. Mineral. Petrol. **18**, 17.
HSU, L. C. (1968) J. Petrol. **9**, 40.
KHLESTOV, V. V. (1964) Dokl. Akad. Nauk SSSR **154**, 842.
LUTS, B. G. and L. N. KOPANEVA (1968) Dokl. Akad. Nauk SSSR **179**, 1200.
MARAKUSHEV, A. A. and KUDRYAVTSEV (1965) Dokl. Akad. Nauk SSSR **164**, 179.
NEWTON, R. C. and J. V. SMITH (1967) J. Geol. **75**, 268.
REINHARDT, E. W. (1968) Can. J. Earth Sci. **5**, 455.
RICHARDSON, S. W. (1968) J. Petrol. **9**, 467.
RICHARDSON, S. W., P. M. BELL and M. C. GILBERT (1968) Amer. J. Sci. **266**, 513.
SCHREYER, W. and J. F. SCHAIRER (1961) J. Petrol. **2**, 324.
SCHREYER, W. and H. S. YODER (1960) Carnegie Inst. Wash. Yearbook **59**, 90.
SCHREYER, W. and H. S. YODER (1964) Neues Jahrb. Mineral. Abhandl. **101**, 271.
WYNNE-EDWARDS, H. R. and P. W. HAY (1963) Can. Mineralogist **7**, 453.

HIGH PRESSURE EXPERIMENTAL STUDIES ON THE MINERALOGICAL CONSTITUTION OF THE LOWER CRUST

TREVOR H. GREEN

Hoffman Laboratory, Harvard University, Cambridge, Mass. 02138, U.S.A.

Diorite and gabbroic anorthosite have been proposed as two possible overall compositions constituting the lower crust. The mineral assemblages stable in these compositions under anhydrous conditions at temperatures of 900–1200 °C and pressures of up to 36 kb have been determined. The low pressure mineralogy is dominated by plagioclase, with subordinate pyroxene and minor quartz. With increasing pressure garnet appears, and garnet, quartz and clinopyroxene form at the expense of plagioclase. Finally at pressures greater than 20 kb at 900–1200 °C plagioclase disappears and the high pressure assemblage consists of clinopyroxene+quartz (coesite) + garnet + K-feldspar ± kyanite(?). Extrapolating the experimental results to P–T conditions predicted for a stable, anhydrous lower crust and calculation of compressional wave velocities for these compositions supports the models of a lower crust composed of diorite or gabbroic anorthosite, where the mineralogy of these compositions consists of clinopyroxene, sodic plagioclase and subordinate quartz, garnet.

1. Introduction

Previous experimental investigations on natural rock systems have established the mineral assemblages formed in basaltic and granitic (adamellite) rocks at high pressures and temperatures under anhydrous conditions (GREEN and RINGWOOD, 1967; GREEN and LAMBERT, 1965). These results, complemented by data of other workers on relevant simple systems (BIRCH and LECOMTE, 1960; KUSHIRO and YODER, 1966), have been used to interpret the mineralogy expected in natural rocks under the pressure–temperature conditions predicted for the lower crust. The present investigation provides data on high pressure mineral assemblages in diorite and gabbroic anorthosite compositions, which fall between the extremes of basaltic and granitic rocks previously studied.

Recently proposed crustal models indicate that the earth's continental crust has an overall composition approximating to diorite or andesite (TAYLOR and WHITE, 1965). In crustal areas where there has been a long history of metamorphism and igneous activity (e.g. Precambrian shield) the crust may be differentiated into an upper granodioritic fraction underlain by a more basic fraction (TAYLOR, 1968; HARRIS, 1967). This basic fraction would approximate to gabbroic anorthosite if the initial overall composition was dioritic or andesitic (GREEN, 1968). Thus rocks of diorite or gabbroic anorthosite composition may well be major constituents of the lower crust, and determination of their mineralogies under lower crustal pressure–temperature conditions is important in enabling comparison of the properties of these compositions with the geophysically determined data for the lower crust. Accordingly this paper describes an experimental investigation of the mineral assemblages found in diorite and gabbroic anorthosite with increasing pressure, followed by estimates of the geophysical properties of these assemblages and finally a comparison with the available geophysical data for the lower crust in order to qualitatively evaluate the merits of these models.

It should be emphasized that from a detailed petrological point of view it is unlikely that the lower crust consists of a homogeneous layer of diorite or gabbroic anorthosite. Rather it is more likely to be composed of a mixture of rock types (cf. DEN TEX, 1965) among which diorite and gabbroic anorthosite may dominate. Alternatively, when viewed on a gross scale (e.g. by geophysical methods) the varied rock types combine to give characteristics expected of a rock of uniform intermediate composition (RINGWOOD and GREEN, 1966).

2. Experimental

The high pressure experimental work has been conducted using a solid medium piston-cylinder high pressure apparatus of the type described by BOYD and ENGLAND (1960a, 1963). The detailed experimental procedure has followed that of BOYD and ENGLAND (1960b, 1963) and GREEN and RINGWOOD (1967). Loss of pressure transmitted to the sample due to friction effects and non-uniform distribution of pressure in the solid medium pressure cell is allowed for by applying a -10% correction to the nominal pressure values for a single stage, instroke run (i.e. approaching the desired P–T values with the piston moving in), and the resulting pressures are believed to be accurate to $\pm 3\%$ for the range 15–40 kb (GREEN et al., 1966). For pressures lower than 15 kb the results are probably accurate to $\pm 5\%$. In two-stage runs where the piston has been retracted a -4.5% pressure correction has been applied, based on the results of GREEN et al. (1966) for similar two-stage runs, and the probable accuracy is $\pm 5\%$. Temperature is measured with a Pt/Pt 10% Rh thermocouple and is believed accurate to $\pm 15\,°C$. No corrections for any pressure effect on the e.m.f. of the thermocouple have been made.

The compositions were carefully prepared by thoroughly mixing highest purity Fisher chemical compounds in the requisite proportions (table 1). The gabbroic anorthosite composition is based on the average of 7 analyses of gabbroic anorthosite from the Adirondack anorthosite complex (BUDDINGTON, 1939). A crushed glass of this composition was held in a sealed, evacuated silica tube with an iron pellet at 900 °C for 24 hours, then chemically analyzed for FeO and Fe_2O_3 to check the oxidation state. This procedure resulted in devitrification of the glass to a finely crystalline mix of feldspar and pyroxene which was used for the experimental work. The diorite composition is based on the average andesite composition proposed by TAYLOR (1968). It was prepared as a reacted mix, held in an evacuated silica tube with an iron pellet at 900 °C for 24 hours. Subsequent chemical analysis of the diorite indicated that the iron was present as FeO, with negligible Fe_2O_3.

In this project it was necessary to carry out runs at as low temperature as possible while still allowing equilibrium to be obtained in a reasonable time, in order to simulate as closely as possible the P–T conditions of the lower crust. Experience showed that reaction rates were extremely slow under anhydrous conditions at temperatures less than 1000 °C. The presence of water provides a catalytic effect for silicate reactions, but only a low water content as a catalyst could be tolerated in the present work otherwise hydrous phases crystallized, and it is unlikely that in general the activity of water in the lower crust is sufficient to produce such phases (RINGWOOD and GREEN, 1966). Thus the experiments were carried out under conditions of low water activity in the subsolidus fields of the two compositions for the temperature range 900–1200 °C at pressures up to 36 kb. Even at these temperatures, times of up to 72 hours were needed to enhance attainment of equilibrium. The results of the runs were then extrapolated to P–T conditions of the lower crust.

At 900–1000 °C unsealed gold capsules were generally used, though in a few cases silver–palladium capsules

TABLE 1

Composition and CIPW norms of synthetic diorite and gabbroic anorthosite used in the experimental work.

	Diorite	Gabbroic anorthosite
SiO_2	59.9	53.5*
TiO_2	0.7	1.0*
Al_2O_3	17.3	22.5*
Fe_2O_3	—	0.9†
FeO	6.3†	4.7†
MnO	—	0.1
MgO	3.4	2.1*
CaO	7.1	9.9*
Na_2O	3.7	3.7*
K_2O	1.6	1.1*

* denotes content determined by electron microprobe analysis of a glass fragment.
† denotes content chemically determined.

CIPW norms	Diorite	Gabbroic anorthosite
Qz	9.2	2.1
Or	9.4	6.5
Ab	31.3	31.3
An	25.8	41.5
Diop	7.8	6.3
Hyp	15.0	8.5
Mt	—	1.3
Ilm	1.3	1.9
Density (g/cm³)	2.84	2.84
V_P (km/s) approx.	6.6	7.0

TABLE 2

Results of experimental runs on the diorite composition at 900 °C

Pressure (kb)	Temperature (°C)	Time (hrs)	Type of sample capsule	Phases present*				Comments
27 then 4	900 900	48 67	Au	px	plag	qtz	—	Well crystallized; uncommon laths of orthopyroxene distinguishable; plag ≫ px > qtz
27 then 6.5	900 900	48 64	Au	px	plag	qtz	—	Medium grainsize; plag ≫ px > qtz
27 then 9	900 900	48 62	Au	px	plag	qtz	<u>ga</u>	Medium grainsize; laths of orthopyroxene distinguishable; plag ≫ px > qtz > ga
9	900	24	Au	px	plag	qtz	—	Fine grained; no evidence for garnet; plag ≫ px > qtz
11.3	900	24	Au	px	plag	qtz	—	Medium grainsize; no evidence for garnet; plag ≫ px > qtz
18 then 11.3	1100 900	6 50	Ag–Pd	px	plag	qtz	ga	Medium grainsize; plag ≫ px > qtz > ga
13.5	900	6	Ag–Pd	px	plag	qtz	—	Fine grained; no garnet evident; plag ≫ px > qtz
13.5	900	20	Ag–Pd	px	plag	qtz	ga?	Fine grained; uncertain, rare garnet; plag ≫ px > qtz
13.5	900	48	Au	px	plag	qtz	ga	Medium grainsize; plag ≫ px > qtz > ga
18	900 ± 50	48	Au	px	plag	qtz	ga	Medium grainsize; plag > px > qtz > ga
27 then 22.1	900 900	48 64	Au	px	plag	qtz	ga	Medium grainsize; px ≫ qtz > plag, ga; definite growth of plag compared with 27 kb 900 °C run
22.5	900	48	Au	px	plag	qtz	ga	Medium grainsize; px ≫ qtz, plag > ga
22.5 then 22.5	1000 900	16 71	Au	px	plag	qtz	ga	Medium grainsize; px ≫ plag, qtz > ga; slightly more garnet than 22.5 kb 900 °C run
24.8	900	48	Au	px	felds	qtz	ga	Medium grainsize; px ≫ qtz > ga > felds; trace of feldspar, probably K-feldspar not plag
27	900	48	Au	px	felds	qtz	ga	Fine grained; px ≫ qtz > ga > felds
31.5	900	48	Au	px	felds	qtz	ga	Fine grained; px ≫ qtz > ga > felds

* Underlines denote phase identified by optical means alone.
px = pyroxene; plag = plagioclase; qtz = quartz; ga = garnet; felds = feldspar (K-rich).

were substituted. Runs at 1100 °C were conducted in silver–palladium capsules and at 1200 °C platinum capsules were used, with reduced experiment time to minimize iron loss from the sample to the platinum capsule. At this higher temperature equilibrium was reached in much shorter times. The pressure cell components were not dried, except for runs on the andesite composition at 1100 °C in the lower pressure range where a dried pyrophyllite spacer was used, in order to prevent excessive melting resulting from access of water to the sample. No boron nitride sleeve was used in the pressure cell. This procedure allowed minor access of water to the sample to promote reaction, but the amount present did not result in observable crystallization of hydrous phases.

Even with runs of 48 hours duration, difficulty in nucleating garnet was experienced in both compositions at 900 °C. Accordingly the incoming of garnet with increasing pressure was determined by two-stage runs. In the first stage the charge was taken into the garnet field and held under conditions where it was known from previous runs that garnet formed in the allowed time, then the P–T conditions were changed to those desired for the particular run and held for 2–3 days. Finally the charge was removed and examined to determine whether garnet remained stable or had reacted away. This procedure was, in effect, a type of reversal of reaction by a two stage experiment. The final disappearance of plagioclase was studied in a similar manner.

TABLE 3

Results of experimental runs on the diorite composition at 1000–1200 °C

Pressure (kb)	Temperature (°C)	Time (hrs)	Type of sample capsule	Phases present†					Comments
{ 18 then 9.3	1000 1000	24 70	Ag–Pd	px	plag	qtz	—	—	Well crystallized; uncommon orthopyroxene laths distinguishable; plag ≫ px > qtz
11	1000	31	Ag–Pd	px	plag	qtz	ga	—	Medium grainsize; plag ≫ px > qtz > ga
22.5	1000	16	Au	px	plag	qtz	ga	—	Medium grainsize; px > plag > qtz > ga
25	1000	10	Au	px	felds	qtz	ga	—	Medium grainsize; px ≫ qtz > ga > felds; possible trace of plag remaining
25	1000	48	Au	px	felds	qtz	ga	—	Medium grainsize; px ≫ qtz > ga > felds; possible trace of plag remaining
36	1000	17½	Au	px	felds	coes	ga	—	Medium grainsize; px ≫ coes > ga > felds
12.2	1100	4	Ag–Pd	px	plag	qtz	—	glass	Well crystallized; minor glass; plag ≫ px ≫ qtz
12.2	1100	24	Ag–Pd	px	plag	qtz	—	glass	Well crystallized; common glass; plag ≫ px > qtz
13.5	1100	4	Pt	px	plag	qtz	ga?	—	Medium grainsize; plag ≫ px > qtz; uncertain, rare garnet
14	1100	4	Ag–Pd	px	plag	qtz	ga	—	Fine grained; plag ≫ px > qtz > ga
14	1100	12	Ag–Pd	px	plag	qtz	ga	glass	Well crystallized, very minor melting; plag ≫ px > qtz > ga
15.8	1100	5	Ag–Pd	px	plag	qtz	ga	glass	Medium grainsize; minor melting; plag ≫ px > qtz > ga
18	1100	6	Ag–Pd	px	plag	qtz	ga	glass	Medium grainsize; minor melting; px > plag > qtz > ga
22.5	1100	10	Ag–Pd	px	plag	qtz	ga	—	Fine grained; px > plag > qtz > ga
24.8	1100	10½	Ag–Pd	px	plag	qtz	ga	—	Fine grained; px > plag, qtz > ga
27	1100 ± 30	24	Ag–Pd	px	felds	qtz	ga	—	Fine grained; px ≫ qtz > ga > felds
31.5	1100	24	Ag–Pd	px	felds	qtz	ga	—	Fine grained; px ≫ qtz, ga > felds
36	1100	4½	Ag–Pd	px	felds	qtz	ga	—	Fine grained; px ≫ coes, ga > felds
36	1200	4	Pt	px	felds	coes/qtz	ga	—	Medium grainsize; px ≫ coes, qtz > ga > felds
36	1200	11½	Pt	px	felds	coes	ga	—	Medium grainsize; px ≫ coes > ga > felds

† Underlines denote phases identified by optical means alone.
px = pyroxene; plag = plagioclase; qtz = quartz; ga = garnet; felds = feldspar (K-rich); coes = coesite.

3. Results

3.1. *Diorite*

The detailed experimental results for this composition are given in tables 2 and 3 and are summarized in fig. 1.

At 900 °C garnet first appeared stable at 6.5 kb after a two-stage run, but was not obtained in a single stage run until 13.5 kb (48 hour run), pointing to the difficulty in nucleating garnet in this composition at 900 °C. Garnet, quartz and clinopyroxene increased in amount with increasing pressure, while plagioclase decreased, until at 24.8 kb only a trace of feldspar remained. This amount of feldspar remained unchanged with further increase in pressure and was probably potash feldspar.

At 1000 °C garnet first appeared at 11 kb in a single stage run and was unstable in a two-stage run at 9.3 kb. In a similar fashion to the 900 °C series of runs, garnet, quartz and pyroxene increased in amount with increasing pressure while plagioclase decreased, until at 25 kb only a trace of feldspar (K-rich) remained. In the experiments at 1100 °C minor glass occurred in the lower pressure runs. Garnet first appeared at 14 kb and plagioclase disappeared by 27 kb. Kyanite was not observed in any of the runs.

3.2. *Gabbroic anorthosite*

The detailed experimental results for this composition are given in table 4 and are summarized in fig. 2.

In a series of two-stage runs at 900 °C garnet was

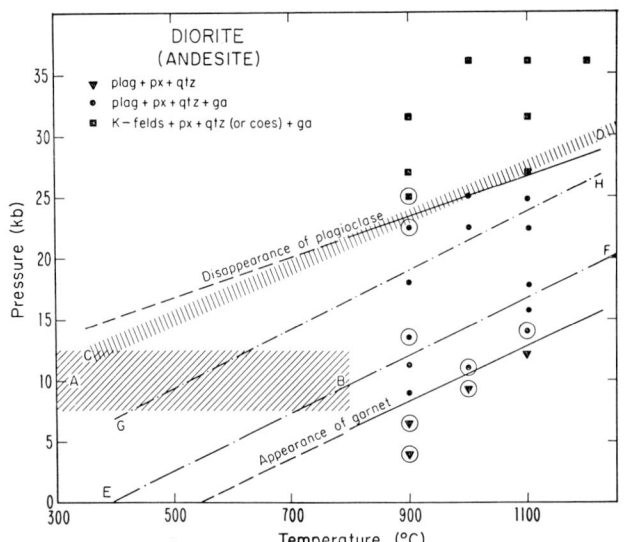

Fig. 1. Results of high pressure experimental runs on the diorite composition.

Shaded area AB: Postulated P–T conditions for the lower crust (BIRCH, 1955; CLARK, 1961, 1962; CLARK and RINGWOOD, 1964).
Shaded area CD: Experimental determinations of the albite ⇌ jadeite + quartz equilibrium plot in this field (BIRCH and LECOMTE, 1961; BOETTCHER and WYLLIE, 1968; NEWTON and KENNEDY, 1968; NEWTON and SMITH, 1967).
EF: Kyanite–sillimanite equilibrium (RICHARDSON et al., 1968).
GH: Experimental determination of the anorthite ⇌ grossular + kyanite + quartz equilibrium (HAYS, 1966) — extrapolated.

Encircled points represent experiments where a low pressure assemblage has reacted to form a high pressure assemblage, or alternatively a high pressure assemblage has reacted to a low pressure assemblage after a two-stage run.

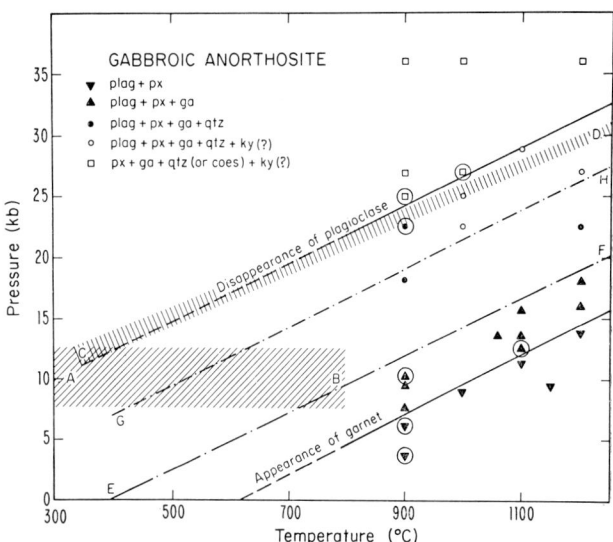

Fig. 2. Results of high pressure experimental runs on the gabbroic anorthosite composition.

Shaded area AB: Postulated P–T conditions for the lower crust (BIRCH, 1955; CLARK, 1961, 1962; CLARK and RINGWOOD, 1964).
Shaded area CD: Experimental determinations of the albite ⇌ jadeite + quartz equilibrium plot in this field (BIRCH and LECOMTE, 1961; BOETTCHER and WYLLIE, 1968; NEWTON and KENNEDY, 1968; NEWTON and SMITH, 1967).
EF: Kyanite–sillimanite equilibrium (RICHARDSON et al., 1968).
GH: Experimental determination of the anorthite ⇌ grossular + kyanite + quartz equilibrium (HAYS, 1966) — extrapolated.

Encircled points represent experiments where a low pressure assemblage has reacted to form a high pressure assemblage, or alternatively a high pressure assemblage has reacted to a low pressure assemblage after a two-stage run.

unstable at 4.3 and 6.6 kb but became stable at 7.7 kb. It first appeared at 10.1 kb in a single stage run, again pointing to the difficulty in nucleating garnet in experimental runs at temperatures as low as 900 °C. Only a trace of feldspar (probably K-feldspar) was present at 25 kb and higher pressures, but plagioclase was stable to 22.5 kb. At 1000 °C plagioclase disappeared between 25 and 27 kb. In a similar fashion to the diorite composition garnet, quartz and clinopyroxene appeared at the expense of plagioclase with increasing pressure at temperatures of 900–1200 °C. As reported previously (GREEN, 1967) kyanite occurred in some runs at 1200 °C but it was not positively identified at 900–1100 °C in the present work, though it may have been present in amounts of the order of 5%. Overlapping peaks on the diffractometer charts from other phases in the runs preclude definite identification of small amounts of kyanite. The garnet cell size increased significantly with increasing pressure ($a = 11.66$ Å at 18 kb to $a = 11.76$ Å at 36 kb) reflecting a marked increase in grossular content (cf. GREEN, 1967).

In the highest pressure runs the silica phase identified was coesite, rather than quartz.

4. Discussion and application of results

4.1. Mineralogical changes and reactions

The mineralogical changes observed result from a series of pressure dependent reactions. These reactions are complex, involving members of feldspar, pyroxene and garnet solid solution series. Prior to the first appearance of garnet there is a small decrease in the proportion of plagioclase relative to pyroxene, particularly in the gabbroic anorthosite composition. This effect is attributed to reactions of the type (i) and (ii) resulting in the formation of aluminous pyroxenes and quartz, with the co-existing plagioclase becoming more sodic:

(i) $m(Mg,Fe)SiO_3 + CaAl_2Si_2O_8 \rightleftharpoons Ca(Mg,Fe)Si_2O_6 + (m-2)(Mg,Fe)SiO_3 \cdot MgAl_2SiO_6 + SiO_2$,
  orthopyroxene   anorthite   clinopyroxene   aluminous orthopyroxene   quartz

(ii) $mCa(Mg,Fe)Si_2O_6 + CaAl_2Si_2O_8 \rightleftharpoons mCa(Mg,Fe)Si_2O_6 \cdot CaAl_2SiO_6 + SiO_2$.
  clinopyroxene   anorthite   aluminous clinopyroxene   quartz

The first appearance of garnet, and its subsequent increase in amount with increasing pressure may result from the reaction of pyroxene with the anorthite component of plagioclase, and also in the later stages at higher pressure, from the breakdown of aluminous pyroxene, e.g.

(iii) $4(Mg,Fe)SiO_3 + CaAl_2Si_2O_8 \rightleftharpoons (MgFe)_3Al_2Si_3O_{12} + Ca(Mg,Fe)Si_2O_6 + SiO_2$,
  orthopyroxene   anorthite   pyrope-almandine   clinopyroxene   quartz

(iv) $Ca(Mg,Fe)Si_2O_6 + CaAl_2Si_2O_8 \rightleftharpoons Ca_2(Mg,Fe)Al_2Si_3O_{12} + SiO_2$,
  clinopyroxene   anorthite   grossular-almandine-pyrope   quartz

(v) $m(Mg,Fe)SiO_3 \cdot MgAl_2SiO_6 \rightleftharpoons (Mg,Fe)_3Al_2Si_3O_{12} + (m-2)(Mg,Fe)SiO_3$,
  aluminous orthopyroxene   pyrope-almandine   orthopyroxene

(vi) $mCa(MgFe)Si_2O_6 \cdot CaAl_2SiO_6 \rightleftharpoons Ca_2(Mg,Fe)Al_2Si_3O_{12} + (m-1)Ca(Mg,Fe)Si_2O_6$.
  aluminous clinopyroxene   grossular-pyrope-almandine   clinopyroxene

The pyroxene content also increases from the breakdown of the albite component of the feldspar solid solution at moderately high pressures, according to reactions (vii) and (viii):

(vii) $NaAlSi_3O_8 + mCa(Mg,Fe)Si_2O_6 \rightleftharpoons mCa(Mg,Fe)Si_2O_6 \cdot NaAlSi_2O_6 + SiO_2$,
  albite   clinopyroxene   omphacite   quartz

(viii) $NaAlSi_3O_8 \rightleftharpoons NaAlSi_2O_6 + SiO_2$.
  albite   jadeite   quartz

At intermediate pressures, where the anorthite molecule of plagioclase is involved in reactions with pyroxene, the pyroxene solid solution becomes rich in $CaAl_2SiO_6$. At higher pressures, the breakdown of aluminous pyroxenes to give garnet, combined with the breakdown of the albite component of the plagioclase to form jadeite and quartz will result in a pyroxene solid solution characterized by a high jadeite content, rather than a high $CaAl_2SiO_6$ content.

At higher pressures than those at which final breakdown of plagioclase occurs, the only change in the pyroxene–garnet–quartz(–kyanite?) assemblage is in the gabbroic anorthosite composition where there is an increase in grossular content of the garnet, probably resulting from the further breakdown of $CaAl_2SiO_6$-rich pyroxene with increasing pressure.

4.2. Comparison of results with natural rock mineralogies

As pointed out by GREEN and LAMBERT (1965) and RINGWOOD and GREEN (1966) the pressure–temperature conditions under which basaltic compositions transform to eclogitic assemblages (garnet + clinopyroxene ± quartz ± kyanite, but no plagioclase) correspond to conditions at which clinopyroxene, garnet, quartz and sodic plagioclase form in more acid compositions. This holds for both the diorite and gabbroic anorthosite compositions investigated in the present work, since plagioclase is still stable above 25 kb at 1100 °C in these two compositions but it has disappeared by 25 kb at 1100 °C in each basaltic composition studied at high pressure (GREEN and RINGWOOD, 1967; GREEN, 1967).

The pyroxene–garnet–quartz–sodic plagioclase (± kyanite?) assemblage obtained at high pressure corresponds to some garnet granulites found in high grade metamorphic terranes. There are several recorded occurrences of assemblages analogous to those obtained in the experimental work in natural rocks of overall andesite or gabbroic anorthosite composition. Thus BUDDINGTON (1939, 1952) describes garnet-bearing gabbroic anorthosite (garnet moderately rich in grossular,

TABLE 4

Results of experimental runs on the gabbroic anorthosite composition at 900–1200 °C (note: additional results depicted in fig. 2 are taken from Green, 1967)

Pressure (kb)	Temperature (°C)	Time (hrs)	Type of sample capsule	Phases present†						Comments
11.3 then 4.3	900 900	13 24	Ag–Pd	px	plag	—	—	—	—	Fine grained; plag ≫ px
11.3 then 6.2	900 900	11½ 24	Ag–Pd	px	plag	—	—	—	—	Fine grained; plag ≫ px
11.3 then 7.7	900 900	12 24	Ag–Pd	px	plag	ga	qtz?	—	—	Medium grainsize; plag > px > ga, qtz? uncertain X-ray evidence for quartz
9	900	4	Ag–Pd	px	plag	—	—	—	—	Fine grained; plag ≫ px; no garnet observed
11.3 then 9.6	900 900	12½ 25	Ag–Pd	px	plag	ga	qtz?	—	—	Fine grained; plag ≫ px > ga, qtz?; uncertain X-ray evidence for quartz
10.1	900	4	Ag–Pd	px	plag	ga	—	—	—	Fine grained; plag ≫ px ≫ trace garnet
10.1	900	12	Ag–Pd	px	plag	ga	qtz?	—	—	Fine grained; plag ≫ px > ga, qtz? uncertain X-ray evidence for quartz
18	900	48	Au	px	plag	ga	qtz	—	—	Medium grainsize; plag > px > ga > qtz; $a_{ga} = 11.66 \pm 0.01$ Å
22.5	900	48	Au	px	plag	ga	qtz	—	—	Medium grainsize; px, plag ≫ ga > qtz; $a_{ga} = 11.76 \pm 0.02$ Å
22.5 then 22.5	1000 900	48 64	Au	px	plag	ga	qtz	—	—	Medium grainsize; px > plag ≫ ga > qtz (c.w. 22.5 kb 1000 °C run, plag content lower; ga, qtz higher)
36 then 22.5	1000 900	48 72	Au	px	plag	ga	qtz	—	—	Well crystallized; px > plag > ga > qtz; plag definitely grown compared with 36 kb 1000 °C run
25	900	48	Au	px	felds	ga	qtz	ky?	(amph)	Medium grainsize; px ≫ ga > qtz > felds > ky? minor amphibole also present; $a_{ga} \approx 11.72 \pm 0.02$ Å
27	900	30½	Ag–Pd	px	felds	ga	qtz	ky?	—	Fine grained; px ≫ ga > qtz > felds > ky?; evidence for kyanite not definitive, felds (trace only) probably K-felds, no plag
27	900	48	Au	px	felds	ga	qtz	ky?	—	Medium grainsize; px ≫ ga > qtz > felds > ky?; felds trace amount only; $a_{ga} \approx 11.76 \pm 0.02$ Å
36	900	48	Au	px	felds	ga	coes	ky?	—	Medium grainsize; px ≫ ga > coes > felds > ky?; felds trace amount only; $a_{ga} \approx 11.76 \pm 0.02$ Å
9	1000	11½	Ag–Pd	px	plag	—	—	—	glass	Fine grained; minor glass; plag ≫ px
22.5	1000	48	Au	px	plag	ga	qtz	ky?	—	Medium grainsize; px > plag, ga > qtz > ky?; $a_{ga} = 11.71 \pm 0.01$ Å
25	1000	24	Au	px	plag	ga	qtz	ky?	—	Medium grainsize; px ≫ plag, ga > qtz > ky?
27	1000	48	Au	px	felds	ga	qtz	ky?	—	Medium grainsize; px ≫ ga > qtz > felds, ky?; felds trace amount only; $a_{ga} = 11.74 \pm 0.01$ Å
36	1000	48	Au	px	felds	ga	coes	ky?	—	Medium grainsize; px ≫ ga > coes > felds, ky?; felds trace amount only; $a_{ga} = 11.76 \pm 0.02$ Å
13.5	1060	3	Au	px	plag	ga	—	—	—	Fine grained; plag ≫ px > ga
11.3	1100	4	Ag–Pd	px	plag	—	—	—	—	Fine grained; plag ≫ px
12.4	1100	4	Ag–Pd	px	plag	ga	—	—	—	Medium grainsize; plag ≫ px > ga
13.5	1100	2	Ag–Pd	px	plag	ga	qtz?	—	—	Medium grainsize; plag ≫ px > ga > qtz?; uncertain X-ray evidence for quartz
28.8	1100	23	Ag–Pd	px	plag	ga	qtz	ky?	—	Medium grainsize; px ≫ ga > qtz ≫ plag > ky?
16.0	1200	2½	Pt	px	plag	ga	—	—	glass	Medium grainsize; minor glass, plag ≫ px > ga

† Underlines denote phases identified by optical means only.
px = pyroxene; plag = plagioclase; qtz = quartz; ga = garnet; felds = feldspar (K-rich); coes = coesite; ky = kyanite; amph = amphibole.

>20 mol%) from the Adirondacks. Dioritic gneisses of similar chemical composition to andesite also occur, but contain little or no garnet. The garnet tends to be lowest or absent in rocks with the highest quartz content, the rocks most closely approaching the diorite of this paper in composition. Both the gabbroic anorthosite and diorite assemblages are dominated by plagioclase with subordinate clinopyroxene and minor garnet. Comparison with the experimental results indicates probable $P-T$ conditions of about 750 °C and 5 kb. DE WAARD (1966) suggests pressures as high as 10 kb and temperatures up to 800 °C for the metamorphism in the Adirondacks, in which case garnet would be expected in rocks of intermediate composition under anhydrous conditions.

DAVIDSON (1943) refers to a garnet-bearing anorthositic gneiss (plagioclase dominant, with subordinate garnet and minor clinopyroxene and hornblende) on South Harris. In the same area DEARNLEY (1963) describes tonalites (approximately equivalent to andesite in composition) with the assemblage orthopyroxene–clinopyroxene–plagioclase. Also in basic rocks there is a reaction relation between olivine and plagioclase (producing garnet and quartz). Conditions satisfying these 3 features would be 3–5 kb at 700–800 °C as indicated by the present work and work of GREEN and RINGWOOD (1967).

DEN TEX and VOGEL (1962) and ESKOLA (1952) record intermediate rocks with assemblages orthopyroxene–clinopyroxene–plagioclase–quartz indicating $P-T$ conditions in the field below the incoming of garnet in fig. 1. QUENSEL (1951) and GROVES (1935) refer to intermediate composition rocks with the assemblage: –2 pyroxenes–plagioclase–garnet–quartz–hornblende, indicating metamorphism under $P-T$ conditions within the garnet field of fig. 1. GROVES (1935) also notes reaction rims involving hypersthene–plagioclase reacting to form diopside and garnet and indicates that the pyroxene becomes increasingly omphacitic in composition. Similarly KOZLOWSKI (1958) reports an intermediate rock with the assemblage garnet (grossular 22%), omphacitic pyroxene, plagioclase (oligoclase), quartz and minor microcline, biotite, amphibole and rutile. Kozlowski's conclusion that this assemblage was formed under conditions which would have produced eclogites from basalts is consistent with the present experimental results.

4.3. *Lower crustal mineralogy*

Assuming a linear extrapolation of the incoming of garnet and the final disappearance of plagioclase to probable temperatures in the stable lower crust (300–700 °C; BIRCH, 1955; CLARK, 1961, 1962) an indication of the mineralogy of an anhydrous lower crust of overall andesite and/or gabbroic anorthosite composition may be obtained. As mentioned in section 2, RINGWOOD and GREEN (1966) have argued that large areas of stable continental crust are essentially dry, and it is emphasized that the present results are only applicable to a consideration of the mineralogy of a dry crust.

For an andesitic composition the highest pressure assemblage obtainable below the quartz to coesite transition (coesite is only known to have formed naturally in association with meteorite impact) consists of clinopyroxene, quartz, garnet and K-feldspar, but for likely $P-T$ conditions in the lower crust, the final breakdown of plagioclase in this composition is not attained, so that some sodic plagioclase will occur as well. This is similar to the assemblage proposed by RINGWOOD and GREEN (1966) except for the absence of kyanite. This phase was not identified in any of the experimental runs on andesite, but it may have occurred in amounts up to about 5% (see section 3.2), but not 15% as calculated by Ringwood and Green. This suggests that the high pressure assemblage obtained experimentally is probably richer in aluminous pyroxenes, quartz and garnet than the assemblage calculated.

Similarly, in the case of the gabbroic anorthosite, the mineralogy expected under stable lower crustal $P-T$ conditions would consist of plagioclase, pyroxene, minor garnet, minor quartz and possibly minor kyanite. The grossular content of the garnet would be significant, and for any given bulk composition, increase with increasing pressure of formation.

4.4. *Experimental results and geophysical models for the lower crust*

Seismologists generally recognize compressional wave velocities of 5.8–6.3 km/s as characteristic of the upper crust, increasing to 6.6–7.4 km/s in the lower crust (GUTENBERG, 1955, 1959; RICHARDS and WALKER, 1959; STEINHART and MEYER, 1961; JAMES and STEINHART, 1966). The nature of the downward increase in velocity is controversial (for a summary see JAMES and

STEINHART, 1966); in some regions it is argued to be discontinuous at the Conrad discontinuity (e.g. RICHARDS and WALKER, 1959), whereas in other areas it is considered to be continuous (e.g. TATEL and TUVE, 1955). Early crustal models proposed to explain this increase have attributed it to a granitic upper crust overlying a basic lower crust. PAKISER and ROBINSON (1966) have given estimates of the average composition of three super-provinces of the North American continental crust, based on seismic evidence. Their conclusion is that the overall composition is intermediate, though they assume that there is a composition change from granitic to basic with depth, rather than a mineralogical change in a single bulk composition.

However RINGWOOD and GREEN (1966) have argued, using experimental results on high pressure assemblages in basic compositions, that an anhydrous lower crust of basic composition could not give rise to the observed seismic properties of this region. RINGWOOD and GREEN (1966) further suggested that rocks approaching diorite ($\approx$ andesite) in average chemical composition would best fit the physical properties of the lower crust. The present experimental work supports these conclusions. Garnet is stable in diorite under conditions equivalent to high-pressure granulite facies and eclogite facies (i.e. conditions expected in the lower crust) and so in an anhydrous lower crust, the presence of relatively dense phases garnet and aluminous pyroxene may be responsible for increasing seismic velocity with depth, rather than a compositional change (cf. GREEN and LAMBERT, 1965). A gabbroic anorthosite composition would behave in a similar fashion to diorite.

The density of the high pressure mineral assemblage has been measured directly on two experimental runs using a Berman balance. The gabbroic anorthosite composition at 25 kb, 900 °C has a density of 3.15 g/cm^3 while the diorite composition at 18 kb, 900 °C has a density of 2.88 g/cm^3. Using solutions 6 and 7 of the equation given by BIRCH (1961) ($V_P = a + b\rho$ where V_P is the compressional wave velocity in km/s at 10 kb and ρ the density in g/cm^3) the calculated compressional wave velocities at 10 kb confining pressure for these assemblages are 7.4 km/s (gabbroic anorthosite) and 7.0 km/s (diorite). The particular gabbroic anorthosite run represents a mineral assemblage stable at P–T conditions greater than those generally considered likely in the lower crust, so that the calculated seismic velocities may be regarded as the upper limit for a lower crust of this composition. The diorite run represents a mineral assemblage likely for intermediate P–T conditions predicted in the lower crust (e.g. extrapolated to 550 °C at 10 kb), so that the calculated seismic velocity may be regarded as a median value for a lower crust of dioritic composition. The compressional wave velocities should be somewhere between 6.8 and 7.4 km/s for gabbroic anorthosite or 6.6 and 7.2 km/s for diorite, where the lower values are those predicted for the low pressure mineralogy (table 1).

Acknowledgements

This work has been conducted under the auspices of the Committee on Experimental Geology and Geophysics, Harvard University and this support is gratefully acknowledged. Dr. J. F. Hays critically read the manuscript and gave much helpful advice during the course of the project.

References

BIRCH, F. (1955) Physics of the crust, in: A. Poldervaart, ed., *Crust of the Earth*, Geol. Soc. Am. Spec. Papers **62**, 101.
BIRCH, F. (1961) J. Geophys. Res. **66**, 2199.
BIRCH, F. and P. LE COMTE (1960) Am. J. Sci. **258**, 209.
BOETTCHER, A. L. and P. J. WYLLIE (1968) Geochim. Cosmochim. Acta **32**, 99.
BOYD, F. R. and J. L. ENGLAND (1960a) J. Geophys. Res. **65**, 741.
BOYD, F. R. and J. L. ENGLAND (1960b) J. Geophys. Res. **65**, 749.
BOYD, F. R. and J. L. ENGLAND (1963) J. Geophys. Res. **68**, 311.
BUDDINGTON, A. F. (1939) Geol. Soc. Am. Mem. **7**.
BUDDINGTON, A. F. (1952) Am. J. Sci. Bowen Vol., 37.
CLARK, JR., S. P. (1961) Carnegie Inst. Wash. Yearbook **60**, 185.
CLARK, JR., S. P. (1962) Temperatures in the continental crust, in: C. M. Herzfeld, ed., *Temperature: Its measurement and control in science and industry* **3** (Reinhold Publ. Co., New York) 779.
DAVIDSON, C. F. (1943) Trans. Roy. Soc. Edinburgh **61**, 71.
DEARNLEY, R. (1963) Quart. J. Geol. Soc. London **119**, 243.
DEN TEX, E. (1965) Geol. Mijnbouw **44^e**, 105.
DEN TEX, E. and D. E. VOGEL (1962) Geol. Rundschau **52**, 95.
DE WAARD, D. (1967) J. Petrol. **8**, 210.
ESKOLA, P. (1952) Am. J. Sci. Bowen Vol., 133.
GREEN, D. H. and I. B. LAMBERT (1965) J. Geophys. Res. **70**, 5259.
GREEN, D. H. and A. E. RINGWOOD (1967) Geochim. Cosmochim. Acta **31**, 767.
GREEN, T. H. (1967) Contrib. Mineral. Petrol. **16**, 84.
GREEN, T. H. (1968) Experimental fractional crystallization of quartz diorite and its application to the problem of anorthosite origin, in: Y. Isachsen, ed., Symp. origin of anorthosite (N.Y. State Geol. Surv. Mem.), in press.
GREEN, T. H., A. E. RINGWOOD and A. MAJOR (1966) J. Geophys. Res. **71**, 3589.
GROVES, A. W. (1935) Quart. J. Geol. Soc. London **91**, 150.
GUTENBERG, B. (1955) Wave velocities in the Earth's crust, in:

A. Poldervaart, ed., *Crust of the Earth*, Geol. Soc. Am. Spec. Papers **62**, 19.

GUTENBERG, B. (1959) *Physics of the Earth's interior*, Intern. Geophys. Ser. **1** (Academic Press).

HARRIS, P. G. (1967) Segregation processes in the upper mantle, in: S. K. Runcorn, ed., *Mantles of the Earth and terrestrial planets* (Wiley, London) 305.

HAYS, J. F. (1966) Carnegie Inst. Wash. Yearbook **65**, 234.

JAMES, D. E. and J. S. STEINHART (1966) Structure beneath continents: A critical review of explosion studies 1960–1965, in: J. S. Steinhart and T. J. Smith, eds., *The Earth beneath the continents*, Am. Geophys. Union Geophys. Monograph **10**, 293.

KUSHIRO, I. and H. S. YODER, JR. (1966) J. Petrol. **7**, 337.

KOZLOWSKI, K. (1958) Bull. Acad. Polon. Sci. Ser. Sci. Chim. Geol. Geograph. **6**, 723.

NEWTON, M. S. and G. C. KENNEDY (1968) Am. J. Sci. **266**, 728.

NEWTON, R. C. and J. V. SMITH (1967) J. Geol. **75**, 268.

PAKISER, L. C. and R. ROBINSON (1966) Composition of the continental crust as estimated from seismic observations, in: J. S. Steinhart and T. J. Smith, eds., *The Earth beneath the continents*, Am. Geophys. Union Geophys. Monograph **10**, 620.

QUENSEL, P. (1951) Arkiv Mineral. Geol. **1**, 227.

RICHARDS, T. C. and D. J. WALKER (1959) Geophysics **24**, 262.

RICHARDSON, S. W., P. M. BELL and M. C. GILBERT (1968) Am. J. Sci. **266**, 513.

RINGWOOD, A. E. and D. H. GREEN (1966) Petrological nature of the stable continental crust, in: J. S. Steinhart and T. J. Smith, eds., *The Earth beneath the continents*, Am. Geophys. Union Geophys. Monograph **10**, 611.

STEINHART, J. S. and R. P. MEYER (1961) Carnegie Inst. Wash. Publ. **622**.

TATEL, H. A. and M. A. TUVE (1955) Seismic exploration of a continental crust, in: A. Poldervaart, ed., *Crust of the Earth*, Geol. Soc. Am. Spec. Papers **62**, 35.

TAYLOR, S. R. (1968) Geochemistry of andesites, Proc. Intern. Symp. Origin and distribution of elements V, Intern. Assoc. Geochem. Cosmochem., in press.

TAYLOR, S. R. and A. J. R. WHITE (1965) Nature **208**, 271.

STABILITY AND STRUCTURAL RELATIONS OF (Mg, Fe) METASILICATES

TH. ERNST and R. SCHWAB

Mineralogisches Institut, Universität Erlangen-Nürnberg, Erlangen, West Germany

The polymorphism of $MgSiO_3$ and $MgSiO_3$–$FeSiO_3$ solid solutions, which are widespread in nature, has been the subject of numerous investigations which have obtained very contradictory results as to the number of modifications, their fields of stability and their structures (fig. 1).

As we now know, these diverse findings can be explained mainly on two facts, namely the use of starting materials of different origin and chemical composition, and the difficulties in identifying the polymorphs on the basis of their similar X-ray powder patterns.

Following HARALDSEN (1930) we shall interpret as protoenstatite the phase which is formed when talc is calcined, particularly the technical product which contains excess silica and some iron. To find out the stability relations it seemed necessary to use pure $MgSiO_3$, a compound which can be synthetized from the respective oxides MgO and SiO_2 in a ratio one to one.

On heating, the primary products are olivine and cristobalite, which transform slowly to the metasilicate at temperatures between 1450 and 1550 °C. It is necessary to grind the material several times. Subsequent annealing at about 1000 °C will give pure $MgSiO_3$ in which olivine or cristobalite can no longer be detected by X-ray or optical methods. The material synthetized in this way is monoclinic and corresponds both optically and in its X-ray pattern to clinoenstatite (MORIMOTO, *et al.*, 1960). On heating, clinoenstatite can be transformed to protoenstatite at 1260 ± 10 °C, and the protoenstatite remains unchanged up to the point of incongruent melting at 1555 °C. The powder patterns within this region are identical with those of protoenstatite formed from talc. Thus it is clear that the term protoenstatite is correct for the high-temperature modification of $MgSiO_3$.

Protoenstatite, formed from pure $MgSiO_3$ at tem-

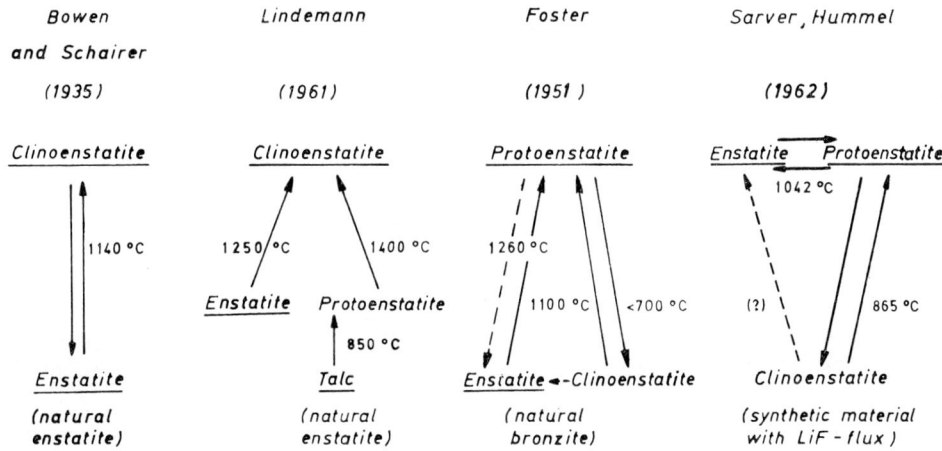

Fig. 1. Polymorphism of $MgSiO_3$ according to BOWEN and SCHAIRER (1935), LINDEMANN (1960), FOSTER (1951) and SARVER and HUMMEL (1962). Modifications regarded as stable are underlined. The starting materials are given in parentheses.

451

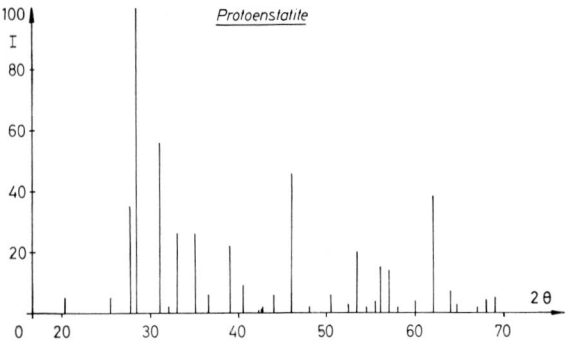
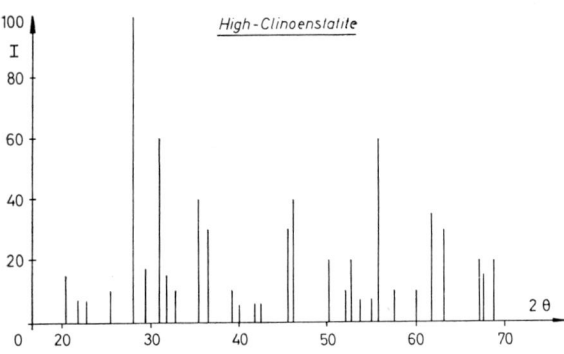
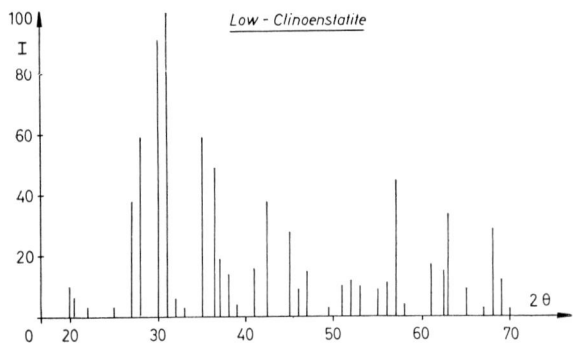

Fig. 2. X-ray powder patterns of low-clinoenstatite, high-clinoenstatite (1145 °C) and protoenstatie (from talc) (CuK$_\alpha$ radiation).

peratures above 1260 °C, can be quenched only partially, and commonly transforms spontaneously to clinoenstatite at room temperature. This contrasts with the behaviour of protoenstatites formed from talc, which are stabilized by excess silica.

We may remember now that protoenstatite is formed reversibly at 1260 ± 10 °C. A detailed investigation, however, reveals that this phase transformation does not originate from unchanged clinoenstatite but from a transitional phase which comes into existence at 1080 ± 10 °C. Clinoenstatite, henceforth more exactly called low-clinoenstatite, transforms at 1080 °C to a phase whose X-ray powder pattern is very similar to that of protoenstatite, but also exhibits some analogies to the diagram of low-clinoenstatite, as fig. 2 demonstrates. Nevertheless, there exist some characteristic differences which make an unequivocal identification possible: for example, the transitional phase has an additional reflection at $d = 3.06$ Å ($T = 1145$ °C), which never appears in protoenstatite (SMITH, 1959). Moreover there are some differences also at higher diffraction angles, like splitting of reflections and differences in intensities, which will not be discussed further; in general the X-ray powder patterns point to a close structural relationship between the three phases.

The interpretation that there might exist a really polymorphic phase between 1080 and 1260 °C and that the diagrams may not be interpreted as a result of order–disorder transitions or of the metastable coexistence of two or more phases, is supported by thermal analyses. Besides normal DTA-methods we applied the "dynamical differential-calorimetry" from *Schwiete* (DDK) and found that the transitions could be fixed surprisingly well in this way. In order to get distinct heat effects it was necessary to take pure and well crystallized samples which had been stored for several months. In this connection it is important to point out that the polymorphic behaviour of the substances is influenced by their chemical composition and thermal history.

The field of the transitional phase is separated by reversible and discontinuous heat effects from the fields of low-clinoenstatite and protoenstatite. The phase therefore has to be regarded as another high-temperature polymorph of MgSiO$_3$. For the present it will be called high-clinoenstatite, mainly because of its physicochemical behaviour. Unlike protoenstatite, high-clinoenstatite is not quenchable, and we are not yet able to stabilize it in any way. We believe that some previous investigators may have detected this phase, too, but unfortunately did not carry on their studies (BROWN and SMITH, 1963) or perhaps may have regarded it as protoenstatite or as low-clinoenstatite (PEROTTA and STEPHENSON, 1965; LINDEMANN, 1960). Commonly high-clinoenstatite can easily be taken as protoenstatite, especially if the identification is confined to lower diffraction angles and to the stronger reflections.

We have not yet discussed the stability of the rhombic

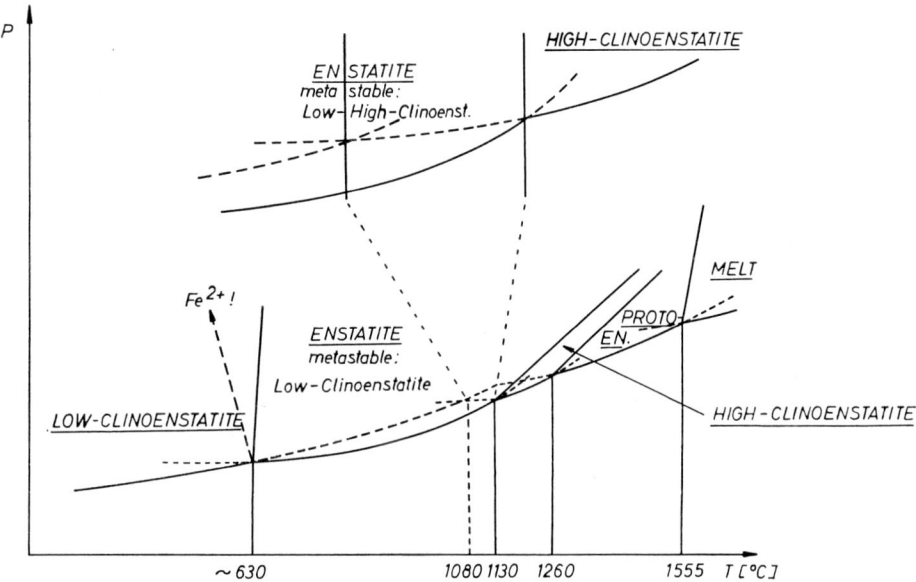

Fig. 3. Schematic P–T diagram of MgSiO$_3$.

enstatite, which is commonly regarded as the stable low-temperature polymorph of MgSiO$_3$. The synthesis of pure rhombic MgSiO$_3$ is not possible at 1 atm. Therefore it is necessary to apply either hydrothermal conditions or high pressures. The synthesis of the pure rhombic enstatite described in this study, was performed at 25 kb and 1300 °C, starting with low-clinoenstatite.

If large quantities are needed, it is more convenient to synthetize the rhombic phase at 1 atm at about 960 °C with LiF or LiOH as a flux. These fluxes cause a marked lowering of the transition temperatures and an increase of the transformation enthalpies.

The calorimetric investigations revealed that the rhombic enstatite forms the phase of higher enthalpy and entropy with respect to low-clinoenstatite. Low-clinoenstatite should be stable below 600 °C, which is in good agreement with the transformation temperature of 630 °C, extrapolated by BOYD and ENGLAND (1965) from high pressure experiments. Because enstatite is the less dense phase (as the lattice constants and the optical constants in table 1 demonstrate), high pressures will lead to an extension of the field of low-clinoenstatite to higher temperatures at the expense of the field of the orthopyroxene.

However, the latter will extend its field of stability

TABLE 1

Pure synthetic enstatite:	Pure synthetic low-clinoenstatite:
optics:	
$n_\alpha = 1.647$	$n_\alpha = 1.650$
$n_\beta = 1.649$	$n_\beta = 1.653$
$n_\gamma = 1.657$ (R. Schwab)	$n_\gamma = 1.660$ (R. Schwab)
lattice constants:	
$a_0 = 18.203$ Å	$a_0 = 9.604$ Å
$b_0 = 8.807$ Å	$b_0 = 8.815$ Å
$c_0 = 5.202$ Å	$c_0 = 5.170$ Å $\beta = 71.65°$
$\frac{1}{2}V = 416.97$ Å^3	$\frac{1}{2}V = 415.47$ Å^3 (PEROTTA and
(R. SCHWAB)	STEPHENSON, 1965)

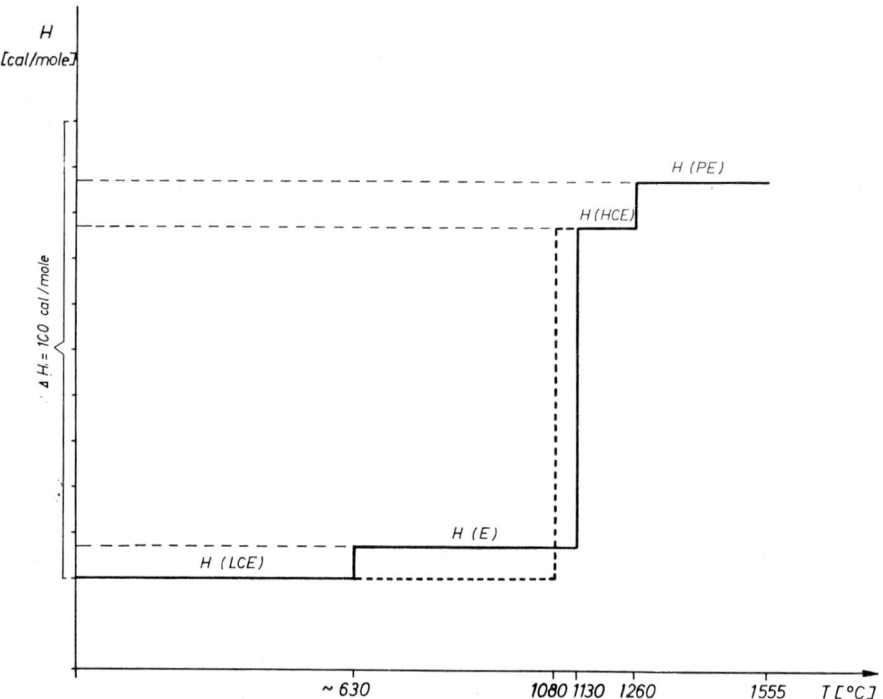

Fig. 4. The transition enthalpies for pure $MgSiO_3$. All data are referred to the enthalpy of the α–β-quartz transition (= 86 cal/mole) according to MAJUMDAR et al. (1964).

against the fields of the high-volume phases high-clinoenstatite and protoenstatite which finally will vanish. These relations are presented schematically in fig. 3. Fig. 4 gives the transition enthalpies for pure $MgSiO_3$.

The presence of ferrous iron affects the stability of $MgSiO_3$ in the same way as increasing pressure, causing a marked increase in thermal stability of the rhombic enstatite relative to the high-temperature phases high-clinoenstatite and protoenstatite. At a content of 25 mole % of $FeSiO_3$ the orthopyroxene melts without any previous polymorphic transition (SCHWAB, 1967). However, there are some indications from the measured transformation enthalpies, that the field of the rhombic enstatite might also be extended against the field of low-clinoenstatite to lower temperatures. This assumption is supported by the high pressure experiments of AKIMOTO et al. (1965) in pure $FeSiO_3$.

The synthesis of orthopyroxenes at 1 atm is easily achieved in the system $MgSiO_3$–$FeSiO_3$ if it is performed under partial pressures of oxygen which preserve all iron in the bivalent state. It is also possible (as in LiF-containing systems) to demonstrate the reversibility of the transition from high-clinoenstatite to enstatite: slow cooling will lead to enstatite in an exothermal transformation, rapid cooling or quenching will lead to low-clinoenstatite in an exothermal, but at first metastable reaction (fig. 3).

Conclusions

Contrary to previous results the existence has to be assumed of at least one more high-temperature modification of $MgSiO_3$. This polymorph is not identical with the high-clinoenstatite described by PEROTTA and STEPHENSON (1965) especially in Ca-containing products.

Moreover there is the very important result, that Fe^{2+} will influence the polymorphism of $MgSiO_3$ in the same sense as pressure in the high temperature region. It is remarkable that the maximum stability of the rhombic phase exists at 1 atm at a content of about 25 mole % $FeSiO_3$ in the system $MgSiO_3$–$FeSiO_3$.

The structure determination of the high-clinoenstatite modification described here will probably be still more difficult than with protoenstatite, where it is possible to synthetize stabilized single crystals out of talc (LINDEMANN, 1951). Because high-clinoenstatite has its position in between the fields of protoenstatite and low-clinoenstatite on one hand and protoenstatite and enstatite on the other hand, it would be very informative to have an exact knowledge about the structures of the respective end members. In the case of the rhombic enstatite exact and concordant structure analyses are available (WARREN and MODELL, 1930; LINDEMANN, 1961), but there is a controversy on the structures of low-clinoenstatite and protoenstatite between MORIMOTO et al. (1960) and SMITH (1959) on one side and LINDEMANN (1960) on the other side. Thus a new structural investigation of the polymorphs of $MgSiO_3$ seems to be necessary. These investigations have been started.

References

AKIMOTO, S., T. KATURA, Y. SYONO, H. FUJISAWA and E. KOMADA (1965) Polymorphic transitions of pyroxenes $FeSiO_3$ and $CoSiO_3$ at high pressures and temperatures, J. Geophys. Res. **70**, 5269–5278.

BOWEN, N. L. and J. F. SCHAIRER (1935) The system $MgO-FeO-SiO_2$, Am. J. Sci. **29**, 151–217.

BOYD, F. R. and J. L. ENGLAND (1965) The rhombic enstatite-clinoenstatite inversion, Ann. Rept. Geophys. Lab. Carnegie Inst. Wash. 1964–1965, 117–120.

BROWN, W. L. and J. V. SMITH (1963) High-temperature X-ray studies on the polymorphism of $MgSiO_3$, Z. Krist. **118**, 186–212.

FOSTER, W. R. (1951) High-temperature X-ray diffraction study of the polymorphism of $MgSiO_3$, J. Am. Ceram. Soc. **34**, 255–259.

HARALDSEN, H. (1930) Beiträge zur Kenntnis der thermischen Umbildung des Talks, Neues Jahrb. Mineral., Abhandl., Beilage Band **61 A**, 139–164.

LINDEMANN, W. (1960) Strukturuntersuchung der Kettensilikate unter besonderer Berücksichtigung der Mg-Pyroxene, Habilitationsschrift, Universität Erlangen.

LINDEMANN, W. (1961) Beitrag zur Enstatitstruktur (Verfeinerung der Parameterwerte), Neues Jahrb. Mineral., Monatsh. (10), 226–233.

LINDEMANN, W. (1961) Gitterkonstanten, Raumgruppe und Parameter des γ-$MgSiO_3$. Naturwissenschaften **48** (11), 428–429.

LINDEMANN, W. (1951) Darstellung und Kristallstruktur des γ-$MgSiO_3$, Dissertation (Univ. Erlangen).

MAJUMDAR, A. J., H. A. MCKINSTRY and RUSTUM ROY (1964) Thermodynamic parameters for the α-β-quartz and α-β-cristobalite transitions, J. Phys. Chem. Solids **25**, 1487–1489.

MORIMOTO, N., D. E. APPLEMAN and H. T. EVANS, JR. (1960) The crystal structure of clinoenstatite and pigeonite, Z. Krist. **114**, 120–147.

PEROTTA, A. J. and D. A. STEPHENSON (1965) Clinoenstatite: High-low inversion, AAAS **148**, 1090–1091.

SARVER, J. F. and F. A. HUMMEL (1962) Stability relations of magnesium metasilicate polymorphs, J. Am. Ceram. Soc. **45**, 152–156.

SCHWAB, R. G. (1967) Die Bedeutung und die experimentelle Beherrschung des Sauerstoffpartialdruckes bei der Synthese und Untersuchung Fe^{2+}-haltiger Silikate, Neues Jahrb. Mineral., Monatsh. (7/8), 244–254.

SMITH, J. V. (1959) The crystal structure of protoenstatite $MgSiO_3$, Acta Cryst. **12**, 515–519.

WARREN, B. E. and D. I. MODELL (1930) The structure of enstatite $MgSiO_3$, Z. Krist. **75**, 1–14.

ANALCITE-JADEITE PHASE BOUNDARY*

MURLI H. MANGHNANI

Hawaii Institute of Geophysics, University of Hawaii, Honolulu, Hawaii 96822, U.S.A.

Thirty-one P–T runs were made, using a piston-cylinder apparatus, to study the phase boundary of the reaction analcite ⇌ jadeite+vapor (H_2O). The starting material was a stoichiometric mixture of the products and reactants. Reversal of the reaction was obtained. The phase boundary is given by the equation, $P = 4000$ bars $+ 10.5\ T$ (°C). At about 10 kb and 600 °C the phase boundary is intersected by the melt curve, jadeite+vapor ⇌ liquid which has a steeper slope. The previously determined phase boundary of the reaction, nepheline+albite ⇌ 2 jadeite, (ROBERTSON et al., 1957) also passes through this invariant point. The above experimental method was also adopted to extend further the studies by YODER (1954) and GREENWOOD (1961) of the reaction, ½ nepheline+½ albite+H_2O ⇌ analcite. The results of the thirteen runs for this reaction fit well with their data as well as the data of BOETTCHER and WYLLIE (1967, 1969).

1. Introduction

Jadeite ($NaAlSi_2O_6$), a common component of the omphacitic pyroxenes, is often found closely associated with eclogites and with certain metamorphic rocks. The fact that jadeite is one of the minerals that form at high pressures makes it important in petrologic and geophysical studies of the Earth's interior.

Much discussion on the stability fields of jadeite is found in solid-state phase-transition studies involving the breakdown of sodium-aluminum silicates such as albite (BIRCH and LECOMTE, 1960), and albite and nepheline (ROBERTSON et al., 1957):

$$NaAlSi_3O_8 \rightleftharpoons NaAlSi_2O_6 + SiO_2, \quad (1)$$
$$\text{Albite} \qquad \text{Jadeite} \qquad \text{Quartz}$$

$$NaAlSi_3O_8 + NaAlSiO_4 \rightleftharpoons 2\ NaAlSi_2O_6. \quad (2)$$
$$\text{Albite} \qquad \text{Nepheline} \qquad \text{Jadeite}$$

Jadeite may also form by dehydration of analcite ($NaAlSi_2O_6 \cdot H_2O$):

$$NaAlSi_2O_6 \cdot H_2O \rightleftharpoons NaAlSi_2O_6 + H_2O. \quad (3)$$
$$\text{Analcite} \qquad \text{Jadeite} \qquad \text{Vapor}$$

The reaction involves removal of H_2O from the zeolitic structure of analcite and the formation of the denser polymorph, jadeite. This transformation has significant bearing on dehydration reactions in general,

* Hawaii Institute of Geophysics Contribution No. 291.

and particularly on the genesis of rocks containing analcite and jadeite.

Another reaction which should be considered here in connection with reaction (3), is:

$$NaAlSi_2O_6 \cdot H_2O \rightleftharpoons \tfrac{1}{2} NaAlSi_3O_8 + \tfrac{1}{2} NaAlSiO_4 + H_2O. \quad (4)$$
$$\text{Analcite} \qquad \text{Albite} \qquad \text{Nepheline} \qquad \text{Vapor}$$

The purpose of this paper is to present the experimental results obtained for reaction (3), and to compare them with the theoretical curve of FYFE and VALPY (1959). Another purpose is to present additional experimental data for reaction (4), and discuss reaction (3) in the light of these findings.

2. Previous investigation

2.1. *System: analcite ⇌ jadeite+water*

The first experimental investigation of this reaction reported is that of GRIGGS and KENNEDY (1956) who employed a simple opposed-anvil squeezer to determine the phase boundary of the above reaction (see also the abstract of GRIGGS et al., 1955). Reproduced in fig. 1 is their phase relationship which shows that the slope dP/dT for this reaction is negative (approximately equal to -17 b/°C).

FYFE and VALPY (1959) pointed out that the initial dP/dT slope for reaction (3) should be positive since in this transformation both the volume and the entropy

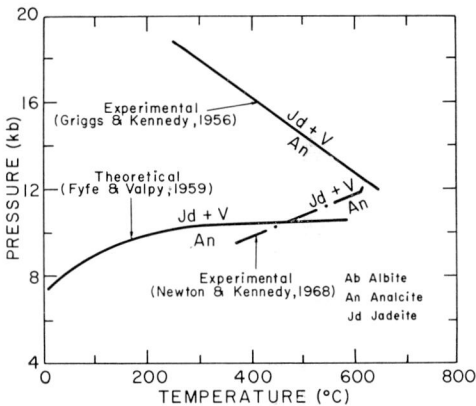

Fig. 1. Previously reported experimental and theoretical equilibrium boundaries of the reaction, analcite ⇌ jadeite+vapor.

changes (ΔV and ΔS) are negative. They deduced the phase boundary for the reaction by combining the thermochemical data of reaction (2) and the experimental data of reaction (4), since the addition of (2) and (4) leads into (3):

$$\tfrac{1}{2}\text{NaAlSi}_3\text{O}_8 + \tfrac{1}{2}\text{NaAlSiO}_4 \rightleftharpoons \text{NaAlSi}_2\text{O}_6 \quad (2)$$
$$\quad\text{Albite} \qquad\quad \text{Nepheline} \qquad \text{Jadeite}$$

$$\text{NaAlSi}_2\text{O}_6 \cdot \text{H}_2\text{O} \rightleftharpoons \tfrac{1}{2}\text{NaAlSi}_3\text{O}_8 + \tfrac{1}{2}\text{NaAlSiO}_4 + \text{H}_2\text{O}$$
$$\quad\text{Analcite} \qquad\qquad \text{Albite} \qquad\quad \text{Nepheline} \quad \text{Vapor}$$
$$(4)$$

$$\text{NaAlSi}_2\text{O}_6 \rightleftharpoons \text{NaAlSi}_2\text{O}_6 + \text{H}_2\text{O}.$$
$$\quad\text{Analcite} \qquad \text{Jadeite} \qquad \text{Vapor}$$

Their basis for deriving the curve was that at pressures and temperatures along the univariant reaction curve of (4), the ΔF values for (3) are equivalent to those for (2), since ΔF for (4) along this curve is zero. In their calculation of the equilibrium curve (see fig. 1) an average density of water vapor of 0.82 g/cm³ at equilibrium conditions was assumed.

More recently, NEWTON and KENNEDY (1968) have experimentally determined the phase relationship in the analcite-jadeite system (see fig. 1).

2.2. Reaction: analcite ⇌ ½ albite + ½ nepheline + vapor

The equilibrium relationships of this reaction to 2.8 kb (total pressure = water pressure) are fairly well established through the detailed studies of YODER (1954) and GREENWOOD (1961). The phase boundaries from their investigations are shown in fig. 5.

Recently, APPS (1968) theoretically derived an equilibrium curve for the above univariant reaction by

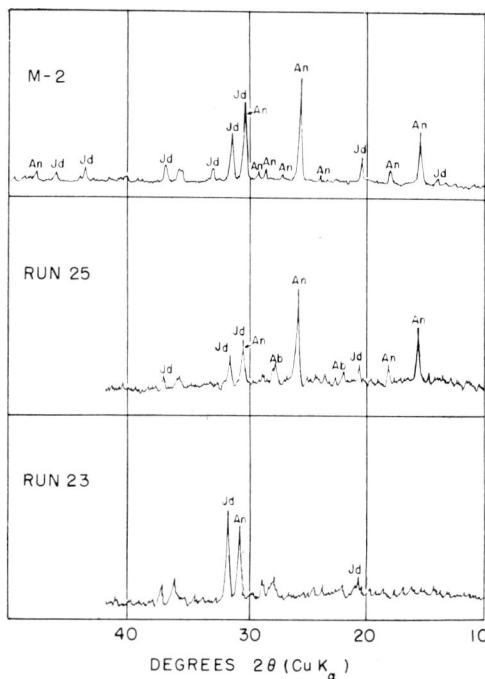

Fig. 2. X-ray diffractograms of the starting stoichiometric mixture M-2 (upper part) and the end products of the two runs. In run no. 25 (7 kb, 700 °C) analcite grew; in run no. 23 (11 kb, 600 °C), jadeite grew.

evaluating the change of volume of the vapor phase in terms of the free energy values, based on the thermodynamic tables for water (SHARP, 1962). APPS's (1969, personal communication, unpublished) curve, reproduced in fig. 5, is in good agreement with YODER's (1954) and GREENWOOD's (1961) results.

3. Experimental procedures

The pressure-temperature experiments in this study were carried out in a piston-cylinder apparatus based on the design of BOYD and ENGLAND (1960), and described in NEWTON and KENNEDY (1963). The diameter of the tungsten carbide pressure chamber was 1 inch, chosen in order to minimize the error due to friction between chamber wall and piston (see NEWTON, 1966).

The starting material in the investigation of reaction (3) was a stoichiometric mixture (M-2) of natural specimens of analcite (from Paterson, New Jersey) and jadeite (from Clear Creek Area, California). A stoichiometric mixture (M-4) of analcite, albite (Amelia, Virginia), and nepheline (Ontario, Canada) was used for reaction (4). The chemical analyses of the specimens used

Table 1

Chemical analyses (in weight %) of the natural specimens used in the experiments (Analyst: H. Asari)

	Analcite	Jadeite	Albite	Nepheline
SiO_2	56.10	58.82	68.35	43.41
Al_2O_3	22.48	24.91	20.02	34.72
Na_2O	13.35	14.65	11.40	15.79
K_2O	—	—	0.03	4.91
CaO	0.01	0.31	0.12	0.85
MgO	0.03	0.19	0.02	0.07
FeO	0.01	0.22	0.08	0.10
Fe_2O_3	0.04	0.37	0.03	0.07
TiO_2	0.05	0.18	trace	0.06
MnO	—	0.17	0.05	trace
P_2O_5	0.03	0.01	0.01	trace
H_2O	8.16	0.10	0.09	0.05
Total	100.26	99.93	100.20	100.03

are given in table 1. The mixture was prepared by intimately mixing and grinding the powdered minerals in acetone. The purity of the specimens as well as the homogeneity of the starting materials was checked by X-ray diffraction analysis; diffractograms of the starting mixtures, M-2 and M-4, are shown in figs. 2 and 3, respectively. No extraneous mineral impurities were detected.

In each experimental pressure-temperature run, 8 to 10 mg of the starting material and some distilled water were sealed in a platinum tubing of 0.0625-inch diameter and of 0.005-inch wall thickness. Each experiment was carried out in a talc pressure-cell assembly with graphite internal furnace. Lead foil of 0.0002-inch thickness was wrapped around the talc–graphite furnace assembly. "Molykote" lubricant was applied around the wall of the pressure chamber to further reduce the friction between the chamber wall and the moving piston.

The duration of the runs made is indicated in tables 2 and 3. Those in which water was retained, were considered successful. The friction correction (i.e., the difference between the pressure indicated by the ram pressure and the actual pressure on the furnace assembly) was estimated from the in-stroke and out-stroke cyclic displacement of the piston versus ram pressure (NEWTON and KENNEDY, 1963). The friction correction was estimated to be 1.5, 1.6, and 1.8 kb at pressures of 4, 5, and 10 kb, respectively. The correction found was approximately of the same order of magnitude as that determined by NEWTON (1966) for a 1-inch pressure

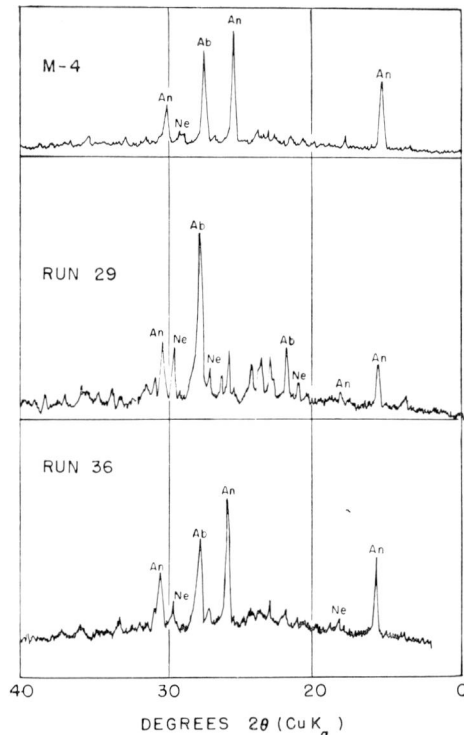

Fig. 3. X-ray diffractograms of the starting stoichiometric mixture M-4 (upper part) and the end products of the two runs. In run no. 29 (4.5 kb, 625 °C), albite and nepheline grew; in run no. 36 (2.5 kb, 600 °C), analcite grew.

assembly. The error in the pressure values reported here are considered to be within ±5% of the reported values.

Identification of the stable phases was made by X-ray diffraction analysis. The peak-heights of the resulting phases were compared with those for the phases in the starting material, and the growth of one phase (say, analcite) at the cost of another (say, jadeite plus water) was indicated by an increase in the peak-height of analcite and a decrease in the peak-height of jadeite, and vice versa (see fig. 2). Fig. 3 similarly shows the X-ray diffraction patterns related to the reaction (4). Where no transformation took place (e.g., run no. 21b, table 2), no change in peak-height was observable. In some runs of reaction (3), nepheline and albite phases were also formed; the significance of this will be discussed later.

4. Results and discussion

Table 2 lists the conditions of the experiments conducted, and the results obtained for reaction (3). Fig. 4,

TABLE 2

Data for the experiments conducted on the system: analcite ⇌ jadeite+vapor. The starting material was a stoichiometric mixture of analcite and jadeite, with an excess of water.

Run no.	Pressure (kb)	Temperature (°C)	Duration (hr)	Results
21b	10	22	12.0	No reaction
20	5	350	5.8	No observable reaction
19	6.4	350	6.0	Analcite grew
48	7	350	8.5	Analcite grew
18	10	350	6.1	Jadeite grew
46	6	400	6.0	Analcite grew
13	6	450	6.0	Analcite grew; dry run
16	7.5	450	6.2	Little analcite growth
14	8	450	12.0	Jadeite grew
4	6	500	6.0	Jadeite grew
25	7	500	6.7	Analcite grew; trace of albite present
3	8	500	5.8	Little analcite growth
17	8.5	500	6.0	Jadeite grew
2	12	500	6.0	Jadeite grew
21	9	535	6.0	Jadeite grew; traces of nepheline and albite also present
22	8	540	5.7	Analcite grew; traces of nepheline and albite also present
26	12	540	5.7	Jadeite grew
45	9	550	6.0	Analcite grew
6	8	575	5.8	Analcite grew
9	9	575	5.5	Analcite grew
12	10	575	6.0	Jadeite grew
5	5	600	5.7	Analcite grew
27b	9	600	5.0	Analcite grew
23	11	600	6.0	Jadeite grew; dry run
44	12	600	5.5	Jadeite grew; nepheline present
1	16	620	5.7	Jadeite grew
8	9	625	9.5	Albite and nepheline (?) grew
7	10	630	6.0	Glass formed
24	11	650	5.5	Glass formed
28	14	650	6.0	Jadeite grew
43a	15	700	6.1	Glass formed
47	12	700	6.2	Glass formed; nepheline and albite present

TABLE 3

Data for the experiments conducted on the system: analcite ⇌ ½ albite+½ nepheline+vapor. The starting material was a stoichiometric mixture of analcite, albite, and nepheline, with an excess of water.

Run no.	Pressure (kb)	Temperature (°C)	Duration (hr)	Results
39	2.0	500	4.0	Analcite grew
37	2.0	500	6.0	Analcite grew
36	2.5	500	4.0	Analcite grew
38	3.0	550	4.0	Analcite grew
42	4.5	570	4.3	Analcite grew
31	2.0	600	4.5	Albite and nepheline grew
30	2.5	600	4.0	Albite and nepheline grew
41	5.5	600	4.0	Analcite grew
29	4.5	625	4.5	Albite and nepheline grew
32	3.5	650	4.5	Albite and nepheline grew
33	2.5	700	4.5	Glass formed
43b*	12.5	700	4.5	Glass formed
34	5.5	750	4.0	Glass formed

* Not plotted on fig. 4.

reaction (3) is estimated as:

$$dP/dT \approx \Delta S/\Delta V \approx 16.6 \text{ b/°C}.$$

This is higher than the above values.

A theoretical curve for reaction (3) was obtained following the method of NEWTON and KENNEDY (1963) in which one takes into account the compressibility effect of vapor phase and ignores the effects of thermal expansion and of compressibility of the solid phases. The resulting expression (NEWTON and KENNEDY, 1963) for reaction (3) is:

$$0 = \Delta V(P-P_0) - \Delta S(T-T_0) - \tfrac{1}{2}(V_0\beta)_{H_2O}(P-P_0)^2 +$$
$$- (\Delta C_p/2T_0)(T-T_0)^2 + (V_0\alpha)_{H_2O}(P-P_0)(T-T_0),$$

a plot of the results, delimits the equilibrium boundary of this reaction. The dP/dT slope between 6.5 kb and 350 °C, and 9.8 kb and 600 °C (invariant point I) is found to be about 10.5 b/°C. This value is slightly higher than the value of 9.5 b/°C in the temperature range of 500–600 °C reported by NEWTON and KENNEDY (1968). The difference between the two slopes, however, is within the experimental error.

Using the thermochemical data for analcite, jadeite and water (table 4), dP/dT at ambient conditions for

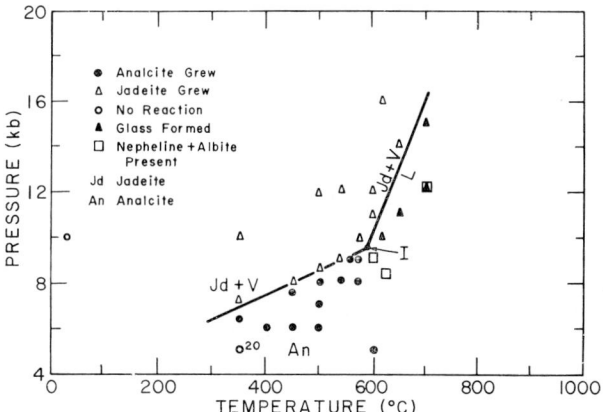

Fig. 4. Experimental results of the reaction analcite ⇌ jadeite+vapor.

TABLE 4

Thermodynamic properties of analcite, jadeite, and water at 1 bar 25 °C

Mineral	Composition	Molar Volume (cm³/mole)	Density (g/cm³)	ΔS (cal/°C·mole)	ΔH^0 (kcal/mole)	ΔF^0 (kcal/mole)	References
Analcite	$NaAlSi_2O_6 \cdot H_2O$	97.5	2.258	56 ± 0.60	-32.75	-34.08	Kelley and King (1961); Barany (1962)
Jadeite	$NaAlSi_2O_6$	60.98	3.315	31.90 ± 0.30	-36.50	-34.62	Kelley and King (1961); Kracek (1953); Barany (1962)
Water	H_2O	18.00	1.00	16.72 ± 0.03			

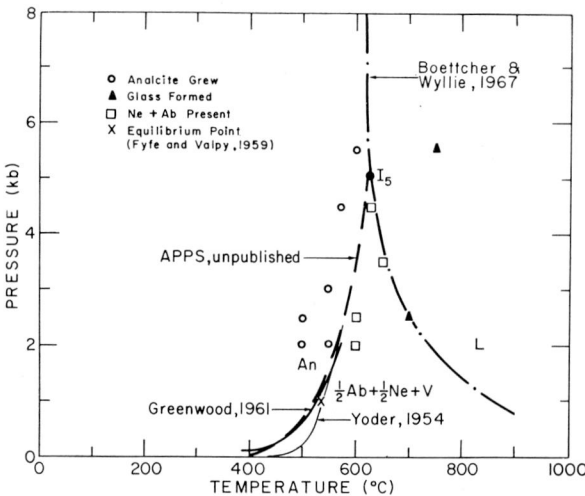

Fig. 5. Experimental results of the reaction, analcite ⇌ ½ nepheline + ½ albite + vapor compared with the theoretical curve (Apps, unpublished). I_5 is the intersection point of the reactions analcite ⇌ ½ albite + ½ nepheline + H_2O and ½ albite + ½ nepheline + H_2O ⇌ liquid determined by Boettcher and Wyllie (1967).

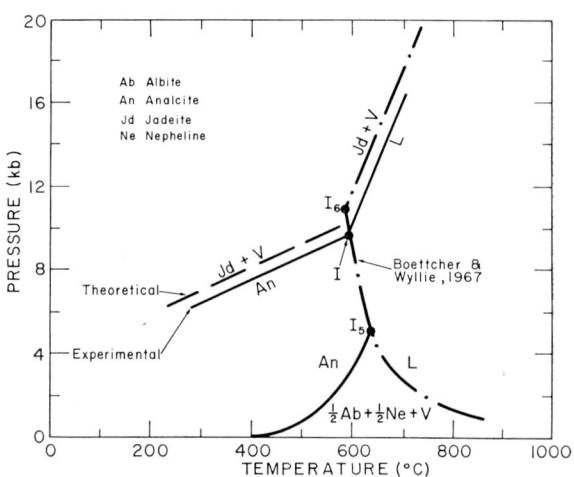

Fig. 6. Summarized diagram of the phase relationship of various fields in the system of total composition of analcite.

where P_0, T_0 is the invariant point, β is compressibility, α is thermal expansion. The invariant point was chosen as 10 kb and 600 °C. In solving the above equation for P and T (i.e., equilibrium conditions), ΔS was assumed constant. The values used were $V_0\beta = 5.6 \times 10^{-4}$ cm³/mole·atm, $V_0\alpha = 1.0 \times 10^{-3}$ cm³/mole·°C and $\Delta C_p/2T_0 = 0.0213$ cal/(°C)²·mole, obtained by extrapolating the specific heat data for analcite and jadeite (Kelley et al., 1953).

The experiments for the reaction (4) are listed in table 3 and the results are plotted in fig. 5 (dashed line). The experimental data fit well with the theoretical curve of Apps (unpublished) and the previously established curves of Yoder (1950) and Greenwood (1961). From fig. 5, it is reasonable to say that the equilibrium curve of reaction (4) intersects the melting curve of nepheline + albite + vapor ⇌ liquid at 5 kb and about 625 °C. This is in good agreement with the point (I_5) determined by Boettcher and Wyllie (1967, 1969). Peters et al. (1966) have reported the invariant point at 4.75 ± 0.25 kb and 665 ± 5 °C in the system.

The phase boundary for the reaction analcite ⇌ jadeite + water is somewhat masked in that nepheline and albite appear in the analcite field (run numbers 8, 21 and 22, table 2, fig. 4). The presence of nepheline and albite may be explained by the fact that the solid solution of analcite, depending on the kinetics, goes through a metastable phase consisting of nepheline solid solution plus albite and vapor (Saha, 1961) during transformation to jadeite. This is further complicated by the melting relationships in silicate–water systems (Kennedy, 1961). Peters et al. (1966) have shown that analcite melts incongruently to albite, nepheline and liquid above the invariant point. They have also analyzed, in detail, various phases and the invariant point in this reaction. BOETTCHER and WYLLIE (1967, 1969) determined the melting relationships in the reactions: albite+nepheline+vapor ⇌ liquid; albite+analcite+vapor ⇌ liquid; and albite+analcite ⇌ jadeite+vapor. They found the invariant point between the first two reactions to lie at 5 kb and 635 °C (point I_5, fig. 2, p. 573, BOETTCHER and WYLLIE, 1967) where analcite becomes stable, and that between the second and the third reactions at 11 kb and 580 °C (point I_6 in the afore-referenced figure) where jadeite becomes stable (see also fig. 6). Their melting curve, jadeite+vapor ⇌ liquid, which has a positive dP/dT slope, passes through I_6. Fig. 6 shows the phase relationship between the reactions (3) and (4). The slopes of the experimental (solid line) and theoretical (dashed line) for reaction (3) are similar. Above I (10 ± 0.3 kb, 600 ± 25 °C) the slope of the experimental curve is found to increase. The results of the present study (e.g. the location of the invariant point I and the melting curve), jadeite+vapor ⇌ liquid, are in good agreement with the results obtained by BOETTCHER and WYLLIE (1967, 1969).

The reaction (3) is masked by the melting relationships in that glass, consisting either of nepheline+albite, or of jadeite, is formed at high temperatures (>600 °C) above 10 kb and that the reaction (3) terminates at I (corresponding point I_6 of BOETTCHER and WYLLIE, 1967). It is interesting to note that the phase boundary of the reaction, nepheline+albite ⇌ jadeite (ROBERTSON et al., 1957), passes just below I_6.

Perhaps the most important findings of the present study are first, that the reversibility of the reaction, analcite ⇌ jadeite+vapor, has been achieved (fig. 4); and second, when both the reactant (analcite) and the products (jadeite+water) are present, this reaction can proceed to the right at pressures and temperatures encountered in the Earth's crust. The previously reported pressures required for the formation of jadeite were higher than reported here. Finally, it should be of interest to experimentally determine the phase relationship of reaction (3) at lower temperatures.

Acknowledgments

The author expresses appreciation to Professor G. C. Kennedy, University of California at Los Angeles, for providing facilities to do the preliminary work in his laboratory. Thanks are due Dr. J. A. Apps for the use of his unpublished data, and Professors R. C. Newton, A. L. Boettcher and H. J. Greenwood for critically reviewing the paper. The author has benefited from his discussion with Dr. H. S. Yoder, JR. on this problem. The help of Mr. M. Newton in the experimental work is gratefully acknowledged. Partial support of this investigation was obtained from NSF Grant GP-5584 and ONR contract nonr 3748(05).

References

APPS, J. A. (1968) Program Ann. Meeting Geol. Soc. Am., 1968, 8.
BARANY, R. (1962) U.S. Bur. Mines, Rept. Invest. No. **599**, 17 pp.
BIRCH, F. and P. LECOMTE (1960) Am. J. Sci. **258**, 209.
BOETTCHER, A. L. and P. J. WYLLIE (1967) Nature **216**, 572.
BOETTCHER, A. L. and P. J. WYLLIE (1969) Am. J. Sci., in press.
BOYD, F. R. and J. L. ENGLAND (1960) J. Geophys. Res. **65**, 741.
FYFE, W. S. and G. W. VALPY (1959) Am. J. Sci. **257**, 316.
GREENWOOD, H. J. (1961) J. Geophys. Res. **66**, 3923.
GRIGGS, D. T. and G. C. KENNEDY (1956) Am. J. Sci. **254**, 722.
GRIGGS, D. T., W. S. FYFE and G. C. KENNEDY (1955) Bull. Geol. Soc. Am. **66**, 1569.
KELLEY, K. K. and E. G. KING (1961) U.S. Bur. Mines Bull. **592**, 149 pp.
KELLEY, K. K., S. S. TODD, R. L. ORR, E. G. KING and K. R. BONNICKSON (1953) U.S. Bur. Mines Rept. of Invest. No. **4955**, 21 pp.
KENNEDY, G. C. (1961) in: Landsberg and Van Mieghem, eds., Advances in Geophysics 7 (Academic Press) 303.
KRACEK, F. C. (1953) Ann. Rept. Dir. Geophys. Lab. Carnegie Inst. Wash. **52**, 69.
NEWTON, M. S. and G. C. KENNEDY (1968) Am. J. Sci. **266**, 728.
NEWTON, R. C. (1966) Am. J. Sci. **264**, 204.
NEWTON, R. C. and G. C. KENNEDY (1963) J. Geophys. Res. **68**, 2967.
PETERS, TJ., W. C. LUTH and O. F. TUTTLE (1966) Am. Mineralogist **51**, 736.
ROBERTSON, E. C., F. BIRCH and G. J. F. MACDONALD (1957) Am. J. Sci. **255**, 115.
SAHA, P. (1959) Am. Mineralogist **44**, 300.
SAHA, P. (1961) Am. Mineralogist **46**, 859.
SHARP, W. E. (1962) University of California, Lawrence Radiation Laboratory Report, UCRL-7118, 51 pp.
YODER, H. S., JR. (1950) Am. J. Sci. **248**, 225, 312.
YODER, H. S., JR. (1954) Analcite, Carnegie Inst. Washington Year Book **53**, p. 121.

ECLOGITES FROM METAMORPHIC COMPLEXES OF THE USSR

N. L. DOBRETSOV and N. V. SOBOLEV

Institute of Geology and Geophysics,
Siberian Branch of the USSR Academy of Sciences, Novosibirsk, USSR

The review investigation of eclogites and associated rocks under complex geological environment should provide new evidence for solving the problem of eclogite origin. As an example of extremely varied and complex geological environment and rock compositions we may consider eclogites from the Precambrian metamorphic terranes associated with different facies of metamorphism over the area of the USSR.

In considering eclogites from the metamorphic complexes, several authors classify them into two large groups with respect to the specific composition of the garnets and pyroxenes, e.g.: eclogites from gneisses and those associated with glaucophane schists (COLEMAN et al., 1965; SMULIKOWSKI, 1964). As an example of eclogites from gneisses one usually takes the eclogites from Norway, Münchberg massif, Eastern Sudeten region, Scotland etc., whereas as reference eclogites of the second group one usually takes the eclogites from California.

In the present paper we consider the peculiarities of the geological environment and the composition of eclogites from the metamorphic complexes of the Urals, Kazakhstan and Tien-Shan areas which have many similar features. In considering these eclogites, we distinguish among them eclogites from gneiss terranes, eclogites from the average-temperature schists and those coexisting with glaucophane schists. The eclogite terranes are separated into three independent metamorphic formations (DOBRETSOV et al., 1969): eclogite-kyanite gneiss, eclogite-kyanite schists and eclogite-glaucophane schists.

All the examples of eclogites occurring in the USSR which will be discussed below have been studied by the authors as well as by a number of other investigators.

The most typical example of widespread eclogites and coexisting rocks is the zone of metamorphic rocks extending along the main Urals fault. These metamorphic assemblages with eclogites constitute the marginal part of the metamorphic basement of the Urals eugeosyncline. The eclogites are known here as well as in gneiss and schist complexes in numerous sites of the Polar, Northern and Middle Urals area (Marun-Keu range, Pai-Er range, Ufalei massif) and also in the Southern Urals (Maksyutov complex) and in Mugodjar complex as well. A number of assemblages is scattered within the Kokchetav-Ulutau-Makbal-Aktyuz massifs, which are the marginal part of the metamorphic basement of the Riphean Lower Paleozoic geosyncline (DOBRETSOV et al., 1966). These geographically remote metamorphic rocks reveal some definite relations and genetic affinities. Besides the above mentioned eclogites discussed in the present paper, similar formations in the metamorphic assemblages over the USSR area are known in the gneiss terranes of the Kola Peninsula.

As typical well-studied examples of eclogites in the gneiss terranes of eclogite-kyanite-gneiss formations we shall treat the metamorphic complex of Kokchetav. The structure of this complex and the distribution of eclogites in it is shown in fig. 1. The eclogitic formation relating to the lower part of the section is possibly of Archean or Proterozoic age, being composed by kyanite-biotite gneisses and schists, and by massifs of later granitic gneisses with eclogite bodies (EFIMOV, 1964).

Eclogites compose both single, large isometric bodies 2.5 km long and chains of extending lenses of the boudinage type. In the granite gneisses and the sites of intense diaphtoresis there are also observed irregular "swarm-like" bodies. Their formation is connected with diaphtoresis and "dissolving" of the eclogite bodies.

Apart from typical eclogites, the Enbek-Berlyk site

Fig. 1. Geological sketch map of the Kokchetav massif eclogite-gneiss formation. Eclogite formation with eclogite bodies (1); granitegneisses (2); "overeclogitic" formation (3); quartzite schists (4); Pz granites (5); acidic dikes with almandine (6); Cenozoic sediments (7); faults (8).

is typical of the series of eclogitized rocks with transformations from weakly eclogitized gabbro-diabases with druse-like structures to typical eclogites (DOBRETSOV and SOBOLEV, 1969). Near lake Kumdy-Kul there are present specific pyrope-bearing rocks with titan-bearing olivine (EFIMOV, 1961) occurring as inclusions in the large eclogite body. The garnet of these rocks is pyrope with increased $FeO/(FeO+MgO)$ ratio (31%) and is distinguished from typical pyrope peridotites from kimberlite pipes and the Chekh massif (see table 2). Of wide distribution here are different types of diaphtorites after eclogites, such as garnet amphibolites and amphibolites, garnet-jedrite rocks, as well as different products of lixiviation of eclogites (quartz-kyanite-almandine and other rocks). They are developed both as rims around the eclogites and as entirely replacing the latter.

The age of the eclogites is determined by the isochron method as being 1.5 to 2.0×10^9 yr old, by the Rb–Sr method $900–1300 \times 10^6$ yr, and by the K–Ar method $550–750 \times 10^6$ yr, both for eclogites and for the country rocks.

An exact counterpart of the Kokchetav complex is the Aktyuz metamorphic complex with eclogites. The analyses of garnet and clinopyroxene from these eclogites are listed in table 2.

In the Polar Urals in the Marun-Keu range, composed of the Precambrian and ortho- and para-gneisses alternating with basic rocks and mica schists, eclogites are represented as separate small bodies with frequently gradual transformation of carinthine and kyanite-zoisite eclogites to plagioclase peridotites and olivine gabbronorites (UDOVKINA, 1964). Here are known garnetized peridotites as well as rocks with the reaction development of garnet-pyroxene assemblage due to hypersthene and plagioclase, just as in the Kokchetav massif. The eclogites are distributed over the Middle Urals area within the Ufalei metamorphic complex (VINOGRADSKAYA, 1964). They can be encountered along with rutile amphibolites both in the lower structural stage rocks of this complex composed of gneisses and amphibolites of Lower Proterozoic age, and in the new rocks of the upper structural stage composed of Ordovician metamorphic schists.

Examples of eclogite-kyanite schists formations are the rocks from the Makbal dome of the Kirgiz range (Tien-Shan range). In fig. 2 is shown the northern part of this dome. Here occur the lenses of garnet-chlorite schists which were likely to form as a result of diaphtoresis of the large bodies of eclogites extending to 2–3 km. The country rocks of these are quartzites, marbles, graphitized schists and some other schists of apparent sedimentary origin. In the lower part of the

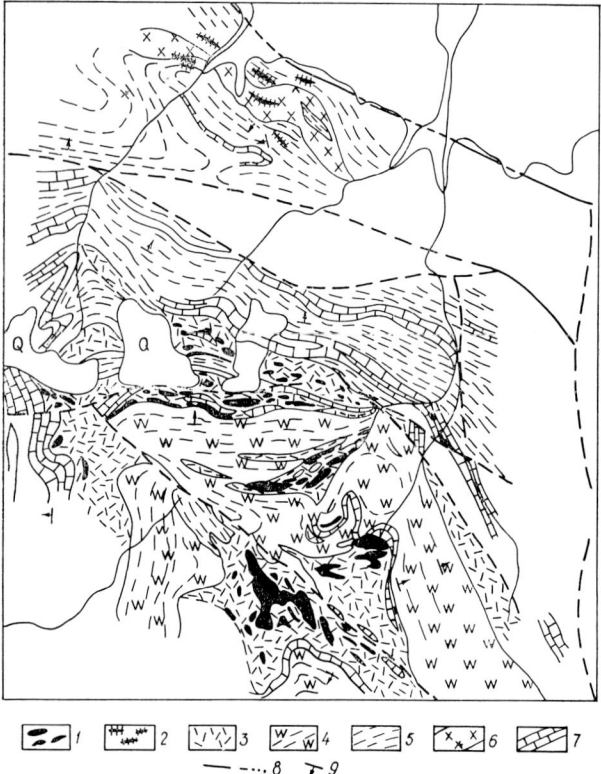

Fig. 2. Geological sketch map of the northern part of the Makbal dome (Kirgizsk mountain range), showing the location of eclogite-schists formation. Eclogites (1); diabase dikes (2); apoeclogitic (?) garnetiferous schists (3); quartzites (4); carbonaceous and other schists (5); gabbro-amphibolites (6); marbles (7); faults (8).

section there occur characteristic minerals such as staurolite and kyanite, while in the upper part one encounters assemblages of the green schist facies. Eclogite bodies are lacking here, but instead there are greenstone altered metagabbroids similar to the lenses of apoeclogite rocks as to shape and size. In several sites there occur intersecting contacts of eclogites, in particular apophyses and the dikes. The eclogites here are almost entirely transformed to diaphtorites, the most typical of which are garnet-amphibole-zoisite rocks, rarely garnet-glaucophane rocks and different products of lixiviation such as quartz-almandine rocks, etc.

An example of eclogite-glaucophane schist formation is the very striking and peculiar Maksyutov complex of rocks in the Southern Urals. The structure of the Maksyutov complex and the distribution of eclogites in it is shown in fig. 3. This complex of metamorphic rocks is confined to a sutural anticlinorium of the

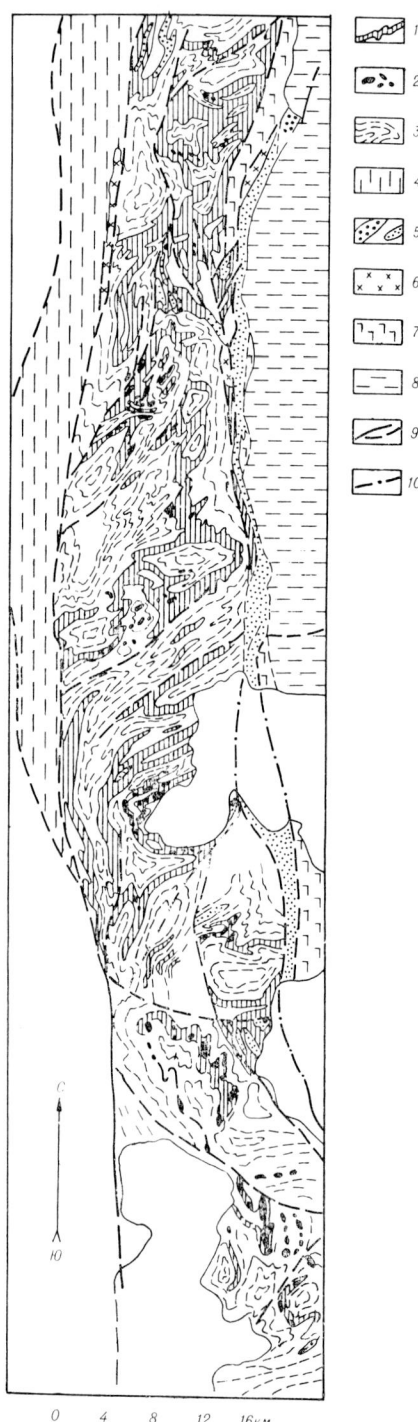

Fig. 3. Geological sketch map showing the location of eclogite-glaucophane formation of the Maksyutov complex of the Southern Urals. Maksyutov complex-Rt_{1-2} (1–3): rock bands with metabasites predominance (1), eclogites and apoeclogite bodes (2), other metamorphic schists (3); Suvanyak complex-Pt_3 (4); serpentinites (5); Magnitogorsk synclinorium-Pz (6–8): gabbro (6), effusives (7), sediments (8); faults under Cenozoic sediments (9).

Urals-Tau range, here being most ancient. From the east this complex is separated by a hyperbasite belt from Magnitogorsk sinclinorium composed of Ordovician, Silurian and Devonian rocks. Westward there extends the Suvanyaksky complex of weakly metamorphosed (greenschist facies), initially sedimentary rocks overlying and perhaps stratigraphically above the Maksyutov complex of the Upper Riphean age.

Among the metamorphic rocks of the Maksyutov complex there are rock bands with predominantly greenstone rocks associated with eclogites, which are confined to eclogites and apoeclogitic rocks. All the other rocks of this complex are composed of quartzites and metapelite (mica, graphite and other schists) rocks which contain glaucophane, jadeite pyroxene, lawsonite, garnet, and rarely kyanite. These schists are undoubtedly of initially sedimentary origin. The peculiarity of these rocks is the presence of complex reversed folds.

The eclogite distribution shown in fig. 3 shows their confinement to the central part of the Maksyutov complex band and their interrelation with the greenstone rock horizons. Such a localization of eclogites and higher temperature schists in the centre of the zone is related mainly to the zonal occurrence of later greenschist diaphtoresis. However, it is not impossible that primary metamorphic zonality takes place.

The shape of the eclogite bodies and their interrelation with the rocks of the Maksyutov complex are shown in the Sakmara river section (fig. 4). Here we have massive eclogite bodies up to 100 m thick, as well as rock bands saturated with small relic lenses of eclogites, resulting probably from diaphtoresis of similar eclogite bodies. These apoeclogite rocks and eclogites alternate with common graphite-bearing rocks, schists and quartzites. Inside the eclogite bodies and apoeclogitic rocks there occur enstatite rocks of radiated texture. The presence of enstatite rocks as well as variations of the body composition is likely to provide some evidence on their primary magmatic inhomogeneity.

Formations analogous to the Maksyutov complex occur in the Charsk ultrabasic belt (Eastern Kazakhstan), as blocks in the ultrabasic rocks, as well as in several small outcrops in the Southern Tien-Shan range.

Apart from the lenses of enstatite rocks in the Maksyutov complex there occur serpentinites which are associated with specific rocks rich of Ca, containing almandine-grossularite garnet and lawsonite as well as albite-jadeite and quartz-almandine-jadeite rocks.

In several regions studied, two eclogite-bearing formations occur simultaneously. Thus, in the Marun-Keu range, eclogite-kyanite-gneiss formations alter to garnet-glaucophane-schists in the northern part. In the Ufalei massif, eclogites are contained in the gneiss complex of the basement and in the schists overlying this complex.

By their chemical composition all the eclogites studied are very similar. This is demonstrated in table 1,

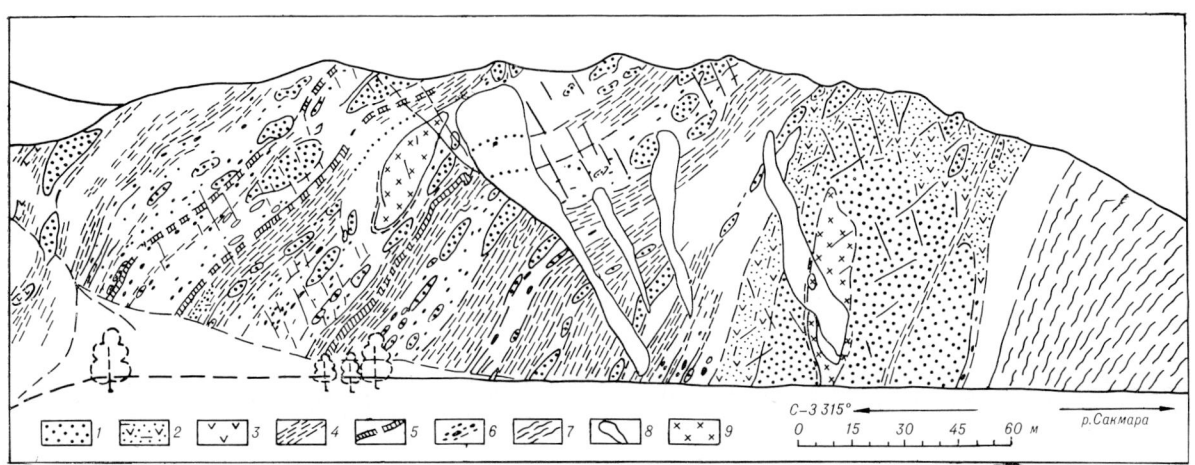

Fig. 4. The Maksyutov complexes rock section on the Sakmara river in the South Urals. Eclogites (1); apoeclogitic rocks (2); greenstones (3); metasedimentary rocks often with graphite (4); quartzites (5); apoeclogitic schists with small eclogite bodies (6); mica-quartzite schists of Yumaguzin rock series (7); Cenozoic (8); enstatite rocks (9); р. Саκмara is the Sakmara river.

TABLE 1

Chemical composition of eclogites and related rocks from different metamorphic complexes of the USSR

	n		SiO_2	TiO_2	Al_2O_3	Fe_2O_3	FeO	MnO	MgO	CaO	Na_2O	K_2O
Tholeite basalts of oceanic olivine basalt formation (Kutolin, 1968)	110	$\bar{x}$	49.15	2.09	15.09	3.35	7.56	0.17	7.75	10.61	2.23	0.30
		s	1.33	0.52	1.04	1.36	1.28	0.02	1.23	0.70	0.40	0.15
Eclogites of the Kokchetav complex (Efimov, 1964)	6	$\bar{x}$	48.00	2.42	13.28	2.33	13.30	0.22	5.32	11.49	2.08	0.31
Eclogites of the Makbal complex (Efimov, 1964)	4	$\bar{x}$	48.53	2.94	14.69	3.75	11.91	0.19	6.03	7.95	2.54	0.80
Eclogites of the Maksyutov complex	18	$\bar{x}$	48.80	2.27	15.17	3.57	8.53	0.16	6.55	9.59	3.34	0.60
"Eclogitic schist", Maksyutov complex (Chesnokov, 1963)	1	$\bar{x}$	61.40	0.85	16.56	4.85	5.49	0.17	1.04	0.82	8.33	0.04
Quartz-garnet-jadeite rock, Maksyutov complex (Lennykh, 1968)	1	$\bar{x}$	71.90	0.57	11.80	1.94	2.76	0.06	1.41	2.45	7.04	0.18

n – number of analyses, $\bar{x}$ – arithmetical mean, s – standard deviation.

in which are listed the mean values of eclogite compositions from different metamorphic complexes of the USSR. Eclogite compositions approach the mean composition of several basalts, in particular tholeitic ones (table 1).

Apart from eclogites in the series of metamorphic complexes studied here, in particular in the Maksyutov complex, there are known specific rocks intercalated with eclogite rocks. First, there are garnet-bearing rocks, rich of Ca, with garnet of almandine-grossularite composition, clinopyroxene, lawsonite, muscovite, albite and zoisite. The garnet from these rocks contains 55–70% of grossularite ($n = 1.775$–1.780; $a_0 = 11.675$–11.720 Å).

In the regions studied, different ultrabasites are associated with eclogites as inclusions in the eclogite (see above) though typical pyrope peridotites are not known here.

The very rare so-called "eclogite schists" (Chesnokov, 1963) and quartz-garnet-jadeite rocks (Lennykh, 1968) similar to those described in Columbia (Green et al., 1968) are of stimulating interest in the Maksyutov complex. By their chemical composition they differ highly from eclogites, chiefly in the sharp increase in SiO_2 and Na_2O content (table 1). The garnet of such rocks is almandine with a high Ca-content and the clinopyroxene is almost pure jadeite (table 2). "Eclogite schists", judging by their chemical composition (table 1), also contain jadeite pyroxene.

Thus, among the rocks of eclogite type studied above one can distinguish:
1) Eclogite proper, of gabbro composition.
2) Eclogitized metagabbroids.
3) Garnet-pyroxene rocks rich of Ca.
4) Ultrabasites associated with eclogites.
5) Quartz-almandine-jadeite rocks of acidic composition.

The points corresponding to the f-value (FeO/(FeO + MgO) ratios) of garnets and pyroxenes of the eclogites studied, as well as of eclogites from several metamorphic complexes, are plotted on the diagram of the f-value distribution (fig. 5). From the plot one can see that all the garnets, except those from eclogites of the Polar Urals, fall into the region of more increased f-values, from 64% to 90%. As cited in the literature, the f-value of the eclogite garnets roughly corresponds to their type: decreased for eclogites from gneisses and increased for eclogites from glaucophane schists. For the former the

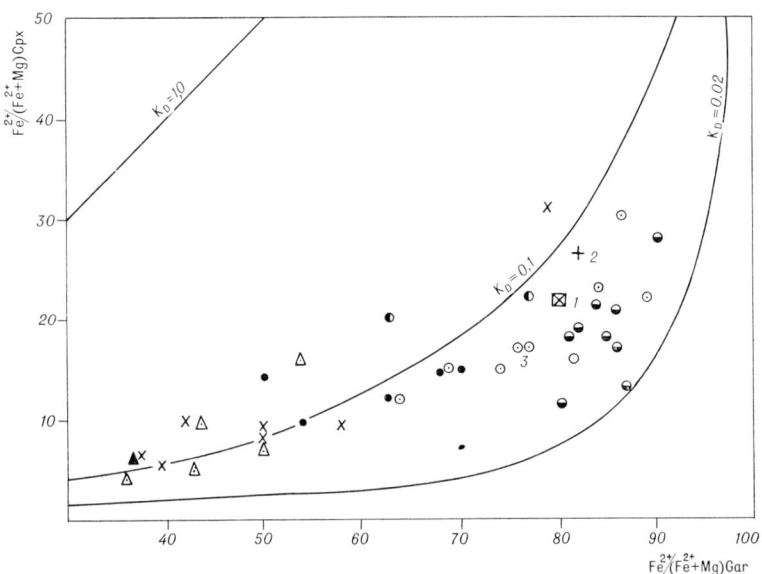

Fig. 5. The distribution diagram of the FeO/FeO+MgO ratios for garnet-clinopyroxene pairs from eclogites of different metamorphic complexes. From Norwegian eclogites – crosses; from Sudetes eclogites – triangles; from Polar Urals eclogites – solid circles; from the Maksyutov complex – open circles; from the Kokchetav complex – half open circles; from California – half open circles; from Aktyuz – crossed square marked by 1; the plot of some Na-rich pyroxenes from the Maksyutov complex are marked by 3.

variability range of the f-value is 35–75%, for the latter it is 75–90% (COLEMAN et al., 1965; SMULIKOWSKY, 1965).

However, the garnets from eclogites from the Maksyutov complex show a wider range of composition with respect to the f-value; they fall in part into the field of composition corresponding to garnets from gneisses.

As for pyroxenes, their composition varies widely in the eclogites from the Maksyutov complex. Here one has established both omphacitic and chloremelanite pyroxenes with 6.5% Na_2O (see table 2) and typical diopside–jadeite pyroxenes having 9.0% Na_2O.

By considering the distribution diagram of the f-value for garnet-pyroxene pairs, we observe that K_D of this ratio is lower for the pairs from the Maksyutov complex (open circles in the diagram) than those for eclogites from the gneiss complexes.

This may in part be connected with the conditions of their formation, i.e. K_D can be affected both by temperature and pressure. However, one can also assume the effect of Na concentration which is noticed also for corresponding data concerning eclogites from the kimberlite pipes. Re-estimation of Fe^{3+} in pyroxenes rich of Fe_2O_3 is also probable for the chloremelanite pyroxenes, whose points generally lie on diagrams lower than those of some other pyroxenes.

In the eclogite problem, in our opinion, two main questions can be singled out:

1) Finding the eclogites of similar composition in very differing metamorphic complexes.

2) General diaphtoresis of eclogites and consistency between the diaphtoresis and the degree of metamorphism of country rocks.

Along with the earlier assumptions on the regional distribution of eclogite facies (TRUSOVA, 1956; MEDVEDEVA, 1960) and on the eclogites of the metamorphic complexes as being extruded from the upper mantle (EFIMOV, 1964), evidence has recently been obtained from the interpretation of experimental data (GREEN and RINGWOOD, 1966) on the stability of eclogites.

We cannot agree either with the first or with the second assumption. We should first stress the magmatic origin of the majority of the eclogites studied. Then in some cases one can assume a direct crystallization of eclogites from the melt and in others a gabbro crystallization followed by an eclogitization, altered by intense diaphtoresis.

In fig. 6 are shown conditions of metamorphism of rocks and their correlation with the conditions of eclogite formation. The position of the $P-T$ lines on the diagram corresponds to earlier data published by us for the scheme of facies (SOBOLEV et al., 1966, 1967), in which account is taken of the experimental data and an estimate of P_{H_2O} was given for different facies (SOBOLEV et al., 1966). For the low temperature field of the diagram, P_{H_2O} is roughly $0.8-0.9 \times P_{total}$, for high temperatures approximately $0.2-0.3 \times P_{total}$. The points stand for the field characterizing the $P-T$ conditions under metamorphism of eclogite-bearing terranes. As seen from fig. 6, this region is situated between three extreme types of metamorphism, corresponding to: 1) Eclogite-glaucophane complexes of the California and Southern Urals type (the lowest temperature and the highest pressure); 2) Eclogite-kyanite schists complexes of the Makbal type; 3) High temperature gneiss complexes of the Norway type. All the remaining metamorphic complexes with eclogites take intermediate positions.

As to the eclogites, we have some reasons to anticipate that eclogites from all terranes studied were initially more high temperature formations. This is supported by their intense diaphtoresis and the absence of weak manifestation of it in the country rocks. Taking into account all the above characteristics of the magmatic origin of eclogites, one can assume that eclogites are specific magmatic formation intruded and crystallized at the moment of metamorphism of country rocks, under increased pressures. Relatively high temperature and the rarity of volatiles in eclogites is determined by the deep nature of intruded magma. Such an origin and

TABLE 2

Chemical analyses of garnets and clinopyroxenes from eclogites and related rocks

	Garnets					Clinopyroxenes			
	K-84	534	K61-1	85/2	516-2	534	K61-1	85/2	516-2
SiO_2	40.99	38.08	40.04	38.11	41.26	53.32	56.42	55.37	57.80
TiO_2	0.19	0.60	0.50	0.72	0.38	1.12	0.40	0.42	0.67
Al_2O_3	22.44	20.72	21.22	20.77	19.92	8.69	11.84	11.38	19.77
Cr_2O_3	0.01	0.10	—	0.01	—	0.19	—	0.05	—
Fe_2O_3	1.41	2.25	2.45	0.80	2.00	7.27	0.88	2.19	2.11
FeO	12.78	25.13	20.45	25.58	24.38	3.95	2.15	2.85	1.48
MnO	0.12	0.37	0.22	0.76	0.77	0.05	tr.	0.01	—
MgO	16.16	3.61	6.90	1.83	2.22	7.82	8.50	7.51	2.02
CaO	5.59	9.31	8.06	11.21	8.17	11.73	12.79	13.14	3.33
Na_2O	—	—	—	—	—	5.36	6.82	6.85	12.72
K_2O	—	—	—	—	—	0.19	0.12	0.11	—
loss ign.	—	—	—	—	—	0.60	0.34	—	0.20
Total	99.69	100.29	99.94	99.79	99.10	100.19	100.26	99.85	100.20
Si	2.991	2.962	2.972	2.957	2.974	1.940	1.988	1.974	1.990
Ti	0.013	0.038	0.028	0.043	0.026	0.028	0.011	0.011	0.017
Al^{iv}	1.930	1.927	1.958	1.974	1.992	0.060	0.012	0.026	0.010
Al^{vi}	—	—	—	—	—	0.312	0.479	0.453	0.793
Cr	—	0.009	—	—	—	0.004	—	—	—
Fe^{3+}	0.070	0.064	0.042	0.026	0.008	0.201	0.025	0.060	0.054
Fe^{3+}	0.009	0.069	0.109	0.022	0.124	—	—	—	—
Fe^{2+}	0.781	1.661	1.341	1.722	1.736	0.120	0.064	0.086	0.046
Mn	0.004	0.024	0.014	0.053	0.056	0.002	—	—	—
Mg	1.754	0.422	0.805	0.223	0.281	0.422	0.447	0.398	0.103
Ca	0.439	0.788	0.678	0.968	0.740	0.457	0.483	0.501	0.124
Na	—	—	—	—	—	0.381	0.466	0.472	0.848
K	—	—	—	—	—	0.009	0.004	0.004	—
Exc. Si	—	0.047	0.162	0.110	0.534^+	—	—	—	—
f (FeO/(FeO+MgO))	31	80	64	89	87	22	13	18	31

+ quartz admixture.

Samples analyzed: K-84 – from pyrope-titan-olivine rock, Kokchetav massif, anal. by I. K. Kuznetsova; 534 – from eclogites of the Aktyuz complex, anal. by I. K. Kuznetsova; K61-1 – from eclogites of the Maksyutov compl., (after LENNYKH, 1968); 85/2 – ibid., anal. I. K. Kuznetsova; 516-2 – from alm.-quartz-jadeite rock, Maksyutov complex (after LENNYKH, 1968).

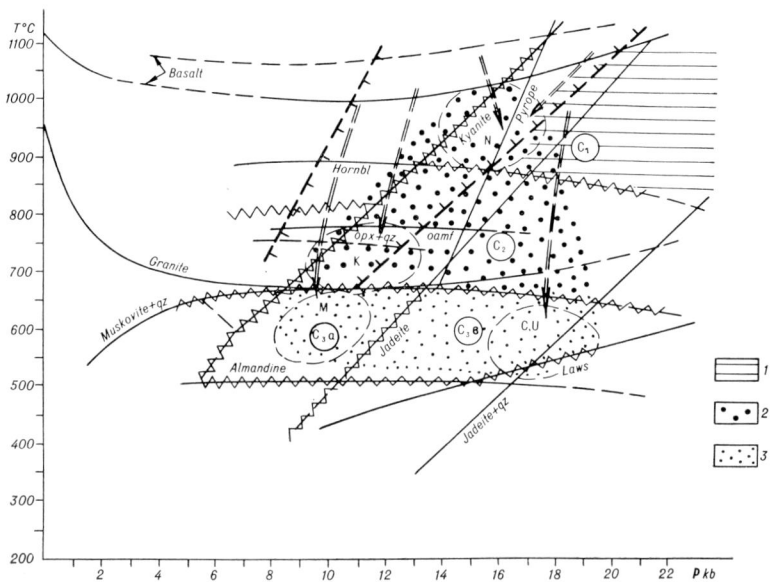

Fig. 6. *P–T* scheme of metamorphic facies. The field of eclogites origin (1); *P–T*-conditions of metamorphism of eclogite gneiss formation (2): of Norway type (N-), of Kasakhstan type (K); *P–T*-conditions of metamorphism of eclogite schist formations (3): of Makbal type (M), of California and South Urals type (C, U).

the successive diaphtoresis of eclogites in their cooling from magmatic temperatures to those of the enclosing rocks is schematically shown by the arrows in the diagram of fig. 6. It is clear that in some cases they were crystallized just as eclogites; in others they were crystallized as gabbro rocks or eclogite-like rocks.

Due to the sloping nature of the transformation field of the gabbro-eclogite (GREEN and RINGWOOD, 1966), they could first have been crystallized as gabbro and then become eclogitized.

From the regions studied we have drawn the conclusion that there are indications of two ways in which eclogites were formed, though they may hardly be distinguished in most eclogites. This origin can explain the fact that eclogitization itself is of an evident regressive nature.

References

CHESNOKOV, B. V. (1963) Eclogites of Southern Urals and their practical significance, in: *Magmatism, metamorphism and metallogeny of the Urals*, 257–263 (in Russian).

COLEMAN, R. G., D. E. LEE, L. B. BEATTY and W. W. BRANNOCK (1965) Eclogites and eclogites: their differences and similarities, Bull. Geol. Soc. Am. **76**, 483–508.

DOBRETSOV, N. L., V. V. REVERDATTO, V. S. SOBOLEV, N. V. SOBOLEV, E. N. USHAKOVA and V. V. KHLESTOV (1966) *Facies of regional metamorphism of the USSR* (Nauka) (in Russian).

DOBRETSOV, N. L. and N. V. SOBOLEV (1969) Eclogites in metamorphic complexes of Kazakhstan, Tien-Shan, Southern Urals and their genesis, in Sobolev Vol., part 2 (in Russian).

DOBRETSOV, N. L., V. S. SOBOLEV and V. V. KHLESTOV (1969) Principles of subdivision of regional metamorphic formations, Geol. i Geofiz. Akad. Nauk SSSR Sibirsk. Otd. **3**, 3–16 (in Russian).

EFIMOV, I. A. (1961) On the find of pyrope serpentinites in Precambrian deposits of Kokchetav massif, Trans. KazIMS **5** (in Russian).

EFIMOV, I. A. (1964) Eclogite formation of the Precambrian of Northern and Southern Kazakhstan, in: *Petrographical formations and petrogenesis problems* (in Russian).

ESKOLA, P. (1921) On the eclogites of Norway, Vid. Skr. 1, Mat. Nat. Kl. N8.

KUTOLIN, V. A. (1968) Petrochemical peculiarities of various formational types of basalts and the composition of the Earth's upper mantle, Repts. Sov. Geol. XXIII IGC, Probl. 1 (in Russian).

LENNYKH, V. I. (1968) Regional Metamorphism of Precambrian terranes of the western slope of the Southern Urals and Ural-Tau range, Ufa. p. 67 (in Russian).

MEDVEDEVA, I. E. (1960) The origin of eclogites of Makbal uplift (Nothern Tien-Shan), Proc. of High school, Geol. Prosp. N 11 (in Russian).

SOBOLEV, V. S., N. L. DOBRETSOV and V. V. KHLESTOV (1966) Regime of H_2O and CO_2 by the progressive regional metamorphism, Dokl. Akad. Nauk SSSR **166**, N2 (in Russian).

SOBOLEV, V. S., N. L. DOBRETSOV, V. V. REVERDATTO, N. V. SOBOLEV, E. N. USHAKOVA and V. V. KHLESTOV (1967) Metamorphic facies and series of facies in the USSR, Medd. Dansk Geol. Foren. **17** (4).

SMULIKOWSKI, K. (1964) An attempt of eclogite classification, Bull. Acad. Pol. Sci., Ser. Sci. Geol. Geogr. **12**, N1.

TRUSOVA, I. F. (1957) Paragenetic analysis of crystalline schists of the Lower Archean of Kokchetav massif, Soviet Geol. **51** (in Russian).

UDOVKINA, N. G. (1964) Eclogites and eclogitization processes in Polar Urals region (exampliﬁed by Marun-Keu range), in: *Physicochemical conditions of magmatism and metasomatosis* (Nauka) (in Russian).

VINOGRADSKAYA, G. M. (1964) Petrology of granitoids of Ufalei region in Southern Urals, Trans VSEGEI (new series), 162–214 (in Russian).

A CRUSTAL-UPPER MANTLE MODEL FOR THE COLORADO PLATEAU BASED ON OBSERVATIONS OF CRYSTALLINE ROCK FRAGMENTS IN A KIMBERLITE DIKE

THOMAS R. MCGETCHIN and LEON T. SILVER

Air Force Institute of Technology Wright-Patterson Air Force Base, Ohio, U.S.A.
California Institute of Technology, Pasadena, California, U.S.A.

Electron microprobe investigations of minerals from kimberlite and dense crystalline rock fragments in the Moses Rock Dike, San Juan County, Utah, suggest equilibration at upper mantle pressures. Crystalline fragments constitute 3% of the breccia within the dike; only 300 ppm are dense types of possible mantle origin; 97% are fragments derived from near-surface sedimentary strata.

Field studies of fragment size of the sedimentary clasts show a decrease in fragment size with distance of upward transport from original locations in walls. Applied to the crystalline fragment populations, this provides an empirical basis for reconstruction of the vertical stratigraphy of crystalline rocks in the crust and upper mantle.

Inferred on the basis of size, abundance, and petrographic character: metabasalt, granite, and granite gneiss are abundant in the upper part of the crust, along the dike walls; diorite, gabbro, mafic amphibolite constitute intermediate crustal layers, and mafic granulite and possibly hydrated ultramafic rocks, the lower crust. The suite of crustal rocks is predominantly metavolcanic or metaplutonic, not metasedimentary.

Dense and ultramafic fragments (possibly mantle rocks) include antigorite-tremolite schist, jadeite-rich clinopyroxenite, eclogite, spinel-websterite, and spinel-lherzolite. The presence of garnet lherzolite is inferred from mineral inclusions observed in pyropic garnets.

The Mohorovicic discontinuity apparently occurs within a petrologically complex region and may coincide with phase *and* compositional transitions, including hydration. A compositional transition within the mantle between spinel and garnet peridotite (lherzolite) is consistent with observations. The variety and abundance of ultramafic and dense types, together with the complexity of their textures suggests the mantle may be as complicated as the crust in composition and history.

Observations suggest that the upper mantle may contain significant volatile material (water and/or CO_2) at modeset temperatures (950 °C at 150 km).

TECTONOPHYSICS

MANTLE BENEATH THE JAPANESE ARC

HIROO KANAMORI*

Earthquake Research Institute, University of Tokyo, Japan

Seismic velocity structure of the mantle beneath Japan has been determined on the basis of teleseismic explosion, long-period surface wave, and $d\Delta/dt$ data. A marked difference in structure is found across the deep seismic plane which dips from the Pacific Ocean side towards the continent. The mantle on the continental side of the seismic plane has such low velocity and Q (≈ 80) as to require a partial melting in addition to a possible compositional change. The low velocity seems to be associated with the underlying earthquake activity. It is found that the lower bound of the low velocity zone is relatively abrupt; this also provides a favorable evidence for the partial melting. A clear later phase branch observed at $15° < \Delta < 20°$ suggests a rapid velocity increase at depths 375 to 400 km. This rapid velocity increase can be attributed to the onset of the olivine–spinel phase change of $(MgFe)_2SiO_4$, if the temperature at 400 km depth is about 1300 °C. This low temperature may call for a hydrous state in the upper mantle in order to cause the partial melting. The earthquake energy release has a notable maximum around 400 km depth in Japan region. The coincidence of this depth with that of the rapid velocity increase suggests that the olivine–spinel phase change is in some way related to the generation of deep focus earthquakes.

1. Introduction

A number of studies have shown that the seismic velocity in the mantle beneath Japan is considerably low and that a large lateral variation of seismic velocity and attenuation exists (see e.g. JEFFREYS, 1966; ANONYMOUS, 1966; AKI, 1961; KATSUMATA, 1960; HISAMOTO, 1965; and SAITO and TAKEUCHI, 1966). Recent studies by UTSU (1966, 1967) pointed out some of the important features of the mantle beneath Japan; he showed that the seismic velocity and Q are significantly lower in the mantle on the continental side of the deep seismic zone than in the mantle on the ocean side. This paper attempts to construct more detailed and quantitative mantle models for the Japanese arc based upon accurate teleseismic explosion, surface wave, and $d\Delta/dt$ data.

2. Teleseismic explosion data

Four explosions, Longshot in 1965, and three Nevada explosions in 1968, provided good opportunities for delineating the gross characteristic of mantle structure beneath the Japanese arc. Fig. 1 shows the distribution of high-sensitivity seismograph stations. Also shown

* Now visiting at Department of Earth and Planetary Sciences, Massachusetts Institute of Technology, Cambridge, Mass. 02139.

are the contours of the deep seismic plane in the Japanese region. The arrow indicates the direction of incidence of the wave paths from the Longshot (Aleutian Is.) and Nevada. We see that the direction is almost normal to the strike of the southern portion of seismic plane where the deep earthquake activity is highest. The cross section is schematically shown in the insert. The path crosses the seismic plane, travels through the mantle, and arrives at a station. Readings from short-period vertical component seismograms are used for the analysis. The seismograms for one of the Nevada explosions are reproduced in fig. 2. The onsets are clear and the readings are probably accurate to ±0.2 sec. The detailed data will be published elsewhere. The travel-time residuals from the Jeffreys–Bullen table were calculated. It was found that the eastern stations registered early arrivals while the western stations registered late arrivals. This observation can be interpreted in terms of a structural anomaly associated with the deep seismic plane. In order to see this we take the length of the path above the deep seismic plane and denote it by L (see fig. 1). The value of L is small at the eastern stations and large at the western stations. Fig. 3 shows the Longshot travel-time residual from the Jeffreys–Bullen time at each station plotted against

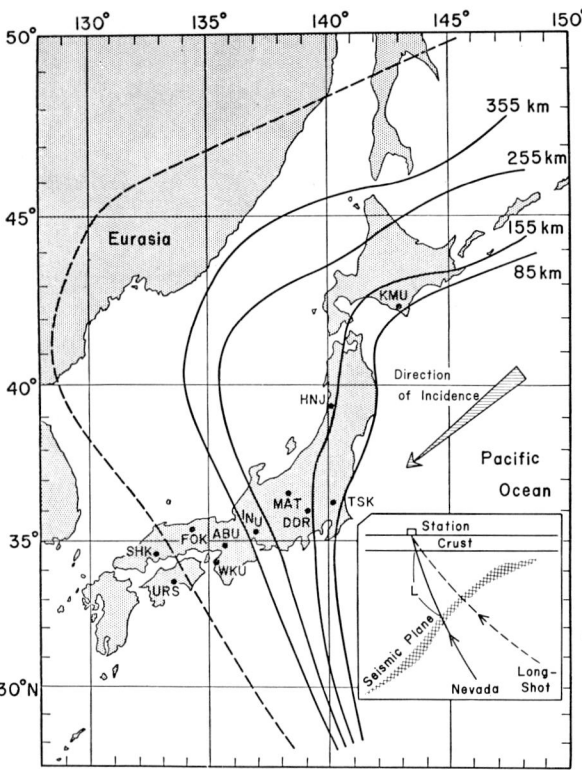

Fig. 1. Location of high-sensitivity seismograph stations (dots with three-letter abbreviations), the contours of the deep seismic plane (WADATI, 1935; and SUGIMURA and UYEDA, 1968), and the direction of incidence from the Longshot and Nevada explosions. The insert shows the cross section and the ray paths.

L. It is quite evident that the larger the value of L, that is, the longer the path above the seismic plane, the later the arrival. This observation can be explained if the average velocity in the mantle on the continental side of the seismic plane is smaller than that on the oceanic side. The dashed lines give theoretical slopes for various velocity differences. We see that a velocity difference of 0.2 to 0.4 km/s can explain the observation.

In addition to the travel times, the value of Q has been determined from the amplitude of the P and PcP phases at the stations SHK and TSK (see KANAMORI, 1968). It was found that the value of Q on the continental side of the deep seismic plane is very low, about 80. However, since the amplitude of body waves may be affected by many other factors, this value may be subject to a considerable uncertainty.

Figs. 4 and 5 show similar plots for two of the Nevada explosions. The result for the other Nevada explosion is almost identical with this. The general

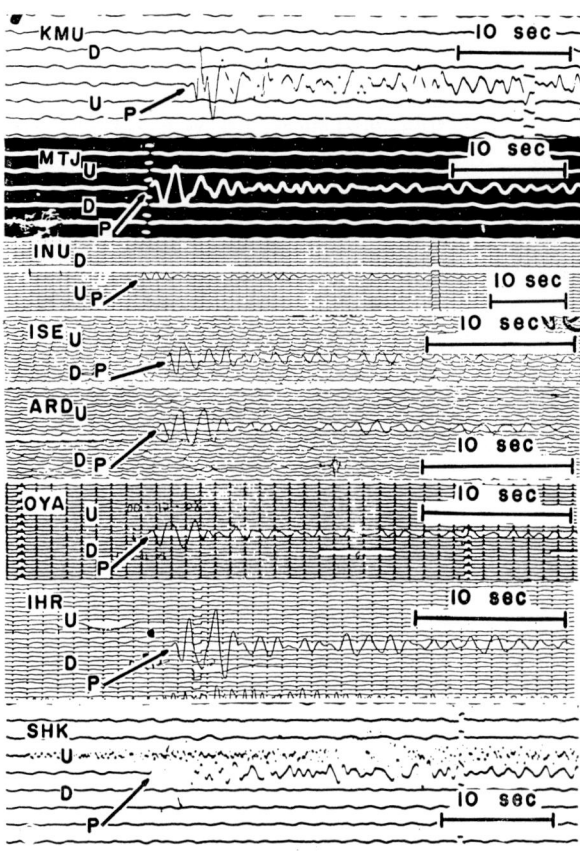

Fig. 2. Seismograms of the Nevada explosion of April 26, 1968 recorded at Japanese stations.

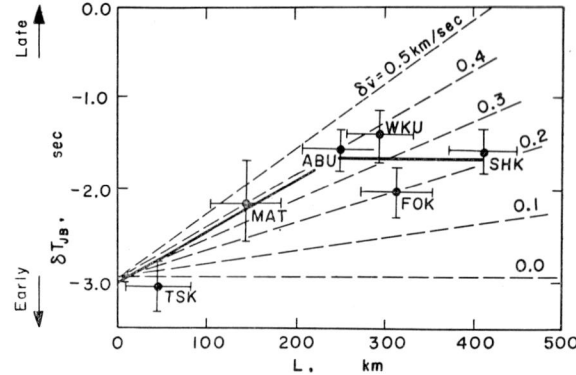

Fig. 3. The J-B residuals for the Longshot plotted against the path length above the seismic plane. Dashed lines show the slopes for various velocity contrast across the seismic plane. Corrections are made for ellipticity and station elevation.

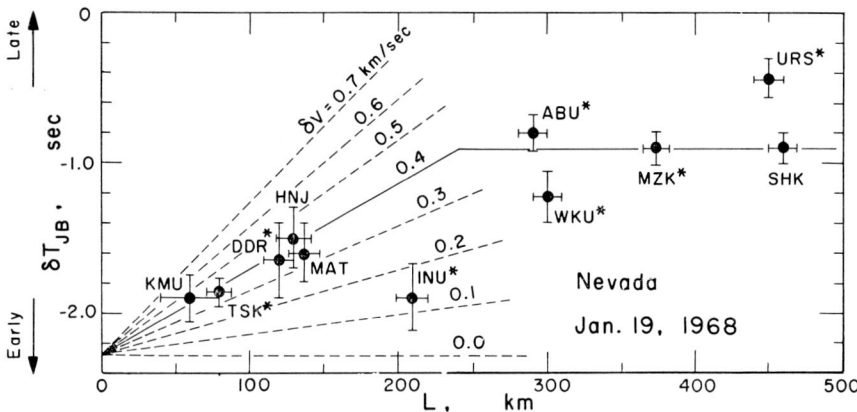

Fig. 4. The J-B residuals for the Nevada explosion of January 19, 1968. Dashed lines give the slopes for various velocity contrast. The solid line shows the general trend of the travel-time residual. The asterisks attached to the station names indicate that the travel-time residual represents the average of the residuals at neighboring stations. Station MZK is close to FOK.

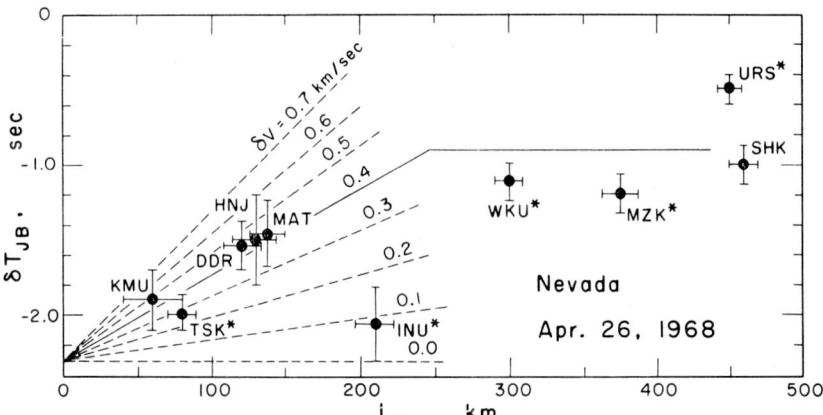

Fig. 5. The J-B residuals for the Nevada explosion of April 26, 1968.

trend is about the same as that for the Longshot. However, there is an indication that δT_{JB} levels off for L longer than 250 km. This means either that the low-velocity region is bounded at a depth around 200 to 250 km or that the velocity contrast becomes small towards the continent. Fig. 6 shows a schematic mantle model beneath Japan thus derived. The mantle above the seismic plane is anomalous in that it has extremely low-Q and low-velocity which may call for a partial melting.

It is to be noted that this kind of observation does not constitute an evidence either for or against the existence of the lithosphere slab as was suggested by OLIVER and ISACKS (1967), since the travel times to all the stations are equally affected by that slab. Fig. 6

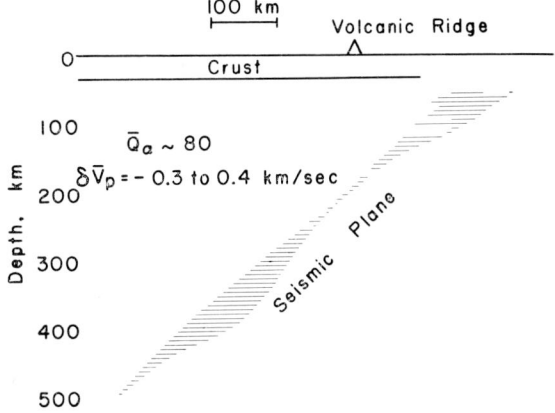

Fig. 6. Schematic cross section of the mantle structure beneath the Japanese arc.

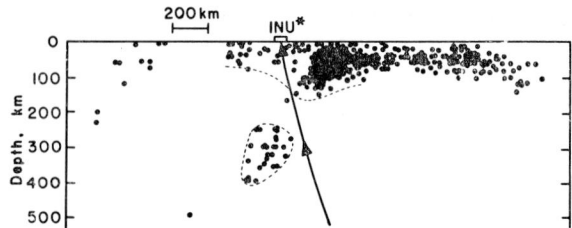

Fig. 7. Ray path to INU and vertical cross section of seismicity (KATSUMATA, 1956).

should be considered as giving a vertically-averaged structure.

Another important feature in these results is that the station INU always registered relatively early arrivals for the Nevada explosions. The station was not operated at the time of the Longshot. A closer inspection of the seismicity of Japan reveals that the path to INU actually passes through a gap of seismic activity. Fig. 7, which shows an appropriate cross section given by KATSUMATA (1956), demonstrates this situation. This observation suggests that the low velocity is actually associated with the underlying earthquake activity; the mantle without underlying earthquake activity does not show the anomalous low velocity.

In order to compare Japanese mantles with those elsewhere, the travel-time residuals at Japanese stations are compared with those at world-wide stations in figs. 8 and 9. We see that the range for the Japanese stations is about the same as that for the world-wide stations at the corresponding distance. The earliest arrivals at Japanese stations are comparable to those at stations on shields, and the latest to those at orogenic belts.

The fact that the general trend of the J–B residual is the same for explosions at different distances (Longshot, $\Delta \approx 35°$, Nevada explosions, $\Delta \approx 80°$), and that the incidence angle at stations is very small for Nevada explosions may preclude the possibility that the observed variation of the J–B residual is due to the systematic error of the J–B table.

3. Surface wave data

Group velocities of long-period surface waves have been measured to study regional characteristics of the mantle beneath the Japanese arc (KANAMORI and ABE, 1968). Fig. 10 shows the great circle paths from epicenters to Japan in relation to various tectonic features such as trench, volcanic ridge and seismic plane. As

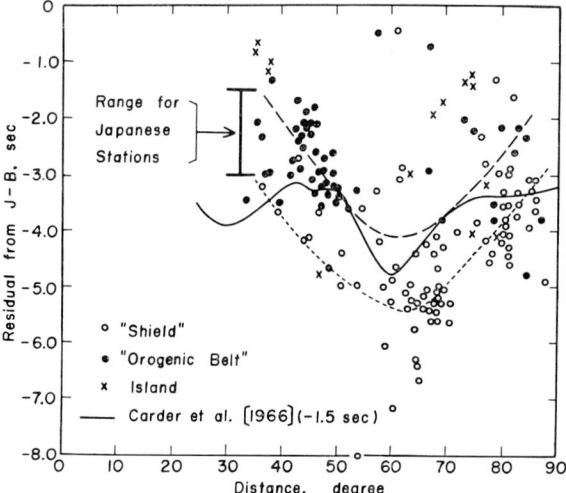

Fig. 8. Travel-time residuals at world-wide stations for the Longshot (from KANAMORI, 1968).

shown in the inserts A and B, the paths from the Aleutian Is. and New Zealand travel on the oceanic side of the arc, while the paths from the Flores Is., Banda sea, W. Caroline Is. and New Britain Is. travel on the continental side of the arc. Further, the former paths are below the seismic plane while the latter paths are above it. We tried to see whether or not any difference exists in the dispersion character between these two groups of path.

For an accurate determination of group velocities for relatively short paths, we used seismograms recorded

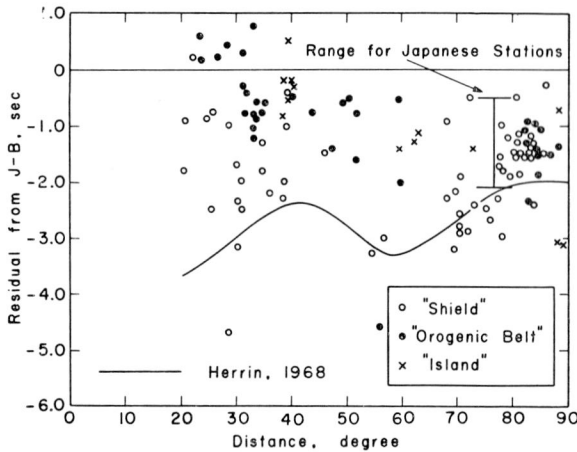

Fig. 9. Travel-time residuals at world-wide stations for the Nevada explosion of April 26, 1968. The classification of shield and orogenic belt is based on UMBGROVE (1947). The data are based upon the Earthquake Data Rep. of U.S. Coast and Geodetic Survey.

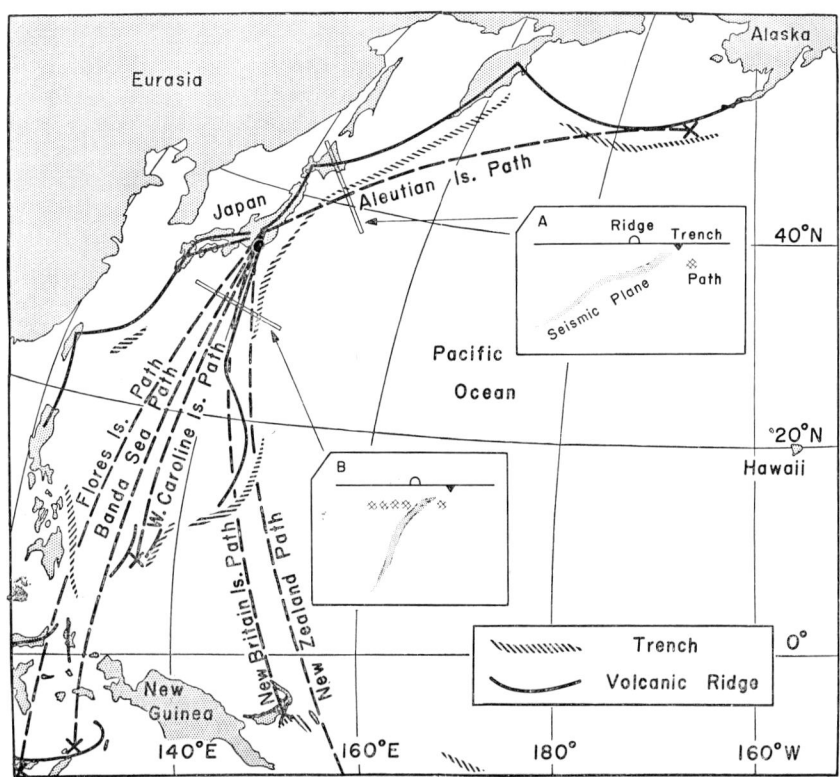

Fig. 10. Great circle paths from epicenters to Japan. The inserts show the vertical cross section of the paths relative to volcanic ridge, seismic plane, and trench. (From Kanamori and Abe, 1968).

on magnetic tapes and applied the band-pass filtering technique. Group velocities of Love waves to a period of 120s are shown in fig. 11. It is clear that, at long periods, the group velocities for the paths on the continental side (Banda sea, W. Caroline, and Flores Is. paths) are about 0.1 km/s lower than those on the oceanic side (Aleutian Is. and New Zealand paths). The standard oceanic mantle models such as the 8099 and CIT11A models fit the data for the latter paths reasonably well. To fit the data for the former paths the model designated as ARC-1 is constructed. Fig. 12 compares the model with the standard models. The major difference of the ARC-1 model from the standard models is that the S-velocity is reduced by 0.3 to 0.4 km/s over the depth range 30 to 80 km. These models can explain the Rayleigh wave data too. Fig. 13 shows the results for Rayleigh waves. The group velocities for the Aleutian path can be fitted reasonably well by the CIT11A model, the fastest of all. The ARC-1 model fits the data for the paths on the continental side of the arc.

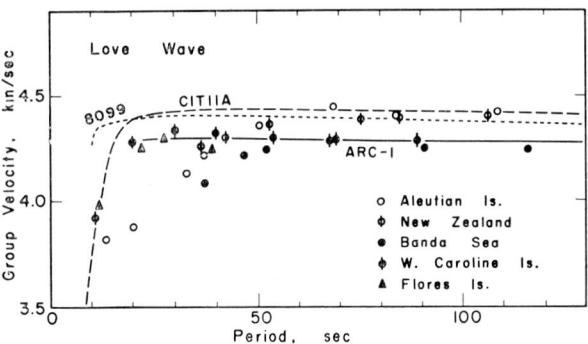

Fig. 11. Group velocities of Love waves for various oceanic paths. Three curves are the group velocities for three structures shown in fig. 12. (From Kanamori and Abe, 1968.)

Thus we can conclude that the shear velocity is 0.3 to 0.4 km/s smaller over the depth range 30 to 80 km on the continental side of the arc than on the oceanic side. This feature is consistent, at least qualitatively, with the teleseismic explosion data.

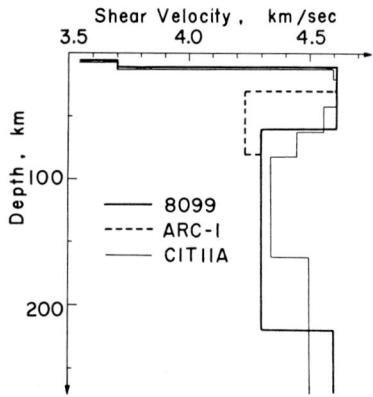

Fig. 12. Shear velocity distributions for three oceanic models. (From KANAMORI and ABE, 1968.)

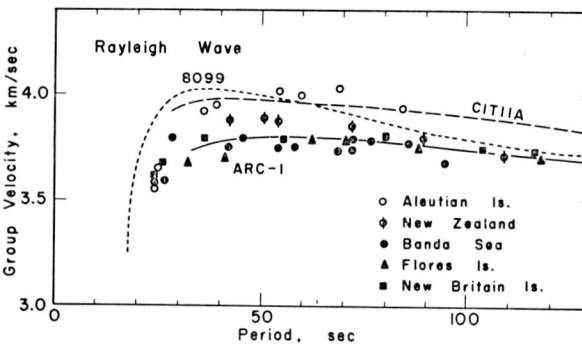

Fig. 13. Group velocities of Rayleigh waves for various oceanic paths. (From KANAMORI and ABE, 1968.)

4. dΔ/dt data

The measurements of dΔ/dt of P-waves were made at an array station in Japan (KANAMORI, 1967). The locations of the station and the epicenters of earthquakes used for the analysis are shown in fig. 14. From the location of the stations relative to the epicenters we see that, above 300 km, most of the propagation paths are on the continental side of the seismic plane.

The dΔ/dt data show five important features (see fig. 15):

(1) The P_n velocity is almost constant to $\Delta = 12°$.

(2) A relatively abrupt change of dΔ/dt from about 8 to 8.6 km/s was observed at $\Delta = 13°$.

(3) At $\Delta = 19.3°$, dΔ/dt abruptly increases from 8.8 to 9.8 km/s. This increase is followed by a rapid but almost continuous increase of dΔ/dt to $\Delta = 23°$.

(4) A clear later phase branch was observed over the

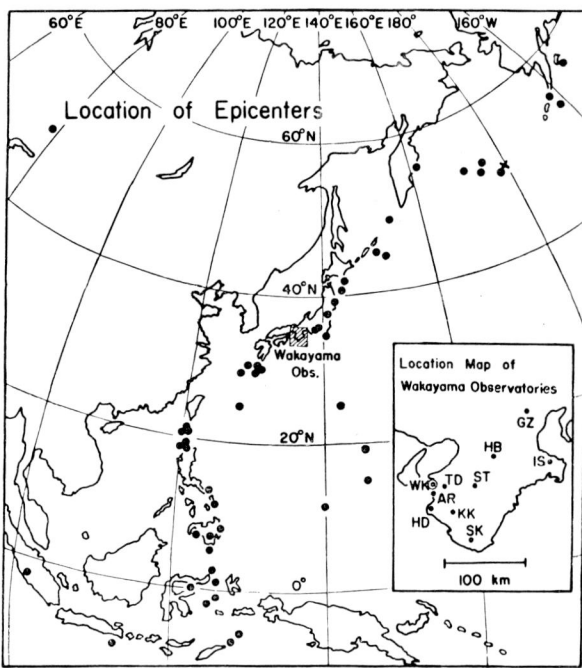

Fig. 14. Location of epicenters of earthquakes used for dΔ/dt measurements. The insert shows the location of stations. (From KANAMORI, 1967.)

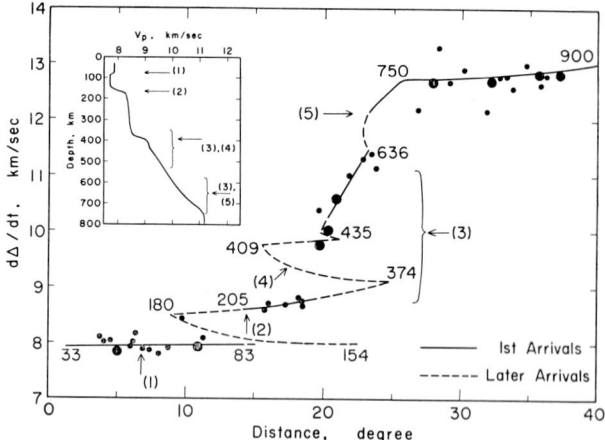

Fig. 15. The apparent velocities, dΔ/dt, of P-waves as a function of distance. The numbers on the curve indicate the depth of penetration in km. The curve is calculated for the model given in the insert.

range from $\Delta = 16°$ to $22°$ (dΔ/dt was not measured for this branch).

(5) A complicated behavior in dΔ/dt versus Δ curve was found at distances between $23°$ to $27°$ (not shown in this figure).

From (1) and (2) we can construct a model in which

the P-velocity is relatively constant at 7.9 km/s down to a depth of about 180 km where it abruptly increases to 8.5 km/s. By (3) and (4), a rapid velocity increase starting from the depth of 375 km is suggested. This rapid velocity increase can be attributed primarily to the onset of the olivine-spinel transition of $(MgFe)_2SiO_4$. No detailed velocity structure could be determined for the depth range from 450 to 700 km. But it is evident that the velocity increase is very rapid over this depth range. Judging from the complicated nature of the seismograms for the distance range from 23° to 27°, it is possible that this rapid velocity increase consists of a number of minor discontinuities. To detect these discontinuities, however, is beyond the resolving power of the present method. It is to be noted that no positive evidence for any other major discontinuity at depths below 500 km was found. In this respect, the present model differs from the continental model such as the one found by JOHNSON (1967) (see also ANDERSON and TOKSOZ, 1963; NIAZI and ANDERSON, 1965; JULIAN and ANDERSON, 1968; ARCHAMBEAU et al., 1966; and HALES et al., 1968). In other respects, the present model is in a remarkable accord with the continental model. Fig. 16a shows the comparison and ANDERSON's (1967) two olivine mantle models, one for the $(Mg_{0.8}Fe_{0.2})_2SiO_4$ mantle, and the other for the $(Mg_{0.6}Fe_{0.4})_2SiO_4$ mantle. It is seen that Anderson's interpretation which was originally made on the continental model also applies on the whole to the island-arc model. However, the major difference between the continental and the island-arc models exists at depths from 420 to 550 km. This difference can be attributed to the difference in the tectonic nature of the two mantles. The continental model represents the mantle where no deep earthquake activity exists, while the island-arc model represents the mantle with high deep seismic activity. Fig. 16b shows a schematic cross section of the island-arc mantle sampled here. Fig. 16c is the curve of the earthquake energy release as a function of depth in the Japanese region (MIYAMURA, 1968; and MIZOUE, 1967). Note the coincidence between the depths of the rapid velocity increase and the peaks of the energy release. It is suggested that the phase transition is in some way related to the deep seismic activity. The marked difference of the mantle structure across the seismic plane suggests that a large scale mantle movement as proposed by recent multi-disciplinary studies may also exist beneath the Japanese arc as indicated by the arrow in fig. 16b. The olivine in the down-going mantle material will break down to spinel when it passes through the depth

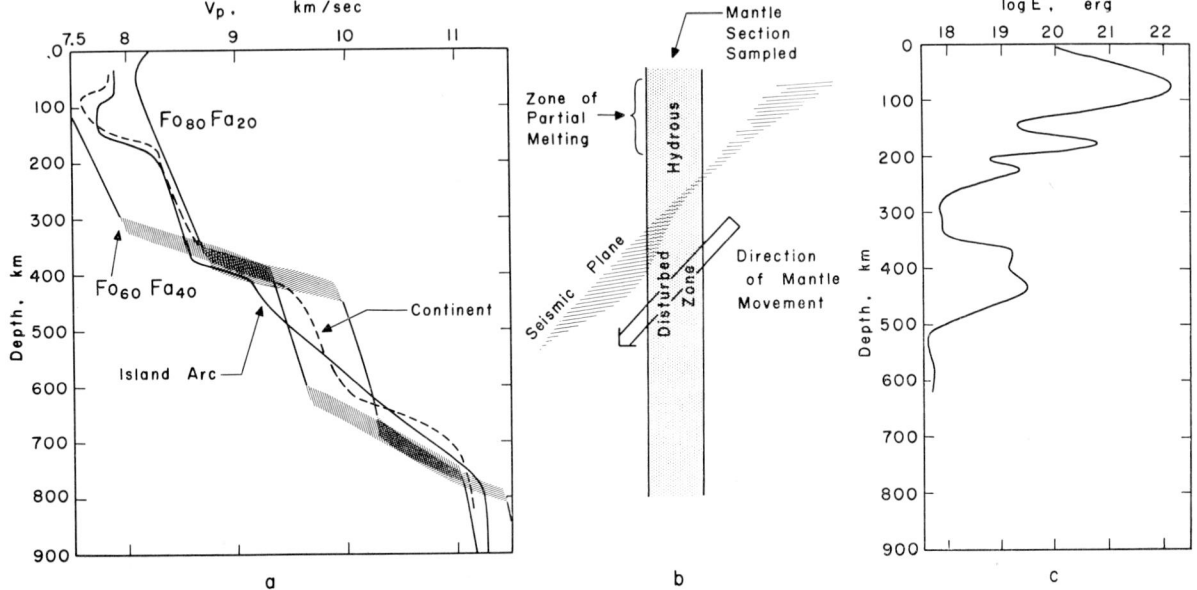

Fig. 16. a. Comparison of velocity structures beneath the continent (after JOHNSON, 1967) and beneath Japan. Two olivine mantle models according to ANDERSON (1967) are shown. b. Mantle section sampled in $d\varDelta/dt$ study and schematic mantle model beneath Japan. c. Earthquake energy release curve in the Japanese region according to MIZOUE (1967) and MIYAMURA (1968).

of about 400 km. Since the olivine–spinel transition is an exothermic reaction, the above process may liberate a significant amount of heat which in turn may cause the low-velocity in the overlying mantle. The volume change and the possible reduction of the strength associated with this reaction may cause a stress concentration at this part of the mantle, thereby causing a high earthquake activity. In order to make more detailed discussions about these processes we need accurate thermochemical data of the olivine–spinel transition and a thermodynamical treatment of the pertinent fracture mechanism.

It is to be noted that the depth range where we found a large difference between the continental and island arc mantles is just below the deep seismic plane. It is possible that the composition and the temperature conditions in this part of the mantle beneath Japan are so disturbed by the mantle movement that the clearly defined discontinuities as found in the continental mantle do not exist. RINGWOOD (1969) suggested that the discontinuity at 650 to 700 km may represent a combined effect of several phase transformations such as spinel → ilmenite + MgO, garnet → ilmenite etc. If the pressure, temperature and composition conditions beneath Japan are such that these transformations occur at considerably different depths, thereby smearing out the discontinuity over a wide interval, the lack of clear discontinuity at depths between 600 and 700 km is not unreasonable.

If we use the phase diagram of $(MgFe)_2SiO_4$ system given by AKIMOTO and FUJISAWA (1968), the temperature at 400 km depth is no higher than 1500 °C.

We also note that the velocity increase around 200 km depth is abrupt; this may be considered as a positive evidence for a partial melting in the overlying mantle. Experimental works (SPETZLER and ANDERSON, 1968; and MIZUTANI and KANAMORI, 1964) have shown that a major change of velocity occurs at the onset of melting. If the mantle above 200 km is partially molten and this partially molten zone is terminated at the depth of 200 km, a sharp velocity change will be most naturally expected.

5. Conclusion

Because of the difference in the sensitivity of the methods and in the regions sampled, the structures derived from different sets of data differ slightly in detail. However, they all seem to point out that the mantle on the continental side of the Japanese arc is, on the whole, characterized by extremely low-Q and low-velocity; they are probably lower than those in any other tectonically active area. The low velocity is associated with earthquake activities which in turn seem to be related to sharp velocity changes.

A relatively low temperature at the depth of 400 km has been suggested while a partial fusion may be required to explain the low-velocity, low-Q and the sharp velocity increase at the depth of around 200 km. Thus our results demand that the mantle material be partially molten at relatively low temperatures, around 1300 °C. This requirement is hardly satisfied by a dry peridotite. However, recent work by KUSHIRO et al. (1968) shows that the melting temperature of lherzolite drops markedly under a hydrous condition; the solidus can be lowered to 1000 °C. Thus, if we assume that the mantle beneath Japan is in a slightly hydrous condition this difficulty is removed.

It is probable that the down-going mantle movement generates heat and causes earthquakes through the olivine–spinel transition.

The partial melting may greatly increase the efficiency of the heat transfer in the mantle on the continental side of the arc. This will explain, at least qualitatively, the large heat flows observed on the continental side of the Japanese arc (VACQUIER et al., 1966).

Acknowledgments

I am grateful to the staff of the following observatories for the seismograms used in the present study: Urakawa, Aobayama, Honjo, Tsukuba, Dodaira, Matsushiro, Inuyama, Abuyama, Wakayama, Shiraki, Tottori and Kochi Observatories. A part of this paper was written while I was a visiting professor in the Department of Earth and Planetary Sciences at the Massachusetts Institute of Technology.

References

AKI, K. (1961) Bull. Earthquake Res. Inst. Tokyo Univ. **39**, 255.
AKIMOTO, S. and H. FUJISAWA (1968) J. Geophys. Res. **73**, 1467.
ANDERSON, D. L. (1967) Science **157**, 1165.
ANDERSON, D. L. and M. N. TOKSÖZ (1963) J. Geophys. Res. **68**, 3483.
ANONYMOUS (1966) Research group for explosion seismology, Explosion seismological research in Japan, in: J. S. Steinhart et al., eds., *The Earth beneath the continents* (Am. Geophys. Union, Washington, D.C.) 334.

ARCHAMBEAU, C. B., E. A. FLINN and D. C. LAMBERT (1966) J. Geophys. Res. **71**, 3483.

CARDER, D. S., D. W. GORDON and J. N. JORDAN (1966) Bull. Seismol. Soc. Am. **56**, 815.

HALES, A. L., J. CLEARY, H. DOYLE, R. GREEN and J. ROBERTS (1968) J. Geophys. Res. **73**, 3885.

HERRIN, E. (1968) Bull. Seismol. Soc. Am. **58**, 1193.

HISAMOTO, S. (1965) J. Seismol. Soc. Japan **18**, 142, 195 (in Japanese).

JEFFREYS, H. (1966) Geophys. J. **11**, 5.

JOHNSON, L. R. (1967) J. Geophys. Res. **72**, 6309.

JULIAN, B. R. and D. L. ANDERSON (1968) Bull. Seismol. Soc. Am. **58**, 339.

KANAMORI, H. (1967) Bull. Earthquake Res. Inst. Tokyo Univ. **45**, 657.

KANAMORI, H. (1968) Bull. Earthquake Res. Inst. Tokyo Univ. **46**, 841.

KANAMORI, H. and K. ABE (1968) J. Phys. Earth Tokyo **16**, 137.

KATSUMATA, M. (1956) Geophys. Mag. **27**, 483.

KATSUMATA, M. (1960) Kenshin Ziho **25**, 19 (in Japanese).

KUSHIRO, I., Y. SYONO and S. AKIMOTO (1968) J. Geophys. Res. **73**, 6023.

MIYAMURA, S. (1968) Seismicity of island arcs and other arc tectonic regions of the Circum-Pacific zone, in: L. Knopoff et al., eds., *The crust and upper mantle of the Pacific area* (Am. Geophys. Union, Washington, D.C.) 60.

MIZOUE, M. (1967) Bull. Earthquake Res. Inst. Tokyo Univ. **45**, 679.

MIZUTANI, H. and H. KANAMORI (1964) J. Phys. Earth Tokyo **12**, 43.

NIAZI, M. and D. L. ANDERSON (1965) J. Geophys. Res. **70**, 4633.

OLIVER, J. and B. ISACKS (1967) J. Geophys. Res. **72**, 4259.

RINGWOOD, A. E. (1969) Dept. Geophys. Geochem. Australian Nat. Univ. Canberra Publ. **666**, 1.

SAITO, M. and H. TAKEUCHI (1966) Bull. Seismol. Soc. Am. **56**, 1067.

SPETZLER, H. and D. L. ANDERSON (1968) J. Geophys. Res. **73**, 6051.

SUGIMURA, A. and S. UYEDA (1968) Japan and its environ (in preparation).

UMBGROVE, J. H. F. (1947) *The pulse of the Earth*, 2nd ed. (Nijhoff, The Hague).

UTSU, T. (1966) J. Fac. Sci. Hokkaido Univ. Ser. VII **2**, 359.

UTSU, T. (1967) J. Fac. Sci. Hokkaido Univ. Ser. VII **3**, 1.

VACQUIER, V., S. UYEDA, M. YASUI, J. SCLATER, C. CORRY and T. WATANABE (1966) Bull. Earthquake Res. Inst. Tokyo Univ. **44**, 1519.

WADATI, K. (1935) Geophys. Mag. **8**, 305.

EVALUATION OF THE ISOSTATIC MECHANISM AND ROLE OF MINERALOGIC TRANSFORMATIONS FROM SEISMIC AND GRAVITY DATA

G. P. WOOLLARD

University of Hawaii, Honolulu, Hawaii 96822, U.S.A.

1. Introduction

The concept of mineralogic transformations being an important factor in determining the physical parameters of the crust and upper mantle and hence, important from the standpoint of maintaining isostatic equilibrium is not new. KENNEDY (1959), in examining possible explanations for the marked difference in seismically defined crustal thickness found beneath ocean basins and continents, concluded that a reversible transformation between eclogite and basalt controlled by differences in thermal gradient could explain the observed relations. HESS (1955) in examining the problem of crustal uplift associated with plateaus concluded that serpentenization of olivine through the action of water could be an important factor in both determining crustal uplift and maintaining isostasy. There are many other mineralogic transformations that could play an active role in the isostatic mechanism and in determining changes in surface elevation and the physical parameters of the crust and upper mantle. For the most part, these transformations require high pressures and temperatures, and therefore are generally assumed to be deep-seated. As such they offer a logical explanation for the seismic velocity reversal at about 110 km and the deeper lying seismic velocity discontinuities in the upper mantle. However, it is also possible that some of these transformations can occur at relatively shallow depths through the presence of water (see SCLAR, 1970).

It, however, is not our purpose here to review possible mechanisms that might contribute to the phenomenon of isostasy so much as to call attention to seismic and gravitational evidence for mineralogic changes under relatively low temperature and pressure conditions that appear to be both continuously active and of a reversible nature.

2. Isostatic considerations and the nature of the problem

As shown by HEISKANEN and VENING MEINESZ (1958), a random selection of gravity values on a worldwide basis from both ocean basins and continental areas shows that on average isostasy is approached within 5 mgal. However, these authors note that certain geographic areas show apparent deviations from isostatic equilibrium of about 20 mgal on average. If we examine the areas that appear to be out of isostatic equilibrium, it is found that most have a relatively narrow width with a half wave length of 100 km or less, and that they can be related to specific geologic features such as island arc trenches, batholiths as the Sierra Nevada in California, horsts as the Harz Mts. in Europe, and grabens as the Rift Valleys in Africa. However, there are some regions of 5° to 10° in width, as peninsula India, that also appear to depart significantly from isostatic equilibrium. Whereas it is easy to conceive of relatively narrow short wave length geologic features being out of isostatic equilibrium due to crustal deformation, or intrusion, it is difficult to conceive of broad areas of the earth's crust exceeding 3° in width not being compensated unless there is evidence, as in eastern Canada, and Fenno-Scandinavia of recent regional crustal loading occasioned by a former ice cap, or evidence of vertical convective movement in the underlying mantle such as appears to underly the Mid-Atlantic Ridge. Apparent regional departures from isostasy over other areas, particularly the tectonically stable shield areas, are more likely to be related to regional changes in the mean density of the crust and upper mantle, and a consequence of mass distribution and departure of the actual crust from the isostatic crustal model rather than any real departure in isostatic equilibrium.

The point being made here is that it makes no difference from the standpoint of isostatic equilibrium how the density and thickness of the crust varies. As long as the crust is not restrained by external forces, the mass of the mantle displaced ($R \times \sigma_m$) will always equal that of the crustal column ($H \times \sigma_c$), and as a result, there will be equal mass above some level at depth. However, it does not follow that there will be no change in gravity when there are regional changes in crustal parameters since the force of gravity (g) depends on the relation $M_1 M_2 / r^2$ and, furthermore, only the vertical component of gravity (gz) = $g \cos \theta$ is measured. Therefore, the dimensions and spatial distribution of crustal and mantle blocks having differences in density as well as earth curvature become important in defining observed gravity. It is to be noted that the basic relation $\Delta h \times \sigma_c = \Delta R (\sigma_m - \sigma_c)$ which is assumed in the isostatic reduction in relating the topographic mass effect for any change in surface elevation (Δh) to its compensation effect through a crustal root increment of ΔR thickness does not hold when there are changes in crustal density. This is because the density of the intervening crustal column has also changed. Instead of $\Delta R = \Delta h \times \sigma_c / (\sigma_m - \sigma_c)$, when σ_c differs from σ_s (the density of the standard sea column of H_s thickness),

$$\Delta R = \frac{\Delta h \sigma_c}{(\sigma_m - \sigma_c)} + \frac{H_s (\sigma_c - \sigma_s)}{(\sigma_m - \sigma_c)}$$

and

$$\Delta h \sigma_c = \Delta R (\sigma_m - \sigma_c) - H_s (\sigma_c - \sigma_s).$$

There are thus two gravitational contributions that have to be considered in the compensation for the topographic mass under these conditions: one from the crustal root increment and one from the difference in density of the intervening crustal column corresponding to H_s. Because of the above, it is found that there is excess gravity (positive free air and isostatic anomalies) where seismic velocity values suggest the crust has an abnormal density. Also the seismic data define an abnormal value of crustal thickness for the surface elevation, and frequently there is geologic evidence of subsidence in such areas (basins). Similarly it is found that there is a deficiency in gravity (negative free air and isostatic anomalies) where the seismic velocity values suggest the crust has a subnormal density. The data also define a subnormal crustal thickness for the surface elevation, and frequently there is geologic evidence of uplift in such areas. There is also good evidence (WOOLLARD 1968) that σ_c as reflected in the mean seismic velocity of the crust as well as abnormality in crustal thickness, and free air and isostatic gravity anomalies also vary directly with the seismic velocity of the upper mantle. It therefore appears there is an intimate relationship between the parameters of the crust and those of the upper mantle. As it is evident that where the mantle velocity is abnormal, the crust has taken on mass (an increase in both density and thickness), and that where the mantle velocity is subnormal the crust has experienced a loss of mass (decrease in density and thickness), it appears that there is a reversible process at the "M" discontinuity that involves a transfer of mass between the crust and mantle. This is most evident in the presence or absence of a high velocity (6.8–7.3 km/s) layer at the base of the crust and in the overall mean velocity values for the crust. However, in order to maintain isostatic equilibrium, the transfer of mass between the crust and mantle has to be equal, and the implication of the corresponding change in sign for the velocity values for the crust and mantle with thickening and thinning of the crust as well as subsidence and uplift and change in gravity is that the density significance for the change in mantle velocity is opposite from the normal relation whereby density increases with increase in velocity. The only other explanation is that there is a related deeper seated transformation, as yet not defined, which maintains equilibrium conditions and governs the movement of the overlying rock column. This would still require a vertical transfer of mass to satisfy the gravity relations, and whereas it is probable that where the mantle and crustal velocity increases a deeper low velocity zone would not be detected; it is reasonable to expect that where the crust and mantle velocity decreases any deeper lying high velocity zone, if it exists, would be found. What fragmentary evidence we have about velocity layering between the "M" discontinuity and the low velocity zone at about 110 km, suggests that there is no discrimination between the two types of areas under consideration as regards sub-Moho layering. In the Williston Basin area of eastern Montana and western North Dakota in the United States where the thickest crust and highest mantle and crustal velocity values are found, there is a well defined sub-Moho layer with a velocity of 9 km/s. In the Basin and Range area of Utah in the United States where the mantle velocity reaches its lowest value (7.6–7.7 km/s) and the crust is abnormally thin (25 km) for the surface elevation (1500 km), there is a suggestion of a possible normal mantle velocity of around 8.2 km/s at a depth of about 75 km, but nothing as pronounced or as well defined as the 9 km/s layer beneath the Williston Basin, which shows up at depths ranging from 68 to 80 km below sea level.

We therefore must conclude that there is either reversible vertical transport of mass throughout the rock column above the depth of compensation and that the compensation is achieved through mass changes near the base of the column, or alternatively that there is mass exchange between the crust and mantle, and perhaps also at greater depths. Conceivably, every seismic discontinuity below the depth of compressibility effect on velocity could represent a zone of either past or present mass exchange defined originally by a given threshold pressure and temperature. The observed inter-relation of gravity, crustal movement and crustal and upper mantle parameters suggest that if this is the case, then there is probably a transfer of iron involved which, as observed with changes in fosterite-fayalite ratio, raises density and reduces velocity when the Fe/Mg ratio increases, and lowers density and raises velocity when the Fe/Mg ratio decreases. Also, as brought out by SIMMONS (1964) the percentage of CaO and Al_2O_3 can also have a marked anomalous effect on the inter-relationship of velocity and density.

In its simplest terms the phenomenon involves in one direction, an increase in gravity which suggests both near surface control and an increase in crustal density; crustal subsidence which implies a decrease in density contrast between the crust and mantle; an increase in crustal thickness through addition of a high velocity basal layer which suggests conversion of mantle material to crustal material, and a raise in mantle velocity which normally would be taken as evidence of an increase in mantle density, but which can also be interpreted as a decrease in density if there is a migration of iron or calcium. In the other direction there is a decrease in gravity, a decrease in crustal thickness, crustal uplift and a decrease in mantle velocity. That the two phenomenon operated simultaneously in adjacent areas is suggested by the geologic record and gravity field over adjacent uplifts and basins.

This then is the statement of the problem, and the balance of this paper is devoted to the presentation of substantiating evidence.

3. Gravity relations

That departures in isostasy because of density changes are far more extensive than suggested by the data used by HEISKANEN and VENING MEINESZ (1958) is indicated by fig. 1 which shows a comparison of average Bouguer anomaly values as a function of average elevation values for $3° \times 3°$ size areas on the different continents. As the Bouguer anomaly is computed on the basis of the surface elevation and the included mass effect and only varies slightly with change in crustal density for any given elevation, the marked spread of about 80 mgal in Bouguer anomaly values for any elevation shown in fig. 1 have to originate from the mass distribution below sea level. Although the implication is that there are significant regional changes in mean crustal density and upper mantle density, they could alternatively be caused by significant mass inhomogeneities at depth that bias the gravity field of the different continental areas. Another alternative is that the actual form of the earth departs significantly from

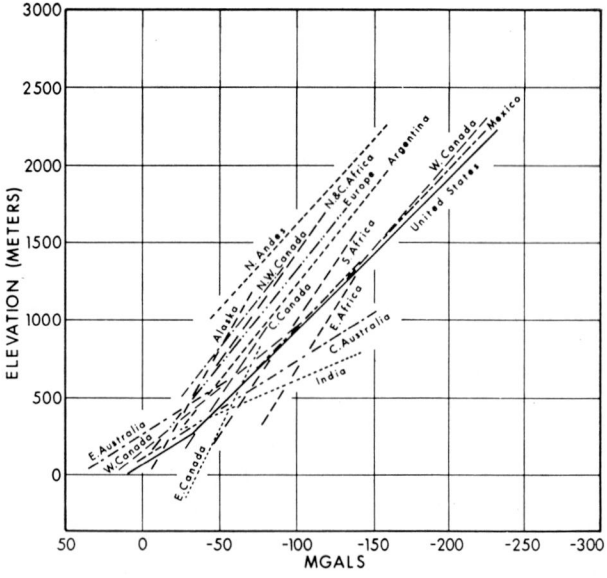

Fig. 1. Relation of average Bouguer anomalies to surface elevation for $3° \times 3°$ size areas on the different continents.

the simple biaxial ellipsoid of revolution used in the international gravity formula for defining the theoretical value of gravity at sea level. The principal evidence bearing upon the latter two explanations is obtained from the analyses that have been made of artificial earth satellite orbital perturbations. An 8th degree spherical harmonic representation of the satellite defined free air gravity anomaly field (KHAN and WOOLLARD, 1968) with a polar flattening of 1/298.25 and using the zonal coefficients of KOZAI (1964) and the tesseral harmonic coefficients of GAPOSHKIN (1966) does show that on a relative basis the relations shown in fig. 1 are evident in the satellite derived gravity anomaly field. In analyzing the global gravity pattern defined by satellite data, KHAN and WOOLLARD (1968) show that on the basis of the half-width values ($X\frac{1}{2}$ = width of the anomaly at half amplitude) of the individual anomalies which can be related to the depth of origin, few anomalies have an apparent depth as great as 1000 km below the Earth's surface, and most, no greater than 250 km. These writers also show that $10° \times 10°$ average values of surface data give a similar anomaly pattern to that derived from satellite data, and conclude that no firm case can be made for localized deep-seated mass inequalities as the integration of the gravitational field from near-surface mass inequalities at satellite height has an expression equivalent to that of a deep-seated source. They also note that while there is evidence in the satellite data of triaxiality in the northern hemisphere at about 0° longitude and at about 30° east longitude in the southern hemisphere, there is no continuity in pattern between the northern and southern hemispheres in crossing the equatorial region, but instead a reversal in sign for the geoid at these longitudes. The explanation for the observed surface anomalous gravity field therefore cannot be resolved at this time by satellite data. The fact that they do corroborate the anomalous surface field, however, does verify the reality of significant regional changes in gravity.

4. Seismic determinations of crustal and upper mantle abnormality

In gross terms it is known from seismic refraction studies that the thickness of the Earth's crust as defined by the Mohorovicic discontinuity is directly related to surface elevation. Laboratory studies of the physical properties of rocks at confining pressures comparable to those encountered over the depth range from the surface down to the "M" discontinuity suggest the density of the crust is in the range of 2.87 to 2.9 gm/cm³ and that the upper mantle has a density in the range of 3.28 to 3.32 gm/cm³. The density differential between the crust and mantle thus lies in the range, 0.38 to 0.45 gm/cm³. It is clear from these values that there is a significant density contrast between the crust and mantle and a high probability that the buoyant support of the crust by the mantle is a significant factor in maintaining isostasy. Whether there are deeper mass distributions of equal significance as the "M" discontinuity is problematical. Deeper seismic discontinuities above the low velocity zone at about 110 km, while detected in some areas, appear to be exceptions rather than the rule. The only significant argument lies in whether there is an overall systematic relation between changes in surface elevation and changes in mantle depth and whether the pattern of change in Bouguer anomaly values with surface elevation which on the basis of fig. 1 can be written as $BA = -0.84 \Delta h + 0$ where Δh is in meters can be satisfied by the depth of the "M" discontinuity.

Once a standard model is established it then becomes possible to examine departures from the model in terms of their cause such as phase transformations or changes in chemical composition.

If there is hydrostatic equilibrium between the crust and upper mantle, a plot of mantle depth values determined seismically as a function of elevation should give a linear relation in the form $y = mx + b$ where the slope factor (m) is the ratio of crustal root increment (ΔR) to surface elevation increment (Δh), and, b is the sea level intercept value for the mean thickness of the crust at sea level. This follows from Archimedes principle whereby a floating body displaces its own mass of the supporting medium. The fundamental relation as expressed earlier is $R = H\sigma_c/\sigma_m$ where for the case under consideration R is the column of mantle material displaced by the crust; H is the thickness of the crust; σ_c is the density of the crust, and σ_m is the density of the underlying mantle. As the freeboard $(F) = H - R$, any change in surface elevation (Δh) under equilibrium conditions requires a corresponding change in R that depends on the ratio $\sigma_c/(\sigma_m - \sigma_c)$ since $\Delta R = \Delta h \sigma_c/(\sigma_m - \sigma_c)$. Because $\Delta R/\Delta h = \sigma_c/(\sigma_m - \sigma_c)$, if $\Delta R/\Delta h$ can be shown emperically to approximate a constant, it can be concluded that the depth of the "M" discontinuity is a primary factor in controlling Δh and that there is an average constant value of $(\sigma_m - \sigma_c)$. Under these conditions the value of σ_m can be derived if a reasonable estimate of σ_c can be determined from laboratory studies of crustal rocks.

That there will be a considerable spread in seismic values for the depth of the mantle at any elevation is to be expected on the basis of the spread in Bouguer anomaly values for any elevation shown in fig. 1. The magnitude of the spread can be estimated from the data of table 1 based on equilibrium conditions.

TABLE 1

Gravity effect of change of 1 km in crustal root for different density contrasts between the crust and mantle

$\Delta\sigma = (\sigma_m - \sigma_c)$	Mgal/km change in R
0.5 gm/cm³	20.92 mgal
0.45	18.83
0.4	16.74
0.35	14.65
0.3	12.56

The above is based on the relation Δg for $M\Delta h = \Delta g$ for $M\Delta R$ with Δg for $M\Delta h = 2\pi G \Delta h \sigma_c$ being the Bouguer anomaly increment for Δh. Therefore $\Delta R = \Delta$Bouguer anomaly$/2\pi G(\sigma_m - \sigma_c)$.

If $\Delta\sigma = 0.4$ gm/cm³ the observed gravity relations shown in fig. 1 would suggest that on average there will be a spread of about 4.8 km in the depth of the mantle for any elevation.

In dealing with the actual seismic data the ocean data can be incorporated by using a synthetic elevation $(\Delta h') = \Delta h - \Delta d$ where Δh is the depth of water and $\Delta d = \Delta h \sigma_{H_2O}/\sigma_c$, the equivalent rock column for the water column. However, since there is no identifiable granitic layer in the oceans comparable to that on the continents, the value assumed for σ_c was 2.745 gm/cm³ which is the average density of the basement crystalline rock complex on the continents (WOOLLARD, 1962). The effect on the synthetic elevation $(\Delta h')$ in using this density rather than a value of 2.85 or 2.9 gm/cm³ which as will be seen later are nearer the apparent value of crustal density is shown in table 2.

As the maximum difference in derived synthetic elevation using 2.745 gm/cm³ rather than an approximate value of the mean crustal density is of the order of 0.15 km for a depth of water of 6000 meters and greater, the choice of density used is not critical.

Fig. 2 shows the plot of land data for elevations up to 2500 m, and ocean data for all depths in terms of the actual depth of water (Δh). This plot shows: 1) that while there is a general trend suggested for the oceans as a whole, there are few data for the transition zone between the edge of the continental shelf and the deep ocean floor; 2) that the oceanic trenches constitute a separate cross-cutting relationship that is similar to that for the oceanic rises; and 3) that the oceanic islands and ridges represent a third independent relationship. If it is assumed there is a regular transition in crustal thickness with elevation across the continental slope where there are essentially no data between the continental boundary and the deep ocean, the following average relations are described for the depth of the mantle (H_m) with change in depth of water (Δh) in km.

General relation $H_m = -33.2 + 4.64 \Delta h$
Oceanic trenches $H_m = -(0.72 + 2.14 \Delta h)$
Oceanic rises $H_m = -(1.9 + 2.14 \Delta h)$
Oceanic islands $H_m = -17.8 + 3.8 \Delta h$

If the general relation for the depth of water (Δh) is replaced by one using the synthetic elevation ($\Delta h'$) based on condensing the water column to an equivalent layer of rock so that these data can be incorporated with continental data the relation is $H_m = -33.2 + 7.4 \Delta h'$.

On the continents the data similarly fall into discrete packages; the most significant ones besides the general relation are: a) the Basin and Range values in the United States where the mantle velocity is markedly subnormal as mentioned earlier, and b) those areas where the crust and mantle have an abnormal velocity and thickness. On an overall basis the continental data alone suggest the ratio $\Delta R/\Delta h$ lies between 7.5 and 8.0

TABLE 2

Synthetic elevation values in the ocen for different values of σ_c

H₂O depth km	$\sigma_c = 2.745$		$\sigma_c = 2.85$		$\sigma_c = 2.9$	
	Equiv rock km	Syn. elv km	Equiv rock km	Syn. elv km	Equiv rock km	Syn. elv km
1.0	0.375	0.625	0.362	0.638	0.356	0.644
2.0	0.750	1.250	0.724	1.276	0.712	1.288
3.0	1.125	1.875	1.086	1.914	1.068	1.932
4.0	1.500	2.500	1.448	2.552	1.424	2.576
5.0	1.875	3.125	1.810	3.190	1.780	3.220
6.0	2.250	3.750	2.172	3.828	2.136	3.864
7.0	2.625	4.375	2.534	4.466	2.492	4.508
8.0	3.000	5.000	2.896	5.104	2.848	5.152

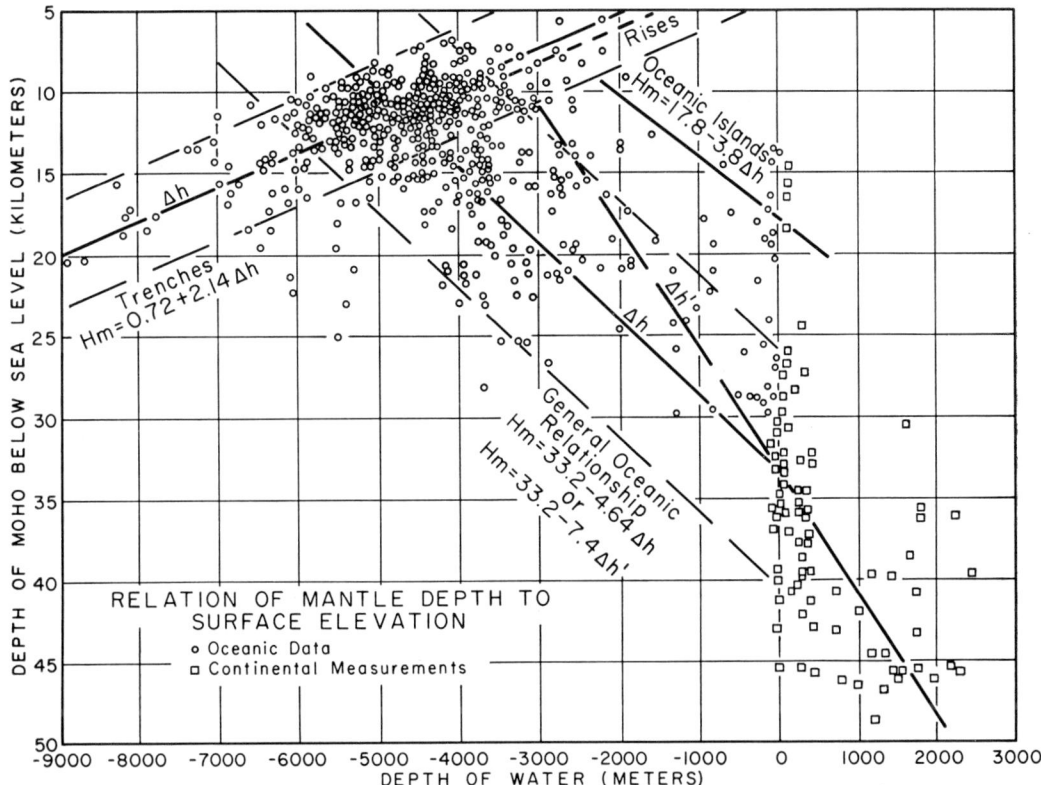

Fig. 2. Relation of depth of "M" discontinuity to ocean depth on world-wide basis.

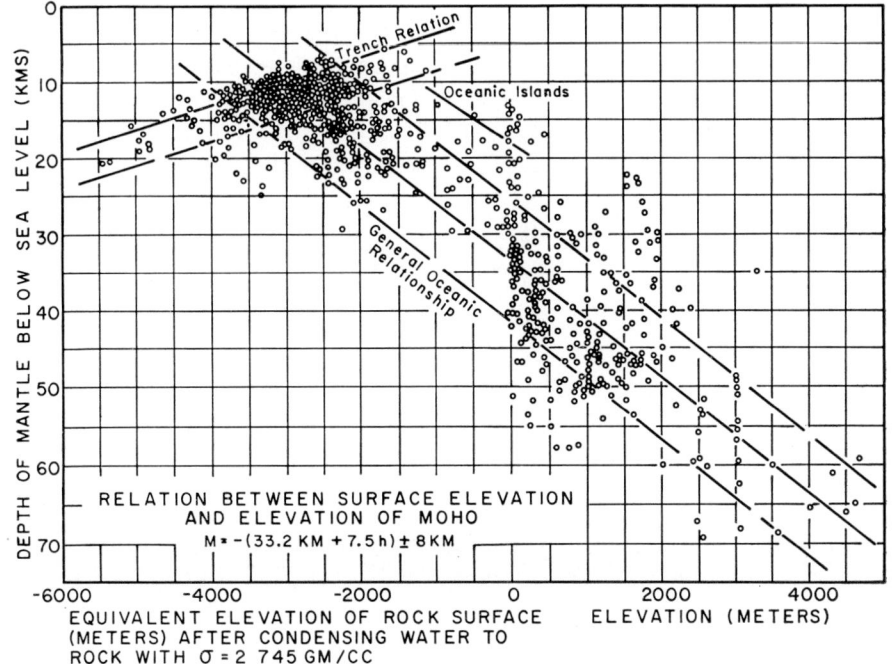

Fig. 3. Relation of depth of "M" discontinuity to surface elevation of continents and condensed water column equivalent rock elevation in the oceans.

with a sea level intercept value corresponding to 33.2 km and 34.0 km respectively. A combined plot of both land and sea data (fig. 3) using a density of 2.745 gm/cm^3 in condensing the water column to an equivalent rock column shows that the relation could be either $H_m = -(33.2+7.5\Delta h)$ or $H_m = -(34+7.7\Delta h)$. If the water column is condensed to equivalent rock material of 2.90 gm/cm^3, the relations could be either $H_m = -(34.5+7.55\Delta h)$ or $H_m = -(34.0+7.3\Delta h)$ depending on the weight given the high altitude observations. On any basis it appears the ratio of $\Delta R/\Delta h$ lies between 7.3 and 7.7 and the sea level intercept value between 33.2 and 34.5 km.

As the mean crustal velocity excluding sediments ($V < 4$ km/sec) for some 260 seismic measurements at sea is 6.35 km/sec and that for some 160 continental measurements is 6.51 km/sec, (WOOLLARD, 1962) there is no marked difference in overall crustal characteristics other than thickness and structural composition. An average of the two sets of data gives a velocity value of approximately 6.43 km/sec. Whether the Nafe-Drake curve relating seismic velocity to density (TALWANI et al., 1959) or that of the writer (WOOLLARD, 1968) is used, the crustal density indicated is in the range 2.87–2.89 gm/cm^3.

If we use the ratio $\Delta R/\Delta h = \sigma_c/(\sigma_m - \sigma_c)$ varying from 7.0 to 7.7 with σ_c in the range 2.87–2.89 gm/cm^3, the corresponding values of σ_m are as shown in table 3.

It is clear from the above table that if isostatic equilibrium is accomplished at the crust-mantle interface defined seismically, and the experimental determination of $\Delta R/\Delta h$ and the relations used for defining the density equivalence of seismic velocity values are valid, the upper mantle cannot, within the limits of probable experimental error, have a density lower than 3.243 gm/cm^3 or higher than 3.315 gm/cm^3. If the median mantle velocity from all seismic crustal measurements is also considered, it is found to be approximately 8.15 km/sec (8.08 km/sec on the continents and 8.20 km/sec in the oceans). The most probable rock type satisfying these average parameters of density and seismic velocity under the pressure and probable temperature environment of the mantle is an olivine-rich peridotite or pyrolite. The mantle immediately beneath the crust could also be an eclogite with a density of around 3.5 gm/cm^3 if we choose only those eclogites having a seismic velocity of around 8.15 km/sec under mantle environment conditions. However, in this case, the density contrast between the crust and mantle would be around 0.5 gm/cm^3 and the ratio $\sigma_c/(\sigma_m - \sigma_c)$ would be about 5.8 to 6.0 rather than 7.3 to 7.5. A principal argument for their not being an eclogite mantle therefore lies in the empirical relation of ΔR to Δh.

If we examine the relation of Bouguer anomaly changes to surface elevation (fig. 1) which on the average can be expressed as BA $= -0.084\Delta h + 0$ when $\sigma_c = 2.67$ gm/cm^3 as plotted in fig. 1 or BA $= -0.074\Delta h + 0$ when $\sigma_c = 2.9$ gm/cm^3, it is seen that the average change in Bouguer anomaly for 1 km change in surface elevation (Δh) is 74 mgal. This should equal the compensation effect defined by a plot of Bouguer anomaly values versus mantle depth values if compensation is achieved at the crust-mantle interface. Fig. 4 is a plot of the data for those seismic sites shown in fig. 2 for which there are gravity values. This plot also includes phase velocity depth determinations to the "M" discontinuity to both expand the sample of data and permit a comparison of these data with explosive

TABLE 3

Value of mantle density (σ_m) for values of crustal density (σ_c) = 2.87 to 2.9 gm/cm^3 and $\Delta R/\Delta h$ = 7.0 to 7.7 under equilibrium conditions

$\Delta R/\Delta h$	σ_m for $\sigma_c = 2.9$	$\Delta\sigma$	σ_m for $\sigma_c = 2.89$	$\Delta\sigma$	σ_m for $\sigma_c = 2.88$	$\Delta\sigma$	σ_m for $\sigma_c = 2.87$	$\Delta\sigma$
7.7	3.277	0.377	3.266	0.376	3.254	0.374	3.243	0.373
7.6	3.281	0.381	3.270	0.380	3.259	0.379	3.247	0.377
7.5	3.286	0.386	3.275	0.385	3.264	0.384	3.252	0.382
7.4	3.292	0.392	3.280	0.390	3.269	0.389	3.257	0.389
7.3	3.297	0.397	3.286	0.396	3.275	0.395	3.263	0.393
7.2	3.303	0.403	3.292	0.402	3.280	0.400	3.269	0.399
7.1	3.309	0.409	3.298	0.408	3.286	0.406	3.275	0.405
7.0	3.315	0.415	3.304	0.414	3.292	0.412	3.281	0.411

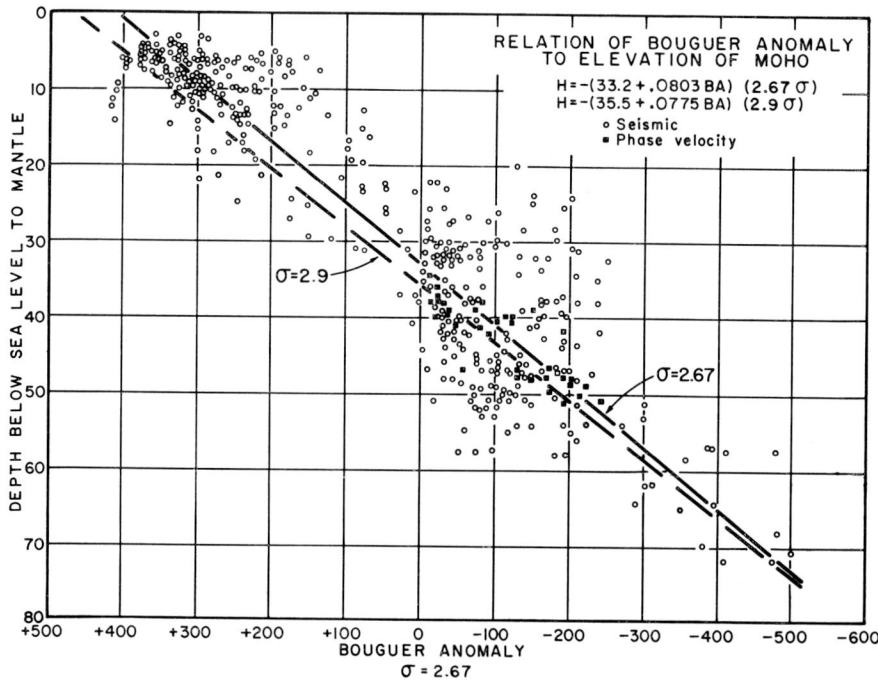

Fig. 4. Relation of depth of "M" discontinuity to Bouguer gravity anomaly values.

refraction results. A best fit to these data, which are based on Bouguer anomaly values computed with a density of 2.67 gm/cm³, gives the expression $H_m = -(33.2+0.803\,BA)$ where H_m is the depth of the mantle below sea level. If a crustal density of 2.9 gm/cm³ had been used, the expression would be $H_m = -(35.5+0.775\,BA)$. The respective change in mgal per kilometer of root increment (ΔR) is 12.5 mgal and 12.9 mgal. For a value of $\Delta R/\Delta h = 7.5$ as suggested empirically by fig. 3, when $\sigma_c = 2.9$, the slope of the Bouguer anomaly curve for change in elevation should therefore be about 97 mgal per 1000 meters. This is in poor agreement with the average value for fig. 1 but is in close agreement with the value for the United States (93 mgal/1000 m using $\sigma_c = 2.9$ gm/cm³) which is the principal source of the land seismic measurements used in fig. 4. Complete agreement with the United States values could be obtained by changing the value of $\Delta R/\Delta h$ from 7.5 to 7.2, or alternately using a slightly lower crustal density, or by making a slight adjustment in slope in fitting the Bouguer anomaly values to the seismic depths.

In connection with the Bouguer anomaly relations it should be noted that the magnitude of the observed change in Bouguer anomaly with elevation does not agree with the theoretical slab effect of the crustal root increment for 1 km change in surface elevation. If all the compensation is achieved at the "M" discontinuity, the theoretical Bouguer anomaly should be $BA = 2\pi G \times \Delta R \times (\sigma_m - \sigma_c)$. As shown in table 3 when $\Delta R/\Delta h = 7.2$ to 7.5, $(\sigma_m - \sigma_c) = 0.4$ to 0.386 gm/cm³. Assuming either value, BA should be about 120 mgal. There is a logical explanation for this apparent discrepancy of about 25 mgal between the theoretical and observed Bouguer anomaly for 1 km change in elevation. It can be attributed in part to the depth of the compensation mass distribution which is normally below the depth of the base of the sea level column on the continents, and in part, to Earth curvature. As a result, the compensation effect only equals about 75% of the topographic mass effect out to $R = 167$ km (zone 0), the distance at which Earth curvature becomes important. The gravitational effect of topography is therefore not matched by that of its crustal root counterpart locally, and the topographic effect will always be more positive. As the Bouguer anomaly has a negative sign, this reduces the anomaly. Therefore if the theoretical root effect is 120 mgal per kilometer change in surface elevation, the expected Bouguer anomaly will only be about 75% of this value. In this case, the expected value is about –90 mgal per km change in elevation which agrees closely with the empirical relations established.

No firm case can therefore be made in general for mass distributions deeper than the "M" discontinuity being important in maintaining isostatic equilibrium. It is also clear that the upper mantle, on the average, is more like a dunite or pyrolite than an eclogite, and the most probable value of $\Delta R/\Delta h$ appears to be around 7.5 which corresponds to a value of $(\sigma_m - \sigma_c) = 0.385$ gm/cm^3 when σ_c is in the range 2.88 to 2.90 gm/cm^3.

5. Dependence of crustal parameters on mantle velocity

That there is a dependence of crustal parameters, as expressed in the depth to the "M" horizon, and the mean velocity of the crust has been brought out earlier by the writer (WOOLLARD, 1968). The dependence of

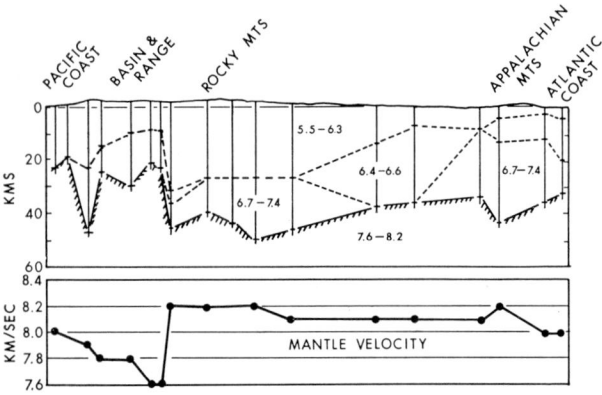

Fig. 5. Composite crustal and mantle velocity profile across the United States.

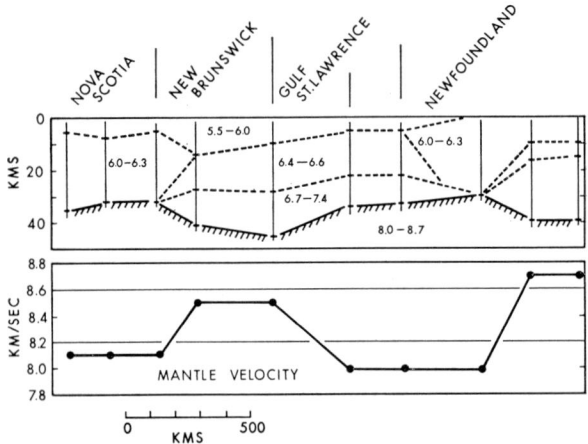

Fig. 6. Crustal and mantle velocity section in Canadian Maritime Provinces based on EWING et al. (1966).

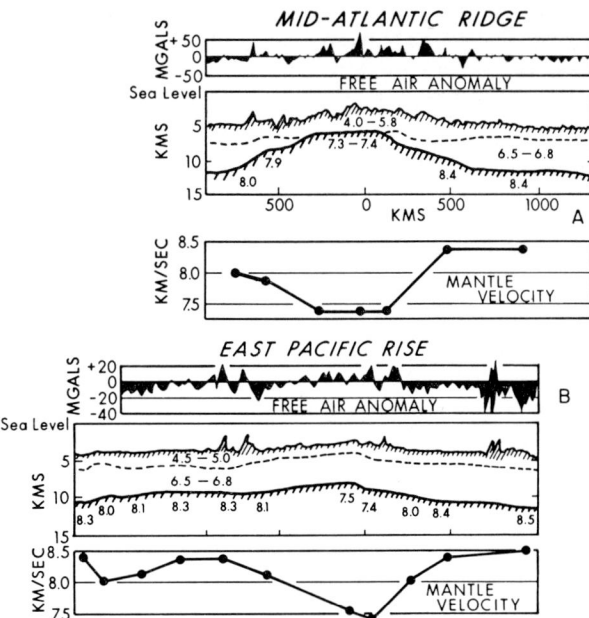

Fig. 7. Comparative crustal, gravity and mantle velocity profiles across the Mid-Atlantic Ridge and East Pacific Rise.

mantle depth on mantle velocity is evident from the typical profiles shown in figs. 5–7. Fig. 5 is a regional profile across continental North America; fig. 6 is a more restricted profile in the Appalachian tectonic province of Newfoundland and the adjacent Maritime Provinces. Fig. 7 shows the relations observed in crossing the Mid-Atlantic Ridge and the East Pacific Rise.

If all the data for North America (over 100 data points) are looked at, it is found that although there is a considerable spread in values, average values do define a systematic relationship between mantle depth (H_m) and mantle velocity (V_m). This plot is shown in fig. 8, and the relation defined can be written $H_m = -[34 + (V_m - 7.95)33.2]$ km.

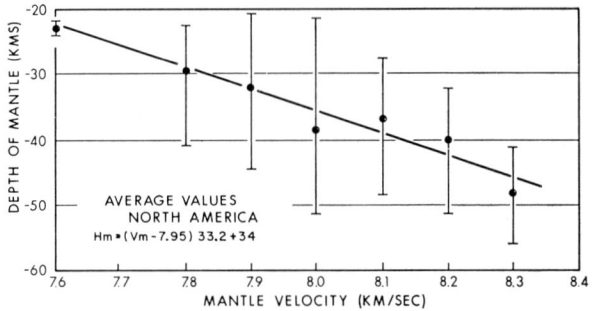

Fig. 8. Relation of depth of "M" discontinuity to mantle velocity in North America.

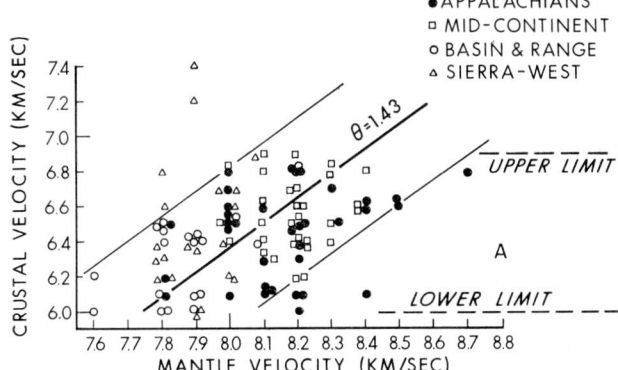

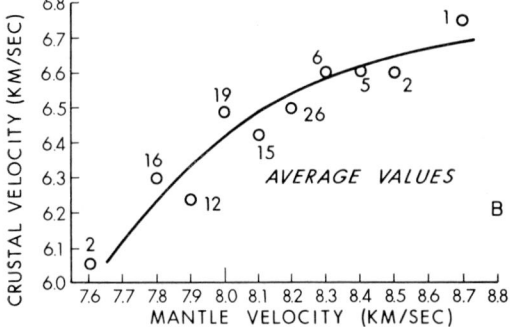

Fig. 9. Relation of mean crustal velocity to mantle velocity in North America.

sults in an exponential-type relationship with the curve becoming asymptotic at the upper velocity boundary as shown in fig. 9B.

That the changes in crustal parameters are related to an increase in density of the crust where the crust thickens is indicated by the plot of mean isostatic anomaly values as a function of mantle velocity. This plot (fig. 10) can be expressed as $IA = (V_m - 8.05)$ 85.6 mgal. The change of ≈ 8.6 mgal per 0.1 km/sec change in isostatic anomaly corresponds therefore to 3.4 km change in mantle depth. The gravity is anomalous because when H_m increases, the negative gravity effect of the additional crustal root increment is more than offset by a positive gravity effect. This can only originate through an increase in crustal density. That 0.1 km/sec change in mantle velocity results in the equivalent of 0.03 gm/cm³ change in crustal density is evident from fig. 11 which is a plot of theoretical changes to be expected in mantle depth for differences in $(\sigma_m - \sigma_c)$ using $\sigma_c = 2.9$ gm/cm³ and $(\sigma_m - \sigma_c) = 0.4$

The relation of crustal velocity (V_c) to mantle velocity is not a simple one because in general the mean velocity of the crust is bounded by discrete limits of 6.0 km/sec and 6.9 km/sec. Although crustal velocity values of over 7.0 km per second appear to characterize the Olympic Peninsula area of the state of Washington in the United States, this is an exceptional occurrence. In general, it is found that although the velocity of the crust varies over the full range of values for any value of mantle velocity as shown in figure 9A, there is on the average a dependence on mantle velocity that re-

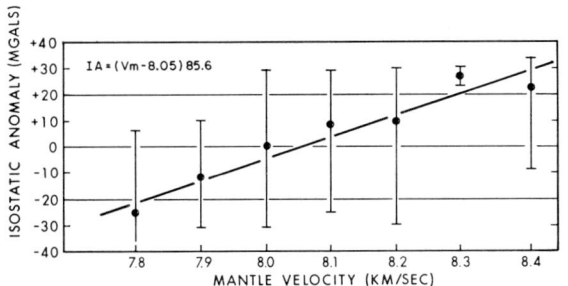

Fig. 10. Relation of isostatic gravity anomaly to mantle velocity in North America.

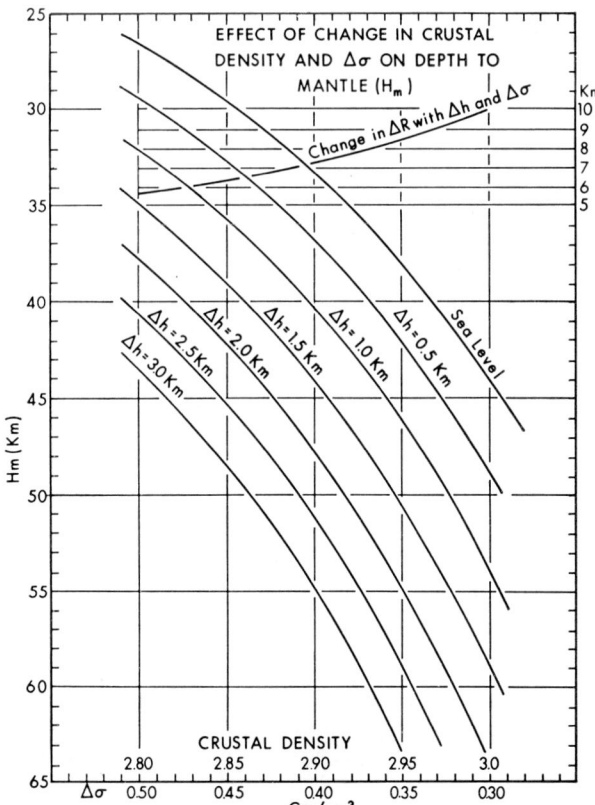

Fig. 11. Theoretical changes in mantle depth for changes in the density contrast between the crust and mantle for different values of surface elevation.

gm/cm³ as approximations of normal conditions.

The general relations portrayed are therefore those set forth in the beginning of this paper defining the problem. As the available heat flow data suggest the mantle velocity is subnormal (<8.15 km/sec) and that the crust has a subnormal mean velocity (<6.43 km/sec) and a subnormal thickness for the surface elevation based on $H_m = 33.2 + 7.4 \Delta h$ where there is high heat flow, and that the relations are opposite where there is abnormal heat flow, it appears the relations are not static but temperature controlled. A dynamic system is also suggested by the geologic evidence of contemporaneous crustal uplift and subsidence in adjacent areas over long periods of time in the tectonically stable shield area. Certainly the seismic data indicate that the crustal and upper mantle relations beneath the basins and uplifts studied to date fit the above general relations and that there has been thickening and densification of the crust with a decrease in $(\sigma_m - \sigma_c)$ beneath basins with an increase in gravity, and that uplifts are areas where the process has operated in reverse with a decrease in gravity.

In the oceans we also find evidence for extensive changes in crust and mantle relations. One of the most pronounced ones of opposite sign to the relations shown in fig. 7 for the oceanic rises being oceanic ridges formed over fracture systems. In contrast to the rises these are not related to crustal spreading. Fig. 12 shows the relations across the Hawaiian Ridge. It is quite evident from this section, that if the original oceanic crust conformed to that defined by RAITT (1963) for the abyssal portion of the Pacific Ocean with a water depth of 5 to 5.5 km, an upper crustal layer 1.71 km thick with a velocity of 5.07 km/sec, and a basal layer 4.86 km thick having a velocity of 6.69 km/sec, there have been extensive changes both beneath the ridge and to the south of the ridge. To the north of the ridge the velocity structure and thickness of the crust and the velocity of the upper mantle conform closely to Raitt's average model. Although there has been tectonic uplift to form an arch which places the Moho at a depth of only 10 km below sea level at the crest of the arch, there has been no distortion of crustal parameters or change in mantle velocity. Presumably the arch is a consequence of crustal subsidence beneath the ridge after it was originally built up above the sea floor by outpourings of basaltic lava that now extend up to over 4000 meters above sea level on the island of Hawaii. If the interface between the 5.0 km/sec and 6.8 km/sec layers beneath the ridge (site B) is used as a marker horizon, the subsidence that has taken place cannot have been much more than 2.5 km. This would account for the tectonic uplift of about 2 km beneath the arch. However, it is clear that the basal layer of the crust beneath the ridge with a velocity of 6.8 km/sec has thickened from approximately 5 km to over 11 km at the expense of former mantle material. It is also to be noted that the mantle velocity here is abnormally high (8.5 km/sec). Site A is the section over an intrusive area (rift), and as shown by FURUMOTO et al. (1965), seismic studies over one of the two primary volcanic pipes that led to the formation of the island of Oahu show 7.6 km/sec material at a depth of only 2 km beneath the surface. South of the Hawaiian Ridge, the entire velocity structure and thickness of the crust is abnormal, and as seen, the mantle velocity is markedly high (8.6 km/sec). The general trend in relationships found on the continent therefore are also evident here and beneath the rises.

6. Relations where there is sub-Moho layering

Although sub-Moho layering above the velocity reversal at about −110 km has not received much attention, there is some indication that where this layer is found, there is a connection between the depth of the "M" discontinuity, the velocity of the "M" horizon, and the velocity and depth of the sub-"M" layer. In other words, a downward extension of the pattern is observed between the parameters of the crust and the velocity of the mantle associated with the "M" dis-

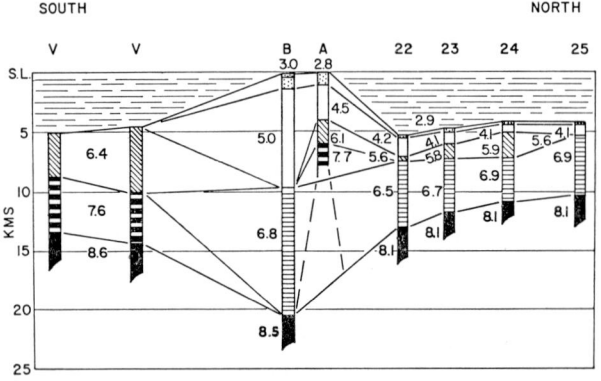

Fig. 12. Composite crustal section across the Hawaiian Ridge. A = normal ridge section, B = section over center of volcanic eruption.

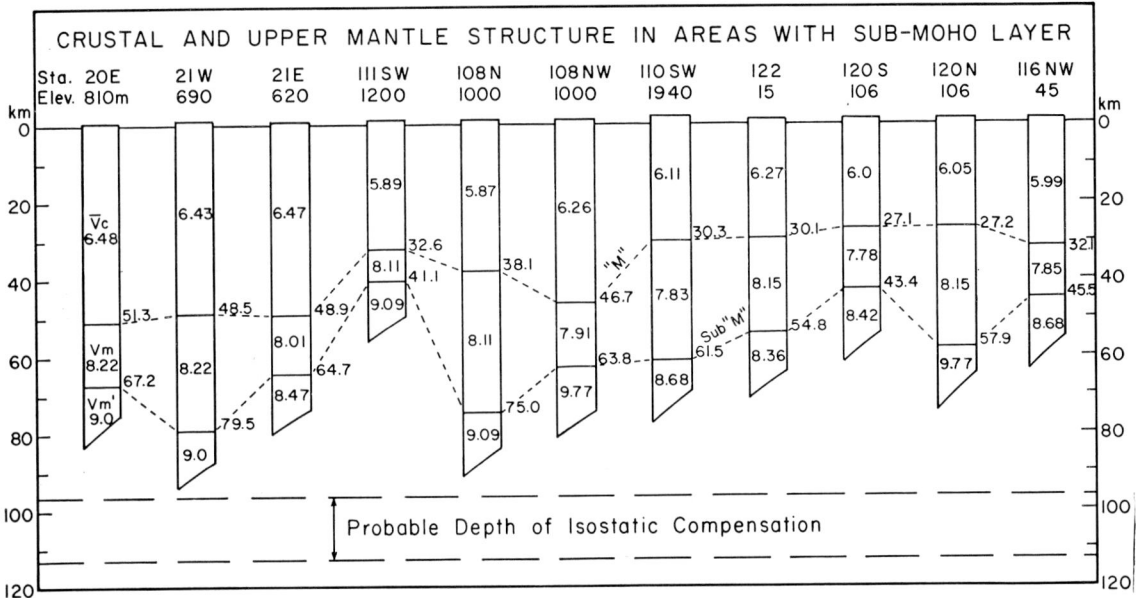

Fig. 13. Seismic sections in areas where there is good evidence of sub-"M" layering.

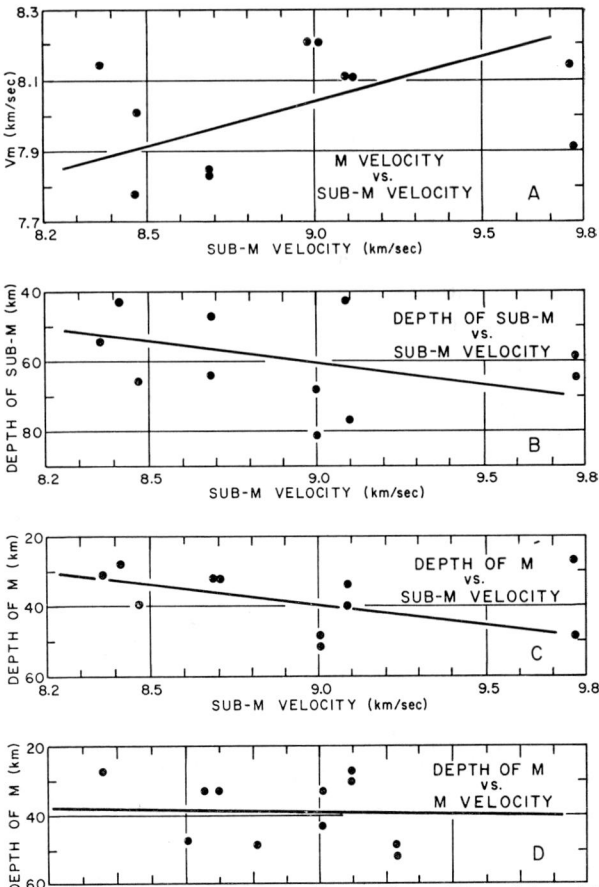

continuity. That sub-"M" layering is not general, but of a local nature, is suggested by the fact that despite long-range recording from nuclear blasts it has not been reported. All of the evidence that appears valid is from local high explosive blasts, and this is only for four areas (eastern Montana, California, Minnesota, and Utah). The available evidence on the sub-"M" layering is summarized in fig. 13. Stations 20 E, 21 W, and 21 E are in eastern Montana, and the balance are in California. The data for Utah and Minnesota were not included because there is only a single set of observations for each of these areas.

In order to see if there were any systematic relations, the several parameters that might have significance were plotted as a function of sub-"M" velocity. The plots shown in fig. 14 are:

A) the dependence of the "M" velocity on the sub-"M" velocity;

B) the dependence of the depth of the sub-"M" horizon on the sub-"M" velocity;

C) the dependence of the depth of the "M" horizon on the sub-"M" velocity;

Fig. 14. Relation of sub-"M" and "M" parameters to the velocity beneath sub-"M" layering. A – relation of mantle velocity. B – relation of depth of sub-"M" layer. C – relation of depth of "M" discontinuity. D – relation of depth of "M" discontinuity to "M" velocity.

D) the dependence of the depth of the "M" horizon on the velocity of the "M" discontinuity.

This last plot was made to see if compensation was indicated at the "M" horizon as is normal, or whether it was apparently achieved at the depth of the sub-"M" horizon.

An average best fit to each of the above plots suggest:

a) there is an apparent dependence of the "M" velocity value on the sub-"M" velocity value which amounts on the average to about +0.26 km/sec for 1 km/sec increase in sub-"M" velocity. If only one point were a poor observation, the slope factor could be as high as 0.36.

b) there is an apparent dependence of the depth of the sub-"M" horizon on sub-"M" velocity that amount on the average to about +11 km for 1 km/sec increase in sub-"M" velocity.

c) there is an apparent dependence of the depth of the "M" discontinuity on sub-"M" velocity that amounts on the average to about +13 km for 1 km/sec increase in sub-"M" velocity but could be nearer +20 km if one point (the same apparently erratic site as in plot A) was rejected.

d) there is either no, or little apparent dependence of the depth of the "M" horizon on the "M" velocity where there is sub-"M" layering.

We can conclude from the above that whatever controls the formation of sub-Moho layer, its effect extends through to the crust and that the resulting mass distribution down to the sub-"M" layer is that which is in equilibrium with sub-"M" material down to the depth of equal mass. To test this the data for the four sites having a sub-"M" velocity of ≈ 9.0 km/sec were used to define what density would give equal mass at the probable depth of compensation (96–113.6) km, defined originally by BOWIE (1917) as that depth yielding a minimum isostatic anomaly using the Pratt-Hayford concept of isostasy. For present purposes the depth adopted was 105 km.

The data are:

Sta 20E, $h = 810$ m, $H_m = 51.3$ km, $H = 52.1$ km,
  $V_c = 6.48$ k/s, $\sigma_c = 2.91^*$ gm/cm^3,
  "M" to Sub-"M" layer $(M) = 15.9$ km,
  $V_m = 8.22$ km/sec,
  $\sigma_m = 3.335^*$ gm/cm^3,

* Based on velocity-density relation of WOOLLARD (1968).

Depth to Sub-"M" layer $H_{SM} = 67.2$ km,
Sub-"M" layer to 105 km $(SM) = 38.8$ km,
$V_{SM} = 9.0$ km/sec.

Sta 21W, $h = 690$ m, $H_m = 48.5$ km, $H = 49.2$ km,
  $V_c = 6.43$ km/sec, $\sigma_c = 2.885$ gm/cm^3,
  $M = 31.0$ km, $V_m = 8.22$ km/sec,
  $\sigma_m = 3.335$ gm/cm^3,
  $H_{SM} = 79.5$ km,
  $SM = 25.5$ km, $V_{SM} = 9.0$ km/sec.

Sta 111SW, $h = 1200$ m, $H_m = 32.6$ km, $H = 33.8$ km,
  $V_c = 5.89$ km/sec, $\sigma_c = 2.67$ gm/cm^3,
  $M = 8.5$ km, $V_m = 8.11$ km/sec,
  $\sigma_m = 3.322$ gm/cm^3,
  $H_{SM} = 41.1$ km,
  $SM = 63.9$ km, $V_{SM} = 9.09$ km/sec.

Sta 108, $h = 1000$ m, $H_m = 38.1$ km, $H = 39.1$ km,
  $V_c = 5.87$ km/sec, $\sigma_c = 2.665$ gm/cm^3,
  $M = 36.9$ km, $V_m = 8.11$ km/sec,
  $\sigma_m = 3.322$ gm/cm^3,
  $H_{SM} = 75.0$,
  $SM = 30.0$ km, $V_{SM} = 9.09$ km/sec.

If we equate masses above the –105 km level on the basis that the total mass above this level is $H \times \sigma_c + M \times \sigma_m + SM \times \sigma_x$ and solve for σ_x it is found that the apparent density equivalent for $V_{SM} \approx 9.0$ km/sec = 3.08 gm/cm^3 using Montana stations 20E and 21W and 3.19 gm/cm^3 using California stations 111SW and 108. The agreement is therefore close, and the fact that the density value for 9.0 km/sec is less than that for normal upper mantle materials suggests that either the density values used for σ_c and σ_m are too large or that the depth of equal mass is considerably greater. An alternative is that the higher velocity noted below the sub-"M" horizon is occasioned by a decrease in Fe/Mg ratio and an actual decrease in density which would account for the increase in velocity value.

To test the validity of the above result, a plot was made to determine if there is a systematic relationship between the depth of sub-"M" layering and surface elevation since fig. 14-D indicates that where there is a sub-"M" layer the normal dependence of depth to the mantle on V_m is not evident. This plot (fig. 15) shows that except for two sites (111SW and 110SW) whose values are remote from all the other values, a systematic relation is defined with $\Delta R/\Delta h \approx 24$. As

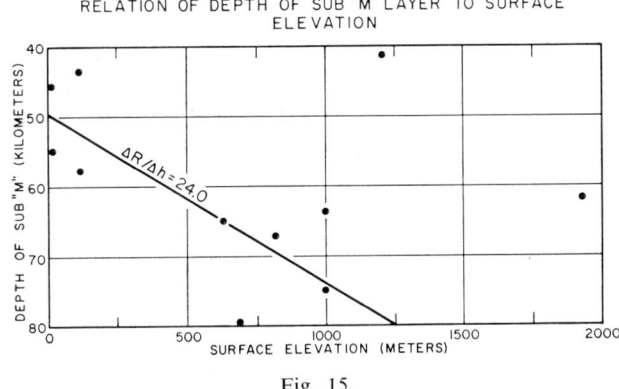

Fig. 15.

7. Evidence for regional changes in crust-mantle relations

That the value of R/F changes both locally and regionally in North America and is not strictly related to changes in elevation can be shown if we substitute density values for seismic velocity values and solve for R and F using the relations $R = H\sigma_c/\sigma_m$ and $F = H - R$. The velocity-density relation used was that described by the writer (WOOLLARD, 1969). If the Nafe-Drake relation published by TALWANI et al. (1959) or the BIRCH (1961) relation for a fixed atomic weight is used, the values of R/F will be somewhat lower than those shown below. The derived R/F values by physiographic provinces along an east-west traverse are as shown in table 4.

The relations shown in table 4 explain the wide scatter in values shown in fig. 3 for the relation of the depth of the mantle to surface elevation. There are several areas where there are significant differences in the ratio R/F as computed using the observed values of H and derived values of R and F and as computed using the ratio $\sigma_c/(\sigma_m-\sigma_c)$. Where there are only one or two observations, the difference in R/F values could be due to poor data. In some cases it could be related to unknown sub-"M" layering in the mantle. In at least two areas (the Appalachian Mountains and Puget Sound) there appears to be a definite departure from isostatic equilibrium. It is clear that in addition to R/F not being constant, changes in R/F are not related to elevation although conceivably one could imagine an

$\Delta R/\Delta h = \sigma_c/(\sigma_m - \sigma_c)$ under equilibrium conditions, it is possible to determine σ_m if σ_c is known. Mean values of σ_c were determined for the column above the sub-"M" layer using the density-velocity plot of the writer (WOOLLARD, 1968), and except for site 111SW, all values are within ± 0.1 gm/cm^3 of a median value of 3.0 gm/cm^3. Using this value for σ_c and $\Delta R/\Delta h = 24$, the value of $(\sigma_m - \sigma_c)$ is 0.125 gm/cm^3. The density indicated for the material beneath the sub-"M" layer is therefore 3.125 gm/cm^3. This is essentially the same value as obtained using the alternate approach for isostatic balance above the -105 km level.

The evidence for isostasy and the maintenance of equilibrium through vertical mass exchange with consequent changes in seismic velocity and density values therefore appears to hold for these areas as well as for the general situation where there is no apparent sub-"M" layering.

TABLE 4

Area	Sample	R/F	Elv (m)	σ_c (gm/cm^3)	σ_m (gm/cm^3)	$\Delta\sigma$ (gm/cm^3)	Equilibrium R/F for σ_m and σ_c
Taconic-Appalachian Piedmont	25	5.3	10	2.86	3.34	0.48	5.95
Appalachian Mts.	2	7.5	525	2.85	3.32	0.47	6.06
Gulf Coast	3	4.9	53	2.88	3.37	0.49	5.85
Interior Shield	17	6.8	270	2.79	3.33	0.54	5.16
High Plains	15	7.6	1040	2.97	3.35	0.38	7.82
Rocky Mts.	4	7.7	1600	2.97	3.34	0.37	8.0
Columbia Plateau	5	7.4	1250	2.91	3.29	0.38	7.65
Colorado Plateau	5	7.1	1940	2.90	3.31	0.41	7.1
Basin and Range	18	5.8	1400	2.78	3.27	0.49	5.65
Sierra Nevada	4	8.3	1680	2.95	3.29	0.34	8.7
Cascade Mts.	1	8.4	1070	2.94	3.30	0.46	6.4
Coastal California	12	6.3	10	2.83	3.29	0.46	6.2
Puget Sound	2	8.6	5	3.11	3.27	0.17	18.2

exponential relationship if the Basin and Range data and that for Puget Sound were rejected. However, neither of these groups of data can be rejected as being poorly substantiated. They do represent extraordinary conditions (in opposite phase) from all the other values and are probably indicative of marked changes in mantle regime. In general aspects, the Basin and Range relations are very similar to those found under the oceanic rises and as mentioned earlier both areas are characterized by high heat flow.

8. Conclusions

Whether the picture presented of crustal and mantle heterogeneity is due to chemical fractionation, undefined mineralogic transformations governed by local variations in heat flow or a combination of effects is not known. However, there appears to be an orderly intimate inter-dependence of physical parameters between the crust and mantle. In most areas the inter-dependence conforms to that to be expected for isostatic equilibrium at the "M" discontinuity. On the basis of probable density equivalent values for seismic velocities, it appears that no specific rock terms can be used to describe the crust or mantle. It does appear that the crust is a derivative of the mantle and that its overall physical parameters, and in many places its internal structure and composition also, are directly related to those of the mantle as defined by the velocity of the mantle. There is a similar inter-dependence of overall physical parameters where there is sub-"M" layering. However, in these cases isostasy appears to be related to the column above the sub-"M" discontinuity rather than that above the "M" discontinuity. Although the orderly vertical mass distribution found in most areas could be the product of vertical chemical fractionation, the evidence of reversals in pattern as defined by changes in crustal and upper mantle parameters in association with crustal uplift and subsidence suggest a reversible process. The fact that abnormalities in heat flow also appear to correlate with abnormalities in crustal and mantle parameters suggests a process that is sensitive to changes in thermal regime. Although seismic velocity values at depth would be sensitive to temperature abnormalities, it is unlikely that temperature alone would change the layered structure of the crust although it could affect the observed velocity values for both the crust and mantle and also influence the derived values of crustal thickness since the thickness value is in part dependent upon the ratio V_c/V_m. The only logical alternative appears to be temperature control of a mineralogic transformation involving a change in density and rigidity.

References

BIRCH, F. (1961) The velocity of compressional waves in rocks to 10 kb, Part 2, J. Geophys. Res. **66**, 2199–2224.

BOWIE, W. (1917) Investigations of gravity and isostasy, U.S. coast and geod. surv., Spec. Publ. **40**, 196 pp.

EWING, G. N., A. M. DAINTY, J. E. BLANCHARD and M. J. KEEN (1966) Seismic studies on the eastern seaboard of Canada, Can. J. Earth Sci. **3**, 89–110.

FURUMOTO, A. S., N. J. THOMPSON and G. P. WOOLLARD (1965) The structure of Koolau volcano from seismic refraction studies, Pacific Sci. **19** (3), 306–314.

GAPOSHKIN, E. M. (1966) A dynamical solution for the tesseral harmonics of the geopotentials for station coordinates, Trans. Am. Geophys. Un. **47**, 47.

HEISKANEN, W. A. and F. A. VENING MEINESZ (1958) *The Earth and its gravity field* (McGraw-Hill) 470 pp.

HESS, H. H. (1955) Serpentines, orogeny and epeirogeny, in: A. Poldervaart, ed., *Crust of the Earth*, Geol. Soc. Am. Spec. paper **62**, 391–408.

KENNEDY, G. C. (1959) The origin of continents, mountain ranges and ocean basins, Am. Scientist **47**, 494–504.

KHAN, M. A. and G. P. WOOLLARD (1968) Methods of analysis and comparison of geophysical data on a plane with specific application to the Solomon Island Area, Hawaii Inst. Geophys. Report 68–17, 76 pp. plus tables.

KOZAI, Y. (1964) New determinations of zonal harmonic coefficients of the Earth's gravitational potential, Smithsonian Astrophys. Obser. Spec. Report 165.

RAITT, R. W. (1963) *The crustal rocks in the sea*, Vol. III, (John Wiley and Sons, N.Y.) pp. 85–102.

SCLAR, C. B. (1970) Phys. Earth Planet. Interiors **3**, 333.

SIMMONS, G. (1964) Velocity of compressional waves in various minerals at pressures to 10 kb, J. Geophys. Res. **69**, 1117–1121.

TALWANI, M., J. L. WORZEL and M. LANDISMAN (1959) Rapid gravity computations for two dimensional bodies with application to the Mendocino Submarine Fracture Zone, J. Geophys. Res. **64**, 49–59.

WOOLLARD, G. P. (1962) The relation of gravity anomalies to surface elevation, crustal structure and geology, Univ. Wisconsin Geophys. Polar Res. Centr. Report 62-9, pp. 292.

WOOLLARD, G. P. (1968) The interrelationship of the crust, the upper mantle and isostatic gravity anomalies in the United States, in: Knopoff, Drake and Hart, eds., *The crust and upper mantle of the Pacific Area*, Am. Geoph. Union Monograph **12**, 312–240.

PHASE TRANSFORMATIONS WITHIN THE EARTH'S MANTLE AS A CAUSE OF CRUSTAL MOVEMENT AND A SOURCE OF CRUSTAL MATERIAL

S. I. SUBBOTIN

University of Kiev, Kiev, USSR

Analysis of the data of investigations into tectogenesis and problems of the crust and mantle physics, carried out by scientists of many countries during last decades and published in the works of ARCHANGELSKY, (1936), BELOUSSOW (1941), BERNAL (1936), BRIDGMAN (1948), BULLEN (1961), GORANSON (1940), GUTENBERG (1951), JEFFREYS (1952), LOVERING (1958), MAGNITSKY (1965), RAMSEY (1950), RIKITAKE (1959), RINGWOOD (1962), SCHATSKY (1955), UFFEN (1959), as well as numerous other authors, makes it possible to outline some ways of solving the problem of the causes of tectogenesis and suggest likely schemes of the formation of the main crustal structures.

The material of the Earth's mantle is subjected to the action of pressures and temperatures whose values are continuous monotonic functions of depth:

$$P = f(H), \quad T = \varphi(H). \tag{1}$$

In terms of the physical properties of the mantle material, the value of its density should also be a generally monotonically increasing function of depth, which,

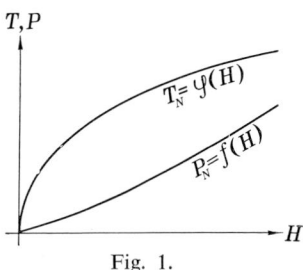

Fig. 1.

however, under certain thermodynamic conditions experiences discontinuous changes at certain depths:

$$\rho = \Psi_1(P,T) = \Psi(H). \tag{2}$$

The above conditions will be termed critical. In that case continuous portions of the curve will be representative of a steady increase in density at the expense of decreasing intermolecular and interatomic distances, while points of discontinuity represent a discontinuous increase of density, resulting from rearrangement in the

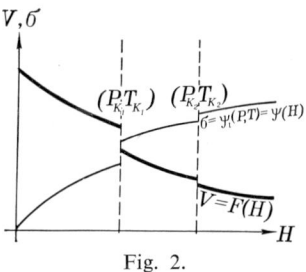

Fig. 2.

structure of the material at the expense of its phase, polymorphic, and electronic conversions. The reverse process, involving decrease in density with relaxation of stresses at shallower depths, follows an analogous pattern.

At points close to critical, a small increment of pressure (either positive or negative), sufficient for such critical conditions to be achieved, may result not only from an increase (decrease) of depth, but also from an increase (or relaxation) of stresses in connection with changes in the rate of rotation of the Earth, or at the expense of the endogenous and exogenous processes occurring in its interior. Spatial distribution of zones, where such additional stresses originate, is related to the inhomogeneities in the material and is generally confined to the upper mantle only. Consequently, it is the upper mantle proper which is responsible for supplying energy for differentiated motions, manifested within the crust in terms of troughs or uplifts of its areas as well as in other structural features.

Crustal structures, besides great diversity of their

forms, vary widely in their horizontal dimensions and amplitudes of vertical displacements. They range from geosynclinal troughs and folded mountain formations, having lateral dimensions of hundreds of kilometers and often extending to many thousands of kilometers, with amplitudes of vertical displacements within the first few tens of kilometers, up to local structural forms whose horizontal dimensions fall within several kilometers or even hundreds of meters only, with amplitudes of vertical displacements amounting to hundreds and tens of meters. The above facts suggest that such a wide range of scales of structural features should be governed by an analogous range of dimensions and depths of distribution of certain mantle volumes where compaction or expansion of its material occurs at the expense of phase and other transformations, which are the cause and source of energy for the formation of crustal structures.

Proceeding from the available petrological schemes of mantle composition, it is believed that phase, polymorphic and electronic transitions may occur at different depths as a function of the material composition and levels of thermodynamic condition.

Two trends of investigations – geological and physical (physics of the Earth) – result in one conclusion, that the processes responsible for tectogenesis occur at different depth levels within the upper mantle of the Earth. And it is natural that shallow horizons of the upper mantle are responsible for the formation of small-scale structural features, while deep horizons control large-scale features. The concepts of a layered non-homogeneous structure of the upper mantle have recently been corroborated by the results of studying the Earth's interior in Middle Asia and the Far East, especially so in the transition zone from the Asiatic continent to the Pacific ocean, and in other places on the globe. The above mentioned results, together with the data of a petrological model of the Earth, suggest an intricate structure of the upper mantle as well as of multiple-stage distribution of energy centres for tectonic movements and sources of the magmatic substance.

The largest structural forms are likely to originate in connection with transformations of the mantle material in the Gutenberg channel zone. It is in this zone that the mantle material is in a mobile, thermally "softened" state, due to a proximity of temperature levels of the mantle material to those of its complete or selective, component by component melting. In such a state when critical levels of thermodynamic conditions are reached, the crystalline lattice of the material is easier destructed and rearranged into a new modification, due to "thermal loosening".

The conditions for phase and other transformations may occasionally occur in the uppermost layers of the mantle, immediately under the crust. Phase conversions in such zones usually result in the formation of small-scale structures.

Thus the cause of downwarping and bulging-up of crustal areas lies in the compaction, compression, i.e. decrease of a certain volume of the mantle material, and in the reverse process – loosening, expansion, i.e. increase of a certain volume of the mantle material, all at the expense of its phase, polymorphic, electronic (and probably also chemical) transitions caused by a change in stress. The compaction brings about sinking of the overlying mantle and crust layers and the formation of troughs; conversely, the expansion causes their swelling up and formation of uplifts.

At the boundaries of contraction or expansion zones of the mantle substance there arise conditions appropriate for the formation of deep fractures.

A deep fracture may originate on the periphery of an area where phase or other transitions of the mantle material at depth occur, in the form of a mobile zone with relaxation (or increase) in stress (a). This is the first link of a deep fracture.

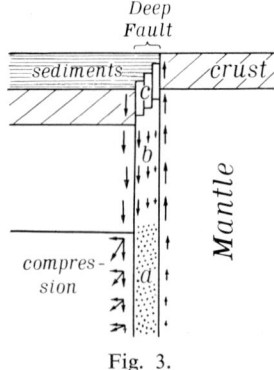

Fig. 3.

Spreading of phase (or other) transformations on to the overlying mantle layers causes appearance of the second link of a deep fracture – the zone of counter stresses and shear deformations (b). A further advance of the process results in the appearance of the third, uppermost link of a deep fracture – the zone of shattering of the Earth's crust (c).

In zones of deep fractures partial melting of the mantle material is possible, as well as concentration and movement up to the Earth's surface of various fluids.

So far as phase, polymorphic, and electronic transitions of the mantle material are considered as the major cause of tectogenesis, the following scheme of crustal downwarping and formation of magmatic chambers may be suggested:

1) Changes in the rotation rate as well as certain endogenous and exogenous processes within the Earth result in creating additional geodynamic and thermoelastic stresses.

2) The additional stresses (positive ones) add up to the normal stresses, and consequently, in some parts of the mantle, thermodynamic conditions reach critical values. The critical conditions promote phase or other transitions accompanied by a decrease in the volume of the material, which process in its turn, due to the gravity effect, causes downward movements of the overlying mantle layers.

3) At the initial stage of the downwarping, as subcrustal mantle layers move downwards, there develops, due to the strength of the "basaltic" and "granitic" crustal layers, a lower pressure zone in the mantle immediately under the crust. Lower pressure results in bringing down the melting temperature of the material, which causes the temperature level to increase; the least

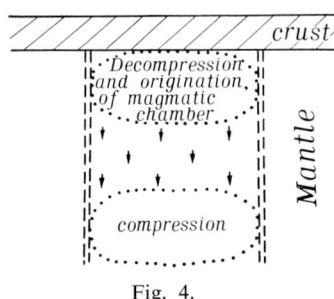

Fig. 4.

high-melting components or the whole mass of the mantle material undergoes melting and a magmatic chamber appears. Depending on thermal conditions and an inhomogeneous composition of the mantle material, a magmatic chamber may originate at some lower depth levels of the mantle, not immediately under the crust.

4) A further course of the process, consisting of the compression of mantle material at depth by way of involving still greater masses in phase or other transformations, brings about a new sinking down of upper mantle layers, which in its turn renders crustal strength incapable of supporting the burden. The area of the crust splits up and collapses, and a depression originates

5) An area of crustal downwarping (depression) be-

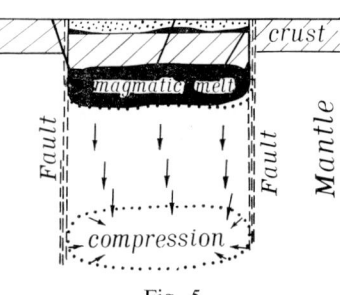

Fig. 5.

gins to be filled up with the magmatic melt supplied through fractures from below, as well as with terrigenous or even chemogenic sediments.

6) As the changes in the rate of rotation of the Earth may be of different sense, and the endogenous and exogenous processes may vary with time, additional geodynamic and thermoelastic stresses should inevitably change their sign. The latter phenomenon should result in a cessation of the compression process and its substitution by a reverse process, i.e. dilatation of the mantle material. The above accounts for the discontinuous character of sedimentation, as well as for partial erosion of the series already accumulated.

A scheme of the formation of crustal uplifts may easily be pictured in much the same way as the downwarping scheme, but with reverse signs of stresses and motions and an expansion of the mantle material instead of its contraction. It is also quite evident that in uplift zones the conditions favourable for the formation of a magmatic chamber are ruled out. Such conditions (decreasing pressure) are likely to arise only beyond an uplift zone, beyond deep fractures bounding

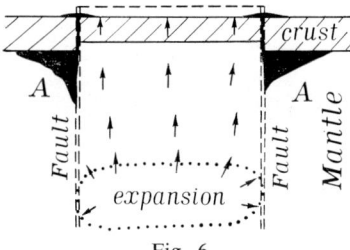

Fig. 6.

an uplifted area of the upper mantle (see zones A, fig. 6). But in the latter case the magmatic chamber will be of much smaller dimensions as compared to that formed with downwarping.

More specifically, the supply of mantle material for formation of the Earth's crust must have taken place in two ways. The first is zonal melting, whose scheme has been adapted to the conditions, prevalent within the Earth, by academician A. P. Vinogradov. The second way implies magmatic phenomena whose scheme is suggested above. It may be assumed that both processes, supplementing one another, have resulted in the accumulation of the material making up the present Earth's crust.

Thus, the Earth's mantle is a source of energy for tectonic movements and responsible for the formation of crustal structure. It is also a source of the material for the crust, for the "basaltic", "granitic", and sedimentary layers within the crust, as well as for mineral resources confined to certain structural forms. It is phase and other transformations of the mantle material which have a dominant role in these processes.

References

ARCHANGELSKY, A. D. (1936) Gravitational anomalies and their geological significance, Sotsrekonstruktsija and Nauka (in Russian).

BELOUSSOV, V. V. (1941) Gravitation and tectogenesis, Izv. Akad. Nauk SSSR, Ser. Geol. Geofiz. (in Russian).

BERNAL, J. D. (1936) Geophysical discussion, Observatory **59**, 268.

BRIDGMAN, P. W. (1948) The compression of 39 substances to 100 000 kg/cm^2, Proc. Am. Acad. Arts Sci. **76**, 3.

BULLEN, K. E. (1961) The deep regions of the earth, in: *The planet Earth* (Foreign Literature Press).

GORANSON, R. W. (1940) Physical effects of extreme pressures, Sci. Monthly **51**.

GUTENBERG, B. (1951) *Internal constitution of the Earth* (New York).

JEFFREYS, H. (1952) *The Earth* (Cambridge Univ. Press, London).

LOVERING, J. F. (1958) The nature of the Mohorovičič discontinuity, Trans. Am. Geophys. Union **39**, 5.

MAGNITSKY, V. A. (1965) *Interior structure and physics of the Earth* (Nedra), Moscov.

RAMSEY, W. H. (1950) On the compressibility of the earth, Monthly Notices Roy. Astron. Soc. **110**.

RIKITAKE, T. (1959) Geophysical evidence of the olivine–spinel transition hypothesis in the Earth's mantle, Bull. Earthquake Res. Inst. Univ. Tokyo **37**, 3.

RINGWOOD, A. E. (1962) A model for the upper mantle, J. Geophysic. Res. **67**, 2.

SCHATSKY, N. S. (1955) Origin of the Patschelm trough; comparative tectonics of ancient platforms, 5. Byull. Mosk. Obstschestva Ispytatelei Prirody, Otd. Geol. (in Russian).

SUBBOTIN, S. I. et al. (1965) Structure of the earth's crust and upper mantle; processes in the upper mantle; Influence of upper mantle processes on the structure of the earth's crust, Tectonophysics **2**, 2–3.

SUBBOTIN, S. I. et al. (1968) *The Earth's mantle and tectogenesis*, (Naukova Dumka, Kiev).

UFFEN, R. J. (1959) On the origin of rock magma, J. Geophys. Res. **64**, 1.

THE MECHANISM OF MANTLE EARTHQUAKES IN RELATION TO PHASE TRANSFORMATION PROCESSES

A. R. RITSEMA

K.N.M.I., de Bilt, Netherlands

A study of the focal mechanism of mantle earthquakes shows that the main phenomenon in the focus is a shear motion along sub-horizontal planes. This shear motion could possibly be triggered by some local process of phase transformation. The direction of shear is uniform over great distances and independent of local deviations in the direction of geological structures at the surface of the earth's crust. The notable gap in earthquake occurrences in depth, often coinciding with the position of the low-velocity layer, marks an abrupt change in sense of motion. Above this level the ocean side seems to be overthrusted by the continent side, below this level the ocean side seems to overthrust the continent side. In all deep-reaching seismic zones the earthquakes seem to be caused by a transfer of mantle material within the low-velocity layer in the direction from the ocean towards the continent.

1. Introduction

First motion studies of the direct longitudinal wave radiating from an earthquake focus show that the focal sphere normally can be divided into four quadrants in which alternatively compressional and dilatational first motions are emitted. This quadrantal pattern, already in the twenties was interpreted by Nakano et al. as being caused either by a sudden deformation of a focal-rock sphere into an ellipsoid, or by a fault movement along one of the two orthogonal nodal planes in the focus where the first motion of the longitudinal wave changes sign (fig. 1). This two-dipole model or double-couple shear model of the focus appeared to be generally valid for both crust and mantle earthquakes.

The use of such a point model for the interpretation of data from mantle earthquakes is justified by the fact that the focal dimensions normally are less than the half wavelength of the recorded waves such as they are used for these studies.

In the course of time, other models have been used to explain the data observed for specific earthquakes, resulting in solutions that are *non-orthogonal*. These models diverge between (see fig. 2):

the single force model of STAUDER (1960),

the single couple model of KEYLIS-BOROK (1957), see also RITSEMA (1957),

the asymmetrical single couple model of SCHÄFFNER (1961),

the single dipole model of BENIOFF (1963),

the double dipole model of unequal strength of RITSEMA (1965),

the explosion or implosion model of EVISON (1963),

the triple dipole model of which two are of equal strength, ISHIMOTO (1932), KAWASUMI (1934), ROBSON et al. (1968),

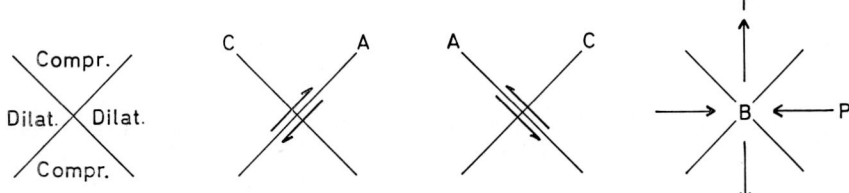

Fig. 1. Distribution of compressions and dilatations in a cross section through the earthquake focus, the directions of possible fault motion (A), and the position of the principal stresses (P, B and T).

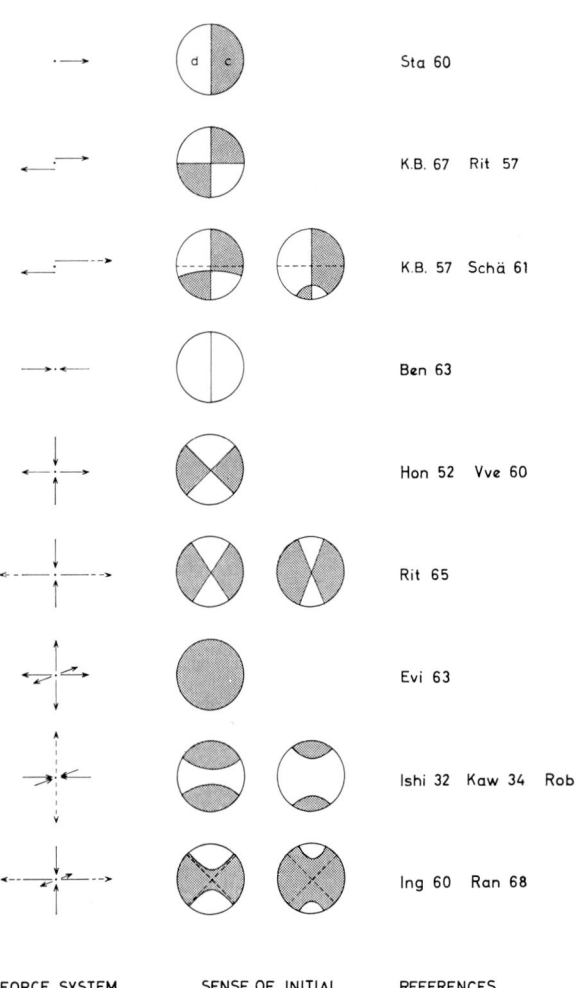

Fig. 2. Force models of the earthquake source, and the distribution of compressions and dilatations on the focal sphere. The following point sources are indicated: single force; single couple; asymmetric single couple or combination of single force and single couple; single dipole; double couple or double dipole; double dipole of unequal strength or combination of double couple and single dipole; radial force, triple dipole of which two are of equal strength or combination of radial force and single dipole; triple dipole of unequal strength or combination of radial force and double dipole.

the most generalized triple dipole model of INGRAM (1960), RANDALL (1967).

A study of such solutions reveals, however, that nearly all of them also, or even better, may be explained by the two-dipole model of the focus.

There are several causes for this ambiguity in interpretation, as for example: a mis-understanding of the orthogonality criterion by the original authors, insufficiency of data, an incorrect focal depth or an incorrect assumption about the P-wave velocity in the focal rock, dipping discontinuities near the source, and of course unreliable data.

Several cases could be cited for each of the above-mentioned invalidities.

Contrary to this relatively loose fit in non-orthogonal solutions published in literature, many tight solutions of the orthogonal type are available in the literature on the subject (see for example HODGSON, 1967; RITSEMA, 1964).

The conclusion then is that if non-orthogonality is real, the part of the seismic P-wave that originated from processes other than the commonly accepted two-dipole, double-couple shear model is small. Under these other processes are included the explosive or implosive source that may be related to a process of a sudden phase transformation (EVISON, 1963; RANDALL, 1968).

2. Materials

A re-evaluation has been made of the basic P and PKP wave data of 373 mantle earthquakes with a depth greater than 60 km having take place in the period 1957 through 1961. Catalogues, literature, ISS and BCIS bulletins and data of our own files have been used.

The criterion used for an earthquake to be considered has not been the number of basic data, but the distribution of the available data on the focal sphere. The data should at least show two consistent groups, one of compressions, the other of dilatations. This means that the earthquakes with many observations of a certain kind and only two or three of the opposite kind, were rejected. Only when a group of at least 5 consistent data is available in addition to a larger number of data of the opposite kind, the earthquake has been considered.

At first, only the earthquakes with a reasonably tight orthogonal solution were considered. The less documented earthquakes then were compared with the tight solutions of the same geographical region.

3. Results

Out of the 71 earthquakes with a reasonably tight solution there are 56 of the dip-slip kind and 15 of the transcurrent kind. A dip-slip fault exhibits a fault motion in the general direction of the dip of the fault plane and about perpendicular to its strike, a transcurrent fault exhibits a fault motion in the strike direction of the fault plane. A comparison with the solutions of this group of 71 earthquakes shows that for 273 of the less documented earthquakes a solution of the dip-slip fault type is feasible. 29 are definitely of the transcurrent fault type.

possible fault motion directions of the 329 earthquakes of the dip-slip fault type shows maxima around 0 and 90°. For the compilation of the figs. 4 systematically the less steep dipping possible fault motion has been used. Consequently only the maximum of around 0° dip is shown. A very much the same figure could be constructed using the steeper of the two possible fault mo-

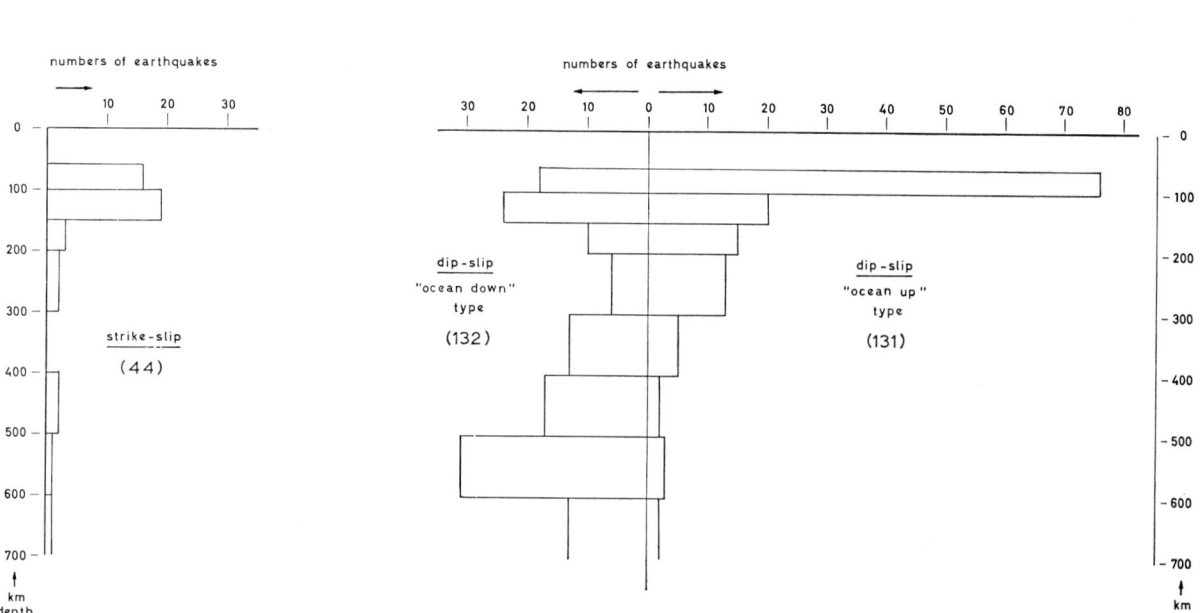

Fig. 3. Frequency distribution in depth of the fault motion types of mantle earthquakes (world sequence 1957–1961). There are 44 strike-slip and 263 dip-slip shocks. 132 Dip-slip earthquakes are of the "ocean down" type, which means that the fault motion is either in ocean down, continent up direction, or in the direction that the ocean block overrides the continental block, continental block underthrusts ocean block. 131 Dip-slip earthquakes are of the "ocean up" type, which means that the fault motion is either in ocean up, continent down direction, or in the direction that the ocean block underthrusts the continental block, continental block overrides ocean block.

A number of 35 out of the total of 44 shocks of the transcurrent fault type is located at a depth smaller than 150 km. Only 4 have a depth greater than 300 km (see fig. 3a).

Evidently, a dip-slip fault motion is the normal reaction of mantle material to earthquake generating processes acting at these depths of 60 to 700 km. This is in complete harmony with the conclusion reached earlier on a smaller number of data (RITSEMA, 1964). It seems likely that the transcurrent motions are related to tangential inhomogeneities occurring mainly in the more shallow layers of the upper mantle.

The frequency distribution of the dip-angle of the

tion directions, and then showing the maximum around 90° dip. The conclusion is that differential sliding does occur preferentially along about horizontal layers, or along about vertical planes.

A study of the more or less horizontal motion directions in the dip-slip fault type earthquakes shows that these directions are uniform over large distances along the strike of the particular seismic zones, and often independent of local strike variations within the zone. Figures 6 and 7 give the examples of the South American mantle earthquakes and those of South-East Asia. Both show this uniform pattern all along the individual seismic zones. It may be concluded that the earthquakes are interrelated and that they are caused by a process of greater dimension than that of the individual seismic arc.

In depth a clear change in sense of the motion direction does occur. In 76 out of 94 earthquakes with depths of 60–100 km either the ocean side moves up with respect to the continental side, or the continental

i.e. which side is the actual active agency in the process, the ocean side, the continental side, the hanging block or the footwall.

In fig. 3b the diametral change in process in the depth range 60–700 km is illustrated. The phenomenon is real and is displayed in all seismic zones where very deep earthquakes do occur.

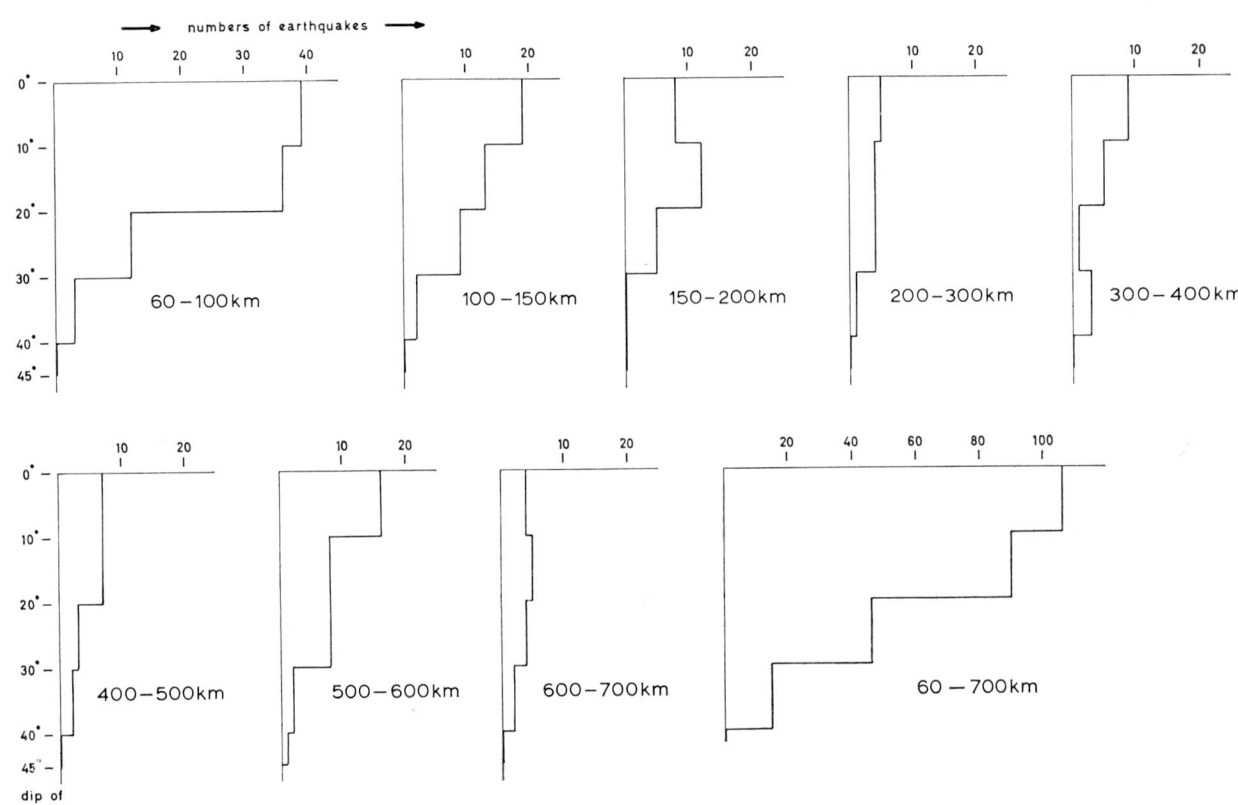

Fig. 4. Frequency distribution in depth of the plunge of the less steep fault motion direction of the earthquakes of the dip-slip fault type (world sequence 1957–1961).

side moves down with respect to the ocean side. The alternative possibilities are that either the ocean side underthrusts the continent, or reversely that the continent overrides the oceanic crust.

For the shocks with a depth of 300 km and more 74 out of 86 give the following picture: either ocean side down with respect to the continent, or the continent side up with respect to the ocean; or the alternative formulation of either the ocean side overthrusting the continental side, or the continental side underthrusting the ocean side. It remains to be decided for each depth level which of the four possibilities is the actual one,

In fig. 5 the percentages of the two different types of mechanism are plotted as a function of depth. It is seen that at depths in between 100 and 300 km the percentages are about equal: 43 are of the ocean up, continent down, ocean underthrusts, continent overthrusts type, a number of 40 of the opposed type. There is some indication for a geographical variation in this particular depth range: in the Kamchatka-Kuriles sector most shocks of these depths seem to accord with the deeper type, in the South American and Sunda arcs they seem to be of the more shallow mantle type.

A link of the change of sign of the dip-slip motions

in depth as illustrated in figs. 3b and 5, with the position of the low-velocity layer in the upper mantle and with the hiatus in the frequency-depth curves for the earthquakes of all deep-reaching seismic zones seems obvious and must evidently be assumed.

The derived horizontal motion directions are in agreement with those of surface shocks compiled by OLIVER, ISACKS and SYKES (1968) in their study on ble, however, to explain the same data by a Pacifico-petal drift of the continents above the low-velocity layer and a same Pacifico-petal flow in the mantle underneath the channel (VAN BEMMELEN, 1965).

4. Conclusion

In summary, it can be stated that differential movements do occur in the upper mantle and that they ap-

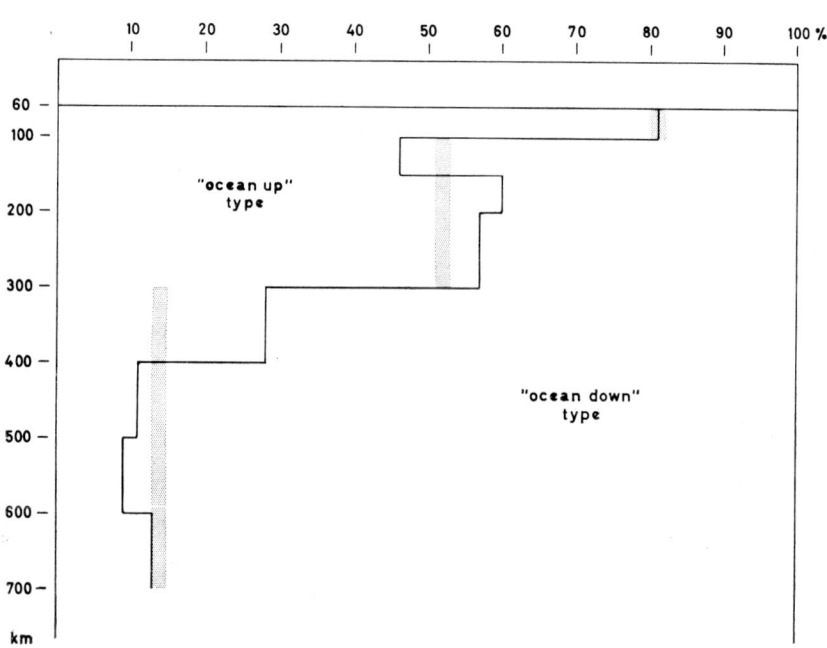

Fig. 5. Distribution in depth of the percentage of the two opposite dip-slip fault motion directions (world sequence 1957–1961) For explanation see caption fig. 3.

Global Tectonics. The observed change of sense of the horizontal motion in depth, however, is not readily explained by Oliver's model of the ocean lithospheric segment plunging down under the continent.

The observed seismological data are in good agreement with the already earlier formulated model of a horizontal flow in the low-velocity layer from ocean towards continent causing sub-horizontal shears with an underthrusting effect at the top and an overthrusting effect at the bottom of the channel (RITSEMA, 1964). In this model (fig. 8) the action comes from the ocean side. For the case of the Pacific area, this means a Pacifico-fugal flow in the low-velocity layer. It is possi- parently are linked on a global scale. Although shear in the mantle earthquakes appears to be the main phenomenon, the seismic data are not in contradiction with the assumption that these shear motions are triggered by local processes of phase transformations taking place where ocean and continent are in active contact with each other. In fact, if such phase transformations take place in the upper mantle layers, it seems straightforward to assume that they should do so in sub-horizontal layers where hydrostatic pressure reaches the crital value. That in such a case the motion itself preferentially should take place along these planes is obvious. Those namely are the planes where an available deviatory stress most easily will be relieved, whether the phase transformation is a sudden or a more gradual process.

The derived facts that should be taken into account in data-interpretations of other geo-disciplines are the following:

levels of the upper mantle, and probably relate to lateral inhomogeneities.

4. Slip in the mantle occurs preferentially either along horizontal planes or vertically.

5. The horizontal component of the motion vectors inside the plane with the more horizontal position are

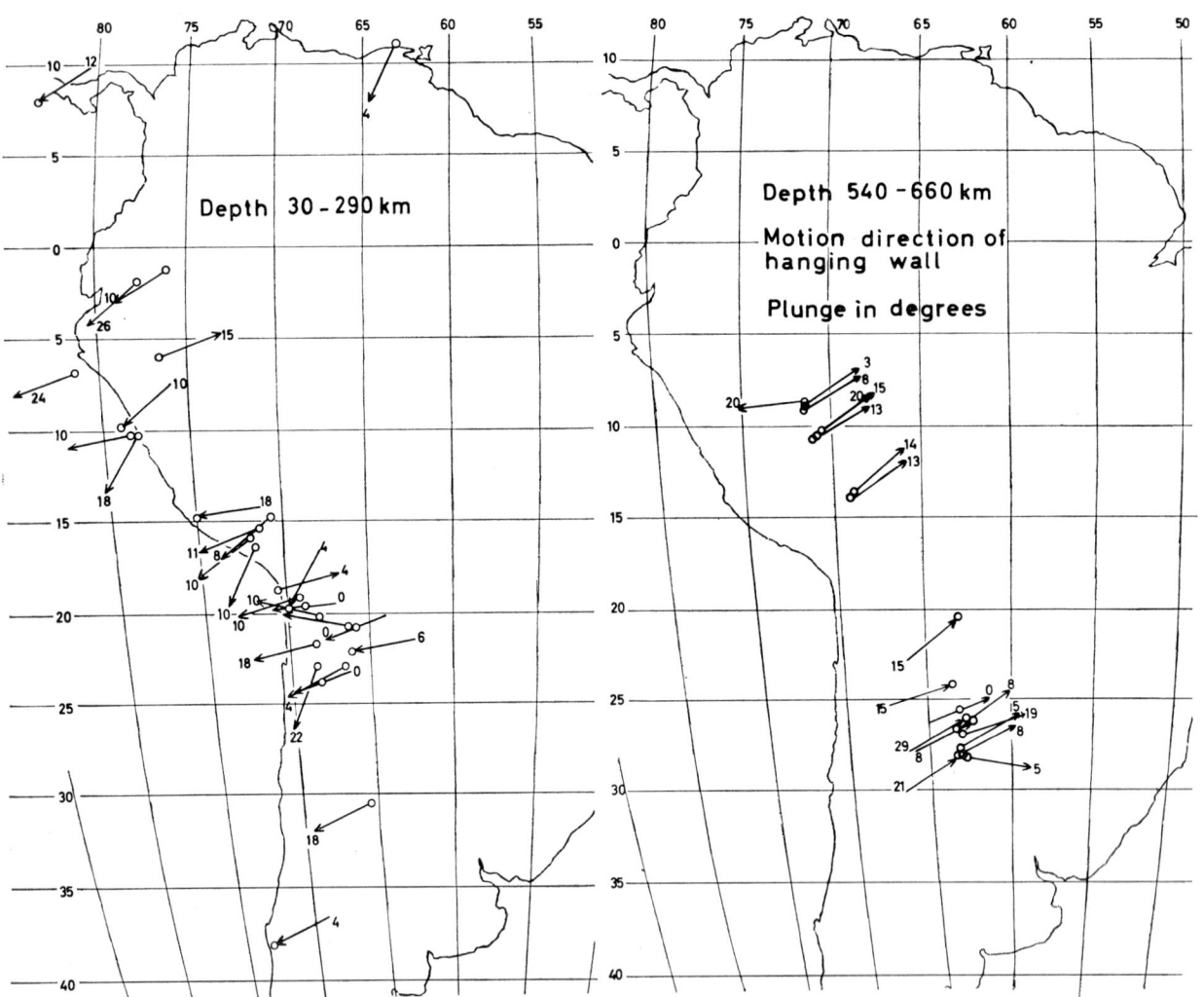

Fig. 6. Fault motion directions in South American mantle earthquakes.

1. Earthquake motions, also at great depth in the mantle, are of the shear-motion type.

2. The normal reaction of mantle material to stress generating processes in the mantle is that of a dip-slip fault motion.

3. The relatively few transcurrent fault motions in the mantle more specifically are bound to the higher

sub-parallel, roughly perpendicular to the tectonic zone in which they occur, but independent of local deviations from this direction.

6. There is a clear and often abrupt change of sense of motion when going to deeper levels in the upper mantle. This 180° change of motion direction seems to be correlated with the notable gap in seismicity that

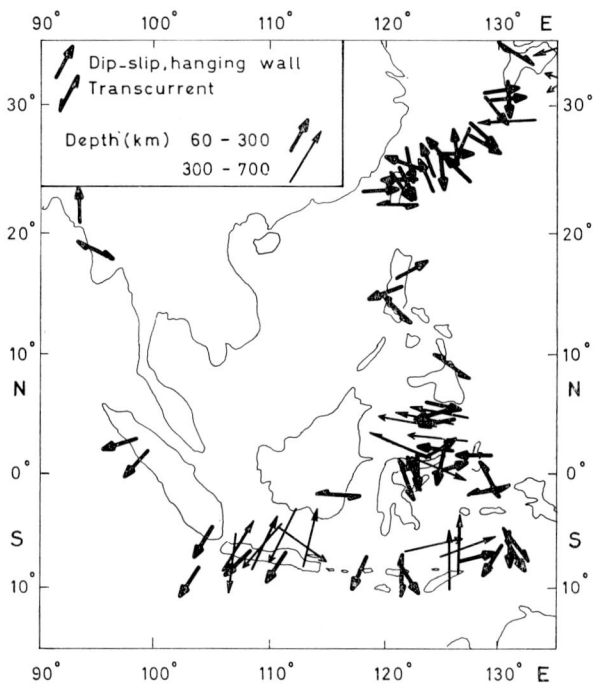

Fig. 7. Fault motion directions in the mantle earthquakes of the Sunda arc, the Celebes–Philippines arc and the Riu Kyu arc (1957–1961).

occurs in all zones with deep earthquakes, and with the position of the low-velocity layer. Always, a transfer of material in the low-velocity layer from ocean towards continent explains satisfactorily the type of motions observed.

References

VAN BEMMELEN, R. W. (1966) On Mega-undations, a new model for the earth's evolution, Tectonophys. **3**, 83.

BENIOFF, H. (1963) Source wave forms of three earthquakes, Bull. Seismol. Soc. Am. **53**, 893.

EVISON, F. F. (1963) Earthquakes and faults, Bull. Seismol. Soc. Am. **53**, 873.

EVISON, F. F. (1966) Polarity of the earthquake source, Nature **211** no. 5046. 273.

HARRINGTON, H. H. (1963) Deep focus earthquakes in South America and their possible relation to continental drift, in: A. C. Munyan, ed., *Polar wandering and continental drift*, Soc. Econ. Paleontologists Mineralogists Spec. Publ. **10**, 55.

HONDA, H. and A. MASATUKA (1952) On the mechanisms of the earthquakes and the stresses producing them in Japan and its vicinity, Sci. Rept. Tohoku Univ. Fifth Ser. **4** (1), 42.

HONDA, H. (1957) The mechanism of the earthquakes, Sci. Rept. Tohoku Univ. Fifth Ser. **9** Suppl., 46 pp.
Also in: Publ. Dominion Obs. Ottawa **20** (2), 295.

INGRAM S.J., R. E. (1960) Generalized focal mechanisms, Publ. Dominion Obs. Ottawa **24** (10), 305.

ISACKS, B., J. OLIVER and L. S. SYKES (1968) Seismology and new global tectonics, J. Geophys. Res. **73** (18) 5855.

ISHIMOTO, M. (1932) Existence d'une source quadruple au foyer sismique d'après l'étude de la distribution des mouvements initiaux des secousses sismiques, Bull. Earthquake Res. Inst. Tokyo Univ. **10**, 449.

ISHIMOTO, M. (1933) La déformation de la croûte terrestre et la production des ondes sismiques au foyer, Bull. Earthquake Res. Inst. Tokyo Univ. **11**, 254.

KAWASUMI, H. (1934) Amplitude of seismic waves with the structure of the earth's crust and mechanisms of their origin, Bull. Earthquake Res. Inst. Tokyo Univ. **12**, 660.

KEYLIS-BOROK, V. I. (1957) Isledovanie mekhanizma zemletryasenii, Tr. Geofiz. Inst. Akdad. Nauk SSSR, Sb. Statei **40**, 166.
Also: Soviet Res. Geophys. **4**, Am. Geoph. Un. (1960).

NAKANO, H. (1923) Notes on the nature of the forces which give

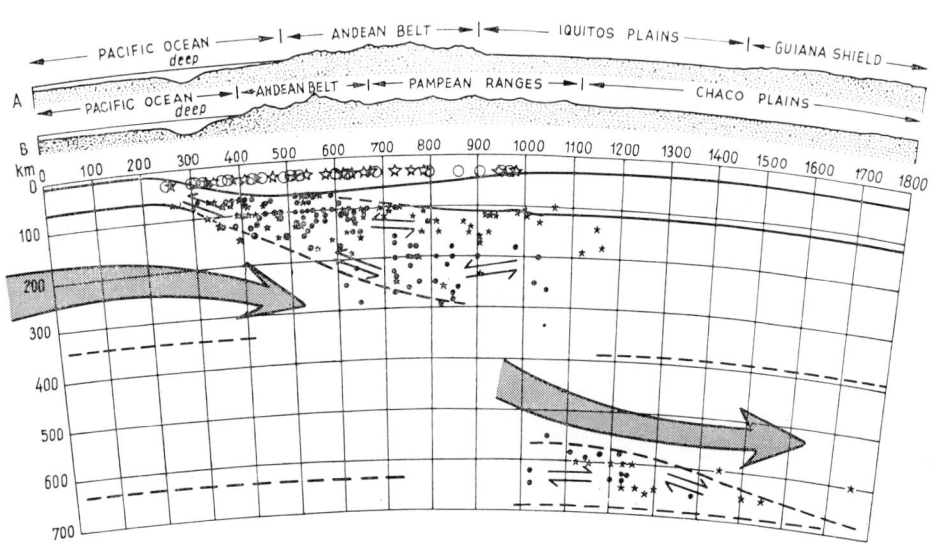

Fig. 8. Cross section of South America (HARRINGTON, 1963) with the location of, and the relative motion direction in the mantle earthquake foci, and the possible causative flow of mantle material from ocean to continent.

rise to the earthquake motions, Seismol. Bull. Centr. Meteorol. Obs. Japan **1**, 92.

OLIVER, J. and B. ISACKS (1967) Deep earthquake zones, anomalous structures in the upper mantle and the lithosphere, J. Geophys. Res. **72**, 4259.

RANDALL, M. J. (1966) Seismic radiation from a sudden phase transition, J. Geophys. Res. **71**, 5297.

RANDALL, M. J. (1968) Relative sizes of multipolar components in deep earthquakes, J. Geophys. Res. **73** (18) 6140.

RITSEMA, A. R. (1957) On the use of transverse waves in earthquake mechanism studies, Verhandel. Lemb. Meteorol. Geofys. Djakarta **52**.

RITSEMA, A. R. (1964) Some reliable fault plane solutions, Pageophys. **59**, 58.

RITSEMA, A. R. (1965) The mechanism of some deep and intermediate earthquakes in the region of Japan, Bull. Earthquake Res. Inst. Tokyo Univ. **43**, 39.

ROBSON, G. R., K. G. BARR and L. C. LUNA (1968) Extension failure: an earthquake mechanism, Nature **218**, 28.

SCHÄFFNER, H. J. (1961) Zur Interpretation von Herdmechanismen durch asymmetrische Dislokationen, Z. Geophys. **27**, 164.

STAUDER S.J., W. (1960) Three Kamchatka earthquakes, Bull. Seismol. Soc. Am. **50**, 347.

VVEDENSKAYA, A. V. (1960) The determination of stresses acting in earthquake foci from observations at seismic stations, Izv. Akad. Nauk SSSR Ser. Geofiz. **4**, 513.

WICKENS, A. J. and J. H. HODGSON (1967) Computer re-evaluation of earthquake mechanism solutions 1922–1962, Publ. Dominion Obs. Ottawa **33** (1).

PRESSURE EFFECT ON THE RATE OF PHASE TRANSITION IN MERCURY TELLURIDE

A. LACAM, B. A. LOMBOS and B. VODAR

Laboratoire des Hautes Pressions, C.N.R.S., 92 Bellevue, France

1. Introduction

This paper presents some new experimental results on the pressure dependence of the phase transition velocity between two polymorphic phases of mercury telluride. Owing to the large resistance variation characterizing this transition, the phenomenon can be conveniently used as a model to study the rates of solid–solid phase transitions, which may be of great interest in geophysics. This is the reason for presenting this paper at this meeting. BRIDGMAN (1940) suggested that there was a solid–solid transition near 12.5 kb at 22 °C with a volume change of approximately 8.4% in this case. A number of observations of velocity phenomena during phase transitions were collected by him in a separate paper (BRIDGMAN, 1916). JAYARAMAN et al. (1963) concluded that the solid–solid transition in mercury telluride at 14.0 kb and at 23 °C is clearly diffusion controlled as indicated by the increasing rapidity of the reaction at higher temperatures. BLAIR and SMITH (1961) observed an abrupt change in the resistivity of mercury telluride at a pressure of about 15 kb, at room temperature. This resistivity change by a factor of 10^5 to 10^4 was essentially reversible. They also reported the sluggishness of this solid–solid transition.

We measured the variation of the resistance of the mercury telluride samples during the transition as a function of pressure and time. From this variation of the transition rate with the pressure, monitored by the resistance of the samples, a "volume of activation" has been estimated using the transition state theory.

2. Experimental procedures

Pressures, up to 28 kb, were generated in a piston-cylinder type apparatus. Four electrical leads permitted the monitoring of the resistances of the samples with a four probe method using a sensitive recording potentiometer. A mixture of n.pentane–i.pentane (50 vol %) served as the pressure transmitting medium. This environment was confirmed as being truly hydrostatic by BARNETT and BOSCO (1967) up to 60 kb, under certain conditions. The pressure was measured with the resistance variation of a manganin coil, calibrated with the transition point of KBr (BIRCH, 1966). The samples, $1 \times 2 \times 4$ mm in dimensions, were cut from a single crystal of mercury telluride, supplied by the Solid State Laboratories of the C.N.R.S. at Bellevue. The sample volume being approximately 10^{-4} times smaller than the usable volume of the pressure chamber, the effect of the volume variation of the sample on the pressure, during the phase transition, could be neglected.

3. Results and discussions

The transition in mercury telluride induced (BLAIR and SMITH, 1961; HARMAN, 1967; VERIE and MARTINEZ, 1968) by pressure is a semiconductor–semimetal transition and that is the cause of the large resistivity variation. This polymorphic transition was found to be from the structure of zincblend type to cinnabar type (JAYARAMAN, 1963). The discussion of the phase transition from the point of view of the solid state physics will be presented in another paper. The phase transition started at about 15.3 kb at 21 °C. From this pressure the resistance variation as a function of time, at constant pressure, was recorded at each small step of pressure increase. It could be seen from our data, that after an initiation period of several minutes, a linear dependence of the logarithm of the resistance as a function of time existed for a few hours at each step. This suggested first-order type kinetics. These linear parts of the curves

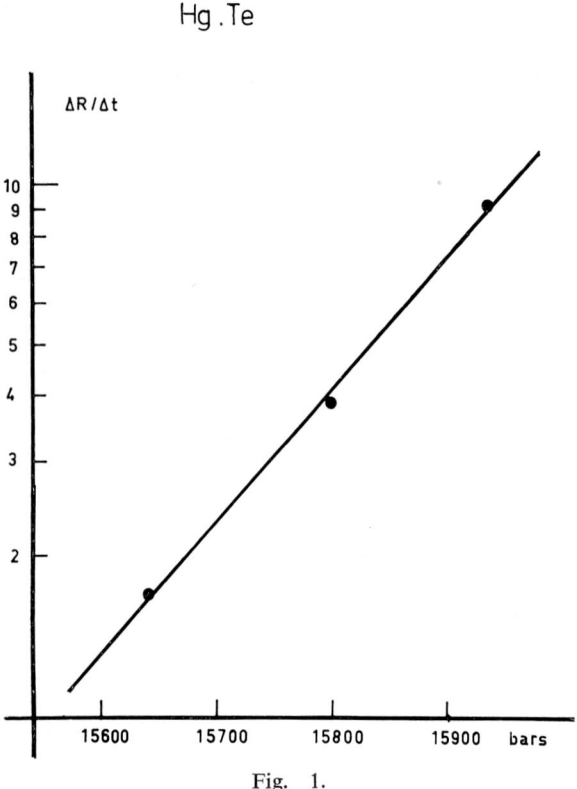

Fig. 1.

were used to determine the rate of variation of the transition as a function of pressure, assuming that the rate of resistance variation represented the variation of the rate of the phase transition, according to Bridgman's discussion on the mono-nucleation of polymorphic transitions (BRIDGMAN, 1916). The results are shown in fig. 1. The linear dependency of the logarithm of the rate variation is clearly apparent.

It has been a common practice to discuss experimental data on the basis of the socalled van 't Hoff equation in a form which was first suggested by EVANS and POLANYI (1935). The value of the "volume of activation" has been tentatively calculated from our experimental results. This was found to be $-126 \, cm^3/mole$. Thus the formation of the transition state involves a net decrease in volume and an increase of the rate constant with increasing pressure.

We do realize that this "volume of activation" is really a composite function, as has been discussed by HAMANN (1963), but its introduction might be useful for the evaluation of rate measurements at high pressure of polymorphic phase transitions. Further quantitative measurements are planned to be carried out to reveal the mechanism of this semimetal to semiconductor transition.

Acknowledgments

The kinetic point of view of the phase transition in mercury telluride is included in a larger study of the pressure effect on this material investigated in collaboration with the Laboratoire de Magnétisme et de Physique du Solide du C.N.R.S. Bellevue. We would like to express our thanks to Dr. Verie and Mr. Weill of this laboratory, respectively for suggesting the phase transition problem in HgTe, and for assistance during the measurements.

References

BARNETT, J. D. and C. D. BOSCO (1967) Rev. Sci. Instr. **38**, 957.
BIRCH, F. (1966) in: S. P. Clarc, Jr., ed., *Handbook of physical constants*, Geol. Soc. Am. Mem. **67**.
BLAIR, J. and A. C. SMITH (1961) Phys. Rev. Letters **7**, 124.
BRIDGMAN, P. W. (1916) Proc. Am. Acad. Arts Sci. **52**, 57.
BRIDGMAN, P. W. (1940) Proc. Am. Acad. Arts Sci. **74**, 21.
EVANS, M. G. and M. POLANYI (1935) Trans. Faraday Soc. **31**, 875.
HAMANN, S. D. (1963) in: R. S. Bradley, ed., *High pressure in physics and chemistry* II, 165.
HARMAN, T. C. (1967) Proc. Int. Conf. II-VI Semicond. Compounds.
JAYARAMAN, A., W. KLEMENT, JR. and G. C. KENNEDY (1963) Phys. Rev. **130**, 2277.
VERIE, C. and G. MARTINEZ (1968) Compt. Rend. Acad. Sci. Paris **266**, 720.

THE EVOLUTION OF THE EARTH'S CORE AND ITS MAGNETIC FIELD

J. A. JACOBS

University of Alberta, Edmonton, Alberta, Canada

The broad spectrum of variations in the Earth's magnetic field is discussed with particular reference to hydromagnetic oscillations in the Earth's core. Palaeo-intensity measurements are reviewed and the effects on the magnetic field of the evolution of the core and of irregular fluctuations in the rate of the Earth's axial rotation are discussed.

1. Introduction

One of the main difficulties in any geophysical investigation is that observations of natural phenomena cover only a very small fraction of the Earth's lifetime – less than about 1 part in 10^6. With such an extremely small sample it is not to be wondered at that our knowledge of many geophysical processes is so rudimentary. Two fields of research, however, – isotopic studies and palaeogmagnetism – have yielded information about the early history of the Earth. The radioactive decay of certain rocks and minerals has enabled geophysicists to establish absolute methods of geochronology and also to construct possible models of the thermal history of the Earth. Palaeomagnetic studies have in the past been mainly concerned with measurements of the direction of the ancient magnetic field and these investigations have had their greatest impact on such problems as continental drift and polar wandering. Palaeomagnetic intensity measurements have received much less attention – mainly because of the difficulty in obtaining reliable results. However increased interest has been shown in such measurements during the last few years; we can hope that in the future it will be possible to learn from them something about the growth and development of the Earth's core. In turn conditions in the Earth's core can provide boundary values for conditions in the mantle.

2. Hydromagnetic oscillations in the Earth's core

The spectrum of variations in the Earth's magnetic

Periods		Origin	Comments
Seconds	Years		
10^{17}	3×10^9	?	?
10^{16}			
10^{15}	3×10^7	Internal and dipolar	Dipole reversals
10^{14}			
10^{13}			
10^{12}			
10^{11}			
10^{10}	300	Internal, non-dipolar	Secular variation
10^{9}	30		
10^{8}			
10^{7}			
10^{6}	3×10^{-2}	External	Magnetic storms
10^{5}	3×10^{-3}	External	Diurnal variation
10^{4}			
10^{3}			
10^{2}			
10^{1}		External	Micropulsations
10^{0}			
10^{-1}		External	Sub-acoustic

Fig. 1. Spectrum of geomagnetic variations

field covers an enormous range – from fractions of a second to millions of years (fig. 1). The spectrum of the variation in the Earth's rotation is another such geophysical phenomenon showing a very broad range. In both cases the causes are probably different for different frequency bands. In the case of the Earth's magnetic field, the short period variations are of external (solar) origin and will not be considered here. The longer period variations may be divided into four groups,
1) Secular variation ($10^2 < T < 10^3$) (T in yr);
2) Historic/Archaeological variations ($T \sim 10^4$);

3) Field reversals ($10^5 < T < 10^6$) (Polarity epochs $T \sim 10^6$) (Polarity events $T \sim 10^5$);
4) Ancient/Geological variations ($T \sim 10^8$).

The frequency band-widths are only approximate – thus during the last few million years there have been a number of polarity events with durations between about $1.5 \sim 7 \times 10^5$ yr, whilst from the Upper Carboniferous $\sim$ Upper Permian (a span of some $40 \sim 50 \times 10^6$ yr, some 3×10^8 yr ago) there are no records of any reversals.

It is now generally accepted that the main geomagnetic field has its origin in magnetohydrodynamic processes in the liquid, metallic core of the Earth. HIDE (1966) has examined free hydromagnetic oscillations of the Earth's core in an effort to account for the secular variation. However a phenomenon as world wide as a reversal can hardly be thought of as a small perturbation of the Earth's field and it may be necessary to investigate large amplitude oscillations of the core. On the other hand the field that "escapes" from the core and is observed at the Earth's surface may be a small fraction ($\sim 1\%$) of the field inside – in fact most theories of the Earth's main field demand a much greater ($\sim 100\Gamma$) toroidal field in the core. Thus it may be possible to regard the dipole field and its reversals as resulting from perturbations of this much larger toroidal field in the core.

Consider plane hydromagnetic waves in a perfectly conducting, incompressible, inviscid, homogeneous fluid of indefinite extent rotating uniformly with angular velocity $\boldsymbol{\Omega}$ and immersed in a uniform magnetic field $\boldsymbol{H}_0$ parallel to $\boldsymbol{\Omega}$. The dispersion relation (LEHNERT, 1954; HIDE and ROBERTS, 1961, 1962) shows that there are two modes (an inertial mode and a magnetic mode) both highly dispersive. The frequency and phase velocity of the magnetic mode are much lower than those of the inertial mode. The angular frequency ω_m of the magnetic mode is given by

$$\omega_m \simeq \pm \frac{V_A^2 k^2}{2\Omega},$$

where k is the wave number and V_A the Alfvén velocity defined by

$$V_A^2 = \mu H_0^2 / 4\pi \rho.$$

Thus the period $T_m = \dfrac{2\pi}{\omega_m} = 16\pi^2 \rho \Omega / \mu k^2 H_0^2$.

For the Earth Ω does not vary much although if the Earth has been slowing down, i.e., if the Earth was rotating faster in the past, then T_m would have decreased, although not by more than a factor of about 5 over geologic time. Taking $\rho \simeq 10$ gm/cm^3, $\Omega \simeq 7 \times 10^{-5}$ rad/sec and $\mu \simeq 1$, $T_m \simeq 1/10\, H_0^2 k^2$, i.e. $\lambda^2 = 10\, H_0^2 T_m$ where $\lambda = 1/k$ is the wave length. Thus T_m increases if H_0 decreases and/or k decreases i.e. as λ increases. The indications are that the Earth's magnetic field has not changed appreciably over geologic time so that T_m essentially depends on λ.

Let us consider the above four different classes of variations of the Earth's main field. Taking $H_0 \simeq 5\Gamma$ (the estimated core value of the poloidal field) and $T_m \simeq 10^3$ yr (for the case of the secular variation) $\lambda \simeq 27.5$ km which appears to be too small. If $T_m \simeq 10^4$ yr (historic/archaeological variations) $\lambda \simeq 85$ km which is again rather small. If $T_m \simeq 10^6$ yr (reversals) $\lambda \simeq 850$ km which is reasonable, while if $T_m \simeq 10^8$ yr (ancient/geological variations) $\lambda \simeq 8500$ km which is too large. The above very crude model merely serves to indicate that hydromagnetic oscillations may well be the cause of some of the variations in the spectrum of the Earth's magnetic field. (Transmission of magnetic energy between different parts of the core is almost certainly mainly due to hydromagnetic waves, diffusive processes being relatively unimportant.) HIDE (1966) has carried out detailed calculations when there is a strong toroidal magnetic field in the Earth's core. He was able to explain many of the observed features of the secular variation (including the slow westward drift of the main magnetic field at the Earth's surface) assuming a toroidal field of 200Γ, $T_m = 720$ yr and $\lambda = 2850$ km.

3. Palaeomagnetic intensity measurements

The results of palaeomagnetic intensity measurements are not conclusive, although certain broad trends have been established. These trends must be reviewed against the background of the general spectrum of variations in the Earth's main field. Since direct field observations were first made some 140 yr ago, the dipole moment has been decreasing, the rate of decrease since 1835 being about 5 per cent per century. Measurements over historical and archaeological time indicate that during the past 2000 yr the geomagnetic dipole moment has decreased by about one-third of its peak value –

prior to the year 0 the dipole moment was increasing. SMITH (1967) has considered the significance of this result and come to the conclusion that it is not the effect of secular variation (the quasi-periodicity is about one order of magnitude higher than that of known non-dipole field effects). Smith also eliminated dipole wobble as an explanation of the archaeological variations. Although the period of the wobble is of the right order ($\sim 10^4$ yr), he rejected any such relationship – if there were such a relationship there should be a correlation between the inclination and field intensity which is not observed, and in any case the field intensity changes are far greater than could arise from the magnitude of the wobble. Smith thus concludes that there is in fact a real variation in the dipole moment – this conclusion also applies to measurements over geologic time.

The variation of the mean dipole moment over the last 4×10^8 yr is shown in fig. 2 (SMITH, 1967). Throughout the Pre-Cambrian it appears that the mean geomagnetic dipole moment was smaller than the Earth's present dipole moment and has been increasing throughout this time. There is an extreme paucity of data in the Cambrian but it would appear that the field increased again as one goes back farther in time (CARMICHAEL, 1967). It is variations over geological time that are of interest here. If the dipole field has been more powerful in the past i.e. if the non-dipole forces have been less effective, can we deduce that the field was more stable then and thus less subject to reversals? All rocks during the Upper Carboniferous ~ Upper Permian, a span of some $40 \sim 50 \times 10^6$ yr some 3×10^8 yr ago, have shown the same polarity (reversed), and unpublished data by Smith indicate a maximum in the field intensity around that time which is not indicated in his earlier paper (fig. 2). Again the oldest field intensity measurement is about 2.8×10^9 yr which presumably indicates that the Earth's core had already then reached an advanced stage in its development. This would preclude the rather slow evolution of the core as advocated by RUNCORN (1962) which did not begin its growth until about that time.

There is one small piece of evidence, however, that may possibly support some of Runcorn's views about the growth of the core. GASTIL (1960), from an examination of a large number of radioactive age determinations, found peaks of world-wide upsurges of orogenic activity. Runcorn has identified these with a reorganization of convection currents in the Earth's mantle, the

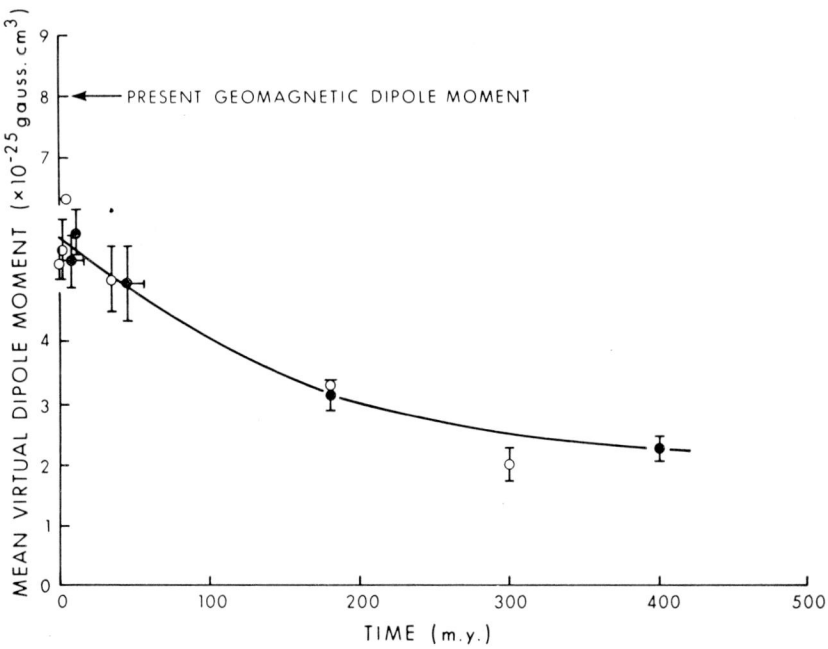

Fig. 2. Mean geological virtual dipole moments (VDM) plotted against time. ●: accurate values, ○: approximate values. Error bars in VDM represent standard errors of the means; error bars in time represent range of potassium-argon ages where appropriate (after SMITH (1967)).

change from the $n = 3$ to the $n = 4$ harmonic occurring 1.1×10^9 yr ago. Such a radical change in the convective pattern might be expected to cause changes in the magnetic regime in the core and it is interesting to note that SCHWARZ and SYMONS (1968) find a high value for the Pre-Cambrian field at precisely 1.1×10^9 yr ago.

HIDE (1967) has also discussed possible correlations between geomagnetic and geological phenomena. He has shown that horizontal variations in properties at the core mantle boundary that could escape detection by modern seismological methods might nevertheless produce pronounced hydrodynamical effects throughout the core, which in turn might produce measurable geomagnetic effects. It is thus possible that motions in the Earth's mantle might affect conditions both at the Earth's surface, where geological events occur, and at the core-mantle boundary where effects on the geomagnetic field may arise. Since only small variations in core motions are necessary to reverse the sign of the dipole field, Hide suggests that perhaps there may be some correlation between reversals and other phenomena that may be affected by motions in the mantle such as tectonic activity, ocean floor spreading, continental drift etc.

4. The rotation of the Earth

The dynamo theory for the generation of a magnetic field from motions in a fluid core predicts the existence of irregularities in the rate of rotation of the planet's mantle. In the case of the Earth astronomical observations have shown that the length of the day changes irregularly. These irregular changes are of the order of 1 in 10^7–10^8 with Fourier components of periods 1–100 yr. The similar time scale in the geomagnetic secular variation which arises from turbulence in the Earth's fluid core shows that induced current systems must exist in the lower mantle which, together with the geomagnetic field, produce varying electromagnetic torques between the mantle and core causing the irregular fluctuations in the length of the day. RUNCORN (1968) has suggested that perhaps Jupiter's rotation rate may show similar irregular fluctuations. He finds a remarkable similarity between the time scale of irregular changes in the Earth's rotation and the rotation period of the Red Spot in Jupiter. The negative results of the attempts to detect main magnetic fields of the Moon and Mars have been explained by the probable absence of large fluid cores and this is to some extent confirmed by the uniformity of their axial rotation. In the case of Venus it is not possible to determine with sufficient accuracy any possible irregular fluctuations in its rate of rotation.

PANNELLA et al. (1968) have obtained values of the length of the synodic month using tidally controlled periodical growth patterns in mollusks and stomatolites for several geologic periods (fig. 3). It would appear

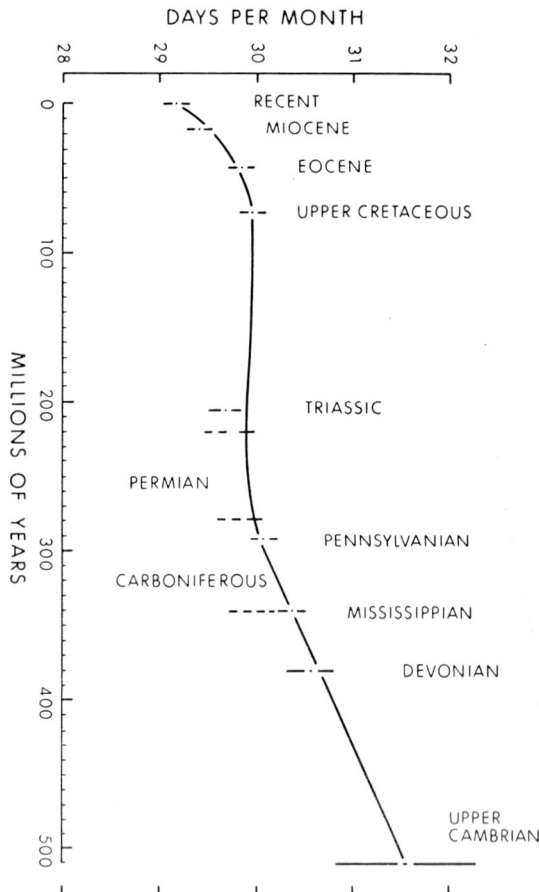

Fig. 3. Variations in the length of the synodic month through geologic time. The error bars show the standard error for each point (after PANNELLA et al., 1968).

that the slowing down of the Earth's rotation has not taken place at a uniform rate – there are two major breaks in the slope of the curve, the slowing down being negligible over a 200×10^6 yr period from the Pennsylvanian to the Upper Cretaceous. Pannella et al. suggest that the changes in the slope may be related to hypothetical events in the Earth's history; e.g. if the slowing down since the late Cambrian is attributed mainly to the loss of energy due to tidal torques in shallow seas, then a redistribution of continents, oceans and shallow seas should affect the amount of energy dissipated and the rotation of the Earth. The high slope between the Cretaceous and the present could be attributed to the rapid drift period of the Upper Cretaceous and to the rise of the Alpine orogenic belt which created widespread shallow seas. However whatever theories are proposed the changes in slope must be related to major events in the Earth's history. We may ask how such events may be reflected in the Earth's core and magnetic field. It is extremely interesting that just after one of these changes in slope, some 300×10^6 yr ago, the Earth's magnetic field maintained a constant polarity (reversed) for about 50×10^6 yr. Does the Earth's field require minimum changes in the rate of rotation in order to experience reversals, becoming relatively "pot-bound" if the rotation rate is sensibly constant? If this should be true we might expect that when the deceleration increased again about the Upper Cretaceous, reversals would become more frequent – this is certainly the case in recent times.

5. Conclusion

The results of space vehicles to detect main magnetic fields of the Moon, Mars and Venus have been negative. Lunik 1 and later Explorer 35 found fluctuating values of about 10γ for the magnetic field near the Moon's surface. The Mariner IV fly by of Mars showed that the surface field at the planet's pole was less than 200γ and that Mars does not possess radiation belts. In the case of Venus the Mariner II measurements in the fly by at a distance of seven radii, showed that the planet's magnetic moment is less than 10% that of the Earth. The dynamo theory indicates that a planet will not possess a general magnetic field unless it has a fluid electrically conducting core in which convective motions can occur. The above results can thus be qualitatively explained since it is generally held that the Moon does not possess an iron core and its long period of rotation will reduce the effectiveness of any dynamo action. Mars like the Moon probably possesses at most a very small core while in the case of Venus, even if it has a liquid core comparable in size to that of the Earth, its slow rate of axial rotation (~ 243 days) would probably prevent the dynamo process from working.

However we still have no answers to such fundamental questions as to how large the liquid core of a planet must be and how fast the planet must rotate in order for a dynamo mechanism to be operative.

It is very difficult to make quantitative estimates of the effect of the size of the Earth's core on the resultant magnetic field. Crude estimates may be made, however, which intuitively seem reasonable. By considering the order of magnitude of the different forces in the equation of motion in the Earth's core, it is easy to show (HIDE, 1956) that viscous and inertial forces can be ignored, and that Coriolis and electromagnetic forces are dominant. The Coriolis term is $\propto \Omega U$ when Ω is the angular velocity of the Earth and U a typical flow velocity. The electromagnetic force is $\propto |\text{curl } \boldsymbol{H} \times \boldsymbol{H}| \propto H^2/L$ where H is the magnetic field strength and L a typical length scale. Thus larger fields can be generated by larger Ω and/or L – which is intuitively obvious. It is unlikely that Ω will have changed sufficiently to have a pronounced effect on H; changes in L (due to the evolution of the core) will be far more profound. The question of the origin of the Moon has some relevance here. If the Moon were captured by the Earth, the Earth's rotation would have been showed by a factor of $\sim 24/4.8 \simeq 5$ with concomitant effects on H. On the other hand other workers (WISE, 1963; O'KEEFE, 1966) believe that the formation of the Earth's core increased the rotational instability of the Earth leading to the origin of the Moon by fission from the Earth. Thus the origin of the Moon and the evolution of the Earth's core may be related.

Finally some additional information may be obtained from ocean floor spreading. If lines of latitude and longitude are drawn, not about the Earth's pole of rotation but about the "spreading pole", the axis of spreading is parallel to the new lines of longitude and the fracture zones are perpendicular to the axis and thus parallel to the lines of latitude. Moreover the fastest rate of spreading occurs along the equator of the new coordinate system, the rate decreasing regularly with distance from the equator. The spreading poles vary for different oceans – for the Pacific and South Atlantic they appear to be very close to the Earth's magnetic poles. Also in Cretaceous times the spreading pole for the North Pacific was near the Cretaceous geomagnetic pole. The relation between the spreading pole and the magnetic pole suggests that the convective motion within the Earth's upper mantle and the Earth's magnetic field may have a common cause – perhaps irregularities in the Earth's orbit. However since the direction of convection does not reverse when the field reverses it is clear that the convective motion does not simply generate the field – nor is it likely that the reversing field could "pump" convection cells. On the other hand there does appear to be yet another link between the Earth's magnetic field and processes occurring in the upper mantle. Further progress is more likely to be made by considering the Earth as a planet, a member of our solar system, than by regarding it as an isolated body.

References

CARMICHAEL, C. M. (1967) An outline of the intensity of the palaeomagnetic field of the Earth, Earth Planet. Sci. Letters **3**, 351–354.

GASTIL, G. (1960) The distribution of mineral dates in time and space, Am. J. Sci. **258**, 1–35.

HIDE, R. (1956) The hydrodynamics of the Earth's core, Phys. Chem. Earth **1**, 94–137.

HIDE, R. (1966) Free hydromagnetic oscillations of the Earth's core and the theory of the geomagnetic secular variation, Phil. Trans. Roy. Soc. London A **259**, 615–650.

HIDE, R. (1967) Motions of the Earth's core and mantle, and variations of the main geomagnetic field, Science **157**, 55–56.

HIDE, R. and P. H. ROBERTS (1961) The origin of the main geomagnetic field, Phys. Chem. Earth **4**, 25–98.

HIDE, R. and P. H. ROBERTS (1962) Some elementary problems in magnetohydrodynamics, Advanc. Appl. Mech. **VII**. 215–316.

LEHNERT, B. (1954) Magneto-hydrodynamic waves under the action of the Coriolis force I, Astrophys. J. **119**, 647–654.

O'KEEFE, J. A. (1966) The origin of the Moon and the core of the Earth, *The Earth-Moon System* (Plenum Press) 224–233.

PANNELLA, G., C. MACCLINTOCK and M. N. THOMPSON (1968) Palaeontological evidence of variations in length of synodic month since late Cambrian, Science **196**, 792–796.

RUNCORN, S. K. (1962) Convection currents in the Earth's mantle, Nature **195**, 1248–1249.

RUNCORN, S. K. (1968) Planetary magnetic fields as a test of the dynamo theory, Geophys. J. **15**, 183–189.

SCHWARZ, E. J. and D. T. A. SYMONS (1968) On the intensity of the palaeomagnetic field 100 million and 2500 million years ago, Phys. Earth Planet. Interiors **1**, 122–128.

SMITH, P. J. (1967) The intensity of the ancient geomagnetic field: a review and analysis, Geophys. J. **12**, 321–362.

WISE, D. U. (1963) An origin of the Moon by fission during formation of the Earth's core, J. Geophys. Res. **68**, 1547–1554.

AUTHOR INDEX

T. J. Ahrens, 205.
S. Akimoto, 186, 189.
D. L. Andersen, 23, 41.
O. L. Anderson, 61.
K.-I. Aoki, 273.
S. Banno, 405.
J. D. Barnett, 54.
W. A. Basset, 51, 54.
R. A. Binns, 156
F. Birch, 178.
A. L. Boettcher, 331.
K. E. Bullen, 36.
C. W. Burnham, 332.
J. L. Carter, 391.
J. A. Cooper, 302.
N. F. Davis, 332.
A. A. Deribas, 348.
N. L. Dobretsov, 348, 462.
S. Endo, 182.
Th. Ernst, 451.
E. J. Essene, 378.
F. A. Frey, 323.
W. S. Fyfe, 196.
E. S. Gaffney, 205.
P. W. Gast, 246.
D. H. Green, 221, 247, 359, 378, 385, 431.
T. H. Green, 441.
D. L. Hamilton, 309.
B. J. Hensen, 378, 431.
W. Hibberson, 247.
A. Irving, 385.
K. Ito, 182.
J. A. Jacobs, 513.
J. C. Jamieson, 201.
T. Jordan, 23.
A. A. Kadik, 343.
H. Kanamori, 475.
N. Kawai, 182.
N. I. Khitarov, 334, 343.
J. D. Kleeman, 302.

K. Koto, 161.
H. Kuno, 273.
A. Lacam, 511.
I. B. Lambert, 316.
R. C. Liebermann, 61.
B. A. Lombos, 511.
J. F. Lovering, 390.
G. J. H. McCall, 255.
Th. R. McGetchin, 471.
I. D. MacGregor, 372, 391.
A. Major, 89.
V. I. Maly, 348.
M. H. Manghnani, 456.
H. K. Mao, 51.
P. B. Moore, 166.
M. Morimoto, 161.
R. W. Nesbitt, 309.
M. J. O'Hara, 236.
F. Press, 3.
V. A. Pugin, 334.
A. F. Reid, 204.
A. E. Ringwood, 89, 109, 359.
A. R. Ritsema, 503.
C. Sammis, 41.
W. Schreyer, 422.
R. Schwab, 451.
C. B. Sclar, 333.
F. Seifert, 422.
L. T. Silver, 471.
A. B. Slutsky, 334.
J. V. Smith, 166.
N. V. Sobolev, 398, 462.
S. J. Subbotin, 499.
I. Syono, 186.
T. Takahashi, 51.
M. Tokonami, 161.
B. Vodar, 511.
C. Y. Wang, 213.
G. P. Woollard, 484.
P. J. Wyllie, 316.

QE
509
P5

JAN 19 1972